ORACLE 10*g* DATABASE ADMINISTRATOR:

IMPLEMENTATION & ADMINISTRATION

By Gavin Powell and Carol McCullough-Dieter

Australia • Brazil • Japan • Korea • Mexico • Singapore • Spain • United Kingdom • United States

Oracle 10*g* Database Administrator: Implementation & Administration
Gavin Powell and Carol McCullough-Dieter

Publisher: Bob Woodbury

Product Manager: Beth Paquin

Acquisitions Editor: Maureen Martin

Marketing Coordinator: Suelaine Frongello

Production Editor: GEX Publishing Services

Cover Designer: Laura Rickenbach

Compositor: GEX Publishing Services

Manufacturing Coordinator: Justin Palmeiro

For product information and technology assistance, contact us at
Cengage Learning Customer & Sales Support, 1-800-354-9706

For permission to use material from this text or product, submit all requests online at **www.cengage.com/permissions**
Further permission questions can be emailed to
permissionrequest@cengage.com

ISBN-13: 978-1-4188-3665-8
ISBN-10: 1-4188-3665-6

Course Technology
20 Channel Center Street
Boston, MA 02210
USA

Cengage Learning is a leading provider of customized learning solutions with office locations around the globe, including Singapore, the United Kingdom, Australia, Mexico, Brazil, and Japan. Locate your local office at: **international.cengage.com/region**

Cengage Learning products are represented in Canada by Nelson Education, Ltd.

To learn more about Course Technology, visit **www.cengage.com/coursetechnology**

To learn more about Cengage Learning, visit **www.cengage.com**

Purchase any of our products at your local college store or at our preferred online store **www.ichapters.com**.

Printed in the United States of America
3 4 5 6 7 8 9 10 09

TABLE OF CONTENTS

PREFACE

This textbook is a detailed guide to basic database administration for the Oracle 10g database. The database administrator (DBA) uses the techniques demonstrated in the book, which are essential for creating the initial database and configuring the storage space, tables, users, and security for a database. The book is a basic guide because it covers some of the first concepts that a DBA needs, omitting more advanced topics such as network configuration and performance tuning.

Intended Audience

This book is intended to support individuals in database courses covering database administration using the Oracle 10g database. It is also intended to support individuals who are preparing for the Oracle Database 10g DBA OCA exam that is required for certification as an Oracle Certified Associate (OCA) or an Oracle Certified Professional (OCP).

Prior knowledge of general relational database terminology and concepts is required. In addition, the reader should have basic knowledge of SQL (Structured Query Language). While it is preferable that the reader have knowledge of Oracle 10g's SQL, the reader's experience using SQL on other databases, such as SQL Server, is acceptable. The reader should be able to write SQL commands for querying, inserting, updating, and deleting data in relational tables.

Oracle Certification Program

This textbook covers the objectives of *Exam 1Z0-042, Oracle Database 10g DBA OCA*. The exam is the first exam required for individuals seeking certification as an Oracle Certified Professional (OCP). OCA certification is a prerequisite to the second exam for OCP certification. Information about registering for these exams can be found at *http://www.oracle.com/education/index.html*.

Approach

The concepts introduced in this textbook are presented in business scenarios. Concepts are introduced and examples of real-life uses for the various concepts are discussed. Then students follow along with hands-on practices to drive home the concepts in every chapter. The case project at the end of each chapter works within the context of a hypothetical real-world business: an online newspaper publishing company named Global Globe News Company. The case studies begin with an empty database, adding tablespaces and adjusting initialization parameters. Each chapter builds on the concepts of the previous chapter, and the case study adds details to the database. By the end of the book, the student has created tables, views, indexes, and objects. The student also creates database users, roles, and profiles that provide the Global Globe's employees with varying security levels of access to the data. This allows students to not only learn the syntax of a command, but also to apply it in a real-world environment. In addition, there are several script files that generate the data. These script files are available to allow students hands-on practice in re-creating the examples and practicing variations of SQL commands to enhance their understanding.

The first thing new database administrators often do is create a database. So, the initial focus of this textbook is on creating a running database. In Chapters 1 and 2, the database software, its components, memory structure, and background processes are introduced. Chapter 3 discusses and uses the commands for creating a new database. Each subsequent chapter zooms in on one area that the DBA must administer, such as tables, indexes, users, and roles. Each chapter covers the clauses of the commands that affect the underlying structure of each component. In this way, the student becomes familiar with the many choices available when creating new database structures, and learns how to select the appropriate choice in each particular situation.

To reinforce the material presented, each chapter includes a chapter summary. In addition, at the end of each chapter, groups of activities are presented that test the student's knowledge, challenging the student to apply what has been learned by solving business problems.

Overview of This Book

The examples, projects, and cases in this book will help students to achieve the following objectives:

- Understand key components of the Oracle 10*g* database architecture and installation options.
- Use DBA tools, specifically the Enterprise Management console and the Database Control, and understand the background processes that perform database operations.
- Create a new database instance using both manual SQL commands and database tools.
- Use data dictionary views to monitor database structures and activities; use SQL to manage control files.
- Understand the concepts and use SQL to create, modify, or remove redo log files, control files, and database-generated diagnostic files.
- Understand the physical and logical structures that make up the database; create, modify, and drop database tablespaces; manage undo data.
- Use SQL to create, modify, and drop many types of tables and indexes using additional clauses that control storage settings, LOB storage, and partitioning.
- Create users with SQL including clauses for storage quotas; modify and drop users.
- Understand and manage system privileges, as well as object privileges, profiles, and roles.

The contents of each chapter build on the concepts of previous chapters. **Chapter 1** describes the key components of the Oracle 10*g* database architecture and installation options. **Chapter 2** introduces DBA tools, including the Enterprise Management console, the Database Control, and describes the background processes that perform database operations. **Chapter 3** shows how to create a database using the Database Configuration Assistant tool, and describes the Oracle instance. **Chapter 4** describes the physical structure of the Oracle database, including datafiles, tablespaces, controlfiles, redo logs, and archive logs. There is also a brief introduction to using the metadata data dictionary views. **Chapter 5** describes physical storage settings for datafiles and tablespaces. **Chapter 6** introduces the very basics of querying an Oracle database. **Chapter 7** describes basic table management, including table structures and how to create several types of tables. **Chapter 8** shows how to create several more complex types of tables and describes how to modify or drop tables. **Chapter 9** examines constraints, indexes, and several specialized database objects, such as sequences, views, and synonyms. **Chapter 10** details basic data changes using Data Manipulation Language (DML) statements. **Chapter 11** looks into advanced data management including coverage of PL/SQL, plus import, export, and SQL Loader utilities. **Chapter 12** examines database security features, including users, privileges, profiles, auditing, and roles. **Chapter 13** gives a brief summary of Oracle 10g performance monitoring. **Chapter 14** examines proactive maintenance, using the Database Control to predict and react automatically, reducing reaction time to resolving potential problems. **Chapter 15** examines backup and recovery strategies, tools, and procedures.

FEATURES

To enhance the learning experience for students, each chapter includes the following elements:

- **Chapter Objectives:** Each chapter begins with a list of the concepts to be mastered by the chapter's conclusion. This list provides a quick overview of chapter contents as well as a useful study aid.
- **Methodology:** As new commands are presented in each chapter, the syntax of the command is presented and then an example illustrates the command in the context of a business operation. This methodology shows the student not only *how* the command is used but also *when* and *why* it is used. The step-by-step instructions in each chapter enable the student to work through the examples in this textbook, engendering a hands-on environment in which the student reinforces his or her knowledge of chapter material.
    - **Tip:** This feature, designated by the *Tip* icon, provides students with practical advice. In some instances, tips explain how a concept applies in the workplace.
    - **Note:** These explanations, designated by the *Note* icon, provide further information about concepts or syntax structures.
- **Chapter Summaries:** Each chapter's text is followed by a summary of chapter concepts, as a helpful recap of chapter contents.
- **Review Questions:** End-of-chapter assessment begins with review questions that reinforce the main ideas introduced in each chapter. These questions ensure that students have mastered the concepts and understand the information presented.
- **Exam Review Questions:** Certification-type questions are included to prepare students for the type of questions that can be expected on the certification exam, as well as to measure the level of understanding.

- **Hands-on Assignments:** Along with conceptual explanations and examples, each chapter provides hands-on assignments related to the chapter's contents. The purpose of these assignments is to provide students with practical experience. In some cases, the assignments are based on tables or other structures created using a script provided with the book that the student runs just before working the assignments.
- **Case Project:** One or two cases are presented at the end of each chapter. All the case projects are based on a fictional business called Global Globe News Company and ask students to create the database structures discussed in the chapter as part of the growing database system for the company. These cases are designed to help students apply what they have learned to real-world situations. The cases give students the opportunity to independently synthesize and evaluate information, examine potential solutions, and make recommendations, much as students will do in an actual business situation.

The Course Technology Kit for Oracle 10*g* software, available when purchased as a bundle with this book, provides the Oracle database software on CDs, so users can install on their own computers all the software needed to complete the chapter examples, hands-on assignments, and cases. The software included in the kit can be used with Microsoft Windows Server or XP operating systems. The installation instructions for Oracle 10*g* and the login procedures are available at *www.course.com/cdkit* on the Web page for this book's title.

Teaching Tools

The following supplemental materials are available when this book is used in a classroom setting. All teaching tools available with this book are provided to the instructor on a single CD-ROM:

- **Electronic Instructor's Manual:** The Instructor's Manual that accompanies this textbook includes the following elements:
  - Additional instructional material to assist in class preparation, including suggestions for lecture topics.
  - Answers to end-of-chapter review questions, exam review questions, hands-on assignments, and case projects (when applicable).
- **ExamView®:** This objective-based test generator lets the instructor create paper, LAN, or Web-based tests from test banks designed specifically for this Course Technology text. Instructors can use the QuickTest Wizard to create tests in less than five minutes by taking advantage of Course Technology's question banks—or create customized exams.
- **PowerPoint Presentations:** Microsoft PowerPoint slides are included for each chapter. Instructors might use the slides in three ways: as teaching aids during classroom presentations, as printed handouts for classroom distribution, or as network-accessible resources for chapter review. Instructors can add their own slides for additional topics introduced to the class.

- **Data Files:** The script file necessary to insert data into the Global Globe database tables is provided through the Course Technology Web site at *www.course.com*, and is also available on the Teaching Tools CD-ROM. Additional script files needed for use in specific chapters are also available through the Web site.
- **Solution Files:** Solutions to the end-of-chapter material are provided on the Teaching Tools CD-ROM. Solutions may also be found on the Course Technology Web site at *www.course.com*. The solutions are password protected.

READ THIS BEFORE YOU BEGIN

To the User

Data Files

Much of the practice you do in the chapters of this book involves creating, modifying, and then dropping a database structure, such as a table, index, or a user. Most of the practices in the chapters and the hands-on exercises at the end of the chapters can be done without running any data files. At certain points in the book, however, you will need to load data files created for this book. Your instructor will provide you with those data files, or you can obtain them electronically from the Course Technology Web site by accessing *www.course.com* and then searching for this book's title. When you reach a point in the book where a data file is needed, the book gives you instructions on how to run each data file. The data files provide you with the same tables and data shown in the chapter examples, so you can have hands-on practice re-creating the practice commands. It is highly recommended that you work through all the examples to reinforce your learning.

Script files for all chapters are found in the **Data** folder under their respective chapter folders (for example **Chapter09** and **Chapter10**) on your data disk and have the filenames that correspond with the instructions in the chapter. If the computer in your school lab—or your own computer—has Oracle 10g database software installed, you can work through the chapter examples and complete the hands-on assignments and case projects. At a minimum, you will need the Oracle 10g Release 2 of the software to complete the examples and assignments in this textbook.

Using Your Own Computer

To use your own computer in working through the chapter examples and completing the hands-on assignments and case projects, you will need the following:

- **Hardware:** A computer capable of using a Microsoft Windows Server operating system (2000 or 2003), or Windows XP. You should have over 256MB of RAM and 5Gb of hard disk space available before installing the software.
- **Software:** Oracle 10g Release 2 Enterprise Edition is best although the Standard or Personal editions will suffice. The Course Technology Kit for Oracle 10g Software contains the database software necessary to perform all the tasks

shown in this textbook. Detailed installation, configuration, and logon information are provided at *www.course.com/cdkit* on the Web page for this title.

- **Data files:** You will not be able to use your own computer to work through the chapter examples and complete the projects in this book unless you have the data files. You can get the data files from your instructor, or you can obtain the data files electronically by accessing the Course Technology Web site at *www.course.com* and then searching for this book's title.

When you download the data files, they should be stored in a directory separate from any other files on your hard drive or diskette. You will need to remember the path or folder containing the files because you will have to locate the file while in SQL*Plus Worksheet in order to execute it. (The SQL*Plus Worksheet is one of the interface tools you can use to interact with the database.)

When you install the Oracle 10*g* software, you will be prompted to supply the database name for the default database being created. Use the name ORACLASS to match the name used in the book. If you prefer a different name, remember that anywhere the book instructs you to type in ORACLASS, you should type in your database name instead. You will be prompted to change the password for the SYS and SYSTEM user accounts. Make certain that you record the names and passwords of the accounts because you will need to log in to the database with one or both of these administrative accounts in some chapters. After you install Oracle 10*g*, you will be required to enter a username and password to access the software. One default username created during the installation process is SCOTT. The default password for the user name is "tiger." As previously mentioned, full instructions for installing and logging in to Oracle 9*i*, Release 2, are provided on the Web site for this textbook at *www.course.com*.

Visit Our World Wide Web Site

Additional materials designed especially for this book might be available at *www.course.com*. Visit this site periodically for more details.

TO THE INSTRUCTOR

To complete the chapters in this book, your students must have access to a set of data files. These files are included in the Instructor's Resource Kit. They may also be obtained electronically by accessing the Course Technology Web site at *www.course.com* and then searching for this book's title.

The set of data files consists of script files that are executed either at the beginning of the chapter or before starting the hands-on exercises. After the files are copied, you should instruct your students in how to copy the files to their own computers or workstations. Maintain the directory structure found in the original data files: Data\Chapter01, Data\Chapter02, and so on.

You will need to provide your students with this information, which is used in many of the chapters:

- The passwords for the SYSTEM and SYS users on their workstation.
- The database name, which is assumed to be ORACLASS. If the database name is not ORACLASS, inform the students that they should substitute the correct name whenever the text tells them to enter ORACLASS.
- The full path for the ORACLE_BASE and ORACLE_HOME directories on their workstations. ORACLE_BASE is a variable name used in Oracle documentation and in this book to refer to the root directory of the Oracle database installation. ORACLE_HOME is the root directory of the Oracle software.
- The computer name of the workstation.
- The full path names for directories where the students can create additional directories and files. The directories must be on the hard drive (not removable media, such as floppy disks). Around 1.5Gb of storage space is required for each student.

The chapters and projects in this book were tested using Microsoft Windows 2000 operating system with Oracle 10*g* Release 2 Enterprise Edition.

Course Technology Data Files

You are granted a license to copy the data files to any computer or computer network used by individuals who have purchased this book.

CHAPTER 1

ORACLE ARCHITECTURE OVERVIEW

LEARNING OBJECTIVES

In this chapter, you will:

- Learn about Oracle 10*g* architecture and key Oracle 10*g* software components
- Look at the ORACLASS database used in exercises throughout the book
- Discover differences between Oracle 10*g* client and server installation options
- Learn how to use the Oracle Universal Installer
- Examine why to use OFA (Optimal Flexible Architecture)

INTRODUCTION

Oracle Corporation offers many products in addition to its Relational Database Management System (RDBMS) software. The core product, the Oracle 10*g* database, contains a suite of software that includes required components, work-saving tools, utilities, and management software that is primarily used for database administration. The Oracle 10*g* database has extensive enhancement with regard to scalability and manageability.

Oracle also offers related software suites, such as: Oracle 10*g* Application Server (a Web server with special plug-ins for the Oracle database); Oracle Financials (a set of software designed for bookkeeping, accounting, inventory, and sales); and Oracle JDeveloper (an application builder that writes Java code using a Windows-like interface). This book covers only the core Oracle 10*g* RDBMS software suite, which includes the key components described in this chapter.

Your journey into the world of Oracle 10*g* begins as you become familiar with the basic components of the Oracle 10*g* RDBMS software suite. This chapter introduces you to the components that make up that suite, and to the Oracle Universal Installer, which is used to install any of Oracle's products. The chapter closes with a discussion of the architectural strategy that Oracle uses to maximize flexibility when implementing its database and related software products.

INTRODUCTION TO ORACLE 10*g* ARCHITECTURE

The Oracle 10*g* RDBMS software suite (referred to as *Oracle 10g* throughout the book) includes everything you need to build and maintain a relational database. The basic software runs the database engine, manages the data storage for all information in the database, and provides tools to manage users, tables, data integrity, backups, and basic data entry. The basic software includes additional tools and utilities that help you monitor the performance and security of the database.

This section analyzes the components of Oracle 10*g* and describes the function of each component. Understanding how the components fit together helps you prepare for software installation. After installation, you will know which components to use for various tasks, such as creating a new user or adjusting storage space.

NOTE

The Oracle 10*g* RDBMS software suite can be enhanced with many extra software components that Oracle sells for an additional cost. This book covers only the components included when you buy the database software. Typically, you purchase additional software to write applications that access the database. Many other vendors offer software that adds on to an Oracle 10*g* database. For example, the company called ESRI offers a complete graphic information system (GIS) software solution for computer-generated maps that stores all its data in your Oracle 10*g* database.

Key Components of the Oracle 10*g*

Before installing Oracle 10*g*, the Installation window lists all the components it plans to install. There are various components listed for the basic installation alone. Fortunately, most of these components are modules that support a smaller set of major components. Figure 1-1 illustrates a conceptual view of the Oracle 10*g* database engine and the core utilities that comprise Oracle 10*g*. These key components make up Oracle 10*g* running on any operating system or hardware. Oracle has standardized its software so that it looks and acts identically at the user interface level on all platforms. Oracle customized its software release on each platform at the lowest levels to optimize performance based on each platform's characteristics. An example is the process that takes data from Oracle 10*g* memory buffers and writes it to a hard disk.

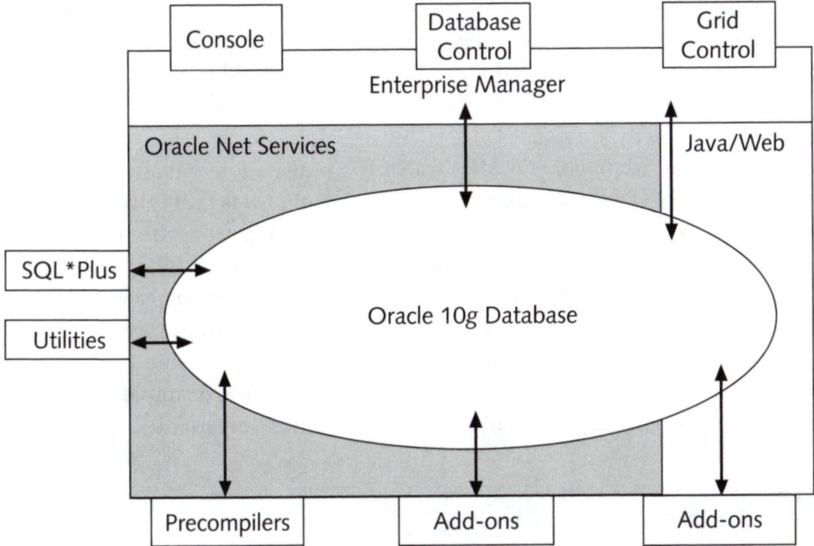

FIGURE 1-1 Oracle 10*g* RDBMS software suite includes several core utilities and tools

Whether you download the software or order CDs (the latest version is on DVD), your software includes the core database suite and several "try and buy" packages. The documentation explains which portions are standard and which require the additional purchase of a license. The following list corresponds to the features shown in Figure 1-1 and briefly describes each of the core utilities and tools. All of these features are part of the Oracle 10*g* RDBMS software suite.

- **Oracle 10*g* database:** If you want, a pre-configured database can be created during installation. When installing an upgrade, this is not needed, but new installations require a database, and using a pre-configured database is convenient. The database itself contains a number of preset features that speed up its use. For example, there are several predefined, high-level users that aid you in creating other users and in monitoring your database statistics. The PL/SQL engine included with the Oracle software provides a specialized programming language you can use for creating triggers, packages, and procedures. The Java Virtual Machine (JVM) engine allows you to store, parse, and execute Java applets, servlets, and stored procedures within the database.
- **Oracle Net Services:** This tool provides the network link between the Oracle 10*g* database and most applications that communicate with the database. Oracle Net Services must be configured to define ports and the protocols allowed to send and receive data for each database in your network. If you are running Enterprise Manager, SQL*Plus, Forms, or non-Oracle products, Oracle Net Services is the gateway into the database.
- **Java/Web:** The other way to access the Oracle 10*g* database is through Java ports. There are a number of different programs and tools installed with your Oracle 10*g* database to support Web and database integration. These tools

include a fully functioning Apache Web server containing the following: modules that hook into the database; database packages to support XML input and output; a pre-compiler for SQLJ (Java code with embedded SQL commands); and a gateway for PL/SQL to allow HTML documents to be delivered directly from the database to the user.

- **Oracle Enterprise Manager (OEM):** Oracle 10*g* builds more enhancements into this Windows-like Database Administrator tool, which was first introduced with Oracle7. The OEM interface is controlled in Oracle 10*g* from the previously seen console, or as a web interface, using the Database Control or Grid Control features. The tool integrates many utilities and monitoring tools into a single interface. Multiple databases across local and remote networks can be mapped and managed within OEM. Figure 1-2 shows the console with its navigation tree displayed. You also get access to wizards (for example, backup, recovery, and SQL analysis wizards) and management packs.

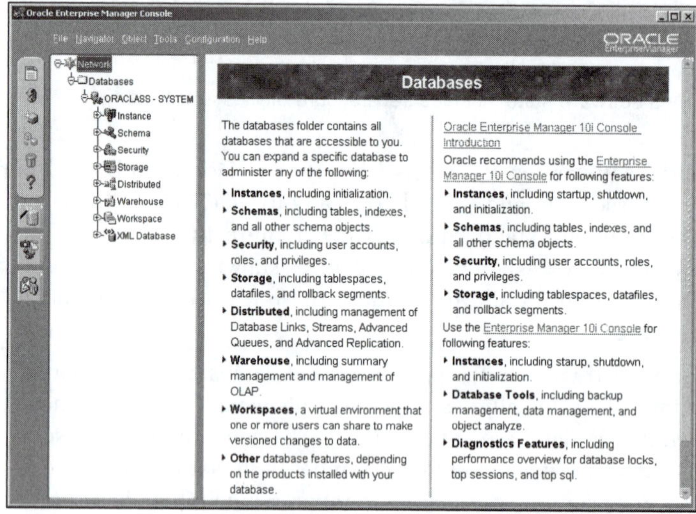

FIGURE 1-2 The Enterprise Manager console provides centralized access to tools

- **OEM wizards:** These are sets of related tools, wizards, and assistants that are added as a group to the Enterprise Manager console. The Standard Pack that comes with your Oracle 10*g* installation includes management tools for creating users, creating tables and objects, and administering storage space. The Database Control and Grid Control provide more sophisticated Enterprise Manager Web-based access in Oracle 10*g*.
- **SQL*Plus:** This tool allows you to create and run queries, add rows, modify data, and write reports using SQL. SQL*Plus accepts standard SQL, with all

the Oracle enhancements to SQL, such as the TO_DATE function. In addition, SQL*Plus provides commands for customizing output, defining variables, and executing files. iSQL*Plus is a Web-based form of SQL*Plus. Web-based implies it runs in a browser like Microsoft Internet Explorer. SQL*Plus Worksheet is a form of SQL*Plus contained within OEM.

- **Utilities:** There is a group of utilities for backup, migration, recovery, and transporting data from one database to another. For example, the Recovery Manager automates database recovery after a failure. Data Pump Export and Import utilities improve on the previous Oracle version EXP (export), and IMP (import) utilities providing a command-line interface to extract tables, schemas, or entire databases into files that can later be reloaded into the same database or into another database. There are additional tools for migration to a different Oracle release, for loading new data from a flat file into a table, and for transferring data from one character set to another. A character set provides special symbols for various languages.
- **Pre-compilers:** Pre-compilers (such as Pro*COBOL and Pro*C) support embedded SQL commands within programs in C, C++, or COBOL. The pre-compiler translates the SQL command into the appropriate set of commands for the program, which is then compiled and ready for executing. Pre-compilers save a great deal of programming time.
- **Add-ons:** Oracle offers many additional components that can enhance the database system. Some of these access the database through Oracle Net Services, whereas others access the database through the Java/Web connectors. A section later in this chapter describes some of the add-ons available.

As you begin working with each of these main components, the supporting modules and the connections between all the components become clearer.

Running the Database

When you install an Oracle 10g database, you install the software components, create database files to store your data, and then start a set of background processes that allocate memory and handle database activities. Figure 1-3 shows a typical installation of the Oracle 10g database.

Oracle defines a database as the collection of operating system files that store your data. The database has three types of files: control files, database files, and redo log files. These file types are described later in this chapter.

The **database software components**, such as Oracle Net Services, Enterprise Manager, and the RDBMS components are installed on the computer. To use the database, you start up a database instance, which allocates **memory** for the database (called the System Global Area, or SGA) and starts up a set of background processes. Background processes are the programs that run on the computer, while an instance of the database is in memory (the database is up and running). The background processes handle user interaction with the database and manage memory, integrity, and I/O for database files. Chapter 2 describes the makeup of the files, memory structures, and background processes of a database instance.

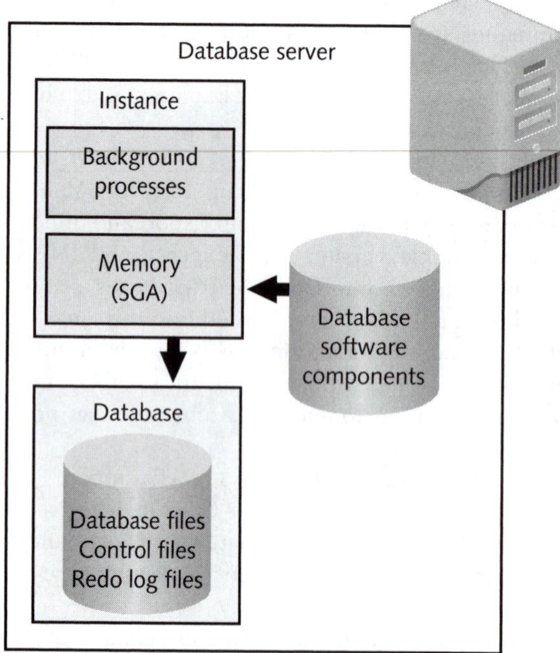

FIGURE 1-3 A computer containing one instance is a single-instance server

The combination of database software, a database (the files), and a database instance (the SGA and the background processes) is called a database server. Figures 1-3, 1-4, and 1-5 show three different configurations used to implement a database server.

- **Single-instance server:** This is the typical installation and contains one computer with one set of database files and one instance that accesses the files. Figure 1-3 shows this configuration.
- **Multiple-instance server:** Another installation possibility for one computer has two instances, each with its own database files. Figure 1-4 shows this configuration. The two instances run independently of one another. This is useful when you want a dedicated instances for each of two development teams, for example, and you have one powerful computer to support both teams.

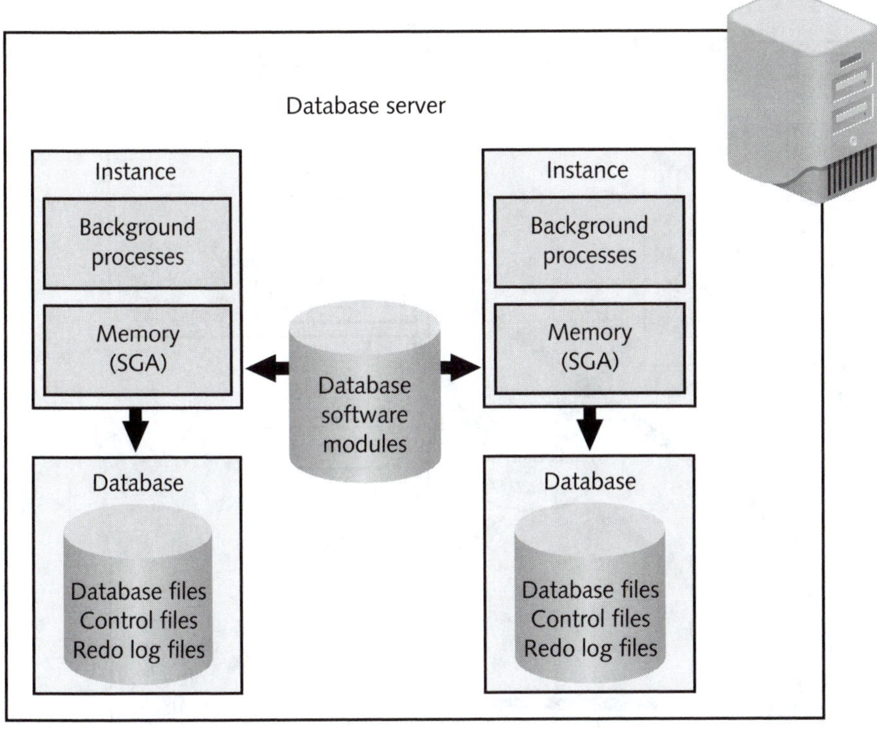

FIGURE 1-4 A multi-instance server hosts two or more instances together on the same computer

- **Clustered servers:** This is known in Oracle 10g as Oracle Clusterware, Oracle Cluster Ready Services, or Oracle Real Application Clusters (RAC). RAC encompasses clustered database instances and grid computing. Figure 1-5 shows a clustered configuration of several computers (database nodes), each one containing a database instance. Another computer (file server), or a specialized storage device, houses the database files, which are shared by all the instances. The cluster manager is a software application that coordinates all the instances and the tasks that each one handles. The cluster manager can reside on a separate computer or on one of the nodes. Clustered servers are complex to manage, but supply increased computing power for applications that require speed for processing massive amounts of data, or complex calculations, such as astronomy or physics applications. One task can be divided into smaller pieces and divided among the instances to complete the task more quickly. Oracle Real Application Clusters is an add-on component that can manage clusters of Oracle 10g instances.

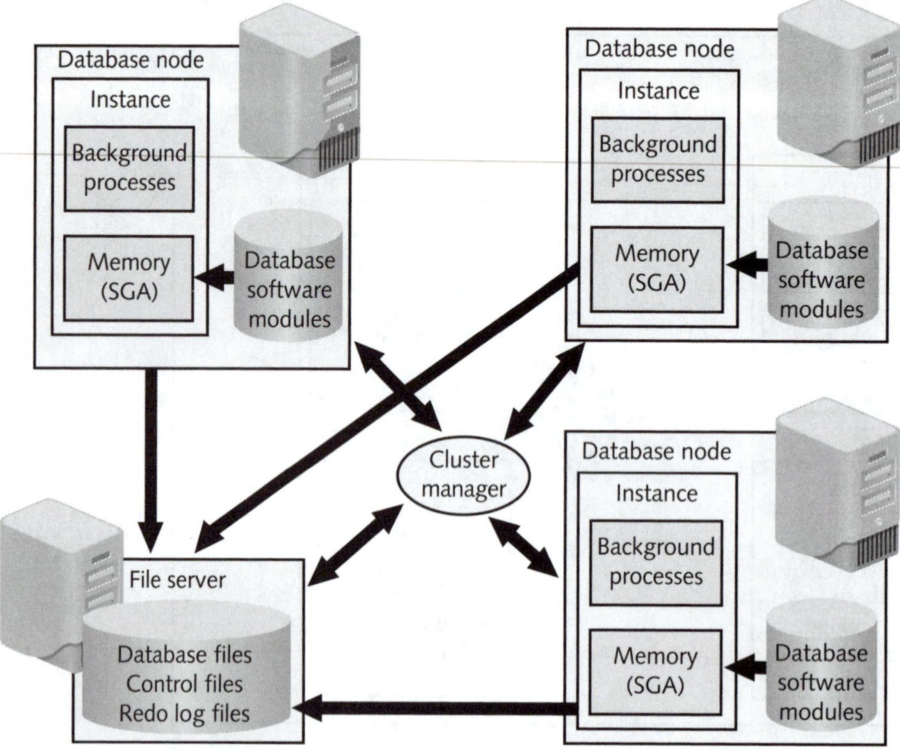

FIGURE 1-5 Many computers can act as one when connected in a cluster

On the single-instance server (the typical configuration for an Oracle 10*g* database), how does the user interact with the database? When a user runs an application that uses the database, the application creates a user process that controls the connection to the database process. On the database side, the process that interacts with the user process is called a server process. There are two basic methods used to connect the user process and the server process:

- **Dedicated server:** This method connects one user process with one server process. This is better for systems with fewer users or with user processes that require a great deal of processing time on the database.
- **Shared Servers:** A minimal number of database server processes are shared with multiple connections, between an application and the database. Shared Servers allows the database to service a larger number of users at once, the same as a web server or application server would do.

If you install Oracle 10*g* Enterprise Edition, Standard Edition, or Personal Edition and choose to have a predefined database created on installation, the database is configured in dedicated server mode.

Some Optional Additions to Oracle 10*g*

Oracle 10*g* can be enhanced by purchasing additional features from Oracle Corporation. Many of these features are included in the base software on a trial basis, and it is sometimes unclear which portions you have purchased and which require an additional license fee. The following list summarizes the optional features that require additional license fees.

TIP

Oracle allows the use of optional features in your test environment without restriction. However, if you begin to use these features in your production environment, you need to purchase the appropriate license.

- **Oracle Partitioning:** Builds the capability to partition tables across multiple tablespaces. High-volume tables, such as historical records in data warehouses, benefit from partitioning by speeding up data retrieval for queries. Data warehouses are storage spaces of data that are used to support management decisions. Online Transaction Processing (OLTP) databases can also benefit substantially from the use of partitioning.
- **Oracle Clusterware:** Provides management tools to support database clusters and computer grids. Database grids can combine multiple databases sharing a central memory area, so that a long-running task can be spread among several databases, speeding up processing time.
- **Oracle Spatial:** This option adds programmed packages to the database to handle spatial objects. Spatial objects store data related to time and space in a way that allows you to calculate distance or time between objects and to draw lines, polygons, points on a map, and similar functions.
- **Oracle Data Mining:** Supports setting up algorithms and functions that search and retrieve data warehouse information. The Java interface for the tool is based on emerging standards for Java Data Mining (JDM).
- **Oracle COM Automation Feature:** PL/SQL developers can manipulate COM components.

TIP

A COM component is typically a Microsoft Windows acronym for a self-contained, executable, compiled binary chunk of code. Typically COM components are Dynamic Link Libraries (DLLs), ActiveX Controls (OCXs), and other executable files (EXEs).

- **Oracle Database Extensions for .NET:** These extensions allow execution of .NET stored procedures against a database.
- **Oracle Advanced Security:** Adds a security layer to Oracle Net Services and

the database. This package encrypts outgoing data before it goes onto the network or across the Internet. It also supports special methods of user authentication such as the programs used with automated tellers, with which you must have a valid bank card in your possession and know the personal identification number (PIN) to access your account data.

- **Oracle Label Security:** Is useful when high security standards must be met. By labeling individual rows with a security profile and matching that with a user's security profile, the database restricts access to rows within a table or view.
- **Oracle Online Analytical Processing (OLAP) Services:** Allows you to combine standard file directory support with database delivery of tables or views. When you view a directory that is controlled by Oracle OLAP services, you see a list of files as you would in any directory. When you open one of these files, Oracle runs an extraction program and creates a file with the extracted data. The DBA predefines the format of the file contents.

Few database installations require all of these options. Typically, data warehouses might use partitioning and data mining to increase performance, whereas banks or accounting firms might use Oracle Advanced Security and OLAP services to control data accessibility.

INTRODUCTION TO ORACLASS DATABASE

This book has a database developed to use throughout all chapters. There are step-by-step instructions in many of the chapters that lead you through the creation of tables, users, indexes, and more. Hands-on assignments at the end of each chapter are intended to run on the database as well. The database is a standalone database instance running on your workstation. This means that you alone have control over the database. All the workstations start out with the same basic database and build on this foundation.

The main features of the ORACLASS database are the following:

- The database has Enterprise Edition software, unless your instructor informs you otherwise.
- The database has a user named CLASSMATE whose password is CLASSPASS. Most of the exercises and step-by-step instructions begin with you logging in as this user. Initially, the CLASSMATE user owns no tables or other objects in the database. You add these as you go along.
- In some chapters, you are asked to run a script to create tables or other objects needed for the chapter. Each chapter contains the necessary instructions for these scripts.
- The database also has the standard users, SYSTEM and SYS, which are installed to provide you with Database Administration rights on the database. The password for the SYSTEM and SYS users will be provided, by your instructor.

Every chapter has an ongoing case project that you develop from beginning to end. The project is the database for the Global Globe newspaper chain.

Global Globe is a national newspaper chain and you have been hired to design and build its database. You are starting from scratch, and the design elements are revealed as you work the case projects at the end of each chapter. You build tables, add indexes, constraints, users, and security roles as you learn about these elements. You monitor the database activity and make adjustments to suit the ever-changing requirements of the Global Globe users and administration.

The database has already been installed and is currently running on your workstation. The following section describes the software components and the database architecture that have been installed, as well as the other options available when installing a new database.

OVERVIEW OF ORACLE 10*g* INSTALLATION OPTIONS

When installing Oracle 10g, you have several high-level choices that determine the components installed on your computer. One important concept to understand before installing Oracle 10g is the multi-tier architecture that Oracle 10g uses. Multi-tier architecture became popular in the 1980s and has grown into an important and lasting design element of many computer systems, not just database systems. The term multi-tier, also referred to as n-tier, means that the data, processing, and user interfaces are divided into separate areas that are fairly independent of one another. Figure 1-6 shows a typical n-tier application system. The advantages of an n-tier system are primarily processing power, as well as isolation of changes to a specific tier. In a typical Oracle 10g application system, the data tier contains the Oracle 10g database server, the middle tier contains the application server, which may or may not be Oracle's application server, and the user tier includes a browser. The user tier may have a custom application instead of a browser, or a combination of browser with Java applet, for example.

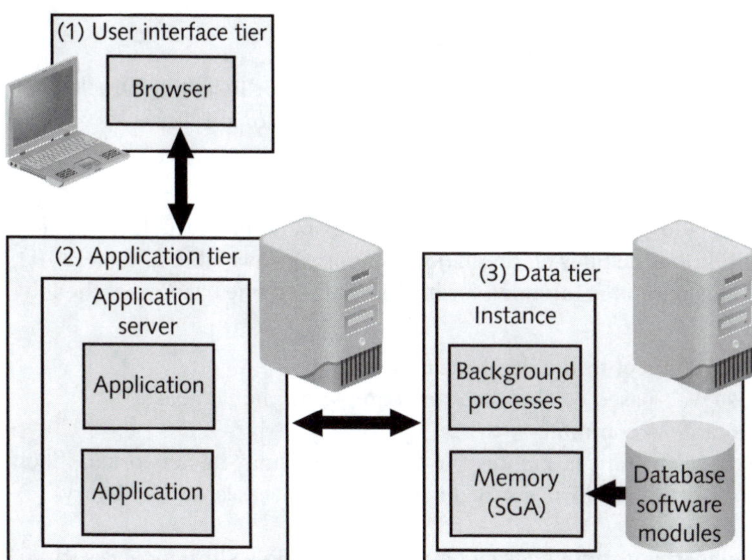

FIGURE 1-6 Three tiers compose a typical database system

This book works almost entirely on the data tier of this architecture. You dig into the processes, memory structures, and data structures that make up the database server. This means that the software that was installed for your use is a database server configuration. The server configuration includes components that contain the database software modules needed to run the database. A second path of installation, the client configuration, is available as well.

To create a database server, you need the server configuration. This configuration installs all the software modules required to run the database and communicate with any application that requires data from the database. In addition, the server configuration provides you with a predefined database (if you want to use it) ready for use.

The client configuration is a combination of the user interface tier and the application server tier. It is a specialized setup for end users running database applications that interact with the database on their own workstations. The client configuration gives the user the communication modules needed to reach the database directly (skipping the application tier). It also has options to install various other components, such as Enterprise Manager, which provides an option for the database administrator to reach the database from a remote computer.

Figure 1-7 shows the client and server paths and the resulting components installed. Looking at Figure 1-7, you can see the decisions you must make when installing Oracle 10g software. First, you choose between client and server installations. Within a client installation, you select either Administrator or Runtime configuration. Within a server installation, you choose between Personal, Standard, and Enterprise editions. In both client and server installations, you can opt for Custom installation, which allows you to select specific components individually.

Comparing Server-Side Installation Options

This section describes in greater detail each of the three options for installing Oracle 10g server components: Enterprise Edition, Standard Edition, and Personal Edition.

Enterprise Edition

When installing the server-side options, the most complete installation is the Enterprise Edition. The Enterprise Edition includes all the major components available in Oracle 10g, as shown in Figure 1-7. This edition is intended for businesses with one or more of the following requirements:

- Multiple users connect to the database concurrently.
- Applications (Web-based or client/server) connect to the database.
- High data volume is common.
- Multiple database instances support the business and may be networked. Each instance consists of its own set of background processes and memory allocation.
- Database replication for duplicating data across multiple sites is necessary.

The Enterprise Edition runs on many different platforms and multiple operating systems, including Unix, Windows, Linux, VMS, and others.

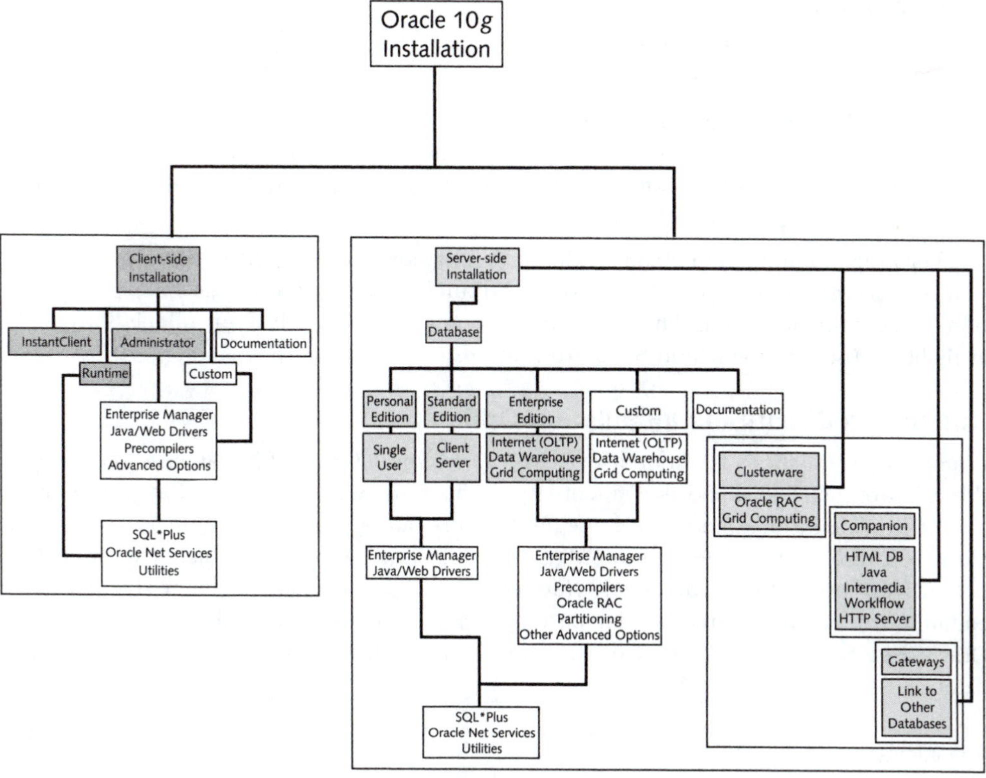

FIGURE 1-7 The components installed depend on which installation path you select

> **NOTE**
>
> To avoid installation errors when using Windows 2000, install Service Pack 3 before installing the Oracle software. When using the Windows operating system, the latest release of Oracle 10*g* is probably best installed on Windows 2003.

Your workstation should have either the Enterprise Edition or the Personal Edition installed.

Standard Edition

The Standard Edition provides basic support for multi-user database applications on a smaller scale than that of the Enterprise Edition. It is intended for use by a smaller number of users, such as those in a single department within a company.

Although replication, security, and Web access are available with the Standard Edition, less emphasis is placed on providing high-volume, high-availability capabilities. The Standard Edition cannot be upgraded with database features, such as partitioning and clustering.

Personal Edition

The Personal Edition allows access by a single user to the database instance. Two primary uses for the Personal Edition are:

- **Programming:** The programmer develops applications that run on the Personal Edition database and then are moved to an Enterprise Edition instance.
- **Deployment:** A company develops and distributes an application/database package for use by single users.

For programming, the Personal Edition is fully compatible with the Enterprise Edition, so software developed using Personal Edition can be ported to the Enterprise Edition without modification. The Personal Edition can be installed on all platforms available to the Enterprise and Standard Editions.

Comparing Client-Side Installation Options

Client-side installations facilitate user access to a remote Oracle 10g database. Therefore, the software installed supports connectivity to this remote database. The four variations on a client-side installation include one for Administrators, one for Runtime use, one for all client software, and a custom option. Figure 1-7 lists the components installed for each option. For both options, the Oracle Net Services component on the client side handles communication with the remote database by sending requests across a network to the Oracle Net Services component on the server side, which passes requests on to the database.

Administrator

The Administrator option provides the user management tools, including the Enterprise Manager, to provide remote management of multiple databases. This option includes many of the same features as the Enterprise Edition option on the server side, except there is no database installed.

Runtime Option

The Runtime option is intended primarily for programmers who are developing applications on their own client machines while using a remote database as the connection to the database. This is often easier to manage than providing a database for each workstation, because the database changes need not be replicated for each programmer. The Runtime option installs basic tools and connectivity features such as Oracle Net Services.

TIP

Both the Administrator and Runtime options support all the operating systems platforms in which you can install Enterprise Edition.

This book primarily covers the Enterprise Edition of the server-side installation, because it is the most complete edition of the database. All the exercises have been tested using Enterprise Edition. There is no reason for errors in the Personal Edition, except for advanced features, such as partitioning.

THE ORACLE UNIVERSAL INSTALLER

Regardless of which installation option you select, Oracle Universal Installer handles the job. The Oracle Universal Installer provides a common user interface on all platforms when installing any Oracle product. Figure 1-8 displays one of the main windows in the Universal Installer. This shows how you use a navigation tree to select and deselect installation options.

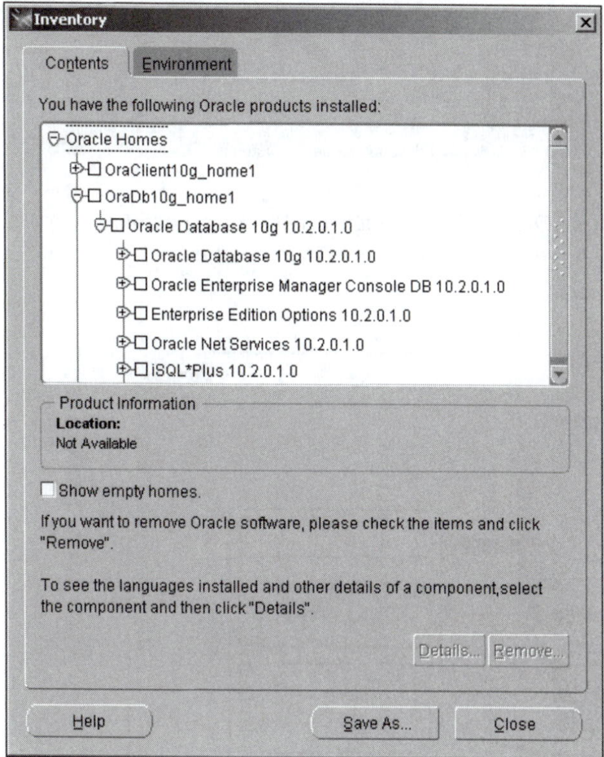

FIGURE 1-8 The Oracle Universal Installer shows options in a navigation tree format

The Oracle Universal Installer keeps log records on your computer of previous installation activity. If you install upgrades or enhancements later, the Oracle Universal Installer skips redundant subcomponents. This saves time during subsequent installations.

The Oracle Universal Installer also enables you to view the Oracle components and subcomponents currently installed and to uninstall any component. Subcomponents are often used in more than one major component, so the Oracle Universal Installer does not allow you to accidentally remove a subcomponent that is still required by installed components.

Oracle Home

A unique feature of the Oracle Universal Installer is its ability to install several versions of Oracle software on a single machine by setting up a separate directory structure for each version and its software components. Oracle executable files are stored in a directory tree referred to as **ORACLE_HOME**. The Oracle Universal Installer defines this directory in the Unix environment variables, or in Windows Registry entries. A single machine can contain more than one value for ORACLE_HOME.

NOTE

On Windows platforms, you assign a name to each Oracle Home directory and use a feature called Home Selector to change the current target. On other platforms, the target is by default the last Oracle Home directory in which the Universal Installer was active.

Here is an example of using multiple Oracle Homes. Figure 1-9 shows a database server with two versions of the database software, and one database instance for each version.

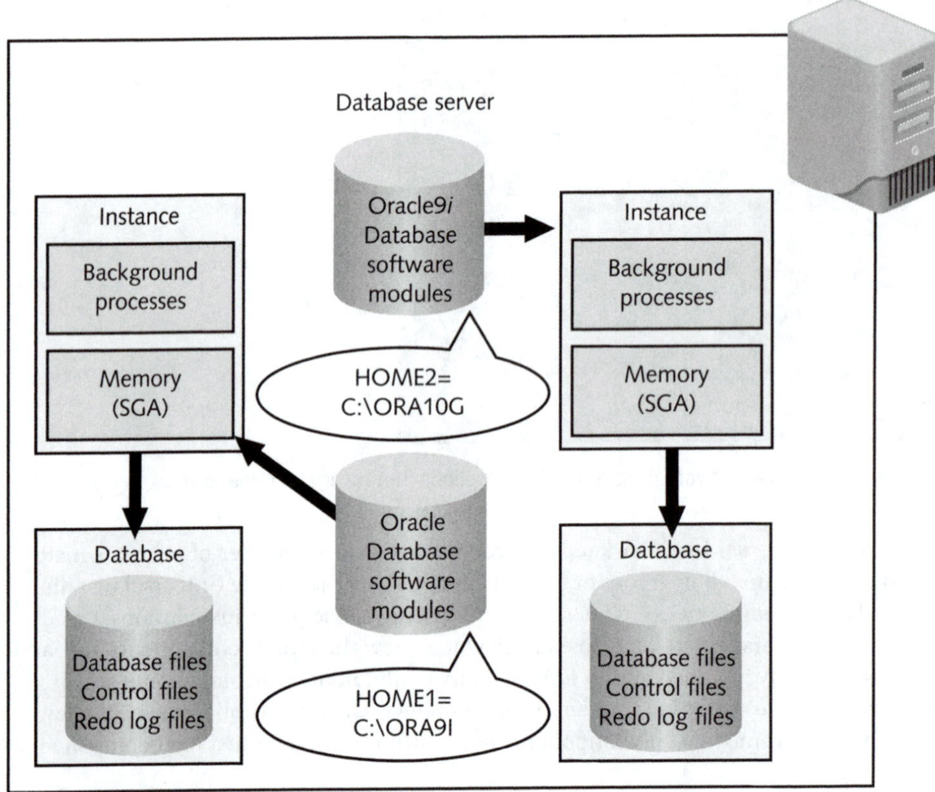

FIGURE 1-9 A database server can handle multiple versions of Oracle software

Looking at Figure 1-9, imagine you are the database administrator (DBA) at an e-commerce Web company. You have Oracle9*i* loaded on your Windows computer. You need that version to support troubleshooting and minor upgrades to your company's production system, which uses Oracle9*i* on a Unix server. Your Oracle9*i* software resides in a directory tree starting at the **C:\ORA9I** directory, which you named **Home1**, when you defined the Oracle Home during installation. Now, you want to install an evaluation copy of Oracle 10*g* without disturbing the Oracle9*i* installation. Early in the installation, you define a new Oracle Home, called **Home2**, with a root directory of **C:\ORA10G**.

You have both database instances up and running on your Windows computer. You want to run SQL*Plus on the Windows computer. How do you determine which instance you reach with SQL*Plus? There are two **sqlplus.exe** executable files in the Windows computer system: one for each Oracle version. You must define a target Oracle Home, which directs any call to Oracle software to only one of the Oracle Home directories. The PATH variable contains both Oracle Home directories. Whichever comes first takes precedence. By running the Oracle Universal Installer feature called the Home Selector, you can easily change the order of the Oracle Home directories in the PATH variable. Let's say you run the Home Selector and select Home2. Now, your system uses the Oracle 10*g* software. Switch to Home1, and your system now uses the Oracle9*i* software.

Silent Install

The Oracle Universal Installer provides a SILENT MODE of installation that enables you to run an installation without any human intervention. The silent install helps when you are making identical installations on multiple machines. With this process, you predefine responses to the questions asked during installation by recording the answers in a response file. Presetting the responses guarantees identical installations. In addition, predefined responses allow you to run several installations at once.

Before installation, start with the response file template provided, and modify it to fit your specifications. Then, instead of using Autostart to run the installation from the CD, use a command line and type the command. On Windows, enter the following command in the directory in which Universal Installer resides:

```
setup.exe -responseFile <filename> -silent
```

Replace <filename> with the actual response filename.

On Unix, enter this command:

```
runInstaller -responseFile <filename> -silent
```

TIP

The Oracle Universal Installer can also be used to install applications that you write by preparing an installation script with the optional Oracle package called the Software Packager.

DESCRIPTION OF OPTIMAL FLEXIBLE ARCHITECTURE (OFA)

The Optimal Flexible Architecture (OFA) brings standard conventions to the directory structure and filenames for your Oracle software. OFA provides standards intended to improve performance of the database by doing the following:

- Spreading I/O functions across separate devices by separating data from software.
- Improving performance by separating products into distinct directories that can be located on separate devices to reduce bottlenecks.
- Speeding up administrative tasks, such as backups, by using naming standards for file types.
- Improving detection and prevention of fragmentation in datafiles by using naming standards that quickly identify which tablespace and datafile are associated with one another.

NOTE

A datafile contains Oracle data in the operating system. A tablespace allows grouping of datafiles within an Oracle database.

The OFA recommends a standard directory structure as well as specific naming conventions.

Directory Structure Standards

The recommended OFA standards for directory structure create order from a myriad of executable files, data storage files, and administration files. The standards allow the storing of multiple versions of the software on a single computer.

The structure begins with a root directory called ORACLE_BASE. All Oracle software files and database storage files reside on subdirectories under ORACLE_BASE. In Windows, the default ORACLE_BASE directory is:

```
C:\oracle\product\10.2.0
```

In Unix or Linux, the default ORACLE_BASE might be something like:

```
/app/oracle/product/10.2.0
```

ORACLE_BASE houses a set of directories, in which all the Oracle software, database data, and administrative files, as well as the ORACLE_HOME directory (or directories) are stored. The following directories are found under ORACLE_BASE:

- **Admin/<database name>:** Stores initialization files and high-level log files.
- **db_1:** Oracle database server installation, storing Oracle software, such as the database engine and Oracle Net Services software.
- **client_1:** Oracle database client installation.
- **oradata/<database name>:** Stores database datafiles, control files, and redo logs.

- **flash_recovery_area:** Oracle database backup and recovery files, including archive log files by default (not current database files). Current database files are those stored in Oradata/<database name>, and are datafiles, control files, and online redo logs.

NOTE

The <database name> is usually stored as the ORACLE_SID environment variable.

As you can see, the database files allow for division by function and by either database instance or software release number. Documentation is stored separately from the executable files. Executable files for different versions of Oracle, such as Oracle 10*g*, Release 2, are stored separately from executable files. File sets for both versions would be installed (by default) below a common directory in the ORACLE_BASE directory. Datafiles for each database instance are separated, although both sets are found below the *oradata* directory. All the previously listed directories contain sets of subdirectories in which actual files are stored. Subdirectories allow you to move sets of related files without losing the standardized directory structure.

The path to all the Oracle software is called ORACLE_HOME containing Oracle binaries. ORACLE_HOME is defined as a variable within your system, so that all Oracle software and subdirectories can be found using a relative address that is prefixed by the value in the ORACLE_HOME variable. For example, you might have a program that calls an executable **runnit.exe** that is found in the **bin** directory. Rather than coding the full path and filename, the program code contains the ORACLE_HOME variable like this: **ORACLE_HOME/ bin/runnit.exe**. The variable keeps the program flexible in case the Oracle software is moved or the directory is renamed. When you have more than one software release installed on a single computer, each has its own ORACLE_HOME. The default ORACLE_HOME of an Oracle 10*g*, Release 2, database server installation on Unix or Linux might be:

```
/app/oracle/product/10.2.0/db 1
```

On Windows, the default ORACLE_HOME for an Oracle 10*g*, Release 2, database server installation is:

```
C:\oracle\product\10.2.0\db 1
```

Similarly, on Windows for a client installation:

```
c:\oracle\product\10.2.0\client 1
```

NOTE

Don't confuse ORACLE_HOME and ORACLE_BASE. ORACLE_HOME contains an installation. ORACLE_BASE can contain one or more different installations, such as two database servers, or a client and a database server installation.

ORACLE_HOME contains a **bin** directory that holds most of the executables. Each subcomponent, such as SQL*Plus and Oracle Net Services, has its own subdirectories that contain additional files used by that subcomponent. For example, Oracle Net Services software has an executable in the **bin** directory and has additional files in the **network** directory. Optional add-on software from Oracle, such as InterMedia, adds its own directory within the ORACLE_HOME structure. Figure 1-10 shows a directory structure for Windows.

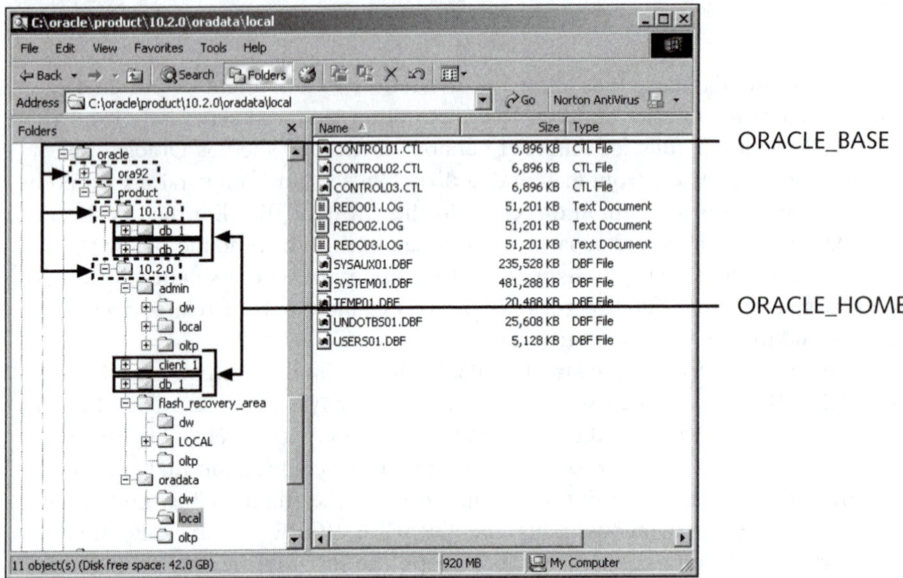

FIGURE 1-10 OFA structure handles complex installations

Looking at Figure 1-10, observe there are three OFA structured Oracle installations for three different releases of Oracle database. The releases are 9.2, 10.1 and 10.2. The 10.1 installation has two installations of Oracle binaries: **db_1** and **db_2**. The 10.2 installation has an Oracle server installation (**db_1**) and an Oracle client installation (**client_1**). Additionally the 10.2 installation contains three databases: DW, LOCAL and OLTP.

File Naming Standards

The OFA standards call for naming files according to their type and sequence of creation. Table 1-1 shows the naming standards recommended. The naming standards are only recommendations, not requirements. If you prefer to use a different naming standard, you can do so without any penalties. You can even rename the files that Oracle creates (for example, the files created when you allow the installation to create a predefined database for you). If you choose to follow the naming standards, when the database grows, you issue Oracle commands to create additional files, telling Oracle to create files that you name according to these conventions. This assures that the original files and all subsequent files follow the same naming standards.

All the files in Table 1-1 are located within the ORACLE_BASE/**Oradata** directory structure in the OFA configuration.

TABLE 1-1 OFA Naming Standards

File Type	Description	Suffix	Examples
Control-file	Administrative file containing up-to-date information on the database structure, log files, checkpoints; critical file for opening the database.	.ctl - The filename should be controlNN, where NN is a sequence number.	Control01.ctl, Control02.ctl
Redo log file	A set of files that record changes to database data. Online redo logs record changes immediately. Arhive redo logs contain copies of current and past redo logs for recovery if needed.	.log - The filename should simply be redoNN, where NN is a sequdnce number.	Redo01.log, Redo02.log
Datafile	A physical file on the computer disk containing data, such as tables and indexes, Each datafile belongs to one tablespace ina database. A tablespace may use nmore than one datafile.	.dbf - The datafile name should include the tablespace name and a sequence number.	System01.dbf, Users01.dbf, Users02.dbf, Temp01.dbf

The primary purpose of these naming standards is to speed up the process of identifying files. Quickly locating files helps during recovery processes and aids in routine tasks, such as backing up files, relocating files to alleviate bottlenecks, and adjusting file sizes.

INSTALLING ORACLE SOFTWARE

It is invaluable for students of Oracle software to actually see the installation process taking place, using the Oracle Universal Installer tool, as shown in the following sequence of steps:

TIP

You are not required to follows these steps as an exercise – JUST READ IT! There are multiple choices to select from, when executing the Oracle Universal Installer, both on the DVD (or CDRom) installation disks. There are also both server and client installations on your computer. All these choices have different screens when executing the Oracle Universal Installer.

1. This is an execution of the Oracle Universal Installer, from the database server menu, on a Windows computer. On a Windows computer start the Oracle Universal Installer by clicking **Start/All Programs/Oracle - OraDb10g_home*n*/Oracle Installation Products/Universal Installer**.

The name of the menu items may vary depending on your Oracle software release (**Oracle - OraDb10g_home*n***), your operating system, and what is already installed on the computer you are presently working on.

The welcome screen, as shown in Figure 1-11 appears first.

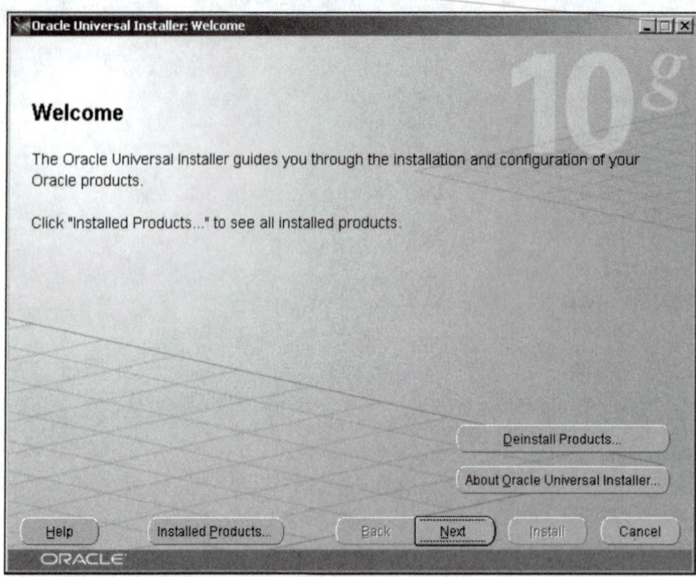

FIGURE 1-11 The Oracle Universal Installer Welcome screen

2. To begin the installation process, click the **Next** button. Figure 1-12 shows the next screen, asking you to specify a source location. The source location specifies where software is being installed from, typically a CD-ROM or DVD drive. In this case my DVD drive is drive F. Click the **Next** button to continue.

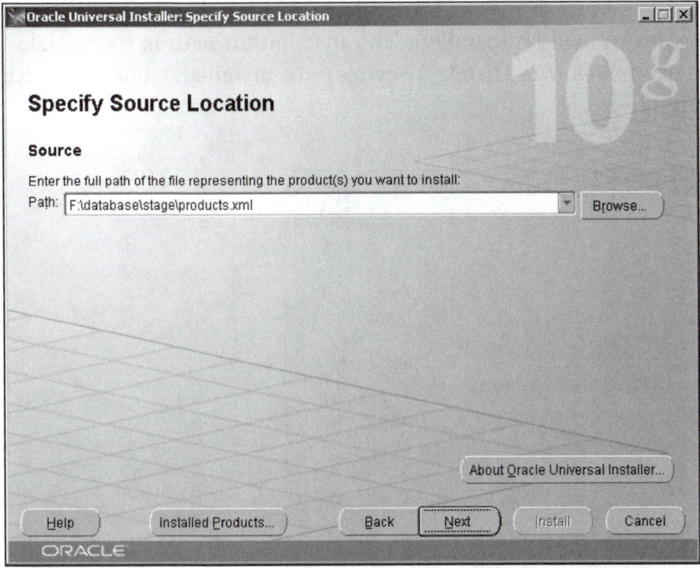

FIGURE 1-12 Specify a source location

3. The next screen as shown in Figure 1-13, gives you the option to install Enterprise edition, Standard edition, Personal edition, or a Custom edition you can make up yourself. Select the Custom option and click the **Next** button to continue.

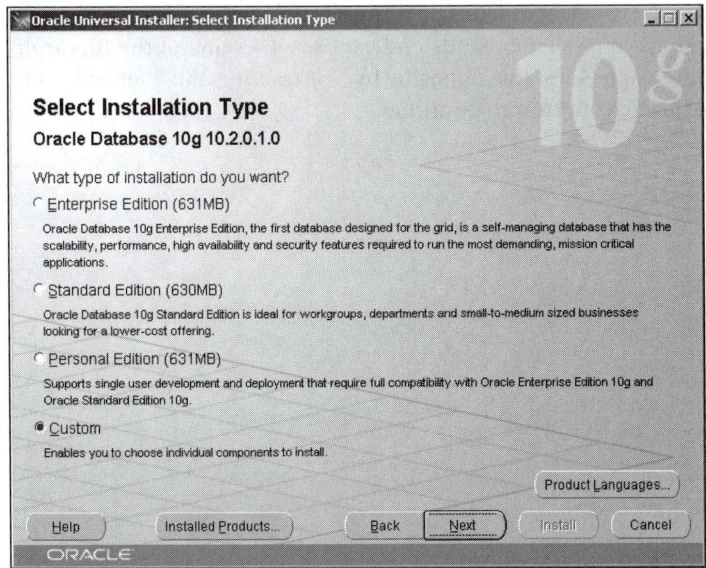

FIGURE 1-13 Select installation type

4. The next screen as shown in Figure 1-14, allows you to select the path into which Oracle software will be installed. The installation path is the Oracle home where executables (*the Oracle binaries*) are installed. Click the **Next** button to continue.

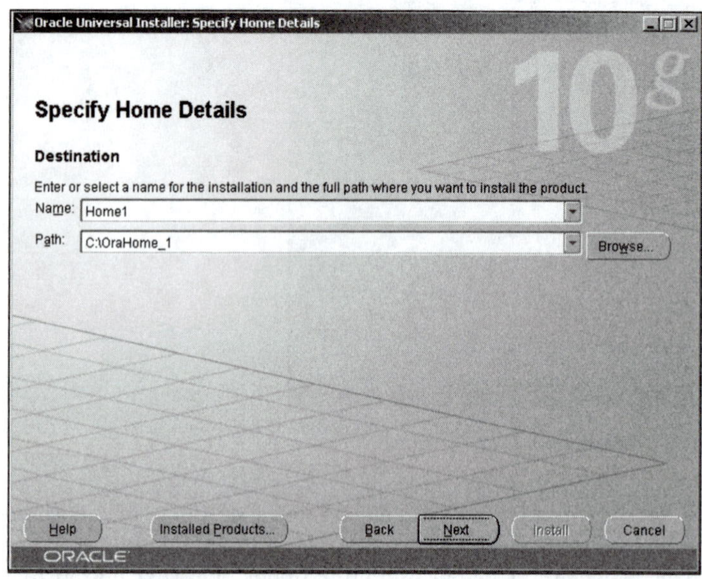

FIGURE 1-14 Specify home details

5. In Figure 1-15 you can select and deselect from a multitude of different options to install. Clicking a plus sign expands and exposes a section of the hierarchy. Clicking a minus sign does the opposite by contracting the hierarchy of options. Click the **Next** button to continue.

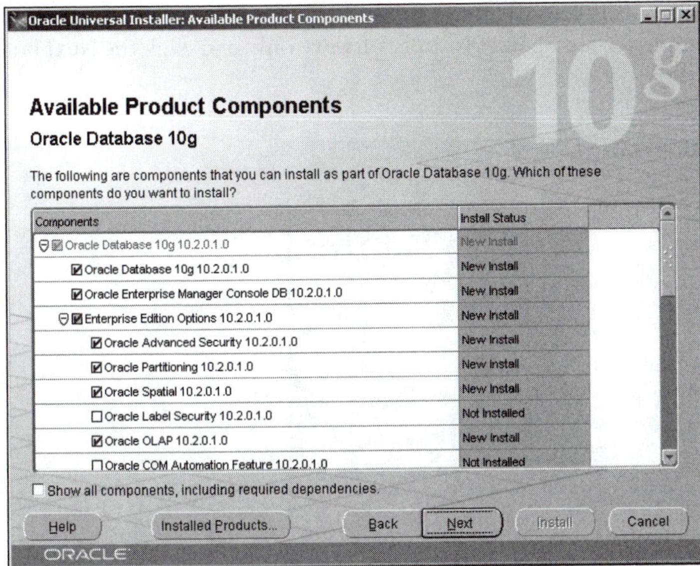

FIGURE 1-15 Available product components

6. Figure 1-16 shows the next screen, showing minimum requirements for Oracle software installation, such as a minimum amount of onboard memory (RAM) required. Click the **Next** button to continue.

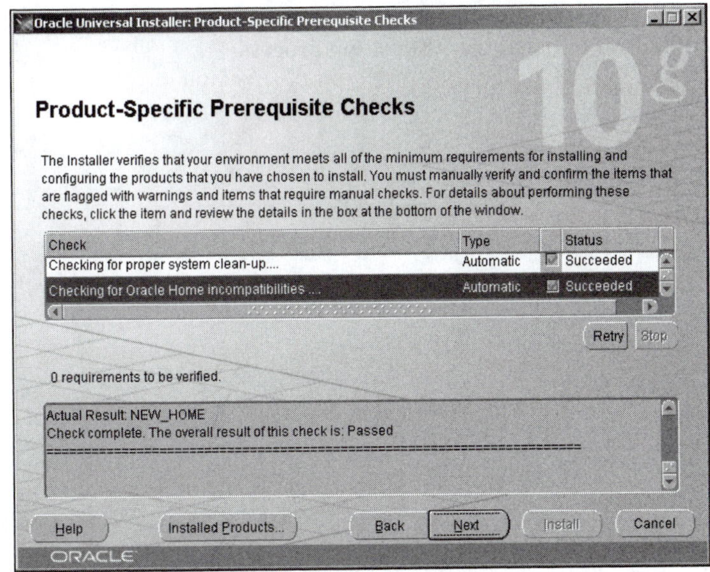

FIGURE 1-16 Product specific prerequisite checks

7. The next screen, as shown in Figure 1-17, allows creation of a database during the installation process. Select **Install Software only** and click the **Next** button to continue.

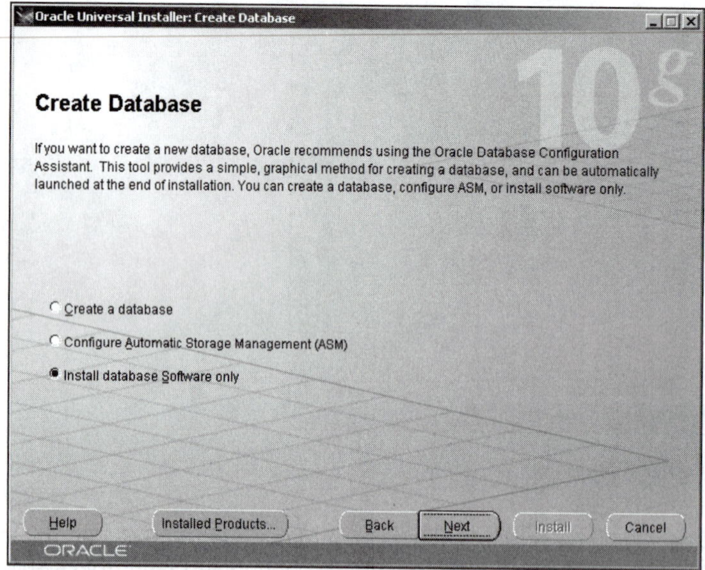

FIGURE 1-17 Create a database

8. Figure 1-18 shows you a screen of all options you have selected for installation. Click the **Install** button to begin the installation process.

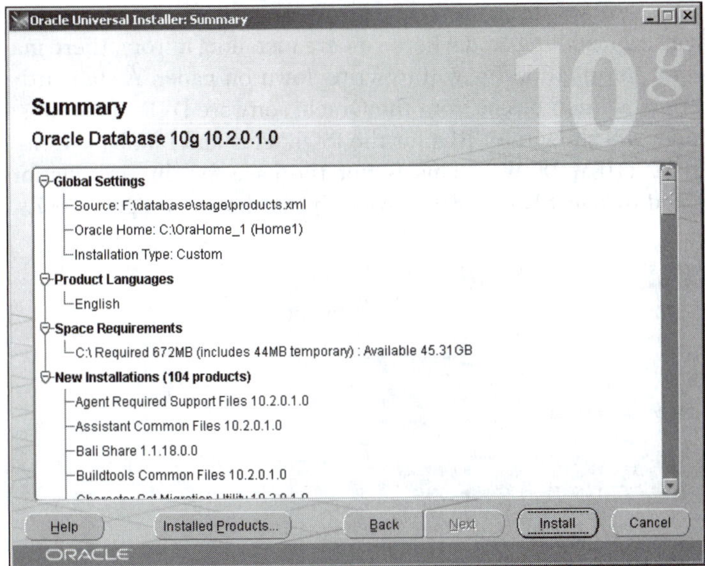

FIGURE 1-18 Summary of options to be installed

9. During the installation process you will be presented with the installation progress percentage bar, as shown in Figure 1-19. The installation process can take some time.

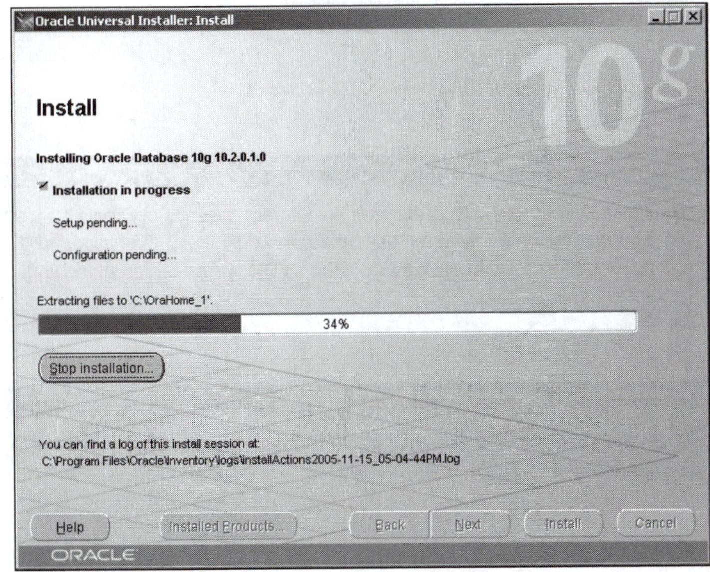

FIGURE 1-19 The progress of installation screen

10. The final screen, as shown in Figure 1-20, is very important. Depending on what version you are installing, and where you are installing it from, there may be very important information for you to write down on paper. A standard database server installation direct from the Oracle software DVD, including a database creation, will show the URLs for the iSQL*Plus and Database Control tools. WRITE THEM DOWN! This is not the case for this installation because you opted to select **Install Software only**, as shown in Figure 1-17.

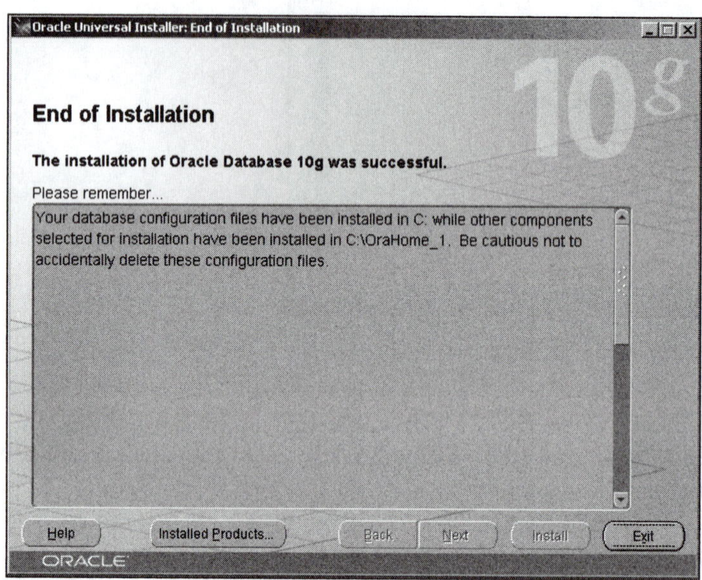

FIGURE 1-20 The End of installation screen

TIP

Please note that the Oracle Universal Installer changes often, even within different releases of Oracle 10*g*. For example, release 10.1.0.1 might be quite different from release 10.2.0.1. Additionally, the Universal Installer will display different screens depending on installation source (the DVD or your Start menu on your computer).

NOTE

For this book you need to install both the Oracle server and the Oracle client installation options.

Chapter Summary

- The core Oracle 10*g* RDBMS software includes a suite of products. The "*g*" stands for Oracle Database 10 Grid. A grid is a group of many computers linked together over a network, acting as a single computer or hardware resource. Thus Oracle Database 10 Grid can be Oracle database installed on a grid of computers.
- The basic installation includes many components and subcomponents.
- The key components of Oracle 10*g* RDBMS software are: Oracle 10*g* database, Oracle Net Services, Java/Web support and tools, Enterprise Manager, SQL*Plus, utilities, and pre-compilers.
- Additional add-on options to Oracle 10*g* can help to increase functionality of the database and add more tools to the basic tool set provided by the basic version.
- A database instance is made up of the memory area (SGA) and the background processes started on a computer.
- A database server contains a database instance and the database files that store the database data.
- A database server can contain a single instance, multiple instances, or can be a combination of multiple servers, called a cluster or grid of computers.
- Some of the key optional components include Oracle Partitioning, Oracle Real Application Clusters (Oracle RAC), Oracle Spatial, Oracle Data Mining, Oracle Advanced Security, Oracle Label Security, and Oracle OLAP services.
- The most complete and robust installation option is the Enterprise Edition for the server side.
- Other choices for the server side are: Standard Edition (for workgroups) and Personal Edition (for individuals).
- On the client side, you can install the Administrator option (for administration of remote databases) or the Runtime option (for running applications that access remote databases).
- The ORACLASS database is provided for running exercises throughout this book.
- A running case project develops a database system for a fictional newspaper chain called Global Globe.
- Multi-tier (n-tier) architecture separates different levels of software components.
- The client-side installation option runs remote applications or remote DBA tasks.
- The server-side installation option runs a database server.
- The client-side installation option has a number of paths: Administrator, Runtime, a custom install, and InstantClient. The InstantClient installation option creates the bare minimum Oracle client installation. InstantClient is for use by OCI (Oracle Call Interface) shared library application use.
- Oracle Universal Installer is a user-friendly interface for installing or uninstalling all Oracle products.
- The Oracle Home Selector feature of the Oracle Universal Installer enables the installation of multiple versions of Oracle software on one computer.

- The Silent Install option pre-builds and runs an installation without any human intervention.
- Optimal Flexible Architecture (OFA) defines standard names for directory structures, database files, control files, and redo log files.
- OFA uses the ORACLE_BASE variable to define the root location for all Oracle subdirectories, including Pradata (where database files reside) and ORACLE_HOME (where Oracle executable software resides).

REVIEW QUESTIONS

1. The Oracle 10*g* Application Server is part of the Oracle 10*g* database software suite. True or False?
2. List the key components that come with Oracle 10*g* RDBMS software.
3. Explain the function of Oracle Net Services in relation to the database.
4. Explain the function of Oracle Net Services in relation to the utilities component.
5. Explain how a pre-compiler works.
6. Describe the difference between the Administrator and Runtime installations.
7. The Personal Edition database (without additional options) is identical to the Enterprise Edition database. True or False?
8. Oracle Universal Installer CANNOT install non-Oracle applications. True or False?
9. List two situations in which you would use the Oracle Home Selector.
10. Write the command to run Universal Installer in SILENT MODE on Windows, in which your response file is named **resp0901.txt**.
11. Describe why an Optimal Flexible Architecture helps improve performance.
12. Oracle executable files are found under the ORACLE_BASE directory. True or False?
13. Oracle administrative files are found under the ORACLE_HOME directory. True or False?
14. The Database Control is used to manage multiple databases on more than a single computer across a network. True or False?
15. Data Pump Import and Export utilities are new and improved versions of the Import and Export utilities? True or False.
16. Shared servers allow connection pooling, and higher concurrency for data warehouse databases? True or False.

Exam Review Questions—Oracle Database 10*g*: Administration (#1Z0-042)

1. Your database server uses shared server processes. You have 50 user sessions. What is the least number of server sessions you can have?
   a. 50
   b. Less than 50
   c. More than 50
   d. 1
2. Your ORACLE_HOME is defined as **D:\oracle\product\10.2.0** and your ORACLE_BASE is defined as **D:\orabase\product**. Which statement is true?
   a. This structure complies with OFA.
   b. This structure does not comply with OFA.
   c. This structure cannot be used because it does not comply with OFA.
   d. This structure is not specified by OFA and, therefore, may or may not comply.
3. Which of these functions is handled by the key components of Oracle 10*g and* can be enhanced with optional software that Oracle sells? (Choose all that apply.)
   a. Query the data
   b. Enforce user security
   c. Do backups
   d. Monitor performance
   e. All of the above
4. Which platforms would commonly be used to install the Personal Edition of Oracle 10*g*, Release 2? (Choose all that apply.)
   a. Windows 2000
   b. Windows XP
   c. Unix
   d. Windows 2003
   e. Linux
5. Oracle Partitioning can aid in which of these applications? (Choose all that apply.)
   a. An accounting application requiring complex calculations
   b. A sales projection application requiring large quantities of historical data
   c. An online store needing fast response time for many users
   d. A small company with a work group of five programmers

6. Which of these optional components enhance the database internally? (Choose all that apply.)
   a. Oracle Advanced Security
   b. Oracle Label Security
   c. Oracle Spatial
   d. All of the above
7. When installing Oracle 10*g* on a computer that currently has Oracle9*i*, which of the following statements are *not* true? (Choose two.)
   a. You must replace the Oracle9*i* software with the Oracle 10*g* software.
   b. Your ORACLE_HOME is permanently set to point to the newest software.
   c. The Oracle Universal Installer handles either release of the software.
   d. A database instance can run on one or the other, but not both releases.
   e. Enterprise Manager is installed with Oracle 10*g* Enterprise Edition.
8. Which components make up a database instance? (Choose two.)
   a. Database software modules
   b. Database files
   c. System Global Area (SGA)
   d. Oracle Net Services
   e. Background processes
9. The Standard Edition of the database is often used for what purposes? (Choose two.)
   a. Testing new versions of the database
   b. Running applications on a client workstation
   c. Developing applications on a programmer workstation
   d. Saving money when operating a low usage website
   e. Running administrative tasks
10. Which component is NOT available on a client-side installation?
    a. Enterprise Manager
    b. Pre-compilers
    c. SQL*Plus
    d. Utilities
    e. None of the above

Hands-on Assignments

1. Looking at only the Windows side of Figure 1-10, list the lowest level directories in the diagram that containing Oracle database binary installation files, for any version of Oracle database. Use full path names in your list. Save your work in a file named **ho0101.txt** in the **Solutions\Chapter01** directory on your student disk.
2. Your office plans to install Oracle 10*g* Enterprise Edition, version 10.2.0 on its server. They want to know ahead of time what the directory structure will look like. The root directory will be **/all_apps/oracle**. Draw a diagram, using Figure 1-10 as a guide, in which you show the location of all the key components and the database datafiles. Save your work in a file named **ho0102.tif** in the **Solutions\Chapter01** directory on your student disk.
3. Using your diagram from Assignment 2, what are the full path values of ORACLE_HOME and ORACLE_BASE? Assume the same directory for both a Windows and a Unix installation, listing values for both Windows and Unix operating systems. Save your work in a file named **ho0103.txt** in the **Solutions\Chapter01** directory on your student disk.
4. You have just installed the Administrator client-side software on your Windows computer. There is another Windows computer on your network that has the Enterprise Edition server-side software installed and running. You start Enterprise Manager console on your client computer and look at the list of users on the remote database. How did the information travel between the two network nodes? Draw a simple diagram showing a client and a server communicating with one another. Save your work in a file named **ho0104.tif** in the **Solutions\Chapter01** directory on your student disk.
5. Your e-commerce business is growing fast. You currently have the Standard Edition of Oracle 10*g* on your server. Your server is having trouble with servicing more than five customers at a time. The server tends to bog down when any user searches the product table. Write a recommendation letter describing three options that help solve one or both of these problems. Save your work in a file named **ho0105.txt** in the **Solutions\Chapter01** directory on your student disk.

Case Project

The Global Globe has an Online Applications branch and a Data Warehouse Applications branch. The Online Applications branch has just moved to a different city. This poses a problem because your central database (Enterprise Edition) was used for programmers in both divisions. Each programmer has a computer with the Standard Edition installed (without a database), with which he or she develops applications. You see that this configuration may not be suitable for a new remote location. Come up with two different strategies for supporting programming at the remote site that *do not* require remote access to the central database. Draw simple diagrams of each plan showing the programmer nodes and any other database nodes at the remote site. Create a list that compares the advantages and disadvantages of each plan. Save your diagram in a file named **case0101.tif** and your list in a file named **case0101.txt** in the **Solutions\Chapter01** directory on your student disk.

CHAPTER 2

TOOLS AND ARCHITECTURE

LEARNING OBJECTIVES

In this chapter, you will:

- Identify the main DBA tools in the Oracle 10*g* software suite
- Configure Oracle Net Services to connect to the database
- Examine Oracle database instance architecture
- Examine Oracle database memory architecture
- Examine Oracle database process architecture
- Examine Oracle database connection management architecture
- Start using the Enterprise Manager
- Go through a brief introduction to the Database Control

INTRODUCTION

As you discovered in the previous chapter, Oracle 10*g* brings with it a host of tools and utilities to aid you in working with the database. Two of the most important tools are Oracle Net Services and Enterprise Manager. In this chapter, you examine the many tools that are available to assist you as a database administrator (DBA). Then, you learn how to set up and configure Oracle Net Services on a network. After you have Oracle Net Services running properly, you go through the steps to configure Enterprise Manager, and then tour the main DBA tools included in Enterprise Manager.

Before going on, if you have not already installed the Oracle client software, in addition to the Oracle database server software on your computer, then please do so now. This book requires use of both Oracle server and Oracle client software on the computer you are working on. In industry it is extremely likely that a database administrator will prefer to work as much as possible using client software. Staying off the database server as much as possible is prudent and helps to avoid costly errors. The Oracle client installation is very similar to, and much simpler to perform, than the Oracle server installation. Please refer to Chapter 1 if you get stuck, or ask your instructor for help. Also, you can use both an Oracle server and an Oracle client installation on the same computer.

OVERVIEW OF DBA TOOLS

In the previous chapter, you were introduced to the key components of the Oracle 10g software suite. Within these key components, you find many tools designed to streamline the work of the DBA. Many of these tools become integrated in the central workspace, the Enterprise Manager console, the Database Control, or the Grid Control. There is a hands-on tour of the console later in this chapter. Some tools are standalone, some are reached from inside the Enterprise Manager console, and some are available from the operating system using a command-line execution. Table 2-1 lists the DBA tools and their uses.

All of the tools that can be run from the operating system are found on the Start menu in Windows and in some branch of the Oracle tree of applications. In Unix and Linux you may need to start tools from a command-line prompt (in a shell). In Windows, you can also start the tools from a command prompt if you want. Table 2-2 shows the operating system commands for starting some of the tools in operating system shells, such as Windows, Unix, and Linux.

TABLE 2-1 DBA tools and their uses

Tool Name	Function
Analyze Wizard	Collects statistics about tables or objects, validates tables, finds continuing rows
Backup and Restore Wizards	Backup and recovery planning and maintenance
Data Upgrade Assistant	Migrates data from older versions to newer versions of the database
Database Conrfiguration Assistant	Creates new database instances. Also called DBCA
Enterprise manager console	Monitors databases and provides access to database management tools
Export and Import Wizards	Exports and import data using navigator-based export tool

TABLE 2-1 DBA tools and their uses (continued)

Tool Name	Function
Instance Manager	Access to configuration, sessions and resource consumer group configuration. Also includes summary advisory performance metrics
Log Miner	Queries redo log files and otherwise. Also called Data Miner)
Net Configuration Assistant	Configures Oracle Net with step-by-step instructions
Net Manager	Configures Oracle Net
Schema Manager	Views and modifies data; views and modifies table structures, indexes, views, objects, procedures, and packages
Security Manager	Views and modifies users, roles, and profiles; creates new users; changes passwords
SQL*Plus	Executes SQL commands; runs reports and queries; starts and stops the database
iSQL*Plus	A browser version of SQL*Plus, used mostly for reporting
SQL*Plus Worksheet	Executes SQL commands; runs queries
Storage Manager	Views and modifies database storage: tablespaces, datafiles, controlfiles, redo logs, archive logs
Universal Installer	Use to install Oracle software, or change an existing installation
Grid Control	The Grid Computing (clustered) version of the Database Control
Database Control	Browser based tool for monitoring, performance, adminstration and maintenance
Enterprise Security Manager	Directory service security management
Warehouse Summary Management	Access to dimensions and materialized views
Workspace Manager	Database versioning workspaces
XML Database Manager	XML database access and maintenance

Whether you start the tools from the command line, console, or a Windows type menu, these tools give you a way to work on the database in a Windows-style environment, in which the actual Oracle commands may be generated for you. However, it is important to be familiar with both automated (using a tool) and manual (using a command) methods for some tasks. Sometimes, you do not have a Windows-like interface for diagnosing and correcting a problem with the database. In these cases, you must understand how to work directly from the command line. Other times, as you become familiar with both methods, you may find that certain tasks are more quickly accomplished using manual methods. For example, changing a user's password can be done with a single command line in SQL*Plus, as shown in Figure 2-1. On the other hand, changing a user's password in Security Manager requires opening Windows, navigating through lists of users, and then typing the password twice, as shown in Figure 2-2. Of course, you must know the command by heart before the manual version is faster than the tool. Later chapters teach both manual and tool-based methods for many tasks.

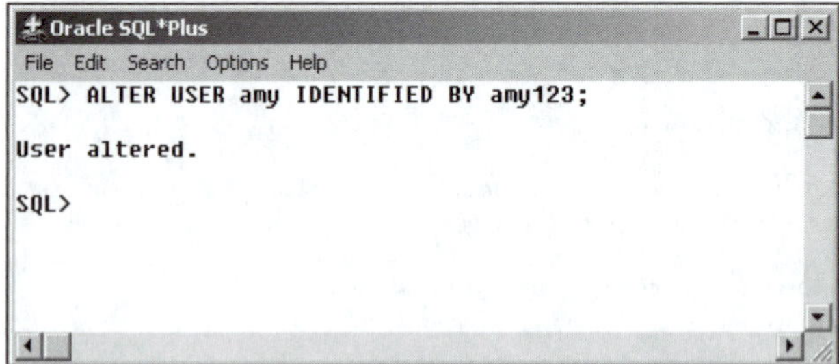

FIGURE 2-1 Changing a user's password manually requires one command

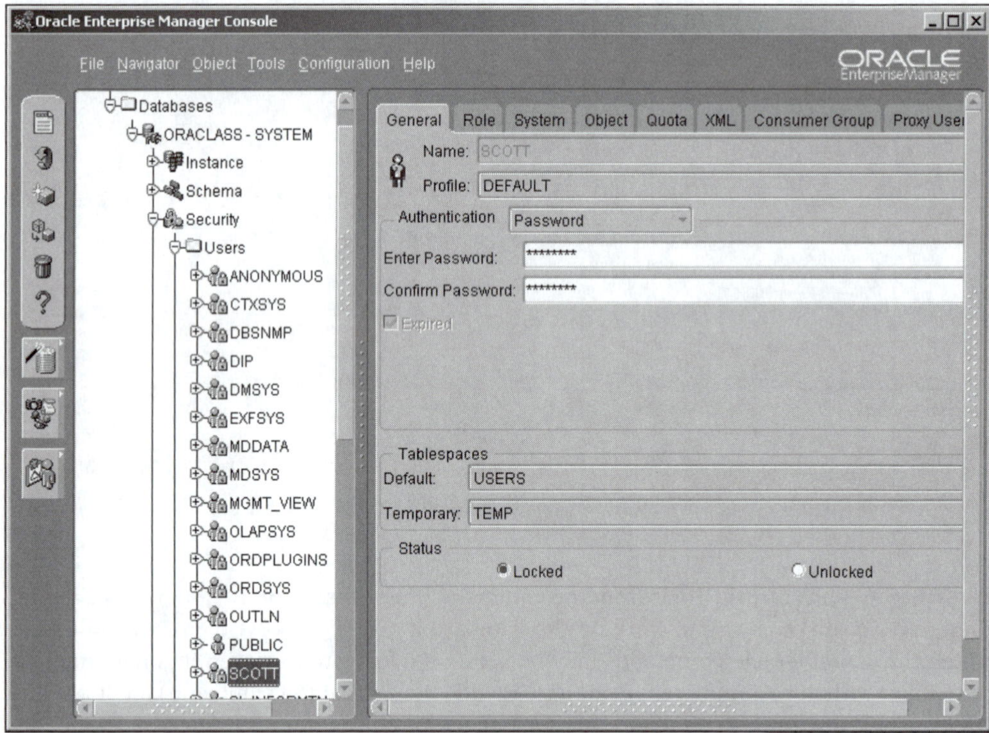

FIGURE 2-2 Changing a user's password in Security Manager takes more time

TABLE 2-2 Operating system commands for starting tools

Tool Name	Windows Command	Unix/Linux Command
Database Upgrade Assistant	dbma	dbma
Database Configuration Assistant	dbca	dbca
Enterprise Manager Console	oemapp.bat console	oemapp console
Net Configuration Assistant	netca	netca
Net Manager	netmgr	netmgr
SQL*Plus	sqlplus <username>[/<password]	sqlplus <username>[/<password]
SQL*Plus Worksheet	oemapp.bat worksheet	oemapp worksheet
iSQL*Plus	http://2600client:5560/isqlplus	http://2600client:5560/isqlplus
Database Control	http://2600client:5500/em	http://2600client:5500/em

Some tasks can be handled by more than one tool. For example, you can stop the database using SQL*Plus or iSQL*Plus. To view data in a table, you can use the Schema Manager or write a query in SQL*Plus, iSQL*Plus or SQL*Plus Worksheet.

Oracle Net Services is a common denominator for these tools. In every case, you must connect to the database via Oracle Net Services. For tools that run inside the Enterprise Manager console, a single logon gives you access to all the tools. When running a tool directly from the operating system, you must log on with a user name, a password, and a database or network name known to Oracle Net Services. The next section describes how to configure and test Oracle Net Services connections.

CONFIGURING ORACLE NET SERVICES TO CONNECT TO THE DATABASE

Nearly every time you access Oracle 10g, you go through Oracle Net Services. Therefore, a critical step in setting up your working environment involves properly configuring Oracle Net Services to access the database on which you want to work.

Overview of Oracle Net Services Architecture

Oracle Net Services is made up of several subcomponents that work together to translate your requests, such as SQL queries, into network packages for the local, or Internet, network. Each network package has a destination and an identifying code that tells the receiving end when it has collected all the network packages for a single request. Figure 2-3 shows a basic diagram of a user with Enterprise Manager on his desktop computer, connecting to an Oracle 10g database on a database server across a local network.

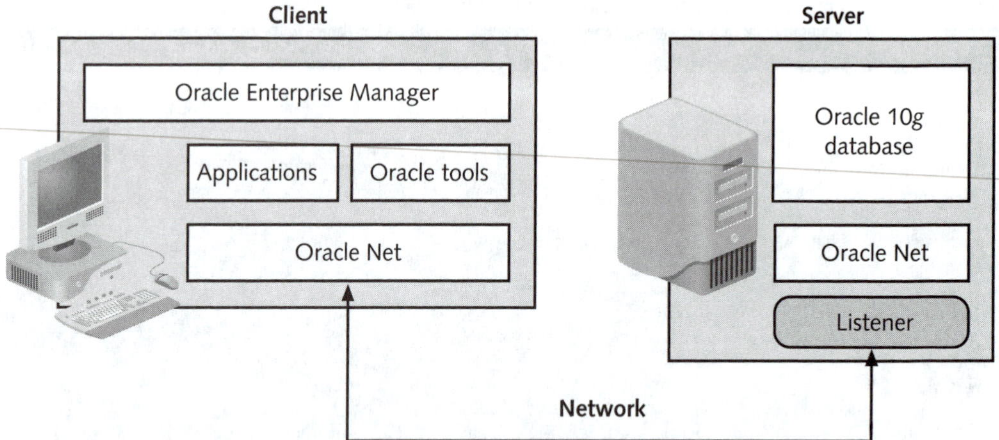

FIGURE 2-3 Communication between a user and the database using Oracle Net Services

As you see in Figure 2-3, Oracle Net Services resides on both the client and the server side of the network. This is necessary when using applications (such as a tax preparation application that uses Oracle to store its data), and when using Oracle tools, such as Enterprise Manager or SQL*Plus, installed on the client side. Both client and server installations of Oracle Net Services must be configured so that they are synchronized to the target database. On the client side, Oracle Net Services accepts requests from the Oracle tool, translates them into the local network protocol (usually TCP/IP), and sends them across the network as network packages. The configuration is stored in the **tnsnames.ora** configuration file on the client computer. On the server side, the computer has its own copy of the **tnsnames.ora** configuration file, defining the database that runs on the computer, or even remote databases on other computers if needed. Oracle Net Services receives the request from the network, translates it into an Oracle protocol, and sends it to the database. The database has a service called the listener process. The listener process waits for requests and responds to requests as needed. A request occurs when a user connection makes a "request" for information from the database. Oracle Net Services must know not only what computer contains the database, but also what the database's name is, and to which port number its listener process is tuned. All these details about a database are stored in Oracle Net Services as the service name of the database. A service name is the set of information (configuration) that Oracle Net Services uses to locate and communicate with an Oracle database. After a service name is defined on both the client and server side, any tool on the client side can reach the server-side database by using the service name combined with a valid user name and password. Typically, a tool prompts the user to log on by requesting provision of all three of these vital pieces of information. Figure 2-4 shows the logon screen for SQL*Plus as an example. The service name goes in the box called Host String.

When you are using tools that reside on the same computer as the database, you might assume that no Oracle Net Services connection is needed. You have the choice to use either Oracle Net Services or the bequeath protocol. A bequeath protocol is allowed only when you are logged on to the database machine. Either method works equally well. A bequeath

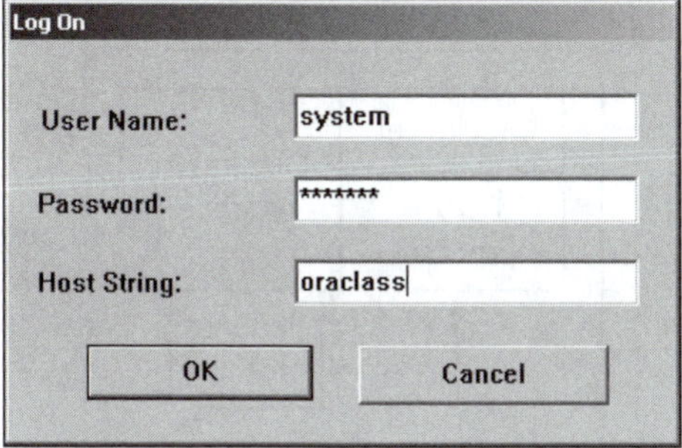

FIGURE 2-4 Specify the Oracle Net service name in the service box when logging in

protocol allows a direct connection to a database, on a database server computer, without going through Oracle Net services and without requiring a network name.

NOTE

A network name is a label given to a specific network configuration, usually a specific computer, at a specific TCP/IP address—in the case of an Oracle database server, housing a particular Oracle database.

Figure 2-5 shows the two primary access points for reaching the database server from a client computer. In addition, it shows the path of communication used when you access the database while logged on to the server. The three methods are:

- **Client with Oracle Net:** When a client computer runs applications such as those written in C or COBOL, the client computer must install the client-side version of Oracle 10g. This provides Oracle Net Services and its tools, so the client can configure a service name to reach the database server over the network.
- **Client with JDBC driver:** A client running a browser with a Java applet can use the JDBC thin driver to access a remote database via the Internet. The JDBC thin driver can be included in the applet, so that no additional software needs to be installed. On the server side, the database only requires a standard TCP/IP protocol program (which is generally already installed and used for most Internet communication).
- **Terminal with direct connection:** When you log directly on to the database server (without using a network name), you access the database directly via the bequeath protocol. To use this, you simply leave the service name blank when logging on to any tools or utilities. You are automatically directed to the database that is running on the machine, based on the environmental configuration of the database server. This technique also supports a workstation computer that has its own copy of the database running. By default, you reach the database on your computer when you omit the service name during logon.

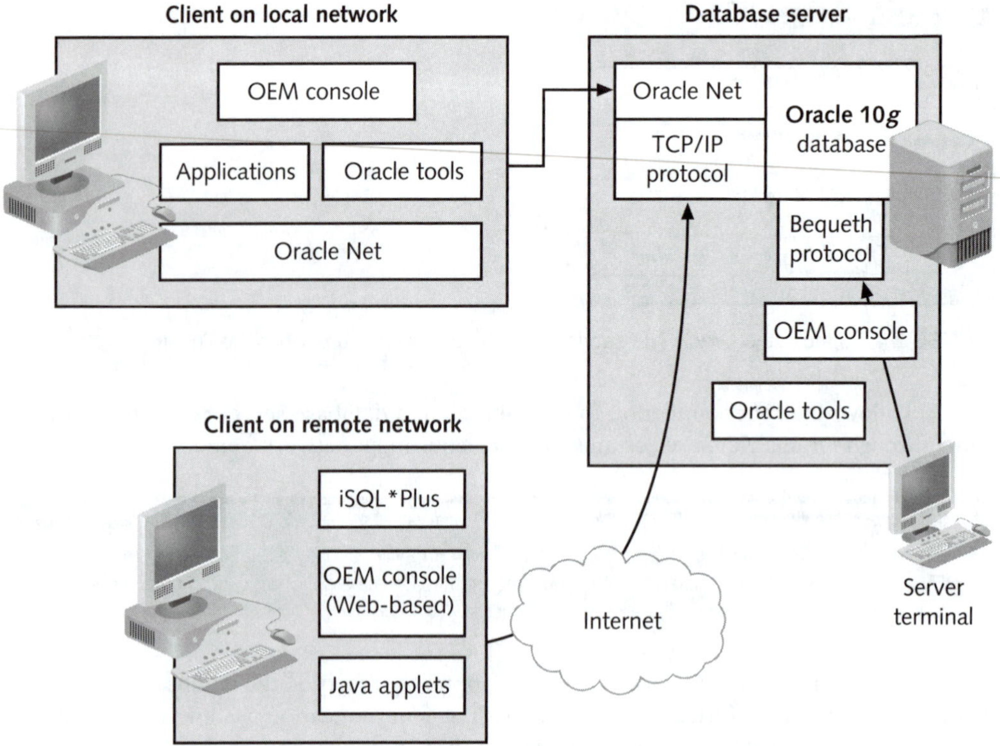

FIGURE 2-5 Three methods of connecting to the database

Before examining specific steps of how to configure an Oracle Net Services connection between client and database server machine, you need to understand there are various methods of connecting to a database.

Network Naming Methods

Computers on a network use network connections to communicate with each other. A network connection is established over a network protocol, such as TCP/IP. TCP/IP uses what is called an IP address, such as 123.456.789.012, where the naming method allows the allocation of a simple name, such as **yahoo.com**, to the route across the Internet to that website.

Oracle database can use various different naming methods. What is a naming method? A naming method is a way in which a node name on a network will be translated into a form understandable and routable by hardware and network software on that network. Different naming methods apply under different circumstances:

- **Local naming:** Use a configuration file called **tnsnames.ora**, commonly called TNS (Transparent Network Substrate). TNS is the default method for connecting to a database.
- **Directory naming:** Service names and their mappings are stored in a lightweight directory access protocol (LDAP) server, much like a DNS server.

- **Host naming:** This method uses an operating system-based IP address to host-name mapping host files. Host files on Solaris Unix are placed in the /etc/system directory and on Windows in the c:\windows\system32\drivers\etc directory.
- **External naming**: Third-party naming services involve third-party software applications that are not part of Oracle software.
- **Easy connect**: No naming, such as TNS lookup in the **tnsnames.ora** is required, allowing direct access to a database server. This option allows a direct connection to a database server using a connection string as shown in this syntax diagram:

```
CONNECT <user>/<password>@host:port/service
```

And in the following example, connecting to an Oracle database on a host (a host is a computer on a network) called 2000server, through port 1521, to a database called OLTP:

```
CONNECT <user>/<password>@2000server:1521/OLTP
```

Those are the different naming methods. The most commonly used is local naming with TNS names. The following section leads you through the steps to set up an Oracle Net Services connection on a client machine, using the Net Manager tool, with the local naming method.

Step-By-Step Configuration of Oracle Net Services Using Net Manager

Your first hands-on assignment is to configure Oracle Net Services on your workstation so that you can use the Enterprise Manager, SQL*Plus, and other tools needed for the rest of the book. Your instructor provides you with appropriate user names, passwords, and database names.

Follow these steps to configure Oracle Net Services:

1. To start Net Manager in Windows, on the taskbar, click **Start/All Programs/Oracle.../Configuration and Migration Tools/Net Manager**. In Unix or Linux, type **netmgr** on the command line of a shell. The initial screen for Net Manager appears, as shown in Figure 2-6.
2. Expand the Local node by clicking the **plus sign** to the left of the Local icon.
3. Expand the Service Naming node by double-clicking the **folder icon**. This is shown in Figure 2-7. The only service name listed is that called extproc_connection_data. This service name is used to execute procedures created externally to Oracle database, such as C programmed procedures.

NOTE

The EXTPROC_CONNECTION_DATA network name is used to call external programs (external to Oracle software), such as DLL's. It should be generated into the Oracle Net Services configuration when a database is created using the Database Configuration Assistant (DBCA tool). If it is missing, it was either deleted, or you created a database manually, or your database was migrated from an older version of Oracle, or you are using a version of Oracle 10*g* that may not be functioning properly.

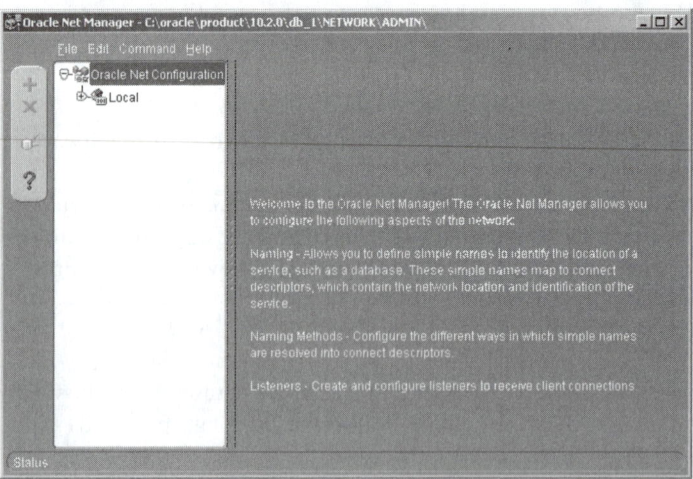

FIGURE 2-6 No login is required for Oracle Net Manager

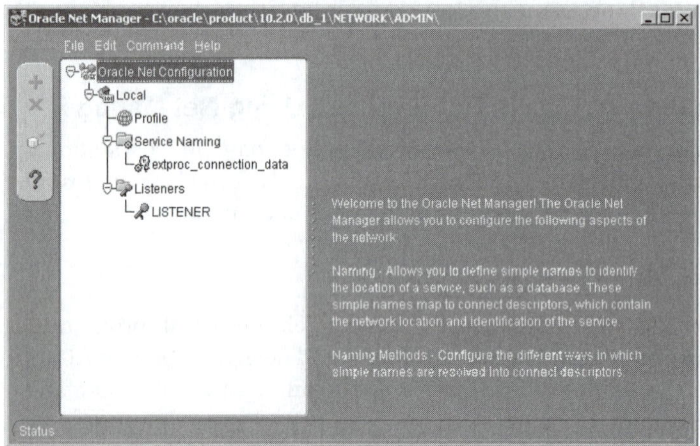

FIGURE 2-7 No services exist to begin with

4. Now highlight Service Naming, and click the big **green plus sign** in the left margin of the window, as shown in Figure 2-7. The Net Service Name Wizard starts. Type the new service name, **ORACLASS**, in the box as shown in Figure 2-8, and click **Next.**
5. For the connection, select **TCP/IP** (Internet Protocol) as the protocol, as shown in Figure 2-9, and click **Next**.
6. Type the computer name on which the database resides in the Host Name box. *Your instructor provides this name.* Accept the default port setting of **1521**. This port setting is the default port setting for most Oracle database listeners. Figure 2-10 shows the window filled in with my database server computer name, as **1300server**, and the port default setting of 1521. Click **Next**.

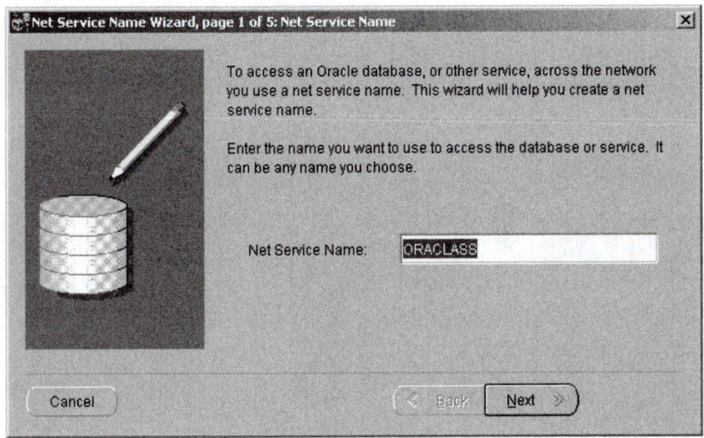

FIGURE 2-8 Name the new service here

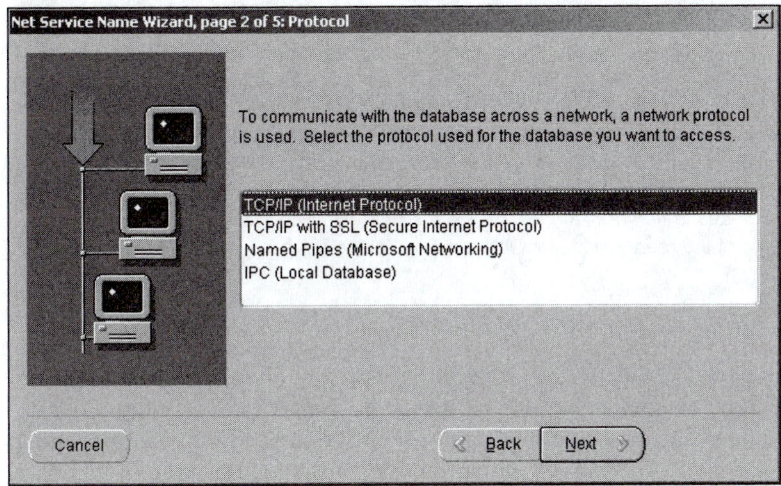

FIGURE 2-9 TCP/IP is the typical protocol for networks

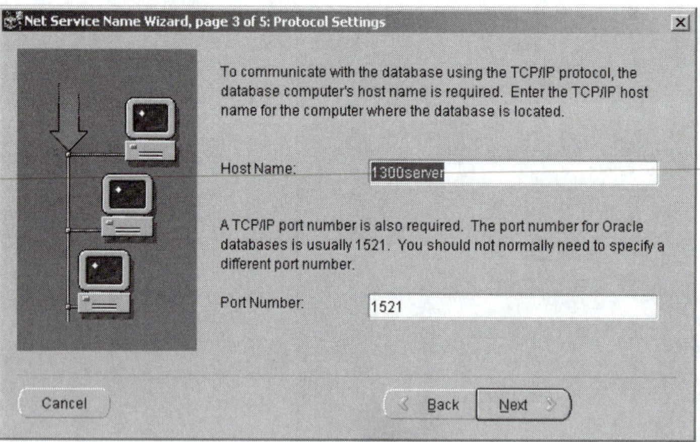

FIGURE 2-10 Identify the host computer on which the database runs

NOTE

To discover the name of the Windows computer you are working on, click **Start/Control Panel** (this will be **Start/Settings/Control Panel** in older versions of Windows). Then double-click **System**. Click the **Network Identification** tab. In Windows XP, click the **Computer Name** tab. The computer name appears listed in the Full computer name field. You can also find the computer name of a Windows computer by typing the command SET into a command-line shell and searching for the COMPUTERNAME environment variable. This will only tell you the name of the computer you are working on. If you are on a client computer, and not on a database server, you will not find the host name of the database server computer – only the host name of the client computer. On a Unix or Linux machine use the ENV command in a shell.

7. Accept the default selection of Oracle8*i* or later, and fill in the Oracle 10*g* service name for the database. *Your instructor provides this information.* Figure 2-11 shows the screen with a service name of **class.edu**. If you are setting up a service name for Oracle8 and lower, the database **SID** (System Identifier) is used. The SID has been replaced by the service name in Oracle8*i* and higher. This change gives more flexibility when implementing complex database configurations, such as clustered or parallel databases. Leave the Connection Type selection as its default "Database Default." The connect type can be "Database Default," "Dedicated Server," or "Shared Server." Choosing "Database Default" allows Oracle Net Services to connect according to the database setup, rather than dictating which mode to use. Click **Next** to proceed to the next window.

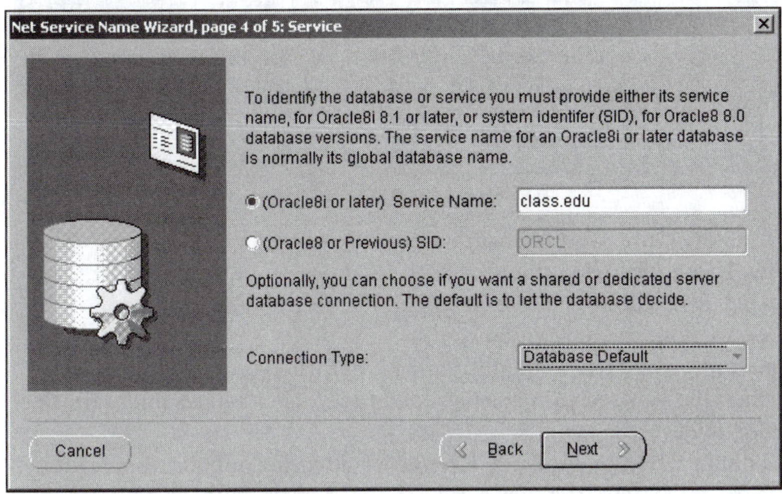

FIGURE 2-11 Identify the database instance by its service name

8. Click **Test**. This starts an automatic connection to the database using the new service name you created. The test logs on as the default user *scott*, with password *tiger*. Figure 2-12 shows the test. This is the default user and password installed in Oracle databases to store demonstration tables. If you have removed *scott* or changed the password, the automated test fails.

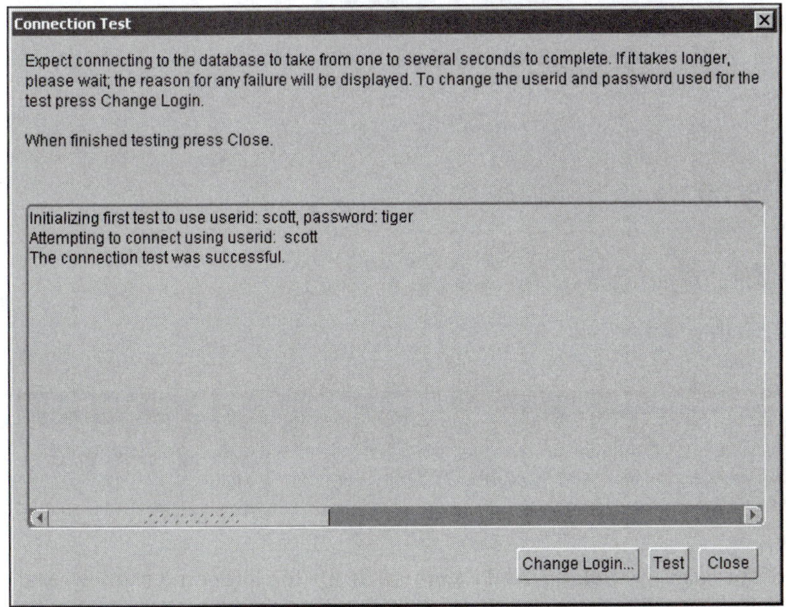

FIGURE 2-12 Test the new connection

NOTE

When creating a database using the Oracle Installer or the DBCA tool the user *scott* will remain a locked account by default. Oops! Yes you can go and unlock the account, but don't bother. You will unlikely ever find another use for the *scott* schema. I usually delete it after first creating a new database, and usually with the glee of a first act of plunder! It's useless!

You can adjust the user/password very easily using any valid user by clicking **Change Login**, changing the user name and password, clicking **OK**, and then clicking **Test** again as SYS or SYSTEM. It's that easy.

9. Click **Close** to close the test window.
10. Click **Finish** to complete the definition. This returns you to the main window, in which the new service name is added to the list of names under the Service Naming folder.
11. Save the configuration by selecting **File/Save Network Configuration** from the menu. Figure 2-13 shows the new service name and its configuration.

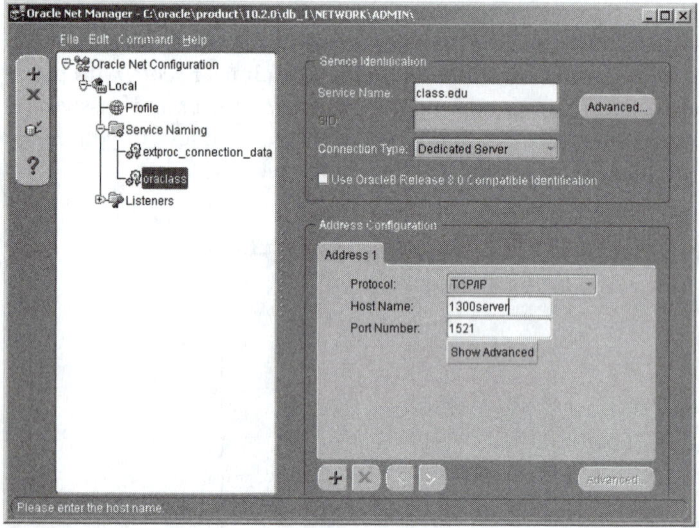

FIGURE 2-13 LSNRCTL utility help options

NOTE

Make changes to any service name configuration by typing over the current settings. Remember to test and save your changes. Test changes by selecting **Command/Test Service** from the menu.

12. Close Net Manager by clicking the **File** option at the top left of the window, and then selecting **Exit** from the menu.

The file **tnsnames.ora** stores the Oracle Net Services configuration that you work on when using Net Manager. This file is found in the **ORACLE_HOME/network/admin** directory on Unix or Linux, and C:\ORACLE_HOME\network\admin directory on a Windows computer.

NOTE

Use caution when editing this file. You must use a plain-text editor, such as Notepad or vi, to avoid adding any characters to the file that would corrupt it.

The next step is to briefly examine the TNS configuration files. There are actually two files to look at. These are the **tnsnames.ora** file on the client computer, and the listener configuration file on the database server computer.

The ORACLE_HOME and ORACLE_BASE Variables

ORACLE_HOME and ORACLE_BASE are variables used throughout this book to represent registry entries in Windows, and $<named> variables in Unix or Linux. The ORACLE_HOME variable refers to the directory where Oracle installs its executable files (commonly referred to as the Oracle binaries). In Windows, ORACLE_HOME will typically be something such as **C:\oracle\product\10.2.0\db_1** (varies by Oracle version and release). The ORACLE_BASE variable is the directory upward from the ORACLE_HOME variable, and typically something like **C:\oracle\product\10.2.0** in Windows. In Unix or Linux ORACLE_HOME and ORACLE_BASE variables will be stored as profile settings, generally or locally, and are usually referred to extensively from scripting as $ORACLE_HOME and $ORACLE_BASE—in other words, accessible variables.

TNS Configuration Files

The **listener.ora** file is stored on the database server, providing configuration for how the listener process listens over the network, for connection requests. The **listener.ora** file is stored in the **$ORACLE_HOME/network/admin** directory:

```
LISTENER =
  (DESCRIPTION_LIST =
    (DESCRIPTION =
      (ADDRESS_LIST =
        (ADDRESS = (PROTOCOL = TCP)(HOST = 1300server)(PORT = 1521))
      )
    )
  )
```

The **tnsnames.ora** file, as shown below, is placed onto the client machine, allowing communication between the client machine and the listener on the database sever:

```
ORACLASS =
  (DESCRIPTION =
    (ADDRESS_LIST =
      (ADDRESS = (PROTOCOL = TCP)(HOST = 1300server)(PORT = 1521))
    )
    (CONNECT_DATA =
      (SERVICE_NAME = class.edu)
    )
  )
```

NOTE

The **tnsnames.ora** configuration file is used to translate a connection string typed by the user into the network and database, addressing information needed to locate the database on the network.

Working with the Listener

The listener listens for requests made by user connections. When a user submits a query to the database, the user's session connection sends a request to the database server. The database server computer responds to that request with the listener process. The listener process then allocates a server process, directly (dedicated servers) or indirectly (shared servers using a dispatcher process). The listener listens for requests and passes them on to the appropriate database server process for execution in the database.

The first thing you need to know about the listener is how to stop and start it. On a Windows computer you can go to the Services window and stop and start the listener service (**Start/Control Panel/Administrative Tools/Services**, or **Start/Settings/Control Panel/Administrative Tools/Services** in older versions of Windows). You can also use the listener control utility (lsnrctl). Options can be executed as command-line parameters into the lsnrctl as in the following example:

```
lsnrctl stop
lsnrctl start
```

Type the word help within lsnrctl to see all options, as shown in Figure 2-14.

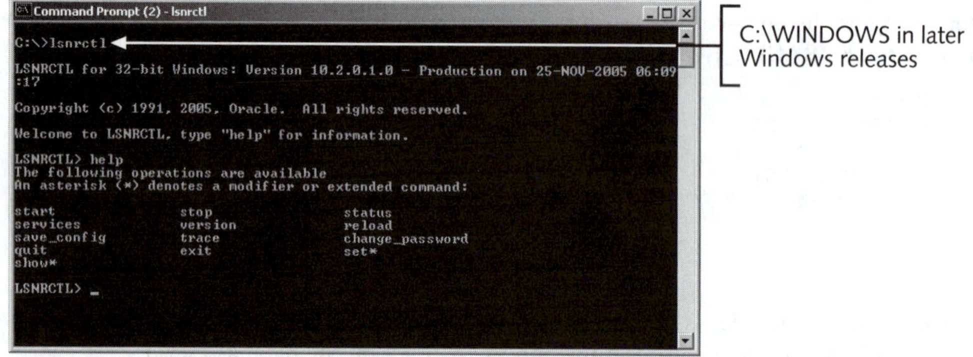

FIGURE 2-14 A new service must be saved to take effect

Listener features include the ability to change the listener queue size and set listener logging and tracing. You can also create multiple listeners and load balance between those separate listener processes. Increasing the listener queue size allows a larger number of listener requests to be serviced by allowing requests to wait in a larger queue. Listener logging and tracing can be a drain on resources and is defaulted as off. Set the queue size, logging, and tracing in the listener configuration file as shown in the following **listener.ora** listener process configuration file on the database server.

```
LISTENER =
  (DESCRIPTION_LIST =
    (DESCRIPTION =
      (ADDRESS_LIST =
        (ADDRESS =
           (PROTOCOL = TCP)
           (HOST = 1300server)
           (PORT = 1521)
           (QUEUESIZE = 50)
        )
      )
    )
  )
LOGGING_LISTENER = OFF
TRACE_LEVEL_LISTENER = OFF
```

Multiple listeners can be load balanced as randomized load balancing between multiple listeners, pointing at the same database. Load balancing can help performance by providing multiple listener connection points to a database server. If one listener is busy, another listener process can be deferred to, thus reducing load on the first listener. The configuration file below adds a listener. The listener called LISTENER2 uses a different port number to the listener called LISTENER. Both of the two listeners, LISTENER and LISTENER2, allow connections to the same database.

```
LISTENER =
  (DESCRIPTION_LIST =
    (DESCRIPTION =
      (ADDRESS = (PROTOCOL = TCP) (HOST = 1300server) (PORT = 1521))
    )
  )
LISTENER2 =
  (DESCRIPTION_LIST =
    (DESCRIPTION =
      (ADDRESS = (PROTOCOL = TCP) (HOST = 1300server) (PORT = 1522))
    )
  )
```

Load balancing between multiple listener processes is placed in the **tnsnames.ora** configuration file on the client computer. The **tnsnames.ora** configuration file demonstrates this with two addresses, one pointing at each listener as indicated by the two different port numbers.

```
ORACLASS =
  (DESCRIPTION =
    (ADDRESS_LIST =
      (LOAD_BALANCE = YES)
      (ADDRESS = (PROTOCOL = TCP) (HOST = 1300server) (PORT = 1521))
      (ADDRESS = (PROTOCOL = TCP) (HOST = 1300server) (PORT = 1522))
    )
    (CONNECT_DATA =(SERVICE_NAME = class.edu))
)
```

As a final note, you cannot use Net Manager to start and stop the listener but you can configure listener processes using Oracle Net Manager.

Using the Net Configuration Assistant

The Net Configuration Assistant allows definition of all possible individual sections of configuration for Oracle Net Services. As shown in Figure 2-15, listeners, naming methods, local names, and directory services can be configured. The Net Configuration Assistant is more of a wizard when compared to Net Manager. Where Net Manager gives a Windows Explorer-like picture of Oracle Net Services, the Net Configuration Assistant provides, quite literally, assistance in the form of a step-by-step wizard.

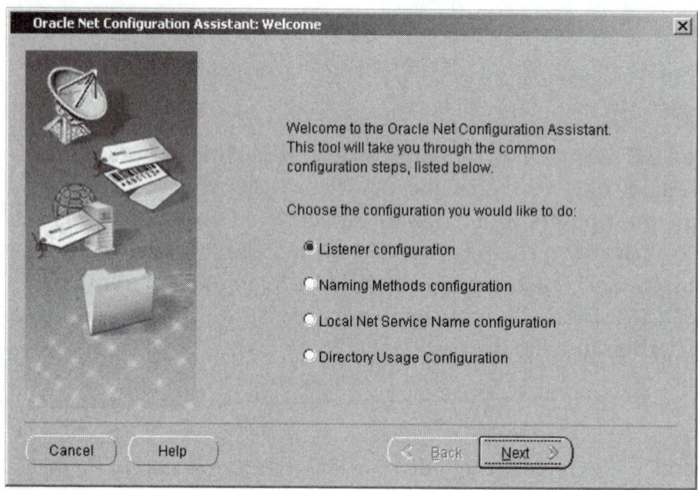

FIGURE 2-15 The Net Configuration Assistant is a step-by-step wizard

Using SQL*Plus, SQL*Plus Worksheet, and iSQL*Plus

If you can't connect to the database using SQL*Plus it could be that the configuration of the listener on the server is incorrect, or the client configuration of the **tnsnames.ora** file is incorrect. An easy way to validate configuration is to use an Oracle version of the TCP/IP ping utility, called **tnsping**. The **tnsping** utility attempts to contact the database server through Oracle Net Services, using both the client TNS configuration, and the database server listener process configuration, as shown in Figure 2-16.

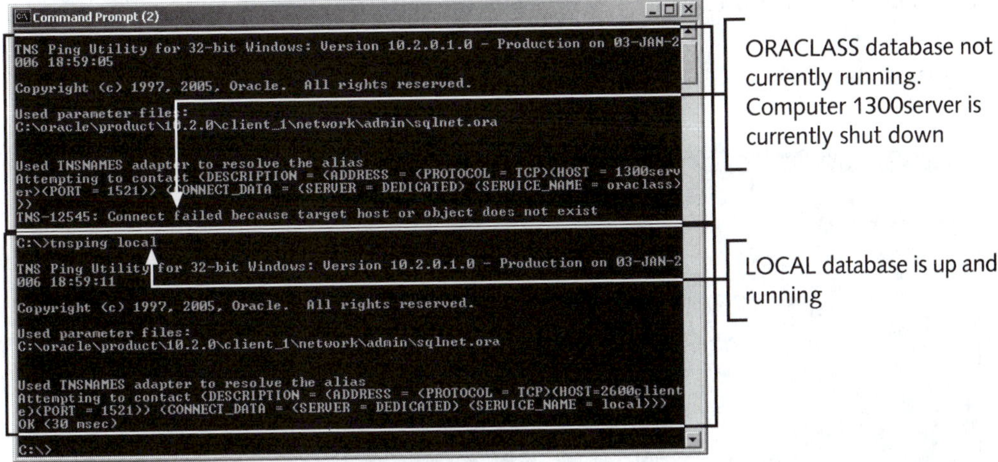

FIGURE 2-16 TNSPING pings a database server

To execute SQL*Plus, select **Start/All Programs/Oracle .../Application Development/ SQL Plus** from the menu. As shown in Figure 2-17 a user name, password and host string (network name) are required to connect.

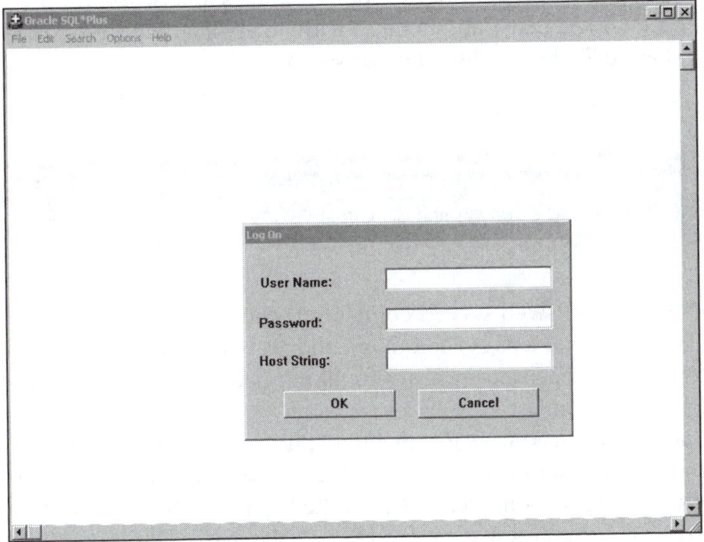

FIGURE 2-17 Connecting to SQL*Plus

To execute SQL*Plus Worksheet, select **Start/Programs/Oracle/Application Development/SQLPlus Worksheet** from the menu. Figure 2-18 shows the connection screen for SQL*Plus Worksheet. Note how SQL*Plus Worksheet allows a connection as a database super user (SYSDBA and SYSOPER), and SQL*Plus does not. Connect to

SQL*Plus as a super user with the command-line version of SQL*Plus, as in the following example code:

```
sqlplus <username>/<password>@<tns network name>
```

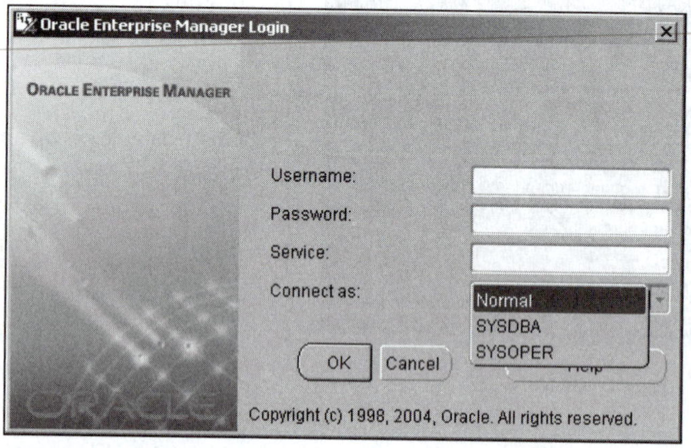

FIGURE 2-18 Connecting to SQL*Plus Worksheet

iSQL*Plus (Internet SQL*Plus) is a Web-based version of SQL*Plus. It gives you a way to write queries and other SQL commands via the Internet, or across a network, returning results in a Web browser. An Oracle installation comes with an application server called iSQL*Plus, a service in Windows called something like this:

```
OracleOraDb10g_home<n>iSQL*Plus
```

NOTE

The name of the service depends on the Oracle version and release in use.

The application server is an HTTP Web server, configured when Oracle software is installed. On a Unix or Linux system, run a **ps –ef** command and **grep** with the Oracle installation user name. Search for the **isqlplus** process. You should not need to start or stop the isqlplus process on a Unix or Linux computer.

Start up a Web browser and type the address for the iSQL*Plus service on the iSQL*Plus application server. The browser address to execute iSQL*Plus is as shown in Figure 2-19. Log in using user name, password, network name, and click the login button. The LOCAL database used here is added to the **tnsnames.ora** file as follows, a database called `LOCAL` running on a host (computer) called `2600client`:

```
LOCAL =
  (DESCRIPTION =
    (ADDRESS_LIST =
      (ADDRESS =
         (PROTOCOL = TCP)
         (HOST = 2600client)
         (PORT = 1521))
```

```
    )
    (CONNECT_DATA =
        (SERVICE_NAME = local)
    )
  )
```

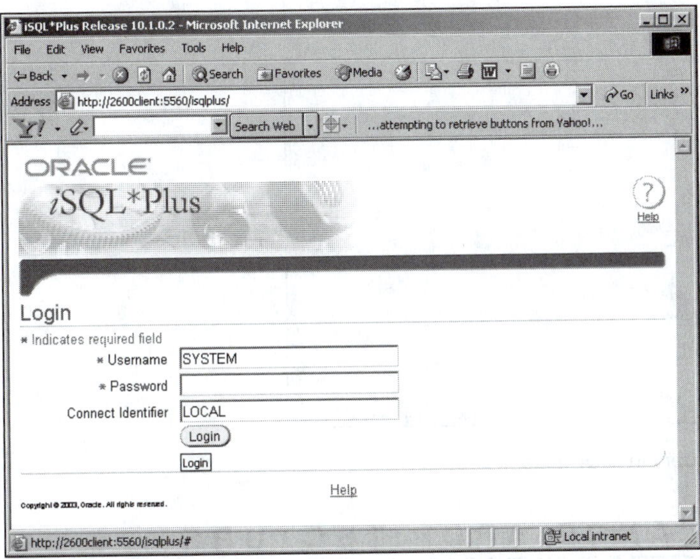

FIGURE 2-19 Executing iSQL*Plus

iSQL*Plus output looks like an HTML table, as shown in Figure 2-20. It may be helpful to think of the tools you have been working with so far as the outer shell of the database system. The next section takes you "under the hood" to look at the internal workings of the database.

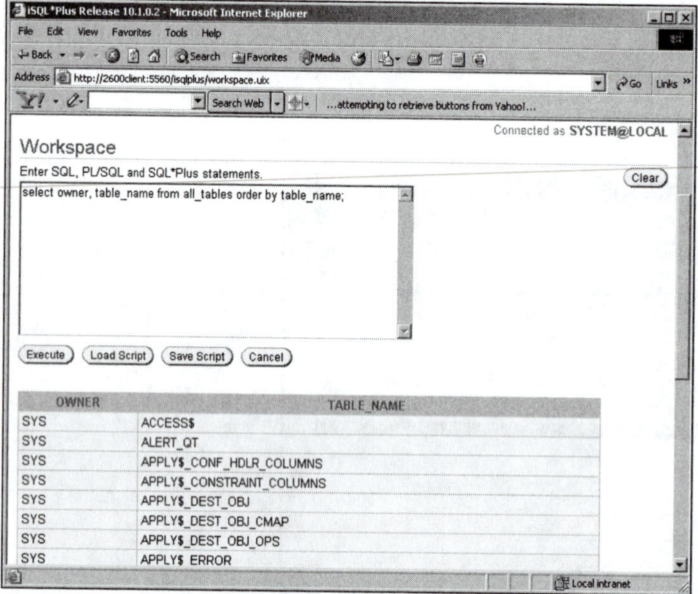

FIGURE 2-20 Executing a query in iSQL*Plus

ORACLE INSTANCE ARCHITECTURE

To start this section, remember that in Chapter 1 you learned that the database instance runs on a database server and uses data inside the database. The Oracle instance (instantiation of Oracle Database) is the part of an Oracle database executing in memory, when a database is started. That instance is made up of processes (running in the background) and memory structures. What actually occurs, internally in the database, when a user connects with that database? The next few sections describe the constituent parts of the Oracle instance in detail.

Shared and Dedicated Server Processes

The processes that support a user's activities on the database begin the moment the user logs on to the database. There are two configurations for this connection: the shared server mode and the dedicated server mode. Figure 2-21 shows an overview of the processes involved when the database is in shared server mode.

The user's application might be a C program, a Java applet, or an Oracle tool, such as SQL*Plus. In any case, after a request to connect to the database is received by the database server, Oracle creates a user session. The user session handles the communication with the user's application.

Then, Oracle creates a server process. In the case of a **dedicated server**, every user session (connection) has its own server process. With a shared server, several user sessions may share a single server process. The dispatcher background process distributes the connections from user sessions to server processes. A **shared server** uses CPU and memory more efficiently than a dedicated server by swapping out user sessions during idle time. This

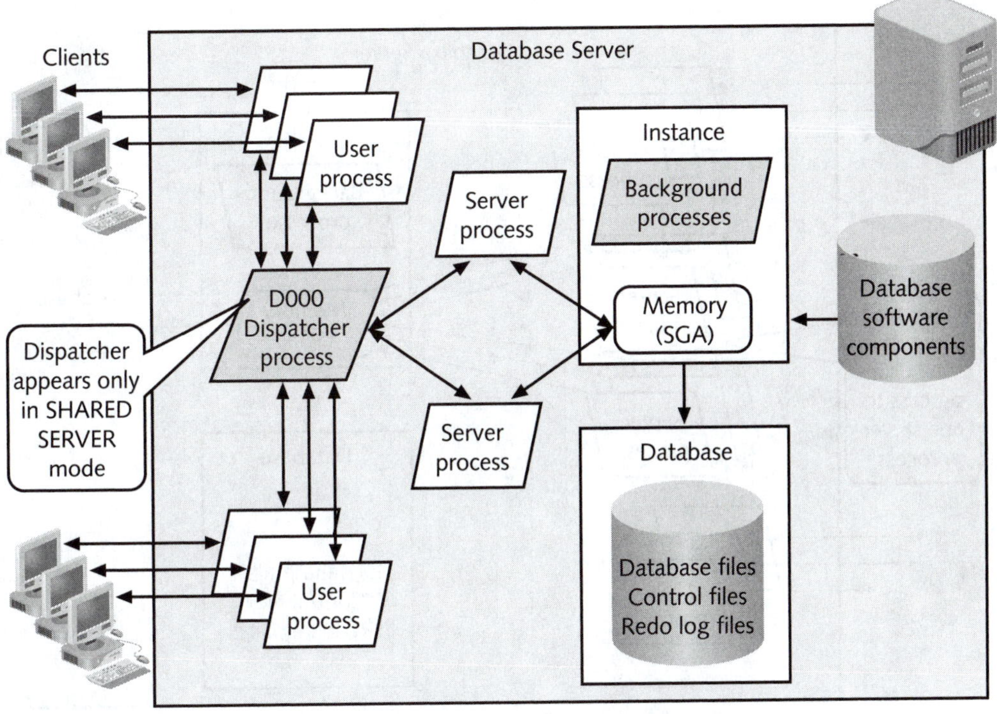

FIGURE 2-21 Shared server mode uses a dispatcher to distribute connections

is the best configuration for an OLTP database system with a large number of users (typically more than 1,000 concurrent user connections), especially if they are online users, because there is usually plenty of idle time between activity for sharing the server process.

Figure 2-22 shows the user and server processes for a database in dedicated server mode. Everything is the same, except there is no dispatcher to be the intermediary between the user sessions and the server process. Every user session has its own server process, which is why they are termed dedicated. This is the best configuration for a database system with large amounts of memory and CPU power because user sessions never have to wait for a server process to be available.

The link from the user session, through the server session, and to the database instance is called a connection. When you log on to SQL*Plus, for example, the tool tells you that you are connected. This means that the tool has established a connection with the database instance through the user session and server process on the database server. A session lasts from the time you make a connection until you end the connection. When you log off, for example, or exit an application, you end your session.

Setting a user connection as being dedicated or shared can be controlled in the client-side configuration of Oracle Net Services, in the **tnsnames.ora** file, for shared server processes, as shown in the following configuration:

```
ORACLASS =
  (DESCRIPTION =
    (ADDRESS_LIST =
```

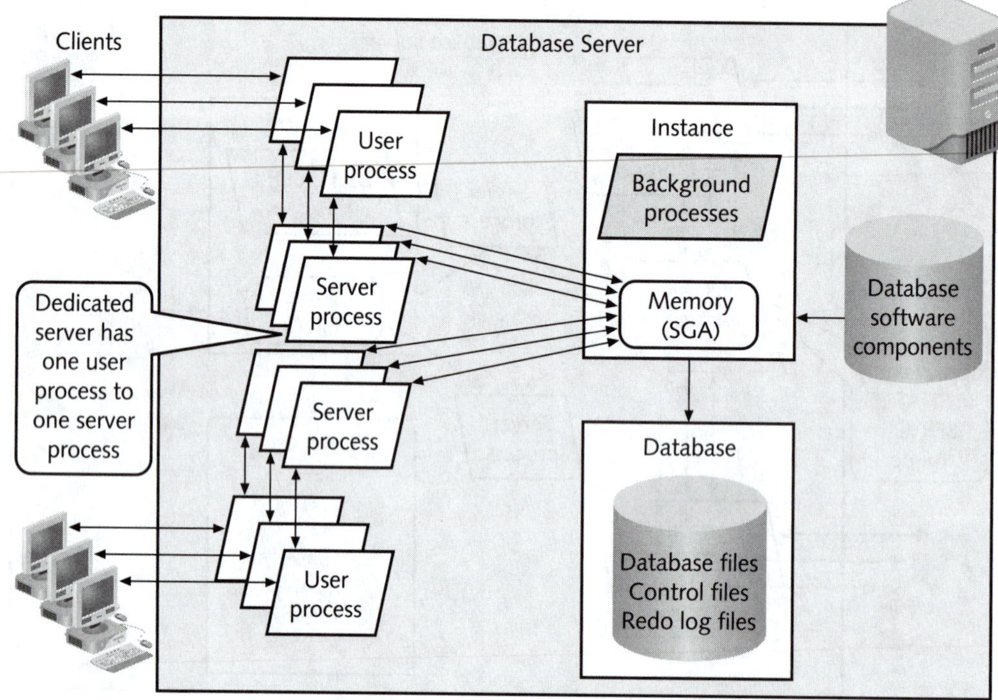

FIGURE 2-22 A single dedicated server process can service many user connections

```
        (ADDRESS = (PROTOCOL = TCP)(HOST = 1300server)(PORT = 1521))
      )
      (CONNECT_DATA =
        (SERVICE_NAME = oraclass)
        (SERVER = SHARED)
      )
    )
```

And for dedicated server processes, as shown in the following configuration:

```
ORACLASS =
  (DESCRIPTION =
    (ADDRESS_LIST =
      (ADDRESS = (PROTOCOL = TCP)(HOST = 1300server)(PORT = 1521))
    )
      (CONNECT_DATA =
        (SERVICE_NAME = oraclass)
        (SERVER = DEDICATED)
    )
  )
```

Now you have established a connection. The next step occurs when you perform a task that requires the database to interact with your user session. For example, you have logged on to SQL*Plus and you now execute a query. This is where the database's memory and background processes come in to play.

Background Processes

The background processes support and monitor the server processes and handle database management tasks to keep the database running efficiently and to help maintain fast performance. Figure 2-23 shows most of the background processes and how they interact with the System Global Area (SGA) and the datafiles. The more obscure background processes are excluded from Figure 2-23 to avoid cluttering the diagram. Snnn, CJQn, QMNn are a little too obscure. And those under the section describing other background processes are even more obscure than obscure.

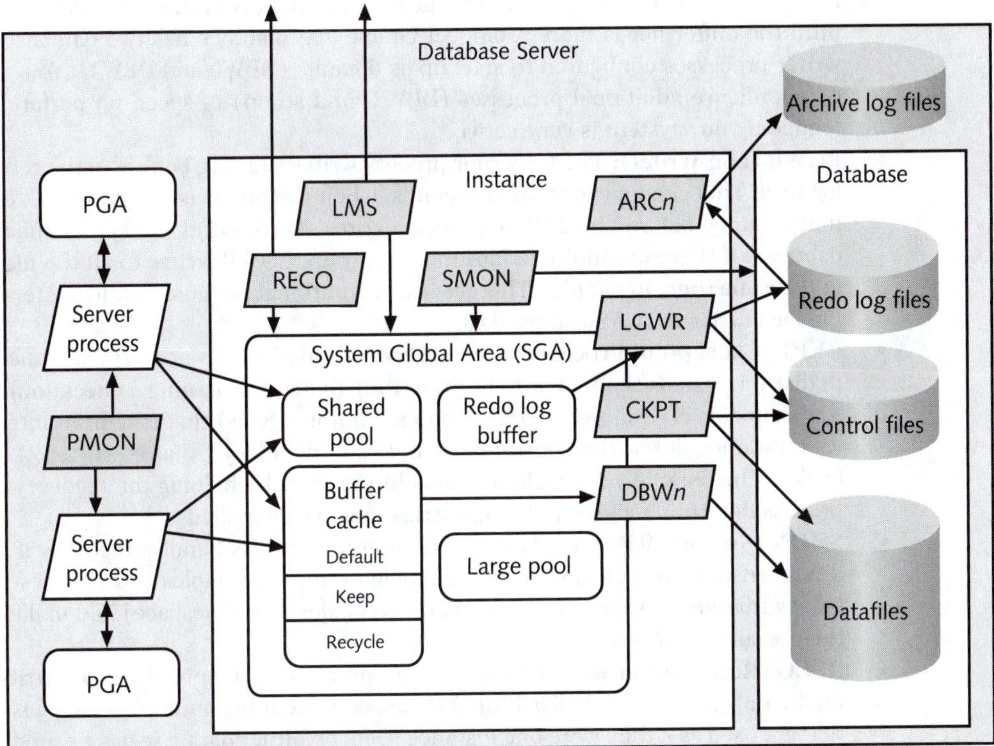

FIGURE 2-23 The background processes handle activity behind the scenes

> **NOTE**
> The SGA contains many of the memory buffer components for an up and running (online) Oracle instance.

The following list describes how the background processes pictured in Figure 2-23 work within the database.

- **PMON (Process Monitor):** The process monitor process cleans up after user sessions are finished. Any resources the user sessions were using are restored. PMON also monitors the server and user sessions and tries to restart them if they have stopped unexpectedly.
- **DBW*n* (Database Writer):** The database writer process writes buffers to datafiles. The only buffers that are written back to the datafiles are buffers that have been modified. For example, you issue an UPDATE command that modifies ten rows. The buffer containing the changed data is in the buffer cache, until the buffer needs more room. An Oracle 10g instance has two database writer processes configured to start up as defaults (DBW0 and DBW1). You can configure additional processes (DBW2, and so on) to speed up performance if your system is very active.
- **LGWR (Log Writer):** The log writer process writes redo log buffers to the redo log files. This occurs every three seconds, when the buffer is one-third full, or immediately before the DBW*n* process writes its changed buffers to the datafiles. If there are multiple files in a log group, LGWR writes to all the files at the same time, in parallel. This acts as an additional safeguard for logs, storing log entries in two separate files.
- **CKPT (Checkpoint Process):** The checkpoint process does not write any data to disks. It signals the DBW*n* to begin writing to disk, by issuing a **checkpoint**. A checkpoint is assigned a **system change number (SCN)** that is written into each redo log entry, the control files, and each data block that is written back to the datafiles. A checkpoint aids in recovery by helping the recovery process determine which redo log entries must be reapplied.
- **SMON (System MONitor):** The system monitor process handles recovery if it is needed. It also cleans up any unneeded temporary tables and restores blocks that are released (such as when you coalesce a tablespace) and makes them available for use.
- **RECO (Recoverer Process):** The recoverer process is only present in a distributed database system. A **distributed database system** has multiple instances that are used as if they were one instance. Data on either database can be modified by either instance. The recoverer process communicates with other databases when the databases are connected in a distributed system. Distributed systems allow data changes to span more than one database. The recoverer process looks for distributed processes and helps fix errors in changes that have failed because of communication problems. The recoverer process keeps trying to make the connections that are needed, making the distributed system better at recovering from communication problems.
- **ARC*n* (Archiver):** The archiver process is only present when a database is in ARCHIVELOG mode. The archiver process copies the redo log files to a separate location outside the database. Redo log files are cyclic (they are rotated and reused). This preserves the redo log file data as archived log files as historical changes to the database because redo log files are reused for further database changes. This allows database changes no longer retained in redo log

files to be used for recovery at a later date, if needed. An Oracle 10g instance has two archive log writer processes configured to start up by default (ARC0 and ARC1). You can create up to 10 (ARC0 through ARC9) archiver processes.

- **Dnnn (Dispatcher Process):** Dispatcher processes are shown in Figure 2-21, numbered as D000, D001, and so on. Dispatcher processes are only present in shared server mode to distribute user connection requests, between server processes. In a dedicated server environment, no dispatcher processes are started.
- **Snnn (Server Process):** Server processes service database connections (between database server and users) both with dedicated server configuration and with shared server configuration (using dispatchers). Multiple server processes can be executed, such as S000, S001, and so on.
- **CJQ*n* (Job Queue Process):** The job queue processor runs scheduled jobs, submitted in background, using the DBMS_JOBS package, as scheduled batch job operations. An Oracle 10g instance has a single job queue process configured to start by default (CJQ0). There can be more than one job queue process running at the same time, if load requires it (CJQ1, CJQ2, and so on). Job queue processes also spawn job queue slave processes numbered as J000, J001, and so on.
- **QMN*n* (Queue Monitor Process):** This process is optional and is utilized when running Oracle Streams Advanced Queuing. Up to 10 queue monitor processes can be configured (QMN0, QMN1, and so on). By default only one process is started (QMN0).
- **Other background processes:** MMON does various manageability-related background tasks. MMNL does light-weight manageability-related tasks. MMAN covers internal database tasks. RBAL, one or more ORBn, and OSMB processes are present in an Automatic Storage Management instance.

Memory Components

The two main sections of memory for the Oracle instance are the System Global Area (SGA) and the Program Global Area (PGA). Figure 2-24 shows a map of the PGA, the SGA, and most of the components within each. The java pool and the streams pool are left out of Figure 2-24 so as to make the diagram easier to read.

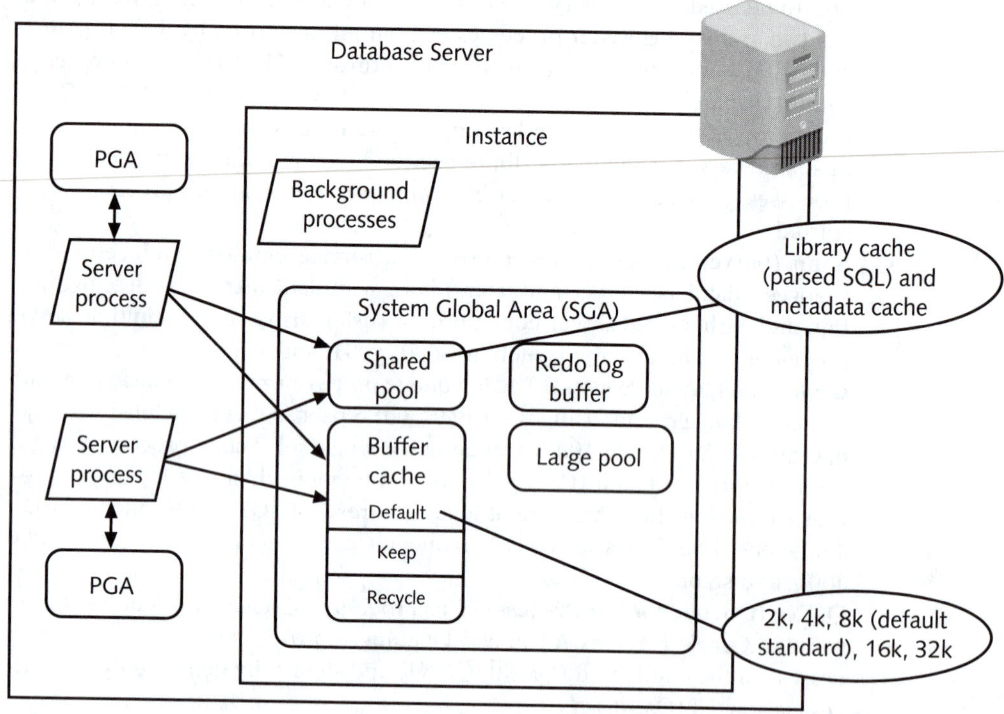

FIGURE 2-24 Memory has two primary components, SGA and PGA

The SGA is allocated when an instance is started and is deallocated when the instance is shut down. The memory components within the SGA are:

- **Buffer cache:** Data blocks stored on disk that have been read or modified are stored in memory (in the buffer cache) for quick retrieval from random access memory (RAM). The blocks remain in memory as long as there is room in the buffer cache. The more frequently a block is used, the longer it stays in the buffer. The buffer blocks are reused as needed to make room for newer data blocks. The buffer cache is populated when a server process requests data (for example, when you execute a query) from the Oracle instance. If the buffer already has the block requested, the server process reads the block from the buffer. If not, the server process reads the block from disk to the buffer, and then uses the block. The block stays in the buffer until the buffer needs to reuse the space. In Oracle 10*g* the buffer cache can consist of multiple buffer caches, of different sizes. Those sizes can be 2K, 4K, 8K, 16K, and 32K—depending on the operating system. The default size is 8K. When creating a tablespace with a block size that is not the default size for the database, then a matching buffer cache must be created. The keep and recycle pools are an alternative to multiple buffer caches of varying block sizes, if a little out of date. The keep pool retains data in the buffer for longer. The recycle pool removes data from the buffer more rapidly.

NOTE

Blocks can be read from disk to buffer cache more than one block at a time. Typically an I/O heavy database environment, such as a data warehouse, will perform better when reading lots of contiguous blocks at once. The term contiguous means physically "next to" on disk.

- **Shared pool:** This area stores parsed SQL in memory, in the library cache section of the shared pool, when a server process sends a request, such as a query. In the shared pool, Oracle stores results of parsing (interpreting the SQL into machine language) and the results of the Optimizer's path selection process. Oracle compares any new requests coming in with requests already in the shared pool. If possible, Oracle reuses old requests to save time. Oracle has a list of comparison rules, which it uses to decide whether to reuse a stored request. If it cannot find a match, the new request is parsed, and the Optimizer determines the path. The other part of the shared pool is the metadata cache, or row cache. The metadata cache stores dictionary objects or metadata. Metadata is the data about the data, or all the tables, indexes, and so on, in the database.
- **Redo log buffer:** Whenever a change is made to data, the redo buffer stores a copy of the changed data and the original data, in case it is needed.
- **Large pool:** This optional memory area improves response time for background processes and for backup and recovery processes.
- **Java pool:** Used for java code buffering specific to user sessions.
- **Streams pool:** This is used for Oracle Streams. An Oracle Stream is like a direct pipeline between different databases, often used to manage replication (duplication) of data across multiple databases.

The PGA is effectively used in session connection memory and is broken into private chunks for each server process. This is the work area for the application code that works with the data for the application. The application's compiled code resides here along with copies of data blocks that the application is working on.

This section introduced you to all the internal workings of the database instance (the background processes and SGA) and the networking component of the database (Oracle Net Services). Next, you explore the front door of the database where it is easy to decipher the contents of your database: Enterprise Manager and the Database Control.

INTRODUCING ENTERPRISE MANAGER

In previous versions of Oracle, executing the Enterprise Manager console (or simply console) was complicated, and occasionally problematic at best. The reason was that too much power was placed into the console software. In Oracle 10g that power is divided between the console and the Database Control (Grid Control for clustered servers). The Database Control is a browser-based Web GUI (graphical user interface), containing much of the more complex functionality of database administration. Let's begin with the console.

Running the Enterprise Manager Console

Enterprise Manager and its console can be used as a centralized control center for many of the available DBA tools. Use it to find your way around the database quickly without getting bogged down in SQL commands and complex data dictionary views. Much of what you need to know easily can be discovered using the Enterprise Manager console. For example, perhaps you are wondering who is working on the database right now. Find out by opening the Enterprise Manager console and viewing the active sessions listed under a database instance. Not only the current activity, but the database initialization parameters, the storage size, and the structural details about tables, views, and other database objects are easily visible in a navigation tree structure.

The Enterprise Manager console is only available in the Oracle client installation, for Oracle 10g. To start the Enterprise Manager console (referred to as the console throughout this book) in standalone mode, follow these steps:

1. Click **Start/All Programs/Oracle .../Enterprise Manager Console**. The Enterprise Manager Console Login screen appears.
2. If there are no databases defined within the console Navigator window, then the window shown in Figure 2-25 will be displayed. Figure 2-25 presents the option of adding connections to the console, for all databases existing as network names, in the **tnsnames.ora** file. Databases not to be added are deselected.

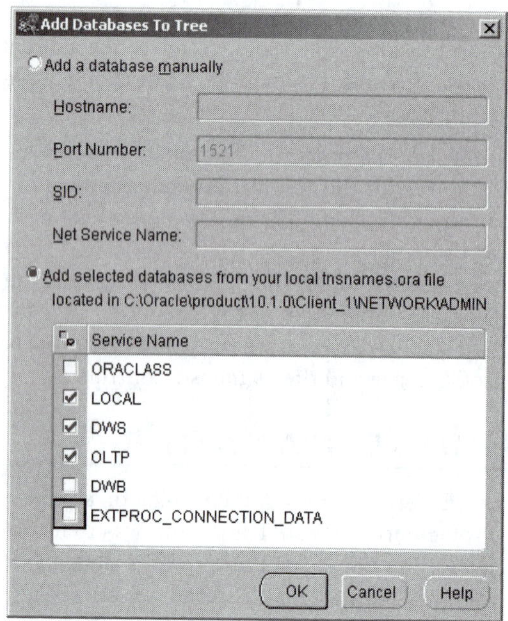

FIGURE 2-25 The Console adds network named databases to its Navigator tree

3. In the main window of Enterprise Manager console, double-click the **Databases** folder. This lists all available databases.

NOTE

The ORACLASS service name should already be listed. If it is not, perform steps 4 and 5. Otherwise, skip steps 4 and 5. *The next two steps are needed only when a new service name must be added to the console.*

4. You can also click the **Navigator** menu, and then click **Add Database to Tree**. This starts a wizard that adds the new database service to the console.
5. As shown in Figure 2-26, click the second radio button labeled "Add selected databases from your local **tnsnames.ora** file." Then adjust the check boxes so that only the **ORACLASS** database is checked. Click **OK** to complete the task.

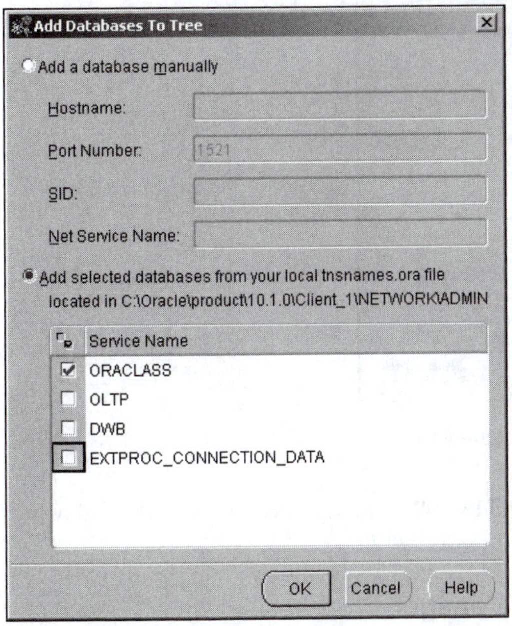

FIGURE 2-26 Add the new database to the console

The top radio button allows you to add databases that are not defined in your local Oracle Net Services configuration. This is a shortcut method to add new databases, because it updates the local **tnsnames.ora** file, adding the database to the console. It is limited, however, to the TCP/IP protocol only.

6. Expand the **ORACLASS** database node by either clicking the **plus sign** to the left of the node or double-clicking the **database name**. A Database Connect Information window appears.

7. Log on to the database as **SYSTEM** as shown in Figure 2-27: SYSTEM is a predefined user in the Oracle 10g database with DBA privileges. When using the console, you should log on as SYSTEM, so you have appropriate privileges to perform tasks (such as creating new users and tables) within the console. *Type the password provided by the instructor*. Click the check box so that the information is saved in the console's preferred credentials for this database. Clicking the check box initiates settings that make it unnecessary for you to type the user name and password; they are automatically entered when you expand the database node. Click **OK** to continue. A window opens explaining that Oracle stores encrypted passwords. Click **OK** to continue.

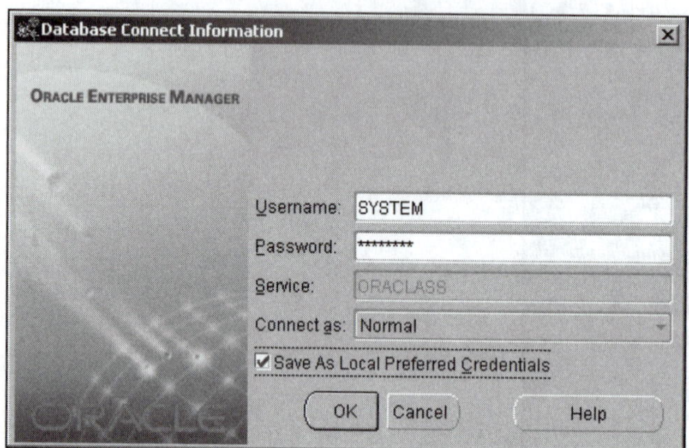

FIGURE 2-27 Set up the login once, saving it for the future

With the console open, begin the tutorial by following the directions in the following sections.

Viewing the Features of Enterprise Manager

Start the console as described previously and begin a tour of the Enterprise Manager's primary DBA tools. These tools aid in the creation and maintenance of database users, files, and tables.

There are four primary tools and many additional tools found in the console. The four primary tools are:

- **Instance Manager:** Monitors activities in the database. It does not allow starting and stopping of the database when running the console from a client machine (even logged in as SYS with the SYSDBA privilege).
- **Schema Manager:** Views table structures, creates new tables, indexes, views, and any other type of object stored in the database.
- **Security Manager:** Creates new users, allocates storage resources to users, and changes passwords.
- **Storage Manager:** Monitors storage use (tablespaces and datafiles) and adds more space as needed or adjusts settings on existing storage units.

- **Other Management Tools:** Other less frequently utilized tools are Distributed Database Manager, Warehouse Manager (data warehouse database), Workspace Manager (database versioning of database objects), and finally XML Database Manager.

See Table 2-1 for a list of other tools and utilities found in the console.

The next sections are tutorials that guide you through each of the four major console tools.

Instance Manager

You are already in the console, so follow these steps to examine the Instance Manager.

1. Double-click the **Instance** icon (or click the **plus sign**). The window on the right displays a description of the Instance Manager tool, with access to help screens and descriptions. All the top-level icons in the console have an introductory window similar to this one.
2. Click the **Configuration** icon. This brings up a status window on the right side of the console. This section of the Instance Manager helps with these tasks:
   - Setting initialization parameters
   - Monitoring and adjusting memory usage
   - Locating archive log files
   - Monitoring of resource sharing
   - Automated Undo management
3. Click the **All Initialization Parameters** button. The list displays all initialization parameters available and their current settings. Changes to parameters can be made in this list. The method of applying a change to a parameter depends on the type of parameter. Dynamic parameters (those with a check in the Dynamic column) can be reset immediately by simply clicking the Apply button that executes the change immediately. Static parameters can be changed here, but the change must be saved in a configuration parameter, initialization file (executed on database startup, **init.ora)**, and then the database must be shut down and restarted to apply any changes.
4. Select **audit_trail**, and then click **Description**. As you see in Figure 2-28, a short definition of the use of this parameter displays in the bottom of the window. This feature enables you to quickly determine the effect that changing a parameter might have on your database performance.
5. Click the **Category** column heading, resorting the list by category, and scroll down to the **Pools** category and then to the **SGA Memory** category. As you can see, there are a large number of parameters that affect the memory usage of the database instance.

NOTE

The ability to sort lists by column heading appears throughout the tools in the console.

6. Click the **Cancel** button to return to the main console window.

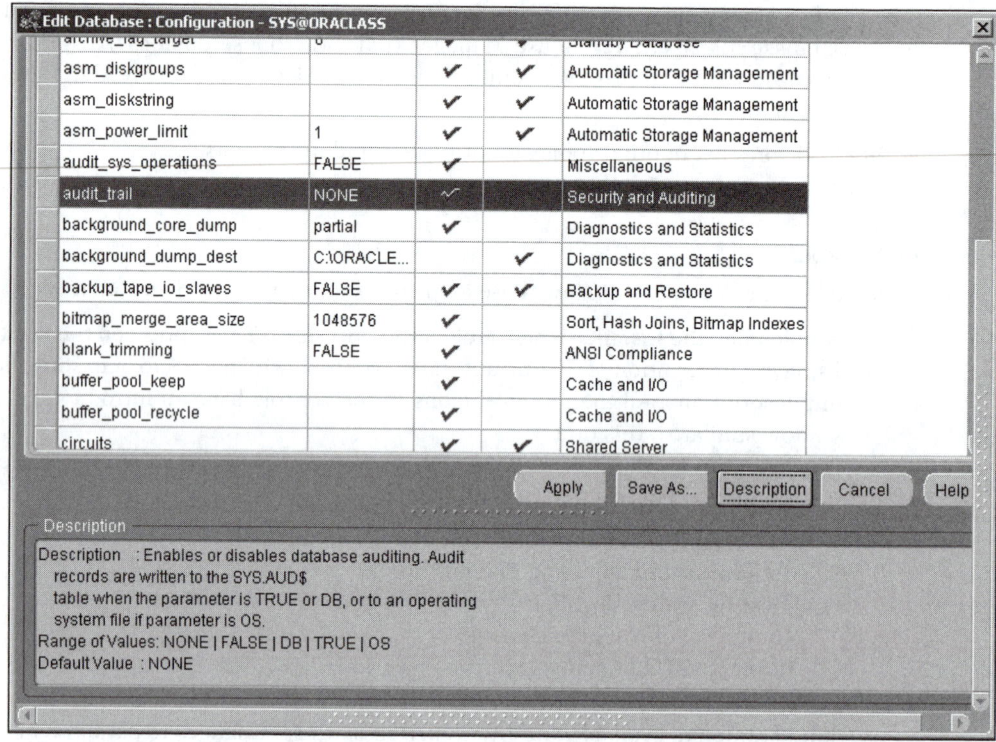

FIGURE 2-28 Initialization parameters have default and customized settings

NOTE

Changes made in the console are applied to the database through SQL commands that are built and then executed in the background. To see the commands, click **Navigator/Application SQL History** in the menu.

7. Double-click **Sessions**. This reveals a list of users with sessions started in the database. At least six background processes are also listed. The number of background processes varies depending on your database configuration.
8. Click **SYSTEM** under Sessions. The right side displays details about the session as shown in Figure 2-29. This is the session you are running while in the console. (Recall that you logged on to the database with the user name SYSTEM.) Notice the **Kill Session** button. This button allows you to end a user session from the console. For example, a new programmer might have created and run a query that has incorrect selection criteria. Because of the error, the query has run for 20 minutes and has slowed down the database enough to cause other users to call the DBA and complain. You, as the DBA, find the user session and stop it using the **Kill Session** button.
9. Collapse the Instance Manager node by clicking the **minus sign** next to the Instance icon.

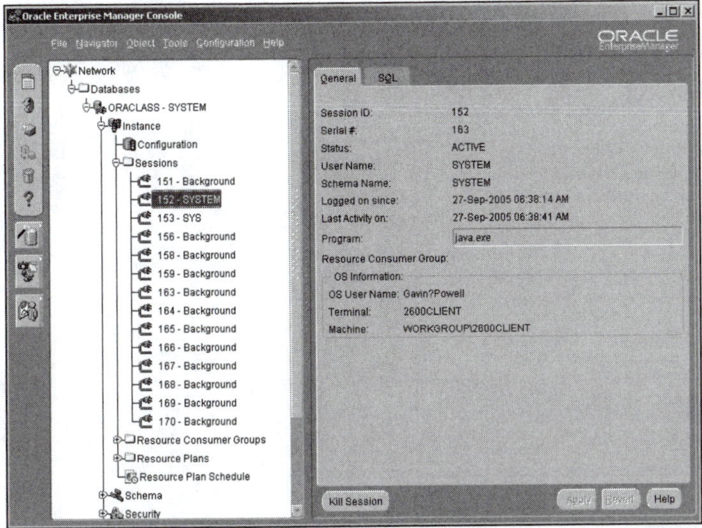

FIGURE 2-29 Individual users show up here when they use the database

Leave the console open, and continue to the next section to examine the Schema Manager.

Schema Manager

The Schema Manager provides a way to manipulate anything in the database that is in a schema. A schema is a set of database structures, such as tables, indexes, user-defined attributes, and procedures. Each schema is owned by one Oracle user and is identified by the user's name. For example, the user named Sidney creates two tables and a view. These three objects are collectively called the Sidney schema. A database object is not only an object table, but also any other structure held in the database, such as a relational table, an index, a PL/SQL procedure, a Java servlet, or even a customized attribute with methods and rules attached.

Begin using the Schema Manager by completing the following steps:

1. Double-click the **Schema** icon in the console. The list you see below the Schema icon, as shown in Figure 2-30, are the users that own database objects stored in the Oracle 10g database.
2. Scroll down in the left window and double-click the **SYSTEM** schema. All the objects in the SYSTEM schema are listed in the right side of the console. All object types are listed in the left side.
3. Double-click the **Tables** folder. A list of tables appears below the folder. In the right window, a sorted list displays all the tables owned by SYSTEM, their name, tablespace, and so on.

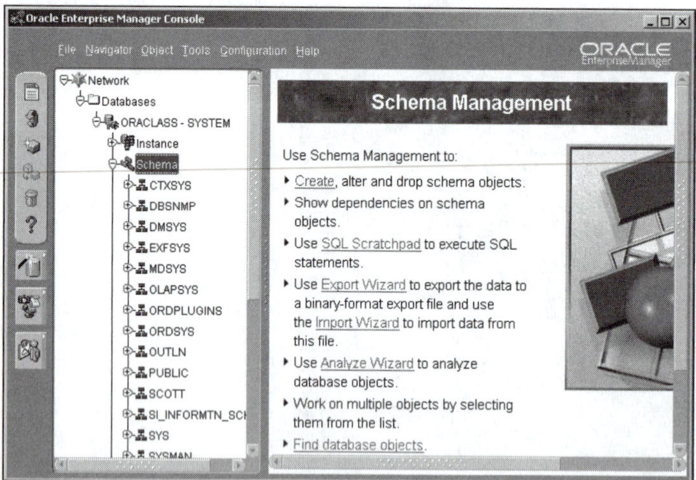

FIGURE 2-30 Schemas contain many types of database objects

4. Scroll down and double-click the **HELP** table. Now, a Property window appears in the right side, as shown in Figure 2-31. The Property window shows the columns contained in the table, and much more. Notice the tabs along the top named Constraints, Storage, and so on.

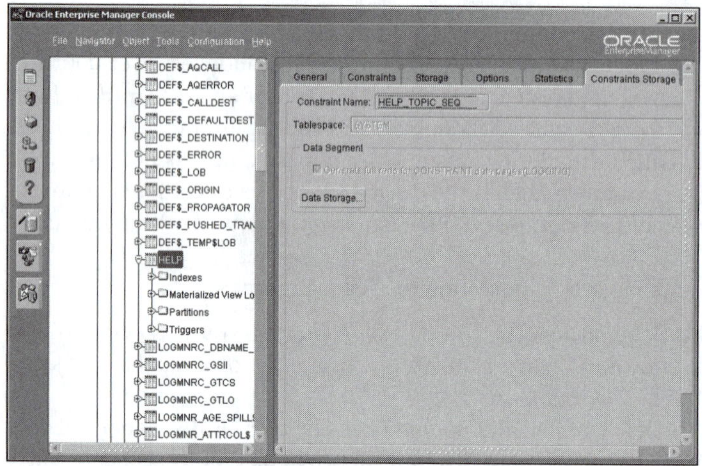

FIGURE 2-31 The table property sheet displays many details about a selected table

NOTE

If a Constraints Storage tab appears at the top right of the window, as in Figure 2-31, after clicking the Constraints tab, and then doesn't disappear after selecting a different tab; don't worry about it. It's probably a little interface bug. It's not going to cause a problem.

The Property window and its related tabs allow you to perform these tasks:

- Add a new column
- Remove an existing column
- Change the characteristics of a column, such as its data type or size
- Add and remove constraints, such as primary and foreign keys
- Adjust storage allocation
- View current statistics such as number of rows

5. Click the **Indexes** folder below the HELP table in the left side of the console. This displays a list of indexes created on this table on the right side of the window. You can also view all indexes under the Indexes folder listed on the same tree level as the Tables folder.
6. Right-click the **HELP** table. A pop-up menu appears as shown in Figure 2-32. This menu displays more tasks that you can perform on a table using Schema Manager. For example, one of the menu items is Remove, which drops the table from the database. Another menu item is View/Edit Contents, which displays a spreadsheet for editing the table's data.

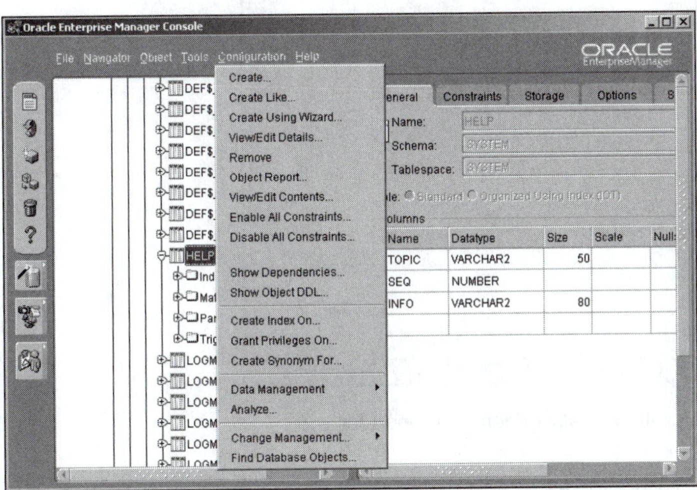

FIGURE 2-32 The pop-up menu lists many specialized tasks for tables

7. Select **Show Object DDL** from the pop-up menu. DDL stands for Data Definition Language. After selecting Show Object DDL, Oracle 10*g* generates the SQL command used to create the object and displays it in a window. Viewing the SQL code for tables, indexes, and other objects can be a useful learning aid. You have the option of saving the DDL to a file as well.
8. Click **Close** to return to the main console window.
9. Scroll down and right-click the **Views** folder. A new pop-up menu with different selections appears.
10. Select **Save List** in the pop-up window. A window appears giving you some choices on how your list will be generated and saved. The Save List function

can be used in many areas of the console to generate a list of the items you are viewing and save them to a text file or HTML document. Figure 2-33 shows an example of the list of views saved in HTML format.

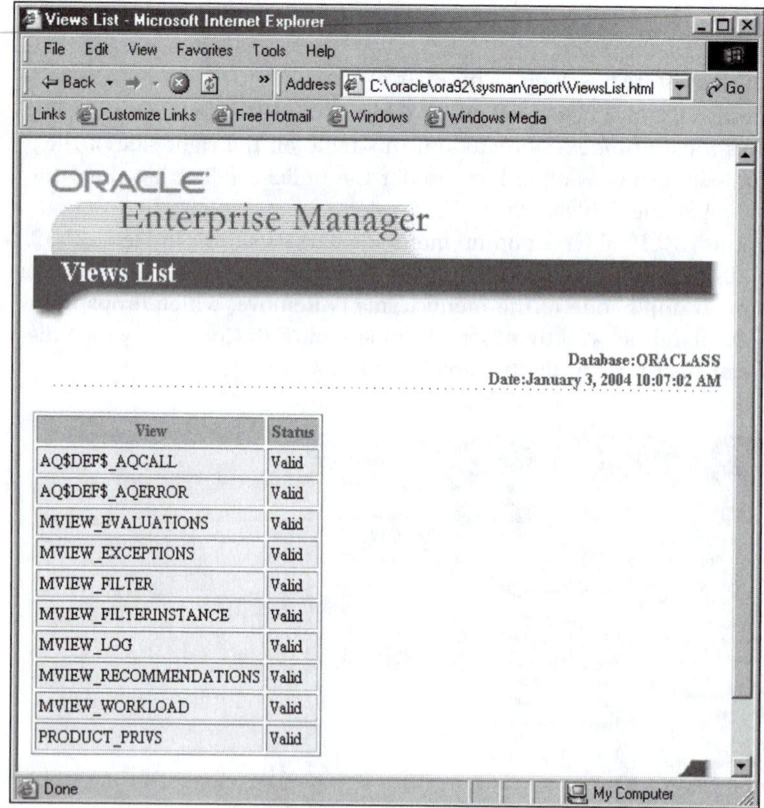

FIGURE 2-33 The console can automatically generate a report for you

11. Click **Cancel** to return to the console window.

Leave the console open to continue the tutorial in the next section, which covers the Security Manager.

Security Manager

The Security Manager's primary tasks involve setting up and maintaining users in Oracle 10g. Follow these steps to examine Security Manager in the console.

1. In the console, double-click the **Security** icon. The introductory window of the Security Manager appears, displaying an overview of its purpose.
2. Double-click the **Users** folder. This displays a list of users.

3. Scroll down and select the **SYSTEM user**. A Property window appears for the SYSTEM user on the right side of the console, shown in Figure 2-34. Here are some of the tasks you can perform in this property sheet:
   - Change the password
   - Expire the password (forcing the user to change her password)
   - Modify the default tablespace where all objects created by a user are stored
   - Assign or revoke roles and privileges
   - Adjust storage limits (quotas)

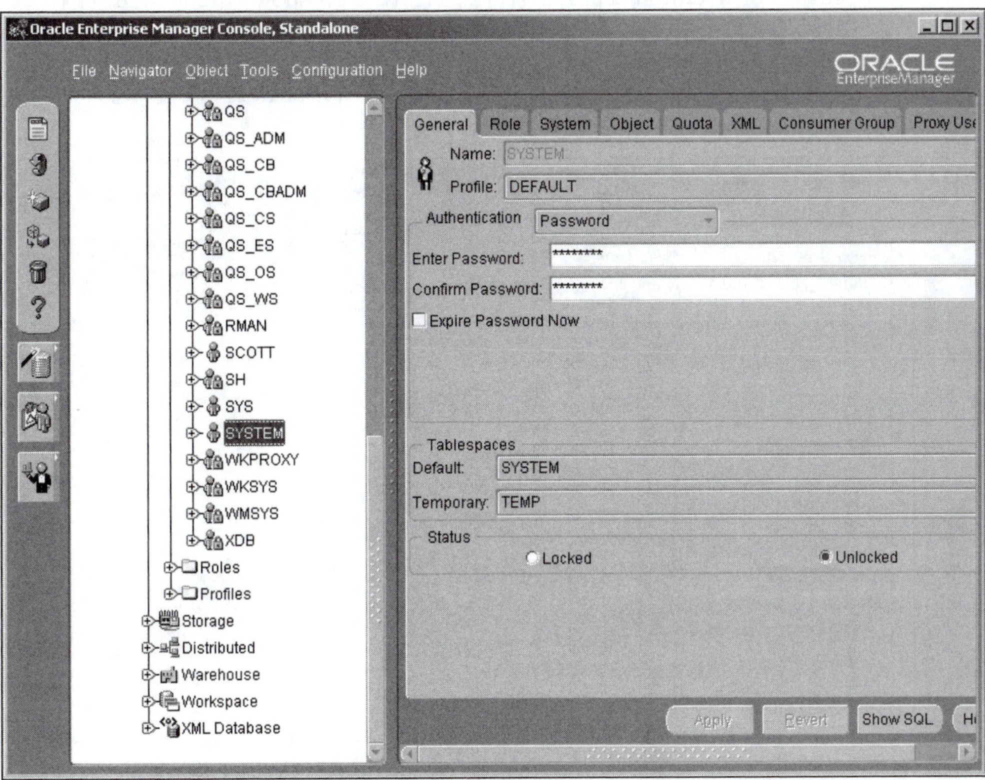

FIGURE 2-34 The user property sheet allows changes to passwords, among other tasks

4. Double-click the **Roles** folder. This displays a long list of roles that are predefined in the Oracle 10g database to help organize all the different permissions needed for various tasks. A **role** is a set of related privileges that can be granted as a set to another role or to a database user.
5. Scroll down on the left side of the console and select the **RESOURCE** role. The Property window for the role appears in the right side of the console. The tabs across the top represent the types of privileges that can be assigned to a role:
   - **Role**: Embed the privileges of the selected role into this role
   - **System:** Allows system-related tasks, such as creating users or tables

- **Object:** Allows access (select, insert, update, delete, or a combination of these) to data in tables, views, or other database objects
- **XML:** Allows access to XML attributes
- **Consumer Group:** Sets the resource allocations that are inherited by any user assigned the role

6. Click the **System** tab. The privileges listed at the bottom of the window are those granted to the RESOURCE role. In previous versions of the Oracle database the RESOURCE role was assigned to any user allowed to create objects in the database. Oracle is now trying to discourage use of the RESOURCE role because of potential security issues. Figure 2-35 shows the content of the RESOURCE role, allowing allocation of group resources to users. (The RESOURCE role was typically used for developers, not end users.) For example, the CREATE TABLE privilege allows a user to create a table.

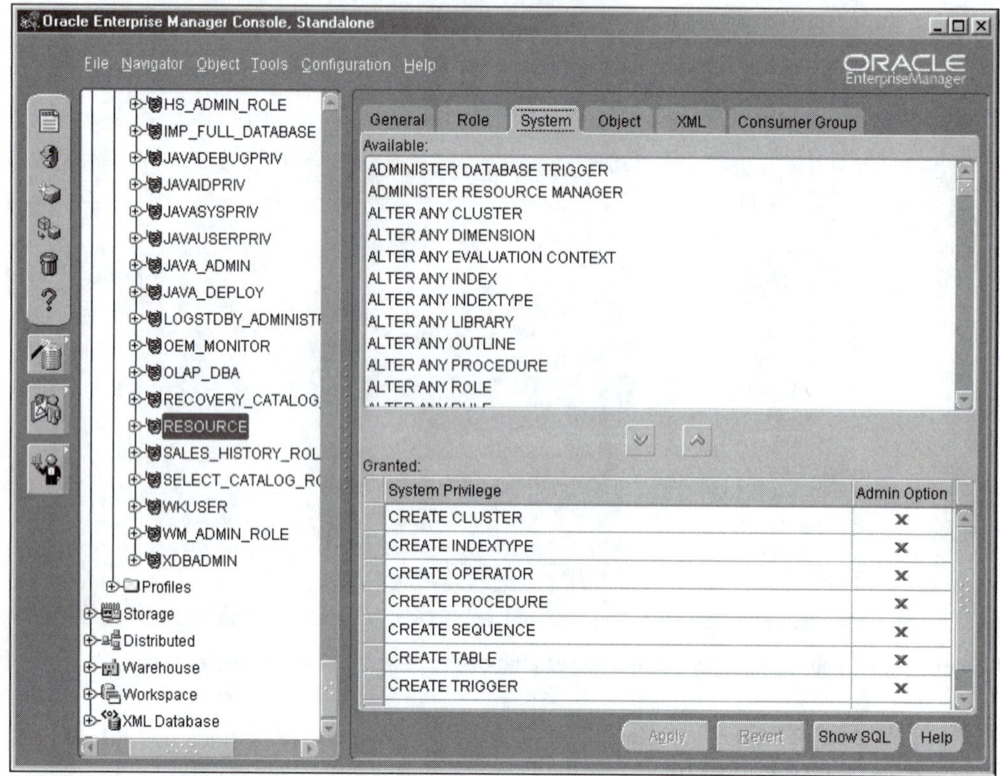

FIGURE 2-35 Roles are assigned sets of privileges

Keep the console as it is to continue to the next section, which describes the Storage Manager.

Storage Manager

The final primary tool you are to examine is the Storage Manager. This tool, as its name implies, performs storage-related tasks in the Oracle 10g database. Follow the steps to see the main points of interest in the Storage Manager.

1. Double-click the **Storage** icon on the left side of the console. This starts the Storage Manager.
2. Select **Tablespaces** under the Storage icon. The right side displays data about all the tablespaces in the database. In Figure 2-36, observe the column that illustrates the percentage of storage used in each tablespace.

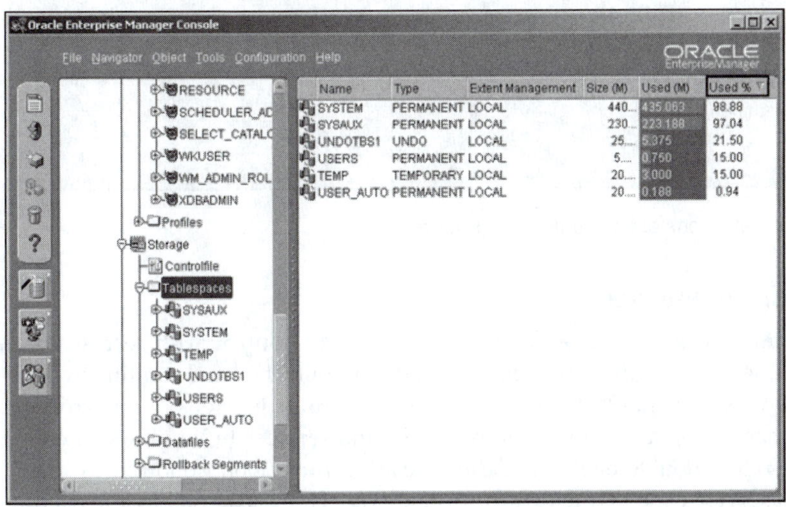

FIGURE 2-36 Tablespace storage information can easily be read at a glance

3. Double-click the **Datafiles** folder. A similar display of storage information appears in the right side.
4. Click the datafile with the name **TEMP01.DBF**. A property sheet displays details about this datafile.
5. Click the **Storage** tab. The storage screen displays as shown in Figure 2-37. Notice the check box that indicates the datafiles can be automatically extended. This feature saves a great deal of human intervention. It allows Oracle 10g to add more space, in the predefined increments shown in this window, to the datafiles when datafiles runs out of allocated space. You can also set a cap on the maximum size of datafiles to prevent them from consuming too much disk storage.
6. Close the console by clicking the **X** in the top-right corner of the window. However, it is better practice to go to the File menu at the top left, left click, and click exit. Traditionally a window closed using the **X** in the top-right corner, aborts all processing in a brutal fashion.

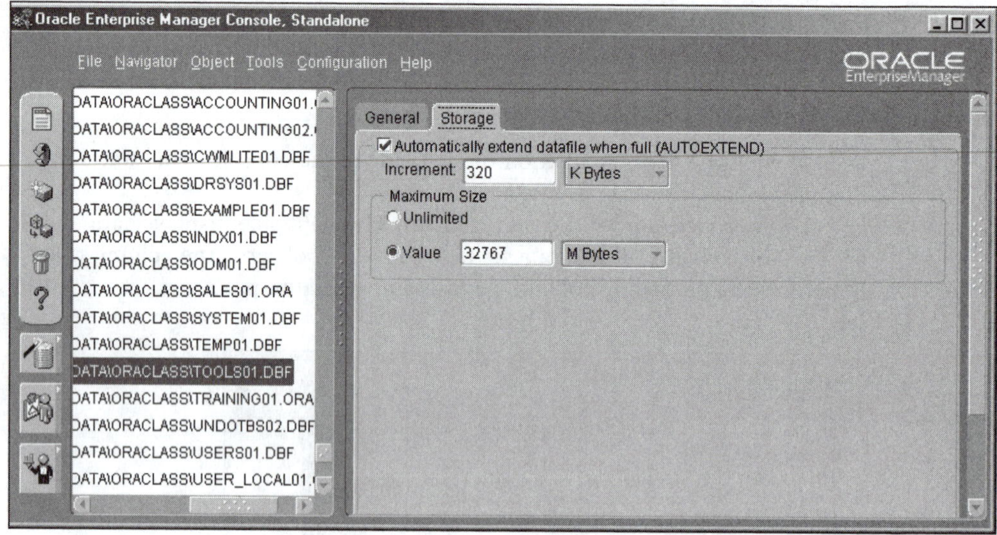

FIGURE 2-37 Storage definitions allow datafiles to grow in size

The Database Control Interface

Now let's briefly examine what can be done using the Database Control, with respect to database administration. Let's begin by getting the Database Control up and running by typing this URL into a browser: **http://2600client:1158/em**. If nothing happens or an error is returned, check the service on a Windows computer. To find services in Windows, click **Start/Control Panel**. Then double-click the **Administrative Tools** icon, followed by the **Services** icon. Find a service called **OracleDBConsole<SID>**. The **<SID>** postfix means the service is named for the database name within which its metadata is stored.

NOTE

To verify the Database Control is running on Unix or Linux, use a **ps** command, such as **ps –ef grep | oracle**, finding all processes running under the Oracle operating system user name.

NOTE

The Database Control requires a login.

NOTE

Port numbers in the Database Control URL vary with different versions of Oracle.

Much of what is contained within the Database Control is beyond the scope of this book. However, the sheer scope of its power and versatility should be mentioned. When the Database Control is executing, there are four tabs across the top left of the tool: Home, Performance, Administration, and Maintenance:

- **Home:** This screen provides general information about the database and the Oracle installation. Included is general information about database CPU usage, active database sessions, recovery capabilities, space usage efficiency, and diagnostics, all shown in Figure 2-38. Additionally, the Home screen contains alerts (both critical alerts and less significant warnings) and various related links and drill-down options, all shown in Figure 2-39.

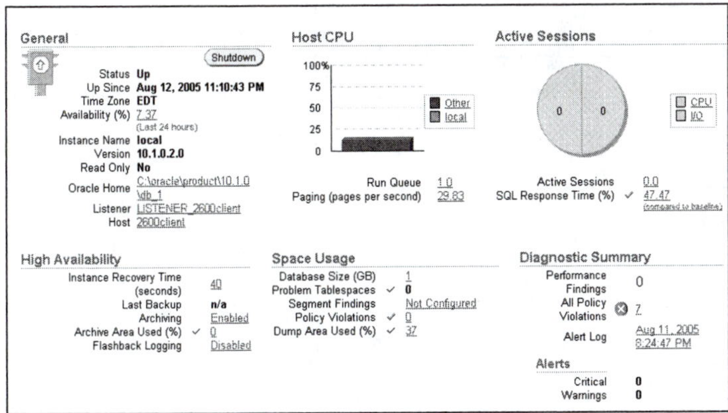

FIGURE 2-38 Database Control Home screen activity indicators

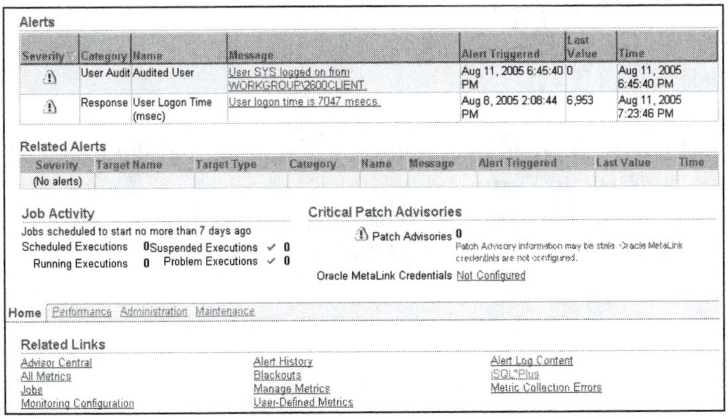

FIGURE 2-39 Database Control Home screen alerts, warnings, and links

- **Performance:** This screen shows performance information, both good and bad, as shown in Figure 2-40. The objective is to indicate problem areas and allow actions, to define a problem further, using drill-down links (as shown in Figure 2-41), such as Top Sessions, Top SQL, Database Locks, and others.

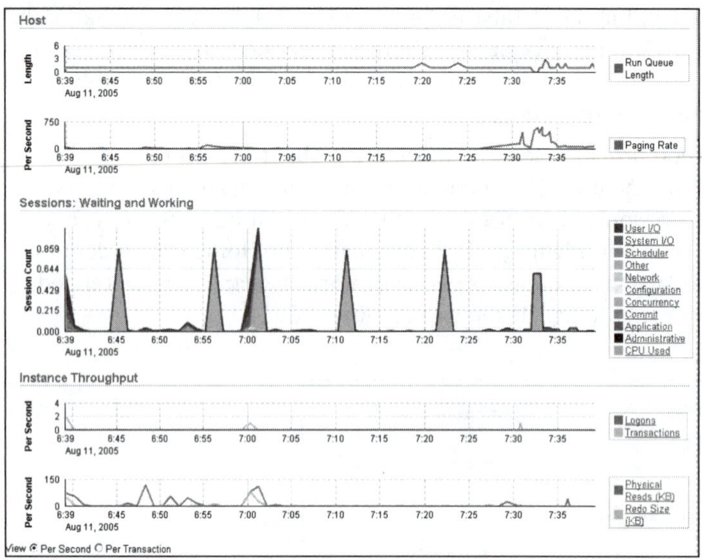

FIGURE 2-40 Database Control Performance screen graphical displays

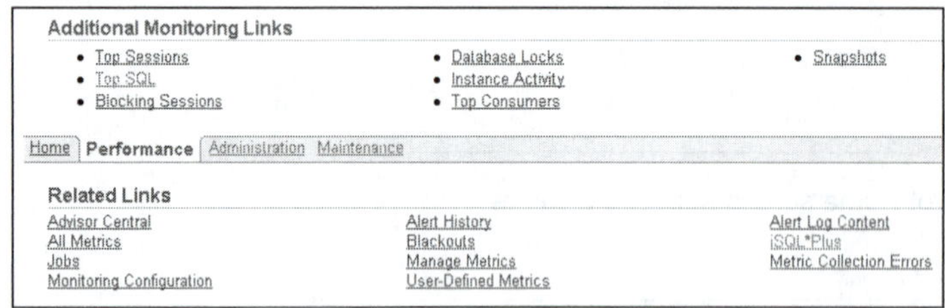

FIGURE 2-41 Database Control Performance screen alerts, warnings, and links (drill-downs included)

- **Administration:** This screen provides a multitude of options for administration functionality, as shown in Figure 2-42.
- **Maintenance:** This screen provides comprehensive maintenance access to database utilities, backup/recovery, and deployment activities, as shown in Figure 2-43.

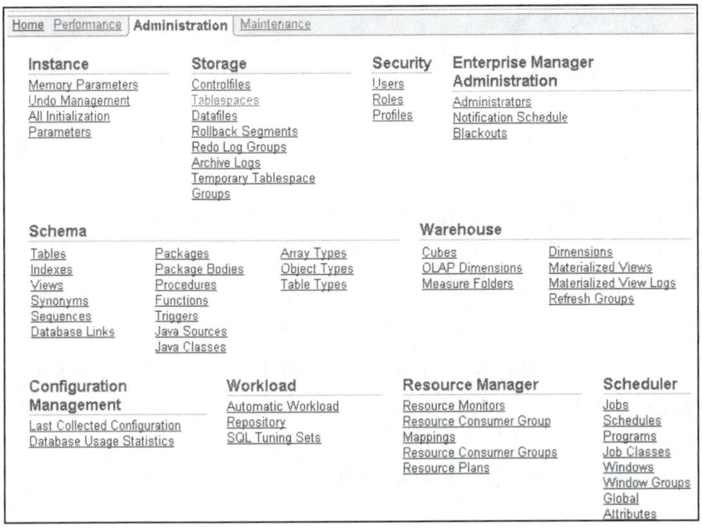

Home Performance **Administration** Maintenance

Instance
Memory Parameters
Undo Management
All Initialization Parameters

Storage
Controlfiles
Tablespaces
Datafiles
Rollback Segments
Redo Log Groups
Archive Logs
Temporary Tablespace Groups

Security
Users
Roles
Profiles

Enterprise Manager Administration
Administrators
Notification Schedule
Blackouts

Schema
Tables
Indexes
Views
Synonyms
Sequences
Database Links
Packages
Package Bodies
Procedures
Functions
Triggers
Java Sources
Java Classes
Array Types
Object Types
Table Types

Warehouse
Cubes
OLAP Dimensions
Measure Folders
Dimensions
Materialized Views
Materialized View Logs
Refresh Groups

Configuration Management
Last Collected Configuration
Database Usage Statistics

Workload
Automatic Workload Repository
SQL Tuning Sets

Resource Manager
Resource Monitors
Resource Consumer Group Mappings
Resource Consumer Groups
Resource Plans

Scheduler
Jobs
Schedules
Programs
Job Classes
Windows
Window Groups
Global Attributes

FIGURE 2-42 The Database Control Administration screen provides extensive administrative functionality

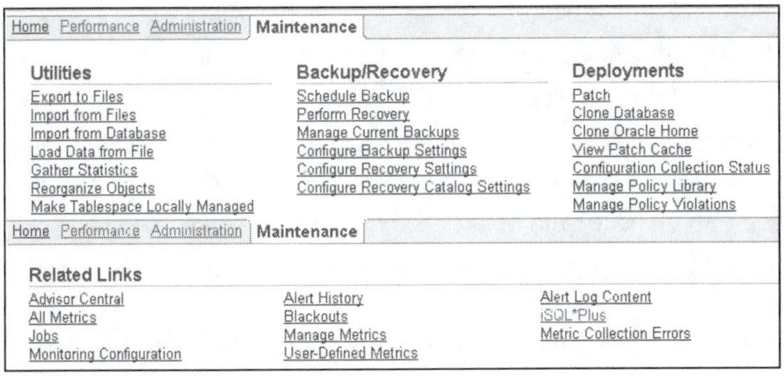

Home Performance Administration **Maintenance**

Utilities
Export to Files
Import from Files
Import from Database
Load Data from File
Gather Statistics
Reorganize Objects
Make Tablespace Locally Managed

Backup/Recovery
Schedule Backup
Perform Recovery
Manage Current Backups
Configure Backup Settings
Configure Recovery Settings
Configure Recovery Catalog Settings

Deployments
Patch
Clone Database
Clone Oracle Home
View Patch Cache
Configuration Collection Status
Manage Policy Library
Manage Policy Violations

Home Performance Administration **Maintenance**

Related Links
Advisor Central
All Metrics
Jobs
Monitoring Configuration
Alert History
Blackouts
Manage Metrics
User-Defined Metrics
Alert Log Content
iSQL*Plus
Metric Collection Errors

FIGURE 2-43 The Database Control Maintenance screen provides excellent access to database utilities, backup/recovery, and deployment activities

This concludes the overview of the four primary tools in the Enterprise Manager and basic use of the Database Control. These tools are revisited in later chapters as related subjects are covered, such as creating a table.

Chapter Summary

- Using command-line interfaces to manually execute commands may be necessary, even if a Windows-like tool can do the job.
- Oracle Net Services is a connection architecture that allows communication between a tool and the database.
- Oracle Net Services must be configured on both the client and the server when using Oracle tools on the client side.
- The Oracle service called the Listener waits to receive requests sent to the database.
- A service name defines the database's name, location, and listening port.
- Use Net Manager and the Net Configuration Assistant to configure Oracle Net Services.
- Net Manager contains a wizard to guide you through the steps of configuring a new service name.
- The file **tnsnames.ora** stores Oracle Net Services configuration settings on the client side.
- The file **listener.ora** stores Oracle Net Services configuration settings on the server side.
- In SHARED SERVER mode, a dispatcher process determines which user session connects to each server process.
- SHARED SERVER mode has one server process for many user connections.
- DEDICATED SERVER mode has one server process for each user connection.
- A server process reads data from the datafiles and places it in the buffer cache.
- The shared pool stores parsed SQL commands for possible reuse (in the library cache), and metadata objects, such as tables and indexes (in the metadata or row cache).
- The redo buffer stores all changes to data before the changes are written to a disk.
- The large pool adds more memory to improve response time, under specific circumstances.
- The PGA stores shared connection memory allocations, separately for each application.
- The background processes are named PMON, DBW*n*, LGWR, CKPT, SMON, RECO, ARC*n*, Dnnn, LMS, and CJQn, among others.
- The Instance Manager tracks database activity, such as user sessions, initialization parameters, and memory usage.
- Use the Navigator selection from the menu to add a database to the console.
- When logging on to a database in the console, set the user and password as preferred credentials, and you need not enter them again.
- Help files within the console provide details on the various tools contained within Enterprise Manager.
- The Schema Manager provides details on schema objects, such as tables, views, and triggers.
- Use Schema Manager to add or drop columns, change storage, and view the data of tables.

- The Security Manager is focused on users and privileges.
- Roles group privileges into related sets.
- Use the Security Manager to change passwords and adjust quotas of users.
- The Storage Manager displays information about datafiles and tablespaces.

Review Questions

1. The Log Miner reads log files. True or False?
2. The dispatcher process distributes processes among the available server processes. True or False?
3. Using a tool is always preferable to using a manual method. True or False?
4. The file that stores Oracle Net Services configuration data in a client computer is called ________________.
5. The primary purpose of the Enterprise Manager console is to
   a. Gather data about remote databases
   b. Centralize access to tools and utilities
   c. Share user and server connections
   d. Automate backups
6. You have issued a query. Which memory area stores the parsed SQL command?
   a. Redo log buffer
   b. Shared pool
   c. PGA
   d. Buffer cache
7. The ________________ Manager displays the full path name of datafiles.
8. Which of these tasks can be performed using the Instance Manager?
   a. Killing a user session
   b. Dropping a table
   c. Creating a new user
   d. Running a backup
9. The ________________ process writes buffers to the datafiles.
10. Which of the following are naming methods that can be used by Oracle Net Services? (Choose two.)
   a. Local naming
   b. Localized naming
   c. Host naming
   d. Proxy naming

11. Which line below has an error?

```
1. LISTENER =
2.   (DESCRIPTION_LIST =
3.     (DESCRIPTION =
4.       ADDRESS_LIST =
5.         (ADDRESS = (PROTOCOL = TCP)(HOST = 1300server)(PORT = 1521))
6.       )
7.     )
8.   )
```

   a. 1
   b. 3
   c. 4
   d. 5

12. Multiple listeners are load balanced in a randomized manner. True or False?

Exam Review Questions—Oracle Database 10*g*: Administration (#1Z0-042)

1. A procedure created by the user named HAROLDK, logged in as SYSTEM, is in which schema?
   a. SYSTEM
   b. PUBLIC
   c. The database instance
   d. HAROLDK
   e. None of the above
2. The following tasks can be handled in the Security Manager. (Choose three.)
   a. Assign users the ability to create tables
   b. Change a user's session
   c. Limit the resources a user can use on the database
   d. Change a table's owner
   e. Create a new user
3. A datafile should not be removed from the operating system without removing from the database first. True or False?
4. Which of the following tools can be used to establish a connection between tools and the database? (Choose two.)
   a. Net Manager
   b. SQL*Plus
   c. Net Configuration Assistant
   d. Instance Manager
   e. All of the above

5. When logging on to a tool, which items are typically required, using appropriate configuration files? (Choose all that apply.)
   a. User name
   b. Port number
   c. Service name
   d. Password
   e. All of the above
6. You are configuring a new service name in Oracle Net Manager. The Test step has failed. What might be wrong? (Choose three.)
   a. Invalid date and time on database server
   b. Incorrect port number specified
   c. Database instance not running
   d. User SCOTT was removed or password was changed
   e. Missing or invalid NET_CONNECTION parameter
7. You log in to SQL*Plus Worksheet and run a query. Which parts of the database instance are involved? (Choose three.)
   a. ARC*n*
   b. PGA
   c. Server process
   d. DBW*n*
   e. LGWR
8. Which of the following best describes the role of the ARC*n* process?
   a. Copying redo log files to a remote storage area
   b. Monitoring user sessions
   c. Connecting to distributed databases to resolve connections
   d. Distributing user sessions among server processes.
9. You are running the Enterprise Manager console. Your computer shuts off suddenly. Which background process cleans up the memory used by your user session?
   a. SMON
   b. PMON
   c. DBW*n*
   d. LMS
10. Which of the following make up the database instance? (Choose three.)
    a. SGA
    b. PGA
    c. Datafiles

d. Background processes
e. Oracle Net Services
f. Database software

11. Which part of the SGA contains a record of changes made to data blocks?
    a. Shared pool
    b. Redo log buffer
    c. Large pool
    d. Program global area
12. Which of these operating system non-editable files are used by the Oracle 10*g* database? (Choose three.)
    a. Redo log file
    b. listener.ora file
    c. Archived log file
    d. Control file
    e. tnsnames.ora files
    f. init.ora files
13. Which of these commands will start the listener?
    a. lsnrctl stop
    b. lsnrctl go
    c. lsnctl start
    d. lsnrctl start
    e. None of the above

Hands-on Assignments

Run the setup script to prepare the following assignments. To run the script, follow these steps:

a. Start the Enterprise Manager console.
b. Double-click the **Databases** folder.
c. Double-click the **ORACLASS** database icon.
d. Select **Tools** from the top menu, choose **Database Applications**, and then choose **SQL*Plus Worksheet**. This opens the SQL*Plus Worksheet.
e. Select **Worksheet/Run Local Script** from the top menu.
f. Navigate to the **Data\Chapter02** directory on your student disk, select **prech02.sql**, and click **Open**. This runs the script.
g. Close SQL*Plus Worksheet by clicking the **X** in the top-right corner.

You are now ready to work on the hands-on assignments and case projects.

1. Using the Schema Manager, find the WANT_AD table in the CLASSMATE schema and display the DDL command that created the table. Save this as the file named **ho0201.sql** in the **Solutions\Chapter02** directory on your student disk and print it. (*Hint*: Use the right mouse click.)
2. Run the Analyze Wizard on the CLIENT table in the CLASSMATE schema by right-clicking over the CLIENT table to bring up the menu. Then scroll down almost to the bottom of the menu, and left-click the **Analyze ...** option. Then view the results and answer these questions. Save your answers in a file named **ho0202.txt** in the **Solutions\Chapter02** directory on your student disk.
   a. How many rows are in the table?
   b. What is the average length of a row?
   c. How many empty blocks are there for the table?
3. Using the console, look at the view named WANT_AD_COMPLETE_VIEW in the CLASSMATE schema. List all the dependencies of this view. (*Hint*: Right-click the **view name** to start.) Generate a report in HTML format and save the report in a file named **ho0203.html** in the **Solutions\Chapter02** directory.
4. Look at the view named CLIENT_VIEW in the CLASSMATE schema and answer these questions:
   a. Is this view valid? If not, list the Oracle error number(s) and message(s) you find.
   b. How can this view be corrected?
   c. Fix the view.
   d. Save the DDL for the corrected view in your solutions directory in a file named **ho0204.sql** in the **Solutions\Chapter02** directory on your student disk.

      (*Hint:* Compile the view to enable the Show Errors button.) Save the answers to (a) and (b) in a file named **ho0204.txt** in the **Solutions\Chapter02** directory on your student disk.
5. Practice saving a password.
   a. Change the password of the **CLASSMATE** user to **STUDENT** using the console. Test your change by connecting to **CLASSMATE** in the SQL*Plus worksheet. Explain what tool you used to change the password. Save your explanation in a file named **ho0205.txt** in the **Solutions\Chapter02** directory on your student disk.
   b. Change the password of the **CLASSMATE** user back to **CLASSPASS** using the SQL*Plus worksheet. Save the SQL statement that you used in a file named **ho0205.sql** in the **Solutions\Chapter02** directory on your student disk.

Case Project

Using the information in the console, draw a diagram of the CLASSMATE schema. Save the diagram in a file named **case0201.tif** in the **Solutions\Chapter02** directory on your student disk. Include the following features in your diagram:

a. All tables with the following information for each table: table name, primary key name, and all columns (name, data type, size, null/not null, and default value).

b. All relationships between tables. For each relationship show whether it is one-to-many, one-to-one, or many-to-many. Label with a verb, and also list the constraint name that enforces the relationship.

c. Mark columns that are indexed with an asterisk for non-unique indexes, and with a plus sign for unique indexes.

CHAPTER 3

CREATING AN ORACLE INSTANCE

LEARNING OBJECTIVES

In this chapter, you will:

- Learn the steps for creating a database
- Understand the prerequisites for creating a database
- Configure initial settings for database creation
- Create, start, and stop a database instance
- Learn the basics of managing configuration parameter files
- Learn the purpose and location of the alert log and trace files

INTRODUCTION

Database administrators often work with existing databases for some time before actually creating a new database. As a result, the DBA often focuses on activities, such as backup and tuning, which often require a great deal of time and effort. Nonetheless, a competent Oracle 10*g* DBA must be familiar with database creation. A new database may be the best way to install a new version of the Oracle software or to set up a replicated site, for example. A few parameter settings for the database cannot be altered after database creation, making preparation and planning a critical part of database creation. This chapter describes the steps involved in planning, configuring, creating a new Oracle 10*g* database instance, and starting and stopping that database instance. Hands-on exercises give you a chance to experiment with creating new database instances on your own.

STEPS FOR CREATING A DATABASE

Creating a new database helps you understand how Oracle 10g works and gives you experience working with parameter settings, with several of the assistant tools provided by Oracle 10g, and with manual commands as well. Table 3-1 lays out the steps for database creation in a checklist format.

TABLE 3-1 Checklist for creating a database

Main Step	Detailed Steps	Comment
1. Install Oracle software. Can be performed by the Oracle Installer	Install Oracle 10g software	Choose Enterprise, Standard, or Personal Edtion
	Install Oracle Net Services software	Needed for all editions (usually automatically installed)
	Install Oracle 10g client software	Install on client machine needing remote access to a database
2. Establish user (for Unix only; automatic for the Windows operating systems). Performed by the Oracle Installer	Create operating system user with privileges needed to create, start and stop the database	The user may or may not be the same user who installed the software (see the specific guide to your operating system for details)
	Create operating system environment variables for the user	ORACLE_HOME and ORACLE_BASE are required, and more may be needed (see the specific guide to your operating system for more information)
3. Confirm memory and storage availability. Can be performed by the Database Configuration Assistant (DBCA) if so required	Plan the disk storage size and location	Determine how many disks are available and design the distribution of data
	Decide on the DBA authentication method	Choose between operating system and password file authentication
4. Determine initial settings in the initialization configuration parameter file. Can be performed by the DBCA if so required	Decide on the file management method	Choose between the manual and Oracle Managed Files (OMF) database methods
	Set the initial parameters	Decide the database's name and other settings, and place them in the initialization configuration parameters file (**init.ora**)
5. Choose the type of database installation	Decide on assisted or manual creation	Use the DBCA or the CREATE DATABASE command
	Run the DBCA	If using the DBCA, follow the instructions on the tool to create the database

TABLE 3-1 Checklist for creating a database (continued)

Main Step	Detailed Steps	Comment
6. Create the database. This step is for manual database creation only. The DBCA handles this step for you	Write and execute the CREATE DATABASE command	Using decisions made prior to this step, write an appropriate CREATE DATABASE command, and execute that command
	Run scripts to create data dictionary (metadata) views and built-in PL/SQL procedures	Standard scripts for this purpose are provided within the Oracle software installation. These scripts can be found in the ORACLE_HOME\rdbms\admin directory, such as **catalog.sql** and **catproc.sql**
	Add recommended tablespaces and one for user-created tables	If there are no manual additions, the DBCA creates SYSTEM, SYSAUX, USERS, TEMP and UNDOTBS
	Edit the Windows Registry, if using Windows, using **regedit.exe**	Add the new database name to the Windows Registry
7. Test the database	Start up the database	Use the SQL*Plus **startup** command
	Shut down the database	Use the SQL*Plus **shutdown** command

Most of these steps are covered in this chapter, and you can try out some of the options yourself with the Hands-on Assignments at the end of this chapter.

The next section discusses the prerequisites for creating a database, which include Steps 1–5 listed in Table 3-1.

OVERVIEW OF PREREQUISITES FOR CREATING A DATABASE

Creating a database is actually a separate process that occurs after the database software has been installed. You have the option, during the software installation, to have a predefined database created included within the installation process. You may, however, install the software only and create the database at a later time, which gives you some time to think about how you might want to customize your database.

There are several prerequisites you must fulfill before creating a new database. First, the Oracle software must be installed on the computer. With distributed systems, the software may actually reside on a different machine than the database; however, most often, the two reside on the same machine. Second, you must be able to log on as a user with installation privileges and with the correct set of environmental variables in place. Third, the machine must have enough memory and enough disk space to install and start the database. Table 3-1 details each of these three requirements.

NOTE

The details outlined in Table 3-1 are an overview of the steps for installing the Oracle 10*g* software and creating an Oracle instance. Outside the classroom, you should always refer to and follow the documentation provided with your software. Oracle provides an installation guide for each operating system with the in-depth information you need to complete the tasks before, during, and after installation. The guides are found online at the Oracle Technology Network site (*otn.oracle.com*), as well as on the installation CDs, and installed with Oracle software as Oracle documentation.

Unix system pre-installation tasks require more manual steps than Windows systems. For example, in Unix, you must set the environmental variables manually and create an account for database administrators, an account for upgrades, and an account for the Universal Installer. In Windows, these tasks are either not required, or are handled automatically by the Universal Installer during installation of the software. See your operating system installation guide for detailed setup steps for Unix.

The installation guide for your platform contains information specific to the operating system for the minimum storage and memory. For example, if you install Oracle 10g Enterprise Edition on Windows 2000, the minimum requirements for memory and storage are:

- RAM: 512 megabytes minimum, Oracle recommends 1024 megabytes
- Virtual memory: double-up the amount of RAM. Thus 1024 megabytes of RAM requires 2048 megabytes of virtual memory
- Temp space: 100 megabytes
- Storage space: ORACLE_HOME drive for Oracle binary files (system drive) of at least 100 megabytes
- Start database size: At least 800 megabytes
- Total space: At least 1.5 gigabytes are recommended in total
- Video adapter: Greater than 256 colors
- Processor speed: Greater than 450 Mhz processor at the bare minimum

CHOOSING CONFIGURATION

There are a number of important configuration tasks that you must be familiar with before creating the database. These tasks determine the initial settings, file management, and security for your database. In addition, permanent settings (those that cannot be altered after database creation) must be set prior to the database creation. Each of the next sections discusses one of these tasks, which are:

1. Choose a database type. Databases in modern commercial environments are usually transactional (Online Transaction Processing (OLTP)), data warehouses, or a hybrid of those two. Which is it to be?
2. How should the database be managed? Oracle databases can be managed from a central location using the OEM Grid Control, or locally (on the server where the database resides) using the OEM Database Control. You can select to manage with or without Enterprise Manager, and then select the Grid Control, assuming an agent process is running on the server that is running the Grid Control.

3. Decide on the DBA authentication method. The most common choice is to set up a password file for authentication.
4. Select a storage mechanism. The storage mechanism can be in the operating system file system, Automatic Storage Management (ASM), or raw devices. ASM optimizes layout of database files and storage for the best I/O performance. ASM is vaguely similar to a RAID array in which a set of disks, or groups of storage areas are utilized. Raw devices can be used where operating system formatting is not required, such as in some Oracle RAC environments.
5. Decide on the file management method. The regular setup allows specification of Oracle database files, such as datafiles, into specific disk locations. Oracle Managed Files (OMF) requires parameter settings that allow the database to handle creation and maintenance of control files, log files, and datafiles. When using OMF Oracle decides where to put files, and what the files are called.
6. Set the initial parameters. These settings are stored in the **init.ora** file and include settings for the two tasks listed previously, plus many more settings.

Database Type

Database types are generally application driven, or determined by the kinds of applications that users are utilizing against a database. An OLTP database is used to service Internet applications. A data warehouse database is used to service very large data warehouses, creating long running forecasting reporting. Figure 3-1 shows the Database Configuration Assistant (DBCA tool) allows selection of four types of databases: Custom, General Purpose, Data Warehouse, and Transaction Processing. A custom database obviously allows you to define anything and everything – this one is for the experts. A general purpose database combines aspects of both the data warehouse and transaction processing options. The Data Warehouse option will change some configuration, generally allowing low concurrency and very high I/O activity. The Transaction Processing option allows for a lot of change activity, raising concurrency requirements, and lowering I/O activity requirements. An OLTP database is the equivalent of a transaction processing database.

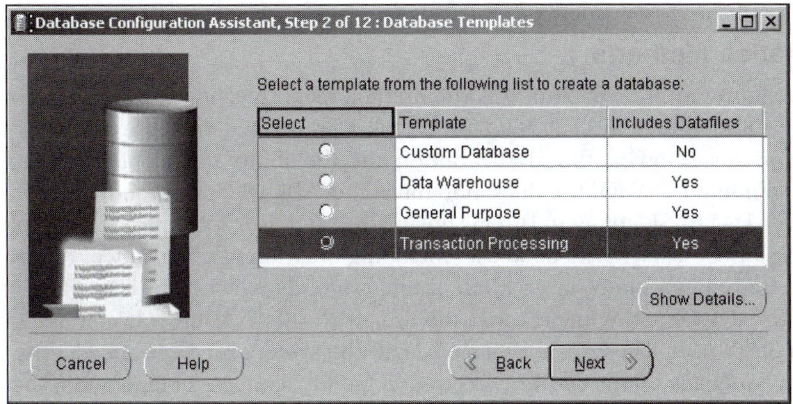

FIGURE 3-1 Choose between different database type templates

Database Management Tool

When creating a database, one of the screens in the DBCA tool allows configuration of a database using Oracle Enterprise Manager, the Database Control, or the Grid Control. The Grid Control allows management of multiple, possibly local and remote databases. The Database Control allows management of a single database, local to the database server computer only. This is shown in Figure 3-2. When managing with the Database Control you can also configure the system to utilize automated email notifications (used to preempt potential problems), and you can also schedule regular backups.

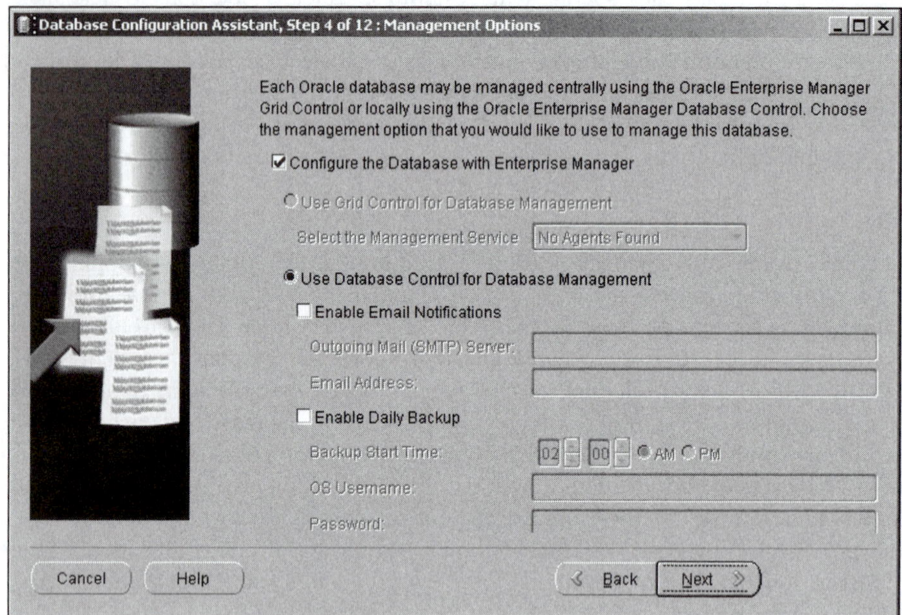

FIGURE 3-2 Manage a database with Enterprise Manager and the Database Control

DBA Authentication Methods

The DBA authentication method encompasses the method used to validate logon of users with the SYSDBA or SYSOPER role. The SYSDBA role is a predefined database role containing all system privileges having the ADMIN option and the ability to execute the CREATE DATABASE command (SYSOPER does not have the ADMIN option and cannot execute the CREATE DATBASE command). This means that any user with the SYSDBA role can perform any task necessary in the database, including creation of an instance, recovery of an instance, and maintenance or creation of tables and data owned by any user in the database. The ADMIN option means that the user with the SYSDBA role can assign any privilege to another user. The SYSOPER role has all the system privileges needed to start up, shut down, and back up the database, as well as modify database components, such as datafiles and tablespaces. The security level of users with this authority is higher than that of normal users. Oracle offers two methods of security. Both methods authenticate the

DBA users logging on with SYSDBA or SYSOPER roles prior to entering the database. The two methods are called operating system (OS) authentication and password file authentication.

Operating System (OS) Authentication

When using Operating System (OS) authentication, the user logs on without specifying an Oracle user name and password. The database retrieves the user's name from the operating system. Use this method when you only allow local access (direct logon to the machine that has the database running), or secured remote network access to your database machine.

The two operating system groups specifically set up for DBAs are referred to as OSDBA (for SYSDBA privileges) and OSOPER (for SYSOPER privileges) in the Oracle documentation. The actual names of these groups depend on your operating system. For example, in Windows, the Universal Installer automatically creates the group for SYSDBA privileges and names it ORA_DBA. Before using the Universal Installer with Unix, you must first manually create a group named DBA for SYSDBA privileges. On both operating systems, you must manually set up the OSOPER group if you want to use it.

To set up OS authentication, follow these steps:

1. Create an operating system user for the DBA.
2. Unix only: Create an OSDBA group.
3. Optional: Create an OSOPER group.
4. Set the initialization parameter REMOTE_LOGIN_PASSWORDFILE to NONE.
5. Assign the operating system user to the OSDBA group or the OSOPER group.
6. Create the corresponding Oracle user in the database with the same name (see the following note).

To log on to SQL*Plus with SYSDBA privileges using the OS authentication method, type the following command on the command line:

```
sqlplus /nolog
```

SQL*Plus starts up without logging you onto the database. Type:

```
CONNECT /@ORACLASS AS SYSDBA
```

The slash tells the database to use OS authentication. The "@ORACLASS" identifies the database to log on to. With OS authentication, the AS SYSDBA parameter tells the database to check that your operating system name is in the OSDBA group and if so, logs you on with SYSDBA privileges. If you are not in the group, the logon fails.

NOTE

The Oracle 10*g* user name and the operating system user name must be identical, except when you have specified that a standard prefix be added to the OS user name. For example, by setting the OS_AUTHENT_PREFIX parameter to OPS$ (the default setting), a user who logs on to the operating system as MARTY corresponds to the database user name of OPS$MARTY. Change the parameter to a null value to eliminate a prefix.

The OS authentication method can be used for other Oracle users as well. One advantage of this method is that user names and passwords are not typed in when logging on to the database. This can be convenient for users. For example, local users at an insurance office all have their own workstations, and log on to them at the beginning of the day. Using OS authentication, they are not required to log on a second time to run an Oracle-based application that calculates auto insurance estimates.

The OS authentication method can be set up for remote database access over the Internet. However, it requires more steps and should only be used when the users reach the database over a secured line, such as one with the secure sockets layer (SSL) protocol. Access via an unsecured line opens the database to security risks, because the database and the local operating system do not have control over the user's logon procedures.

Password File Authentication

As stated previously, Oracle offers two methods of security: Operating System (OS) authentication and password file authentication. The previous section covered OS authentication, and this section covers password file authentication. Password file authentication requires you to set up a special encrypted file that contains user names and passwords of authorized DBA users. Use this method when the DBA must log on from a remote site over an insecure line. Typical TCP/IP networks, such as a standard local area network or the Internet, are not secure lines. Password file authentication is the proper method to use when, for example, your DBA uses the Web-based Enterprise Manager console to start or stop the database. The password file contains user names and passwords for all users that are granted the SYSDBA or SYSOPER privileges. Other users must log on directly with their own user names and passwords.

To set up password file authentication, follow these steps:

Create a new password file. The Oracle command, **orapwd** (Unix and Linux are case sensitive), creates the password file. The location and filename are predefined by Oracle for Windows operating systems. The file is named PWD<sid>.ORA (<sid> is the Oracle instance name) and the file is located in the ORACLE_HOME\database directory. In Unix systems, the filename must be orapw<sid>.ora and you must specify the full path in the name parameter to define the location. Typically, the password file in Unix systems is stored in the ORACLE_HOME/dbs directory.

1. Set the REMOTE_LOGIN_PASSWORD FILE initialization parameter to **EXCLUSIVE**.
2. Log on to the database as SYS with SYSDBA privileges (or another user with SYSDBA privileges). The SYS user is predefined in the Oracle database upon creation.
3. Create the new DBA user name if needed.
4. Grant the SYSDBA privilege or the SYSOPER privilege to the user. This action in the database automatically updates the password file to include this user name and password.

The syntax for the **orapwd** command is:

```
orapwd file=<filename> password=<pwd> entries=<maxusers>
```

Replace <filename> with the filename of the password file. Include the full path if you are using Unix. Replace <pwd> with the actual password for the SYS user. Replace <max-users> with the maximum number of distinct users that can be logged on as SYSDBA or SYSOPER at one time. For example, in a Windows system, you are about to create a password file named **PWDora.ora** which will be located in the **D:\oracle\ora\database directory**. The password for SYS is GOFORIT1209, and you plan to allow no more than ten users at a time to be logged on as SYSDBA or SYSOPER. The command looks like this:


```
orapwd file=D:\oracle\ora\database\PWDora.ora
password=GOFORIT1209 entries=10
```

NOTE

Whenever you change the password of a user with SYSDBA or SYSOPER roles, the change is automatically written to the password file.

You can log on to the database with SYSDBA or SYSOPER roles in Enterprise Manager as well as in SQL*Plus. After logging on to the console as SYSDBA, you can start or stop the database, adjust initialization parameters, and grant SYSDBA or SYSOPER privileges to other users. You can also perform most other DBA functions, such as creating a new user, changing passwords, creating tables, and so on. To log on as SYSDBA, follow these steps:

1. Start the Enterprise Management console.
2. Double-click the **Databases folder** to view databases recognized by the console.
3. Right-click the **Database name** you want to enter, and select **Connect** from the shortcut menu. This brings up a log on window.
4. Type **SYS** in the Username box, the current password in the Password box, and select **SYSDBA** from the list in the Connect as box. Figure 3-3 shows the box filled in correctly.

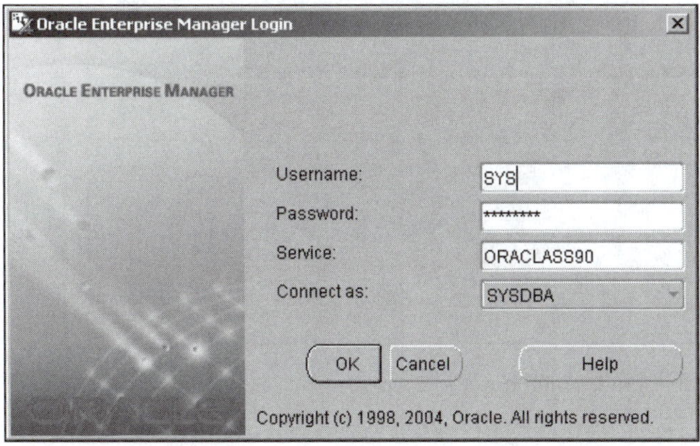

FIGURE 3-3 Log on as SYSDBA by reconnecting to the database in the console

5. Click **OK** to return to the Main Console window. You are now logged on as SYS with SYSDBA privileges.

After you have the password file in place and users authorized, your system has a secure method of preventing unauthorized access to the SYSDBA role. Oracle cautions that the location and name of the password file should be protected from unauthorized viewing to prevent possible attempts to hack the user names and passwords stored in the file. The file is automatically marked read-only and encrypted by Oracle.

Next you shall examine various different storage methods.

Storage Management Methods

Clearly shown in Figure 3-4 are the various storage management options available when creating a database using the DBCA tool. The File System option simply uses the operating system file system to store database files, such as datafiles, control files, and log files. This method allows direct access to files in a shell tool in Unix or Linux, and in Windows through a DOS shell, or a tool such as Windows Explorer. ASM optimizes the layout of a database, favoring I/O performance. ASM requires specialized structures in the form of disk sets and groupings of Oracle database physical structures. Raw devices essentially allow the use of unformatted disk partitions to physical parts of an Oracle database, such as datafiles, control files, and redo log files. Typically raw devices are used with ASM and Real Application Cluster (RAC) installations.

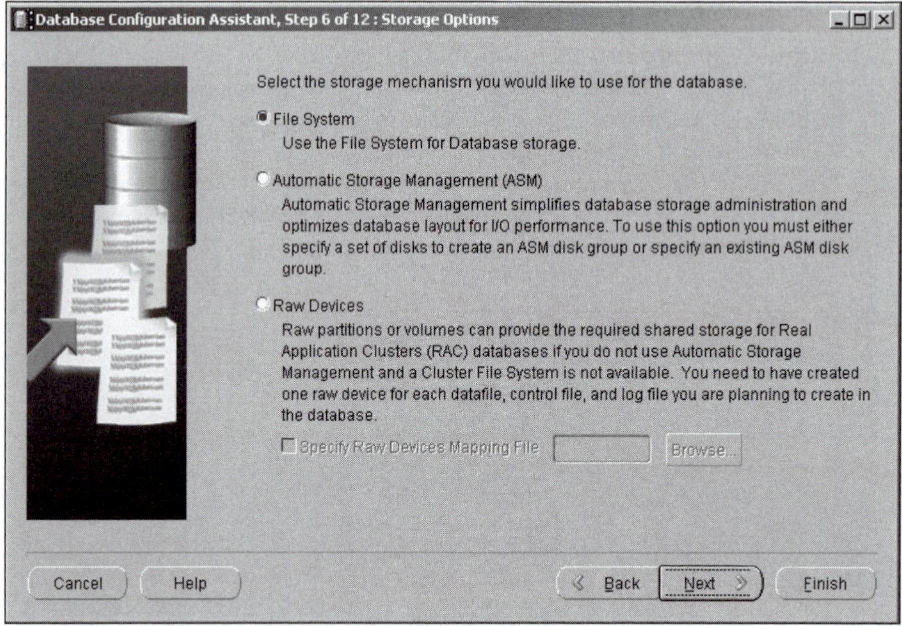

FIGURE 3-4 Choose a storage management method

Before creating your database, it is also important to decide how much control you need over the locations and names of datafiles, control files, and log files maintained by the database. The next section presents guidance on these decisions.

File Management Methods

An Oracle 10*g* database uses several different types of files to run. As described in Chapter 1, the three types of files are control files, redo log files, and datafiles. Each type of file has unique requirements for storage, size, location, and redundancy. The two primary tasks in file management are:

- **Location of files:** A control file stores up-to-the-minute information that is critical to your database's functioning. Oracle recommends that you multiplex your control file. Multiplexing involves setting up at least one identical copy of the control file, stored on a different physical device. Oracle handles multiplexed control files by automatically updating both copies simultaneously, so both are always kept up to date. The location of redo log files is also critical for database performance, because these files record each transaction as it is executed in the database itself. As with control files, it is recommended that you multiplex redo log files. Datafiles should be located on a different physical device than the database software to reduce bottlenecks in accessing data and database functions.
- **Addition, expansion, and deletion of files:** The data in all the tables, objects, indexes, and other database objects is stored in tablespaces. A tablespace is a logical data storage space. Each tablespace is mapped to one or more datafiles. A datafile is a physical file in the OS file system. Storage requirements grow and shrink according to the activity in the database. For example, a table that stores survey results grows every day as surveys are received and entered into the database. Later, a new survey replaces the old one, and old data is purged, and new data added to the table. As data storage needs change, the datafiles storing the data fill up and empty. New datafiles may be added, or old ones may be expanded. Likewise, if the volume of activity grows, the size or number of redo log files must be adjusted. Additional control files might be added to reduce the risk of losing the all important control file, making recovery more likely in the case of a database failure, caused by possible physical disk damage.

There are two basic file management methods available for a new database: user-managed and Oracle Managed Files. Each has advantages and disadvantages. The Oracle Managed Files method automates more of the DBA file-related tasks, but is reputed to slow performance. And many DBA's prefer to retain access control at this level of detail, rather than hand over structural decisions to something like Oracle database software. Figure 3-5 shows the various options available when creating a database using the DBCA tool.

The next two sections describe the two choices of file management offered in Oracle 10*g*.

User-Managed File Management

Earlier versions of Oracle had no additional options for file management. The DBA was responsible for designing the file management strategy, from file locations and names to monitoring file size and allocating more files as needed. The primary advantage of this method is the administrator's total control of file management. The primary disadvantage is that many tasks involve manual intervention. The Oracle8 Database release began to

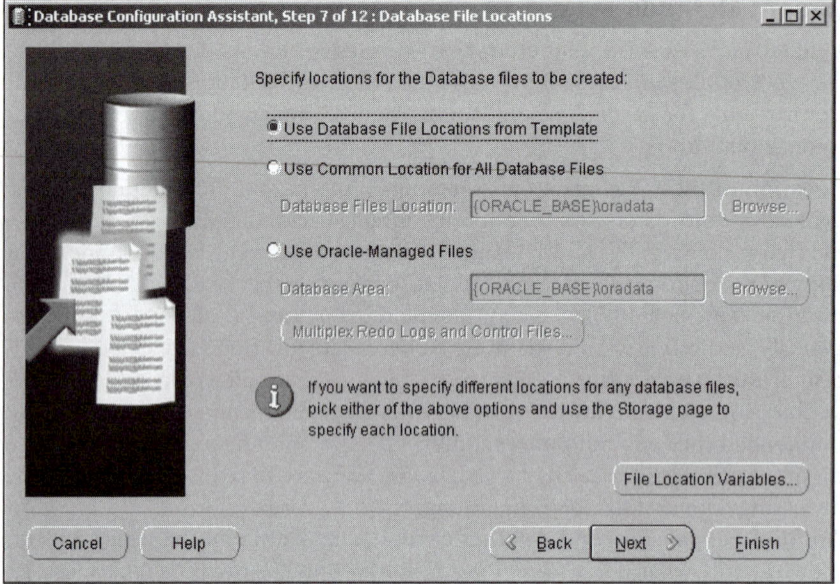

FIGURE 3-5 Choose a file management method

attempt to improve this situation by adding the capability of extending the datafile sizes automatically. However, monitoring a datafile's use versus unused space still requires manual intervention.

A good reason for using the user-managed method of file management is to continue with a customized file management standard that was in place for earlier versions of the database. The consistency of continuing with an established methodology may outweigh the disadvantage of having to assign the name, size, and location of your datafiles manually.

Control files are created when the database is created, based on the CONTROL_FILES parameter in the initialization parameter file. Here is an example of the parameter settings in a Unix system that create two control files located on two different physical devices named d1 and d2:

```
CONTROL_FILES = (/d1/oracle/control01.ctl, /d2/oracle/control01.ctl)
```

Redo log files are also created when the database is created. To implement user-managed redo log files, use the LOGFILES clause in the CREATE DATABASE command itself. Omitting the LOGFILES clause means that the redo log files are created as Oracle Managed Files.

The SYSTEM tablespace and its corresponding datafiles are created when the database is created. Omitting a fully qualified DATAFILE clause in the CREATE DATABASE command causes Oracle 10*g* to create Oracle Managed Files as the datafiles for the SYSTEM tablespace. The DATAFILE clause must include the file's name, size, and location.

Oracle Managed Files

Oracle Managed Files (OMF) automates most of the menial tasks of file management, leaving more important decisions in the hands of the DBA. OMF handles the creation, expansion, and deletion of files as database size changes. You, as DBA, provide the Oracle 10g with the information it needs to know what disks to use for storage.

The advantages of using Oracle Managed Files include:

- Adherence to Optimal Flexible Architecture (OFA) naming standards
- Automatic removal of dependent datafiles when a tablespace is dropped
- Simplified syntax for the CREATE DATABASE command
- Automated expansion and addition of datafiles as storage requirements change

NOTE

When you remove a tablespace, you get an error message, if there are any tables or other database objects stored in the tablespace, unless you specify INCLUDING CONTENTS in the command.

The main disadvantage of using OFA is the inability to control the exact size and name of datafiles, control files, and log files. These tasks are automated for you. You can, however, choose to create some files as user-managed files and leave others as Oracle Managed Files in the same database.

To implement Oracle Managed Files, specify values in these initialization parameters:

- DB_CREATE_FILE_DEST to define the directory where all datafiles and tempfiles (files used for storage of temporary data during sorting and other operations) are to be created.
- DB_CREATE_ONLINE_LOG_DEST_ to define one or more directories where control files and redo log files are to be created. The "n" on the end is replaced by a 1, 2, and so on to indicate the location of each control file and log file. By specifying more than one directory, Oracle 10g automatically creates multiplexing control files and redo log files. This parameter is optional. If omitted, Oracle 10g uses the same directory for control files and redo log files that you specify for datafiles.

Here is an example of settings for these parameters. Your new database is on Windows and you have three physical disk drives. You have placed all the Oracle software on the C drive. You want to use the D drive for the datafiles and one set of the control and redo log files. The E drive will contain the multiplexed copies of the control and redo log files. The two initialization parameters might look like this:

```
DB_CREATE_FILE_DEST = D:\oracle\product\10.2.0
DB_CREATE_ONLINE_LOG_DEST_1 = D:\oracle\product\10.2.0
DB_CREATE_ONLINE_LOG_DEST_2 = E:\oracle\product\10.2.0
```

The directories that you specify must already exist when you create the database. When executing the CREATE DATABASE command, you must also omit the parameters that specify names for the control file, redo log file, and SYSTEM tablespace datafiles.

NOTE

Oracle 10*g* creates new database files as Oracle Managed Files by default, as long as the DB_CREATE_FILE_DEST parameter has been defined and you leave out the filename when creating any new files.

Oracle Managed Files have names that comply with OFA (Optimal Flexible Architecture). Standard naming patterns are used for each type of file, and unique character strings are added to guarantee unique names. For example, the third datafiles for the USERS tablespace might be named **ora_users_2ixhgl0q.dbf**.

Oracle Managed Files can be viewed in the Storage Manager like any other file. Oracle Managed Files also display in the standard Data Dictionary views that display datafiles, such as V$DATAFILE or DBA_DATA_FILES. Data Dictionary views provide system-related information by querying the database's internal management tables and presenting the data as views (metadata views).

The final area of preparation for creating a database involves setting the critical initialization parameters.

Set the Initialization Parameters

The previous sections described choices to make before creating a new database. Most of these choices require you to set initialization parameters. In addition to the parameters already discussed, there are several other initialization parameters that must be set before launching into database creation.

Initialization parameters are placed into a file usually called **init<sid>.ora** and located in the ORACLE_BASE/admin directory structure. Replace <sid> with the Oracle instance name. The **init<sid>.ora** file contains a list of all the initialization parameter values that the database uses for creation, and also for subsequent start ups. There are well over 200 initialization parameters. Any parameters not listed in the initialization file are automatically set to their default values. When a database is created using the DBCA tool, a parameter file is generated. Figure 3-6 shows part of the file. The format of the file is simple: Parameter names are followed by an equal sign, and then the value of the parameter follows; comment lines are marked with a leading # sign; blank lines are ignored.

In Oracle 10g configuration parameters are most generally separated into basic parameters, and all other parameters. Most database installations can be managed by changing only the basic parameters, shown in Table 3-2. Other parameters are more obscure, and should not necessarily be tampered with.

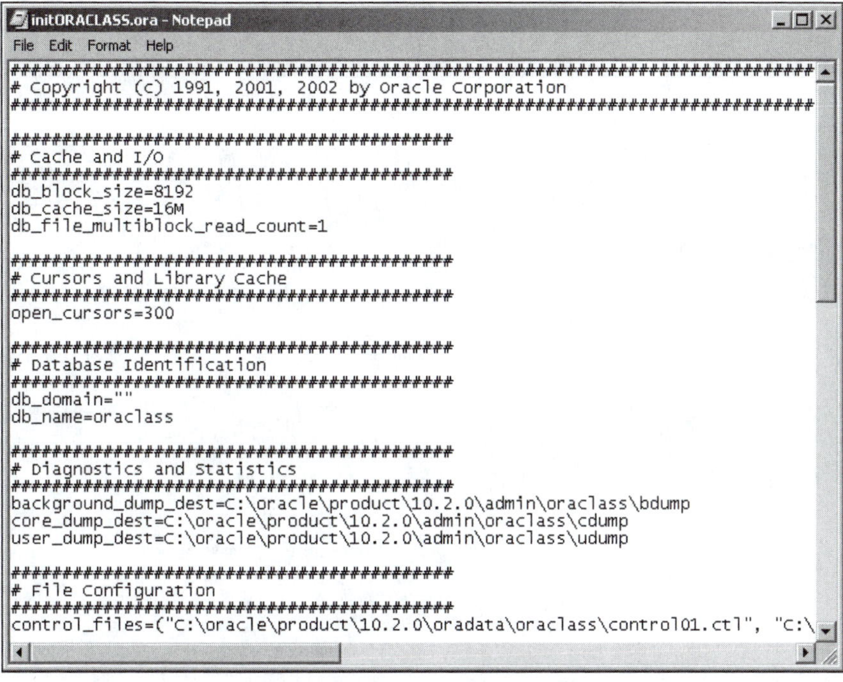


```
###########################################################################
# Copyright (c) 1991, 2001, 2002 by Oracle Corporation
###########################################################################

###########################################
# Cache and I/O
###########################################
db_block_size=8192
db_cache_size=16M
db_file_multiblock_read_count=1

###########################################
# Cursors and Library Cache
###########################################
open_cursors=300

###########################################
# Database Identification
###########################################
db_domain=""
db_name=oraclass

###########################################
# Diagnostics and Statistics
###########################################
background_dump_dest=C:\oracle\product\10.2.0\admin\oraclass\bdump
core_dump_dest=C:\oracle\product\10.2.0\admin\oraclass\cdump
user_dump_dest=C:\oracle\product\10.2.0\admin\oraclass\udump

###########################################
# File Configuration
###########################################
control_files=("C:\oracle\product\10.2.0\oradata\oraclass\control01.ctl", "C:\
```

FIGURE 3-6 The initialization configuration file has a plain text format

TABLE 3-2 The basic initialization parameters

Parameter	Description	Example
COMPATIBLE	Setting to the latest release of your installation alows Oracle software to take best advantage of the latest features	10.2.0.2
CONTROL_FILES	Set the location and name of one or more control files	(/d1/oracle/control01.ctl, /d2/oracle/control01.ctl) defines two files. Set to Null to use OMF (Oracle Managed Files)
DB_BLOCK_SIZE	The block size used by Oracle 10g to allocate space as files grow in size. The block size cannot be chaged after a database is created	Most commonly set to 8K (8192 bytes)
DB_CREATE_FILE_DEST	Defines the location of OMF datafiles and tempfiles	/db1/oradata/oraclass

TABLE 3-2 The basic initialization parameters (continued)

Parameter	Description	Example
DB_CREATE_ONLINE_LOG_DEST_n	Defines the location of OMF control files and redo log files	/db1/oradata/oraclass
DB_DOMAIN	The network domain name location of a database on a network, often distributed database environments	ohio.toystore.net
DB_NAME	The name of the database	oraclass
DB_RECOVERY_FILE_DEST	Flash recovery area containing control files and log files, both redo and archive log files	/db1/oracle/product/10.2.0/flash_recovery_area
DB_UNIQUE_NAME	DB_DOMAIN (network domain name) + DB_NAME (database name), making for a name gloablly unique to a network of distributed database servers	ohio.toystore.net/oraclass
JOB_QUEUE_PROCESSES	Maximum number of job executing processes running at once, determing how many scheduled jobs can be executed concurrently	Defaulted at 10
LOG_ARCHIVE_DEST_n	Path of destination when set to LOCATION, of up 10 separate archive log set copies	LOCATION="C:\oracle\product\10.1.0\oradata\oraclass\archive"
OPEN_CURSORS	Stop an individual database connection from opening too many cursors at once	Defaulted at 300
PGA_AGGREGATE_TARGET	Private database connection memory allocated to all server processes	
PROCESSES	Maximum number of processes allowed including all background and network servicing processes	

TABLE 3-2 The basic initialization parameters (continued)

Parameter	Description	Example
REMOTE_LOGIN_ PASSWORDFILE	Uses an operating system based password file	Defaulted at EXCLUSIVE implying password file used by local database only
SESSIONS	Maximum number of allowable database connections	
SHARED_SERVERS	Server processes used by shared database connections	
UNDO_MANAGEMENT	Automated undo is the default in favor of manual rollback segments	AUTO

Another important factor with respect to initialization parameters is the ability to change some of those numerous parameters with the database up and running. Figure 3-7 shows a picture of the V$PARAMETER view in an Oracle database. The ISSES_ MODIFIABLE and ISSYS_MODIFIABLE columns are highlighted.

- **ISSES_MODIFIABLE**. A parameter can be changed at the session level, for a particular session, for the life of that session (while that session remains connected to the database). The ALTER SESSION command can be used.
- **ISSYS_MODIFIABLE**. A parameter can be changed for all current and future sessions. The ALTER SYSTEM command is used. This column is set to either IMMEDIATE, DEFERRED, or FALSE. All three options apply to the action of the ALTER SYSTEM command when changing a parameter. IMMEDIATE means any change will take effect immediately. DEFERRED takes effect for all subsequent connections to the database. FALSE means that the parameter can only be changed when using a binary parameter file, and the change will only come into effect when the database is restarted.

Figure 3-8 shows a brief picture of the settings for the ISSES_MODIFIABLE and ISSYS_ MODIFIABLE columns. Changing some of the parameters with the database online (up and running) requires use of a binary parameter file, taking effect at the next database restart. Managing text and binary parameter files is described briefly later in this chapter.

You have now reviewed all the prerequisites for preparing to create a database. Continue to the next section for some hands-on practice in creating an actual database.

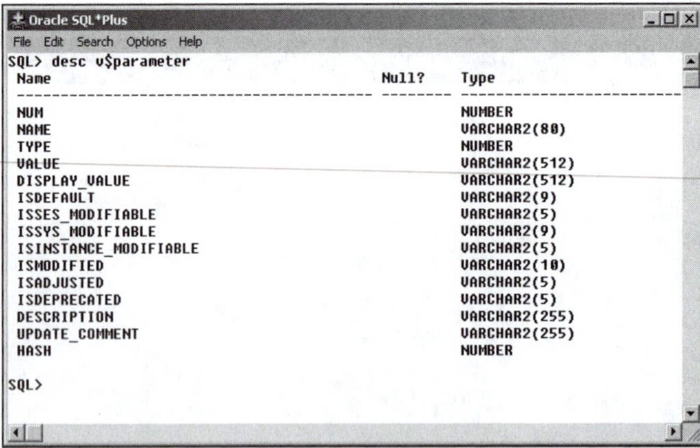

FIGURE 3-7 Changing configuration initialization parameters with the database online

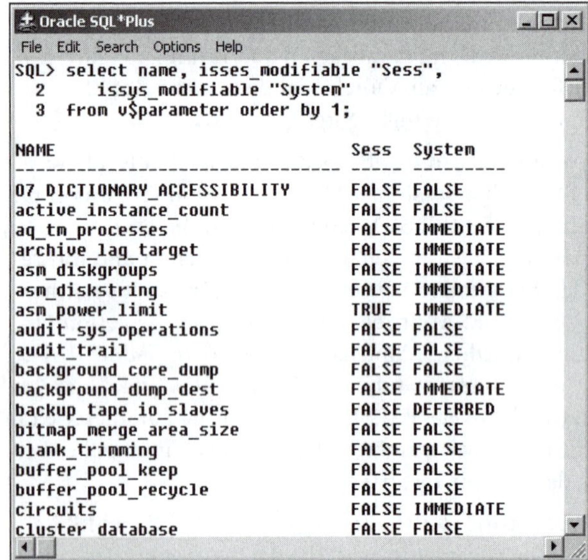

FIGURE 3-8 Examining session and system-level modifiable parameters

CREATING A DATABASE

After you have determined all the settings and reviewed the prerequisites for your operating system, it is time to create the new database. You have two techniques from which to choose: the Database Configuration Assistant (DBCA tool) and the CREATE DATABASE command. The CREATE DATABASE command will be presented, but you are not expected to actually execute it for this book.

The DBCA tool guides you through database setup and creation using windows with instructions for each step. This is a good choice for newer DBAs and tends to create a more standardized database setup. You can also use the DBCA tool to generate scripts for creating a database manually, by far the easiest method.

NOTE

The Database Configuration Assistant is automatically started during the installation of Oracle 10*g* software if you opted to have a seed database created as part of your installation process. A seed database is a database with predefined size, initialization parameters, and datafiles. Using this option saves time and allows for a fast start after software installation.

The CREATE DATABASE command gives you greater flexibility but unnecessary complexity with settings. However, you do need to be familiar with the syntax of the command. Even seasoned DBAs will use the DBCA tool to generate scripts when using a new version of Oracle, just to see what the latest version does. Subsequently, you can use the generated scripts if you have multiple consistent databases to create at different sites because you can save your settings and scripts.

The next sections guide you through the steps for creating a database, with some syntax details.

Create a New Database Using the Database Configuration Assistant

Follow these steps to create a new database.

1. Start up the Database Configuration Assistant. In Windows, click **Start/All Programs/Oracle ... /Configuration and Migration Tools/Database Configuration Assistant**. In Unix or Linux, type **dbca** in a shell and press **Enter**. The Welcome window displays (not shown). Click **Next** to continue.
2. The Step 1 window displays as shown in Figure 3-9, where a database can be created. Additionally the configuration of an existing database can be changed, a database can be deleted, and a template configuration can be generated. A template allows creation of the same parameter structured customized database in the future.
3. Select the **Create a Database** radio button and click **Next** to continue.
4. Figure 3-10 shows the options you can choose for creating database types. The four options are:
   - **Custom Database:** The initial settings of the database are the same as those of the General Purpose database; however, you can customize the tablespaces and datafiles before creating the database.
   - **Data Warehouse:** The database is optimized for larger I/O intensive queries and storing static data. Previous versions of Oracle changed some memory parameters for different database types. This is not the case for the DBCA tool in Oracle 10g.

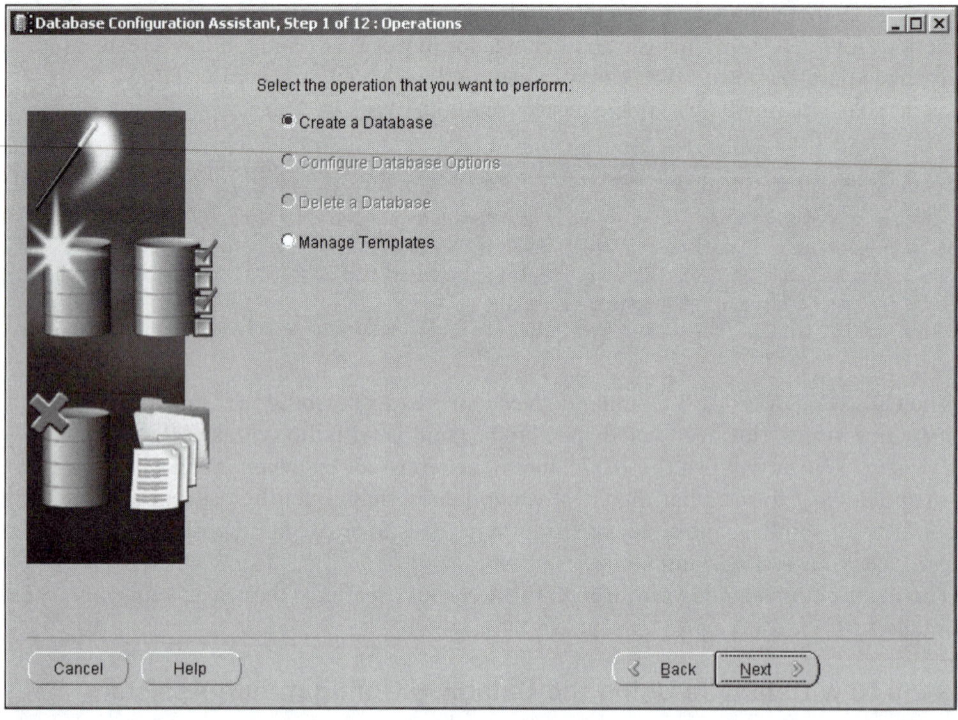

FIGURE 3-9 Choose to create a new database

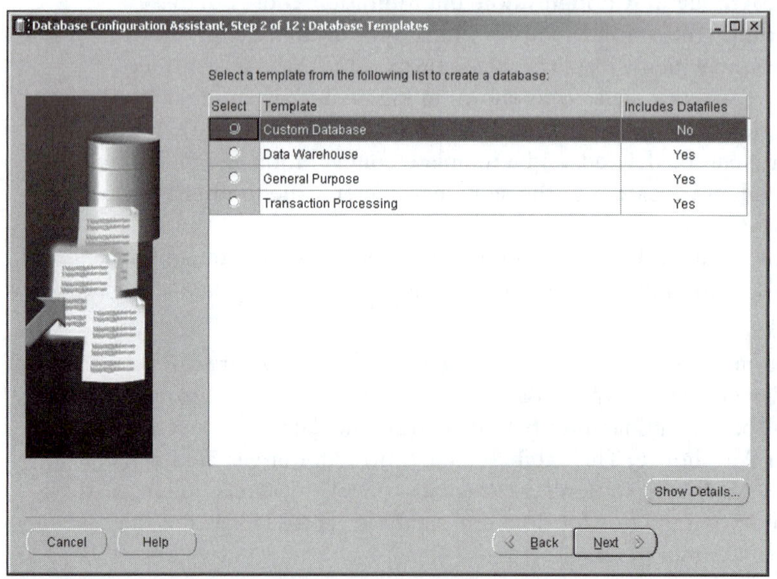

FIGURE 3-10 Choose variations on basic database settings

- **General Purpose:** The database is geared toward a mixture of static data and changing data – a hybrid database.
- **Transaction Processing:** The database is most efficient for transactions in which users update, insert, or delete data – an OLTP (Online Transaction Process, or Internet database).

5. Select **Custom Database**, and click **Next** to continue.
6. You will call this database **trial01.classroom**. So, type **trial01.classroom** in the Global Database Name text box. (The SID (System Identifier) is automatically filled in as **trial01**, as you see in Figure 3-11.) Click **Next** to continue.

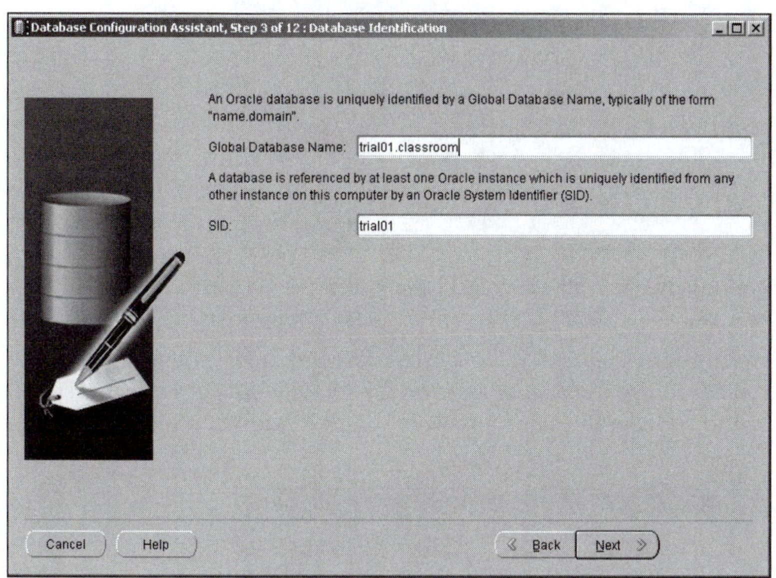

FIGURE 3-11 Give your database a unique name

7. The next screen is as shown in Figure 3-12, allowing configuration of a database with the Database Control. Leave the **Configure the Database with Enterprise Manager** check box checked. Click **Next** to continue.

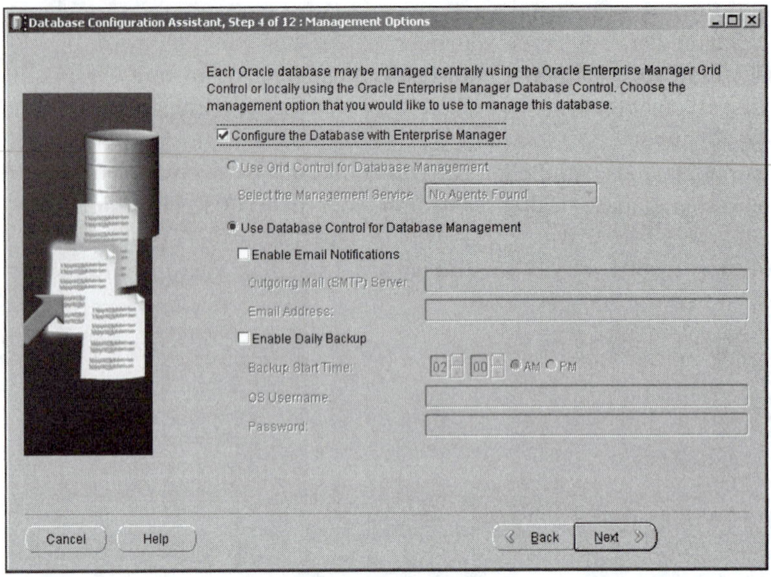

FIGURE 3-12 Your progress is charted as the assistant goes to work

8. The screen shown in Figure 3-13 concerns password management. Retain the radio button called **Use the Same Password for All Accounts** in a selected state, then enter and confirm the same password. Click **Next** to continue.

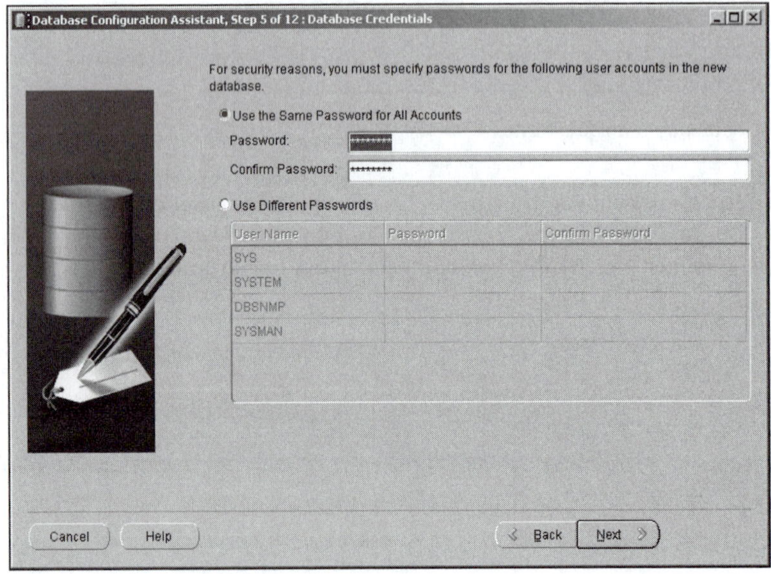

FIGURE 3-13 Password management screen

9. The next screen is also shown in Figure 3-4. It allows the selection of storage options, including the operating system file system, ASM, and raw devices. Typically the file system is most commonly used and the other two are for RAC installations. Leave the **File System** radio button selected. Click **Next** to continue.
10. Once again, the next screen is also shown in Figure 3-5. This step allows for file location selection, including OMF. OMF allows Oracle to control file names and locations. Leave the **Use Database File Locations from Template** radio button selected. Click **Next** to continue.
11. The screen shown in Figure 3-14 allows the selection of recovery options, including flash recovery and enabling of archiving. The **Enable Archiving** check box will not be checked by default. Check it to allow for copying of filled redo log files to archive log files. Flash recovery has the same effect as archive log storage, but only for a limited period of time. However, archive log files will gradually use up disk space, so you need to monitor how much disk space remains. Click **Next** to continue.

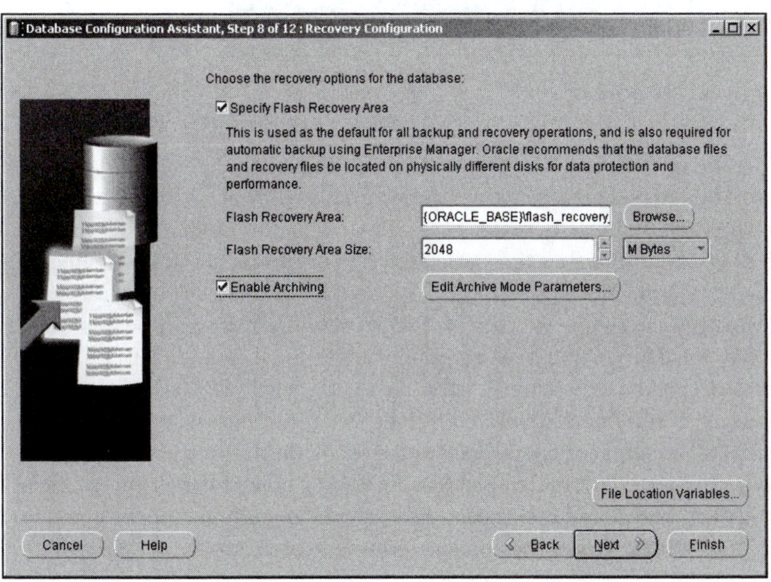

FIGURE 3-14 Setting recovery options

12. In Figure 3-15 you will be presented with various check boxes, generating built-in Oracle options into a new database. Uncheck all the boxes, but leave the **Enterprise Manager Repository** check box checked. Click **Next** to continue.

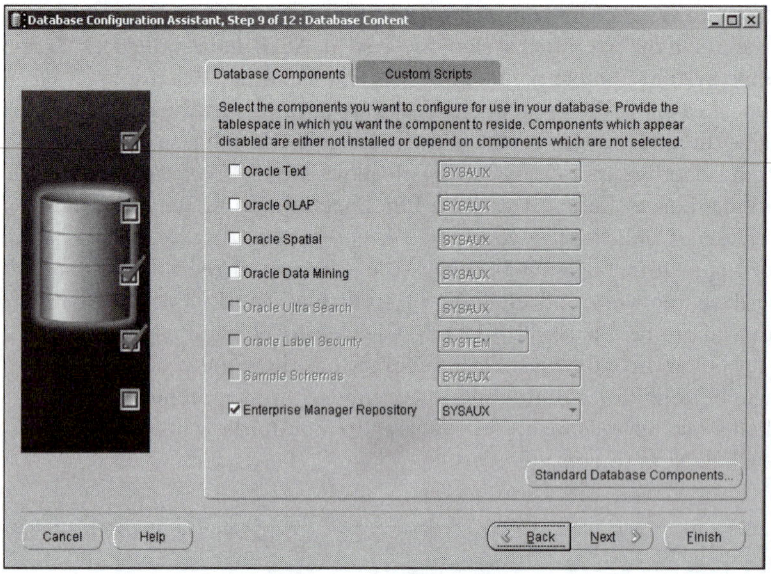

FIGURE 3-15 Allocating database components

13. Figure 3-16 shows options for setting memory, sizing, character set, and connection mode parameters. Leave all parameter settings as they are, viewing each page using the **Memory**, **Sizing**, **Character Sets** and **Connection Mode** tabs at the top of the window. Make sure that the **Dedicated Server Mode** option is selected under the Connection Mode screen tab. The Dedicated Server Mode works well for small workgroups, giving each user a distinct set of resources while working in the database. In reality dedicated servers generally work well for OLTP Internet databases (highly concurrent) with up to as many as 1000 concurrent connections. The other option, Shared Server Mode, can help when there are more than 1000 concurrent users. Shared Server Mode shares the resources to make the best use of idle time and to reuse memory buffers whenever possible. However, typically web and application servers can manage connection pooling better than shared servers, which run in an Oracle database.

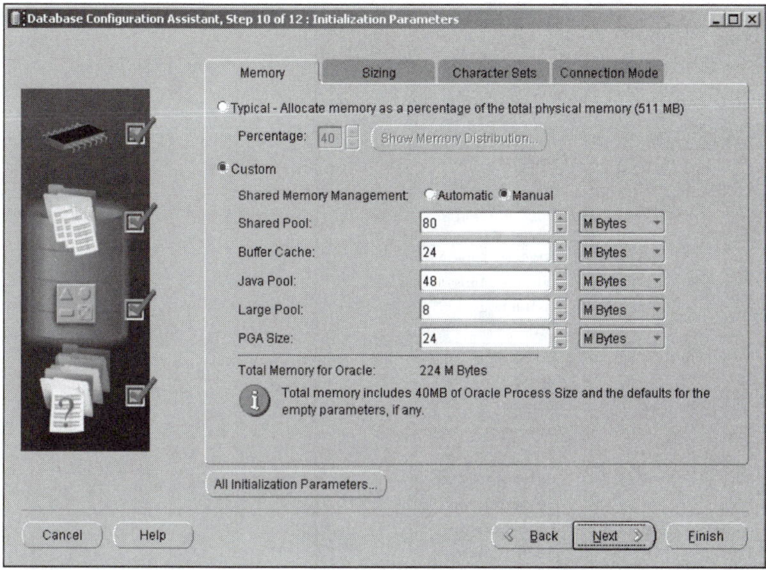

FIGURE 3-16 Setting configuration initialization parameters

14. Click the **All Initialization Parameters** button. This brings up the window shown in Figure 3-17, with many of the parameters. Click the **Show Advanced Parameters** button to show all parameters. You can sort the parameters by the column headings as you could within Instance Manager. For example, to view all the parameters based on categories, click the **Category** tab at the top of its column. The **Show Description** button gives a brief description of a highlighted parameter.
15. Click the **Name** column to re-sort the parameters in alphabetical order. Then scroll down to the **open_cursors** parameter. Change the parameter to 350 by highlighting the box that contains **300,** and typing in 350.

NOTE

Parameters that are not checked and still show values in the window are displayed with their default values.

16. Click **Close** to return to the main window. Click **Next** to continue.

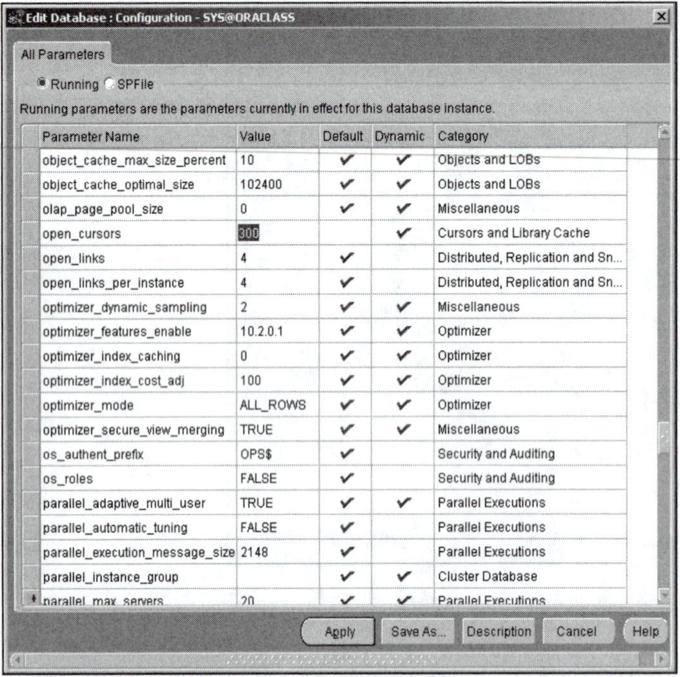

FIGURE 3-17 Initialization parameters are set or changed by typing into the boxes

17. The next step, as shown in Figure 3-18, allows changes to all underlying database files, including control files, datafiles, tablespaces (containing datafiles), and redo log files. Click the **Datafiles** folder. On the right side you find a list of datafiles and their directory locations, as shown in Figure 3-18.
18. The directory names use variables to specify the exact directory to be used. Recall that ORACLE_BASE and ORACLE_HOME are standard variables set up when the database software is installed. It is possible to change the names and locations of the datafiles by typing over the existing values. For this example, accept the default values as they are displayed. Click **Next** to continue.

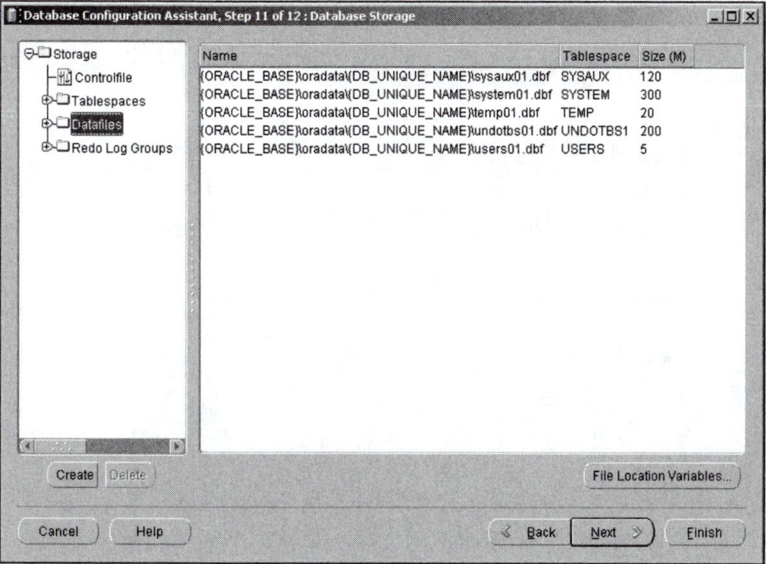

FIGURE 3-18 Datafile locations contain variables

19. You come to the final screen before the database is created. Here you can create the database, create a template that can be used as the beginning point for another database to be created later, or do both. If you choose to save a database template, a DBCA template is created that contains all the changes you have made to the standard selections. This template is added to the list of templates available the next time you use the Database Configuration Assistant. Accept the default, in which only the Create Database check box is marked as in Figure 3-19.

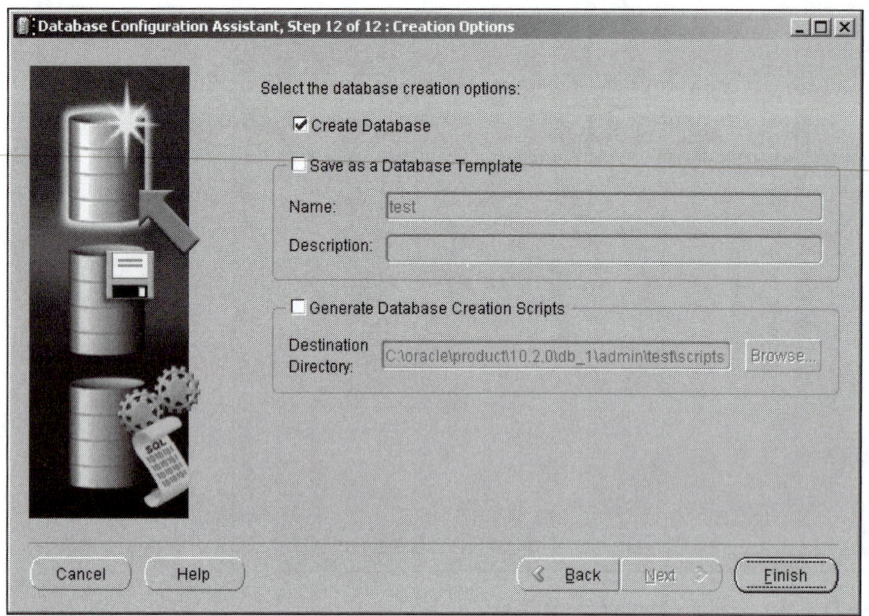

FIGURE 3-19 The assistant is ready to create the database

20. Click **Finish** to begin the creation process. The Summary window displays, displaying all the settings to be used for the new database.
21. Click **OK** to continue. The Assistant starts up a background SQL process and executes the commands it needs. The process takes 10 or 15 minutes to complete and sometimes longer on an older computer.
22. A progress window appears to display the steps, as shown in Figure 3-20. Each step gets checked off as it is completed.
23. Various windows will appear, prompting for Oracle Net Services configuration options. Accept all the defaults and follow the prompts until completed.
24. Click **Exit** to end the Database Configuration Assistant session.

You have completed the creation of a new database. A new database created using the DBCA tool can take up even as much as 2Gb, depending on options selected in Figure 3-15. Oracle 10*g* is one of the most powerful databases in its class, handling massive amounts of data at high speeds. For example, the Amazon.com Web site, which has thousands of concurrent users searching its book database, is run by Oracle Database software.

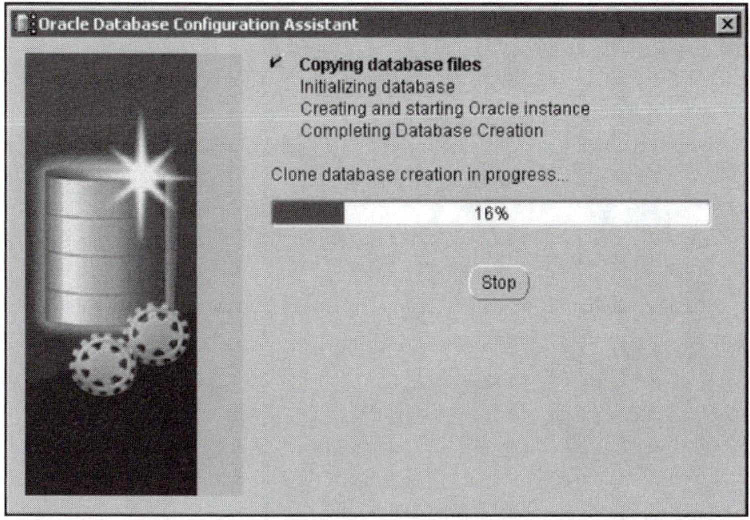

FIGURE 3-20 Your progress is charted as the Assistant goes to work

Connecting to a New Database

One of the last steps that the DBCA tool performs when creating a new database is starting up the database. Now it is ready to use. Or is it? Recall that in order to access a database, the Oracle Net Services connection must be made between your software (such as SQL*Plus or the console) and the database. This means that you must handle the Oracle Net Services configuration before you can access the database easily.

The DBCA tool actually executes the Net Manager for you, in order to configure the new database's Named Service.

The steps for configuring a database in Net Manager were covered in Chapter 2. Review the material again if needed and configure the new database. This is what you need to know:

- Name the Net Service Name **trial01**.
- The Database Service Name is **trial01.classroom**.
- Use the default port and protocol settings of **1521** and **TCP/IP**.
- Ask your instructor for the correct settings for Host Name.

Figure 3-21 shows the Net Manager with the new database configured as a Named Service. Your settings will be similar.

Now that Oracle Net Services is properly configured, the database must be added to the console. You did this in Chapter 2 as well, so refer to the steps in the section titled "Running the Enterprise Manager Console," and repeat the process for the **trial01** database.

In the console, follow these steps to view the status of the database.

1. Click the plus sign for the **trial01** database, as shown in Figure 3-22. The Database Connection Information window is displayed.

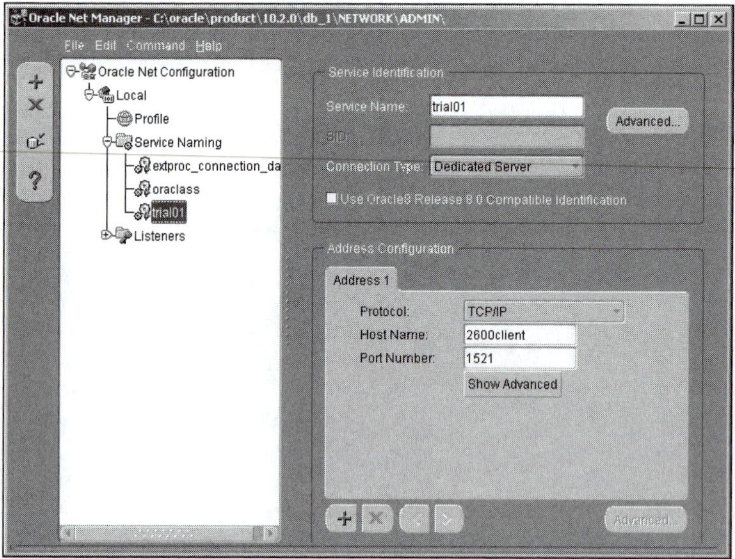

FIGURE 3-21 Net Manager after configuring the trial01 database

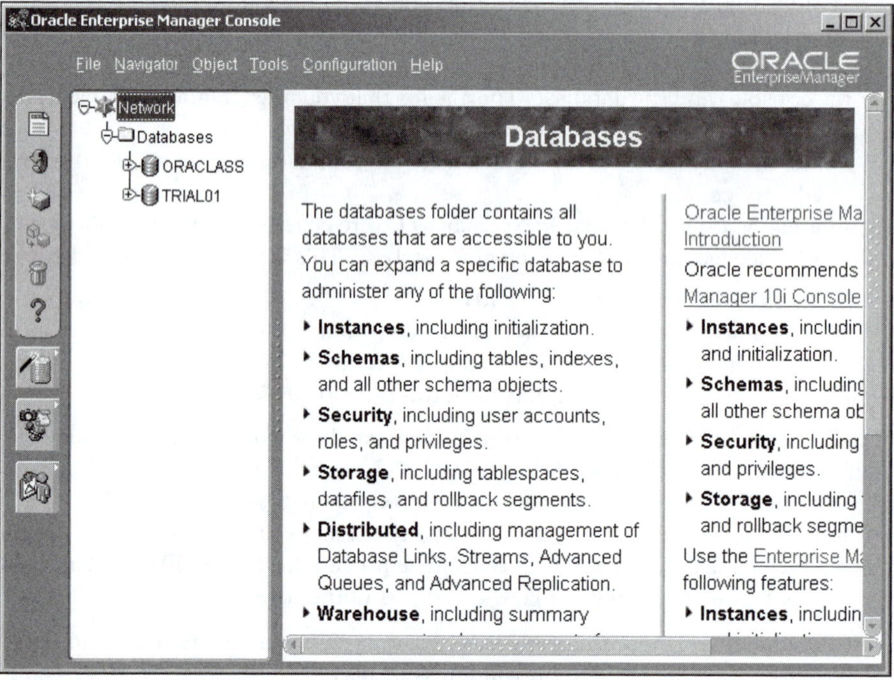

FIGURE 3-22 The console recognizes the new database

2. Type **SYS** in the Username box and the current password for SYS into the Password box. If the first S you type does not appear, type it again. This is a little-bitty bug that just doesn't seem to want to go away. Next select **SYSDBA** in the Connect as box. Figure 3-23 shows the window at this point. Click **OK** to continue. This logs you on to the database as the SYS user with SYSDBA privileges.

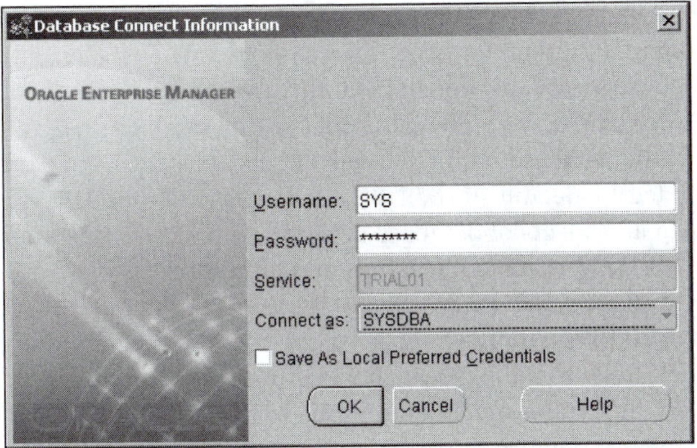

FIGURE 3-23 Do not save these settings as default because they are only for special tasks

3. Double-click the **Instance icon**, and then click the **Configuration icon** to view the status of the database, as you see in Figure 3-24.

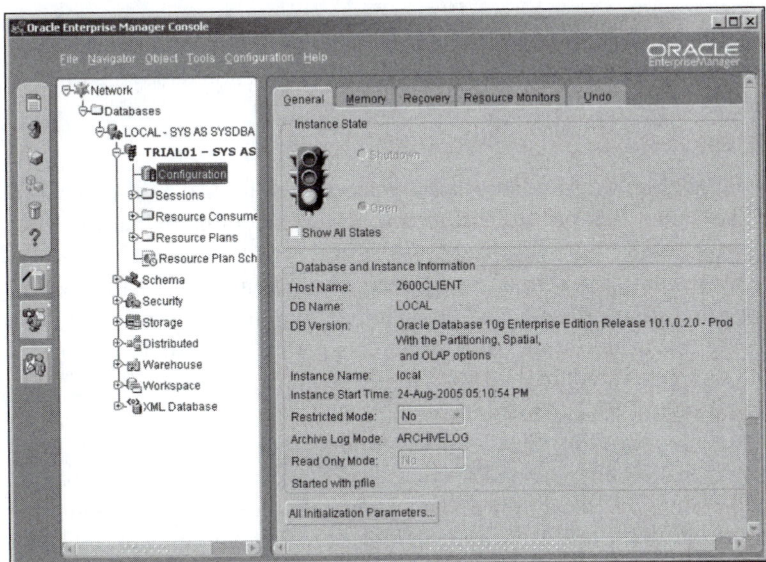

FIGURE 3-24 A green light signifies that the database is running

4. Exit the console by clicking the **X** in the top-right corner of the window.

TIP

The automated method of creating a database is convenient and highly interactive but it can be somewhat restrictive. Editing the scripts generated by the DBCA tool allow for some additional flexibility in settings, as well as the option to reduce the amount of hands-on interaction needed.

Creating a Database Manually

You have worked through the automated method for creating a database. The DBCA tool is convenient. However, it does not always provide enough flexibility for some advanced and highly complex database needs. For example, for a company that stores satellite images, adding 500Mb a day on to the size of the database requires much larger tablespace and datafile extension sizes, than that provided by default in the DBCA tool. Although the DBCA tool does allow explicit alterations to all aspects of database creation, sometimes it is more prudent to manually build scripts, allowing visual comparison and alteration.

Creating a database manually involves several steps. Some steps are the same as those you used to create a database with the DBCA tool, whereas others take the place of the steps that are automated by the tool. Review Table 3-1 at the beginning of this chapter to see what steps come before creating the database manually.

You will not actually perform a manual database creation in this section, but you will see how to do a manual database creation very easily. Perform the following steps:

1. Start up the DBCA tool again. Create a new database and call it **trial02**. Accept all default settings in the tool, or as otherwise changed and described in the previous section entitled Create a New Database Using the Database Configuration Assistant. Go all the way through, clicking Next, until you get to one of the last screens, as shown in Figure 3-19.
2. Deselect the Create Database check box, and select the Generate Database Creation Scripts check box. The destination directory should be set to the following:

   ```
   ORACLE_HOME\admin\trial02\scripts
   ```

3. Click the Finish button. Another screen appears containing details of what the scripts will create. Click the OK button, and select the option to exit the DBCA tool when requested.
4. You need to examine the script files that have been generated by the DBCA tool script generation process. The files can be found in the destination directory, as described in step 2 above. Figure 3-25 shows a Windows Explorer picture of the files generated by the DBCA tool. You may have to use a shell and cd plus ls commands in Unix or Linux.
5. The files include numerous SQL*Plus executable .SQL files, a configuration parameter file called init.ora, and a batch file for executing all the scripts in a single shell command. A batch file is an executable file in a Windows shell.
6. Open up the batch file and have a quick look at it. The contents of the batch file trial02.bat are shown in Figure 3-26. The batch file creates all the directory structures, sets the ORACLE_SID variable, creates a Windows service, starts the Windows service, and starts up the execution of the scripts by executing the script trial02.sql into SQL*Plus.

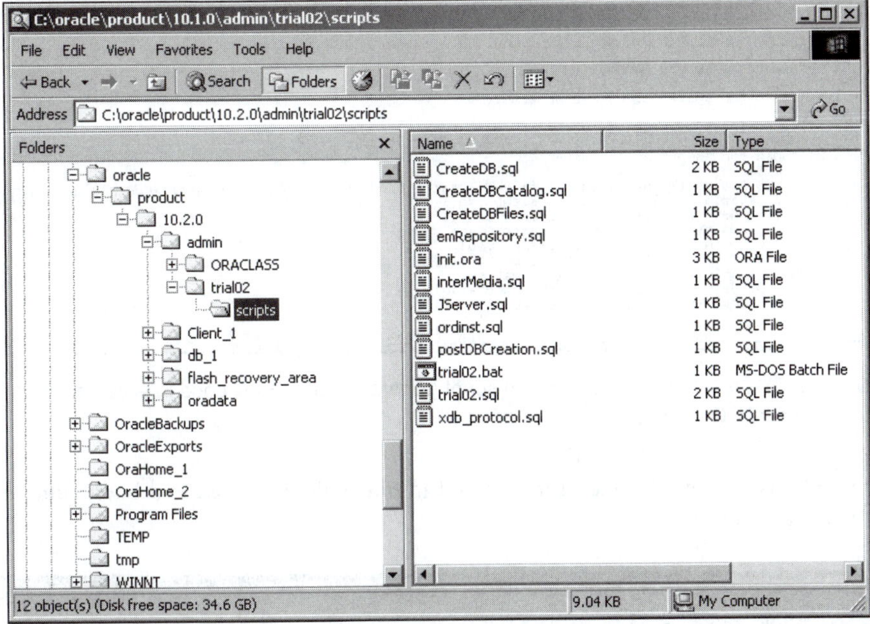

FIGURE 3-25 The Database Configuration Assistant (DBCA tool) can generate database creation scripts

trial02.bat - Notepad

File Edit Format Help

```
mkdir C:\oracle\product\10.2.0\admin\trial02\adump
mkdir C:\oracle\product\10.2.0\admin\trial02\bdump
mkdir C:\oracle\product\10.2.0\admin\trial02\cdump
mkdir C:\oracle\product\10.2.0\admin\trial02\dpdump
mkdir C:\oracle\product\10.2.0\admin\trial02\pfile
mkdir C:\oracle\product\10.2.0\admin\trial02\udump
mkdir C:\oracle\product\10.2.0\db_1\cfgtoollogs\dbca\trial02
mkdir C:\oracle\product\10.2.0\db_1\database
mkdir C:\oracle\product\10.2.0\flash_recovery_area
mkdir C:\oracle\product\10.2.0\oradata\trial02
set ORACLE_SID=trial02
C:\oracle\product\10.2.0\db_1\bin\oradim.exe -new -sid TRIAL02 -startmode manual -spfile
C:\oracle\product\10.2.0\db_1\bin\oradim.exe -edit -sid TRIAL02 -startmode auto -srvcstart system
C:\oracle\product\10.2.0\db_1\bin\sqlplus /nolog @C:\oracle\product\10.2.0\admin\trial02\scripts\trial02.sql
```

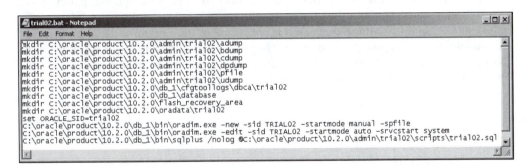

FIGURE 3-26 The DBCA tool database creation script generation is executed from a central batch file

7. Examine the trial02.sql script as shown in Figure 3-27. The script prompts for all the necessary system level passwords, placing password values into variables. You could change the way this works by hard-coding all the passwords throughout all the scripts if you wanted.

The trial02.sql script also executes all the database creation scripts within it. The most significant scripts are CreateDB.sql, CreateDBFiles.sql, CreateDBCatalog.sql, and postDBCreation.sql. Examine the contents of these four files and you should get a fair idea of what is entailed in the internal processes when creating a new database. Creating a new

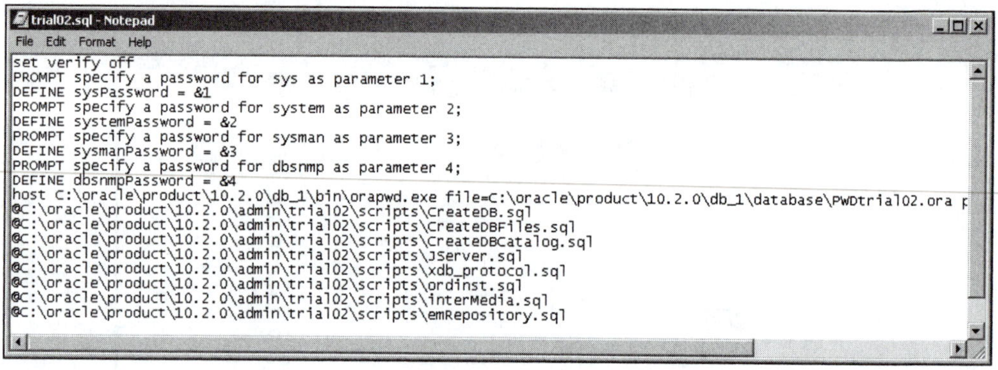

trial02.sql - Notepad

File Edit Format Help

```
set verify off
PROMPT specify a password for sys as parameter 1;
DEFINE sysPassword = &1
PROMPT specify a password for system as parameter 2;
DEFINE systemPassword = &2
PROMPT specify a password for sysman as parameter 3;
DEFINE sysmanPassword = &3
PROMPT specify a password for dbsnmp as parameter 4;
DEFINE dbsnmpPassword = &4
host C:\oracle\product\10.2.0\db_1\bin\orapwd.exe file=C:\oracle\product\10.2.0\db_1\database\PWDtrial02.ora p
@C:\oracle\product\10.2.0\admin\trial02\scripts\CreateDB.sql
@C:\oracle\product\10.2.0\admin\trial02\scripts\CreateDBFiles.sql
@C:\oracle\product\10.2.0\admin\trial02\scripts\CreateDBCatalog.sql
@C:\oracle\product\10.2.0\admin\trial02\scripts\JServer.sql
@C:\oracle\product\10.2.0\admin\trial02\scripts\xdb_protocol.sql
@C:\oracle\product\10.2.0\admin\trial02\scripts\ordinst.sql
@C:\oracle\product\10.2.0\admin\trial02\scripts\interMedia.sql
@C:\oracle\product\10.2.0\admin\trial02\scripts\emRepository.sql
```

FIGURE 3-27 The DBCA tool database creation script generation SQL*Plus code is executed from a central SQL script

database manually is a very advanced topic area and not really required in a book such as this one in any great detail.

NOTE

Use / on Unix or Linux, and \ on Windows.

The syntax of the CREATE DATABASE file contains numerous, optional parameters. The following syntax contains the primary parameters needed for creating the database. The following CREATE DATABASE command is contained in the CreateDB.sql script file, for the trial02 database script generations you just executed. 3 spaces have been added at the beginning of lines to exaggerate code actually on the same line in the script file. And the ORACLE_BASE variable is substituted for actual path names:

```
CREATE DATABASE "trial02"
   MAXINSTANCES 8
   MAXLOGHISTORY 1
   MAXLOGFILES 16
   MAXLOGMEMBERS 3
   MAXDATAFILES 100
   DATAFILE 'ORACLE_BASE\oradata\trial02\system01.dbf' SIZE 300M REUSE
   AUTOEXTEND ON NEXT 10240K MAXSIZE UNLIMITED
   EXTENT MANAGEMENT LOCAL
   SYSAUX DATAFILE 'ORACLE_BASE\oradata\trial02\sysaux01.dbf' SIZE 120M
REUSE
   AUTOEXTEND ON NEXT 10240K MAXSIZE UNLIMITED
   DEFAULT TEMPORARY TABLESPACE TEMP TEMPFILE
   'ORACLE_BASE\oradata\trial02\temp01.dbf' SIZE 20M REUSE
   AUTOEXTEND ON NEXT 640K MAXSIZE UNLIMITED
   UNDO TABLESPACE "UNDOTBS1" DATAFILE
   'ORACLE_BASE\oradata\trial02\undotbs01.dbf' SIZE 200M REUSE
   AUTOEXTEND ON NEXT 5120K MAXSIZE UNLIMITED
   CHARACTER SET WE8MSWIN1252
   NATIONAL CHARACTER SET AL16UTF16
   LOGFILE GROUP 1 ('ORACLE_BASE\oradata\trial02\redo01.log') SIZE 10240K,
```

```
GROUP 2 ('ORACLE_BASE\oradata\trial02\redo02.log') SIZE 10240K,
GROUP 3 ('ORACLE_BASE\oradata\trial02\redo03.log') SIZE 10240K
USER SYS IDENTIFIED BY "&&sysPassword"
USER SYSTEM IDENTIFIED BY "&&systemPassword"
;
```


NOTE

To see the complete set of parameters, go to the Oracle 10*g* documentation installed with the Oracle software. Look in the book entitled SQL Reference, and search for the CREATE DATABASE command.

The different syntax options are briefly described as follows:

- **CREATE DATABASE <databasename>:** The name of the database; must be identical to the value of DB_NAME in the initialization parameters.
- **MAXINSTANCES <n>:** The maximum number of instances of this database that can be open at the same time; the minimum is 1 and the maximum depends on your operating system.
- **MAXLOGFILES <n>:** The maximum number of redo log groups. Redo log groups can contain one or more copies of the same redo log file.
- **MAXLOGMEMBERS <n>:** The maximum number of members (redo log files) in each redo log group.
- **MAXDATAFILES <n>:** An optional setting that determines the maximum number of datafiles that can be open when the database is open. This determines the size of the control file (where space for the names of all open files is reserved).
- **DATAFILE ...:** The path and filename of the datafiles for the SYSTEM tablespace datafile. You can specify more than one datafile for the SYSTEM tablespace. The <storage_settings> option contains the datafile's storage parameters and generally includes SIZE and EXTENT values for a datafile.
- **SYSAUX DATAFILE ...:** Places optional Oracle add-on functionality into a tablespace (datafile) separated from the very high usage metadata stored in the SYSTEM tablespace. The objective is better overall database performance.
- **DEFAULT TEMPORARY TABLESPACE ...:** The temporary tablespace is a specially formatted datafile, used for sorting data. Every user is assigned a temporary sorting tablespace. This command ensures all users are allocated temporary sort space automatically, rather than defaulting to the SYSTEM tablespace and thus conflicting with the heavy usage SYSTEM tablespace.
- **UNDO TABLESPACE ...:** The name of the tablespace Oracle 10g uses to store undo transaction data for rollback of erroneous application data change operations. Be identical to the value of the UNDO_TABLESPACE initialization parameter.

- **CHARACTER SET** and **NATIONAL CHARACTER SET**. These options are beyond the scope of this book.
- **LOGFILE GROUP ('redo log file member 1', 'redo log file member 2',), GROUP 2 ...:** The group number for log files; usually sequential starting with 1. Each group has at least one file associated with it; repeat the GROUP phrase once for each log file group. You must have at least two groups.

The CREATE DATABASE command becomes elegantly simple when you use OMF settings. When the DB_FILE_DEST initialization parameter is set, Oracle 10*g* knows where to locate all the datafiles. Adding the DB_ONLINE_LOGFILE_DEST_n parameter tells Oracle 10*g* where to place the control files and redo log files. In fact, if you omit both of these parameters from the **init<sid>.ora** file, Oracle 10*g* can still use OMF, placing files in a default directory. To invoke OMF, simply omit all of the filenames and locations in the CREATE DATABASE command. For example, the following command is valid, and results in the creation of the SYSTEM tablespace, online redo logs, and control files (the number of redo logs and control files depends on the initialization parameters).

```
CREATE DATABASE testOFM
MAXINSTANCES 1
CHARACTER SET WE8MSWIN1252
NATIONAL CHARACTER SET AL16UTF16;
```

Use the DBCA tool to generate a CREATE DATABASE command for the database you wish to create. Use the script generation option shown in Figure 3-19 by checking the Generate Database Creation Scripts checkbox and unchecking the Create Database checkbox. The previous edition of this book went through the entire process of creating a database manually without using the DBCA tool first. This is unnecessary for Oracle 10*g*. Oracle 10*g* is more GUI (Graphical User Interface) tool driven.

To execute the scripts generated by the DBCA tool scripts generation option, you have to execute the batch file created in the ORACLE_BASE/admin/trial02/scripts directory. It's as easy as that.

STARTING AND STOPPING THE INSTANCE AND DATABASE

Sometimes it is necessary to use shell commands rather than a Windows-like interface. For example, you are at home and have no choice but to connect to your Unix server using telnet or an ssh connection, over a slow telephone line. Telnet only gives you command-line access. What can you do without the comforts of a navigation tree and the console's handy code generators? It is important to be prepared for unusual situations by becoming familiar with the manual methods of starting and stopping the database.

NOTE

Even if the database was created with the Database Configuration Assistant, it can be started or stopped manually, and vice versa. Therefore, you may use whichever method is convenient at the time.

Follow these steps to shut down a running database using SQL*Plus:

1. To start a Command Prompt window, click **Start**, point to **All Programs**, point to **Accessories**, and then click **Command Prompt** in Windows. In Unix or Linux open up a shell. Make sure your environment variables include the ORACLE_HOME path.
2. Start up SQL*Plus without logging on by typing this command and pressing Enter:

```
sqlplus /nolog
```

   You should see the SQL> prompt awaiting your next command.
3. Connect as SYS with SYSDBA privileges by typing the following command using the current password for SYS and pressing **Enter**.

```
CONNECT SYS/<password>@trial01 AS SYSDBA
```

4. Type SHUTDOWN IMMEDIATE and press **Enter**. SQL*Plus displays the following lines:

```
Database closed.
Database dismounted.
ORACLE instance shut down.
```

There are four options for shutdown:

- **NORMAL:** Waits for all users to log off, prohibits new connections, forcibly rolls back currently running transactions, and then shuts down the database. SHUTDOWN NORMAL is the most prudent method of rapidly shutting down an Oracle database and should always be used if possible. It may take a little longer than other more drastic options, but ensures a smooth and invisible shutdown. If you start this type of shutdown and decide you don't want to wait, cancel the command by holding the **Control** key and typing **C**. This cancels the shutdown and returns you to the SQL> prompt. Then you can type SHUTDOWN IMMEDIATE to shut down without waiting.
- **TRANSACTIONAL:** Automatically completes and commits all pending transactions, then forcibly disconnects all connected users, and shuts down the database.
- **IMMEDIATE:** Disconnects all users, rolls back pending transactions, and prohibits new connections and new transactions even for already connected users. This option may require recovery on database startup if an inconsistent shutdown occurred. This is faster and convenient for environments in which you are confident that you won't disturb users. It can also help you if you have tried SHUTDOWN NORMAL or SHUTDOWN TRANSACTIONAL, and are waiting too long because of some outstanding transactions.
- **ABORT:** Abort simply stops all Oracle processes, and releases all memory buffers. It shuts down without saving any pending transactions or rolling back any unsaved changes. All users are forcibly disconnected. This method is the fastest. However, you should reserve its use for situations in which the database is not functioning properly and you are unable to use SHUTDOWN IMMEDIATE successfully. SHUTDOWN ABORT can corrupt the database! A SHUTDOWN ABORT will require recovery on database startup. This is the most risky option.

Clean up processing required by SHUTDOWN NORMAL during the shutdown process is just as fast as recovery processing caused to the subsequent startup as a result of a SHUTDOWN ABORT. You will not save any time using SHUTDOWN ABORT. Aborting a database causes similar recovery processing on startup. And as already stated, a SHUTDOWN ABORT can corrupt your Oracle database. A similar situation applies to SHUTDOWN IMMEDIATE, although SHUTDOWN IMMEDIATE is not quite as brutal as SHUTDOWN ABORT.

To perform a basic database start up, follow the previous Steps 1 through 3, type this command, and then press **Enter**. Remember to replace ORACLE_BASE with the appropriate path of your Oracle database software installation. In my case this is C:\oracle\product\10.2.0 :

```
STARTUP PFILE=ORACLE_BASE\admin\trial01\pfile\inittrial01.ora
```

The PFILE parameter is needed, if you have not as yet created the SPFILE (server binary parameter file, which is also known as the binary parameter file). The default for the command

```
STARTUP
```

If starting up a database using the text parameter file (PFILE) then you must explicitly indicate use of the text parameter file. The binary parameter file is used in the STARTUP command by default (**spfileTRIAL01.ora**), and is located by default in the ORACLE_HOME/database directory. So when starting up the database without specifying PFILE, the SPFILE in the ORACLE_HOME/database directory will be used to retrieve parameters. If you execute STARTUP with no PFILE parameter and the SPFILE does not exist, then you will get an error.

When starting up the database, SQL*Plus responds with status information about the database and ends with the statement: Database started. Figure 3-28 shows an example of the status messages you see when issuing the STARTUP command.

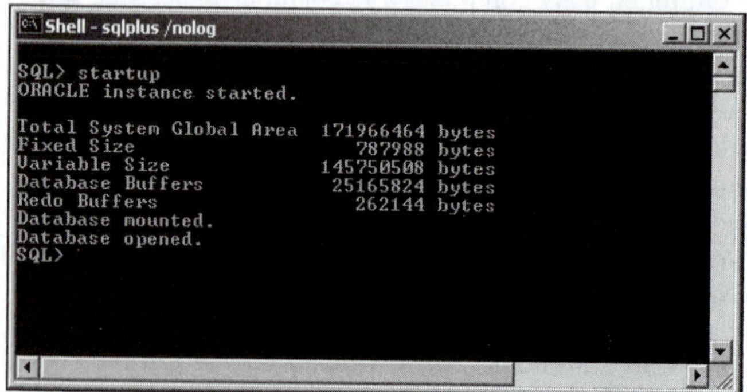

FIGURE 3-28 The STARTUP command displays memory storage size

Like the SHUTDOWN command, there are various options for starting up a database:

- **NOMOUNT:** STARTUP NOMOUNT only starts up all background processes, and allocates space in memory for all Oracle database instance memory buffers. This mode is used for database administration only.
- **MOUNT:** STARTUP MOUNT mounts the database files, essentially creating the connection between file pointers in the control file and the files themselves, such as datafiles and redo log files. As for the NOMOUNT mode, MOUNT mode is used for database administration only. STARTUP MOUNT first executes STARTUP NOMOUNT before mounting the database instance. After executing a STARTUP MOUNT operation, the ALTER DATABASE OPEN command can be executed to open the database for general use. ALTER DATABASE MOUNT similarly applies to STARTUP NOMOUNT.
- **OPEN:** The default method of starting up the database (the STARTUP command with no additional options) will open the database in read/write, unrestricted mode. Additionally, the STARTUP command executes the steps NOMOUNT, MOUNT, and OPEN – in that order. Other options are STARTUP OPEN FORCE, which forces a SHUTDOWN ABORT, followed by a STARTUP OPEN command. STARTUP OPEN RESTRICT starts up the database in restricted mode allowing database administrator access only. The STARTUP OPEN RECOVER option places a database into recovery. STARTUP OPEN PARALLEL is used for Oracle RAC installations.
- **PFILE:** This option has already been examined above, allowing database startup using an explicitly defined and located text-based configuration initialization parameter file.

After the database has been opened, it is ready for use.

MANAGING TEXT AND BINARY (SERVER) PARAMETER FILES

The text parameter file is a text editable form of the configuration initialization parameter file and is by default stored in the ORACLE_BASE\admin\trial01\pfile directory. For a database named as **trial01**, it is traditionally called **inittrial01.ora**. Starting up an Oracle 10g database using a text parameter file requires an explicit sequence of commands, restarting the database as follows:

```
SHUTDOWN;
STARTUP MOUNT PFILE='C:\oracle\product\10.2.0
0\admin\tria01\pfile\inittrial01.ora';
ALTER DATBASE OPEN;
```

The binary parameter file is by default stored in the ORACLE_HOME\database directory and traditionally called **spfiletrial01.ora** (for a database named as **trial01**). The binary parameter file can be used to create an editable text parameter file or visa versa using the CREATE PFILE and CREATE SPFILE commands, as in the following commands respectively:


```
CREATE PFILE='C:\oracle\product\10.2.0\admin\trial01\pfile\inittrial01.
ora' FROM SPFILE='C:\oracle\product\10.2.0\db_1\database\spfiletrial01.
ora';
CREATE SPFILE='C:\oracle\product\10.2.0\db_1\database\spfiletrial01.
ora' FROM PFILE='C:\oracle\product\10.2.0\admin\trial01\pfile\inittrial01.
ora';
```

THE ALERT LOG AND TRACE FILES

Another part of the Oracle database that you need to know about now is the alert log and the trace files. The alert log is essential because it will contain all the essential information for the smooth running of your database. All critical errors will be written to the alert log. The alert log is a last resort, and sometimes a first resort, when trying to solve a problem. For example, if the database crashes because a datafile is corrupt about missing, it is very likely that the only place you will find this information is in the alert log for the database. The alert log is contained within the ORACLE_BASE\admin\trial01\bdump directory and for your **trial01** database, will be called **alert_trial01.log**.

Trace files contain more detailed log and tracing information about general processing. Trace files can be used to track down problems not causing critical failures, generally using special tools. However, direct interpretation of trace files using those special tools, such as TKPROF, is more a tuning and analysis issue, as opposed to database administration, and is beyond the scope of this book. For the purposes of this book, it is important to know the trace files exist and that the alert log can be used to track down the reason for a critical problem, as already mentioned.

Expanding OFA Just a Little More

The OFA or Oracle's Optimal Flexible Architecture needs to be expanded from Chapter 1 to include various other factors introduced in this chapter. Chapter 1 described the OFA as requiring a directory structure as follows:

- **Admin/<database name>:** Stores initialization files and high-level log files. These files are stored within a directory structure contained within the admin directory as follows:
    - **bdump**: Contains background dump trace files, including the alert log and background process trace files.
    - **cdump**: The core dump directory containing memory state core dump files at the time of a database crash.
    - **create**: Contains scripting used during database creation.
    - **pfile**: Contains text-based parameter files.
    - **udump**: User dump destination containing trace files of user sessions connections to the database.
- **db_1:** This is the ORACLE_HOME directory containing the Oracle database software executable files.
- **client_1:** Oracle database client installation.

- **oradata/<database name>:** Stores database datafiles, control files, and redo logs.
- **flash_recovery_area/<database name>:** Backup and recovery files. Archive logs are by default contained within the flash_recovery_area/<database name>/archivelog directory. Within the archivelog directory are date stamped directories containing archive log files for each day.

The next chapter discusses more details you need in preparation for using your new database.

Chapter Summary

- Installing the Oracle 10*g* database software is a separate process from that of creating a database, although often both occur during the same installation session.
- Databases can be created using the Database Configuration Assistant (DBCA tool) or manually using the CREATE DATABASE command.
- When creating a database manually it is best to generate scripts using the DBCA tool first, and then to edit those scripts accordingly. Scripts are created by default in the ORACLE_BASE/admin/<databasename>/create directory.
- The operating system-specific installation guide describes minimum requirements for installing a new database.
- The DBA authentication method determines how Oracle 10*g* validates users logging on with SYSDBA or SYSOPER privileges.
- Operating system (OS) authentication relies on the operating system's security to validate the user's name and password, and authorization group.
- The REMOTE_LOGIN_PASSWORDFILE parameter is set to NONE for OS authentication.
- Password file authentication stores user names and passwords and group membership in an encrypted file in the operating system.
- Set REMOTE_LOGIN_PASSWORDFILE to EXCLUSIVE for password file authentication.
- The ORAPWD utility generates the password file for SYSDBA and SYSOPER and then the database maintains it with changes to passwords.
- Control files can be multiplexed, meaning that each subsequent control file is an exact copy of the first control file.
- Because control files are important, multiplexed copies of those control files should be located on different physical devices to guard against damage.
- You can prevent bottlenecks in data access by placing data on several physical devices, which spreads the demand across more resources.
- Oracle Managed Files eases the DBA's ongoing problem of monitoring and controlling the growth of datafiles.
- User-managed file management offers more detailed control over datafiles than Oracle Managed Files, but also requires more manual maintenance tasks.
- The OMF method automates removal of dependent datafiles when a tablespace is dropped.
- OMF handles creation, naming, and sizing of datafiles.
- The parameter of DB_CREATE_FILE_DEST initialization sets the location of datafiles when using OMF.
- The DB_CREATE_ONLINE_LOG_DEST_ initialization parameters set the location of control files and redo log files when using OMF.
- OMF uses OFA (Optimal Flexible Architecture) as its file-naming standard.

- Initialization parameters are divided into two major types: basic and advanced. In short, Oracle is stating that basic parameters can be altered easily. However, when considering altering advanced parameters, some planning is the prudent approach. Research the precise meanings of advanced parameters before changing.
- When using a binary parameter file, initialization parameters can be changed at the session level for the life of a database connection. Initialization parameters can also be changed system wide. Some parameters require a database restart for changes to be effective.
- The Database Configuration Assistant (DBCA tool) leads you through several steps to create a new database.
- You can choose from four types of database configurations in the DBCA tool, including Custom, Data Warehouse, Transaction Processing, and General Purpose.
- Dedicated Server mode does not work well for very large OLTP databases. Shared Server mode works better for very large groups that have many users accessing the database simultaneously, typically over 1000 concurrent database connections.
- The DBCA tool provides an opportunity to customize memory size and initialization parameters, if you choose to do so.
- Adjusting of tablespace and datafile sizes and locations depends on the database type you choose when using the DBCA tool.
- After creating a new database, use Net Manager to set up a Net Service name for the database.
- To create a database manually, your first step is to set up a directory structure for the files that are to be created. However, prudence dictates that it is best to generate scripts first using the DBCA tool.
- Create a password file to implement password file authentication when the new database is created.
- A database service must be started if you are using Windows, but is not required if you are using Unix or Linux.
- The CREATE DATABASE command generates the datafiles, control files, and so on.
- All manually created databases should also have tablespaces called SYSTEM for metadata, SYSAUX for Oracle add-on options, temporary storage (usually called TEMP), an undo tablespace, and a tablespace for other schemas (usually called USERS).
- All manually created databases should include the minimum scripting generation options, as created by the DBCA tool script generation option. At a bare minimum **catalog.sql** creates metadata, data dictionary views, and **catproc.sql** basic built-in PL/SQL procedures.
- The SHUTDOWN command in SQL*Plus requires you to log on as SYSDBA.
- SHUTDOWN IMMEDIATE is faster than SHUTDOWN NORMAL. SHUTDOWN NORMAL is more prudent.
- SHUTDOWN ABORT is used only when the database has errors and does not shut down with NORMAL, IMMEDIATE, or TRANSACTIONAL.

- A database can be started up with a text or binary configuration initialization parameter file. Starting up a database by default using the STARTUP command starts up a database using the binary parameter file.
- Using a binary configuration initialization parameter file allows changing of most parameters with the database up and running. An up and running database is known as an online database.
- The alert log contains critical errors when something goes wrong.

Review Questions

1. All initialization parameters can be modified after database creation. True or False?
2. List the tools you can use to create a database.
3. ORACLE_BASE and ORACLE_HOME are used only in Windows operating systems. True or False?
4. The ________________ describes system-specific details such as memory requirements.
5. List these steps in chronological order from start to finish.
   a. Write and execute the CREATE DATABASE command.
   b. Start up the database.
   c. Set the initial parameters.
   d. Install Oracle 10*g* RDBMS software.
   e. Shut down the database.
6. Which of the following steps occurs while using the Database Configuration Assistant?
   a. Setting MAXINSTANCES
   b. Selecting database type
   c. Logging onto SQL*Plus
   d. Saving old data
7. When using OS authentication, the OS user must be the same as the Oracle user name when the ________________ initialization parameter is null.
8. When in the console, you must log on with SYSDBA privileges to start the database. True or False?
9. Which of these are primary tasks in file management? Choose all that apply.
   a. Choosing the location of files
   b. Selecting the tablespace names
   c. Deleting old files and adding new ones
   d. Reviewing log files
   e. All of the above
10. Explain how user-managed file management gives more control to the DBA than the Oracle Managed Files method.

11. The DB_CREATE_FILE_DEST defines the OFA standard for the database. True or False?
12. To add a remark to the **init.ora** file, begin the line with this symbol:.
13. List the two types of initialization parameters.
14. You must include the CONTROL_FILES parameter in the **init.ora** file when creating a database manually. True or False?
15. The following utility creates a database service in Windows.
    a. ORAPWD
    b. DBSTART
    c. ORADIM
    d. EXP
16. Data Dictionary views are created by running the ________________ script, and PL/SQL procedures are created by running the ________________ script.
17. In Windows, you can set the value of ORACLE_SID in the Windows Registry. True or False?

Exam Review Questions—Oracle Database 10*g*: Administration (#1Z0-042)

1. Choose the phrase that best describes a DBA authentication method.
    a. Sets the maximum number of instances to run simultaneously
    b. Determines which users are allowed to log on
    c. Specifies how to validate users who log on with special privileges
    d. Enables remote access to the database
2. Which of the following initialization parameters affect DBA authentication? Choose all that apply.
    a. REMOTE_LOGIN_PASSWORDFILE
    b. COMPATIBILITY
    c. OS_AUTHENT_PREFIX
    d. ORAPWD
    e. All of the above
3. Which of the following database components can be defined in the CREATE DATABASE statement? (Choose three.)
    a. **init.ora** file
    b. Undo tablespace
    c. Temporary tablespace
    d. SYSTEM tablespace
    e. Log groups
    f. SYSTEM password

4. When creating a database manually, which of the following statements is true regarding the control file? (Choose one.)
   a. The control file is multiplexed by default.
   b. The directory in which the control file is created is automatically created.
   c. The CONTROL FILE clause must be included in the CREATE DATABASE statement.
   d. None of the above.
5. Examine the following statement.

```
1 CREATE DATABASE PROD02
2 MAXLOGFILES 5
3 MAXDATAFILES 100
4 SYSTEM TABLESPACE DATAFILE
' C:\oracle\product\10.1.0\oradata\trial02\sys01.dbf' SIZE 325M
5 UNDO TABLESPACE UNDOTBS
6 DATAFILE
' C:\oracle\product\10.1.0\oradata\trial02\\undo01.dbf' SIZE 25M
7 LOGFILE GROUP 1
('C:\oracle\product\10.1.0\oradata\trial02\redo01.
log') SIZE 50M,
8 GROUP 2 (' C:\oracle\product\10.1.0\oradata\trial02\redo02.
log') SIZE 50M;
```

   Which line has an error?
   a. Line 7
   b. Line 8
   c. Line 4
   d. Line 5
   e. None of the above
6. Examine the following statement.

```
CREATE DATABASE PROD02;
```

   Which of the following statements are true? Choose the best answer.
   a. The statement will fail.
   b. The SYSTEM tablespace will be 200 M in size.
   c. The statement will create Oracle-managed datafiles.
   d. The control file may or may not be multiplexed.
   e. The control file will be 200 M in size.
7. You are administering an Oracle 10*g* database. Which one of these tools should you use if you need to shut down the database using a command line?
   a. Oracle Database Configuration Assistant
   b. SQL*Plus
   c. Telnet
   d. Enterprise Manager

8. You are configuring the password file for the PROD01 database. Which of these commands creates a password file named orapwPROD01.ora located in the /usr/pwd directory and allows no more than four additional DBAs to be assigned the SYSDBA role?
   a. orapwd file=/usr/pwd/orapwPROD01.ora password=trueblue2002 entries=5
   b. orapwd file=/usr/pwd/orapwPROD01.ora password=trueblue2002 entries=4
   c. orapwd pwdfile=/usr/pwd/orapwPROD01.ora entries=5
   d. orapwd file=/usr/pwd/orapwPROD01.ora maxpassword=5
9. You are preparing to create a second database instance. You copy the init.ora file and prepare to use it for the second database instance. Which initialization parameter must be changed so that it does not conflict with the first database instance?
   a. REMOTE_LOGIN_PASSWORDFILE
   b. DB_CREATE_FILE_DEST
   c. COMPATIBLE
   d. DB_NAME

Hands-on Assignments

1. You are about to create a new database for the Statistics Department of a large insurance firm. The database functions primarily as a storage area for data on past insurance clients and their claim history. Each month, any closed accounts are loaded into the database. There are 25 offices around the country that plan to load data and use the database for gathering statistics. You want to use the Database Configuration Assistant. Using the figures and steps in the chapter to help remind you of the choices available, what selections would you make while running the DBCA tool? Explain your reasoning briefly. Save your work in a file named ho0301.txt in the Solutions\Chapter03 directory on your student disk.
2. You are the DBA at a factory. A new Oracle 10*g* database will soon replace the current database. The old database has serious performance problems that seemed to be caused by an overload on the disk drive's I/O channel. Your analysis has also shown that two important and large tables are frequently accessed: the CAR_PART table and the ENGINE_PART table. You have a new computer with three large-capacity disk drives: F, G, and H. Describe how you plan to alleviate the I/O problem in the new Oracle 10*g* database using the user-managed file management method. Save your work in a file named ho0302.txt in the Solutions\Chapter03 directory on your student disk.
3. Using the situation described in Hands-on Assignment 2, describe how you would solve the same problem using the Oracle Managed Files method. Save your work in a file named ho0303.txt in the Solutions\Chapter03 directory on your student disk.
4. Look at the following lines from an **init.ora** file and list the errors you find; then rewrite the code with corrections and place it in the **Solutions/Chapter03** directory on your student disk with the name **ho0304init.ora**.

```
#
Cache and I/O
#
db_block_size=4096, db_domain="detroit.usa"
remote_login_passwordfile=EXCLUSIVE
```

```
control_files=(" C:\oracle\product\10.1.
0\oradata\trial02\control99.ctl")
maxinstances=2
compatible=1001
#
Database Name
#
db_name=prod1001.detroit.usa
instance_name=trial02
```


5. Using either Oracle documentation (the Oracle 10*g* Database Reference book) or the Initialization Parameter window of the Instance Manager, complete Table 3-3. The first row is filled in as an example.

TABLE 3-3 Initialization Parameters

Name	ALTER SYSTEM?	ALTER SESSION?	Default value
BLANK_TRIMMING	No	No	False
DB_BLOCK_BUFFERS			
GLOBAL_NAMES			
OPEN_CURSORS			
FIXED_DATE			
JAVA_POOL_SIZE			

Store your results in the Solutions\Chapter03 directory on your student disk. Store the query to retrieve parameters in ho0305.sql, and the results of your query in ho0305.txt.

6. You created a database named trial01 in this chapter. Change the initialization parameter PGA_AGGREGATE_TARGET to **25M** using the Enterprise Manager console. (Recall from Chapter 2 that you can modify initialization parameters in the Instance Manager under the Configuration icon.) To test your results, log on to SQL*Plus and type the SQL*Plus command SHOW PARAMETERS before and after your change.
7. Again using the trial01 database, change the JOB_QUEUE_PROCESSES to **12** using the ALTER SYSTEM command. Use SHOW PARAMETERS to test your results before and after the change. Save your SQL script in a file named **ho0307.sql** in your **Solutions/Chapter03** directory.
8. Add a new tablespace named INDX to the trial01 database. Make it an Oracle Managed file. Save the command you used to create the tablespace in the Solutions/Chapter03 directory in a file named **ho0308.sql**. What is the name of the datafile that was created? Where is the datafile located?
9. Look at the following CREATE DATABASE command, and find the errors. Then write a corrected version and save it in the Chapter03 solutions directory in a file named **ho0309.sql**.

```
CREATE DATABASE ultradb FOR UPDATE OF paralleldb
MAXINSTANCES 25
DATAFILE TABLESPACE 'ORACLE_BASE\oradata\system01.dbf'
CHARACTER_SET US7ASCII
```

```
LOGFILE GROUP A ('D:\ORACLE_BASE\oradata\redoA.log') SIZE 50M,
GROUP B ('D:\ORACLE_BASE\oradata\redoB.log') SIZE 50M;
```

10. Using the Database Configuration Assistant (DBCA tool), compare the settings for the Data Warehouse model and the Transaction Processing model. What is different and why? Save your work in a file named ho0310.txt in the Solutions\Chapter03 directory on your student disk.

Case Project

When you arrive at the Global Globe office, you find that your boss, the Senior DBA has left for Hawaii on vacation. You have one day to create a new database on the 50 programmer workstations in your office. You need help. You decide that the only way to get it all done is to create a script that each programmer can run on his or her own workstation. You know, however, that if the instructions for running the script get longer than a few short sentences, no one will do it correctly! Your job is to set up a foolproof database creation script. Earlier in this chapter, you used the DBCA tool (Database Configuration tool) to generate scripts for a database called trial02. Adjust the parameters and the scripts, so that all the tablespaces, control files, and redo log files are Oracle Managed Files. Then create a batch file that runs both scripts. (*Hint*: To start SQL*Plus and run a file inside SQL*Plus, use this command format: SQLPLUS <user>/<password> @<dbname> @filename.sql.) Change the database name to trial03. The script must run without any user intervention. (You may, for the sake of this exercise, assume that the programmer has administrator authority on his or her own workstation, and that any commands to adjust the Windows Registry are not required, because you have not seen how to revise the Windows registry from the command line.) *Hint*: You should have to change the contents of, and even some of the names of the script files CreateDB.sql, CreateDBFiles.sql, init.ora, postDBCreation.sql, trial02.bat, and trial02.sql.

Save the changed scripts and batch file only in the **Solutions/Chapter03** directory.

Clean up your workstation by using Oracle Database Configuration Assistant to delete any work databases created in this chapter, such as the trial01, trial02, and trial03 databases. The remaining chapters work with the ORACLASS database instance that was installed before you started the course.

CHAPTER 4

ORACLE PHYSICAL ARCHITECTURE AND DATA DICTIONARY VIEWS

LEARNING OBJECTIVES

In this chapter, you will:

- Examine tablespaces and datafiles
- Understand how the control file, datafiles, redo log files, and archive log files are linked
- Examine advanced database architectures, including OMF, partitioning, replication, standby, and grids
- Manage and multiplex control files
- Use OMF to manage control files
- Create new control files
- View control file data
- Learn to describe redo log files, groups, and members
- Manage redo log groups and members
- List useful dynamic performance views

INTRODUCTION

Upon first inspection, it may seem overwhelming to absorb all the components found in the Oracle 10*g* database. With each chapter, however, you learn to work with new components, building your understanding, little by little. This chapter explores several important areas of the database that you need to understand to manage the Oracle 10*g* database.

TABLESPACES AND DATAFILES

You have already seen some detail covering tablespaces and datafiles in previous chapters. Datafiles are the physical files stored in the underlying operating system. Datafiles contain all the data in an Oracle database. Tablespaces are logical overlays of underlying datafiles, allowing access to data stored in datafiles.

An Oracle 10*g* database essentially consists of various types of tablespaces, all essential to the smooth operation of an Oracle database installation:

- **SYSTEM**: Contains all the metadata, or the data about the data, such as the structures of user application tables (table names, field names, datatypes, and so on). In addition, the SYSTEM tablespace contains Oracle database metadata structures.
- **SYSAUX**: This tablespace is newly created for Oracle 10*g* and contains Oracle optional software metadata and data. In previous versions of Oracle, these structures were either included in the SYSTEM tablespace, or created in a multitude of separate tablespaces.
- **UNDO**: Undo or rollback data can be used to undo non-committed changes in a database.
- **TEMP**: Temporary space is used for very large on-disk sorts, when a sort overflows the size of sorting buffer space in memory allocated to the database server.
- **USERS**: This tablespace is created by default in the Database Configuration Assistant (DBCA tool), and is intended to contain application structures and data. In reality, application data is likely to be split into multiple tablespaces based on different applications. Sometimes performance tuning subdivisions within applications are used, such as partitioning or splitting of table and index data spaces into two separate tablespaces.

Within this tablespace structure, each tablespace can contain multiple datafiles. Those datafiles can even be stored on separate disk drives. However, multiple tablespaces cannot share the same datafile.

Datafiles essentially are the physical layer of an Oracle database, and tablespaces are the logical layer. An Oracle database, as a whole, is made up of a number of integrated parts, as shown in Figure 4-1, otherwise known as the Oracle instance and the Oracle database. All the physical structures shown in Figure 4-1, part of the Oracle database architecture as a whole, are as follows:

- **Datafiles:** Contain all physical data such as tables, indexes, database metadata, procedural code, and anything stored in the database as accessible or usable by an Oracle database user
- **Redo logs:** Transaction records of all database changes
- **Archive logs:** Historical copies of recycled redo logs, maintaining a complete history of all database change activity
- **Control files:** Contains pointers to all datafiles and log files used for synchronization between all of those files
- **Parameter file:** The configuration parameter file is applied to an Oracle database instance whenever the database is started

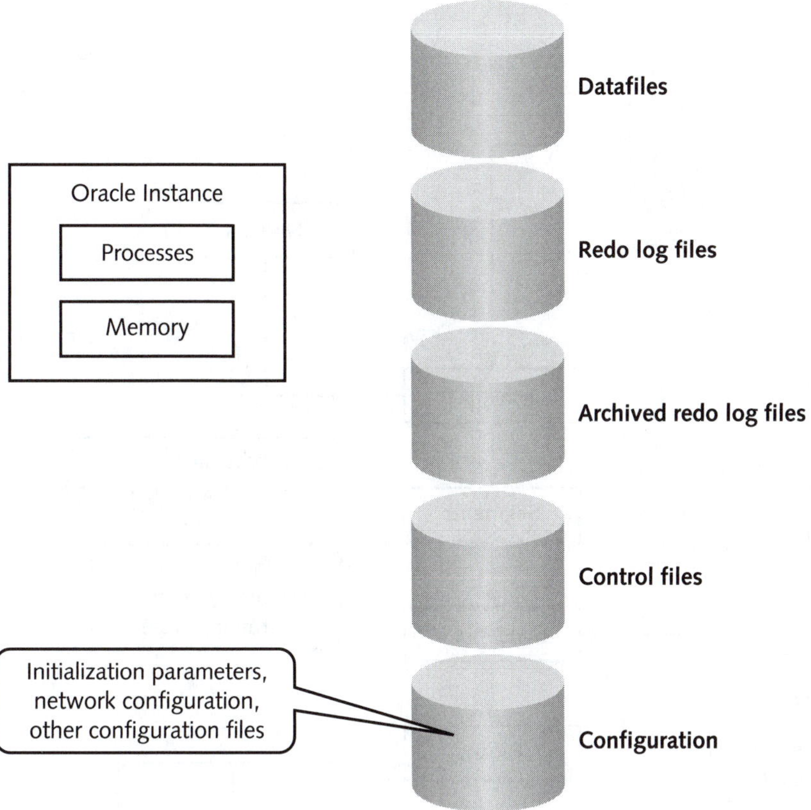

FIGURE 4-1 The Oracle Database and the Oracle Instance

Moving back to the logical structure of an Oracle database, in terms of tablespaces, tables, and indexes, the structure of an Oracle database looks like that shown in Figure 4-2.

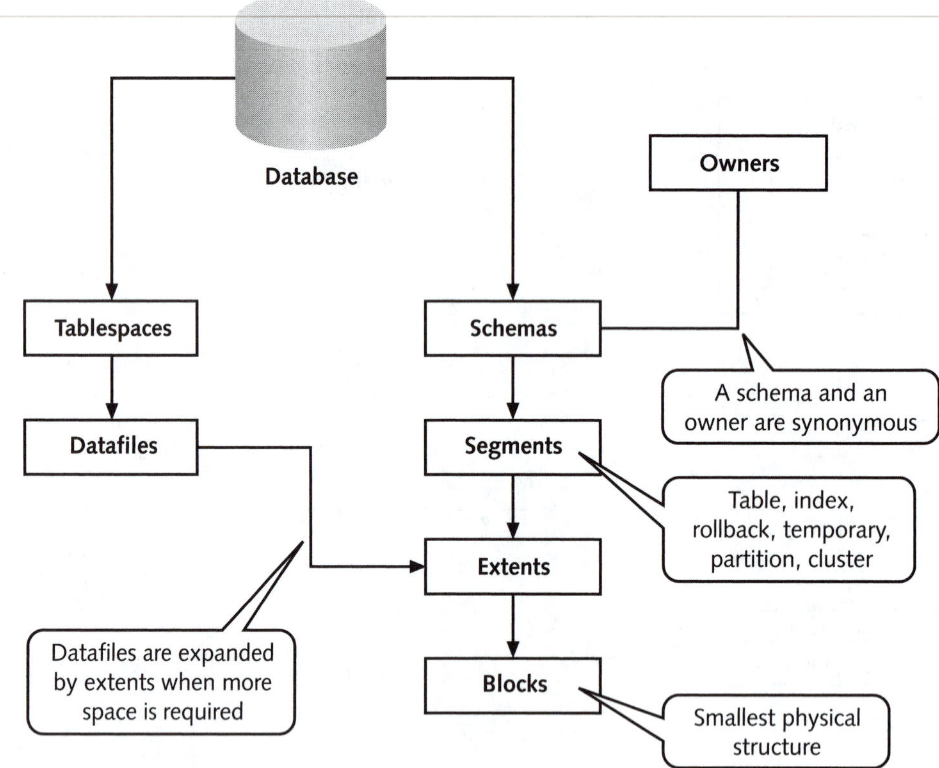

FIGURE 4-2 The logical and physical structure of an Oracle database

NOTE

Ultimately, the entire physical structure of an Oracle database is tied together as shown in Figure 4-3.

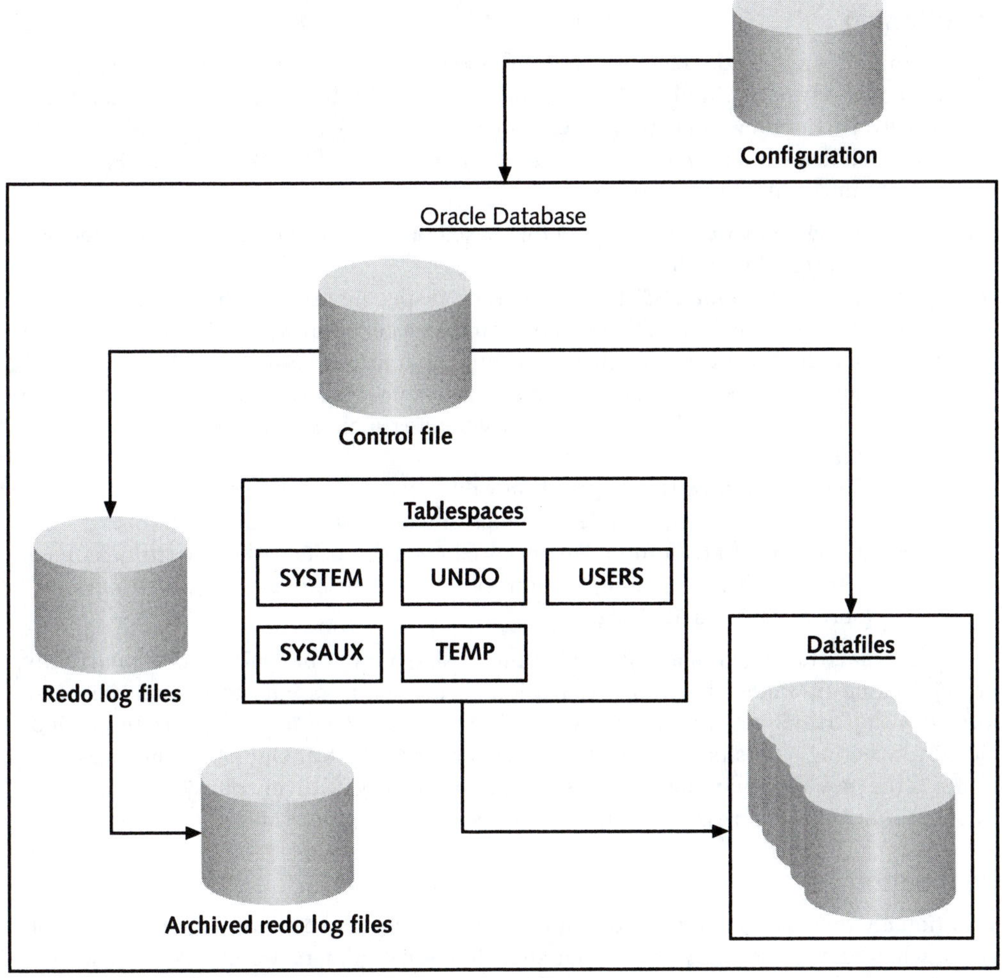

FIGURE 4-3 The flow of physical data and the interrelated structure of an Oracle database

ADVANCED DATABASE ARCHITECTURES

Advanced database architectures do not necessarily apply to the Oracle database alone. You need to know what different types of architecture are, in order to have a brief understanding of what can be achieved with an Oracle database.

Oracle Managed Files (OMF)

OMF allows automated creation and dropping of underlying datafiles, in addition to automated management of both redo log files and control files. OMF has been examined in detail in previous chapters.

Partitioning

Partitioning allows individual tables and their associated indexes to be partitioned (or split) into separate physical chunks. The separate pieces can be executed in parallel or individually. Partitioning can increase performance drastically in very large databases, both for data warehouses and even Online Transaction Processing (OLTP) databases. Partitioning can split tables in a number of ways:

- **Range partitioning**: Split a table based on distinct ranges of values, such as quarters in a year.
- **List partitioning**: Split a table based on specific lists of values, such as state names in the United States. For example, a partition containing data representing the Northeast United States could contain states such as NY, NJ, and PA for New York, New Jersey, and Pennsylvania, respectively. A partition for the West Coast could contain rows from states such as CA (California) and OR (Oregon).
- **Hash partitions**: Divide partition values evenly based on hash values calculated from a column value or values in each row of a table.
- **Composite partitioning**: Partitions can contain other subset partitions or subpartitions, allowing combinations of visibly split ranges or lists, with evenly spread hash partitioning.

Other than the huge impact of parallel processing of multiple partitions concurrently, or eliminating unwanted partitions (partitioning pruning), various tricks can be performed with partitioning. Various types of operations can be performed on partitions individually, affecting only small physical parts of very large tables. One of the most useful things is the moving, renaming of, or destruction of single (effectively physically small) partitions from within otherwise extremely large and cumbersome tables.

Replication

Traditionally replication is intended to link databases distributed over large geographical areas, where data is not only shared but specific chunks of data are exclusive to specific sites. In general, replication occurs in two forms, as either master-to-slave replication or master-to-master replication, as shown in Figure 4-4.

Master-to-slave replication implies data only travels in one direction and master-to-master replication has data traveling in both directions, between any two databases. There can be multiple databases in a set of replicated databases. More specifically using Oracle Replication software, in a master-to-slave database environment, the slave database consists of a database composed solely of materialized views. Effectively, master-to-slave replication does not exist for the Oracle Replication option.

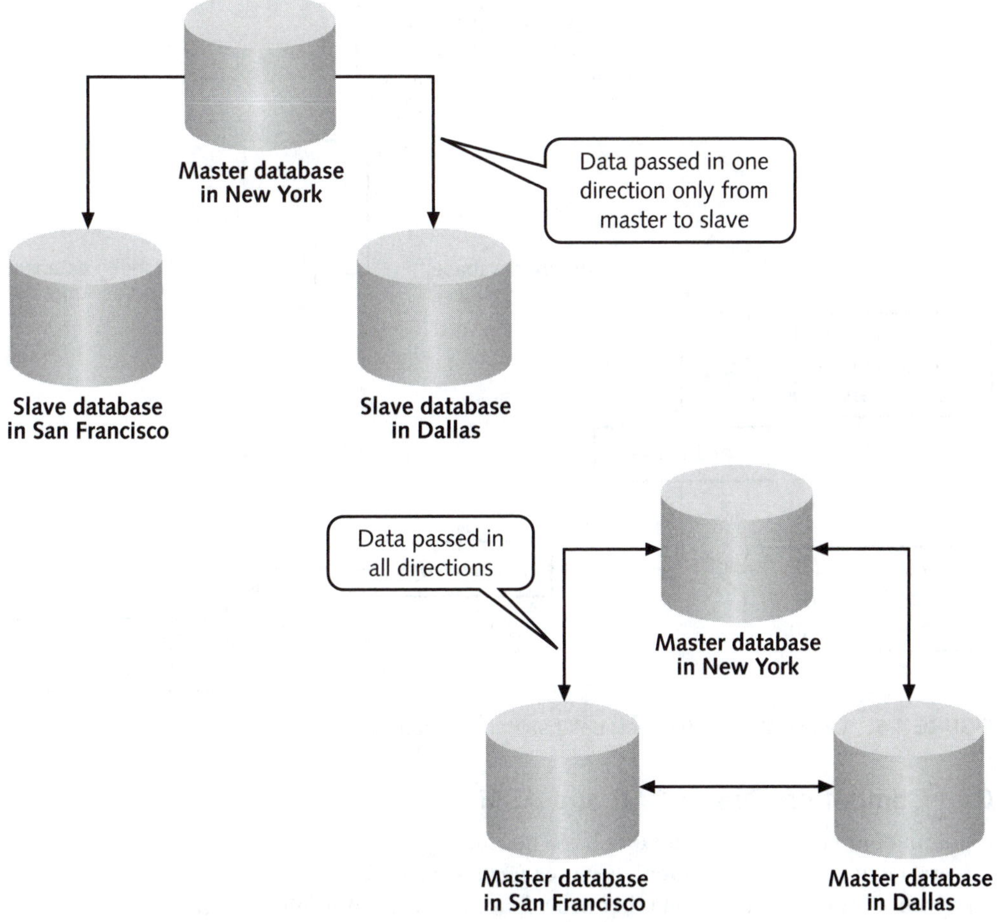

FIGURE 4-4 Master-to-slave and master-to-master replication

One of the most significant problems with the Oracle Replication option is that it is often used inappropriately for backup and failover databases. Oracle Replication has a level of complexity not suited to failover and backup management.

NOTE

Oracle Streams is an alternative, and perhaps much more effective, option for implementing replication—even master-to-master replication.

Standby (Failover) Databases

Standby database architecture is shown in Figure 4-5. There are other more complex ways of implementing standby databases in Oracle 10*g*, but those methods are beyond the scope of this book.

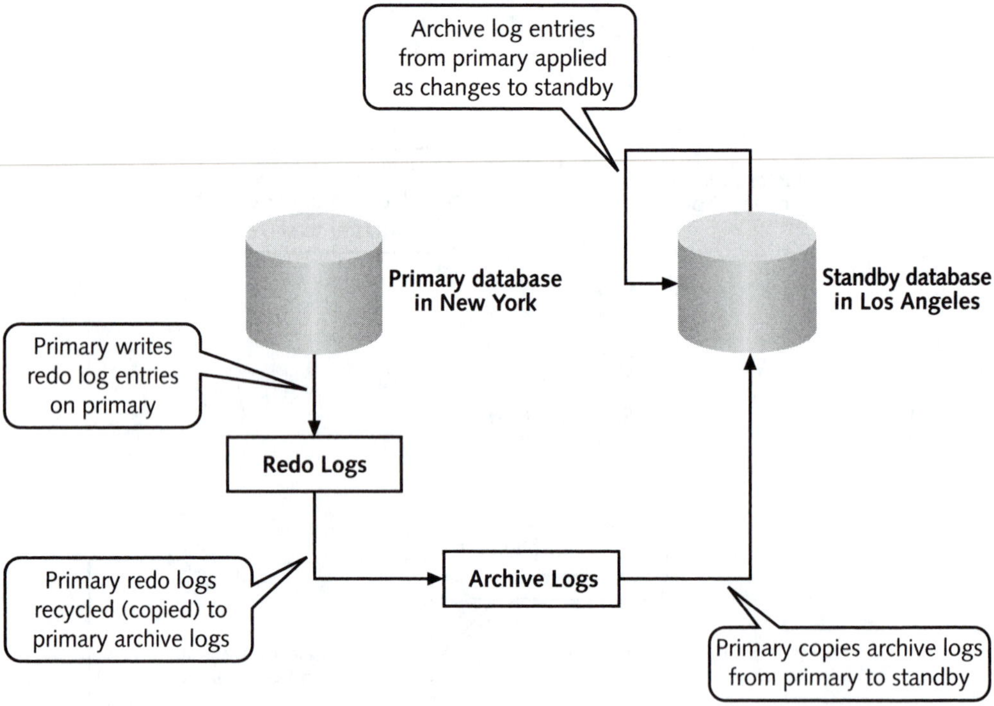

Archive log entries from primary applied as changes to standby

Primary database in New York

Standby database in Los Angeles

Primary writes redo log entries on primary

Redo Logs

Primary redo logs recycled (copied) to primary archive logs

Archive Logs

Primary copies archive logs from primary to standby

FIGURE 4-5 Oracle standby (failover) database architecture

Grid Computing, Oracle RAC and ASM

A grid of computers is made up of a large number of cheaply purchased, simplistic servers—a little a farm of servers, a server farm. Oracle database software contains various features and add-on options that provide for a grid computing architecture. The sum of all these options comprises the capacity for Oracle databases to provide an effective grid computing architecture:

- **Oracle Real Application Clusters (RAC)**: Shares memory (cache) and storage across multiple database instances, running on multiple database server computers. Where a grid contains multiple independent, independently usable computers (containing processors, memory, and disk drive storage), clustering merges the processing and memory powers of multiple workstations, but the disk storage area is shared between the multiple Oracle instances in the cluster.
- **Automated Storage Management (ASM)**: File layout and active load balancing is automated across available disk storage areas using disk groups. ASM simplifies maintenance of storage space and optimizes for I/O performance.
- **Oracle Transportable Tablespaces**: Unplug datafiles from one database, move or copy into another database. Tablespaces can be transported across different platforms and operating systems.

- **Oracle Streams**: Much like opening a direct pipeline between different servers and databases. Streams are update capable and can be used to synchronize different databases, both with a grid of computers and across multiple grids, clusters, or database servers.
- **Oracle Scheduler**: Allows scheduling of batch jobs to be executed repeatedly at specified times.
- **Oracle Enterprise Manager Grid Control**: The Grid Control is the Oracle Enterprise Manager equivalent of the Database Control, but for a grid computing architecture. The Database Control allows management of a single database server. The Grid Control applies to multiple database servers, across a network, or within a grid architectural structure.

The objective of grid computing is to provide high performance, high scalability, automated allocation of resources, information sharing, distribution, and effective security. This is achieved by the sharing of hardware resources, pooling large quantities of basic resources, and automated addition and subtraction of hardware resources based on demand. The overall effect is transparency to the end user. The end user sees a single database server, even though that database server could in reality, be as many as hundreds of very simple individual, cheap blade servers, running Linux.

Now let's examine the control files in detail.

THE CONTROL FILE

An Oracle database is made up of physical structures and logical structures. The physical structures are the datafiles and redo logs. The logical structures are the objects overlaying the datafile structures, such as tablespaces, tables, and indexes.

The control file contains the pointers between datafiles and the redo log, linking them together. A single redo log entry is stamped with a System Change Number (SCN). Whenever a datafile change is made, a redo log file entry is made first—both datafile change and redo log entry get the SCN stamp. If a datafile is ever restored from an old datafile (previously backed up datafile), then changes can be applied from redo log entries to the old datafile, until the datafile reaches the most recent SCN change. In other words, until the datafile and redo log entries become consistent with each other again, such that the datafile is once more up to date with the rest of the entire database.

Additionally, the control file contains up-to-date information on the database structure, log files, and checkpoints. The control file is critical for opening the database as it links the physical and logical structures together. Chapter 3 showed you how to use the CONTROL_FILES configuration initialization parameter to create one or more control files, when creating a new database. The next things to discover are how to manage, multiplex (duplicate), add, rename, relocate, and replace control files.

The control file contains this information about the database:

- **The database name**: Defined either in the CREATE DATABASE command or the DB_NAME initialization parameter.
- **Names and locations of associated datafiles and online redo log files**: Defined by the CREATE DATABASE command and then updated whenever a new tablespace or redo log file is created.

- **The timestamp of the database creation**: Logged when the database is created.
- **The current log sequence number**: Updated whenever the current log group changes. Every log entry is a record of a change in a database. Each entry is assigned a sequence number. If later rebuilding database changes from log file entries, such as in recovery, redo log entries will be applied to datafile backups in the sequence of the log sequence number.
- **Checkpoint information**: Time-stamped redo log entries identifying the most recent changes made to the database, which are matched inside the database structure itself. Redo log entries are also stamped with an SCN, which is a sequential number based on when in the sequence a database change was made. Thus, checkpoints can be made to recover up to a specific time or change number.

The control file is like your credit card: If your credit card is damaged, you still have your account information, but you must replace the credit card before you can access that information. Likewise, if the control file is damaged, the database may be intact, but you cannot open the database until the control file is restored, recovered, or recreated.

Because the control file is so critical, Oracle recommends keeping redundant (duplicate) copies online at all times, in other words multiplexing or simultaneous changes to more than one copy of the control file at the same time. Therefore, multiplexing means making redundant copies of files to guarantee that if the original file is damaged, another can take its place immediately. The next section shows how to manage control files, including maintaining multiple multiplexed copies.

Managing and Multiplexing the Control Files

Now that you have created the database and it is up and running, what else needs to be done with the control files? The first thing you should do is multiplex the control file to ensure against total loss of the file. You can create up to eight copies of the control file. After that, the control file is self-managing unless you make any of these types of changes to the database:

- Add a new control file
- Rename or relocate one or more control files
- Replace a damaged control file

All three of these tasks have very similar procedures. Steps for each task are described in the following sections.

NOTE

It is always a good idea to make a backup of your control files before you begin changing them.

Adding a New Control File

A new control file can be added at any time. For example, you may have created the database with two control files, and then later that day, after reading up on the importance of the control file, decide you really need a third copy on a third device that is more protected from power surges. Or, you may have just arrived on a new job and discovered that the database only has one control file. Your first task for the new assignment is to multiplex the control file. Follow these steps to add a new control file:

1. Shut down the ORACLASS database using SQL*Plus. Shut down by connecting to the database as the SYS user with SYSDBA privileges, then execute a SHUTDOWN IMMEDIATE command.
2. Open the Windows Explorer and locate the control file named **CONTROL1.CTL**. It should be in the **ORACLE_BASE\oradata\ORACLASS** directory.
3. Copy the control file and place the copy in a directory on your hard drive. Ask your instructor which directory to use. Use the Edit menu in the Windows Explorer. For Unix or Linux, use the standard operating system copy command (**cp**) to make a copy of the file to a new location.
4. Rename the new copy to **CONTROL04.CTL**. Use the File menu in the Windows Explorer to complete this task. In Unix or Linux, use the standard operating system command (**mv**).
5. Open the **init.ora** (or **initORACLASS.ora**) file for the ORACLASS database with your Notepad editor. The file is located in **ORACLE_BASE\admin\ORACLASS\pfile** directory.
6. Change the CONTROL_FILES parameter in the database's initialization parameter file by adding the new control file's name to the list of control files. There should be commas separating each filename. The lines should look similar to the following (your actual path may differ, mine is set to drive E on my Windows computer):

   ```
   control_files=
   (
    "C:\oracle\product\10.2.0\oradata\ORACLASS\control01.ctl"
   ,"C:\oracle\product\10.2.0\oradata\ORACLASS\control02.ctl"
   ,"C:\oracle\product\10.2.0\oradata\ORACLASS\control03.ctl"
   ,"E:\oracle\product\10.2.0\oradata\ORACLASS\control04.ctl"
   )
   ```

7. Save the file and close Notepad.
8. Restart the database in SQL*Plus using the STARTUP command, into mounted mode, with the PFILE parameter. At this stage, you can reopen the database, or use the CREATE SPFILE ... FROM PFILE command to generate a binary version of the parameter file, then shut down and restart normally.

NOTE

You don't have to re-create the binary (server) parameter file with the database in mounted mode. However, to avoid problems in a busy environment, it may help to avoid little mishaps. Mount mode restricts access to a database to administration level only.

The database automatically begins multiplexing all the control files listed in the CONTROL_FILES parameter. It verifies that all the files are identical upon startup, which is why you made a copy rather than starting with a blank file.

NOTE

If any of the control files are mismatched, the database will not open.

Renaming or Relocating an Existing Control File

There are a number of practical reasons for renaming or relocating a control file. For example, you may want to change the names of control files to comply with new naming standards your department has implemented. Or you may want to relocate a control file because the database system has grown in size and now has a larger hard drive that is a safer location for one of the control files.

To relocate an existing control file, follow these steps:

1. Shut down the ORACLASS database using SQL*Plus. Once again, shut down by connecting to the database as the SYS user with SYSDBA privileges, then execute a SHUTDOWN IMMEDIATE command.
2. Open the Windows Explorer and locate the control file you just created, which is named **CONTROL04.CTL**.
3. Use the **File** menu in the Windows Explorer to rename the file to **CONTROLX.CTL**. For Unix or Linux, use the standard operating system move command (**mv**) to rename (move) the file.
4. Open the **init.ora (initORACLASS.ora)** file for the ORACLASS database with your Notepad editor. The file is located in the **ORACLE_BASE\admin\oraclass\pfile** directory.
5. Change the CONTROL_FILES parameter in the database's initialization parameter file by modifying the name of the control file to match its current name.
6. Save the file and close Notepad.
7. Next, restart the database using STARTUP in mounted mode with the PFILE parameter. At this stage, you can reopen the database, or use the CREATE SPFILE ... FROM PFILE command to generate a binary version of the parameter file, then shut down and restart normally.

Technically relocating is the same as renaming—all you need to do is to copy (or move) the file, and then change the CONTROL_FILES parameter.

Replacing a Damaged Control File

The loss of one control file can be quickly fixed without any additional recovery steps, as long as you have a second copy. The control file may have been corrupted by a power interruption that unexpectedly shut down the computer. The file also may have been accidentally deleted or damaged when a hard drive failed. Whatever the reason, the database does not start until all control files are restored and available.

If a control file is missing or damaged, Oracle 10g issues this error message when starting up the database:

```
ORA-00205: error in identifying controlfile, check alert log for more info.
```

In the case of a control file error, you must first shut down the database, fix the control file, and then restart the database.

You can fix a control file by replacing the faulty or missing control file with a valid copy of the control file. In this next set of steps, you simulate a missing control file by deliberately deleting one of the control files. Then, you fix the problem by replacing the deleted file with a copy of one of the other control files.

1. Shut down the ORACLASS database using SQL*Plus.
2. Open the Windows Explorer and locate the control file named **CONTROLX.CTL**. This is the control file you renamed in the previous section (Renaming or Relocating an Existing Control File). In Unix or Linux, use the **cd** command to navigate to the directory, and the **ls** command to find the file.
3. Delete the control file. ***This step is performed only to simulate a missing control file.*** Use the File menu in the Windows Explorer to remove the file rather than renaming the file. In Unix or Linux, use the system remove (**rm**) command to delete the file. Be very careful not to destroy anything else. A safer form of the **rm** command is to use the **rm –i** option, which prompts you to indicate what you are deleting.
4. Restart the database again using SQL*Plus. The database startup should fail with the ORA-00205 error message.

   The startup command fails when it tries to mount the database. The Oracle instance will start up in nomount mode, creating processes and allocating memory buffers. The second step of database startup puts a database into mount mode. Mount mode uses the control file to mount logical database objects, such as tablespaces, to the datafiles. A control file is missing and thus the mounting process fails. Opening the database into open mode for general use never occurs in this case because the previous step (mount mode) fails.
5. Shut down the ORACLASS database again in SQL*Plus.
6. Open the Windows Explorer and locate the control file named **control01.ctl**. It should be in the **ORACLE_BASE\oradata\ORACLASS** directory. In Unix or Linux use the **cd** command to navigate to the directory.
7. Copy the **control01.ctl** file and paste it into the directory where you originally placed the **CONTROLX.CTL** file, then rename the copy to **CONTROLX.CTL** to restore the file. In Unix or Linux, use the **cp** command and name the new file **CONTROLX.CTL**. ***These instructions show you how to fix a missing or damaged control file. You replace it with a copy of a valid control file.***
8. Restart the database once again using SQL*Plus.

To remove a control file, follow these steps. When you complete these steps, your database is back to its original configuration of control files.

1. Shut down the ORACLASS database.
2. Open the Windows Explorer and locate the control file named **CONTROLX.CTL**. This is the control file you restored in the previous section. In Unix or Linux use the **cd** command to navigate to the directory.
3. Delete the control file. Use the File menu in the Windows Explorer to remove the file rather than renaming the file. (For Unix or Linux use the **mv** command to relocate rather than rename the file.)
4. Open the **init.ora (initORACLASS.ora)** file for the ORACLASS database with your Notepad editor. The file is located in the **ORACLE_BASE\admin\ORACLASS\pfile** directory.
5. Change the CONTROL_FILES parameter in the database's initialization parameter file by removing the CONTROLX.CTL file from the list of control files.
6. Save the file and close Notepad.
7. Restart the database again using SQL*Plus.

So far, the examples you have been working with involve control files that are managed by traditional methods. With the Oracle Managed File (OMF) option you can accomplish the same tasks. The next section describes the main differences between the traditional and the OMF methods.

Using OMF to Manage Control Files

OMF handles the names and locations of files created in association with the database. Here are the criteria for OMF control files:

- The initialization parameter DB_CREATE_FILE_DEST must be specified.
- The initialization parameter DB_CREATE_ONLINE_LOG_DEST_n can be specified (optional).
- The initialization parameter CONTROL_FILES must be null.

You must set these parameters prior to issuing the CREATE DATABASE command, or during the process of creating the database with the DBCA tool. The DB_CREATE_FILE_DEST parameter specifies the location of OMF managed database files (datafiles, redo log files, and control files).

If you specify a value for DB_CREATE_ONLINE_LOG_DEST_*n*. *n* specifies is 1 to 5 locations, as in: DB_CREATE_ONLINE_LOG_DEST_1, DB_CREATE_ONLINE_LOG_DEST_2, and so on. If DB_CREATE_ONLINE_LOG_DEST_1 is specified then the DB_CREATE_FILE_DEST destination will not contain control files or redo log files. Control files (multiplexed) and redo log files (duplexed) will be created in each of the DB_CREATE_ONLINE_LOG_DEST_1 to DB_CREATE_ONLINE_LOG_DEST_5 directories specified.

NOTE

Multiplexed control files are copies of control files that are written to simultaneously. Only one control file is actually active, others are maintained as copies. Duplexing redo log files means that multiple redo log members, in each redo log group, are written to in parallel.

Regardless of the technique you choose to create the control files, there are times when you must re-create the control file from scratch.

Creating a New Control File

As you have seen, the control file is critical to the health of the database. However, despite best efforts to safeguard files it is a remote possibility that all control files could be damaged or lost. If you have multiplexed the control files across several devices, this rarely happens. Nevertheless, what do you do in this case? Your database cannot function until the control files are restored.

Two other occurrences warrant creating new control files:

- **Changing the value of MAXDATAFILES, MAXLOGFILES,** or **MAXLOGMEMBERS**: MAXDATAFILES, MAXLOGFILES, and MAXLOGMEMBERS are set when a database was created. These clauses affect the size of the control file and can be changed only by re-creating the control file and specifying the changed values of the clauses to be modified. If your system reaches the current maximum value, the only way to change the value is through re-creating the control files with new settings.
- **Change the name of the database**: Make sure that you modify the DB_NAME initialization parameter to match the name you specify in the CREATE_ CONTROLFILE statement while the database is shut down.

NOTE

If your database was damaged, you may have to go through other steps to recover the database.

The following are the general steps for creating a new set of multiplexed control files, assuming the database is undamaged, but the control files are all either missing or damaged. Do not actually perform these steps now. They are listed for information only:

1. Gather a list of all datafiles, including their full paths. There is at least one datafile for every tablespace. You can find this by looking in the directory on which you store your datafiles, such as **ORACLE_BASE/oradata/ORACLASS**. Include the names of Oracle Managed Files as well as traditionally managed files.
2. Gather a list of all redo log files, including their full paths and group number. These files may be in the same directory as the datafiles, or in a separate directory if you specified that during database creation.
3. Build the CREATE CONTROLFILE command and save it in a plain text file. The syntax of the command looks like this:

```
CREATE CONTROLFILE REUSE SET DATABASE dbname
LOGFILE
     GROUP n 'C:\oracle\product\10.2.
0\oradata\ORACLASS\logfilename' SIZE mmm,
```

```
     GROUP 2 'C:\oracle\product\10.2.
0\ORACLASS\logfilename' SIZE <mmm>
NORESETLOGS
DATAFILE
    'C:\oracle\product\10.2.0\oradata\ORACLASS\<filename>', ....
MAXLOGFILES <nn>
MAXLOGMEMBERS <nn>
MAXLOGHISTORY <nn>
MAXDATAFILES <nn>
MAXINSTANCES <nn>
ARCHIVELOG
CHARSET <charsetvalue>;
```


4. Start the database in NOMOUNT mode.
5. Run your CREATE CONTROLFILE command.
6. Start up the database again.

Here are some points to help you understand the syntax:

- The REUSE clause is optional. It tells Oracle 10g to overwrite any existing control files it finds. If Oracle 10g finds existing control files and there is no REUSE clause, Oracle 10g returns an error.
- All the "MAX" phrases should be used to change the original or default setting of any "MAX" phrase. These phrases specify the maximum number of log files, log members, and so on. The MAXLOGHISTORY phrase applies only to Oracle Real Application Clusters.
- You can specify either <dbname> or SET DATABASE <dbname>. Using DATABASE simply identifies the database to connect with the control files. Using SET DATABASE renames the database.
- List all the redo log groups and files that are members of the groups in the LOGFILE phrase.
- The NORESETLOGS tells Oracle 10g to read the log files and save their settings from the last time they were used. An alternative is to specify RESETLOG; however, this is usually used only when the log files must be recovered.
- List all the datafiles in the DATAFILE phrase.
- ARCHIVELOG specifies that the redo log files should be archived before reusing, making recovery easier. The alternative, NOARCHIVELOG, is the default setting.
- Do not specify CHARSET unless your database was originally created in a character set other than the default, US7ASCII (American English).

As an example, here is a CREATE CONTROLFILE command that renames the database to TEST2004 and changes the maximum number of datafiles to 500:

```
CREATE CONTROLFILE REUSE SET DATABASE TEST2004
LOGFILE
     GROUP 1 'D:\oracle\product\10.2.0\oradata\TEST2004\redo01.log',
     GROUP 2 'G:\oracle\product\10.2.0\oradata\TEST2004\redo02.log'
NORESETLOGS
DATAFILE
    'D:\oracle\product\10.2.0\oradata\TEST2004\system01.dbf',
    'D:\oracle\product\10.2.0\oradata\TEST2004\users01.dbf',
```

```
    'D:\oracle\product\10.2.0\oradata\TEST2004\users02.dbf',
    'G:\oracle\product\10.2.0\oradata\TEST2004\accounting01.dbf'
MAXDATAFILES 500;
```


The following command can save time in writing the CREATE CONTROLFILE command for your database. The trick is that you must use it while the database is open. In other words, if you wait until you have lost your control files, this command cannot help you. Follow these steps to try this command now:

1. Start up SQL*Plus by opening a command line and typing this command, replacing <password> with the current password for SYSTEM *provided by your instructor*:

   ```
   sqlplus system/<password>@<ORACLASS>
   ```

2. Type the following command and press **Enter** to execute it:

   ```
   ALTER DATABASE BACKUP CONTROLFILE TO TRACE;
   ```

   SQL*Plus replies, "Database altered." The command writes the CREATE CONTROLFILE command for you into a trace file. Trace files are located in the directory named in the configuration parameter USER_DUMP_DEST initialization parameter. The precise trace file should be the most recently created user dump destination trace file, unless something else is going on with the database at the time.You might have to scroll through the trace file to locate the CREATE CONTROLFILE command. Figure 4-6 shows part of the contents of the trace file. You can use this file, modify any parts that need changing, and save it as your script for restoring the control files.

3. Oracle recommends that you shut down the database at this point and run a backup of all the files, including the control files you just created.
4. Log off SQL*Plus by typing **EXIT** and pressing **Enter**.

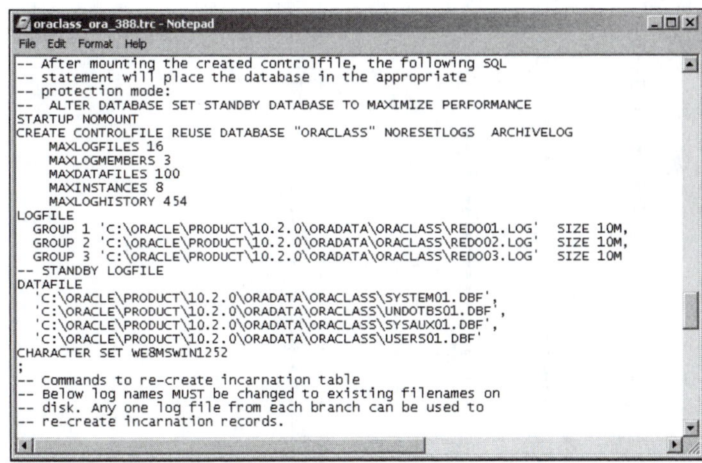

FIGURE 4-6 The trace file contains a complete CREATE CONTROLFILE command

You can use the trace file generated by Oracle 10g to easily write a valid CREATE CONTROLFILE command.

You find out what the control files track in the next section.

Viewing Control File Data

The control file is made up of record sections. Record sections are lists of information by categories within the control file. Some record sections aid in file identification, including records for names and locations of datafiles, tablespaces, temp files, redo log files, and the name of the database. Other record sections aid in recovery activities and contain information on database activity, such as the status of a datafile, details on the most recent backup performed by Recovery Manager, and recovery checkpoints.

Oracle 10g's background processes update the information in the control file whenever certain activities take place. For example, the log writer process (LGWR) updates the control file records whenever a new log sequence number is started.

There are four dynamic performance views that display the high-level contents of the control file.

- V$CONTROLFILE lists the names of the control files in use by the database.
- V$CONTROLFILE_RECORD_SECTION shows data held in the record section, as shown in the query displayed in Figure 4-7.
- V$PARAMETER displays initialization parameter values, including the current value of the CONTROL_FILES parameter.
- V$DATABASE lists current checkpoint numbers and control file sequence numbers.

```
SQL> select * from V$CONTROLFILE_RECORD_SECTION;

TYPE                         RECORD_SIZE RECORDS_TOTAL RECORDS_USED FIRST_INDEX LAST_INDEX LAST_RECID
---------------------------- ----------- ------------- ------------ ----------- ---------- ----------
DATABASE                             192             1            1           0          0          0
CKPT PROGRESS                       8180            11            0           0          0          0
REDO THREAD                          104             8            1           0          0          0
REDO LOG                              72            16            3           0          0          3
DATAFILE                             180           100            4           0          0         16
FILENAME                             524           149            8           0          0          0
TABLESPACE                            68           100            5           0          0          5
TEMPORARY FILENAME                    56           100            1           0          0          1
RMAN CONFIGURATION                  1108            50            0           0          0          0
LOG HISTORY                           36           454           92           1         92         92
OFFLINE RANGE                         56           292            0           0          0          0
ARCHIVED LOG                         584            28            2           1          2          2
BACKUP SET                            40           409            0           0          0          0
BACKUP PIECE                         736           200            0           0          0          0
BACKUP DATAFILE                      116           282            0           0          0          0
BACKUP REDOLOG                        76           215            0           0          0          0
DATAFILE COPY                        660           223            0           0          0          0
BACKUP CORRUPTION                     44           371            0           0          0          0
COPY CORRUPTION                       40           409            0           0          0          0
DELETED OBJECT                        20           818            0           0          0          0
PROXY COPY                           852           249            0           0          0          0
BACKUP SPFILE                         36           454            0           0          0          0
DATABASE INCARNATION                  56           292            1           1          1          1
FLASHBACK LOG                         84          2048            0           0          0          0
RECOVERY DESTINATION                 180             1            1           0          0          0
INSTANCE SPACE RESERVATION            28            63            1           0          0          0
REMOVABLE RECOVERY FILES              32          1000            0           0          0          0
RMAN STATUS                          116           141            0           0          0          0
THREAD INSTANCE NAME MAPPING          80             8            8           0          0          0
MTTR                                 100             8            1           0          0          0
DATAFILE HISTORY                     568            57            0           0          0          0

31 rows selected.
```

FIGURE 4-7 The control file is divided up into separate record sections

Details contained in the record sections are spread out in many V$ dynamic performance views. Table 4-1 shows a list of some of these views.

TABLE 4-1 Control file views

View	Description
V$ARCHIVED_LOG	Archives of redo log entries (copies of recycled redo log files), if the database is in ARCHIVELOG mode
V$DATAFILE	Details about datafiles
V$TABLESPACE	Names and numbers of tablespaces
V$LOG	Online redo log group information
V$LOGFILE	Redo log file information

NOTE

The Recovery Manager's LIST command also displays many details contained in the control file that are used during database recovery.

You can query any of the views listed previously using SQL. For example, the query in Figure 4-8 looks at the datafile information found in the V$DATAFILE view.

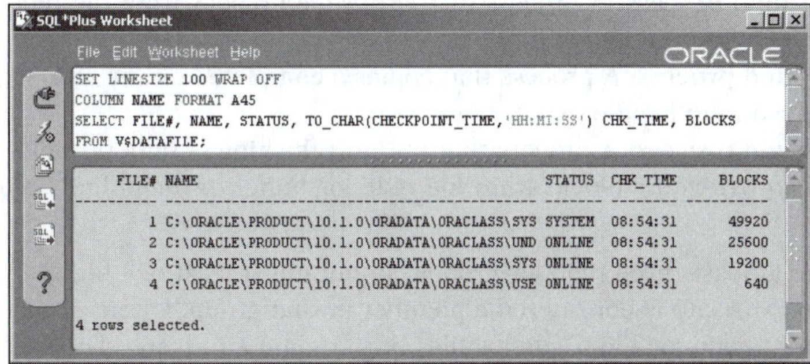

FIGURE 4-8 Use the TO_CHAR function to display times stored in DATE data types

ONLINE REDO LOG FILES AND ARCHIVE LOG FILES

Redo log files record changes to database data. These files are called online redo log files because they are open and available ("online") whenever the database is up and running (online). This section adds to the definition.

Redo log files capture details of database transactions and information about structural or other changes to the database including:

- **Checkpoints:** Checkpoints occur during database shutdown, during a redo log switch (occurs when one redo log file is filled, and the next in the sequence is switched to), and when a checkpoint is forced by an ALTER SYSTEM CHECKPOINT command.
- **Changes:** When a session issues a command changing the database, such as INSERT, UPDATE, or DELETE, redo log files capture information (before the changes are actually committed). The redo log file captures the **Data Manipulation Language (DML)** and **Data Definition Language (DDL)** commands and information on how to undo change operations performed.
- **Datafile changes:** When a new datafile is added or an old one is removed, it is recorded in the redo log file. In addition, when data has changed in the datafile, an entry is made in the redo log file recording that the datafile has changed.

Figure 4-9 shows the components of the database involved with the online redo logs as described in the list that follows:

- **Redo log group:** A set of one or more online redo log files
- **Online redo log member:** A file containing data blocks written by the LGWR process for use in database recovery
- **Archived redo log file:** An offline copy of an online redo log file written by the ARCn process
- **Redo log buffer:** In memory storage of changed data blocks
- **LGWR background process:** A process that writes data blocks from the redo log buffer to the online redo log file
- **ARCn background process:** A process that copies a completed online redo log file to an archived redo log file
- **CKPT background process:** A process that initiates flushing of buffers, causing the LGWR to write data blocks from the redo log buffer to the online redo log file

A database should have at least three redo log groups containing at least one file each. Each redo log file in a redo log group is considered a member of that group. Often, each group contains more than one redo log file. Redo log files in the same group are identical, just as multiple control files are identical. So, when a redo log group is recording data, the data is written to every redo log file in the group, in parallel. If one of the files in the group becomes damaged or unavailable, the other files continue to receive data. As long as one file remains in the current redo log group, the database will continue to function properly. If the LGWR process cannot find any valid redo log files in the current group, the LGWR process instructs the database to shut down immediately. The database also shuts down if the LGWR process cannot find any valid files in the group that will become current after the switch.

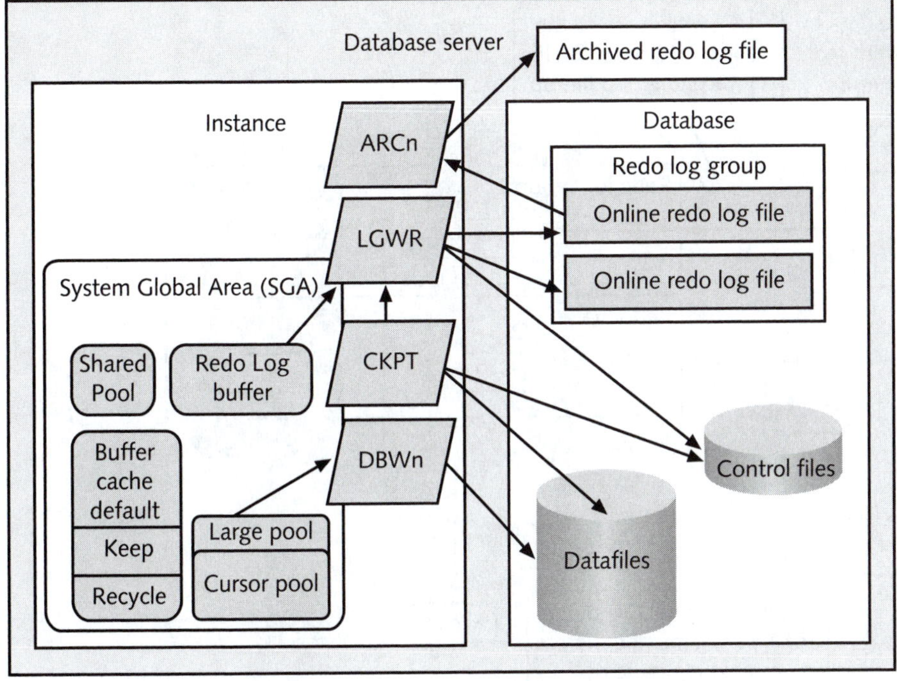

FIGURE 4-9 Redo log components involve memory, files, and background processes

The groups are used in a round-robin fashion, as shown in Figure 4-10. The numbers in Figure 4-10 show these steps:

1. The first redo log group receives data until it is full. All the files in a redo log group are identical in size. They are all written simultaneously, so they contain identical data as well. A redo log group is full when the member files are full.
2. When the first group is full, a redo log switch occurs, and the second group gets data.
3. If the database is in ARCHIVELOG mode (set by an initialization parameter), a file in the first redo log group is copied to an archive file while the second redo log group is being used.
4. When the second group fills up, another log switch occurs, and the first log group is reused, writing over any old data. If the database is in NOARCHIVELOG mode (the default), all the data in the first log group is lost.

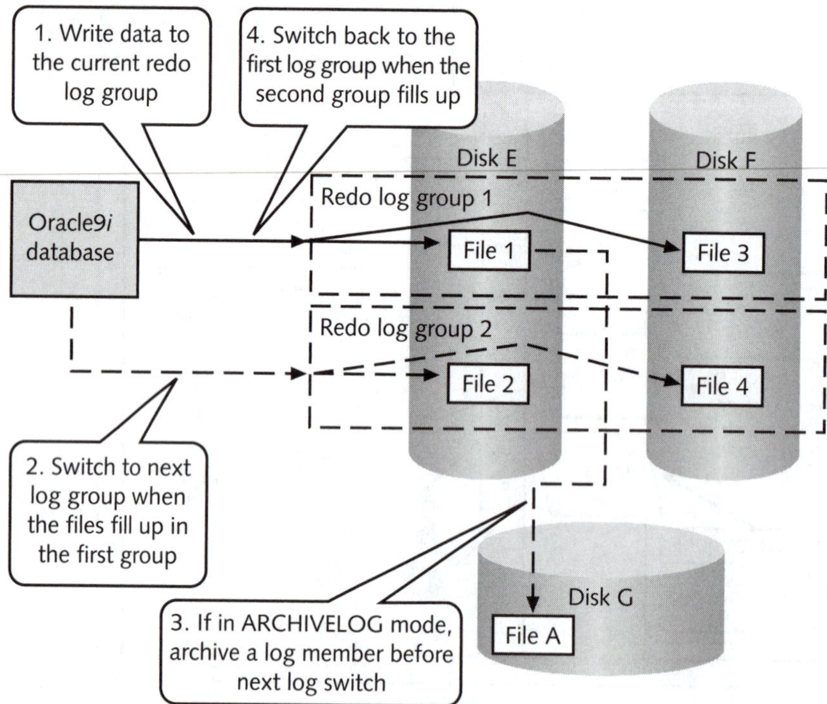

FIGURE 4-10 Redo log groups from one to the next when the members in one group become full

The Purpose of Redo Log Files

The sole purpose of redo log files is to aid in database recovery. The redo log files keep a shorthand list of database changes as they are made. If the database loses some changes because of a power outage, for instance, the recovery process restores the changes by using the redo log files. This is why the redo log files receive the change information before the datafiles are updated. When an unexpected database failure occurs, the process of saving database changes may be interrupted, so the database itself may contain some datafiles that are updated and some that are not. The redo log files contain enough information to restore lost changes.

In minor failures, such as a power outage that lasts a few minutes, the redo log files are automatically checked during database startup, and data is restored, from redo log files into datafiles. However, in major failures, such as the loss of an entire disk due to hardware failure, data would not be saved from the online redo logs alone. You would also need a full backup of the database and archived redo log files that begin after the date of the backup.

TIP

If you want to fully safeguard your database so that you can recover from any failure, you should put your database in ARCHIVELOG mode.

The following steps outline the recovery procedure for problems such as a disk drive crash. Detailed steps that you can use in a real situation are described in Oracle's online documentation on backup and recovery. The steps shown here provide a frame of reference on which you can build.

1. Before database failure, and on a regular basis, perform full database backup procedures.
2. Adjust initialization parameters so that the database starts up in ARCHIVELOG mode as part of routine startup.
3. Upon media failure, such as a disk drive failure, restore the database from the database backup.
4. Apply archived redo log files from the oldest to the newest.
5. Apply online redo log files until all changes have been restored.
6. Apply undo data to remove any uncommitted changes applied by the online redo log files. Undo data is data stored in the undo tablespace or undo segments that identify uncommitted changes.

The result is a database containing every change restored up to the minute that the disk failed.

The Structure of Redo Log Files

The redo log files store a variety of information as a result of database activity. All this information is recorded in the redo log buffer in the System Global Area (SGA). The entire contents of the redo log buffer are written by the Oracle Log Writer background process (LGWR), to the online redo log file, when any of these events occur:

- A transaction issues a COMMIT command.
- The redo log buffer reaches the point of being one-third full.
- Every 3 seconds.
- A checkpoint occurs.

The redo log file contains sets of redo records. A redo record, also called a redo entry, is made up of a related group of change vectors that record a description of the changes to a single block in the database. A single transaction may generate many redo entries (one for each block affected), and each of those redo entries may in turn contain many change vectors (one or more for each record in a block, depending on the change). For example, you write and execute an UPDATE command that looks for all the 10-digit zip codes stored in a CUSTOMER_ADDRESS table. The UPDATE command takes out the hyphen between the first five and the last four digits. This change affects data in many blocks within the table and there are many redo records written to the redo log buffer. Here is another example: When you type the COMMIT command, the redo log buffer's redo records are moved to the redo log files. If the buffer becomes more than one-third full before you commit your transaction, your redo records are written to the log files. In recovery, these changes are applied to the database as if they had been committed and then rolled back using the undo data.

In more detail, it is very easy to surmise what is stored in redo log files in reality. And we can base these facts on what the different commands do, and what already exists in the database, plus what information the redo logs would have to provide to repeat commands previously executed in recovery:

- **INSERT command**: Store the record pointer plus all field values added. Nothing exists so everything must be stored as a single insertion redo log entry.
- **UPDATE command**: Store the pointer to the changed record plus values changed and the field names changed.
- **DELETE command**: Store only the pointer to the row deleted. The row is deleted so the database does not need to know the values deleted, only where to find the row so it can be deleted again in recovery.

Introducing Redo Log File Management

Like control file maintenance, there are several tasks necessary for keeping the redo log files configured and functioning properly. For example, you may notice that the log files fill up very quickly and, therefore, you may decide to change the size of the files, or create more redo log groups.

Before digging into the SQL commands and initialization parameter settings, look into the background processes involved in redo logs.

Figure 4-11 shows a diagram of the redo log processes. In Figure 4-11, a user has issued an UPDATE command followed by a COMMIT command, saving the updated rows. The COMMIT command triggers the LGWR background process to write all the data from the log buffer into the current log group.

The numbers on the figure correspond to these technical notes:

1. When a user submits changes to the database, the changes are stored in the redo log buffer within the SGA memory area of the database.
2. When the user executes a COMMIT command, the user's server process initiates the commit process. The first step in the commit process saves the redo log buffer to the redo log files.
3. The LGWR background process retrieves the buffer and writes it to the current redo log group. It also writes a commit record, identifying the transaction that was committed. In the diagram, Redo log group 1 is the current group. A copy of the buffer and the commit record is written to each file in Redo log group 1. In the diagram, the two members of Redo log group 1 are called **redo01.log** and **redo03.log**, and they reside on separate disk drives.

NOTE

Three redo log groups are recommended but only two are required.

4. After the buffer and the commit record have been written to the redo log group, the buffer contents are deleted. The process of copying the buffer and then deleting its contents is called flushing the buffer.

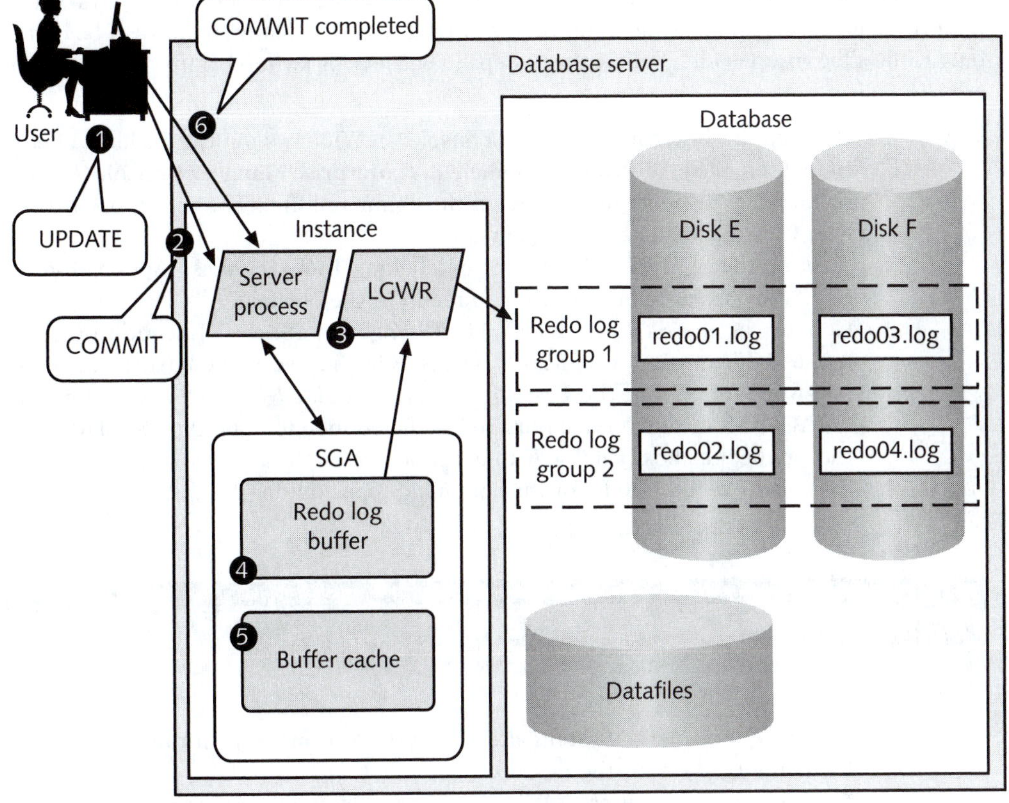

FIGURE 4-11 Committing a change requires writing to the redo log file

5. The changed data blocks are kept in the buffer cache. They are not written to the datafiles at this point. This is called a fast commit. The data is written from the buffer cache to the datafile later.
6. After the redo log buffer data has been written successfully to the redo log group, Oracle tells the server process that the commit succeeded.

The next section describes events that trigger activities involving the redo log files.

Log Switches and Checkpoints

As illustrated in the previous section, a log switch is an event in which the LGWR process stops writing to one log group and begins writing to another log group. A log switch is triggered when a log group fills up with records. Figure 4-10 outlined the steps in a log switch.

> **TIP**
>
> Adding more files to a log group does not change the volume capacity of the log group. Rather, it multiplexes the files, allowing you to store the same redo log data in several locations.

Sometimes, you may need to manually trigger a log switch. Do this when you want to perform maintenance on the log group that is currently active. Maintenance can be done only while a log group is idle. Follow these steps to cause a log switch before the current log group is full:

1. Start up the Enterprise Manager console. In Windows, on the Taskbar, click **Start** then click **All Programs/Oracle ... /Enterprise Manager Console**. In Unix or Linux, type **oemapp console** on the command line. The Enterprise Manager Console login screen appears.
2. Start up the SQL*Plus Worksheet by clicking **Tools/Database Applications/ SQL*Plus Worksheet** from the top menu in the console.
3. Connect as the SYS user. Click **File/Change Database Connection** on the menu. A logon window appears. Type **SYS** in the Username box, the **current password** for SYS in the Password box *as provided by your instructor*, and **ORACLASS** in the Service box. Select the connection and type **SYSDBA** in the Connect as box. Click **OK** to continue.
4. Type and execute the following command to switch log groups:

```
ALTER SYSTEM SWITCH LOGFILE;
```

NOTE

A log file switch can be performed as any DBA user, such as SYSTEM.

5. The system displays "System altered" in the bottom pane of the worksheet. Remain logged on for the next exercise.

   When you execute a COMMIT command, you have seen that it causes the LGWR to write the log buffer to the current log group. One of the details contained in the redo entry is the System Change Number (SCN). Once again, the **SCN** is a sequential number that is incremented for each change that modifies the physical database files. The current SCN is stored in the control file.

A checkpoint is a moment in time when the CKPT background process signals that all used memory buffers are to be written to disk. Used memory buffers are called dirty buffers. In addition, a log switch triggers a checkpoint. The checkpoint process involves incrementing the SCN value and writing the checkpoint information to the control file and the datafile headers. Following this process, all dirty buffers are written to their respective disks. Memory buffers are written to datafiles by the database writer background process (DBWR), and log buffers are written to redo log files by the log writer background process (LGWR).

Checkpoints help in database recovery. If the database shuts down unexpectedly, the automatic recovery process looks at the control file SCN (written by the checkpoint process) and begins rolling forward through changes found in redo log files that are associated with a higher SCN. This speeds up the recovery process by skipping data that has already been written to the physical disk. The redo log files are so critical to recovery that Oracle recommends always duplexing them, as described in the following section.

Duplexing and Other Maintenance

Duplexing of redo logs is the practice of maintaining multiple identical copies of a redo log file to reduce potential risk of loss. When duplexing redo log files, simply add new file members to each group. As illustrated in Figure 4-11, a group can contain many redo log files, and the files can be located on several disk drives. Here are some additional points about redo log groups and duplexing redo log files:

- All the files in a redo log group must be the same size.
- The LGWR process writes concurrently to all redo log files in one redo log group.
- The LGWR process never writes to two different redo log groups at the same time.
- If one or more redo log files are damaged within a redo log group, LGWR writes to the remaining file (or files) and does not stop any database operations.
- If all the files in a redo log group are damaged, LGWR stops all database operations, until a successful log switch occurs. Manual DBA intervention is required to issue a log switch using an ALTER SYSTEM command.
- If a log switch is writing to a pending archive group, LGWR waits until the group is archived (an automatic process), and then continues with the switch.
- If a log switch fails because all the redo log files in the group to which the database is switching are damaged, the database shuts down and must be recovered after restoring the redo log files. Again, manual DBA intervention is required in this case.

The following sections present several important tasks associated with redo log groups and group members.

Adding a Member to a Group

Member files are added to an existing redo log group if the original CREATE DATABASE command failed to add them, or one of the member files was damaged or lost because of accidental deletion or disk failure. In these situations the file should be replaced.

Assume that you have created a new database that has three redo log groups with one file in each group. To duplex the groups by adding one more files to each group, follow these steps:

1. You should be logged on to the Enterprise Manager Console and logged on to the SQL*Plus Worksheet as SYS from a previous section in this chapter.
2. Switch to the console, and double-click the **ORACLASS** database. A list of available tools appears below the database.
3. Double-click the **Storage** icon. A list of storage components appears.
4. Double-click the **Redo Log Groups** folder. Figure 4-12 shows what the screen looks like. Your database should have three redo log groups.
5. Click the first redo log group listed below the Redo Log Groups folder in the left pane. The property sheet appears for this redo log group. Figure 4-13 shows the property sheet. Yours may have a different filename. This illustrates how you can use the console to discover the number of redo log groups, and the

number of files contained in each group. You can also query dynamic performance views to find this information.

6. Switch to the SQL*Plus Worksheet session in preparation for adding a new member to each group.
7. In this example, add a new file named **redo01b.log** to Group 1, a file named **redo02b.log** to Group 2, and a file named **redo03b.log** to Group 3. You will place these new log files in a directory on the hard disk. Ask your instructor for the path of the directory you should use. Type and execute the following three commands to add a new member to each group:

```
ALTER DATABASE ADD LOGFILE MEMBER 'E:\oracle\product\10.2.0\
oradata\ORACLASS\redo01b.log'
TO GROUP 1;
ALTER DATABASE ADD LOGFILE MEMBER 'E:\oracle\product\10.2.0\
oradata\ORACLASS\redo02b.log'
TO GROUP 2;
ALTER DATABASE ADD LOGFILE MEMBER 'E:\oracle\product\10.2.0\
oradata\ORACLASS\redo03b.log'
TO GROUP 3;
```

TIP

No size is specified for the new member, because all members must be identical in size. Therefore, Oracle 10*g* uses the size of the existing members to size the new member.

The files above are created on drive E. The worksheet replies, "Database altered." By adding a second file to each group, You have duplexed the redo logs.

8. Remain logged on to the console and the SQL*Plus Worksheet for the next exercise.

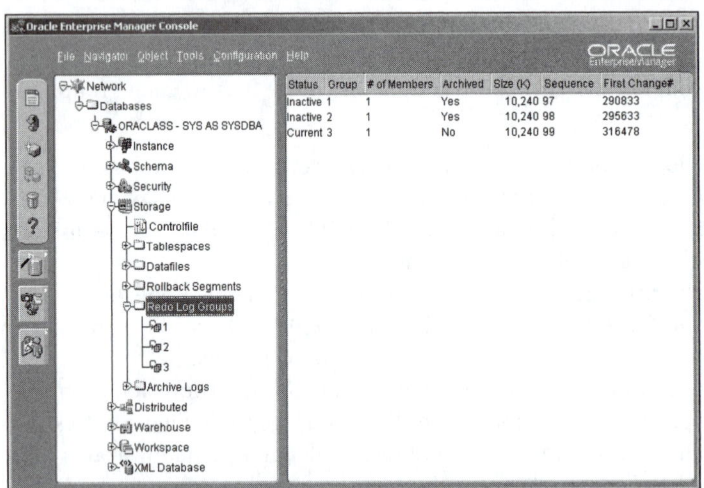

FIGURE 4-12 A database should have at least three redo log groups

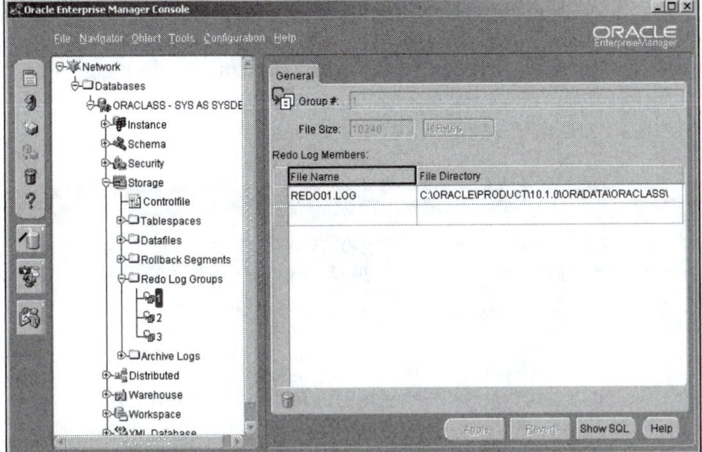

FIGURE 4-13 The redo log group has a single file

Adding a New Group

Continuing with the example database, you now have three groups with two files each. Imagine that the alert log contains warning messages stating that the LGWR process has to wait for the ARCn process. ARCn represents one or more archiver background processes. The first archiver process is named ARC0; the second is ARC1. There can be up to 10 archiver processes. Suppose a warning message is detected several times a day. To correct the problem, you must create a new redo log group so that the archive process has some lead-time to archive the inactive group. To accomplish this, follow these steps:

1. Type and execute the following command in SQL*Plus Worksheet to add a new group. It is not a requirement that all groups contain the same size files or the same number of files. You choose to create a group that matches the other three groups (two files of 20 MB each):

   ```
   ALTER DATABASE
      ADD LOGFILE GROUP 4('C:\oracle\product\10.2.
   0\oradata\ORACLASS\redo04.log',
      'E:\oracle\product\10.2.0\oradata\ORACLASS\redo04b.
   log') SIZE 20M;
   ```

   The files are created on drive C and drive E.
2. Remain logged on for the next exercise.

Renaming or Moving a Redo Log File

If your system acquires an additional disk drive, you may find that moving one member of each redo log group to the extra drive reduces the risk of losing redo logs. This is especially true if two or more members of each group reside on the same physical device. Renaming a redo log file may be done to keep up with new naming standards or to simply make the filenames more meaningful to you.

Renaming or moving a redo log file is similar to working with the control file: You must work with the operating system and with Oracle 10g to accomplish the task. For this exercise, imagine that you choose to rename all the members within the third redo log group that you created in the previous section. The new group has two members; therefore, you are renaming two files. Follow these steps to rename them:

TIP

Oracle recommends that you back up your database before working on the redo log files in the event you need to recover from an error. Oracle also recommends that, after you have successfully completed the change, you back up the control file.

1. Shut down the ORACLASS database by typing the following command in your SQL*Plus Worksheet session and clicking the **Execute** icon. Remember, the SHUTDOWN command can be used only when you are logged on as the SYS user in SYSDBA mode:

   ```
   SHUTDOWN IMMEDIATE
   ```

2. Locate the two members of the new redo log group. In Windows, open Windows Explorer and locate the redo log files you created. They are named **REDO04.LOG** and **REDO04b.LOG**. Remember, these files are created on drive C and drive E. In Unix or Linux, open a command prompt, and use the **cd** command to reach the correct directory.
3. Rename each file to match a new naming pattern. The pattern is:

   ```
   REDO_GRn_Mn.log
   ```

   So, for example, rename **REDO04.LOG** to **REDO_GR4_M0.LOG**, and **REDO04b.LOG** to **REDO_GR4_Mb.LOG**. Use the File menu in Windows Explorer. If you are using Unix or Linux, use the standard operating system copy command (**mv**) to rename the file to its new filename. For example, the following command renames one file:

   ```
   mv REDO04.LOG REDO_GR4_M0.LOG
   ```

4. Switch back to the SQL*Plus Worksheet. Then start up the database in MOUNT mode by typing and executing this:

   ```
   STARTUP MOUNT
   ```

5. Alert Oracle 10g of the renamed files by typing and executing this command

   ```
   ALTER DATABASE
     RENAME FILE 'C:\oracle\product\10.2.0\oradata\ORACLASS\REDO04.
   LOG',
                'C:\oracle\product\10.2.
   0\oradata\ORACLASS\REDO04b.LOG'
               TO 'C:\oracle\product\10.2.0\oradata\ORACLASS\REDO_
   GR4_M0.LOG',
                'E:\oracle\product\10.2.0\oradata\ORACLASS\REDO_
   GR4_Mb.LOG';
   ```

6. Open the database for normal use by typing and executing this:

```
ALTER DATABASE OPEN;
```

7. Remain logged on for the next exercise.

Dropping Redo Log Members or Groups

There are many reasons to drop redo log members or groups, including the following:

- A disk containing redo log members has failed, and you do not want to replace the files.
- Tuning recommendations suggest reducing the number of redo log groups. This is generally an unlikely scenario as parallel writes to duplexed redo logs have a negligible effect on performance, even on the most primitive of hardware architectures.
- Individual log members have become corrupted, and you want to replace them later.

When dropping redo log members, remember that there must be at least one member in every group. In addition, a database must have at least two groups, so you cannot drop a group if it is one of the last two groups.

An online redo log group (or member of a group) can be dropped only when the group is inactive. That means that LGWR process is not using the current group. If the database is in ARCHIVELOG mode, the group must also have been archived. The V$LOG data dictionary view shows group status. The group must have a status of INACTIVE and an archived flag of YES to be dropped.

Follow these steps to check the status of one member, and then drop the file:

1. Type and execute the following query in SQL*Plus Worksheet to determine the status of the third redo log group. Figure 4-14 shows the results. You may have more than two members in each group. As you can see, Group 4 has a status of UNUSED, meaning it has never been used. The archived status is YES. Even though this group is not archived, Oracle 10g initializes a new group by setting the archived status to YES:

```
SELECT GROUP#, STATUS, ARCHIVED, MEMBERS FROM V$LOG;
```

NOTE

Your database may show something different for archived logs. The reason why is because archiving runs in background. In other words, there is no user intervention. Archiving might be running slower on your database server, depending on what else you have running and how big your operating system footprint is. Archiving will catch up eventually.

2. Type and execute the following command to drop the file named **REDO_GR4_Mb.LOG** from Group 4.

```
ALTER DATABASE DROP LOGFILE MEMBER
  'E:\oracle\product\10.2.0\oradata\ORACLASS\REDO_GR4_Mb.LOG';
```

3. Using your operating system, delete the file that you dropped. The database does not drop the file unless the file is an Oracle Managed File (OMF). In Windows, use Windows Explorer to locate and select the file, and then click **File/Delete** from the menu. In Unix or Linux, open a command prompt, and use the **cd** command to navigate to the appropriate directory. Then type **rm –i REDO_GR4_Mb.LOG**, and press **Enter**.

4. Now, drop the entire fourth group by typing the following command in SQL*Plus Worksheet.

   ```
   ALTER DATABASE DROP LOGFILE GROUP 4;
   ```

5. Using your operating system, delete the remaining file that was in Group 4. In Windows, use Windows Explorer to locate and select the file, and then click **File/Delete** from the menu. In UNIX, open a command prompt, and use the **cd** command to navigate to the appropriate directory. Then type **rm –i REDO_GR4_Mb.LOG**, and press **Enter**.

TIP

If the redo log file that you want to drop is in the active group, force a log switch (issues a checkpoint as well) before attempting to drop the file. You can also force archiving if the database is in ARCHIVELOG mode.

6. Type and execute the following command in SQL*Plus Worksheet to force a log switch:

   ```
   ALTER SYSTEM SWITCH LOGFILE;
   ```

7. Repeat the query on V$LOG. There will be three remaining log groups; one will be marked as CURRENT, the other ACTIVE or INACTIVE. An ACTIVE group has files in the group that cannot be dropped. Some transactions that have dirty buffers (changed data not written to the datafile) have redo records in the group, even though a log switch has occurred. To force these outstanding records to be flushed from the buffer, thus changing the status of the group from ACTIVE to INACTIVE, type the following command and execute:

   ```
   ALTER SYSTEM CHECKPOINT;
   ```

8. Close your SQL*Plus worksheet session by clicking **X** in the upper-right corner.
9. Close your console session by clicking **X** in the upper-right corner.

Redo log files occasionally become corrupted and interfere with the proper functioning of a database. If this happens, one option is to drop the group and re-create it. However, in some cases (such as when there are only two groups remaining), a simpler solution is to clear the group. Clearing the group removes all the corrupted data. The syntax of the command is:

```
ALTER DATABASE CLEAR UNARCHIVED LOGFILE GROUP <n>;
```

UNARCHIVED is an optional clause used when the group is not yet archived, and you want to clear it without archiving it. Replace <n> with the actual group number you want to clear.

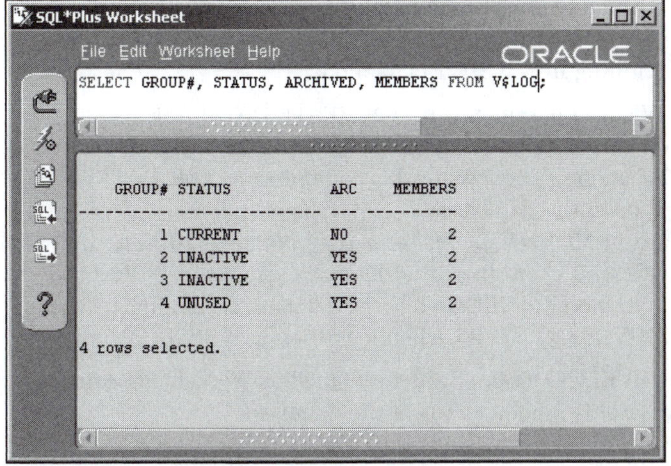

FIGURE 4-14 You can drop a member of a redo group, where the entire group has status of INACTIVE or UNUSED

NOTE

Archive groups may vary in status and may look different on your computer.

Archiving a Redo Log Group

As you know by now, archiving a redo log group ensures that you can perform a complete, up-to-the-minute recovery of your database in the event of catastrophic failure, such as the loss of a disk drive that held your SYSTEM tablespace. Other advantages of archiving include:

- Point-in-time recovery to restore data lost because of user error (such as deleting the contents of an entire table).
- The ability to query archived log files with LogMiner for auditing to find out who performed the DELETE command. With **LogMiner**, you can view and parse the archived redo log files. LogMiner is part of the Oracle database software as an optional feature.
- The ability to update a standby database (replace the failed database immediately with a ready copy).

A standby database is a database clone of the current database. It is kept current with the existing database by applying changes stored in the archived redo logs. If the current database is unrecoverable, the standby database can replace it, speeding up recovery time tremendously.

A database in ARCHIVELOG mode automatically archives redo log files. The opposite (no archiving of redo log files) is a database in NOARCHIVELOG mode. There are several ways to get your database running in ARCHIVELOG mode:

- Specify the ARCHIVELOG parameter in the CREATE DATABASE command.
- Set the initialization parameter LOG_ARCHIVE_START=TRUE and restart the database in MOUNT mode. Then issue the command ALTER DATABASE ARCHIVELOG, and open the database.
- Restart the database in MOUNT mode, issue the command ALTER DATABASE ARCHIVELOG, and then open the database. Finally, set the LOG_ARCHIVE_START parameter to TRUE while the database is running, by issuing the command ARCHIVE LOG START while logged on as a SYSDBA.

Let's experiment with ARCHIVELOG mode. Bring up SQL*Plus Worksheet again, logging in as SYS. Check to see the archive mode status of the database:

```
ARCHIVE LOG LIST;
```

The database log mode should be "No Archive Mode." To put the database into ARCHIVELOG mode do the following steps:

1. Begin by shutting down the database:

   ```
   SHUTDOWN IMMEDIATE;
   ```

2. Now restart in mount mode and change the database to ARCHIVELOG mode:

   ```
   STARTUP MOUNT;
   ALTER DATABASE ARCHIVELOG;
   ALTER DATABASE OPEN;
   ```

3. To force an archive on all CURRENT status redo logs execute the following command:

   ```
   ALTER SYSTEM ARCHIVE LOG CURRENT;
   ```

4. To generate archive logs issue an ALTER SYSTEM command as shown below. Find the archive logs based on the value of the USE_DB_RECOVERY_FILE_DEST parameter:

   ```
   ALTER SYSTEM SWITCH LOGFILE;
   ```

5. In an ARCHIVELOG mode database, you can archive all non-current logs by stating:

   ```
   ALTER SYSTEM ARCHIVE LOG ALL;
   ```

6. Now you switch off ARCHIVELOG mode again by executing the following commands:

   ```
   SHUTDOWN IMMEDIATE;
   STARTUP MOUNT;
   ALTER DATABASE NOARCHIVELOG;
   ALTER DATABASE OPEN;
   ```

There are several initialization parameters that control the archiving processes. Table 4-2 briefly describes each.

TABLE 4-2 Configuration initialization parameters for ARCHIVELOG mode

Parameter	Variants	Description	Example
LOG_ARCHIVE_DEST_n	LOCATION= '<directory>' [MANDATORY | OPTIONAL] [REOPEN = <secs>] OPTIONAL = OK to fail. REOPEN waits before trying again	Destination directory where archived log files are written. The directory must exist. n is the sequence number of the archive process (ARCn). Specify MANDATORY (must succeed)	LOCATION=C:\oracle\product\10.1.0\flash_recovery_area' MANDATORY REOPEN=50
LOG_ARCHIVE_DEST_STATE_n	The state of each destination in LOG_ARCHIVE_DEST_STATE_n. Values can be: enabled (ready for use); defer (do not use); and alternate (use if other destination fails)		Enabled
LOG_ARCHIVE_FORMAT	A standard naming pattern format given to the name of archive log files with variables.%s (log sequence number); %S (log sequence number, fixed length, zero padded); %t (thread number); %T (thread number, fixed length, zero padded)		ar%S.log (becomes, for example, a00125.log). Arch_%t_%s.log (becomes, for example, arch_1_125.log)
LOG_ARCHIVE_MAX_PROCESSES		Maximum archive processes to start on database startup. Values can be 1-10. The processes are named ARC0 to ARC9. The actual number in use is	5

TABLE 4-2 Configuration initialization parameters for ARCHIVELOG mode (continued)

Parameter	Variants	Description	Example
LOG_ARCHIVE_MIN_SUCCEED_DEST		Minimum number of archive destinations written to successfully, for archive to be successfully written	1
LOG_ARCHIVE_START		Start up archiving. TRUE sets database into automatic ARHIVELOG mode on startup, FALSE sets NOARCHIVELOG mode	TRUE

TIP

The only required initialization parameter to achieve successful ARCHIVELOG mode is the LOG_ARCHIVE_DEST_n parameter. (If you are running Personal Oracle, this parameter is called LOG_ARCHIVE_DEST.) The directory must exist.

To see the values currently set for these parameters, you could type **SHOW PARAMETERS LOG** in a SQL*Plus or SQL*Plus Worksheet, as shown in Figure 4-15.

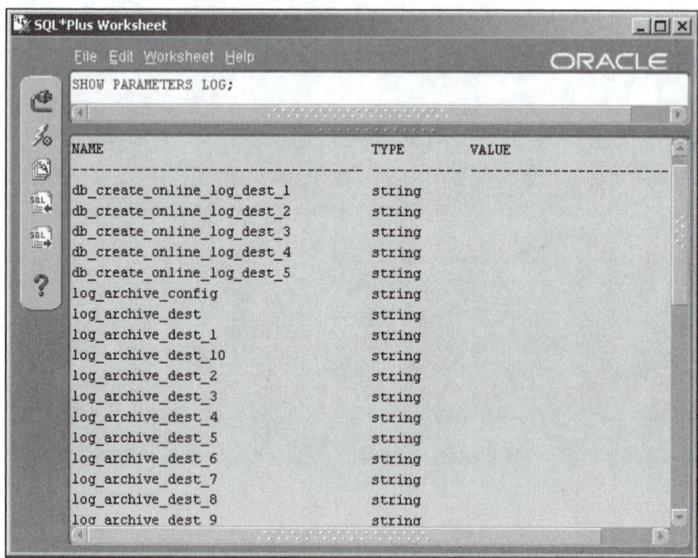

FIGURE 4-15 Display all initialization parameters containing the string "log"

As with control files and datafiles, it is possible to use the Oracle Managed Files method with redo log files. The following section discusses your options.

Using OMF to Manage Online Redo Log Files

When using OMF to manage online redo log files, all you need to do is ensure that the appropriate initialization parameters are set up, and then simply omit the filenames and paths in all commands associated with managing the redo log files.

The same initialization parameters you used with OMF for control files work for the log files. You can store all the files in a single directory (datafiles and all) by specifying DB_CREATE_FILE_DEST. Or, you can spread the control files and redo log files into their own directory or directories by specifying one or more DB_CREATE_ONLINE_LOG_DEST_n. If you have more than one destination, each new destination tells Oracle 10g to create a new duplexed set of redo log files and a multiplexed set of control files.

Here are some examples of initialization parameters, SQL commands, and OMF operations with redo log groups and members.

NOTE

Do not attempt to run the commands shown in this section. They are only examples and will cause errors if run in the ORACLASS database.

Begin by specifying the initialization parameters. For example:

```
DB_CREATE_ONLINE_LOG_DEST_1 = 'C:\oracle\product\10.2.
0\oradata\ORACLASS\logs'
DB_CREATE_ONLINE_LOG_DEST_2 = 'E:\oracle\product\10.2.
0\oradata\ORACLASS\logs'
```

With the initialization parameters complete, the CREATE DATABASE command can easily be used to set up Oracle-managed redo log files. You can omit the LOGFILE parameter completely. The following command creates a user-managed datafile for the SYSTEM tablespace (due to the DATAFILE clause) and creates Oracle-managed log files and control files:

```
CREATE DATABASE NEWDB
DATAFILE 'C:\oracle\product\10.2.0\oradata\NEWDB\system01.dbf' SIZE 250M;
```

Alternatively, you can include the LOGFILE parameter but specify only a file size, as shown in the following example. Do not include any details on groups or datafiles for the log members:

```
CREATE DATABASE NEWDB
DATAFILE 'C:\oracle\product\10.2.0\oradata\NEWDB\system01.dbf'
LOGFILE 10M;
```

The previous statement creates Oracle-managed redo logs, control files, and a user-managed SYSTEM datafile. The default size of OMF log files is 100 MB. Including "LOGFILE 10M" overrides the default size. Figure 4-16 illustrates the redo log groups and members created in this example.

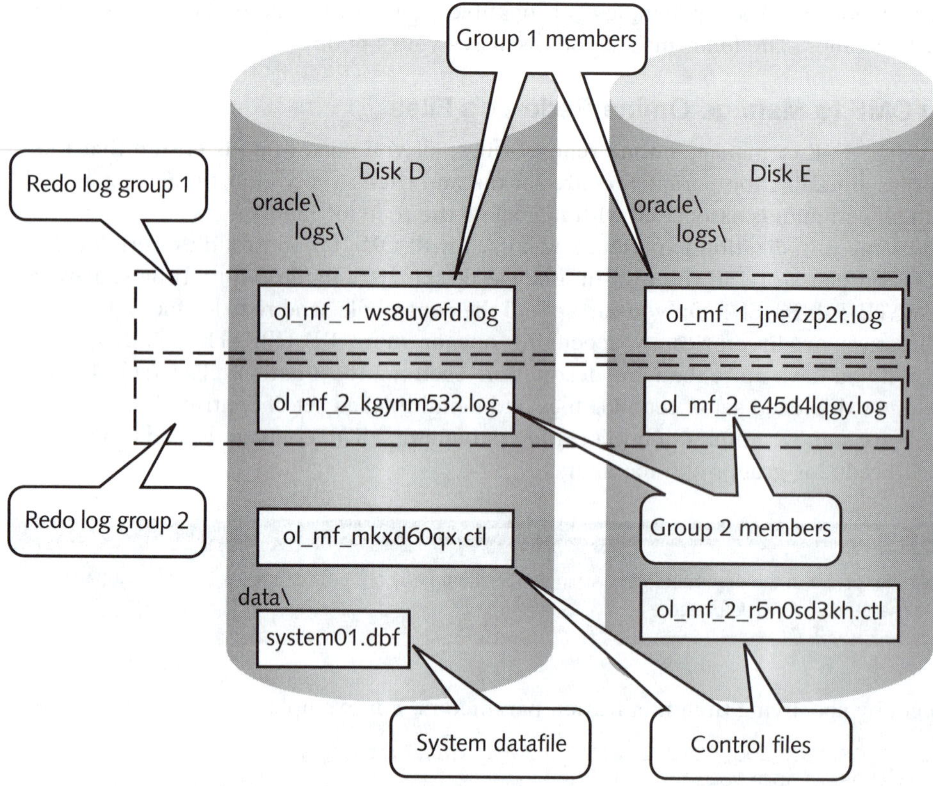

FIGURE 4-16 Oracle-managed redo log files automatically form two groups

TIP

OMFs always begin with a template. The exact template depends on your operating system, but it usually looks like this: "o1_mf_%g_%u_.log", where %g is the group number and %u is a unique eight-character string generated by Oracle 10*g* to ensure that the filename is unique.

Oracle-managed log groups may be add to OMF databases by using the following ALTER DATABASE command. (You must be logged on with SYSDBA privileges to execute the command.)

```
ALTER DATABASE ADD LOGFILE;
```

The database creates a new group and adds one member of the new group into each directory specified in DB_CREATE_ONLINE_LOG_DEST_n. In the example you have been using, in which two parameters were defined, the new group would have two members.

If you define only the DB_CREATE_FILE_DEST and no values for DB_CREATE_ONLINE_LOG_DEST_n in the initialization parameters, all the OMF files are created, including log files, in that one directory. If you omit both the DB_CREATE_FILE_DEST and the DB_CREATE_ONLINE_LOG_DEST_n parameters, Oracle 10g does not create any OMF files.

Viewing Redo Log Information

The dynamic performance views display a variety of data about the online redo log files. Figure 4-14 shows a query of V$LOG dynamic performance view.

The status of a redo log group can be:

- **UNUSED:** Never used, such as just after being added, or after being cleared
- **CURRENT:** Currently in use by LGWR (the CURRENT status implies the group is also ACTIVE)
- **ACTIVE:** Needed for crash recovery, but not currently in use by LGWR
- **CLEARING:** In the process of being re-created after the ALTER DATABASE CLEAR LOGFILE command has been used
- **CLEARING_CURRENT:** Currently being cleared of a closed thread (not a normal state, occurs if there is an error during log switching)
- **INACTIVE:** Not needed for crash recovery

Table 4-3 shows the dynamic performance views that contain details on the redo log groups or their member files.

TABLE 4-3 Dynamic performance view describing redo logs

View Name	Partial Columns List	Description
V$LOGFILE	GROUP#	Log group members, including file names
	STATUS	INVALID: bad file; STALE: incomplete contents; DELETED: no longer available for use; BLANK: file in use
	MEMBER	File name
V$THREAD	THREAD#	In a database with one instance THREAD# always 1
	CURRENT_GROUP#	
	CHECKPOINT_ CHANGE#	System Change Number (SCN), current redo log group number, and other up-to-date details
V$LOG	GROUP#	Group details
	STATUS	Status of each group
V$LOG_HISTORY	SEQUENCE#	Information about archived logs from the control file
	FIRST_CHANGE#	

TABLE 4-3 Dynamic performance view describing redo logs (continued)

View Name	Partial Columns List	Description
	RECID	Each archive log file has a control file RECID
V$ARCHIVED_LOG	RECID	One record for each archived log file
	NAME	Filename
	BLOCKS	
V$ARCHIVE_DEST	DESTINATION	Archive log files location
	BINDING	MANDATORY=successful archive to destination required; OPTIONAL=success not required
	STATUS	
V$ARCHIVE_PROCESSES	PROCESS	ARCn processes status
	STATE	IDLE or BUSY

As always, query the dynamic performance views like any other view.

The next section looks at examining Oracle metadata (tables, indexes, and so on), using data dictionary and dynamic performance views.

THE DATA DICTIONARY

Looking at Data Dictionary Components

Data dictionary views reside in the database like any other views. The views are based on tables owned by the SYS user that are updated automatically by the Oracle 10*g* database. The SYS schema owns both the underlying tables and the data dictionary views. Some of the views are available for anyone to query, whereas others are reserved for DBAs only. A user needs the SELECT_CATALOG_ROLE privilege to be able to query the data dictionary views. To take a quick look at a list of data dictionary views, follow these steps:

1. Start up the Enterprise Manager Console. In Windows, click **Start**, and then click **All Programs/Oracle ... /Enterprise Manager Console**. In Unix or Linux, type **oemapp console** on the command line. The Enterprise Manager Console logon screen appears.
2. Start up the SQL*Plus Worksheet by clicking **Tools/Database Applications/ SQL*Plus Worksheet** from the top menu in the console.

3. Connect as the SYSTEM user. Click **File/Change Database Connection** on the menu. A logon window appears. Type **SYSTEM** in the Username box, the current password in the Password box, and **ORACLASS** in the Service box. Leave the connection type as "Normal." Click **OK** to continue.
4. Type the following commands in the top pane and click the **Execute** icon to run the query. The Execute icon looks like a lightning bolt and is on the left side of the window:

```
SET LINESIZE 100
SET PAGESIZE 60
COLUMN COMMENTS FORMAT A40 WORD_WRAP
SELECT * FROM DICTIONARY
ORDER BY TABLE_NAME;
```

   Figure 4-17 shows part of the results of the query.
5. Remain logged on for the next exercise.

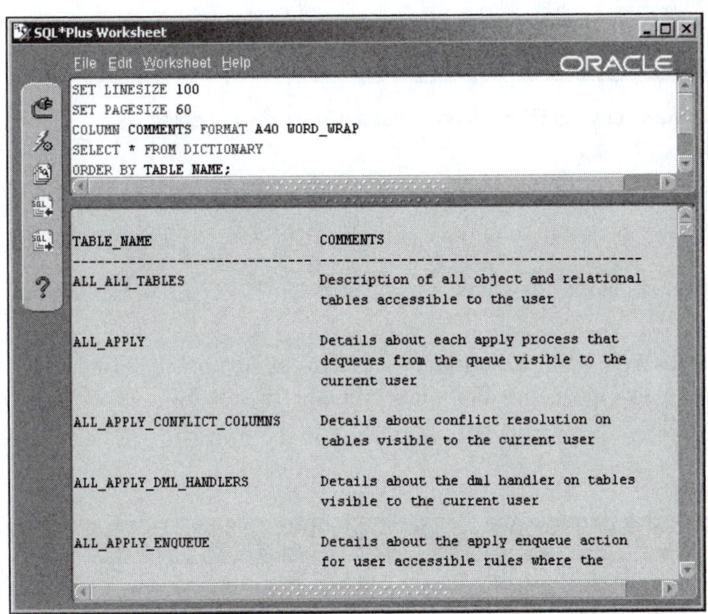

FIGURE 4-17 Find a list of data dictionary views by querying the DICTIONARY view

There are more than 1700 views to learn about. Who can possibly remember all those names? No one! That's why data dictionary view posters are so popular as a quick visual reference. Go to any Oracle convention and you undoubtedly see people handing out posters with lists of hundreds of data dictionary view names, categorized by various headings like "DBA," "Schema," "Performance," "Dynamic," and so on. The data dictionary views are so important to an Oracle DBA that you find these posters plastered in office after office.

In Chapter 3, you saw that data dictionary views provide system-related information by querying the database's internal management tables and presenting the data as views. Like any ordinary view, data dictionary views can be queried. Figure 4-18 shows the results of querying the ALL_TAB_COLUMNS view, which contains the names of all columns in all the tables that the current user owns or is able to view.

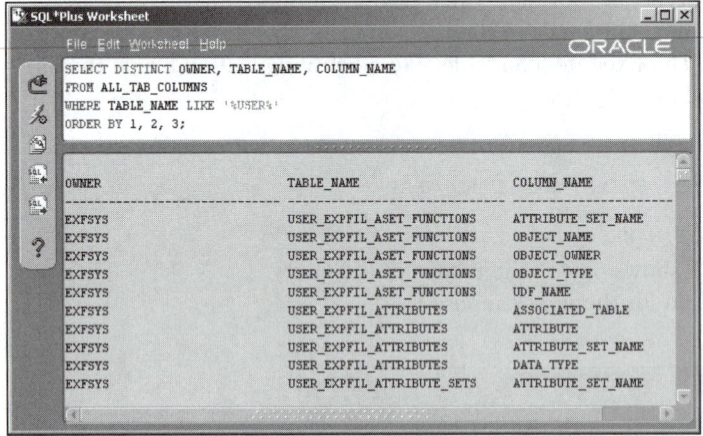

FIGURE 4-18 Data dictionary views contain information about a wide variety of objects

NOTE

Data dictionary views cannot be updated, with one or two exceptions. The internal database system updates the underlying tables.

Oracle's documentation has a list of data dictionary views in the Reference manual. Different types of data dictionary views are defined by their prefixes. Table 4-4 shows the prefixes, descriptions, and examples.

TABLE 4-4 Data dictionary views

Type	Prefix	Description	Examples
Static data dictionary views	USER	Views focused on owned objects	USER_TABLES, USER_VIEWS
	ALL	Views a user can access (could be owned by a different user)	ALL_TABLES
	DBA	Views available to DBAs only, showing information about all objects in a database (more detailed than USER_ and ALL_ versions of the views)	DBA_TABLES, DBA_USERS, DBA_SYNONYMS
Dynamic performance view	V$	Views displaying current database activity	V$LOCK, V$ACCESS
	GV$	Views combining activities across multiple instances for use in RAC environments	GV$LOCK, GV$ARCHIVE

Generally, views with prefixes of USER, ALL, and DBA are in sets, such as USER_TABLES, ALL_TABLES, and DBA_TABLES.

NOTE

USER prefixed views examine everything owned by a user. DBA prefixed views are solely for database administration use, allowing access to everything in the database, regardless of who owns it. ALL prefixed views access all objects accessible by a user, such as a table owned by another user but where the other user has granted access to the connected user.

Each view has nearly identical columns. The only difference is that the USER version omits the OWNER column, because by definition the OWNER is the current user. In addition, the USER version sometimes omits columns to simplify the view. Similarly, V$ and GV$ views are in sets so that every V$ view has its corresponding GV$ view with an additional column (INST_ID) for the instance number. There are also a few stray data dictionary views that do not begin with these prefixes. Most of them are carried over from prior versions of the database and should not be used.

For easier access, all the views except those starting with the DBA prefix have public synonyms and public permission to query. A public synonym is a unique name for an object that allows any user to use the object without prefixing it with the owner name. The views starting with DBA require the owner, SYS, as a prefix. For example, to query the DBA_TAB_COLUMNS table, log on as SYS (the DBA user) and type:

```
SELECT TABLE_NAME, COLUMN_NAME
FROM SYS.DBA_TAB_COLUMNS
WHERE ROWNUM < 200;
```

NOTE

The ROWNUM WHERE clause limits the number of rows returned by the query above to 199 rows in order to prevent a large meaningless query.

A second restriction to views beginning with DBA is that you must have DBA privileges in the database to query the views. The users SYS and SYSTEM are the default DBA users; however, you can create more users as needed by logging on as SYS or SYSTEM and assigning the DBA role to any user.

Using Data Dictionary Views

Data dictionary views give you a quick look into the database and all that it contains. In fact, the Enterprise Manager Console itself uses these views to display information on its screen. Follow these steps to discover how the console uses data dictionary views.

1. Navigate to the **Enterprise Manager Console** window you started up earlier in the chapter.
2. Double-click the **Databases** folder. A list of databases appears below the folder.
3. Double-click **ORACLASS**. The tools appear below the ORACLASS icon.
4. Double-click the **Schema** icon. A list of schemas appears.

5. Double-click the **SYSTEM** schema name. A list of schema object types appears.
6. Double-click the **Tables** folder. A list of tables appears. Figure 4-19 shows the console at this point.

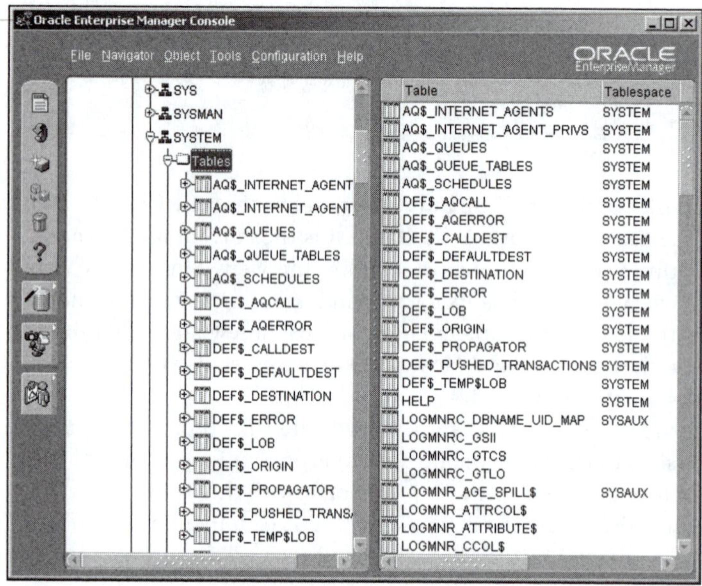

FIGURE 4-19 Navigating in the console is similar to navigating in Windows Explorer

7. Double-click the **HELP** table in the right pane. The property sheet for the table appears. Figure 4-20 shows the Properties window.

   Several data dictionary views were queried to build the Properties window, including DBA_TAB_COLUMNS, DBA_COL_COMMENTS, and DBA_REFS. It is sometimes useful to see the actual queries and other commands executed by the console, especially if you plan to write your own SQL commands to manually perform steps you have used with the console.
8. Close the property sheet by clicking the **Cancel** button in the bottom-center of the box.
9. Click **Navigator/Application SQL History** in the Console menu to view recent SQL commands executed in the console.
10. Scroll to the bottom of the window, and you see the queries that built the Properties window for the HELP table. Figure 4-21 shows the window that appears with SQL commands listed.
11. Close the Application SQL History window by clicking the **Close** button.
12. Remain logged on for the next practice.

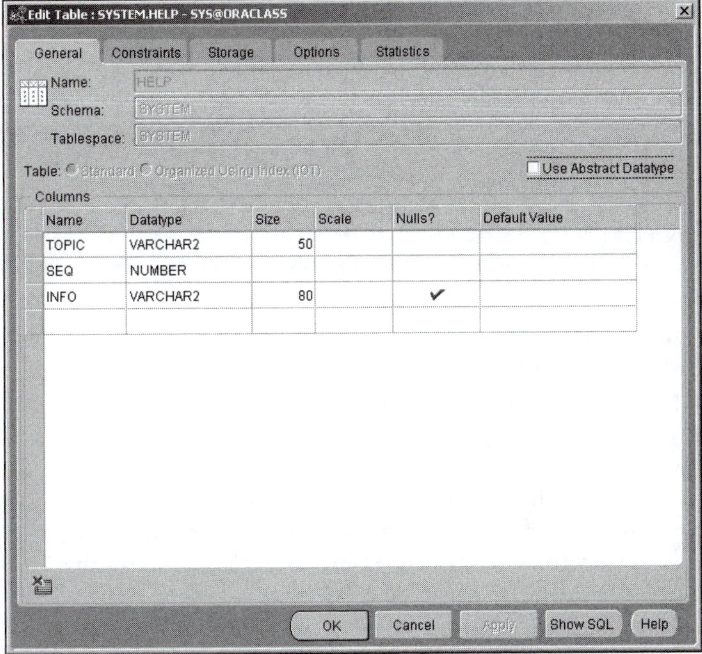

FIGURE 4-20 Property windows like this one are based on queries of the data dictionary views

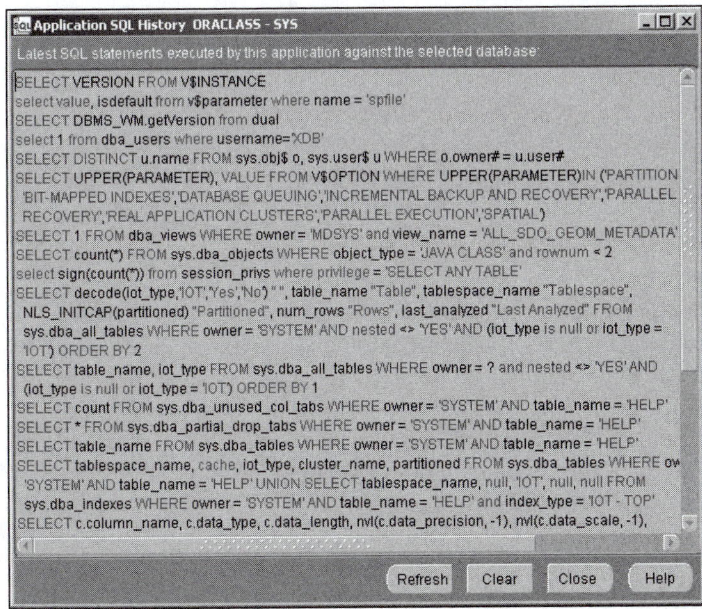

FIGURE 4-21 Scroll to the bottom of the window for the most recent commands

Use data dictionary views to help you build queries, debug SQL problems, and review naming conventions. In the following steps, you will see how to use data dictionary views. The data dictionary views can be queried like any other view in Oracle 10g.

NOTE

Use the SQL*Plus DESCRIBE command (abbreviated to DESC) to retrieve a list of column names and data types for any table or view, including the data dictionary views. For example, use the following steps to sample the DESCRIBE command.

1. Return to your SQL*Plus Worksheet window.
2. Run the DESCRIBE command for the DBA_COL_COMMENTS data dictionary view by typing this command and clicking the **Execute** icon:

```
SET LINESIZE 100
DESC DBA_COL_COMMENTS
```

In the preceding code, the first line adjusts the width of the display to 100 characters. The second line abbreviates the DESCRIBE command to DESC and then names the view to be described. Figure 4-22 shows the results.

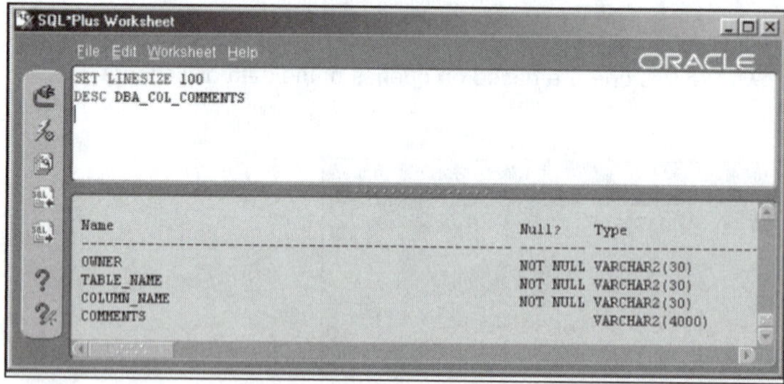

FIGURE 4-22 Refresh your memory of column names with DESC[RIBE]

3. The DESCRIBE command works on many other objects as well, such as tables, object tables, and object types. Imagine that you are writing a query that joins four tables and you have misspelled one of the table names. You are logged on to SQL*Plus Worksheet as the SYSTEM user. You know that the table's schema is QS. Type the following query to retrieve a list of all the tables in the schema. Execute the query by clicking the **Execute** icon.

```
SELECT TABLE_NAME FROM DBA_TABLES
WHERE OWNER = 'QS'
ORDER BY TABLE_NAME;
```

Figure 4-23 shows the results. Now you can return to the query you were writing and use the correct table names.

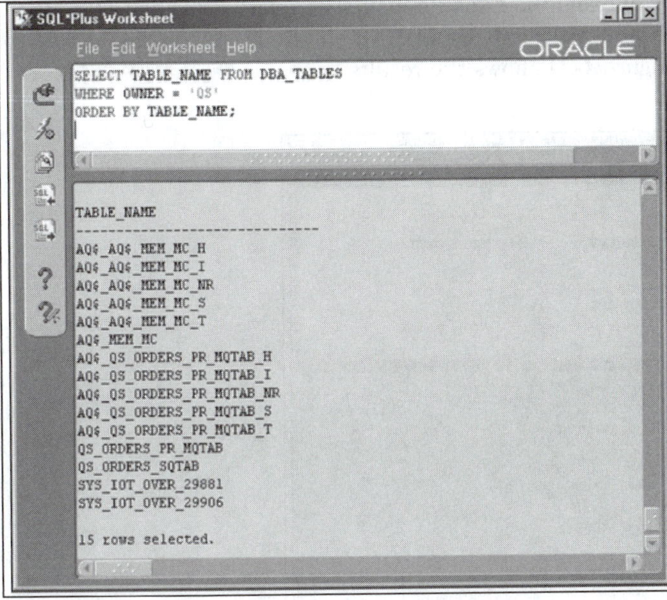

FIGURE 4-23 Use data dictionary views to obtain current table information

4. As a second example, suppose you just made changes to the production database that involved adding and removing columns in existing tables. You know that some views were affected. Type the following query to find a list of all views that are now invalid in the production database. Execute the query by clicking the **Execute** icon. If there are any invalid views, they are displayed.

```
SELECT OWNER, OBJECT_NAME, CREATED
FROM SYS.DBA_OBJECTS
WHERE OBJECT_TYPE = 'VIEW'
AND STATUS = 'INVALID'
ORDER BY OWNER, OBJECT_NAME;
```

5. In a final example, assume that your manager needs a report on the naming standards for database columns, and he wants you to include statistics on how well all the departments that develop database systems are complying with the naming standards. Your naming standards require that columns contain certain suffixes, such as _DATE or _NUM according to their data type. To develop some statistics, you write a query on the data dictionary views that lists all the suffixes found for each data type across all schemas in the database and how many times each suffix is used in a column name. The query is prepared and you are now ready to run it. Click **File/Open** on the SQL*Plus Worksheet menu. A Navigator window appears with the option to select a local file to load into the worksheet.
6. Navigate to the **Data\Chapter04** directory and select the **datatypes.sql** file. Click **Open** to load it into the worksheet.

7. Run the query by clicking the **Execute** icon. The query uses the COLUMN_ NAME column of the USER_TAB_COLUMNS and finds suffixes that start with an underscore. Figure 4-24 shows the results.

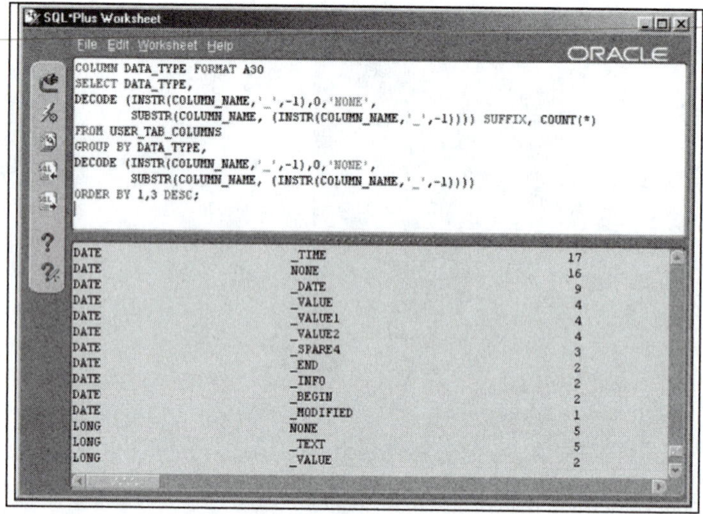

FIGURE 4-24 Columns of the DATE datatype have 11 different suffixes

There are data dictionary views for information about users, tables, objects, indexes, roles, and so on. To help you get started in finding what information is available through the data dictionary views, here is a list of some frequently used views:

- **USER_TABLES**: Tables you own
- **USER_VIEWS**: View name and the query that created the view
- **ALL_DEPENDENCIES**: Dependencies between objects, such as indexes and foreign key constraints
- **USER_ERRORS**: Errors found in views, procedures, and other objects you own
- **USER_INDEXES**: Indexes you own
- **USER_IND_COLUMNS**: Columns in indexes you own
- **DBA_SOURCE**: Source code for all stored objects, such as functions, Java source, and triggers
- **USER_TAB_PRIVS**: Table privileges that were granted to you; for example, you were granted SELECT on the CUSTOMER table by the CUSTACCT schema
- **ALL_TAB_PRIVS_MADE**: All grants given out by you for any table, plus grants by another user for tables you own
- **USER_TAB_PRIVS_MADE**: All grants you issued for tables you own
- **DBA_USERS**: All users in the database and information about them
- **PRODUCT_COMPONENT_VERSION**: Current version numbers and names of all installed components

All these data dictionary views are static because they show information that has been registered in the database and does not change without specific action by a user. For example, a user might execute an ALTER TABLE command to add a new column to a table. The related data dictionary views reflect that change. The next section discusses dynamic performance views, which change each time you query them, because they are based on in-the-moment activities.

Useful Dynamic Performance Views

The dynamic performance views all begin with the prefix V$ and have a counterpart view of the same definition beginning with the prefix GV$. As you remember from Table 4-4, V$ views are of current activities in the database, and GV$ views combine activities across multiple instances. V$ and GV$ views query activity-oriented system tables and display information about the state of your database. These views answer questions such as: What session is using the most CPU time now? How many users are currently logged on? How much memory is being used, and by which transactions? In addition, some of the views contain historical data collected periodically to enable you to assess trends in activities. Documentation of dynamic performance views can be found in Oracle documentation, in the Reference manual.

You will find that the primary use for the dynamic performance views is tuning the database system. Here are some of the key views available for gathering information to help tune the database system:

- **V$SYSSTAT**: Displays the volume of data that is changed and the number of transactions executed, among other important statistics
- **V$SQL**: Shows you the actual SQL commands executed, how many times the commands are reused or shared, and how much CPU each command consumed
- **V$SESSTAT**: Shows the resource consumption by session to help identify long-running or memory-consuming sessions
- **V$SESSION_WAIT**: Records information on sessions that were required to wait for resources (a common cause of slow response time)
- **V$FILESTAT**: Shows the number of reads, writes, and the time used to perform I/O activities on files
- **V$DATAFILE**: Lists the FILE# and FILE_NAME of all control, log, and datafiles

Oracle provides various options for gathering and viewing statistics. Statistics are used to tune a database. Although this book does not cover the details of tuning a database, the options are listed here for your information:

- **ANALYZE command and DBMS_STATS**: The ANALYZE command can be used to compute statistics as follows:

  ```
  ANALYZE <table> COMPUTE STATISTICS;
  ```

  The command above will compute statistics for the table specified plus all attached indexes. The DBMS_STATS package is more sophisticated than the ANALYZE command, allowing among other options, parallel processing on statistics generation.

- **The Database Control and the Oracle Enterprise Manager Console**: Much like STATSPACK, these options save data in a repository set of tables, so that trends can be analyzed over time. Graphs and statistics in easy-to-read panels can be displayed, providing detailed drill-down capabilities.
- **BSTAT/ESTAT scripts**: This method of statistics management is somewhat out of date for Oracle 10*g*. However, you can find these two scripts in the **ORACLE_HOME\rdbms\admin** directory in files **UTLBSTAT.SQL** and **UTLESTAT.SQL**. They produce a simple report of database activity between two points in time. Oracle states that these scripts produce the basic, minimum statistics needed for monitoring a database.
- **STATSPACK**: STATSPACK takes the basic statistics from the BSTAT/ESTAT scripts and adds to them the capability to save the statistics over time. The STATSPACK report provides much more information than BSTAT/ESTAT and is ideal for detecting performance bottlenecks.

To read more about tuning the database, read the Performance Tuning Guide in the Oracle 10*g* database documentation, select *List of books*, and then select the *Performance Tuning Guide*.

The data dictionary views and dynamic performance views provide a wealth of information about the contents and use of the database.

Examining Table Structure Using SQL*Plus and iSQL*Plus

It is important to know how to use SQL*Plus to access data dictionary views. SQL*Plus is the simplest and fastest tool to use for quick lookups in the database.

- **USER_TABLES:** Table structural definitions.
- **USER_TAB_COLS** and **USER_TAB_COLUMNS:** Table column definitions where USER_TAB_COLS includes hidden columns.
- **USER_TAB_COMMENTS** and **USER_COL_COMMENTS:** Comments on table columns.
- **USER_UNUSED_COL_TABS:** This view shows columns in tables marked as SET UNUSED and not physically dropped from tables.
- **USER_OBJECT_TABLES:** Object type table structures.
- **USER_TAB_PARTITIONS** and **USER_TAB_SUBPARTITIONS:** Table partition and subpartition structures.
- **USER_PART_TABLES:** Table partitioning details of tables at the partition rather than the table level as is the case for the USER_TAB_PARTITIONS and USER_TAB_SUBPARTITIONS views.

Chapter Summary

- An Oracle database consists of an Oracle instance and files.
- The Oracle instance consists of processes and memory buffers.
- The physical part of an Oracle database consists of datafiles, redo log files, archive redo log files, control files, and various configuration files.
- An Oracle database is divided into logical structures (tablespaces) and physical structures (datafiles).
- The control file contains pointers to datafiles, redo log entries, and archive log entries. The pointers should match between datafiles and log entries. If pointers in redo logs are ahead of those in datafiles, the database needs recovery.
- Control files track the current datafiles, online redo log files, checkpoints, and log group number.
- By creating more than one control file, you duplicate (multiplex) the control files.
- Add extra control files by copying an existing control file and updating the CONTROL_FILES parameter.
- Rename or relocate an existing control file by moving or renaming the file and updating the CONTROL_FILES parameter.
- Replace a damaged or lost control file by replacing it with a copy of an undamaged control file.
- You can designate control files to be Oracle Managed Files by leaving the CONTROL_FILES parameter null and using the DB_CREATE_FILE_DEST parameter instead.
- Add DB_CREATE_ONLINE_LOG_DEST_*n* parameter values to create multiplexed Oracle managed control files.
- You must create entirely new control files, when all control files are damaged or lost, or when certain database parameters must change.
- Use the CREATE CONTROLFILE command to create all new control files for an existing database.
- Specifying the SET DATABASE clause in the CREATE CONTROLFILE command renames the database.
- Some control file record sections contain information used for database recovery.
- Other control file record sections contain locations and names of files.
- Several V$ views query the control file, such as V$ARCHIVED_LOG.
- Redo log files are also called online redo log files.
- Redo log files primarily contain information on database changes.
- Redo log groups always contain at least one file each.
- At least two redo log groups must exist for a database to function.
- One group at a time is active, and changing to another log group is called a log switch.
- Groups are reused sequentially, and log file data is lost unless the log group is archived.
- Change information is recorded in the redo log before updating the datafile.

- Redo logs can be used to recover from minor failures such as power outages.
- Serious database damage requires archived redo logs, a valid control file, and a database backup for recovery.
- The redo log buffer is part of the SGA.

- Redo log files contain redo records or redo entries made up of change vectors.
- A single transaction may generate many redo entries.
- When a redo log group contains more than one file, it is said to be duplexed.
- The SCN is incremented every time the database changes.
- A checkpoint flushes dirty buffers to be written to disk.
- When redo log files are duplexed, one damaged file does not cause a system error.
- The command ALTER DATABASE ADD LOGFILE MEMBER ... TO GROUP n creates a new redo log file.
- ALTER DATABASE ADD LOGFILE GROUP creates a new redo log group.
- Renaming redo log files must occur with the database in NOMOUNT mode and is done with the ALTER DATABASE RENAME FILE command.
- ALTER DATABASE DROP LOGFILE MEMBER removes a redo log file from the database, but does not remove the physical file (except if it is an Oracle Managed File).
- ALTER DATABASE DROP LOGFILE GROUP removes an entire redo log group.
- ALTER DATABASE CLEAR LOGFILE GROUP removes corrupted data from a group.
- Place a database in ARCHIVELOG mode with the ARCHIVELOG parameter in the CREATE DATABASE command.
- After the database is created, change to ARCHIVELOG mode with the ALTER DATABASE ARCHIVELOG command or by setting LOG_ARCHIVE_START to <TRUE>.
- The only required parameter for ARCHIVELOG mode is LOG_ARCHIVE_DEST_n or the deprecated parameter (now only for Personal Oracle) LOG_ARCHIVE_DEST.
- The LOG_ARCHIVE_FORMAT initialization parameter provides a naming pattern for archived redo log files.
- View current parameter settings with the SHOW PARAMETERS command.
- Use DB_CREATE_FILE_DEST or DB_CREATE_ONLINE_LOG_DEST_n to implement Oracle Managed Files (OMF) with redo log files.
- Specifying more than one DB_CREATE_ONLINE_LOG_DEST_n parameter causes OMF to create multiplexed redo log files.
- The V$LOG dynamic performance view displays redo log group status, with CURRENT meaning the group is in use.
- The V$LOGFILE shows redo log member status, with blank status indicating the file is in use.
- Data dictionary views are owned by the SYS schema and are based on tables owned by SYS.
- Data dictionary views can be queried but never updated.
- Data dictionary views have prefixes of USER, ALL, DBA, V$, and GV$.

- Views prefixed by DBA are for users with DBA privileges in the database.
- Data dictionary views supply information about the structure of the database.
- Dynamic performance views begin with the V$ and GV$ prefixes.
- Dynamic performance views store current activity-oriented data.
- Dynamic performance views are used by STATSPACK and the Enterprise Manager Diagnostic Pack to track performance trends.

Review Questions

1. Describe how to multiplex control files.
2. When you add a datafile, you must manually update the control file as well. True or False?
3. Why would you multiplex control files?
4. List the initialization parameters used for Oracle Managed control files and describe how they are used.
5. The STARTUP command fails if there are no control files. True or False?
6. Describe the difference between ARCHIVELOG mode and NOARCHIVELOG mode.
7. When is the redo log buffer written to the redo log file? (Choose all that apply.)
   a. A transaction changes data.
   b. The redo log buffer becomes one-third full.
   c. A redo log group is archived.
   d. The database starts up.
8. You have determined that the ARCn process is causing the LGWR process to wait while it archives a log group. What should you do?
9. You have two redo log groups. Each group has three members that are 150 megabytes in size. The members reside on the G, E, and F drives. You issue this command:

   ```
   ALTER DATABASE ADD
   LOGFILE MEMBER 'G:\ora\logs\redo0102.log'
        SIZE 200M TO GROUP 1;
   ```

   The statement will fail because:
   a. All redo log groups must have the same number of members.
   b. You cannot add a member to a redo log group.
   c. The size is incorrect.
   d. The size cannot be specified in this command.
10. You have two duplexed redo log groups, and all the members reside on a single disk. Explain why this defeats the purpose of duplexing.
11. When a checkpoint occurs, the background process writes ________________ buffers to disk.
12. Define SCN.
13. If all the members of the current redo log group become damaged, what happens?
14. When is it appropriate to use LOG_ARCHIVE_DUPLEX_DEST?

15. To find the number of the current redo log group, query the ________________ view.
16. The ________________ view displays a list of data dictionary views.
17. The SYSTEM schema owns the data dictionary views. True or False?
18. Which of the following views can a user with DBA privileges query?
    a. USER_COLUMNS
    b. DBA_GRANTS_MADE
    c. V$LOGS
    d. ALL_USERS
    e. All of the above
19. The views prefixed with USER and prefixed with ALL are the same structure except that the ________________ column is missing in the USER views.
20. The views with the ________________ prefix span multiple instances.
21. Explain the difference between the DBA_TABLES view and the USER_TABLES view.
22. Give two examples of when you might query data dictionary views.

Exam Review Questions—Oracle Database 10*g*: Administration (#1Z0-042)

1. The following traits describe V$ views. Choose three.
    a. Cannot be updated
    b. Records database activity
    c. May contain data from multiple instances
    d. Available only to the DBA
    e. Available to any user
2. Control files contain these pieces of information about the database. Choose three.
    a. Timestamp of database creation
    b. Timestamp of last datafile update
    c. Names of datafiles
    d. Checkpoint name
    e. Checkpoint timestamp

3. A database has these initialization parameters:

   CONTROL_FILES=('D:\oracle\oradata\ORACLASS\CONTROL01.CTL')

   DB_CREATE_FILE_DEST='D:\oracle\oradata\datafiles'

   Which statement is true of this database? Choose two.

   a. The control file is multiplexed.
   b. The control file is an Oracle Managed File.
   c. The control file is not multiplexed.
   d. The control file is not an Oracle Managed File.
   e. The control file is in the same directory as the datafiles.

4. Which of the following features are handled by Oracle 10*g* *only* when you use Oracle Managed Files? Choose two.

   a. Location of the control file
   b. Size of the control file
   c. Multiplexing of the control file
   d. Name of the control file
   e. The control file is in the same directory as the datafiles

5. Which of these initialization parameters is used for Oracle Managed Files? Choose two.

   a. DB_CREATE_ONLINE_FILE_DEST
   b. DB_CREATE_ONLINE_LOG_DEST
   c. DB_CREATE_FILE_DEST
   d. DB_CREATE_ONLINE_LOG_DEST_n
   e. CONTROL_FILES

6. Which of these dynamic performance views would you use to view the names of the control files? Choose two.

   a. V$CONTROLFILE_RECORD_SECTION
   b. V$CONTROLFILE
   c. V$PARAMETER
   d. V$DATABASE
   e. V$CONTROLFILE_NAME

7. You want to list all of the indexed columns for objects that you own. Which data dictionary view would you query?

   a. ALL_COL_INDEXES
   b. USER_INDEXES
   c. USER_IND_COLUMNS
   d. USER_TAB_COLUMNS

8. The following lines show the commands and Oracle's responses. What should you enter next?

   SQL>CONNECT SYS/mypwd@PROD AS SYSDBA;

   Connected.

   SQL>STARTUP;

   Total System Global Area 135338868 bytes

   Fixed Size 453492 bytes

   Variable Size 109051904 bytes

   Database Buffers 25165824 bytes

   Redo Buffers 667648 bytes

   ORA-00205: error in identifying controlfile, check alert log

   for more info

   SQL>

   a. STARTUP NOMOUNT;
   b. ALTER DATABASE BACKUP CONTROLFILES TO TRACE;
   c. SHUTDOWN IMMEDIATE;
   d. SHUTDOWN ABORT;

9. What is the best method for moving a control file?
   a. Copy the control file to a new location; shut down the database; modify the CONTROL_FILES parameter; and start up the database.
   b. Issue the ALTER DATABASE RENAME FILE command.
   c. Shut down the database; move the control file; modify the CONTROL_FILES parameter; and start up the database.
   d. Shut down the database; delete the control file; start up the database in NOMOUNT mode; issue the CREATE CONTROLFILE command; and start up the database.

10. What is the greatest advantage of multiplexing the control file?
    a. Less maintenance involved
    b. Less possibility of failure
    c. Faster I/O time
    d. Better archiving

11. Redo log files store information about these types of events. Choose three.
    a. Checkpoints
    b. Archived redo logs
    c. UPDATE commands
    d. Database shutdowns
    e. Log switches

12. A database is in ARCHIVELOG mode, and there are two redo log groups. Group 1 is current and filled. Group 2 was just archived. What happens next? Place these events in order of occurrence.

    1. Archive Group 2
    2. Log switch to Group 2
    3. Log switch to Group 1
    4. Archive Group 1
    5. Fill Group 2

    a. 1, 2, 5, 4, 3
    b. 2, 4, 1, 5, 3
    c. 2, 4, 5, 3, 1
    d. 1, 2, 4, 5, 3

13. A database is in ARCHIVELOG mode, and there are two redo log groups. Group 1 is current and Group 2 has been archived. Then Group 1 becomes full. In what order do the following five events occur?

    1. Archive Group 2
    2. Log switch to Group 2
    3. Log switch to Group 1
    4. Archive Group 1
    5. Fill Group 2

    a. 1, 2, 5, 4, 3
    b. 2, 4, 1, 5, 3
    c. 2, 4, 5, 3, 1
    d. 1, 2, 4, 5, 3

14. Your database is in NOARCHIVELOG mode. You have updated some data in the database. Suddenly, the power fails. You did not have a chance to commit your updates. Which of these statements is true regarding the redo log file?

    a. Your changes were recorded in the redo log buffer but not in the redo log file.
    b. Your changes will be restored only if they were recorded in the redo log file.
    c. The redo log file will be reset, and your changes will be lost.
    d. Your changes will not be restored even if they were recorded in the redo log file.

15. Your database has two redo log groups on one disk. You want to duplex the groups. Which of the following actions is the best strategy?

    a. Create a new redo log group on a separate disk.
    b. Move one group's members to a separate disk.
    c. Move one group to a separate disk.
    d. Create a new member in each group on a separate disk.

16. You issue the following command, which completes successfully:

```
ALTER DATABASE
 ADD LOGFILE GROUP 3 SIZE 500K;
```

What do you know about the database? Choose two.

a. The database is in ARCHIVELOG mode.

b. The database already has two redo log groups.

c. The redo log files are managed by OMF.

d. The database is started and in NOMOUNT mode.

e. The database is started and in MOUNT mode.

17. The following initialization parameters are set:

```
DB_CREATE_FILE_DEST='D:\oracle\data'
DB_CREATE_ONLINE_LOG_DEST_1 = 'D:\oracle\logs'
DB_CREATE_ONLINE_LOG_DEST_2 = 'E:\oracle\logs'
```

You execute the following command:

```
CREATE DATABASE TESTXYZ;
```

Which of the following statements is true?

a. The statement fails because it is missing the LOGFILES clause.

b. The statement succeeds, and the redo log files are duplexed.

c. The statement fails because it is missing the DATAFILE clause.

d. The statement succeeds, and two redo log files are created.

18. It is 11:00 AM on June 24, 2004. Your database has a single instance and is about to archive the redo log file assigned log sequence 21. The LOG_ARCHIVE_FORMAT is **my%S%t.log**. What is the new archive log file named?

a. my210001.log

b. my000240001.log

c. my211100.log

d. my000211.log

19. Which of the following statements correctly describes renaming a log file from '**D:\oralog\redo01.log**' to '**E:\oralog\redo01.log**'?

a. SHUTDOWN IMMEDIATE; move file to new destination; STARTUP MOUNT; RENAME FILE...; STARTUP OPEN;

b. SHUTDOWN IMMEDIATE; move file to new destination; STARTUP NOMOUNT; RENAME FILE...; STARTUP OPEN;

c. SHUTDOWN IMMEDIATE; move file to new destination; STARTUP MOUNT; RENAME FILE...; ALTER DATABASE OPEN;

d. Move file to new destination; SHUTDOWN IMMEDIATE; STARTUP MOUNT; RENAME FILE...; ALTER DATABASE OPEN;

Hands-on Assignments

Before starting the hands-on assignments, prepare the database by running the **setup.sql** script in the **Data\Chapter04** directory on your student disk. You must log on as SYSTEM to run the script. Make sure you open the **setup.sql** script, replacing the path name for the USER_AUTO tablespace's datafile, with the appropriate path name (ask your instructor).

1. You want to make a report of table attributes. This report consists of a series of queries on data dictionary views in which you specify the table name, and the queries return details about that table. Your goal is to have information on the report that is similar (in content, not format) to the information you see when you look at the Schema Manager's table property sheet in the console. Include as much information as you can to match the information you see for the CLASSMATE.WANT_AD table when looking at the General tab of the property sheet (table name, column names, data types, and so on). Review the data dictionary views and select the views you think would produce the best results. Then write a series of queries that display the columns you want displayed on the report.

   Use COLUMN commands to adjust the headings and column width so the report is more readable. Make sure the columns are listed in the same order as they appear on the property sheet. Write the queries to report the details of the WANT_AD table of the CLASSMATE schema. You should log on as CLASSMATE to run the queries.

   Save the script in a file named **ho0401.sql** in the **Solutions\Chapter04** directory on your student disk.

2. Write a query, based on data dictionary views, reporting a table's name, number of rows, average row length, column names (in the order they appear in the table), the average column length, and the high and low values found in each column (the values are in hexadecimal format). When you are logged on as CLASSMATE, run the query. The query should only report on the CLASSMATE schema. Save the query in a file named **ho0402.sql** in the **Solutions\Chapter04** directory on your student disk.

3. Using V$SQL, write a set of queries that display the top ten SQL commands run in the system. Write one query for each of these criteria:
   - Most memory used (add all memory components)
   - Most CPU time
   - Most elapsed time
   - Most rows processed

   Save all the queries in a file named **ho0403.sql** in the **Solutions\Chapter04** directory on your student disk.

4. Your office has new equipment that allows you to multiplex your control files (currently, you only have one control file named **control01.ctl** on a single disk). The database name is TEST01. The SYSTEM user's password is "MYPASS". The current control file is located in the **C:\oracle\product\10.2.0\oradata\TEST01\control1** directory. The new disk drives are labeled D and E. The directory structure for the new drives is the same as the C drive. Write the steps and the commands needed to multiplex the control file and copy it from its current location to the new drives. In other words, the result must be three control files, all on

the three separate drives, leaving the existing control file intact and in place. Save your work in a file named **ho0404.txt** on the **Solutions\Chapter04** directory on your student disk.

5. To prepare for a disaster, you decide that you should have a text form of a control file stored somewhere. Use a form of the ALTER DATABASE command to achieve this objective. (*Hint*: Use the alert log). Save the command to regenerate the control file in a file named **ho0405.txt** in the **Solutions\Chapter04** directory on your student disk. State how you created the command, and where you located the text of the command.
6. List the data dictionary views that you think would be most helpful to a programmer who is writing applications that select, update, and delete rows from tables in the database. Explain briefly why you chose each view. Save your answer in a text file named **ho0406.txt** in the **Solutions\Chapter04** directory on your student disk.
7. Your office database was created by the previous DBA who left for parts unknown. You are reviewing the database setup. You want to know the answers to these questions:
    a. What are the redo log group numbers, and how many members does each group have?
    b. What directory or directories holds the redo log group members?
    c. Are there any archived redo log files? If so, where are they located? How many files are there?

    Write SQL commands or queries to discover all these answers. Save your work script in a file named **ho0407.txt** in the **Solutions\Chapter04** directory on your student disk.
8. You have a database with the following redo log structure that needs a little work:
    - All the members of all redo log groups reside on the same drive, in this directory: **C:\oracle\product\10.2.0\oradata\TEST01**.
    - Groups 1 and 2 have two members and group 3 has three members.
    - Each member should be named redo<nn>[<x>].log. nn is the group number and x represents the drive letter (d or e in this case), representing the member number. However, for group 1, the <x> is excluded (no need to alter this).

    Assuming that the files are not OMF, write all the SQL commands required to spread the members across three drives (drives C, D, and E), in the appropriate manner befitting redo log distribution. And don't forget that all groups should have the same numbers of members. Use the same existing directory structure for all three drives. Save your work in a file named **ho0408.txt** in the **Solutions\Chapter04** directory on your student disk.

Case Projects

1. The Global Globe database has multiplexed control files. You are gathering statistics about the database's current state. Write one or more queries to answer these questions about the ORACLASS database.
   - Which file has had the most physical read activity?
   - List the minimum time for a single I/O process, and the maximum time for a single write process for all datafiles.
   - What is the most recent checkpoint number?
   - When was the database created?

   Save the script in a file named **case0401.sql** in the **Solutions\Chapter04** directory on your student disk.
2. At the Global Globe Newspaper Company, your Enterprise Edition database currently is in NOARCHIVELOG mode. After reading this chapter, you convince your MIS Manager that the database must be made more fail-safe by changing it to ARCHIVELOG mode. Switch your database to ARCHIVELOG mode. Save the script in a file named **case0402.sql** in the **Solutions\Chapter04** directory on your student disk.

CHAPTER 5

BASIC STORAGE CONCEPTS AND SETTINGS

LEARNING OBJECTIVES

In this chapter, you will:

- Differentiate between logical and physical structures
- Create many types of tablespaces
- Configure and view storage for tablespaces and datafiles
- Use and manage undo data
- Learn to describe and configure diagnostic (trace) files

INTRODUCTION

Previous chapters have focused on creating the database and the processes that keep it running smoothly. Beginning with this chapter, the focus shifts toward the internal structure and workings of the database. Topics include the components that you set up to hold user data, as well as a brief look at trace files, their management, and their uses.

This chapter begins with the foundation structures that the database uses to store data. You learn about the logical structures and how they map to the physical structures. You see how to configure these structures and what they are used for. You learn the internal structures needed to speed up performance.

You must also monitor and manage the diagnostic files. In this chapter, you find out what these files contain and how to use them. The chapter ends with a case project in which you set up structures for use in later chapters, as you add actual data, tables, and users to the database.

NOTE

Please remember, as with all the chapters in this book, that Unix, Linux, and Windows path names are represented in order to provide more flexible content. It is assumed that basic knowledge of Windows, Unix and Linux is a given. Paths on your student computers may be different to the paths in this chapter, depending on which operating system you are using.

INTRODUCTION TO STORAGE STRUCTURES

The Oracle 10g database has an internal set of structures that are used to store all the data, users, constraints, data dictionary views, and any other objects you want to create in the database. These structures also contain metadata maintained internally by the database. Metadata is data that tells Oracle 10g about all the structures that store data in the database. For example, the table containing the list of each table's columns, their data types, lengths, and so on, is a table of metadata. The data dictionary views primarily display metadata.

The database server has several components. The database software is installed on the server. The **instance** is made up of the memory (SGA) and the background processes, and the **database** is made up of database files, control files, and redo log files. Figure 5-1 shows the components of the database server.

Previous chapters examined all components of the database. This chapter takes a closer look at previously introduced components, and examines new components. To begin with, the database files are not just ordinary files. Like other files, database files have a physical structure, forming the building blocks of the database system.

Logical Structure Versus Physical Structure

Physical structures are composed of operating system components and have a physical name and location, with direct respect to the underlying operating system. In other words, you can look at the physical parts of an Oracle database within the operating system, such as using Windows Explorer or an **ls** command in Unix or Linux. Physical structures can be seen and manipulated in the computer's operating system. Logical structures are composed of orderly groupings of information that allow you to manipulate and access related data. Logical structures cannot be viewed or modified outside the database (from the operating system). They are part of the integral makeup of the Oracle database, and not the operating system. Logical structures are generally associated with one or more physical structures.

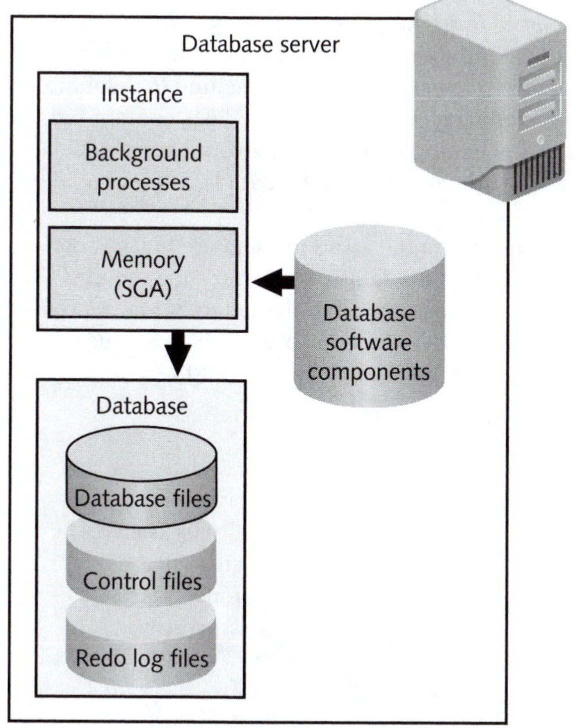

FIGURE 5-1 The database server is a familiar sight by now

Physical structures include:

- **Datafiles:** A datafile is a physical file on the computer. It is sometimes called a database file. It is the primary physical structure that contains all the data you store in the database. When you create a new database, certain datafiles are required. The most important datafiles are those storing the SYSTEM tablespace. A tablespace is a logical structure. Figure 5-2 shows several datafiles in the database. Like other files, datafiles have a size and location you can see in the operating system. A datafile can be made up of several fragmented sections in different locations on the operating system. This allows Oracle 10g to change the size of the datafile. In Figure 5-2, the datafile named **system01.dbf** has two sections. The size of a datafile is defined in bytes, but the smallest unit with which the operating system works is called an operating system block. An operating system block is made up of a group of contiguous bytes (bytes right next to one another) within the file. The operating system block size and the logical database block size (see the bullet on data blocks that follows) are closely related and should be considered when setting the size of a datafile. The operating system block defines the units used by the operating system to handle I/O. For example, when you save a document on a disk, the data is placed onto the physical disk one block at a time, rather than one byte at a time. Handling data in blocks makes I/O more efficient. The operating system block size varies from machine to machine.

- **Redo log files:** As you saw in a previous chapter, redo log files contain redo entries that record changes to the database to allow recovery of changed data. Redo log files are handy when the database loses the data because of a failure of some kind.
- **Control files:** A previous chapter covers the structure and contents of control files. The control files are essential for database startup and recovery. The control files contain the names and locations of all the datafiles and redo log files. Changes to the physical structure of the database, such as adding or dropping a datafile, are automatically updated in the control file.

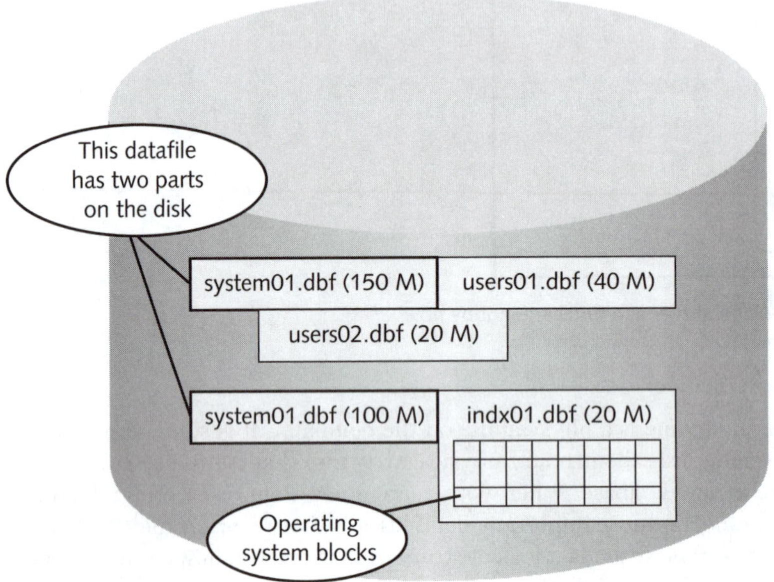

FIGURE 5-2 A datafile is made up of operating system blocks

Logical structures are shown in Figure 5-3 with example names of each component. Logical structures include the following components, from largest to smallest:

- **Tablespace:** A tablespace is the largest logical object. Creating a database requires a number of tablespaces, including the SYSTEM tablespace. A database may contain a large number of tablespaces. Each tablespace contains one or more datafiles. Tables, indexes, and other objects are created within a tablespace. The storage capacity of a tablespace is the sum of the size of all the datafiles assigned to that tablespace. A datafile can be associated with only one tablespace within one database. The contents of the datafiles are stored in encrypted and specialized record structures readable only by the Oracle 10g

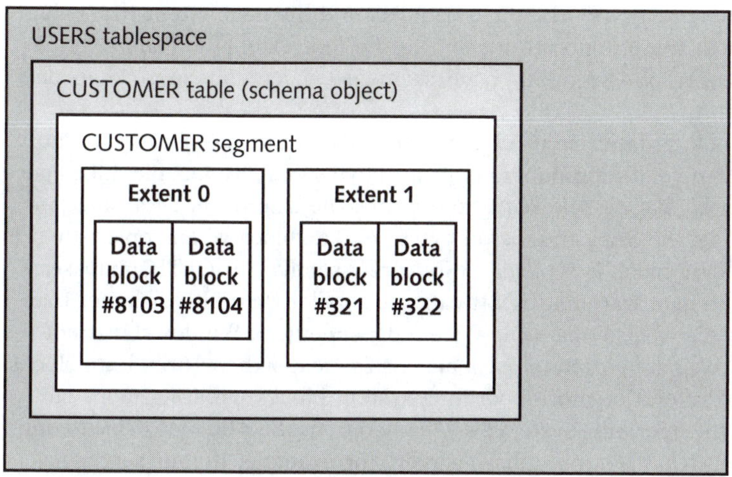

FIGURE 5-3 A tablespace is the largest logical structure and a data block is the smallest

database. You cannot open that file and view the data as you would a regular text file. However, you can view all the data stored in the datafile by using SQL to query the logical components, such as the tables, that are stored in the datafile.

- **Segment:** A segment is the set of extents (see Extents below) that make up one schema object (see Schema object below) within a tablespace. Segments have several different uses, including being the storage holders for a schema object within a tablespace. Segments are discussed in more detail later in this chapter. Each segment belongs to one schema object. Each schema object has only one segment, except for partitioned tables and partitioned indexes, which have one segment per partition.
- **Schema object:** A schema object includes the wide variety of objects that can be created by users in the Oracle 10g database. Tables and indexes are probably the most common types of schema objects. Other types of objects include views, procedures, Java-stored procedures, object tables, and sequences. Each schema object must be contained within one tablespace, with the exception of partitioned tables and partitioned indexes. In the case of partitioned tables and indexes, each partition must be contained within one tablespace, but the partitions for one partitioned table or index may reside in different tablespaces.
- **Extent:** An extent is a contiguous group of data blocks that are assigned to a segment. When more space is needed for an object, such as a table, the space is allocated in the form of one extent. Extents can be different sizes, depending on the storage parameters set for the object or the tablespace (when not specified by the object). For example, you might create a CUSTOMER table and set the storage parameters, so that the first extent (the initial extent that was

allocated upon creation of the table) is 20Mb and the next extent (the extent allocated when there is no more room in the first extent) is 5Mb. Extents always belong to one segment. Each segment can contain more than one extent.

- **Database block:** A database block is the smallest of the logical units. A data block is made up of a set number of physical bytes in a physical file. Like operating system blocks, Oracle reads and writes the size of its data block in chunks. You set the exact number of bytes per data block when you create the database. A data block is typically 8Kb (rarely smaller for OLTP databases; often larger in data warehouses). It can't be smaller than 2Kb or larger than 32Kb. Block size is also operating system dependent. In Windows, it is not recommended or supported to use a block size greater than 16Kb. Data blocks are made up of one or more operating system blocks, creating more efficient use of the operating system for I/O operations. Set the block size to optimize the features of your database server. For example, if your server has plenty of memory and fast disk drives, you could safely set the data block size at four times the operating system size to reduce overall I/O operations and improve response time. On the other hand, if your system has less memory or slower disk drives, set Oracle's data block size exactly equal to the operating system block size to best use the available resources. The initialization parameter DB_BLOCK_SIZE contains the default or standard block size. After a database is created, you cannot change the block size without re-creating the database. However, you can create tablespaces that have a nonstandard block size. Each data block is either assigned to one extent or is available free space in the datafile. Free space is defined as the group of data blocks in the datafile that are available for future allocation to an extent. Each extent contains many data blocks.

Look again at Figure 5-3. The examples listed in the figure illustrate typical examples of names of each logical component. The tablespace example is the USERS tablespace. The schema object example is the CUSTOMER table. Within the CUSTOMER table is the CUSTOMER segment. Segments have the same name as their parent schema object (or the name of their parent partition in the case of partitioned tables and indexes). The CUSTOMER segment has two extents identified as Extent 0 and Extent 1. Extents do not have specific names and are always numbered sequentially starting at zero for each segment. Looking at Extent 0, there are two data blocks: Block 8103 and Block 8104. Each data block is numbered sequentially starting with zero, but the numbering is relative to its position in the datafile, not the extent. This is where the logical and the physical components intersect.

Figure 5-4 shows the connection between the logical components and the physical components of the database. As shown in Figure 5-4, a datafile is always associated with one tablespace. In addition, an extent is always associated with a contiguous group of data blocks within one datafile. There is no direct correlation between a segment and a datafile. A segment can be spread among several datafiles as long as the datafiles belong to the tablespace that contains the segment. Likewise, there is no direct correlation between a schema object and a datafile. Data blocks are made up of one or more operating system blocks, depending on the size of each type of block.

The next section shows you how to create and configure tablespaces and datafiles.

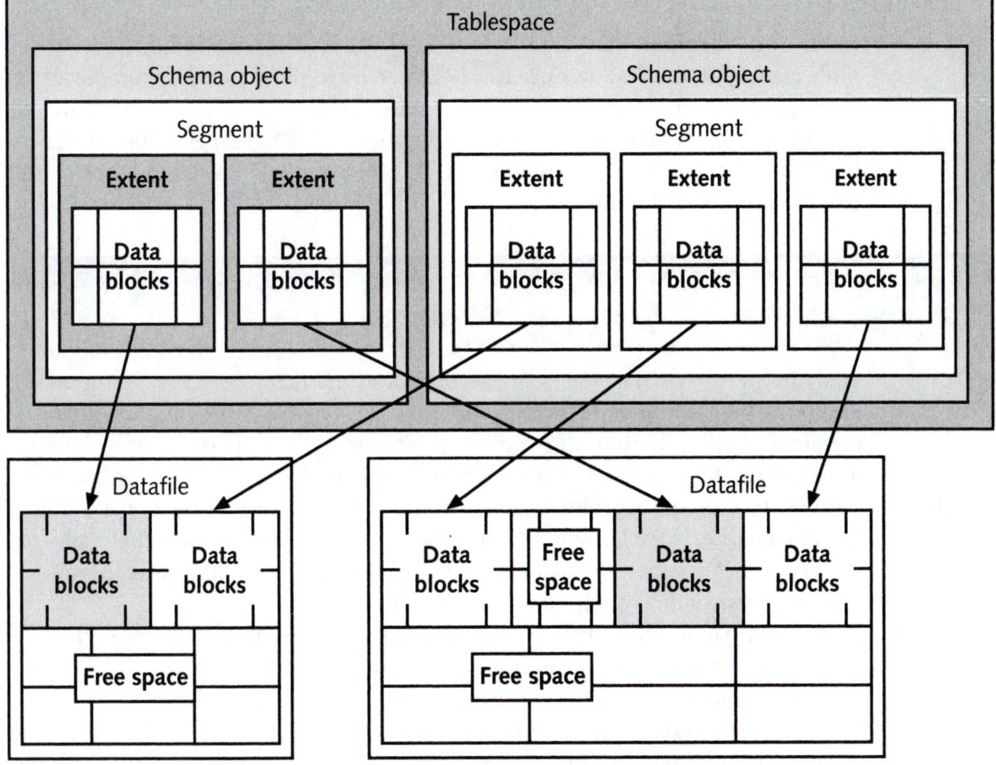

FIGURE 5-4 Ties between the physical and the logical components occur at several levels

Tablespaces and Datafiles

Much of this chapter teaches you about creating, managing, removing, and renaming tablespaces. You also examine viewing metadata about tablespaces and datafiles. Tablespaces are always made up of at least one datafile. Those tablespaces can be of various different types, such as temporary, permanent, undo, and so on. In fact, you cannot create a tablespace without also creating its initial datafile. Similarly, you cannot create a datafile without an associated tablespace. You can, however, create a new datafile to add to an existing tablespace.

Here is the syntax of the CREATE TABLESPACE command that includes clauses that are expanded later in this section.

```
CREATE TABLESPACE [BIGFILE|SMALLFILE] <tablespacename>
DATAFILE <filename> SIZE <nn> AUTOEXTEND ON|OFF
TEMPORARY|PERMANENT
EXTENT MANAGEMENT LOCAL|DICTIONARY
LOGGING|NOLOGGING
ONLINE|OFFLINE
SEGMENT SPACE MANAGEMENT MANUAL|AUTO
```

When you create a tablespace, you specify your choices for:

- **Bigfile or smallfile:** The default is SMALLFILE, the traditional type of tablespace. BIGFILE allows creation of very large permanent or temporary tablespaces, with a single very large datafile. Bigfile tablespaces make management much easier for extremely large databases. A single bigfile datafile can be 32Tb (8Kb block size) and 128Tb (32Kb block size). A smallfile tablespace can contain 1022 datafiles.

NOTE

Bigfile tablespaces must be locally managed.

- **Temporary or permanent data:** Temporary tablespaces store objects during a session only. In addition, they are used for sorting data if you have assigned a temporary tablespace to a user. There are actually two formats for creating temporary tablespaces. See the section titled "Temporary Tablespaces" later in this chapter for the syntax and special considerations for this type of tablespace. Permanent tablespaces are the default type and store permanent objects, such as tables and indexes.
- **Undo data:** Undo tablespaces are specialized to store undo (rollback) data. Undo data is used to store instructions to undo changes when a ROLLBACK (forced or automated) is executed.
- **Local or dictionary-managed tablespaces:** A locally managed tablespace (the default) has a bitmap in the header of the tablespace that stores all the details about free space, used space, and location of extents. A dictionary-managed tablespace stores the details about its free space and other information inside the data dictionary tables, in the SYSTEM tablespace. Locally managed tablespaces are more efficient because less time is spent looking up the locations of data.

NOTE

Dictionary-managed tablespaces are semi-redundant and are allowed in Oracle 10*g* for backward compatibility, and consistency with non-Oracle applications.

- **Storage settings:** The storage settings include details on how Oracle 10g allocates space within the tablespace, its initial size, and more.
- **Datafile information:** You can define one or more datafiles to be created for this tablespace. Specify the initial size of the datafile and whether it is allowed to automatically grow (AUTOEXTEND).
- **Online or offline:** A tablespace that is offline is not available for use, whereas one that is **online** is available (online is the default).

- **Logging or no logging:** This sets the default behavior for logging certain types of changes on objects in the tablespace. LOGGING is the default and specifies that all DDL commands and all bulk INSERT commands, such as those issued by SQL*Loader, are recorded in the redo log buffer.

NOTE

SQL*Loader is a specialized Oracle database tool used for mass loading large amounts of data into a database all in a single operation.

- NOLOGGING means that these two types of commands contain minimal logging. NOLOGGING speeds up the processing time, but in recovery, the changes must be done again manually by rerunning the original commands. A particular object can override the tablespace defaults. Other types of changes to objects, such as updates, inserts other than bulk insertions, and deletions, are always recorded in the redo log buffer, regardless of whether you choose LOGGING or NOLOGGING. The complete syntax for the CREATE TABLESPACE command would fill a whole page. To view all the details, look at Oracle 10*g* online documentation. Then look up CREATE TABLESPACE in the alphabetical list of commands.

The next sections go through examples of creating tablespaces using these parameters and clauses.

The DATAFILE Clause

When creating a tablespace with a user-managed file, specify a datafile name in the command. The syntax of the DATAFILE clause is:

```
DATAFILE '<datafilename>' SIZE <nn>|REUSE
AUTOEXTEND ON MAXSIZE <nn>|UNLIMITED|AUTOEXTEND OFF
```

- You can have more than one datafile by listing multiple files in the DATAFILE clause, separating successive datafiles with a comma delimiter.
- Each datafile must have a specified SIZE. The only exceptions are Oracle managed files (OMF) and named files that already exist. OMF makes files 100Mb by default. If the file already exists, you must also specify REUSE, and then either leave out the SIZE to keep the same size file, or include the SIZE to adjust the existing file's size. All data in an existing file is deleted.
- By adding AUTOEXTEND ON, a datafile can be automatically extended up to MAXSIZE. Omitting AUTOEXTEND and SIZE parameters results in AUTOEXTEND set to ON. You can specify that the datafile has no maximum size by specifying MAXSIZE UNLIMITED.

Follow these steps to create a user-managed tablespace:

1. Type and execute the following command in your current SQL*Plus Worksheet session to create a tablespace named ACCOUNTING that has a user-managed datafile. To make a user-managed datafile, simply include a name in

the DATAFILE clause. Remember to replace ORACLE_BASE with the actual directory provided by your instructor.

```
CREATE TABLESPACE ACCOUNTING
DATAFILE 'ORACLE_BASE\oradata\oraclass\cust_tbs01.
    dbf' SIZE 10M;
```

SQL*Plus Worksheet replies, "Tablespace created."

2. Try these options by typing and executing the following command. Here is an example of a tablespace with two datafiles. The first datafile is set at 20Mb and can be automatically extended in 2Mb increments to 200Mb. The second datafile is initially 2Mb and can expand to an unlimited size. Both files are user-managed files:

```
CREATE TABLESPACE NEWS
DATAFILE '
C:\oracle\product\10.2.0\oradata\oraclass\news01.dbf'
     SIZE 20M AUTOEXTEND ON NEXT 2M MAXSIZE 200M,
     '
C:\oracle\product\10.2.0\oradata\oraclass\news02.dbf'
     SIZE 2M AUTOEXTEND ON MAXSIZE UNLIMITED;
```

3. Remain logged on for the next practice.

TIP

Changes to data are written to datafiles by the DBWn background process. To be more efficient, changes are accumulated in memory and then written later to the physical datafile. If the datafile needs more space, it is allocated when the memory buffers are written to the datafile. Without the AUTOEXTEND ON clause, transactions that cause the datafile to run out of space (such as an INSERT command) cause a critical error.

The next section shows how to create tablespaces with user-managed datafiles.

Implementing Oracle Managed Files (OMF) with Tablespaces

When you use OMF, you allow the database to determine the actual names of datafiles, leave out the DATAFILE clause entirely, and minimize the details you need to supply for a tablespace. Complete the following steps to look at the existing tablespaces in the ORACLASS database and to create a locally managed tablespace that is named USER_TBS:

1. Start up the Enterprise Manager console. In Windows, click **Start** on the Taskbar, and then click **All Programs/Oracle ... /Enterprise Manager Console**. In Unix or Linux type **oemapp console** into a shell. The Enterprise Manager console login screen appears.
2. Double-click the **Databases** folder, and then double-click the **ORACLASS** database icon. A list of tools displays below the database name.
3. Double-click the **Storage** icon to open the Storage Manager.
4. Click the **Tablespaces** folder. The right pane displays a list, as shown in Figure 5-5, showing the type of free space management (LOCAL or DICTIONARY), and the current amount of used and free space in each tablespace. Your screen will show different tablespaces.

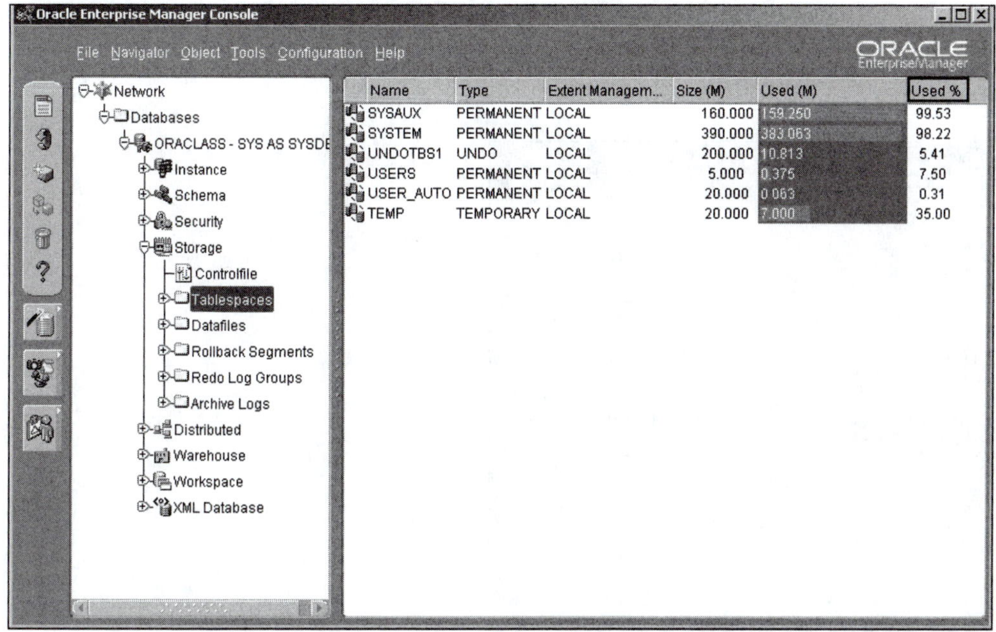

FIGURE 5-5 The used space for each tablespace appears as a bar graph

5. Double-click the **USERS** tablespace. The property sheet for the tablespace displays as shown in Figure 5-6. Here you can see that this is a permanent tablespace that is online and has one datafile.

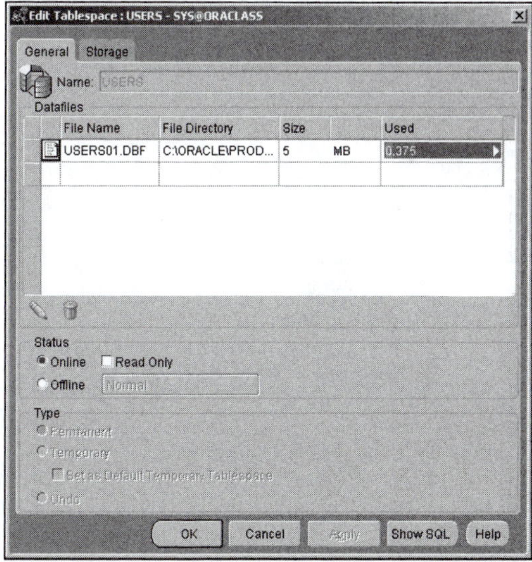

FIGURE 5-6 The USERS tablespace has one datafile

6. Click the **Storage** tab to display the storage settings for the tablespace as you see in Figure 5-7. The brief explanations of each setting are a useful feature of the console.

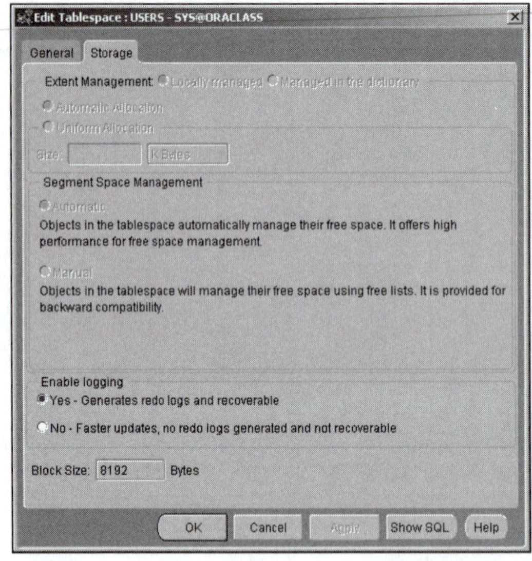

FIGURE 5-7 The USERS tablespace storage settings appear here

7. Click **Cancel** to return to the main console.
8. Start up the SQL*Plus Worksheet by clicking **Tools/Database Applications/SQL*Plus Worksheet** from the top menu in the console.
9. Type and execute the following command to check on the OMF parameters. The DB_CREATE_FILE_DEST must be defined before you can create an Oracle Managed tablespace. Figure 5-8 shows the results.

```
SHOW PARAMETERS DB_CREATE
```

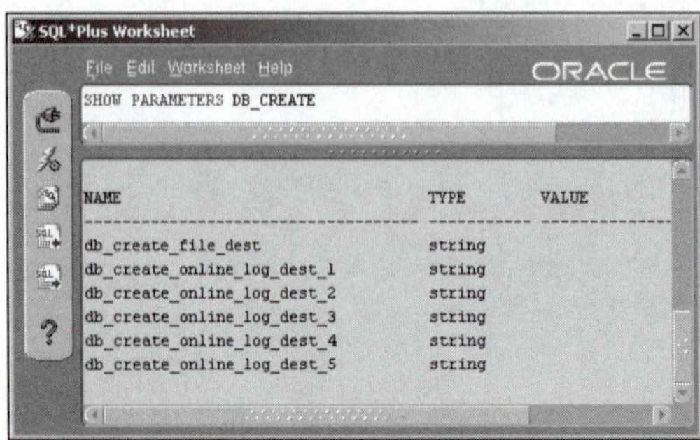

FIGURE 5-8 The DB_CREATE_FILE_DEST parameter is NULL

10. Type and execute the following command to set the DB_CREATE_DEST parameter for your current session. Replace ORACLE_BASE with the actual directory where the new tablespace will be created. Your instructor will provide the exact directory name.

    ```
    ALTER SESSION SET
    DB_CREATE_FILE_DEST='ORACLE_BASE\oradata\oraclass';
    ```

    SQL*Plus Worksheet replies, "Session altered."

11. Type and execute the following command to create an Oracle Managed tablespace. This command contains no additional parameters, so the default settings for OMF are used. The datafile is 100Mb, named by the system, and can automatically extend to an unlimited size.

    ```
    CREATE TABLESPACE USER_OMF1;
    ```

    SQL*Plus Worksheet replies, "Tablespace created."

12. Type and execute this command to create another tablespace using OMF. This time, your datafile is 20Mb in size:

    ```
    CREATE TABLESPACE CUST_TBS
    DATAFILE SIZE 20M;
    ```

    SQL*Plus Worksheet replies, "Tablespace created."

13. Remain logged on to the console and SQL*Plus Worksheet for the next practice.

Next you will examine extent management and segment space management clauses.

The EXTENT MANAGEMENT and SEGMENT SPACE MANAGEMENT Clauses

The EXTENT MANAGEMENT and SEGMENT SPACE MANAGEMENT clauses tell Oracle 10g how to track the usage of blocks within each extent. Recall that the syntax for these two clauses in the CREATE TABLESPACE command is:

```
CREATE TABLESPACE <tablespacename>
...
EXTENT MANAGEMENT LOCAL|DICTIONARY AUTOALLOCATE UNIFORM SIZE
SEGMENT SPACE MANAGEMENT MANUAL|AUTO
...;
```

Blocks are either available for new inserts, available only for updates, or filled. An extent may have thousands of blocks, and every change to a block is tracked whenever data stored in the block changes. The method of tracking the blocks varies dramatically between the two types of extent management. LOCAL is the default for Oracle 10g, although DICTIONARY was the default for earlier releases.

TIP

When the database was created, if the SYSTEM tablespace was created as a locally managed tablespace, no dictionary-managed tablespaces are allowed in the database.

Segment space management applies only to locally managed tablespaces. Specifying AUTO is recommended.

TIP

DICTIONARY was the default for earlier releases; however, Oracle plans to eliminate the ability to create new tablespaces with DICTIONARY management in the near future. Oracle recommends creating all new tablespaces with LOCAL management.

The biggest issue with using dictionary-managed tablespaces is one of performance. A locally managed tablespace stores the map of where extents are in datafiles in a rapid access bitmap, in the header of the datafile. A dictionary-managed tablespace stores the same information in metadata, in the SYSTEM tablespace. This results in extent map searches and expansions competing with all other database metadata operations. In addition, there is a certain amount of space saving using locally managed tablespaces. However, disk space is cheap compared to processing power, and thus space saving is not generally considered relevant. In very rare cases, such as a very busy database server, the downside to using locally managed tablespaces is that there can be localized competition for access to bitmap header extent maps. One solution is to create multiple datafiles. Figure 5-9 illustrates a datafile that belongs to a dictionary-managed tablespace. As new objects are created and then grow, extents are allocated to each object's segment. These extents can be of various sizes. The diagram shows two extents, one 4Mb and one 28Mb. Extents are deallocated (returned to free status) when the object is dropped or truncated. The diagram shows two deallocated extents, one 5Mb and one 2Mb. Deallocated extents return to the free space list in the data dictionary for the tablespace (the list of chunks of free space available in the datafile) as a contiguous chunk of data blocks. For example, the two deallocated extents shown in Figure 5-9 would be listed as two chunks of space in the freelist. For either of those chunks to be usable, the next object that needs an extent must be using that exact size (or smaller) extent. Otherwise, the deallocated data blocks are passed over, and more data blocks at the end of the datafile are used instead. Skipping over unused space in favor of free space at the end of the datafile causes the datafile to grow faster than is necessary.

If there are several deallocated extents next to one another, they are not seen as a single chunk of contiguous free space until they are coalesced. Coalescence describes the combining of multiple adjacent free extents into a single contiguous free extent. For example, in Figure 5-9, even though the two deallocated extents are adjacent to one another, they cannot be listed as a single chunk of 7Mb of free space until they are coalesced. Coalescing occurs periodically through the SMON background process. The SMON process checks regularly for free extents that need coalescing. However, if you have set PCTFREE to 0 (the recommended setting), you must manually coalesce free extents by issuing the ALTER TABLESPACE command. The ALTER TABLESPACE command is discussed in detail later in this chapter.

In previous database releases, the only method of controlling the problem of unused free extents was diligent control over the storage settings for the tablespace and all objects within it. In this release, the DEFAULT STORAGE clause (see examples in the following bulleted list) sets the storage for any object created without specifying its own storage clause. If an object does specify its own storage clause, the settings must be monitored and carefully synchronized to assure that extent sizes stay in multiples of a common size. As the DBA, your job is to set the DEFAULT STORAGE clause for each tablespace and instruct all the developers who

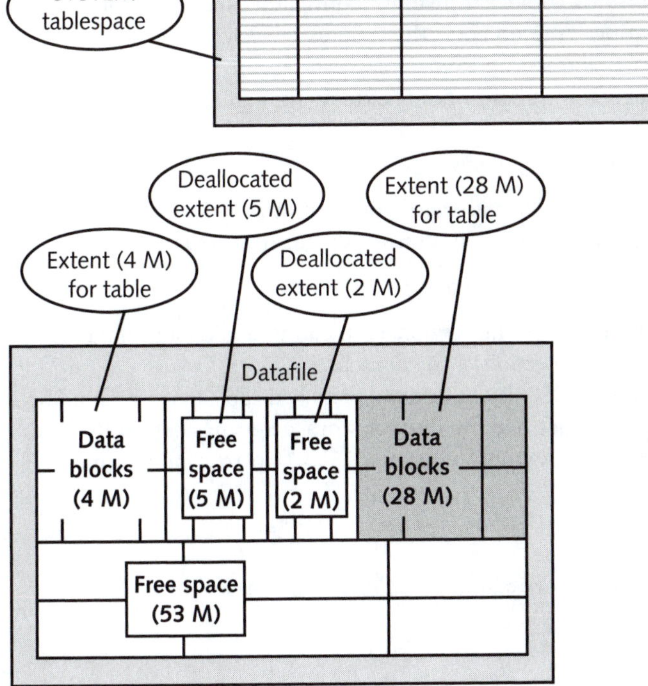

FIGURE 5-9 An example of a datafile in a dictionary-managed tablespace

create tables to omit the STORAGE clause in their CREATE TABLE command so that your default settings are used for all tables.

The syntax for dictionary-managed extents is:

```
EXTENT MANAGEMENT DICTIONARY
MINIMUM EXTENT <nn>
DEFAULT STORAGE (INITIAL <nn> NEXT <nn> PCTINCREASE <nn>
                 MINEXTENTS <nn> MAXEXTENTS <nn>)
```

The MINIMUM EXTENT clause helps keep objects created in the tablespace within similar parameters for extent size. If an object specifies a smaller extent size than the tablespace's MINIMUM EXTENT size, Oracle 10g rounds the number up to the tablespace's MINIMUM EXTENT size.

Within the DEFAULT STORAGE clause, you can set the following parameters:

- **INITIAL:** The initial extent size that is allocated when the object is created.
- **NEXT:** The size of the next extent allocated when the initial extent runs out of room.
- **PCTINCREASE:** The percentage to increase any subsequent extents. For example, if INITIAL is set to 10M (10Mb), NEXT is set to 5M (5Mb), and PCTINCREASE is set to 10, then the first, second, and third extents will be 10M (10Mb), 5M (5Mb), and 5.5M (5.5Mb). Oracle recommends setting PCTINCREASE to zero at all times to keep extent sizes uniform. In reality, this parameter has been set to zero or ignored by experienced DBAs, since Oracle 7.34.

- **MINEXTENTS:** The minimum number of extents allocated when the object is created. This allows the object to use noncontiguous space for the initial extents. When creating large objects, this makes it easier to reuse the allocated storage as space is freed up by deletes.
- **MAXEXTENTS:** The maximum number of extents that an object can grow into.

Creating a Dictionary-Managed Tablespace

Because your SYSTEM tablespace is locally managed, you cannot practice actually creating a dictionary-managed tablespace. If you tried, you would get this error:

```
ORA-12913: Cannot create dictionary managed tablespace
```

You should, however, be familiar with the syntax of creating a dictionary-managed tablespace in case you work on an older version of the database.

The following command creates a 250Mb dictionary-managed tablespace with a MINIMUM EXTENT size of 15Mb. Objects that use the default storage settings get an INITIAL EXTENT of 90Mb, with all subsequent extents equal to 15Mb, up to a maximum of 50 extents.

```
CREATE TABLESPACE USER_TEST
DATAFILE '
C:\oracle\product\10.2.0\oradata\oraclass\user_test01.dbf'
SIZE 250M
AUTOEXTEND ON
EXTENT MANAGEMENT DICTIONARY
MINIMUM EXTENT 15M
DEFAULT STORAGE (INITIAL 90M NEXT 15M PCTINCREASE 0
                 MINEXTENTS 1 MAXEXTENTS 50);
```

Dictionary-managed tablespaces have another problem that causes slower performance of DML commands. All the details of an extent and the data blocks within the extent are stored in data dictionary tables. A freelist is a list of individual blocks within an extent that have room available for inserting new rows. The freelist is not the same as the free space in a datafile. Free space is available for creating new extents. Blocks on the freelist, on the other hand, are blocks inside an existing extent with some extra room. Freelists are stored in the dictionary tables, and therefore the tablespaces are called "dictionary-managed." Updating the dictionary tables causes Oracle 10*g* to create redo log entries. When, for example, you update rows in a table, redo log entries for those updates are recorded in the redo log buffers. In addition, more redo log entries for the updates to the data dictionary tables are also recorded. Undo records are also created for both your updates and for the updates to the data dictionary tables. The amount of work to record a change has doubled because your update actually updates multiple tables behind the scenes.

Locally managed tablespaces reduce and even eliminate the problem of unused gaps of free space by ensuring that all the extents allocated to any object within the tablespace are of the same or multiples of that size as shown in Figure 5-10. In Figure 5-10, there are two extents, one 5Mb and one 25Mb. Both are multiples of 5Mb. There are also two deallocated extents, both 5Mb. Deallocated extents are automatically coalesced with any neighboring free space. In Figure 5-10, rather than two 5Mb deallocated extents, the two would immediately be coalesced into one 10Mb deallocated extent. This reduces the possibility that free space is passed over and left empty over time. No DEFAULT STORAGE clause is used for locally managed tablespaces or the objects within them, so the work of keeping extent sizes uniform is left to the system. You can go further than this to homogenize the extent size. You can tell Oracle 10g that all extents must be exactly the same size (the UNIFORM SIZE clause). If you used the UNIFORM SIZE clause when creating the datafile in Figure 5-10, the 25Mb extent would have been created as five 5Mb extents instead of one 25Mb extent.

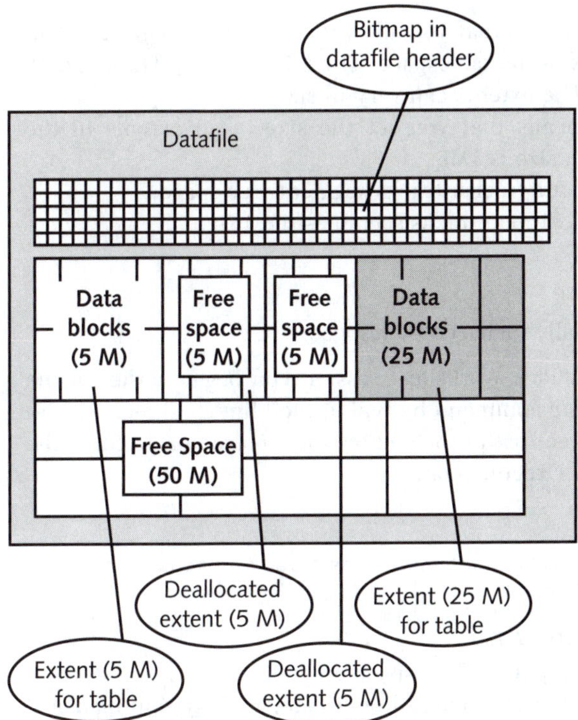

FIGURE 5-10 Locally managed datafiles use a bitmap to track free space

Locally managed tablespaces are more efficient than dictionary-managed tablespaces because the information about storage is stored in a bitmap that is stored with the tablespace, as shown in Figure 5-10. There is a bitmap for each datafile in the tablespace. A bitmap is a small record in the datafile header with one bit for each data block number that marks the beginning of a used group, or a free group, of consecutive blocks in the datafile. Each bit in the bitmap stores information about a block or a group of blocks and tells whether they are used or free. As changes are made to rows stored in the extent, the bitmap updates its record of each changed block. Because the bitmap is stored locally within the tablespace, no redo or undo logs are generated for the bitmap. Therefore, the only redo and undo log records created are for the actual changes made to data. This reduction in redo and undo logging also helps to improve overall database performance.

To further enhance the use of bitmaps and to eliminate the use of freelists, you can specify that the free space of each segment be stored locally as a bitmap as well as the free space of each extent. Do this by setting the SEGMENT SPACE MANAGEMENT to AUTO.

The syntax of the EXTENT MANAGEMENT and SEGMENT SPACE MANAGEMENT portions of the CREATE TABLESPACE command for a locally managed tablespace looks like this:


```
EXTENT MANAGEMENT LOCAL
AUTOALLOCATE|UNIFORM SIZE <nn>
SEGMENT SPACE MANAGEMENT MANUAL|AUTO
```

Here are the meanings of the settings:

- **AUTOALLOCATE**: This means that you allow the system to manage extent size allocation for objects in the tablespace. The MINIMUM EXTENT size the system can set is 64Kb. The extents can vary in size.
- **UNIFORM SIZE**: This means that you set the size of all extents in the tablespace. The minimum size is 1Mb.
- **MANUAL**: This means that a freelist tracks segment-free space.
- **AUTO**: This means that a bitmap tracks segment-free space.

Creating a Locally Managed Tablespace

Use the following steps to create a locally managed tablespace:

1. Continuing with the SQL*Plus Worksheet session you began earlier in the chapter, enter the following command by typing each line as shown. Replace the variable with the correct path provided by your instructor. Execute the command by clicking the **Execute** icon:

   ```
   CREATE TABLESPACE USER_AUTO
   DATAFILE '
   C:\oracle\product\10.2.0\oradata\oraclass\user_auto01.dbf'
   SIZE 20M AUTOEXTEND OFF
   EXTENT MANAGEMENT LOCAL AUTOALLOCATE
   SEGMENT SPACE MANAGEMENT AUTO;
   ```

   SQL*Plus Worksheet replies, "Tablespace created."
2. You have created a 20Mb locally managed tablespace with system-managed extent allocation and automatic segment space management.
3. Remain logged on for the next practice.

The second largest logical structure is a segment. The next section describes the types of segments used in the Oracle 10g database.

Segment Types and Their Uses

So far, the discussion in this chapter has focused on tablespaces and the segments and extents that are created when you create objects within the tablespaces. Tablespaces can contain other types of segments that store different data as well. The types of segments are shown in Table 5-1.

TABLE 5-1 Segment types

Segment Type	Description and Use	Allowed in Which Type of Tablespace
Data segment	Used to store data for objects, such as tables, object tables, and triggers. Each object has one data segment. Exceptions to this rule are: partitioned tables, which have one data segment per partition, and clustered tables, which have one data segment per cluster	Permanent tablespaces (locally or dictionary managed)
Index segment	Used to store all the index data. Each index has one index segment (except partitioned indexes, which have one index segment per partition)	Permanent tablespaces (locally or dictionary managed)
Temporary segment	Created during execution of SQL that needs space to perform sorting or other operations. After the execution is completed, the segment is dropped and its extents return to the free space in the tablespace. Also used for temporary tables and indexes created by a user	Temporary tablespace (usually locally-managed but can be dictionary-managed). Exception: See the following note
Rollback segment	Created automatically by Oracle in an undo tablespace when using automatic undo management. See the section titled "Overview of Undo Data" later this chapter. If using manual undo in management, these segments are created manually	Undo tablespace (if automatic undo management) or any permanent tablespace (if manual undo management)
LOB segment	Created when LOB data is stored out of line from the rest of the table's data	Any tablespace except locally managed with a uniform extent size of two data blocks, or dictionary-managed with extent size of two data blocks. (LOB segment requires a minimum of three data blocks in its first extent)

NOTE

If you have not created any temporary tablespaces, the SYSTEM tablespace is used to store temporary segments. When creating a new database, if the SYSTEM tablespace is locally managed (standard for Oracle 10*g*), a default temporary tablespace should be created during database creation. You can specify the temporary tablespace name, tempfile size, and location in the CREATE DATABASE command. If you do not supply a temporary tablespace name, the SYSTEM tablespace is used, which can degrade performance. Oracle 10*g* creates temporary segments as it needs them. Oracle 10*g* allows you to create the temporary tablespace within the CREATE DATABASE command or during database creation using the Oracle Database Configuration Assistant (DBCA tool). You can add a temporary tablespace to an existing database as well, which you will do in the next section.

In Oracle 10*g*, segments are created automatically when they are needed. For example, when you create a table, the corresponding data segment is created, and when you execute a query that sorts data, the needed temporary segment (for sorting) is created and managed completely behind the scenes. The only exceptions are manual rollback segments, which can be created manually for cross compatibility with older versions of Oracle database.

Manual rollback is redundant, not supported, and not recommended for use in Oracle 10*g*. Use automated undo instead. Manual rollback will ultimately be removed from Oracle database software altogether.

Temporary Tablespace

As you saw in the previous section, temporary segments need a temporary tablespace. A temporary tablespace is almost identical syntactically to a permanent tablespace except that when creating a temporary tablespace, fewer clauses are allowed in the CREATE command. Oracle recommends creating locally managed, temporary tablespaces.

It is possible to create a dictionary-managed, temporary tablespace, although this capability will be unavailable in future releases of Oracle, which will require locally managed, temporary tablespaces.

When creating a locally managed, temporary tablespace, the word TEMPORARY goes before the tablespace name instead of after it. The syntax for creating a locally managed, temporary tablespace is:

```
CREATE TEMPORARY TABLESPACE <tablespacename>
TEMPFILE <filename> SIZE <nn> AUTOEXTEND ON|OFF
EXTENT MANAGEMENT LOCAL UNIFORM SIZE <nn>
```

Here are the meanings of the settings:

- **TEMPFILE**: Specifies a file name and location, which is the same as the DATAFILE clause in the CREATE TABLESPACE command:
- **EXTENT MANAGEMENT LOCAL**: This clause is irrelevant because CREATE TEMPORARY TABLESPACE always creates a locally managed tablespace.
- **UNIFORM SIZE**: Use this clause only when you want to specify the size of the extents. The default size for uniform extents is 1Mb. Extents should always be uniform, especially when there is any amount of deletion activity. Otherwise Oracle manages extent sizes automatically. The accepted practice in Oracle 10*g* is to allow Oracle to manage extent sizes.

In Oracle 10*g* you can also create multiple temporary tablespaces. A single large SQL sorting operation can be split into those multiple temporary sorting spaces, or separate sorting operations can be spread across separate temporary tablespaces. Multiple temporary tablespaces require a tablespace group. The group can be set as the default for the entire database. A user or an operation is assigned temporary sorting space automatically by Oracle 10*g*. If a user is assigned a temporary tablespace group as opposed to just a single temporary tablespace, then Oracle 10*g* decides how different tablespaces, within a group, are assigned as they are required.

A temporary tablespace group is created only for temporary tablespaces, where if the group does not exist it is created, and if it does exist, then another temporary tablespace is added to it. The syntax is as follows:

```
CREATE TEMPORARY TABLESPACE <tablespace name>
  TEMPFILE '...'
  TABLESPACE GROUP <tablespace group name>;
```


Complete the following steps to create a locally managed, temporary tablespace that uses all the default settings and is not an Oracle Managed file:

1. Type and execute the following command in your current SQL*Plus Worksheet session to create a tablespace named TEMP_STARTER that has a user-managed datafile. Replace ORACLE_BASE with the actual directory provided by your instructor.

   ```
   CREATE TEMPORARY TABLESPACE TEMP_STARTER
   TEMPFILE
   'ORACLE_BASE\oradata\oraclasss\temp_starter01.dbf'
   SIZE 2M;
   ```

   The resulting file is 2Mb, with 1Mb uniform extents that can extend to an unlimited number. AUTOEXTEND defaults to ON for temporary tablespaces.
2. Create another temporary tablespace. This one exemplifies using non-default values. Type and execute this command in SQL*Plus Worksheet. Replace ORACLE_BASE with the actual directory provided by your instructor:

   ```
   CREATE TEMPORARY TABLESPACE TEMP_STARTER2
   TEMPFILE 'ORACLE_BASE\oradata\oraclass\temp_starter02.dbf'
   SIZE 6M
   AUTOEXTEND ON NEXT 2M MAXSIZE 50M;
   ```

   This command creates a 6Mb file. The file extends automatically in 2Mb increments to a maximum size of 50Mb.
3. When a user runs a query or other transaction that needs temporary space, a temporary segment is either created or reused in the default temporary tablespace. The default temporary tablespace is the tablespace assigned to the user for all temporary segments. You can assign the default temporary tablespace in two ways:
   - Explicitly assign the **user** a default temporary tablespace using the CREATE USER or ALTER USER command:

     ```
     ALTER USER <username> TEMPORARY TABLESPACE temp;
     ```

   - Explicitly assign the **database** a default temporary tablespace and allow the user to use this tablespace by default. Use the CREATE DATABASE or ALTER DATABASE commands, as described here.
4. If you created the database with the DBCA tool, included the DEFAULT TEMPORARY TABLESPACE clause in your CREATE DATABASE command, a default temporary tablespace has been created for you. Enter and execute the following query to determine the name (if any) of the current, default temporary tablespace.

   ```
   SELECT PROPERTY_VALUE
   FROM DATABASE_PROPERTIES
   WHERE PROPERTY_NAME = 'DEFAULT_TEMP_TABLESPACE';
   ```

The value returned is the current, default temporary tablespace for the database.

5. Occasionally, you may need to change the default that was originally set for the database. Your database may need a larger or smaller size tablespace, or you may want to change from a dictionary-managed tablespace to a locally managed tablespace. Oracle recommends always using locally managed, temporary tablespaces because they are more efficient in handling additions and deletions of segments (a frequent task in the default temporary tablespace). Switch the default temporary tablespace to the new tablespace you just created by typing and running the following command:

```
ALTER DATABASE DEFAULT TEMPORARY TABLESPACE
TEMP_STARTER;
```

6. Remain logged on for the next practice.

TIP

Physical extents are not created in a temporary tablespace datafile until the temporary tablespace is used for something such as a sorting operation. Extents are removed on completion of the same operation. In some operating systems, the temporary tablespace's datafile is not even created until a temporary segment is created the first time.

As mentioned earlier, it is possible to create a tablespace with a data block size different than that set for the database with DB_DATA_BLOCK. The next section describes how this can be done.

Tablespaces with Nonstandard Data Block Size

You can create a special tablespace with its own data block size. One reason for doing this is to help you when transporting tablespaces from one database to another, and the two data block sizes are different. Giving a tablespace a compatible data block size makes transporting the tablespace much more efficient. Another reason is to divide up different types of data to help performance. For example, static data such as your customer names and addresses, rarely changes and is often small in physical size. Thus static data is often better served by smaller block sizes. Transactional data, on the other hand, such as all invoices issued to customers of the last five years, can involve large quantities of data. Thus large block sizes are more appropriate to transactional types of data.

Complete the following steps to find out what is required to create this type of tablespace.

1. First, it is important to know the standard database block size, as defined by the DB_BLOCK_SIZE parameter. Type and execute the following command into your SQL*Plus Worksheet session to find out:

```
SHOW PARAMETERS DB_BLOCK_SIZE
```

The block size is typically 8096 (8Kb). Yours may be different. When you plan to create a tablespace not of standard data block size, you are allowed to use a block size of 2Kb, 4Kb, 8Kb, 16Kb, or 32Kb. You cannot use 8Kb if the database block size is 8Kb. In addition, some operating systems don't allow

32Kb block sizes. Check the specific manual of the operating system to determine any restrictions.

2. The second important piece of information is the current settings for cache memory. Before creating a tablespace with a nonstandard block size, you are required to create a cache space with a matching block size. Recall that Oracle 10g reads and writes data in data block sized chunks. All the memory cache is set up for the standard block size. You must allocate some memory that can work with the new block size. Type and execute this command to view the current memory cache settings:


```
SHOW PARAMETERS CACHE
```

Figure 5-11 shows the results. Your results will be similar. Notice that there are zeros for the db_2k_cache_size, db_4k_cache_size, and so on. The only cache parameter that has space allocated is the db_cache_size, which is calculated by Oracle and is the memory allocation for the standard data block size cache.

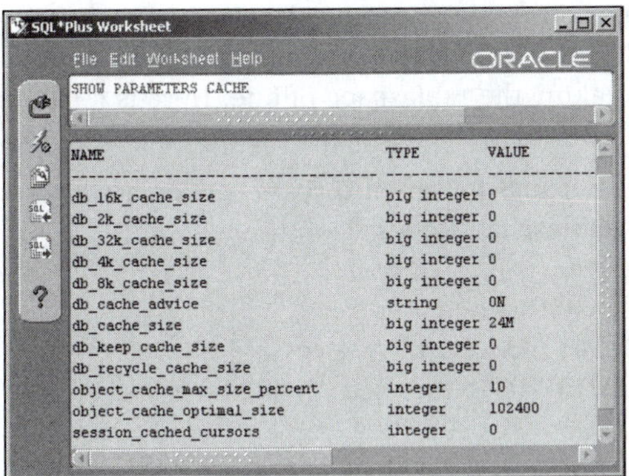

FIGURE 5-11 Alternate cache parameters have no storage allocation at this juncture

3. For this example, you are creating a tablespace with a 2Kb data block size. Therefore, you must allocate storage to the 2Kb cache in Oracle's memory. Enter and execute this command to allocate 4Mb to the 2Kb cache:

```
ALTER SYSTEM SET DB_2K_CACHE_SIZE=4M;
```

4. SQL*Plus Worksheet replies, "System altered." If you receive an error message, reduce the size of the db_cache_size (using ALTER SYSTEM), and then try again.

5. Now you have the buffer prepared, so you can create the tablespace. Execute the following command to create a tablespace with 2Kb block size. Replace the ORACLE_BASE variable with the directory path provided by your instructor.

```
CREATE TABLESPACE TBS_2K
DATAFILE 'ORACLE_BASE\oradata\oraclass\tbs_2k.dbf'
SIZE 4M BLOCKSIZE 2K;
```

SQL*Plus Worksheet replies, "Tablespace created."

6. Remain logged on for the next practice.

TIP

You may not create a temporary tablespace with nonstandard block size.

Now that you have created some tablespaces, how do you modify them? The next section discusses changes to tablespaces.

CONFIGURING AND VIEWING STORAGE

This section describes how to adjust the storage size of existing tablespaces and which data dictionary views can be queried to display information on storage.

Changing the Size, Storage Settings, and Status

After a tablespace has been created, you may want to change its settings to improve performance. For example, if your tablespace keeps running out of space, you may want to modify the AUTOEXTEND setting. Or perhaps your tablespace has plenty of empty space, and you want to coalesce the extents. The ALTER TABLESPACE command handles these types of tasks as well as other tasks, such as taking the tablespace offline. Here is a list of the tasks you can handle using the ALTER TABLESPACE command:

- Change the DEFAULT STORAGE settings for any future objects created in the tablespace (dictionary-managed tablespaces only)
- Change the MINIMUM EXTENT size
- Change from LOGGING to NOLOGGING and vice versa
- Change from PERMANENT to TEMPORARY and vice versa
- Change from READ ONLY to READ WRITE and vice versa
- Coalesce contiguous storage space
- Add a new datafile or temporary file
- Rename a datafile or temporary file
- Begin and end an open backup (a backup of the tablespace while the tablespace is available for use)

TIP

The extent management mode (LOCAL or DICTIONARY) cannot be changed using the ALTER TABLESPACE command. If you want a different extent management mode, you have to create a new tablespace, migrate the data, and then drop the old tablespace.

The syntax for modifying the size and storage settings of the tablespace is:

```
ALTER TABLESPACE <tablespacename>
ADD|RENAME DATAFILE <filename>
       SIZE <nn> AUTOEXTEND ON|OFF REUSE
DEFAULT STORAGE (INITIAL <nn> NEXT <nn> PCTINCREASE <nn>
                 MINEXTENTS <nn> MAXEXTENTS <nn>)
MINIMUM EXTENT <nn>
COALESCE
```

NOTE

Keep in mind that you may only perform one task per ALTER TABLESPACE command. For example, if you want to add a datafile and modify the DEFAULT STORAGE, you must issue two commands.

Let's say that you have created a tablespace that is locally managed and you want to add a new datafile to the tablespace to increase its storage space. Complete the following steps to accomplish this task: 223

1. Enter and execute the following command in SQL*Plus Worksheet by typing each line as shown. Replace the ORACLE_BASE variable with the correct path provided by your instructor.

   ```
   ALTER TABLESPACE USER_AUTO
   ADD DATAFILE 'ORACLE_BASE\oradata\oraclass\user_auto02.dbf'
   SIZE 5M;
   ```

2. SQL*Plus Worksheet replies, "Tablespace altered." You have added a 5Mb datafile to the locally managed tablespace.
3. Remain logged on for the next practice.

Here is an example of changing the DEFAULT STORAGE clause of a dictionary-managed tablespace:

```
ALTER TABLESPACE USER_TEST
DEFAULT STORAGE (INITIAL 120M MINEXTENTS 4 MAXEXTENTS 100);
```

NOTE

The above command does not apply to a locally managed tablespace.

When you modify the DEFAULT STORAGE clause from its original settings, the new settings only affect blocks created after the change. Existing blocks (existing objects) retain their block level settings for INITIAL and MINEXTENTS, as when those objects first used the blocks. This is because these settings are only used during the creation of the object. Both existing and new objects now extend to 100 extents instead of the original 50 extents.

Temporary tablespaces are changed in the same way, except that you use ADD TEMPFILE instead of ADD DATAFILE.

To change an existing datafile's storage, you must use the ALTER DATABASE command rather than the ALTER TABLESPACE command. Follow these steps to try this out:

1. In SQL*Plus Worksheet, change a datafile from AUTOEXTEND OFF to AUTOEXTEND ON by entering and executing the following command. As before, replace the ORACLE_BASE variable with a real directory path provided by your instructor:

   ```
   ALTER DATABASE DATAFILE
   'ORACLE_BASE\oradata\oraclass\user_auto02.dbf'
   AUTOEXTEND ON;
   ```

2. Similarly, you can modify the size, extent settings, and online or offline status of a datafile. Enter and execute this command, replacing the ORACLE_BASE variable as usual:

```
ALTER DATABASE DATAFILE 'ORACLE_BASE\oradata\oraclass
\user_auto02.dbf'
RESIZE 10M;
```


3. Remain logged on for the next practice.

To modify a tempfile, use the ALTER DATABASE TEMPFILE command.

The status of a tablespace defines its availability to end-users and also defines how it is handled during backup and recovery. The two types of status are:

- **ONLINE:** This is the default status for new tablespaces. Online tablespaces are available to users when the database is open.
- **OFFLINE:** The tablespace is not available to users even though the database is open.

You might take a tablespace offline to make sure users do not have access while you perform a backup of the tablespace or while you update the application that uses that tablespace. You can perform an online backup of a tablespace. However, you may want to use the offline backup method because it is faster than the online method.

When you take a tablespace offline, there are three methods you can specify for Oracle 10g's execution of the command:

- **NORMAL:** This is the default, and Oracle 10g performs a checkpoint on each datafile of the tablespace, assuring that all data in memory buffers is successfully written to all the datafiles before the tablespace goes offline. No recovery is needed to put the tablespace back online.
- **TEMPORARY:** Oracle 10g performs a checkpoint on all the available datafiles in the tablespace. Use this mode when a datafile has been damaged, causing the checkpoint attempt while performing a NORMAL offline command to fail. You will have to recover unsaved changes in the bad datafile before putting the tablespace back online.
- **IMMEDIATE:** Oracle 10g does not attempt to do a checkpoint, and you will be required to perform media recovery before the tablespace goes back online. Media recovery is the term used for recovery that requires restoring the database from a backup, rolling forward through archived redo logs, and finally rolling forward through online redo logs. It is called media recovery because the most common reason for this type of recovery is the failure of a disk drive or some other storage media.

Follow these steps to take a tablespace offline, and then restore it to online mode:

1. In your SQL*Plus Worksheet session, enter and execute the following command to take the tablespace offline:

```
ALTER TABLESPACE ACCOUNTING
OFFLINE NORMAL;
```

2. SQL*Plus Worksheet replies, "Tablespace altered." You have taken the tablespace offline. It is possible to take individual datafiles offline by using a similar command.

Type this command to take one datafile of the USER_AUTO tablespace offline, replacing the ORACLE_BASE variable with the correct directory path:

```
ALTER DATABASE DATAFILE
'ORACLE_BASE\oradata\oraclass\user_auto02.dbf'
OFFLINE DROP;
```

The IMMEDIATE, TEMPORARY, and NORMAL options do not apply to taking a datafile offline. In addition, the DROP option is required if you take a datafile offline when your database is in NOARCHIVELOG mode. In NOARCHIVELOG mode it is convenient in some cases when you plan to drop the datafile anyway, because the DROP option takes the datafile offline and drops the operating system file in one command. If, however, you don't want the datafile dropped, you must take the entire tablespace offline or change the archiving mode of the database. When your database is in ARCHIVELOG mode, take a datafile offline without dropping it by removing the DROP keyword.

3. Place the ACCOUNTING tablespace back online by entering and executing the following command:

```
ALTER TABLESPACE ACCOUNTING
ONLINE;
```

4. Remain logged on for the next practice.

Because the operating system file was dropped when you took the datafile offline in the exercise, you cannot return the datafile to online status. If the database is in ARCHIVELOG mode, replacing the ORACLE_BASE variable as usual, the syntax to place a datafile online is:

```
ALTER DATABASE DATAFILE <filename> ONLINE
```

One of the options when creating or altering a tablespace is to make it a read-only tablespace.

Read-only Tablespaces

After you have loaded data into tables, if that data never changes (or rarely changes), such as a table of the states in the United States, you might consider storing all this static data in a read-only tablespace.

The objects in a read-only tablespace, as its name implies, can be queried but not changed. No user can execute an INSERT, UPDATE, or DELETE command on the objects.

The primary purpose of a read-only tablespace is to make large amounts of static data available to the database without having to slow down the database backup and recovery or the need for logging and checkpointing. Back up the tablespace one time, and after that, it never has to be backed up again—unless of course one of those rare changes occurs.

If you need to alter data in the read-only tablespace, simply change it back to read-write status (the default) and make the changes. Then return it to read-only status.

Follow these steps to modify a tablespace to read-only status:

1. In SQL*Plus Worksheet, enter and execute this command to change the ACCOUNTING tablespace to read-only status:

```
ALTER TABLESPACE ACCOUNTING READ ONLY;
```

2. Execute the following query to confirm the status of the tablespace.

   ```
   SELECT TABLESPACE_NAME, STATUS
   FROM DBA_TABLESPACES;
   ```

   Figure 5-12 shows part of the result. Your screen may be different.

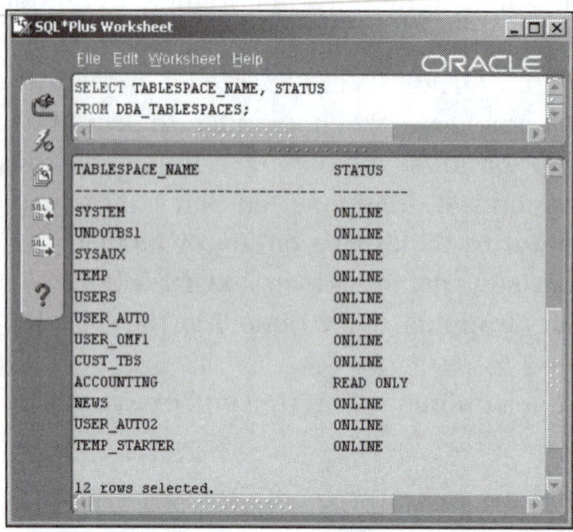

FIGURE 5-12 The DBA_TABLESPACE view shows tablespace status

NOTE

Your result for the query in Figure 5-12 may be slightly different and may even be sorted differently to the example shown. The results depend on your database, its installation, and what you may or may not have cleaned up from previous chapters.

3. To return it to read-write status, use the same command, but specify READ WRITE. Enter and execute this command:

   ```
   ALTER TABLESPACE ACCOUNTING READ WRITE;
   ```

4. Run the query again and notice that the status changed from READ ONLY to ONLINE:
5. Remain logged on for the next practice.

The command for working with tablespaces is for dropping them.

Dropping Tablespaces

A tablespace can be dropped. However, Oracle 10g requires you to use specific clauses to drop tablespaces that contain data. Here is the syntax:

```
DROP TABLESPACE <tablespacename>
INCLUDING CONTENTS
AND DATAFILES
CASCADE CONSTRAINTS;
```

When you drop a tablespace, all references to the tablespace are removed from temporary tablespaces, the data dictionary, and the undo data. If there are transactions running that use data in the tablespace, the tablespace cannot be dropped.

Follow these steps to work with the command:

1. To drop a tablespace that contains no data, you use the basic DROP TABLESPACE command. Enter and execute this command in SQL*Plus Worksheet to drop the ACCOUNTING tablespace, which has no data:

```
DROP TABLESPACE ACCOUNTING;
```

2. To experiment with dropping a tablespace containing data, first you will add some data to one of the tablespaces you created in this chapter. An easy way to do this is to run the prepared script found in the **Data\Chapter05** directory on your student disk. Click the **Worksheet/Run Local Script** on the SQL*Plus Worksheet menu. Navigate to the **Data\Chapter05** directory and select the **add_data.sql** file. Click **Open** to run it in SQL*Plus Worksheet. The script creates tables in three tablespaces.
3. See how much space in each tablespace has been allocated to extents. Type and execute this command:

```
SELECT TABLESPACE_NAME, SEGMENT_TYPE,
EXTENT_ID, BLOCKS
FROM DBA_EXTENTS
WHERE TABLESPACE_NAME IN
('CUST_TBS', 'USER_OMF1', 'USER_AUTO');
```

   Figure 5-13 shows the results. As you can see, each table uses eight blocks, and the index also uses eight blocks.

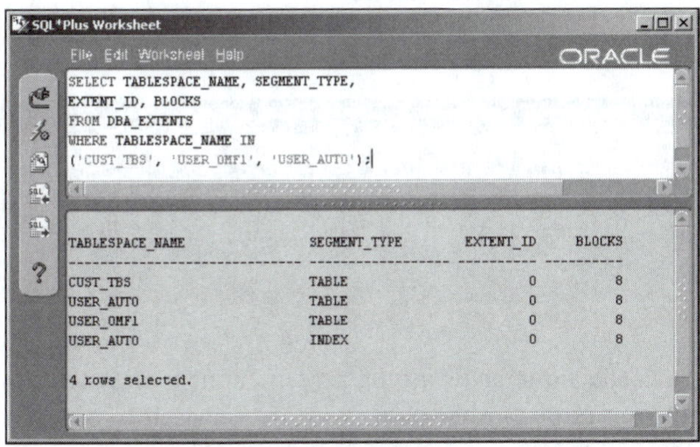

FIGURE 5-13 The first extent of a table has zero as its extent ID number

4. The CUST_TBS contains the CH05NEW_CUSTOMER table with one extent. Drop the CUST_TBS tablespace and all its data with the following command. The command also tells Oracle 10*g* to drop the tablespace's datafiles with the ADD DATAFILES clause:

```
DROP TABLESPACE CUST_TBS INCLUDING CONTENTS
AND DATAFILES;
```

5. The USER_AUTO tablespace contains the CH05PALS table and an index (created for the PRIMARY KEY constraint on CH05PALS). Another table, CH05BID, resides in the USER_OMF1 tablespace. This table has a FOREIGN KEY constraint referencing the CH05PALS table.

NOTE

A primary key uniquely identifies each row in a table. A foreign key contains a copy of a primary key value from a parent table. Primary and foreign keys are both constraints (enforced restrictions) on rows in tables.

To drop the USER_AUTO tablespace, you must use the CASCADE CONSTRAINTS clause. The clause tells Oracle 10*g* to remove any constraints that were created in other tablespaces referencing any table in the USER_AUTO tablespace. Type and execute this command to drop the tablespace and the constraint:

```
DROP TABLESPACE USER_AUTO INCLUDING CONTENTS
CASCADE CONSTRAINTS;
```

6. Oracle Managed datafiles are automatically dropped without using the AND DATAFILES clause; however, you must still use the INCLUDING CONTENTS clause to drop this tablespace because it contains a table. Type and execute this command:

```
DROP TABLESPACE USER_OMF1 INCLUDING CONTENTS;
```

7. Remain logged on for the next practice.

TIP

You can drop a tablespace whether it is online or offline. In fact, it is safer to take a tablespace offline before dropping it. This gives you time to discover any unexpected errors. For example, you may have assigned the tablespace to a developer as his default tablespace. When the tablespace is taken offline (or dropped), he suddenly is unable to create new tables.

Dropping a temporary tablespace is the same as dropping a permanent tablespace. If the tablespace is currently the default temporary tablespace, you must switch to another default tablespace, wait for any outstanding temporary segments remaining in the old temporary tablespace to clear, and then drop the tablespace.

Renaming Tablespaces

You can also rename a tablespace in Oracle 10*g*. Renaming tablespaces can be useful in rare circumstances, such as when switching in and out partitioned tablespaces, or incorporating transported tablespaces into an existing database.

NOTE

A transportable tablespace is an advanced Oracle 10*g* feature used to transport tablespaces between different databases, even different platforms and operating systems.

The following commands create a new tablespace called EMPTYONE, and then rename it to NEWTABLESPACE (replace the ORACLE_BASE variable as usual):


```
CREATE TABLESPACE EMPTYONE
DATAFILE 'ORACLE_BASE\oradata\oraclass\emptyone.dbf' SIZE
10M;

ALTER TABLESPACE EMPTYONE RENAME TO NEWTABLESPACE;
```

And let's clean up after ourselves by dropping the renamed tablespace:

```
DROP TABLESPACE NEWTABLESPACE
INCLUDING CONTENTS AND DATAFILES;
```

You have examined how to create, modify, drop, and rename tablespaces. The next section shows how to review your tablespaces by querying the Data Dictionary.

Querying the Data Dictionary for Storage Data

There are quite a few data dictionary views and dynamic performance views containing information about tablespaces. The most useful views are DBA_TEMP_FILES and DBA_DATA_FILES, because querying these two views shows you all the files connected with tablespaces in the database. Table 5-2 lists many of them.

TABLE 5-2 Views of tablespace information

View	Description
DBA_DATA_FILES	Details on datafiles
DBA_EXTENTS	Details on data extents in all tablespaces
DBA_FREE_SPACE	Free extents within all tablespaces
DBA_FREE_SPACE_COALESCED	Coalesced free space information
DBA_SEGMENTS	Information about all segments in all tablespaces
DBA_TABLESPACES	Descriptions of all tablespaces
DBA_TEMP_FILES	Details on tempfiles
DBA_TS_QUOTAS	Tablespace quotas for all users
DBA_USERS	Default and temporary tablespaces assigned to all users
V$DATAFILE	Datafiles and tablespaces that own them
V$TABLESPACE	All tablespaces from the control file
V$TEMP_EXTENT_MAP	All extents in all locally managed, temporary tablespaces

TABLE 5-2 Views of tablespace information (continued)

View	Description
V$TEMP_SPACE_HEADER	Space used and space free for each tempfile
V$TEMPFILE	Tempfiles and tablespaces that own them

As with any other view, you can query these views.

Complete the following steps to query some data dictionary views:

1. As an example, imagine you want to know whether you should coalesce free extents in the USERS tablespace. Developers use the tablespace to test various database designs. Because of this, many extents would be created and released as the developers create new tables and later drop the tables. Type and execute this query in SQL*Plus Worksheet to review the USERS tablespace:

```
SELECT BLOCK_ID, BLOCK_ID+BLOCKS NEXT_BLOCK_ID, BLOCKS
FROM DBA_FREE_SPACE
WHERE TABLESPACE_NAME = 'USERS'
ORDER BY BLOCK_ID;
```

   An example of the query result is shown on the right side of Figure 5-14. Your results will look similar, but not identical. The goal of the query is to help you determine if any free extents can be coalesced. Recall that free extents must be adjacent to one another to be coalesced. Figure 5-14 illustrates how you interpret the results of the query to determine whether a tablespace needs coalescing. In Figure 5-14, there are two sets of blocks (deallocated extents) in which the NEXT_BLOCK_ID in the first is the same as the BLOCK_ID in the next set of blocks. These sets of blocks are adjacent and can be coalesced. As shown in Figure 5-14, the deallocated extents beginning with BLOCK_ID 2 and 6 can be coalesced, and the extents beginning with BLOCK_ID 47 and 52 can also be coalesced.

2. If your query results indicate that the USERS tablespace needs coalescing, enter and run this command:

```
ALTER TABLESPACE USERS COALESCE;
```

3. Remain logged on for the next practice.

Another type of data that can be stored in tablespaces is undo data. The next section discusses undo data.

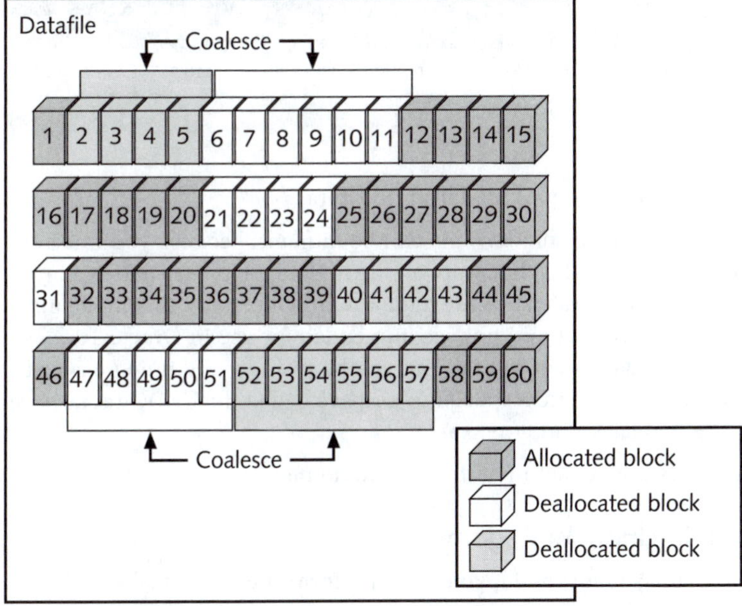

FIGURE 5-14 Adjacent free blocks can be combined into a single free space

OVERVIEW OF UNDO DATA

You have already seen that one of the final steps in database recovery is to roll forward the online redo logs and then apply the undo data. What is undo data and where does it reside in the database?

Undo data is made up of undo blocks. You could think of each undo block containing the before image of the data in the block. For example, when a row in the CUSTOMER table is updated to change Jane's last name to Smith, an exact copy of the block containing Jane's row is copied. It is more complex than that because in reality undo information is stored as pointers, containing references to rows in tables and fields.

Undo copies of data are used to redo the original data if the user issues a ROLLBACK command. The undo data also provides read consistency for users accessing the table between the time the update is pending, and the time the update has been committed. Read consistency means that only the user who issued the update command sees the changes to the data in queries until that user commits the change. At that point, all users see the changed data.

The other use of undo data is during database recovery. The redo log entries are made before data has officially been committed. This occurs when a checkpoint causes the buffers to be flushed. If the database fails and prevents the commit from happening, the data in the redo log is inaccurate. During recovery, the redo log updates the database with the uncommitted change. Then the undo data is applied and replaces the changed data with the original data.

NOTE

You may want to keep undo data in the undo tablespace for a short time after the data has actually been rolled back or committed. This is called **undo data retention** and helps users to complete queries that were begun while the data was in limbo and continued after the data was either rolled back or committed.

There is a hard way and an easy way to manage undo data:

- **Manual undo management mode:** This can be difficult, because you must set up and manage groups of files that store all the undo data in user-managed rollback segments. Earlier versions of the database provided only this mode and it is more or less redundant and not recommended for use in Oracle 10*g*.
- **Automatic undo management mode:** This method is new and much easier to manage. You need only create a special tablespace called the UNDO tablespace and Oracle 10*g* manages undo for you.

The next section describes how to set up automatic undo management.

Implementing Automatic Undo Management

To set up automatic undo management mode, you must perform these two tasks:

1. Set the UNDO_MANAGEMENT initialization parameter to AUTO.
2. Create an undo tablespace.

Using the DBCA tool to create a database will create automated undo by default.

Alternatively, you can set other related initialization parameters, such as UNDO_RETENTION and UNDO_TABLESPACE.

An undo tablespace is an entire tablespace reserved for undo data. Like data and index tablespaces, data in the undo tablespace is added in the form of extents. The extents are called **undo extents**.

You can create the undo tablespace when you create the database. If you set UNDO_MANAGEMENT to AUTO before database creation, you must include the UNDO_TABLESPACE clause in the CREATE DATABASE command. For example, the following command creates the NEWAUTO database with an OMF undo tablespace:

```
CREATE DATABASE newauto
UNDO TABLESPACE UNDO_TBS;
```

You can, of course, use user-managed files for the undo tablespace as well. In this case, supply the datafile name and storage information as you do for any tablespace.

NOTE

Oracle 10*g* uses only one undo tablespace for a database. If you created more than one, Oracle 10*g* uses the first one it finds on startup. Or, you can specify which to use with the UNDO_TABLESPACE initialization parameter.

If you choose to change the undo management mode from user-managed (manual) to automatic, or if you want to switch to a new undo tablespace, you should create the undo tablespace with a CREATE UNDO TABLESPACE command.

TIP

If you start up a database in automatic undo management mode and no undo tablespace exists, Oracle 10*g* creates a default undo tablespace called SYS_UNDOTBS.

As with any other tablespace, drop an undo tablespace by using the DROP TABLESPACE command.

Follow these steps to switch the undo tablespace and adjust the related parameters.

1. Enter and run the follow SQL*Plus command in your SQL*Plus Worksheet session to find information about the undo mode and undo tablespace in your database:

   ```
   SHOW PARAMETERS UNDO
   ```

   Figure 5-15 shows the results. The UNDO_MANAGEMENT parameter is set to AUTO, indicating automatic undo management. The current undo tablespace is UNDOTBS1. (Your results may show a different tablespace name.)

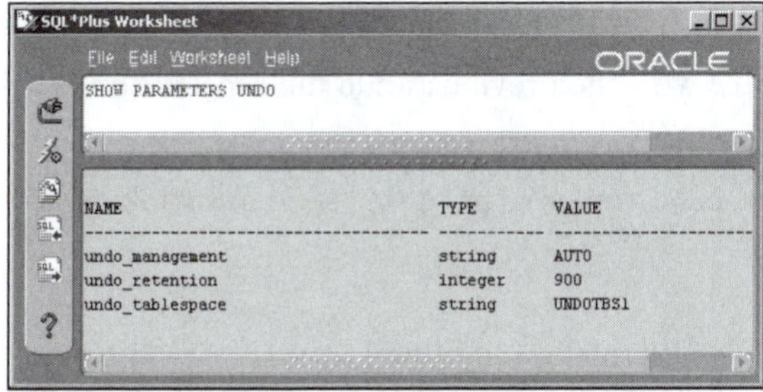

FIGURE 5-15 The three parameters listed define how undo data is managed

2. You want to create a new undo tablespace to replace the current one. The CREATE UNDO TABLESPACE command has fewer options than the CREATE TABLESPACE command. You can only specify the datafile name, size, and autoextend mode. Oracle 10g determines the rest. Enter and execute the following command to create an undo tablespace named UNDO_IT with one 30Mb datafile, with autoextend on. Replace the ORACLE_BASE variable with the directory path provided by your instructor:

   ```
   CREATE UNDO TABLESPACE UNDO_IT
   DATAFILE 'ORACLE_BASE\oradata\oraclass\undo_it.dbf'
   SIZE 30M AUTOEXTEND ON;
   ```

3. Next, adjust the initialization parameters. Consult the following list of parameters to find out what they mean.
   - **UNDO_MANAGEMENT:** This tells Oracle 10g which of the two modes to use for undo data. The two choices are AUTO and MANUAL. This parameter is static. Your database is already in automatic undo management mode. Manual undo management mode is provided for backward compatibility and requires manual creation of undo segments.
   - **UNDO_TABLESPACE:** This specifies the name of the undo tablespace to use for undo data when the database is in automatic undo management mode.
   - **UNDO_RETENTION:** This is the time in seconds to save committed undo data. The default is 900 seconds (15 minutes). Retaining undo data provides support for a new feature: Flashback Query. This feature uses undo data to simulate going back in time and running a query on the old data.

   Switch the current undo tablespace by executing this command:

   ```
   ALTER SYSTEM SET UNDO_TABLESPACE='UNDO_IT';
   ```
4. Lengthen the undo retention time to 30 minutes (1800 seconds) by executing this command:

   ```
   ALTER SYSTEM SET UNDO_RETENTION=1800;
   ```
5. View the changes by executing this command to display the parameters:

   ```
   SHOW PARAMETERS UNDO
   ```
6. Return the database to using the original undo tablespace by executing this command, replacing the variable with the previous undo tablespace name:

   ```
   ALTER SYSTEM SET UNDO_TABLESPACE='<tablespacename>';
   ```
7. Prepare to drop the undo tablespace you created in this session by taking it offline. Execute this command:

   ```
   ALTER TABLESPACE UNDO_IT OFFLINE;
   ```

   You must use caution when dropping an old undo tablespace because, unlike permanent tablespaces, the DROP TABLESPACE command does not return an error message if the tablespace contains data. The old undo tablespace may still contain undo data for some users.
8. Query the data dictionary view DBA_TABLESPACES to check the status of the tablespace. Execute this query:

   ```
   SELECT TABLESPACE_NAME, STATUS
   FROM DBA_TABLESPACES
   WHERE CONTENTS='UNDO';
   ```
9. The status may be ONLINE, OFFLINE, or PENDING OFFLINE. If the status is PENDING OFFLINE, there are undo segments still in use on the old undo tablespace. Wait until the status is OFFLINE before dropping the tablespace.
10. If the status of the UNDO_IT undo tablespace is OFFLINE, execute this command to drop the tablespace:

    ```
    DROP TABLESPACE UNDO_IT;
    ```
11. The SQL*Plus Worksheet replies, "Tablespace dropped."

12. Exit SQL*Plus Worksheet by typing **EXIT** and clicking the **Execute** icon. Click **OK** when the Oracle Enterprise Manager dialog box opens.
13. Exit the console by clicking the **X** in the upper-right corner of the window.

TIP

These are only a few of the many options of the ALTER TABLESPACE command that are allowed for undo tablespaces. You can also add more space by resizing or adding a datafile, renaming a datafile, and starting or stopping an **open backup** (a backup done while the database is running) to an undo tablespace by using the ALTER TABLESPACE command. The other options of the ALTER TABLESPACE command are invalid for undo tablespaces.

Monitoring Undo

Three views in the form of DBA_UNDO_EXTENTS, DBA_ROLLBACK_SEGS, and V$UNDOSTAT. The information is these views is cumbersome, difficult to interpret, and requires access through a SQL*Plus interface. There is a better way using the Database Control. Recall from Chapter 2 that in order to execute the Database Control, execute the URL on your database server: **http://<hostname>:<port>/em**.

TIP

Accessing the Database Control is done by typing a URL into a browser, such as Internet Explorer in Windows. Depending on the release of Oracle 10*g* there are different ways to access the Database Control. In Oracle 10*g*, Release 1 (10.1.0), the URL port number is set at 5500. So if your database server's computer name is abc, then the URL will be *http://abc:5500/em*. For Oracle 10*g*, Release 2, each database on a database server has a distinct port number (ask your instructor). In my case, my ORACLASS database is accessed through the URL *http://piii-450:1158/em*, where my server hostname is piii-450, and the port exclusive to my ORACLASS database is 1158.

Once the Database Control is up and running, click the Administration tab as shown in Figure 5-16. The next step is to click the Undo Management link under the Instance Management group of links at the top left of the screen. The result is as shown in Figure 5-17. From the screen shown in Figure 5-17 you can view and alter the configuration, use the undo advisor, check recent system activity—even graphically as shown in Figure 5-18.

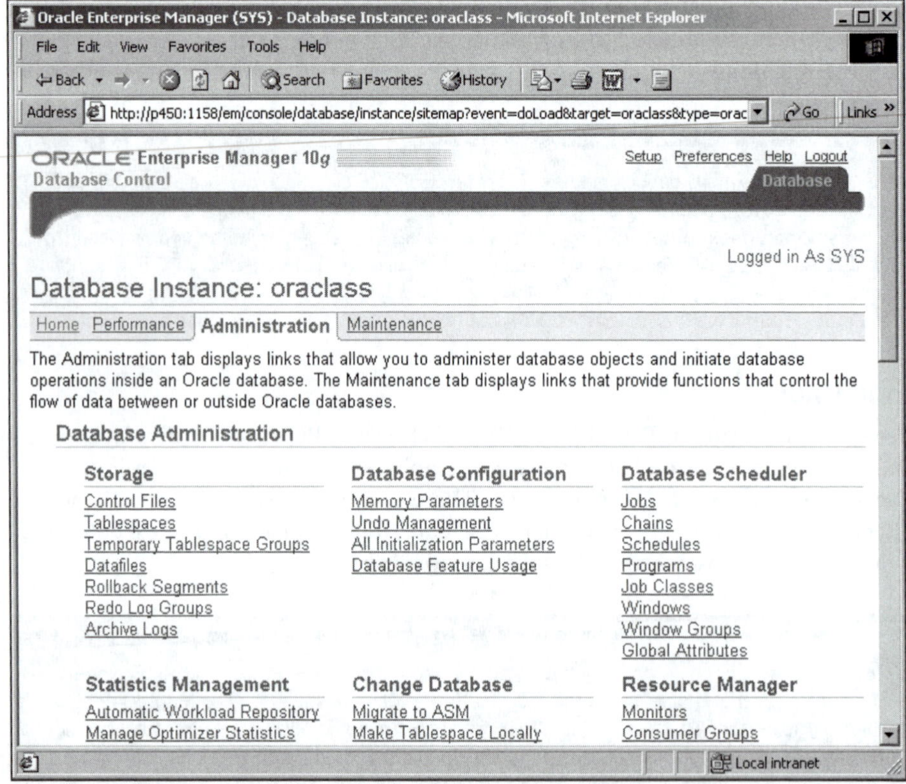

FIGURE 5-16 Click the Administration tab in the Database Control to access Undo Management

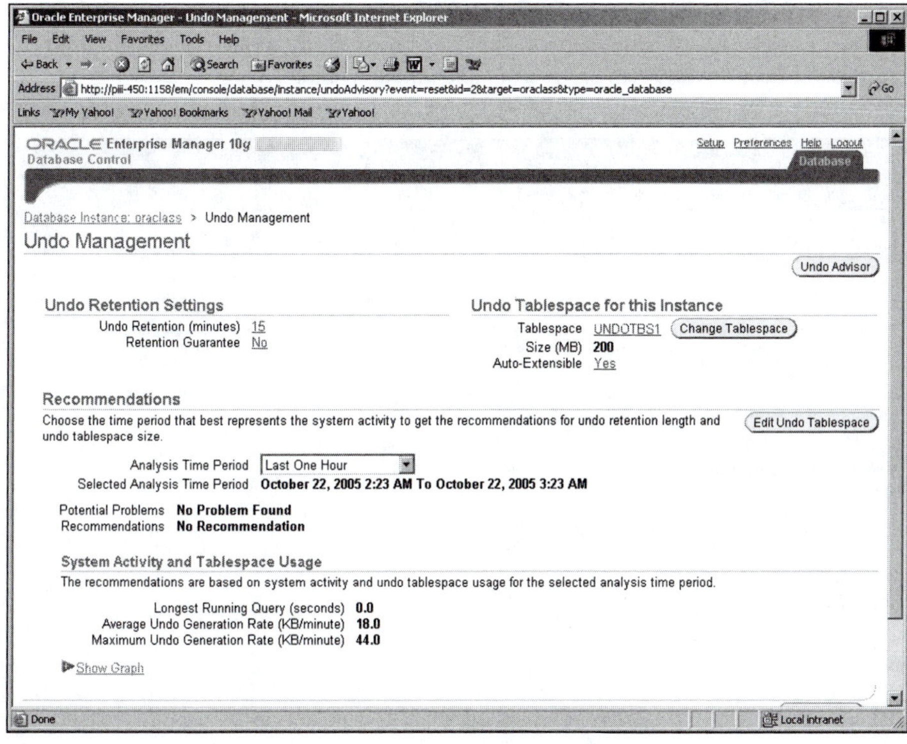

FIGURE 5-17 Undo management from the Database Control

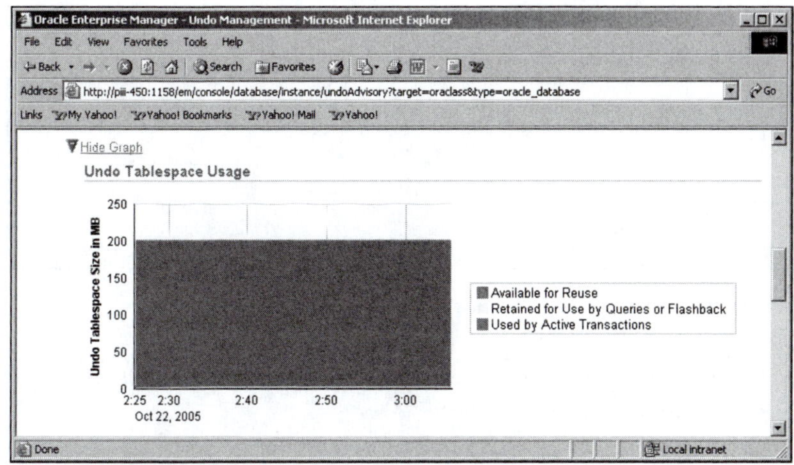

FIGURE 5-18 Undo generation rate and tablespace usage

Click the Undo Advisor button at the top right of the Undo Management screen in the Database Control screen. The result is the screen shown in Figure 5-19, where once again information, settings, and configuration is easily available and adjustable. There is even a graph.

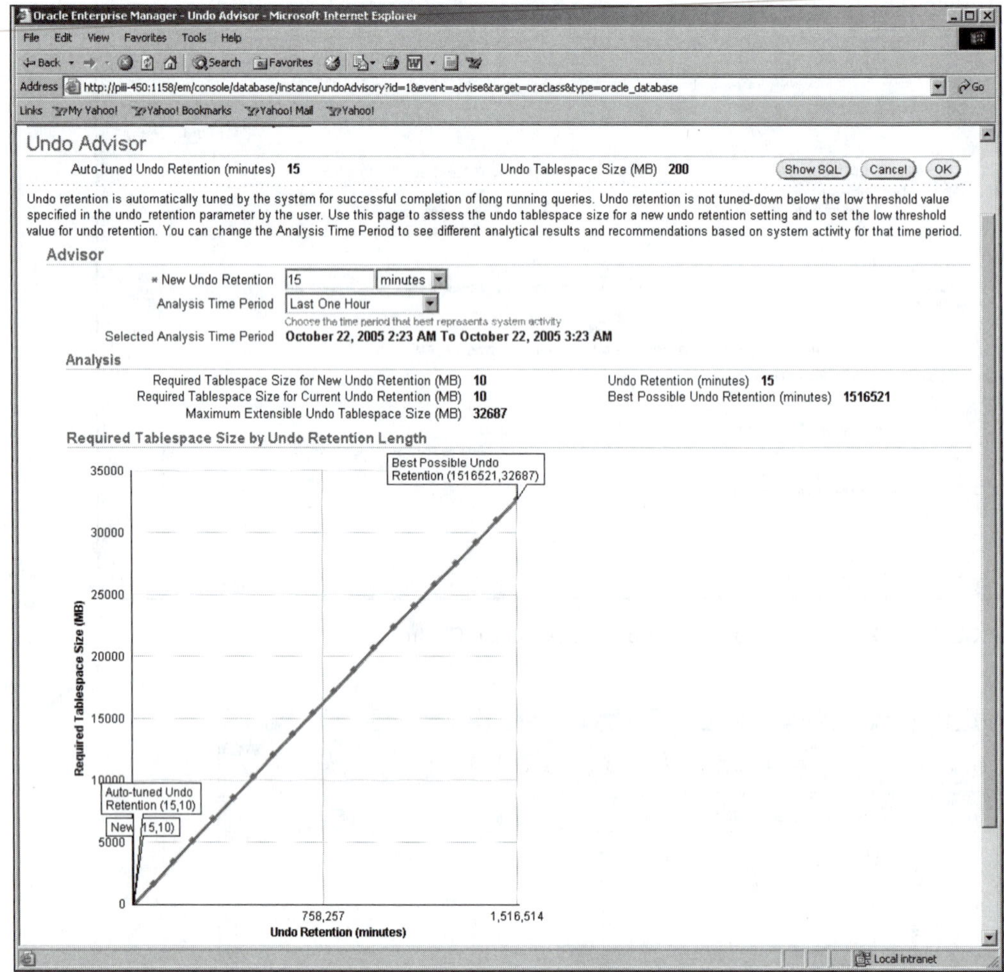

FIGURE 5-19 Using the Undo Advisor in the Database Control

The next section looks at another area in which information about the functioning and health of the database is stored: the diagnostic files.

OVERVIEW OF DIAGNOSTIC (TRACE) FILES

Diagnostic files, or trace files, allow highly detailed resolution of problems. Typically trace files are often used to solve serious problems, such as after a database crash. It is important to note that serious errors are most often written to trace files, and duplicated to the primary trace file—the alert log. Not all critical errors may necessarily be sent to the trace files, and sometimes a database crash may only be sent to a memory core dump and never recorded in the alert log. Let's quickly refresh and expand on some of the admin/<databasename> directories in the ORACLE_BASE directory (ORACLE_BASE/admin):

- **bdump**: Contains background dump trace files, including the alert log and background process trace files.
- **cdump**: The core dump directory containing memory state core dump files, at the time of a database crash.
- **udump**: User dump destination containing trace files of user sessions connections to the database.

Trace files can be found in both the bdump and cdump directories. The values for these directories, and where trace files are sent to from an operating database, are determined by the BACKGROUND_DUMP_DEST and USER_DUMP_DEST parameters, as shown in Figure 5-20. The bdump directory contains trace files produced by background processes. Figure 5-21 shows a Windows Explorer picture of background process trace files. The files are named in relation to the Oracle background processes they represent. Figure 5-22 shows the udump directory, containing trace files, all for specific session server processes.

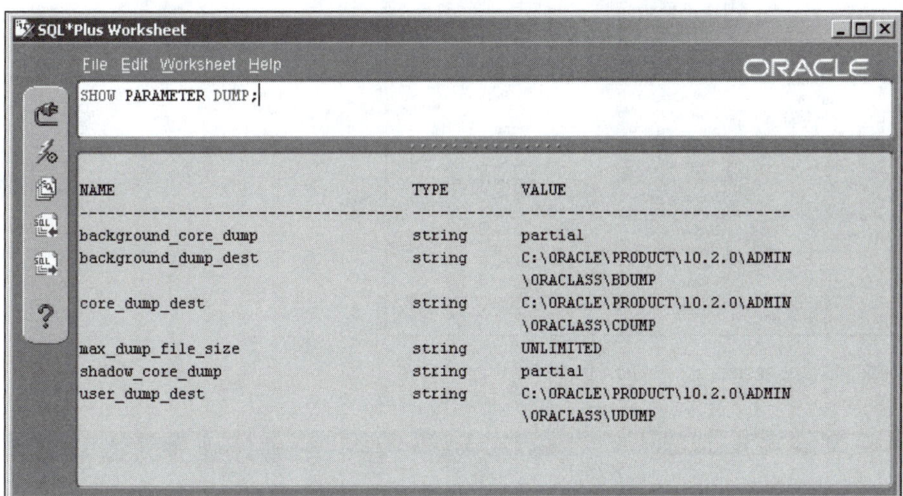

FIGURE 5-20 Trace file locations are determined by configuration parameters

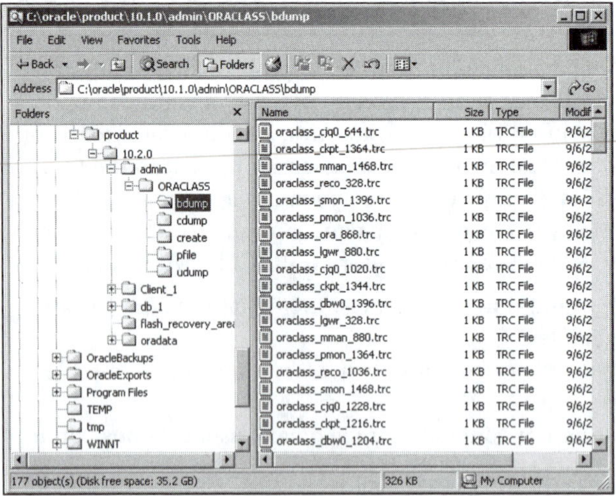

FIGURE 5-21 Background process trace files

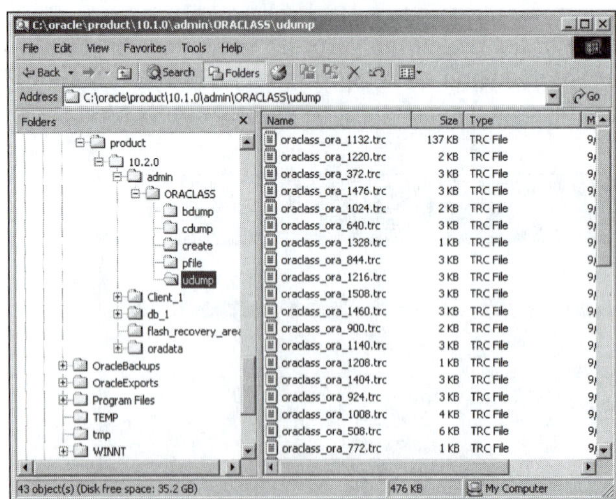

FIGURE 5-22 User session connection trace files

NOTE

I call the udump directory session server process files because in shared server mode, different user connections are shared between different shared server processes. Thus the same connection could have trace information in multiple trace files.

So what is in the trace files and how can all that information be used? When a database is running, it may encounter errors caused by problems in a user's session or problems in the background processes. The database logs errors, warnings, or major events that occur in the database.

Trace files are automatically created and written to by the database. The alert log file is created (if it does not already exist) when the database starts up. The other two types of files are created as events occur. Trace files are not only for logging errors. For example, the alert log contains details of when a database is started up and shut down.

Apart from the BACKGROUND_DUMP_DEST and USER_DUMP_DEST parameters, there are other trace configuration parameters that should be included in this discussion:

- **MAX_DUMP_FILE_SIZE:** Maximum size (in kilobytes, megabytes, or operating system blocks) of trace files. Specify UNLIMITED (the default) to place no limits on trace file size. Change the size for your session with ALTER SESSION or for the running database with ALTER SYSTEM. If a trace file reaches the maximum, it is no longer appended to (tracing is lost). The alter log file is not controlled by this parameter.
- **SQL_TRACE:** Setting of TRUE means that all user sessions have trace files generated with information on their SQL activities. FALSE means no user sessions have trace files generated. The default is FALSE. Be cautious when setting this parameter to TRUE, because it generates numerous files quickly. It is preferable to use the ALTER SESSION command to turn on tracing for specific sessions one at a time.
- **TIMED_STATISTICS:** Setting of TRUE means that SQL trace functions can track the timing of SQL, such as CPU time and elapsed time. FALSE means that timed statistics are not gathered.
- **STATISTICS_LEVEL:** The level of statistics parameter determines the level of statistics gathered for all the trace activities in the database. The BASIC level suppresses all statistics. The default level is TYPICAL, which includes timed statistics and advisories from many processes, such as buffers and shared pool sizing. ALL includes everything in TYPICAL, plus timed operating system statistics and row source execution statistics.

The alert log file is named **alert.log** and resides in the directory that is defined by the **BACKGROUND_DUMP_DEST** parameter. If you do not define BACKGROUND_DUMP_DEST, the **alert.log** file is placed in a default directory that is operating system dependent. The purpose of the alert log is to record messages that you ordinarily see if you are the systems operator watching the server's monitor. The information is recorded in the alert log so it can be reviewed at any time. The information is written in chronological order and consists of informative messages, as well as certain types of serious errors in the alert log:

- Internal errors, deadlock errors, and block corruption errors
- DBA commands, such as ALTER DATABASE, STARTUP, SHUTDOWN, CREATE TABLESPACE, and so on
- Certain error messages about dispatcher processes
- Errors found during automatic refresh of materialized views

Figure 5-23 shows an example of the contents of an alert log file.

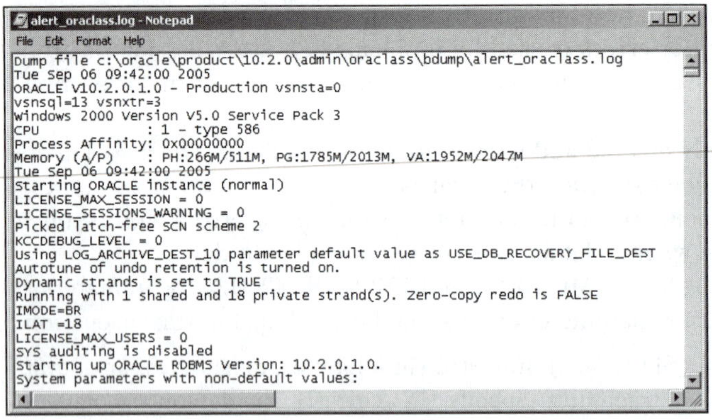

```
alert_oraclass.log - Notepad
File Edit Format Help
Dump file c:\oracle\product\10.2.0\admin\oraclass\bdump\alert_oraclass.log
Tue Sep 06 09:42:00 2005
ORACLE V10.2.0.1.0 - Production vsnsta=0
vsnsql=13 vsnxtr=3
Windows 2000 Version V5.0 Service Pack 3
CPU             : 1 - type 586
Process Affinity: 0x00000000
Memory (A/P)    : PH:266M/511M, PG:1785M/2013M, VA:1952M/2047M
Tue Sep 06 09:42:00 2005
Starting ORACLE instance (normal)
LICENSE_MAX_SESSION = 0
LICENSE_SESSIONS_WARNING = 0
Picked latch-free SCN scheme 2
KCCDEBUG_LEVEL = 0
Using LOG_ARCHIVE_DEST_10 parameter default value as USE_DB_RECOVERY_FILE_DEST
Autotune of undo retention is turned on.
Dynamic strands is set to TRUE
Running with 1 shared and 18 private strand(s). Zero-copy redo is FALSE
IMODE=BR
ILAT =18
LICENSE_MAX_USERS = 0
SYS auditing is disabled
Starting up ORACLE RDBMS Version: 10.2.0.1.0.
System parameters with non-default values:
```

FIGURE 5-23 Statistics and tracing must be adjusted to gather information on SQL tasks

The background trace files are created by each of the background processes if the process encounters an error. The file names always contain the name of the background process. For example, a trace file generated by the LGWR process might be named **oraclass_lgwr_1024.trc**. Trace files are located in the directory you specify in the BACKGROUND_DUMP_DEST initialization parameter. Monitor the trace files to determine if the background processes contain problems. For example, if the log writer background process (LGWR) found a corrupted member of a redo log group, it would write that error to the trace file.

The utility TKPROF reads trace files and loads the data into a readable file or even loads it into a database table so you can query the results. Read about TKPROF in the online documentation book titled *Performance Tuning Guide.*

Perform the following steps to view the current initialization parameters related to the trace files:

1. Start up the SQL*Plus Worksheet. In Windows, click **Start** on the Taskbar, and then click **All Programs/Oracle ... /Application Development/SQLPlus Worksheet**. In UNIX, type **oemapp worksheet** on the command line. The standard Enterprise Manager login screen appears.
2. Type **SYSTEM** in the Username box, the current password for SYSTEM in the Password box, **ORACLASS** in the Service box, and leave the Connect as box displaying "Normal". Click **OK** to continue. The SQL*Plus Worksheet window displays.
3. Type and execute these commands to check the current setting of statistics and the location of trace files:

   ```
   SHOW PARAMETERS STATISTICS
   SHOW PARAMETERS DUMP
   SHOW PARAMETERS TRACE
   ```

   Figure 5-24 shows the results. Your results will be similar, but the directory names may be different.
4. Log off SQL*Plus Worksheet by typing and executing **EXIT**. Then click **OK** in the dialog box.

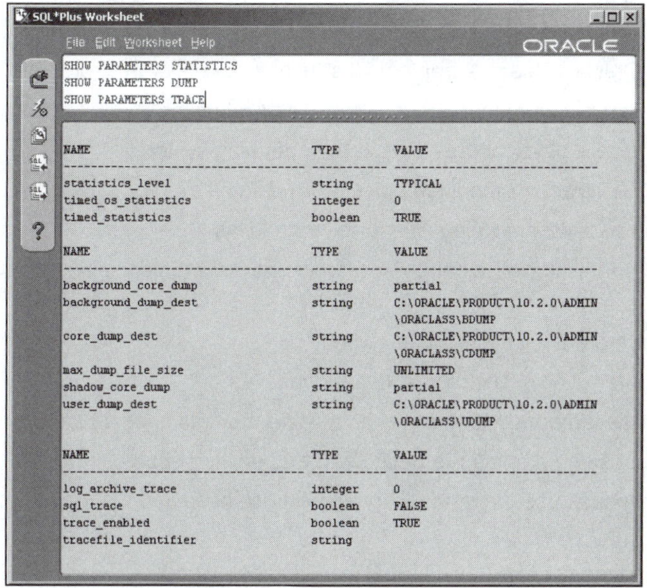

FIGURE 5-24 Statistics, dump, and trace configuration parameters

5. Locate the directory in which background trace files are located. Examine the files in the directory. Note the file names, which tell you which process had errors or warnings.

Both the Enterprise Manager Console and the Database Control contain some excellent tools for gathering and reporting of statistics. Third-party software is also available for diagnostics on Oracle databases.

TIP

Keep an eye on the **alert.log** file for early warnings of problems.

Chapter 6 will examine the basics of querying the database, which is essential to many database administration tasks.

Chapter Summary

- The internal structures of Oracle 10*g* are either logical structures or physical structures.
- Physical structures have a physical presence in the operating system.
- Logical structures do not have a physical presence in the operating system.
- Logical structures cannot be viewed or modified outside the database.
- Physical structures include datafiles, redo log files, and control files.
- Logical structures include data blocks, extents, segments, schema objects, and tablespaces.
- A tablespace has one or more datafiles associated with it.
- Permanent tablespaces store objects, such as tables and indexes.
- Temporary tablespaces store temporary tables and data while the data is being sorted.
- Locally managed tablespaces use a bitmap to track used and unused space.
- Dictionary-managed tablespaces use the data dictionary to track used and unused space.
- The NOLOGGING setting does not log mass INSERT or DDL commands.
- To use OMF, omit the datafile name and (optionally) include the size of the file.
- The REUSE setting allows Oracle 10*g* to reuse an existing file, erasing all its data.
- The AUTOEXTEND ON setting gives a datafile the ability to add to its size automatically.
- Oracle recommends using locally managed tablespaces.
- Dictionary-managed tablespaces tend to waste storage space and perform poorly in comparison to locally managed tablespaces.
- Adjacent, free extents can be manually coalesced in a dictionary-managed tablespace.
- Free extents are automatically coalesced in a locally managed tablespace.
- The MINIMUM EXTENT setting overrides a smaller extent size specified by an object in the tablespace.
- Set PCTINCREASE to zero to keep extent sizes more uniform.
- Dictionary-managed tablespaces use a freelist in the dictionary to track blocks.
- Locally managed tablespaces keep all extents either the same size or a variable size that is controlled by the system.
- The SEGMENT SPACE MANAGEMENT setting is only used for locally managed tablespaces.
- The four types of segments are: data, index, temporary, and rollback.
- Temporary tablespaces should be locally managed, although it is possible to create a dictionary-managed temporary tablespace.
- Before creating a tablespace with a nonstandard data block size, you must create a cache with the corresponding data block size.
- Many of the initial settings of a tablespace can be changed using the ALTER TABLESPACE command.

- You cannot change a tablespace from LOCAL or DICTIONARY mode.
- Changing DEFAULT STORAGE settings does not affect existing blocks, only those created after the change.
- To block access, a tablespace can be changed from ONLINE to OFFLINE.
- Locally managed, temporary tablespaces cannot be taken offline.
- Taking a tablespace offline can be done in NORMAL, TEMPORARY, or IMMEDIATE mode.
- Media recovery is usually needed with any mode but NORMAL.
- A read-only tablespace is not included in regular backups or recoveries.
- Dropping a tablespace with the INCLUDING CONTENTS clause destroys all its data.
- DBA_FREE_EXTENTS tells how much free space is available in a tablespace.
- Undo data allows users to have read consistency, while other users make changes that are not yet committed.
- Manual undo management using manually specified rollback segments is still available but no longer supported in Oracle 10*g*.
- Automatic undo management mode is easier to manage than manual rollbacks, and requires a special undo tablespace.
- Undo blocks reside in undo extents within either an undo tablespace or a rollback segment, depending on the undo management mode.
- The alert log file should be monitored regularly to detect errors.
- The background trace files record errors that occur within the background processes.
- The user trace files log errors from user session connections and transactions executed by users.
- BACKGROUND_DUMP_DEST and USER_DUMP_DEST initialization parameters set the location of the background trace files and the user trace files respectively.

Review Questions

1. Place these in order from smallest to largest.
   1. Segment
   2. Block
   3. Extent
   4. Tablespace

   a. 2, 1, 3, 4

   b. 2, 3, 1, 4

   c. 1, 2, 3, 4

   d. 4, 1, 3, 2
2. Extent management only pertains to locally managed tablespaces. True or False?
3. A logical structure always resides in one physical structure. True or False?

4. Explain the circumstance when the initial storage created for a table would *not* be made up of contiguous blocks.
5. One segment is always created for each. Choose two.
   a. Sequence
   b. Partition
   c. Unpartitioned table
   d. Extent
6. A schema object spans multiple datafiles. Explain how this happens.
7. ________________ extents are automatically coalesced in ________________ managed tablespaces.
8. To create a locally managed tablespace with all extents 3M in size, use the clause EXTENT MANAGEMENT LOCAL ________________ .
9. Undo extents should be stored in an ________________ .
10. Review the following command, and fill in the missing words.

    CREATE TEMPORARY TABLESPACE USERTEMP

    ________________ 'D:\oracle\data\usertemp01.dbf' SIZE 50M EXTENT MANAGEMENT LOCAL ________________ SIZE 2M
11. Any temporary tablespace can be taken offline, at any time. True or False?
12. To find the names of the tempfiles in a temporary tablespace, query the ________________ view.
13. Set the location of the alert log file with the ________________ parameter.

Exam Review Questions—Oracle Database 10*g*: Administration (#1Z0-042)

1. Look at the following command.

```
CREATE TABLESPACE USER_NEWDATA
DATAFILE SIZE 400M
EXTENT MANAGEMENT DICTIONARY
DEFAULT STORAGE (INITIAL 40M NEXT 20M PCTINCREASE 50);
```

   Which one of the following statements is true regarding the previous command?
   a. The tablespace has one Oracle managed datafile.
   b. An object created in this tablespace must have a 40M initial size.
   c. An object created in this tablespace with no storage settings has 80M after three extents.
   d. An object created in this tablespace with no storage settings has 90M after three extents.
   e. The tablespace uses a bitmap to track free space.
   f. None of the above.

2. Study the following command.

```
CREATE TABLESPACE ANYOBJECTS
DATAFILE 'D:\oracle\storage\anyobjects01.dbf' SIZE 500M
SEGMENT SPACE MANAGEMENT MANUAL;
```

Which statement is true about this command?

   a. The datafile is not an Oracle managed file.
   b. The statement will fail, because it is missing the DEFAULT STORAGE clause.
   c. The statement will fail because the EXTENT MANAGEMENT LOCAL clause is missing.
   d. All of the above.

3. The command ALTER TABLESPACE USER_TEST OFFLINE NORMAL has failed. What do you do?
   a. Attempt the command using TEMPORARY instead of NORMAL.
   b. Shut down and restart the database, and then take the tablespace offline.
   c. Drop the tablespace and re-create it.
   d. Back up the tablespace, and then drop and restore it.
4. The following command was issued on the tablespace containing the ZIP_CODE table.

```
DROP TABLESPACE USER_REF;
```

What happens?

   a. The tablespace and the ZIP_CODE table are dropped.
   b. The command fails because the tablespace contains data.
   c. The command fails because the ZIP_CODE table was not listed in the command.
   d. The tablespace is dropped, and all references to the ZIP_CODE table are removed.
5. Which statements are true about dictionary-managed tablespaces? Choose two.
   a. They store data with less wasted space than locally managed tablespaces.
   b. Changes cause dictionary tables to be updated.
   c. Extents are tracked in freelists, in metadata, in the SYSTEM tablespace.
   d. They are commonly utilized in earlier versions of Oracle, and are out of date in favor of locally managed tablespaces.

Hands-on Assignments

1. Open the file on your student disk, in the **Data\Chapter05** directory, named **ho0501.txt**. Replace all variables found in the file with valid values so that the command creates a 5Mb locally managed tablespace called **TEST**. The tablespace has one datafile named **C:\oracle\product\10.2.0\oradata\oraclass\test01.dbf**. Set up automated allocation of extents and automatic segment space management. Save the finished file to **ho0501.sql**, in the **Solutions\Chapter05** directory, on your student disk.

2. Create a SQL script that completes the following tasks successfully. Save the script in a file named **ho0502.sql** in the **Solutions\Chapter05** directory on your student disk.
   a. Create a locally managed tablespace named HANDS_62 with one 10Mb datafile in a temporary directory, such as C:\TEMP in Windows, or /tmp in Unix or Linux.
   b. Add another 5Mb datafile to the tablespace.
   c. Query the Data Dictionary, and display the total available space in the tablespace.
   d. Make the tablespace read only, and take it offline.
   e. Drop the tablespace.

3. Now repeat the same question again, as above, but using an OMF (Oracle Managed Files) configuration, which will delete the datafiles for you when dropping the tablespace. Save the script in a file named ho0502.sql in the Solutions\Chapter05 directory on your student disk.
4. One of your tablespaces seems to have run out of space long before you calculated that it would. There has been a lot of activity in which tables were created, then dropped, and recreated with different storage settings. Explain why your tablespace has a problem, and list two actions you can take to alleviate the problem. Save the answer in a file named **ho0504.txt** in the **Solutions\Chapter05** directory on your student disk.
5. Explain the difference between an extent, a segment, and a table. Give an example of how the three are related. Save your answer in a file named **ho0505.txt** in the **Solutions\Chapter05** directory on your student disk.
6. You have a tablespace that was created with this command:

```
CREATE TABLESPACE CUSTOMERSERVICE
DATAFILE 'C:\temp\cust_serv01.dbf' SIZE 50M
AUTOEXTEND OFF;
```

   Write a SQL script to accomplish these tasks:
   a. Add another file that is identical (except the name) to the original one.
   b. Make the first datafile 25Mb in size.
   c. Make the second datafile expandable forever.

   After you have made the script, write a short explanation of what you believe are the pros and cons of using an unlimited datafile. Save the script and your explanation in a file named **ho0506.txt** in the **Solutions\Chapter05** directory on your student disk.
7. Write a command to create a temporary tablespace with 20Mb storage space. The file must be an Oracle Managed file. Write a query that shows the file name and size. Save the script and the query in a file named **ho0507.sql** in the **Solutions\Chapter05** directory on your student disk.
8. Write a query to determine your current, temporary tablespace when you are logged on to SQL*Plus Worksheet as SYSTEM. Write another query to show the current space used and free space in all temporary tablespaces. Save the script and the query in a file named **ho0508.sql** in the **Data/Chapter05** directory on your student disk.
9. Display the undo-related initialization parameters. Change the retention time to 10 minutes. Write a query using DBA_UNDO_EXTENTS to summarize the used blocks in the undo tablespace. Save your script in a file called **ho0509.sql** in the **Solutions\Chapter05** directory on your student disk.

Case Projects

1. Now that you have read all about locally managed tablespaces, you decide your Global Globe database should use exclusively locally managed tablespaces. Write a query to determine what management mode the SYSTEM tablespace uses. Save the query in your **Solutions\Chapter05** directory in a file named **case501.sql**. Imagine that the SYSTEM tablespace is not locally managed. How will you resolve this?

   Your MIS manager may be dubious about your wish to drop and recreate the database, even though the programming staff has not begun using it yet. Write a paragraph justifying your choice to convince your boss and the Global Globe steering committee that you are right. Save your answer in your **Solutions\Chapter05** directory in a file called **case501.txt**.

2. Create three new tablespaces for the accounting, sales, and training departments. Write SQL commands to create tablespaces as locally managed, automatic extent allocation, and automatic segment space management. Locate datafiles in your ORACLASS database. Do not use OMF datafiles. All tables will have an initial volume of 5Mb. Maximum volume is irrelevant because disk space is inexpensive. Save your script in a file called **case0502.sql** in the **Solutions\Chapter05** directory on your student disk.

CHAPTER 6

THE BASICS OF QUERYING A DATABASE

LEARNING OBJECTIVES

In this chapter, you will:

- Learn about different types of queries
- Cover basic SQL functions and pseudocolumns available in Oracle database
- Discover facts about NULL values, the DUAL table, and the DISTINCT clause
- Learn about filtered, sorted, and aggregated queries
- Discuss advanced queries including joins, subqueries, and other specialized queries

INTRODUCTION

Previous chapters have focused on creating the database itself. This chapter deals exclusively with retrieving both metadata and applications (user) data from a database. The subject matter of this chapter is not included in the first Oracle certification examination for database administration. However, getting information from a database, both from system database and user application structures, is essential for proper database administration. Even database administrators know the basics of how to read tables in an Oracle database. You simply cannot function properly as a database administrator if you do not know some of the most basic SQL functionality availability in Oracle SQL.

Much of what is in this chapter has already been seen in previous chapters of this book. One of the primary purposes of this chapter is to centralize all the necessary facts with respect to querying a database. There is even some repetition across different sections within this chapter itself, solely in the interest of retaining every fact and facet of a SQL command within the scope of each section.

This chapter begins with a foundational introduction to the basic types of queries, building with more complexity as you read through it.

INTRODUCTION TO QUERIES

A query, or database query, uses a special type of SQL command called a SELECT statement. The SELECT statement allows you to specify tables and columns in the tables, from which data is selected. There are numerous different types of queries available in an Oracle database.

Different Types of Queries

The different types of SELECT statement queries allowed in an Oracle database are listed in the following section. To execute these examples yourself, you can start SQL*Plus Worksheet by clicking **Start/All Programs/Oracle ... /Application Development/SQL*Plus Worksheet** (or execute from the console). Connect in SQL*Plus Worksheet as the SYS user.

- **Basic query**: Basic queries simply retrieve rows. This query retrieves all tablespace names:

```
SELECT TABLESPACE_NAME FROM DBA_TABLESPACES;
```

- **Filtered query**: The WHERE clause retains or filters out unwanted rows. This query finds only tablespaces, where the name of the tablespace begins with a capital S:

```
SELECT TABLESPACE_NAME FROM DBA_TABLESPACES
WHERE TABLESPACE_NAME LIKE 'S%';
```

NOTE

Oracle 10*g* uses % to represent any number of characters and _ (an underbar or underscore character) to represent a single character. % and _ (underbar) are known as wildcard characters. For example, 'S%' will find both SYS and SYSTEM but not the USERS or TEMP tablespaces.

- **Sorted query**: The ORDER BY clause returns rows in a specified order. This query returns all the tablespace names, sorted in alphabetical order:

```
SELECT TABLESPACE_NAME FROM DBA_TABLESPACES
ORDER BY TABLESPACE_NAME;
```

- **Aggregated query**: Aggregated queries create groupings or summaries of larger row sets, producing, for example, a sum of values or a number of subtotals from all values. Aggregation is best demonstrated by example. This query returns the total number of tables within each tablespace, rather than one row for every single table, as shown in Figure 6-1.

```
SELECT COUNT(TABLE_NAME), TABLESPACE_NAME
FROM DBA_TABLES GROUP BY TABLESPACE_NAME;
```

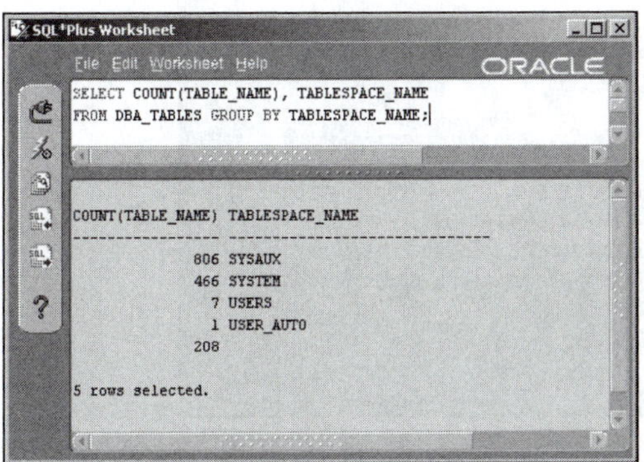

FIGURE 6-1 The GROUP BY clause can aggregate rows into a smaller number of summary rows

- **Join query**: Merge rows from one or more tables, usually linking rows based on related values. This query will return column values from two tables, joined based on a common value, as shown in Figure 6-2:

```
SET WRAP OFF
COL TABLESPACE_NAME FORMAT A10
COL FILE_NAME FORMAT A64
SELECT TABLESPACE_NAME, BLOCK_SIZE, FILE_NAME
FROM DBA_TABLESPACES JOIN DBA_DATA_FILES
USING (TABLESPACE_NAME);
```

NOTE

The WRAP (text wrapping) and COL (column) commands in the query above format columns to 10 and 64 characters in width respectively. Further details can be found for SQL*Plus tools formatting in the SQL*Plus Quick Reference, and the SQL*Plus User's Guide and Reference, in Oracle documentation.

- **Subqueries**: These queries are executed within other queries. One query calls another query, returning results to the calling query (a SELECT statement executed within another SELECT statement). This query is similar to the join query shown in Figure 6-2, except that the name of the datafile is excluded in this query:

```
SELECT TABLESPACE_NAME, BLOCK_SIZE
FROM DBA_TABLESPACES
WHERE TABLESPACE_NAME IN
(SELECT TABLESPACE_NAME FROM DBA_DATA_FILES);
```

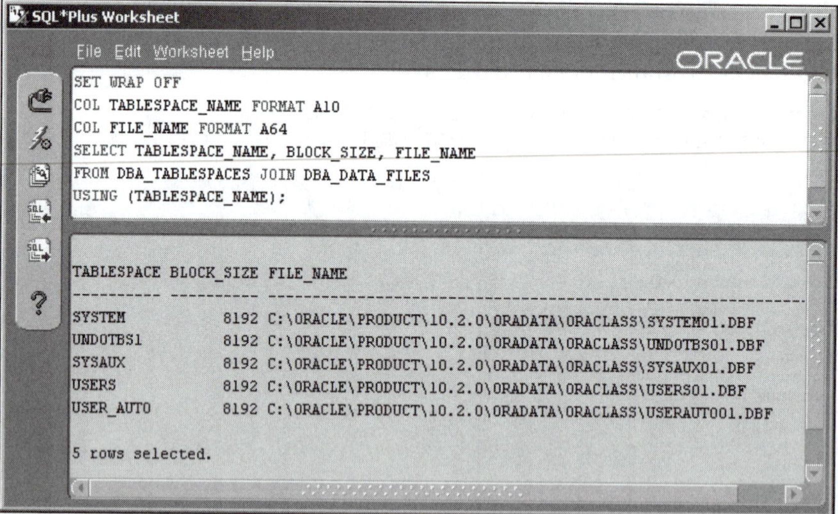

FIGURE 6-2 Tables can be joined using common element columns

NOTE

Certain types of subqueries are also known as semi-joins because they join without returning values from the subquery.

- **Create new table or view**: New tables and views can be created from the results of a query:

```
CREATE VIEW TABLESPACES AS
SELECT TABLESPACE_NAME, BLOCK_SIZE, FILE_NAME
FROM DBA_TABLESPACES JOIN DBA_DATA_FILES
USING (TABLESPACE_NAME);
```

- The view can be accessed just as if it were a table, producing the same result as shown in Figure 6-2:

```
SELECT * FROM TABLESPACES;
```

- **Other specialized queries**: These types of queries include composite queries, hierarchical queries, flashback or version queries, and parallel execution queries. Composite queries use special set operators, such as UNION, to concatenate the rows of two different queries together, into a single resulting set of rows. Hierarchical queries are used to build tree-like hierarchical output row structures from hierarchical data. Flashback or versions queries allow access to data at a previous point in time. Parallel queries execute SQL statements in parallel, preferably using multiple CPU platforms and Oracle partitions.

Oracle SQL Functions and Pseudocolumns

- Oracle SQL allows full expression-based use of both provided (built-in) and user-definable functions. Functions can be loosely classified in the following manner:
  - **Single row functions**: These functions operate on one row at a time. The following example will return the first character for the name of each tablespace:

    ```
    SELECT SUBSTR(TABLESPACE_NAME, 1, 1) FROM DBA_TABLESPACES;
    ```

  - **Datatype conversion functions**: These functions convert values such as numbers to strings. The following example returns a non-standard formatted date as shown in Figure 6-3.

    ```
    SELECT GROUP#, SEQUENCE#,
    TO_CHAR(FIRST_TIME, 'DAY MONTH YEAR') FROM V$LOG;
    ```

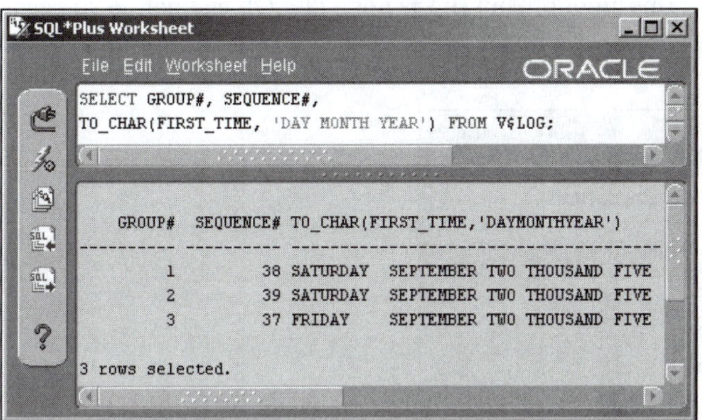

FIGURE 6-3 Oracle database contains many built-in functions for use in queries

  - **Group (aggregate) functions**: These functions apply specific functionality to grouping and summary queries. Some example group functions are:
    - **MIN**: Returns the lowest value from a group of values.
    - **MAX**: Returns the highest value from a group of values.
    - **AVG**: Returns the average of a group of values.
    - **SUM:** Adds values across rows in a group.
    - **COUNT**: Counts the number of rows in a group.
- This query returns a single number as a total (or sum) of all blocks, used by all datafiles in the database:

  ```
  SELECT SUM(BLOCKS) FROM DBA_DATA_FILES;
  ```

- **User-defined functions**: PL/SQL can be used to custom write functions for specific tasks, which are not available in functions provided in an Oracle database.

In the following example, a function is created to convert a phone number to a more readable formatted string:

```
CREATE OR REPLACE FUNCTION GETPHONE(pPHONE IN VARCHAR2)
RETURN VARCHAR2 IS
BEGIN
     RETURN '('||SUBSTR(pPHONE,1,3)||')'||SUBSTR(pPHONE,4,8);
EXCEPTION WHEN OTHERS THEN
     RETURN NULL;
END;
/
```


- Simply copy and paste the above function creation code into SQL*Plus Worksheet and execute. You should get a response back stating "Function Created." Next, execute a query containing the function to convert phone numbers as shown in Figure 6-4. You can be connected to the database as the SYS user to execute the query in Figure 6-4, as the CLASSMATE user is referred to. You can also execute connected as the CLASSMATE user. Not all phone numbers need conversion by this function, but this is not a PL/SQL manual, so you can ignore those details.

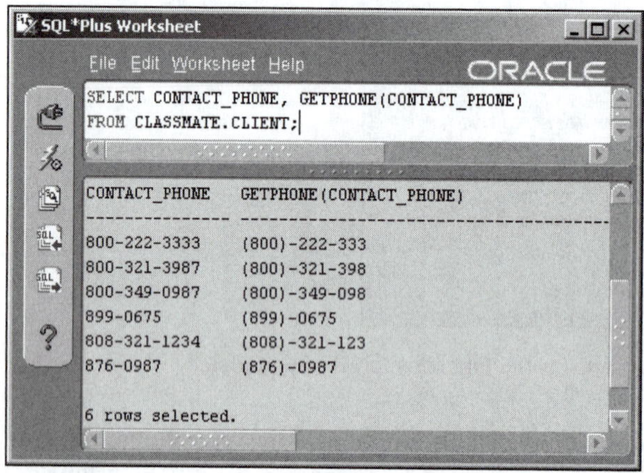

FIGURE 6-4 Users can custom build any function using PL/SQL, for embedding into queries

A pseudocolumn is just that, a "make-believe" column. A pseudocolumn is not a function, but can be added to a table query or into other SQL commands to retrieve values not normally accessed from simple table queries. Pseudocolumns contain values present in an Oracle instance, but not stored in a table. For example, in the query shown in Figure 6-5 the ROWNUM column returns the row number sequence, as each row is retrieved. The ROWID column returns the internal Oracle database row identifier pointer for each row.

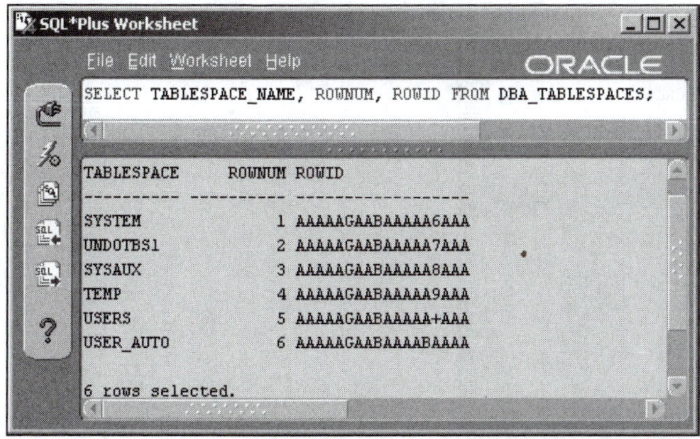

FIGURE 6-5 Pseudocolumns are used to retrieve information from Oracle database, not stored in user tables

What is NULL?

It is very important in any database, and in any programming language, to understand the concept of what a NULL value is. NULL means nothing. NULL is not the same as zero or a space character. Typically a NULL value is an uninitialized or never before accessed variable. Within the realm of a database a NULL valued column, in a row, in a table, is a column, which has never been set to anything (not yet accessed), or has been deliberately set to NULL. Leaving a column value set to NULL saves space.

There are some things that are important to remember about NULL values in an Oracle database:

- NULL represents nothing, not a space, not a zero (0), or even an unknown value; simply nothing.
- Just to reiterate, because it is important, a space character or a 0 value are not NULL.
- NULL values are not included in the most commonly used indexes for an Oracle database, binary tree (BTree) indexes.
- Most Oracle database built-in functions will return a NULL value when passed a NULL value.
- NULL can be tested for using the IS [NOT] NULL conditional operator. In the example shown in Figure 6-6 only rows with NULL valued PCT_INCREASE column values are returned. Locally managed tablespaces do not require an extent PCT_INCREASE increase setting, unless the tablespace is a TEMP (temporary on disk sorting) tablespace; which by default is actually set to zero.
- Any expression containing a NULL value always returns a NULL value.
- The NVL({value}, {replace}) function replaces NULL values in expressions, avoiding SQL errors. The SET NULL environment variable does the same thing in SQL*Plus tools. An example of the NVL function is shown in Figure 6-7 where any NULL valued PCT_INCREASE column values are returned as zero, which is not technically correct but demonstrative of NVL functionality.

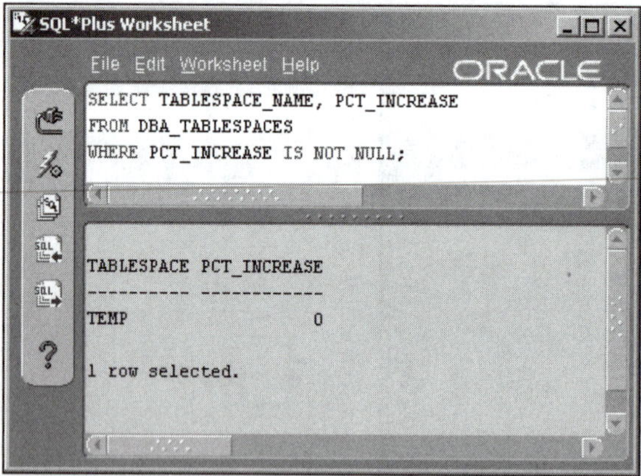

FIGURE 6-6 Filter NULL values with the IS [NOT] NULL operator

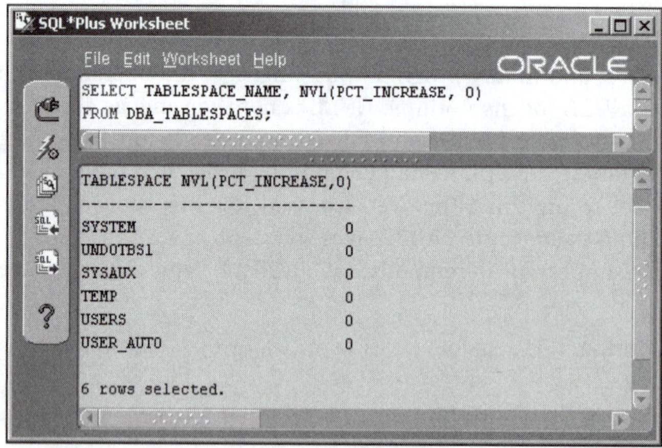

FIGURE 6-7 The NVL function replaces NULL values with a compatible datatype value

- NULL values sort as the highest value by default. Thus in an ascending sort, NULL values will be returned last. It thus follows that in a descending sort NULL values will appear last as shown by the example in Figure 6-8.
- Comparison of a NULL value to any other value will always evaluate to false. For example, if the salary of hourly employees in a table of employees is NULL, then IF SALARY = 0 and IF SALARY <> 0 will both return a false result.

There are some circumstances where NULL values can be used appropriately, for example, in tables where not all field values are known. In a table of employees some employees might be salaried workers and others hourly employees. Thus salaried workers would have a NULL valued hourly rate, and hourly employees a NULL valued salary.

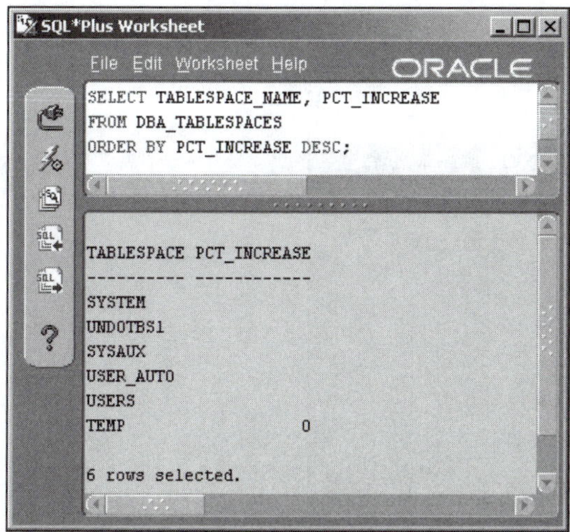

FIGURE 6-8 NULL values appear last on an ascending sort, and first on a descending sort

The DUAL Table and More on Pseudocolumns

The DUAL table is used to request data from an Oracle database, where that data is not in user application tables. The DUAL table is a little like an internal Oracle database cursor. A cursor is a chunk of memory allocated to a query, to contain the results of a query during processing of that query. The DUAL table can only be queried, never updated. The DUAL table is owned by SYS but can be queried by any user.

Figure 6-9 shows two facts about the DUAL table. The first fact is a description of the DUAL table using the DESC command in SQL*Plus Worksheet. You can see that the only column present is called DUMMY (a dummy variable to contain whatever value is retrieved into it). The second fact is a query of the DUAL table showing the DUAL table contains a single row, with the dummy value X, in the DUMMY column.

As already stated, the DUAL table is useful for retrieving values from an Oracle database, where those values are not found in user application tables. For example, to find the current date set within your Oracle database, use the following command:

```
SELECT SYSDATE FROM DUAL;
```

To find the currently logged in user name and his (or her) user identifier use the following command:

```
SELECT USER, UID FROM DUAL;
```

Another simple use of the DUAL table would be to return constant values, as in the following command, whose result is shown in Figure 6-10:

```
SELECT 'This is a string', 999 FROM DUAL;
```

One of the most common uses of the DUAL table is to get the next sequential value from an Oracle sequence object. Sequences are automated counters, counting from a specific value, up to a specified maximum (or an enormous value with no maximum specified). As

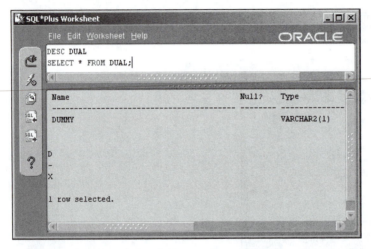

FIGURE 6-9 The DUAL table contains a single column and a single row

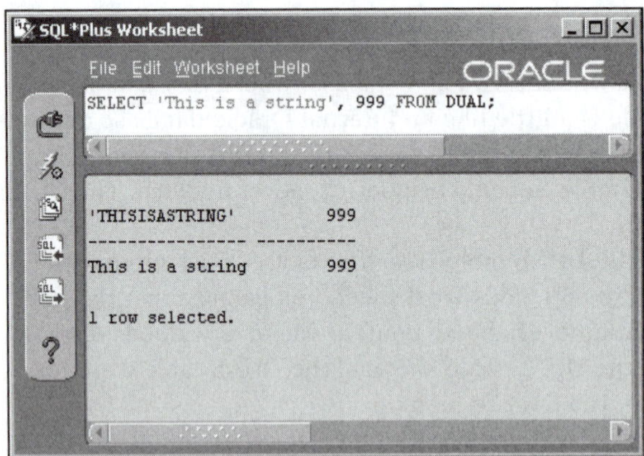

FIGURE 6-10 The DUAL table can be used to return a constant value, literal or numeric

an exercise, create a simple sequence object, whose first value is 1, and each subsequent value will be increased by 1:

```
CREATE SEQUENCE SIMPLE_SEQ START WITH 1 INCREMENT BY 1;
```

Now execute the following DUAL table query a number of times and view the results, as shown in Figure 6-11:

```
SELECT SIMPLE_SEQ.NEXTVAL FROM DUAL;
```

NOTE

A likely use of sequences is for surrogate primary key generation.

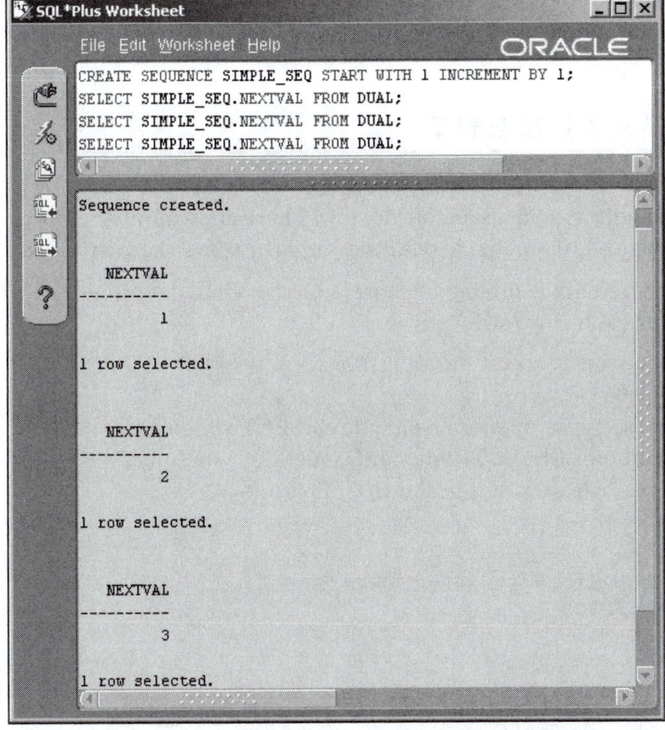

FIGURE 6-11 Get sequence value increments using a DUAL table query

The DISTINCT Clause

The DISTINCT clause is a special SQL clause used to retrieve the first value of each group in a set of duplications. DISTINCT can operate on a single column or multiple columns, where syntax for the DISTINCT clause is as follows:

```
SELECT DISTINCT [(]{column} [, column ... ] ... [)] ...
```

If necessary, connect back to the SYS schema in your SQL*Plus Worksheet window. The DBA_TABLES view under the SYS user finds many tables. If you want to find only user names with tables created in the database, under their user names, you could find that like this:

```
SELECT DISTINCT OWNER FROM DBA_TABLES;
```

Now find the unique tablespace names and owners, with a slightly more complex query:

```
SELECT DISTINCT TABLESPACE_NAME, OWNER FROM DBA_TABLES;
```

You can also use parentheses to find unique values of a single string, as opposed to two separate columns:

```
SELECT DISTINCT (TABLESPACE_NAME||OWNER) FROM DBA_TABLES;
```

The query above gives exactly the same result as the previous two-column query.

That introduces Oracle database querying and some simple tools to help in your quest for the perfect SELECT statement. Next you will examine the SELECT statement in further detail.

THE SELECT STATEMENT

The SELECT statement or clause is the beginning of the SQL command for querying (retrieving) data from a database object, such as a table or view. There are a number of ways that a SELECT statement can be used in an Oracle database, and these are the easy ones:

- **Simple query**: Retrieve data from one or more tables, pulling data from one or more columns for each row retrieved:

```
SELECT TABLESPACE_NAME, BLOCK_SIZE, INITIAL_EXTENT
FROM DBA_TABLESPACES;
```

- **Complex query**: These types of queries include SELECT statements that can be embedded within other SELECT statements, such as a subquery shown in the following example, whose result is shown in Figure 6-12:

```
SELECT
   (SELECT NAME FROM
    V$TABLESPACE WHERE TS# = D.TS#) AS DATAFILE,
    D.NAME AS TABLESPACE
    FROM V$DATAFILE D;
```

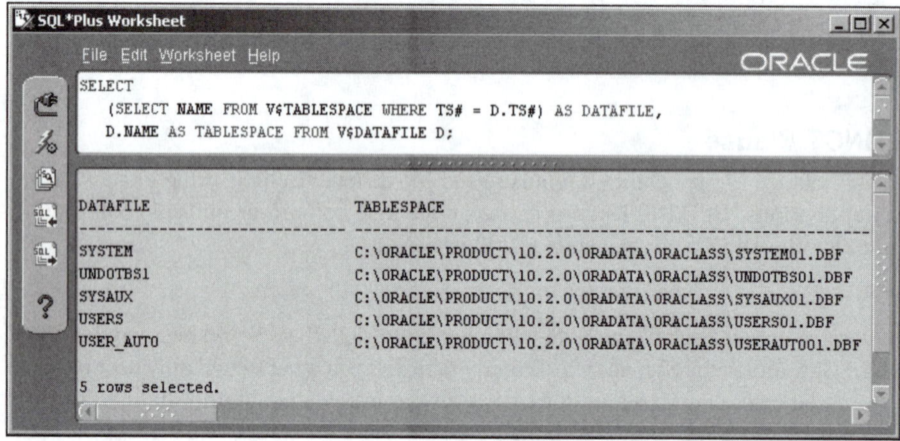

FIGURE 6-12 Complex queries retrieve data from more than one table, in many different ways

- **Create a view or table**: You can create a view or another table using a SELECT statement, which has already been seen in a previous chapter, shown for your convenience in the following two examples:

```
CREATE OR REPLACE VIEW CLASSMATE_TABLES_VIEW AS
SELECT TABLE_NAME FROM DBA_TABLES
WHERE OWNER='CLASSMATE';
```

NOTE

The OR REPLACE clause will overwrite the view if it already exists.

```
CREATE TABLE CLASSMATE_TABLES AS
SELECT TABLE_NAME FROM DBA_TABLES
WHERE OWNER='CLASSMATE';
```

- **Insert, update, or delete data**: As already seen in this chapter, SELECT statements can be embedded in DML commands as subqueries (within INSERT, UPDATE, and DELETE statements).

This is the basic syntax for an Oracle database SELECT statement, or query:

```
SELECT { [alias.]column | expression | [alias.]* [ , ... ] }
FROM [schema.]{table | view|...} [alias];
```

Now you get to see some more examples using more details of the syntax shown above. The easiest way for you to learn is to connect using SQL*Plus Worksheet and execute these examples as you read through to see what they do.

This query finds a single column:

```
SELECT TABLESPACE_NAME FROM DBA_TABLESPACES;
```

This query finds more than one column:

```
SELECT TABLESPACE_NAME, BLOCK_SIZE, INITIAL_EXTENT
FROM DBA_TABLESPACES;
```

This query uses the * (asterisk or star) character to find all columns:

```
SELECT * FROM DBA_TABLESPACES;
```

This query includes expressions on columns:

```
SELECT BYTES, BYTES/1024, BYTES/1024/1024, FILE_NAME
FROM DBA_DATA_FILES;
```

This query makes a little more sense out of the previous query by adding more appropriate headings to the different expression columns; the result is shown in Figure 6-13:

```
SELECT BYTES "Bytes", BYTES/1024 "Kb", BYTES/1024/1024 "Mb",
FILE_NAME "OSFile"
FROM DBA_DATA_FILES;
```

You can achieve the same result using the AS clause for each column:

```
SELECT BYTES AS Bytes, BYTES/1024 AS Kb, BYTES/1024/1024 AS Mb,
FILE_NAME AS OSFile
FROM DBA_DATA_FILES;
```

However, the AS clause can be used in the ORDER BY clause. So following on from the previous two examples, this query will not work because the AS clause should be used for the ORDER BY clause to work using the renamed column:

```
SELECT BYTES/1024 "Kb", FILE_NAME "OSFile"
FROM DBA_DATA_FILES
ORDER BY Kb;
```

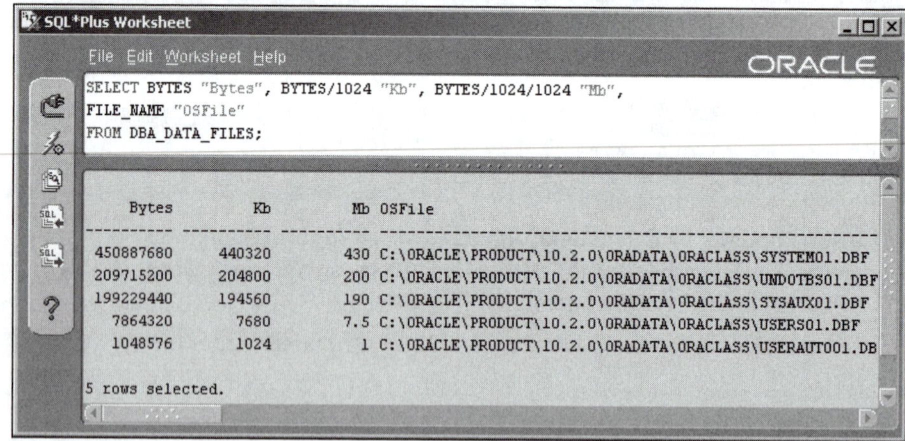

FIGURE 6-13 Column headers can be renamed, useful for expression columns in queries

On the contrary, this query will work using the ORDER BY clause because the AS clause is used to rename the column:

```
SELECT BYTES/1024 AS Kb, FILE_NAME "OSFile"
FROM DBA_DATA_FILES
ORDER BY Kb;
```

An alias can be used to refer to a column, typically in a join or a subquery, as shown in the following two queries. An alias can also act to apply a distinct name to multiple columns with the same name. This query is a join query:

```
SELECT T.NAME, D.NAME
FROM V$TABLESPACE T, V$DATAFILE D
WHERE T.TS# = D.TS#;
```

This query contains a subquery linking calling query and subquery using aliased columns:

```
SELECT T.NAME
FROM V$TABLESPACE T
WHERE EXISTS (SELECT * FROM V$DATAFILE WHERE TS# = T.TS#);
```

And you don't have to use aliases but can use the full table name (in this case a performance view), but this is not recommended as good programming practice:

```
SELECT V$TABLESPACE.NAME, V$DATAFILE.NAME
FROM V$TABLESPACE, V$DATAFILE
WHERE V$TABLESPACE.TS# = V$DATAFILE.TS#;
```

Use of upper- and lowercase is only a factor in queries for strings or expressions enclosed in Oracle database quotation marks. The following three queries are identical:

```
SELECT TABLESPACE_NAME, BLOCK_SIZE FROM DBA_TABLESPACES;
select tablespace_name, block_size from dba_tablespaces;
SELECT tablespace_name, block_size FROM dba_tablespaces;
```

Varying case of characters enclosed in quotation marks can, however, make a big difference, as shown in Figure 6-14.

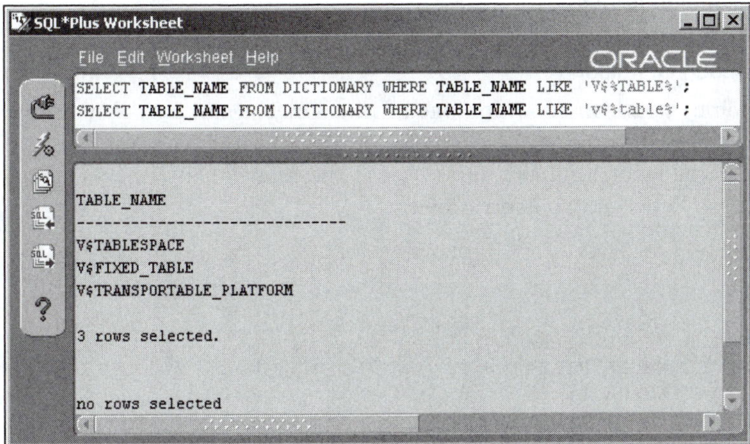

FIGURE 6-14 Upper- and lowercase characters enclosed within Oracle database string quotations can cause varying results

Next you will find out how to filter, sort, and summarize query results.

FILTERING, SORTING, AND SUMMARIZING (AGGREGATIONS) QUERIES

Filtering uses the WHERE clause to filter out unwanted rows, or retain wanted rows. Sorting allows resorting of query results using the ORDER BY clause. Queries can be summarized or aggregated into fewer rows using the GROUP BY and HAVING clauses.

Filtering Queries with the WHERE Clause

Expanding the syntax for the basic SELECT statement here you add the WHERE clause to the equation:

```
SELECT { [alias.]column | expression | [alias.]* [ , ... ] }
FROM [schema.]{table | view|...} [alias]
[
WHERE [schema.]table [alias] { column | expression }
comparison condition
[schema.]table [alias] { column | expression }
[ {AND | OR } [NOT] ... ]
]
Comparison conditions: expression = | > | < | >= | <= | LIKE | EXISTS | ...
 expression
AND and OR are logical operators used to combine multiple sets of expression
comparisons.
```

Once again the best way to learn about the WHERE clause is by example, and you can practice these queries in SQL*Plus Worksheet at the same time, and see the results. The WHERE clause has two distinct facets: (1) comparison conditions, and (2) logical operators. You begin with comparison conditions.

WHERE Clause Comparison Conditions

Here you go through each comparison condition one by one. Once again, try them out in SQL*Plus Worksheet as you read through the examples, just to see what the results are and to help you to absorb the facts. Equi (=), anti (!=, <>) and range (<, >, =<, >=) comparison conditions are used between two expressions:

```
expression [ = | != | > | < | >= | <= ] expression
```

For example:

```
SELECT * FROM V$DATAFILE WHERE FILE# = 1;
SELECT * FROM V$DATAFILE WHERE FILE# <> 1;
SELECT * FROM V$DATAFILE WHERE FILE# >= 5;
```

A subquery is also an expression so the following applies. This comparison condition allows a single row subquery only. A single row subquery returns a single row only:

```
(subquery) [ = | != | > | < | >= | <= ] (subquery)
```

LIKE pattern matches between strings; those strings are expressions. LIKE also requires single row subqueries if used with a subquery:

```
expression LIKE expression
```

For example:

```
SELECT * FROM DICTIONARY WHERE TABLE_NAME LIKE 'V$%A';
```

The percentage sign (%) and the underscore/underbar (_) characters are pattern-matching wildcard characters. A wildcard character can match any character. More specifically, the percentage sign (%) is used as a pattern-matching character representing zero or more characters in a subset of a string. The underscore/underbar character (_) is used to represent one and only one character. In the following example only a single row is returned because only values beginning with V$, end with A, and a total length of five characters are found:

```
SELECT * FROM DICTIONARY WHERE TABLE_NAME LIKE 'V$__A';
```

IN set membership evaluates an expression as being within a set of elements. Because IN implies set membership, IN allows multiple row returning subqueries:

```
expression [ NOT ] IN (expression)
```

IN is often used to check membership of one element in a list of elements. This example checks against a list of literal items:

```
SELECT NAME FROM V$DATAFILE WHERE FILE# IN (1, 2, 3);
```

This example uses a subquery to create the list for IN membership to check against:

```
SELECT NAME FROM V$DATAFILE
WHERE TS# IN (SELECT TS# FROM V$TABLESPACE);
```

EXISTS checks for membership as IN does with a few differences. First, EXISTS only allows an expression on the left. Second, EXISTS is sometimes faster. Third and probably most significantly, both IN and EXISTS allow a correlation between a calling query and a subquery. However, EXISTS executes the correlation row by row, IN searches all rows of the subquery, for every row of the calling query. Thus EXISTS can be much faster. Like the IN condition, EXISTS implies set membership and allows multiple row returning subqueries:

```
[ NOT ] EXISTS (expression)
```


In this example the name of the filename is retrieved whenever the TS# identifier is found in the V$TABLESPACE view:

```
SELECT NAME FROM V$DATAFILE
WHERE EXISTS (SELECT TS# FROM V$TABLESPACE WHERE TS# < 5);
```

This example modifies the previous one such that a correlated value is passed from the calling query into the subquery. This effectively creates a link or correlation between calling query and subquery:

```
SELECT D. NAME FROM V$DATAFILE D
WHERE EXISTS (SELECT TS# FROM V$TABLESPACE WHERE TS# = D.TS#);
```

NOTE

A query using a subquery is also called a semi-join when performing the same or similar function to that of a join.

BETWEEN validates an expression as being between two values, and its inclusive. Also the first value should be less than the second. BETWEEN allows only a single row of returning subqueries:

```
expression BETWEEN expression AND expression

SELECT NAME FROM V$DATAFILE WHERE FILE# BETWEEN 2 AND 4;
```

The next example returns nothing because there is nothing between 2 and 4 when starting the count at 4. In other words, there is nothing greater than or equal to 4, and less than or equal to 2, which is between 2 and 4:

```
SELECT NAME FROM V$DATAFILE WHERE FILE# BETWEEN 4 AND 2;
```

ANY, SOME, and ALL check set membership and allow subqueries that return multiple rows. ANY checks for membership of any element and SOME looks for some elements. ANY and SOME are identical. ALL only returns a result if all elements in both expressions match, and the two sets must be equal in size and content:

```
expression [ = | != | > | < | >= | <= ] [ ANY | SOME | ALL ] expression

SELECT NAME FROM V$DATAFILE
WHERE TS# = ANY (SELECT TS# FROM V$TABLESPACE);
```

Those are the conditional comparisons available in Oracle SQL. Now you examine logical operators.

WHERE Clause Logical Operators

Logical operators in Oracle SQL are AND, OR, and NOT, allowing concatenation of multiple conditional expressions together. Precedence rules apply. Expressions are evaluated from left to right, unless overridden by parenthesized (bracketed) sections. NOT has higher precedence than AND, followed by OR. In the query shown in Figure 6-15, the two STATUS column checks are parenthesized, and thus evaluated first, followed by the CREATION_TIME date field using AND.

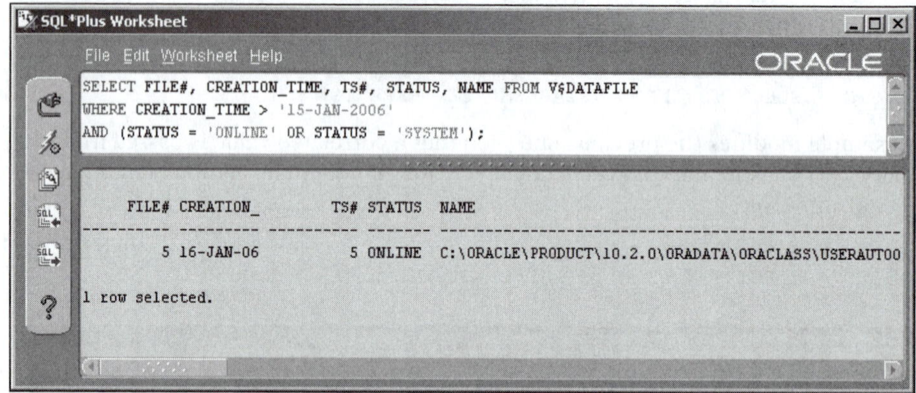

FIGURE 6-15 Parenthesized expressions are evaluated first regardless of the other sequence and precedence operators

Removing parentheses, as shown in Figure 6-16, produces a meaningless or nonsensical result because the first row fails the first comparison (the date check).

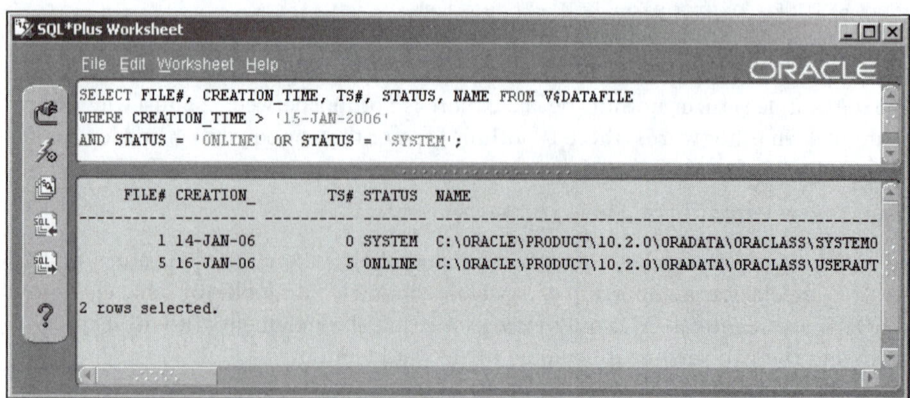

FIGURE 6-16 Omitting parentheses can change the order of execution of operators, producing meaningless results

Top-N Queries

Top-N queries are used to return a small number of rows from a very large query. A Top-N query simply saves time and resources. Top-N queries are executed by filtering against the ROWNUM pseudocolumn. ROWNUM is a pseudocolumn generated for each row, in sequence, as each row is returned from a query. So selecting to return the first 10 rows from a 1,000,000-row query will return only 10 rows, helping performance and minimizing disruption to resources. The following query will find all the objects in a database (logged in as the SYS user). In Oracle 10g this amounts to almost 50,000 different objects. That's a really big query. You don't always want to watch 50,000 rows scroll past you in SQL*Plus Worksheet. If you only want to see the first 10 rows use a query like this:


```
SELECT * FROM DBA_OBJECTS
WHERE ROWNUM <= 10;
```

There is one small quirk when using Top-N queries and that concerns using the ORDER BY clause correctly. The following query will return a senseless result because the ORDER BY clause does not sort all the rows in the query, only the first 10 rows selected by the TOP-N query WHERE clause:

```
SELECT * FROM DBA_OBJECTS
WHERE ROWNUM <= 10
ORDER BY OBJECT_NAME;
```

The only way to resolve this problem is to use a special type of subquery called an inline view. An inline view implies a view created inline within a query. It's a little like a view on the fly. Essentially an inline view is a subquery embedded in the FROM clause of a SELECT statement. The result is that the ORDER BY clause is pushed into the subquery, executing the sort on the entire data set, and the ROWNUM filter is applied to the sorted result:

```
SELECT *
FROM (SELECT * FROM DBA_OBJECTS ORDER BY OBJECT_NAME)
WHERE ROWNUM <= 10;
```

Another little quirk to mention is that selecting values greater than ROWNUM, as opposed to less than, or less than or equal to ROWNUM, will return no rows whatsoever. Thus the following query will return nothing:

```
SELECT * FROM DBA_OBJECTS WHERE ROWNUM > 10;
```

Now you can proceed to sorting query results using the ORDER BY clause.

Sorting Queries with the ORDER BY Clause

The ORDER BY clause is used to sort the results of a query. The syntax for the ORDER BY clause expands the SELECT statement syntax as follows:

```
SELECT { [alias.]column | expression | [alias.]* [ , ... ] }
FROM [schema.]{table | view|...} [alias]
[ WHERE ... ]
[ ORDER BY { { column | expression | position } [, ...] } }
[ ASC | DESC ] [ NULLS { FIRST | LAST } ]
```

NOTE

The ORDER BY clause always appears as the final clause in a SELECT statement.

The best way to learn is by example. You can practice these queries in SQL*Plus Worksheet at the same time, and see the results. The ORDER BY clause is simple. You can sort the results of a query in a number of ways:

- **Sort by one or more columns**: Sort by one or more column names:

```
SELECT BYTES, BLOCKS, FILE_NAME FROM DBA_DATA_FILES
ORDER BY FILE_NAME;
```

- **Positional sort**: Sort by the position of a column, in any order, within the selected list of columns:

```
SELECT BYTES, BLOCKS, FILE_NAME FROM DBA_DATA_FILES
ORDER BY 2, 1;
```

- **Sort by expression**: Sort by one or more expressions and the expression does not have to be in the selected column list:

```
SELECT BYTES, BLOCKS, FILE_NAME FROM DBA_DATA_FILES
ORDER BY BYTES/1024;
```

- **Aliases**: Use aliases to sort with:

```
SELECT BYTES/1024 AS KB, BLOCKS, FILE_NAME FROM DBA_DATA_FILES
ORDER BY KB;
```

- **Ascending and descending sorts**: Queries can be sorted in ascending or descending order, where ascending is the default if ASC or DESC are not specified. The following query will return all rows with the largest Kb value appearing as the first row, and the smallest last:

```
SELECT BYTES/1024 AS Kb, BLOCKS, FILE_NAME FROM DBA_DATA_FILES
ORDER BY KB DESC;
```

- **Sorting NULL values**: In an ascending sort, NULL values are returned last, and in a descending sort NULL values are returned first. The NULLS FIRST clause can be used in conjunction with an ascending sort to return NULL values first, and NULLS LAST returns NULLS last for a descending sort. The following example will return any NULL values in the last rows, despite the descending sort:

```
SELECT TABLESPACE_NAME, NEXT_EXTENT FROM DBA_TABLESPACES
ORDER BY NEXT_EXTENT DESC NULLS LAST;
```

- **Combination sorting**: Various sorting parameters can be combined such that ASC, DESC, NULLS FIRST, and NULLS LAST; apply to each individual column sorted with:

```
SELECT TABLESPACE_NAME, NEXT_EXTENT FROM DBA_TABLESPACES
ORDER BY TABLESPACE_NAME ASC, NEXT_EXTENT DESC NULLS LAST;
```

The ORDER BY clause is not complicated, as you can see. Now you can read about the GROUP BY clause.

Aggregating Queries with the GROUP BY Clause

The GROUP BY clause can become complex when additions of OLAP modeling, grouping sets, rollups, and cube operations are taken into account. However, for the purposes of this book and the art of database administration you only need to know how to create fewer rows from a row set, using a summary or aggregation function, and the GROUP BY clause. The HAVING clause can sometimes be useful as well.

Basic syntax for the GROUP BY clause is as follows:

```
SELECT { [alias.]column | expression | [alias.]* [ , ... ] }
FROM [schema.]{table | view|...} [alias]
[ WHERE ... ]
[ GROUP BY expression [, expression ] [ HAVING condition ] ]
[ ORDER BY ... ]
```


There are a few standard rules to remember about the GROUP BY clause:

- The GROUP BY clause column list must include all columns in the SELECT statement not affected by any aggregate functions.
- The expression for the SELECT statement should include at least one grouping function such as COUNT().
- The GROUP BY clause cannot use the column positional specification like the ORDER BY clause because the result set columns do not exist when the GROUP BY clause is executed and do exist when the ORDER BY clause is executed. The GROUP BY clause summarizes rows for output and the ORDER BY clause sorts the result set of a query.

NOTE

The GROUP BY clause is executed during query execution and the ORDER BY clause runs after retrieval and grouping of all rows. The ORDER BY clause will always add a performance overhead to a query. Implicit or inherent sorting can often be executed in the WHERE and GROUP BY clauses.

Now you can read and practice some easy examples in SQL*Plus Worksheet. The query below will return more than 4,000 rows:

```
SELECT SYS.CLASS, SYS.VALUE, SES.VALUE
FROM V$SYSSTAT SYS JOIN V$SESSTAT SES
ON(SES.STATISTIC# = SYS.STATISTIC#);
```

The GROUP BY clause is added to this query, as shown in Figure 6-17, where averages of two columns are returned for each statistics class. The number of rows returned by the query in Figure 6-17 is now only nine rows using the GROUP BY clause and the AVG (average) functions.

Now you can examine the HAVING clause.

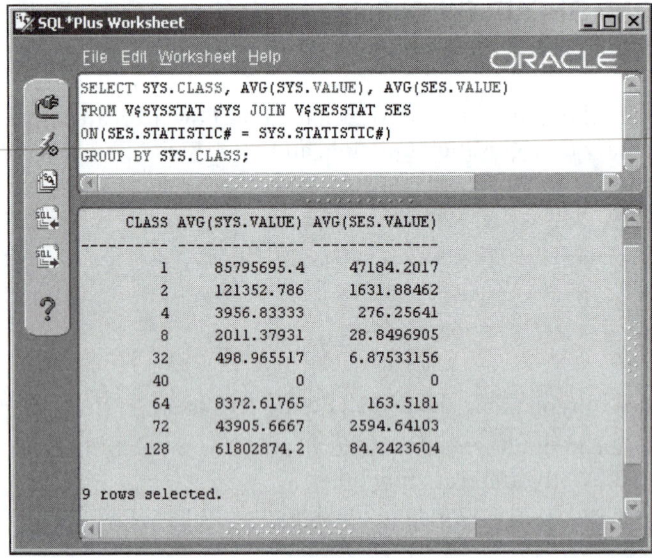

FIGURE 6-17 The GROUP BY clause is used to summarize large numbers of rows into a smaller, aggregated set of rows

Filtering GROUP BY Aggregations with the HAVING Clause

The HAVING clause extends the GROUP BY clause by filtering on resulting grouped rows (returned by the GROUP BY clause). The query below could be used to return a filtered selection of the nine rows produced in the query shown in Figure 6-17:

```
SELECT SYS.CLASS, AVG(SYS.VALUE), AVG(SES.VALUE)
FROM V$SYSSTAT SYS JOIN V$SESSTAT SES
ON(SES.STATISTIC# = SYS.STATISTIC#)
GROUP BY SYS.CLASS HAVING AVG(SYS.VALUE) > 9999;
```

In other words, in Figure 6-17, the HAVING clause can be applied to the nine rows returned after the application of the GROUP BY clause. The HAVING clause is not applied to all 4,000 rows returned by the query.

As already stated, the GROUP BY clause can become much more complex with the addition of OLAP (Online Analytical Processing) functionality, and otherwise. However, for the purposes of database administration, details of OLAP are not required and are out of the scope of this book. Details of this functionality can be found in many other available titles including one of my books called *Oracle SQL Jumpstart with Examples*, DigitalPress, September 2004. And of course there is always the Oracle documentation.

Now you will see how to build some different types of advanced querying in an Oracle database.

ADVANCED QUERIES

So far you have covered the basics of the SELECT statement and its various additional clauses. It is essential that you know the basics of the SELECT statement to be able to use SQL*Plus effectively as a database administrator. There are numerous types of advanced query types available in an Oracle database, including joins, subqueries, and other specialized queries.

Joins

Joins merge columns and rows from more than a single table. The different types of join queries are as follows:

- **Cross-join or Cartesian product**: Merge rows selected from both tables into a single result set, regardless of matching column values in respective rows in either table. Every row in one table is joined with every other row in the second table in the join. In the query shown below and in Figure 6-18 every tablespace is shown as having all the datafile statistics of all datafiles, regardless of tablespace and datafile. This is obviously incorrect:

```
SELECT TS#, PHYRDS, PHYWRTS
FROM V$DATAFILE CROSS JOIN V$FILESTAT
ORDER BY 1, 2;
```

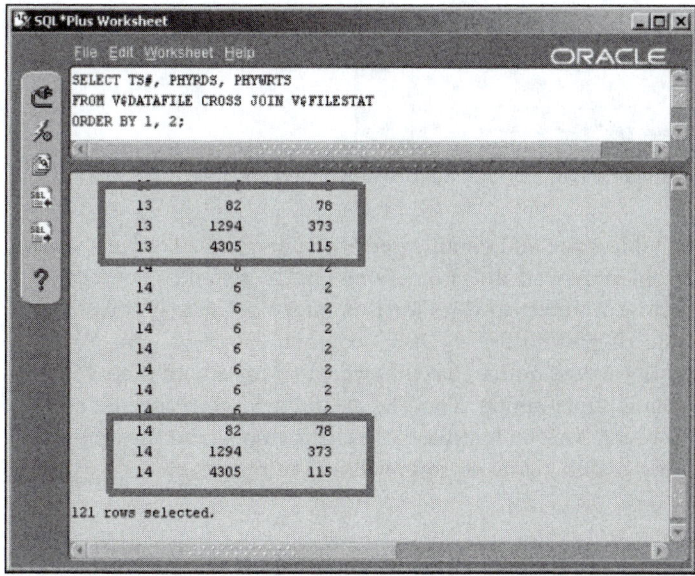

FIGURE 6-18 Cross-join or Cartesian products join everything to everything else

- **Natural or inner join**: Combines rows from both tables using matching column names and column values. It uses the same column names across tables to match rows. The result set includes only rows that match. This is shown in

Figure 6-19 where every tablespace is matched with respectively contained datafiles. Now the query makes some sense:

```
SELECT TS#, PHYRDS, PHYWRTS
FROM V$DATAFILE CROSS NATURAL JOIN V$FILESTAT
ORDER BY 1, 2;
```


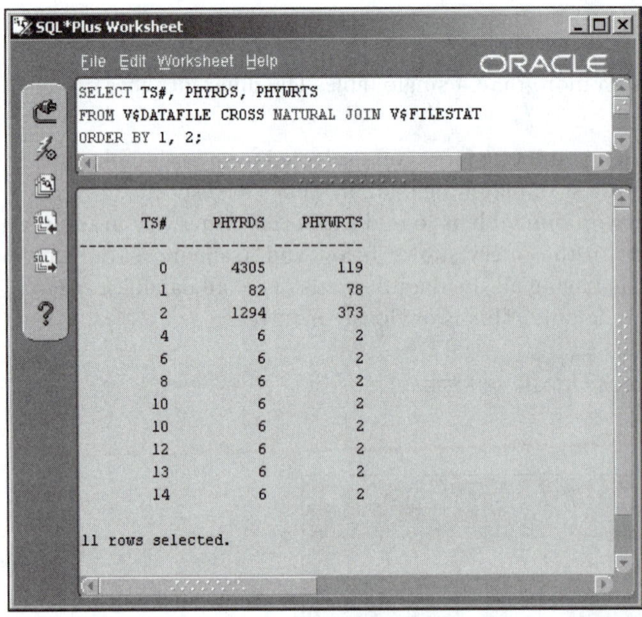

FIGURE 6-19 A natural join uses same named columns to join two data sources

- Now let's join the tablespace and datafile performance views. The next query has an obvious problem in that the two columns in the two data sources are the same. Oracle will produce no rows for this query because it tries to perform a natural join. If you describe (DESC command in SQL*Plus Worksheet) the two views you will notice that the two views have both their TS# and NAME columns named in common. The TS# contains a tablespace identifier in both tables. However, NAME identifies both the filename and the tablespace name—these do not match. Thus no rows will be the result:

```
SELECT TS#, FILE#
FROM V$TABLESPACE NATURAL JOIN V$DATAFILE
ORDER BY 1, 2;
```

- The way to resolve the above query is to utilize the USING clause, forcing a match on the column names specified in the USING clause, as shown in the following query and as shown in Figure 6-20 (the NATURAL keyword is removed because it no longer applies):

```
SELECT TS#, FILE#
FROM V$TABLESPACE JOIN V$DATAFILE USING (TS#)
ORDER BY 1, 2;
```

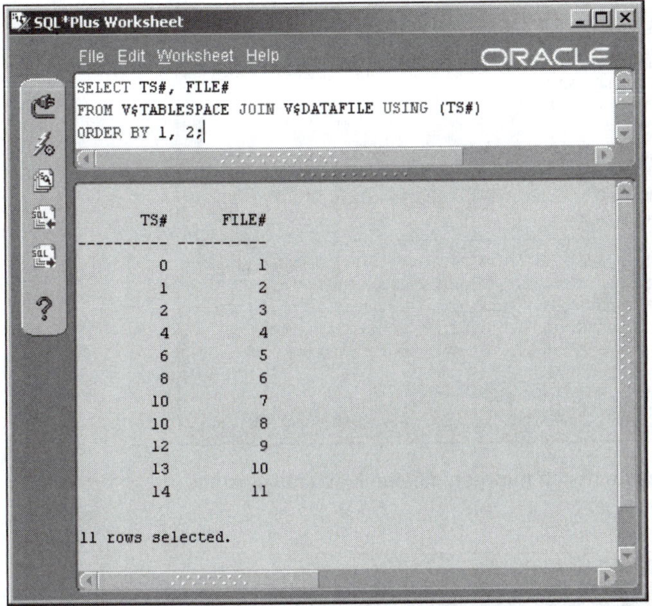
SQL*Plus Worksheet
File Edit Worksheet Help
ORACLE
SELECT TS#, FILE#
FROM V$TABLESPACE JOIN V$DATAFILE USING (TS#)
ORDER BY 1, 2;
TS# FILE#
---------- ----------
0 1
1 2
2 3
4 4
6 5
8 6
10 7
10 8
12 9
13 10
14 11
11 rows selected.

FIGURE 6-20 The USING clause allows a match between specified columns when columns with different values and the same names exist

- The next obvious problem is not same named columns containing different values, but different named columns containing the same values, as shown in the following query and in Figure 6-21. Again a natural join won't work because Oracle is unaware that two differently named columns could have the same values (that can be linked in a join). The solution is the ON clause, as in the query below (note the addition of the column alias qualifiers):

```
SELECT PV.FILE#, MV.FILE_ID, MV.FILE_NAME
FROM V$DATAFILE PV JOIN DBA_DATA_FILES MV
ON (MV.FILE_ID = PV.FILE#)
ORDER BY 3;
```

- **Outer join**: Select rows from two tables as with a natural join but including rows from one or both tables that do not have matching rows in the other table. Missing values are replaced with NULL values.
- **Left outer join**: All rows from the left table, plus all matching rows from the right table. Column values from the right table are replaced with NULL values when the matching right side row does not exist in the left side of the table. In the following query and in Figure 6-22 all tablespaces, regardless of being temporary tablespaces or not, are joined with files for temporary tablespaces. Only temporary tablespaces have temporary datafiles, and thus all non-temporary datafiles are left blank because they are the outer part of the outer join.

```
SELECT TABLESPACE_NAME, FILE_NAME
FROM DBA_TABLESPACES LEFT OUTER JOIN DBA_TEMP_FILES
USING (TABLESPACE_NAME);
```

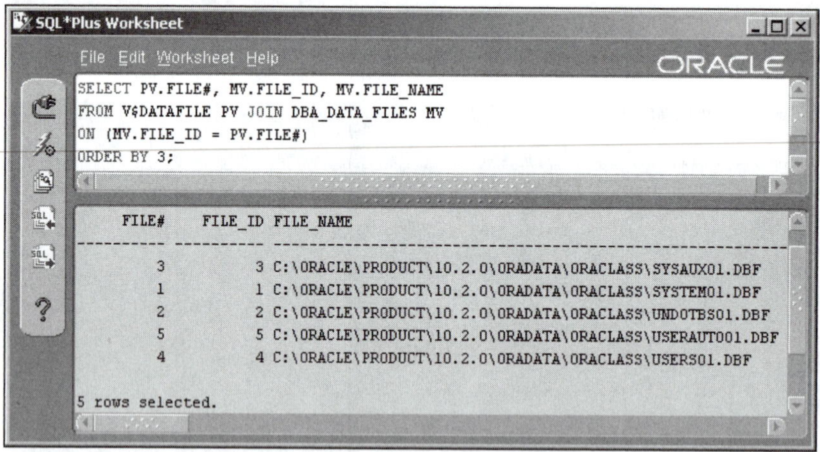

FIGURE 6-21 The ON clause forces a match between differently named columns

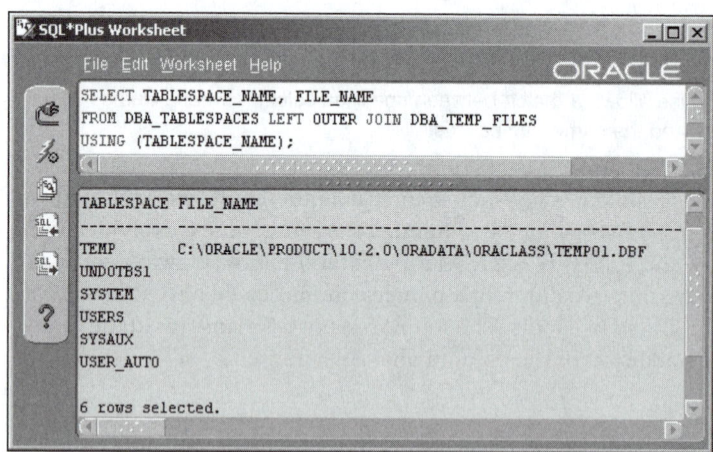

FIGURE 6-22 A left outer join retrieves all matching rows and non-matching rows on the left

- **Right outer join.** All rows from the table on the right plus matching rows from the left table, the opposite of the left outer join. The following query is shown in Figure 6-23:

```
SELECT FILE_NAME, TABLESPACE_NAME
FROM DBA_TEMP_FILES RIGHT OUTER JOIN DBA_TABLESPACES
USING (TABLESPACE_NAME);
```

- **Full outer join**: All rows from both tables with null values replacing missing values. The following query is shown in Figure 6-24:

```
SELECT D.FILE_NAME, T.FILE_NAME
FROM DBA_DATA_FILES D FULL OUTER JOIN DBA_TEMP_FILES T
USING (TABLESPACE_NAME);
```

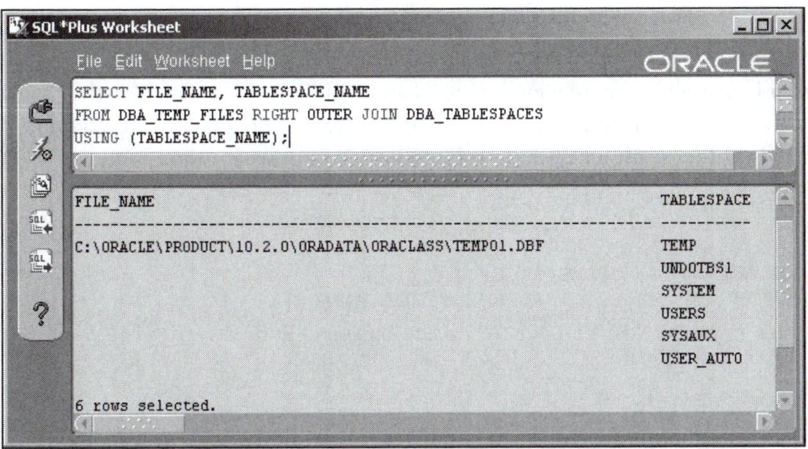

FIGURE 6-23 A right outer join retrieves all matching rows and non-matching rows on the right

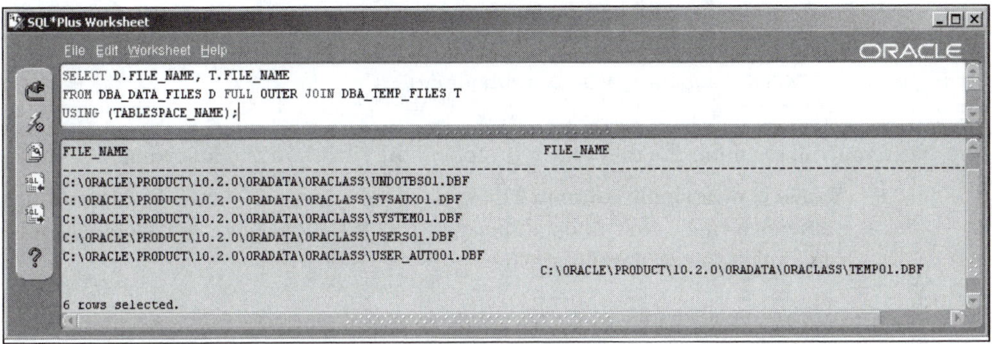

FIGURE 6-24 A full outer join retrieves all matching rows, plus non-matching rows on both the left and the right

- **Self-join**: A self-join does just that—it joins a table to itself. Some tables contain hierarchies of data and are natural candidates for using self-joins.

NOTE

Self-joins are sometimes called fishhook joins.

For example, the DBA_TYPES view contains object type hierarchies where each type (TYPE_NAME) is optionally defined as having a parent type or super type (SUPERTYPE). A super type exists as another type in the same DBA_TYPES view. Run the following query in SQL*Plus Worksheet and examine the results:

```
SELECT P.TYPE_NAME "Parent", C.TYPE_NAME "Child"
FROM DBA_TYPES P LEFT OUTER JOIN DBA_TYPES C
```

```
ON (C.TYPE_NAME = P.SUPERTYPE_NAME)
ORDER BY 1, 2;
```

- **Semi-join**: A vague form of a join using IN and EXISTS operators where the results contained within the subquery are not retrieved as part of the final columns selected for the result join. Both of the following queries are semi-joins:

```
SELECT NAME FROM V$DATAFILE
WHERE EXISTS (SELECT TS# FROM V$TABLESPACE);
```



```
SELECT NAME FROM V$DATAFILE
WHERE TS# IN (SELECT TS# FROM V$TABLESPACE);
```

That's the basics of joins. Now you get to see the basics of subqueries.

Subqueries

Subqueries are queries executed within other queries. One query calls another query, returning results to the calling query (a SELECT statement executed within another SELECT statement). This is a quick summary of subquery syntax where subquery comparison syntax is much the same as WHERE clause comparison syntax:

```
(subquery) { = | != | LIKE | [ NOT ] IN } (subquery)
[ NOT ] EXISTS (subquery)
(subquery) BETWEEN (subquery) AND (subquery)
(subquery) { = | != | > | < | >= | <= } { ANY | SOME | ALL } (subquery)
```

Next you will examine the different types of subqueries as in the following:

- **Single row or single column**: This type of subquery can be used to find a single value or a single row. The subquery in this WHERE clause finds a single column value (a scalar subquery):

```
SELECT * FROM V$DATAFILE
WHERE TS# =
(SELECT TS# FROM V$TABLESPACE WHERE NAME='SYSTEM');
```

The same query, with the WHERE clause removed from the subquery, will not work because you cannot ask if something is equal to multiple values. The subquery in the following query retrieves all tablespaces (multiple values):

```
SELECT * FROM V$DATAFILE
WHERE TS# =
(SELECT TS# FROM V$TABLESPACE);
```

- **Multiple row, single column**: This type of subquery returns a list of values. The following query checks against a list of multiple, single values:

```
SELECT * FROM V$DATAFILE
WHERE TS# IN
(SELECT TS# FROM V$TABLESPACE);
```

- **Multiple column, single or multiple row**: Subqueries can return more than a single column, and their calling queries can validate against those multiple columns, as in the following example:

```
SELECT * FROM DBA_DATA_FILES
WHERE (FILE_NAME, FILE_ID) IN
(SELECT NAME, FILE# FROM V$DATAFILE);
```

- **Regular subquery**: Apart from the single and multiple row and column "stuff," there are regular and correlated subqueries. The following query is commonly known as a regular subquery, simply because there is no direct relationship established by the query itself, between calling query and subquery:

```
SELECT * FROM V$DATAFILE
WHERE TS# IN
(SELECT TS# FROM V$TABLESPACE);
```

- **Correlated subquery**: The word correlation is used to describe a relationship between the calling query and the subquery. The rule to remember with correlated subqueries is that a correlating column must be passed down into the subquery from the calling query, not the other way around. Values cannot be passed from subquery to calling query. Thus a correlated subquery is always dependent on the calling query. The following example changes the previous regular subquery into a correlated query, by matching rows between calling and subquery:

```
SELECT * FROM V$DATAFILE D
WHERE D.TS# IN
(SELECT TS# FROM V$TABLESPACE
```

The point of a correlated subquery is the relationship established between calling query and subquery. The correlation allows a restrictive (finds a matched row for each row in the calling query) rather than exhaustive (finds all rows for every row in the calling query) search within the subquery. In other words, if a subquery is passed a single primary key value, then the subquery will find only a single row, instead finding, for example, 1,000,000 rows in a table accessed by a subquery, for every row found in the calling query. Correlated queries are potentially much faster than non-correlated subqueries.

- The same example now uses the EXISTS rather than the IN operator. EXISTS is traditionally faster than IN because when EXISTS finds a matching value in the subquery, the subquery is supposed to stop executing. The IN operator builds a set first and thus always executes the entire subquery. In other words, if the subquery using EXISTS finds a useful row, at the 10th row out of 1,000,000 rows, the subquery will search 10 rows, not 1,000,000 rows. Tuning experiments seem to indicate that this particular advantage of EXISTS over IN no longer exists as of Oracle 10g. IN is perhaps now just as efficient as EXISTS when used in correlated subqueries.

```
SELECT * FROM V$DATAFILE D
WHERE EXISTS
(SELECT TS# FROM V$TABLESPACE
WHERE TS# = D.TS#);
```

- **Nested subquery**: Subqueries can call other subqueries and so on, ad infinitum. In other words, subqueries can be nested within subqueries, within subqueries. The following example demonstrates this:

```
SELECT * FROM DBA_INDEXES
WHERE TABLE_NAME IN
(
  SELECT TABLE_NAME FROM DBA_TABLES
```

```
WHERE TABLESPACE_NAME IN
 (
  SELECT TABLESPACE_NAME FROM DBA_TABLESPACES
  WHERE TABLESPACE_NAME IN
  (
    SELECT TABLESPACE_NAME FROM DBA_DATA_FILES
  )
 )
);
```


- **Inline View**: An inline view is a subquery embedded in the FROM clause of a calling SELECT statement, which is also a subquery. Values can be passed from the inline view to the calling query, or subquery. This is the only type of subquery where a real join can be created, as opposed to a semi-join:

```
SELECT T.TABLESPACE_NAME, D.FILE_NAME
FROM DBA_TABLESPACES T,
(
SELECT TABLESPACE_NAME, FILE_NAME
FROM DBA_DATA_FILES
) D
WHERE D.TABLESPACE_NAME = T.TABLESPACE_NAME;
```

- **DML subqueries**: Subqueries can be used in all sorts of ways, and in many different forms in Oracle SQL commands. Subqueries are frequently used for embedding with INSERT, UPDATE, and DELETE DML commands. For example, when inserting a new row into a detail record of a two-table master detail relationship, a single row is inserted:

```
INSERT INTO DETAIL_TABLE (DETAIL_PK_ID, MASTER_FK__ID)
VALUES (DETAIL_SEQ,
(SELECT MASTER_ID FROM MASTER_TABLE
WHERE MASTER_NAME = '<a unique master literal value>')
);
```

- In the case of the following UPDATE statement, many rows have their master table foreign keys updated:

```
UPDATE DETAIL_TABLE SET MASTER_PK_ID =
(SELECT MASTER_ID FROM MASTER_TABLE
WHERE MASTER_NAME = '<a unique master literal value>')
WHERE DETAIL_PK_ID = 1;
```

The above two tables do not exist in your database and are used here for demonstrative purposes only.

That's the basic syntax and facts of subqueries. Now let's discuss some more complex (specialized), and perhaps rarely used, types of queries.

Other Specialized Queries

Specialized query types examined in this section are composite queries, hierarchical queries, flashback queries, and parallel queries.

Composite Queries

A composite query concatenates columns and rows of two queries. Set operators (UNION, UNION ALL, INTERSECT, and MINUS) are used to concatenate queries. Restrictions are such that both SELECT column sets in the two queries must have the same number of columns, and data types must be compatible, dependant on SELECT column list position. The following query, also shown in Figure 6-25, is a rather messy application of the UNION operator, merging all rows from two queries:

```
SELECT TABLESPACE_NAME, FILE_NAME FROM DBA_DATA_FILES
UNION
SELECT TABLESPACE_NAME, FILE_NAME FROM DBA_TEMP_FILES;
```


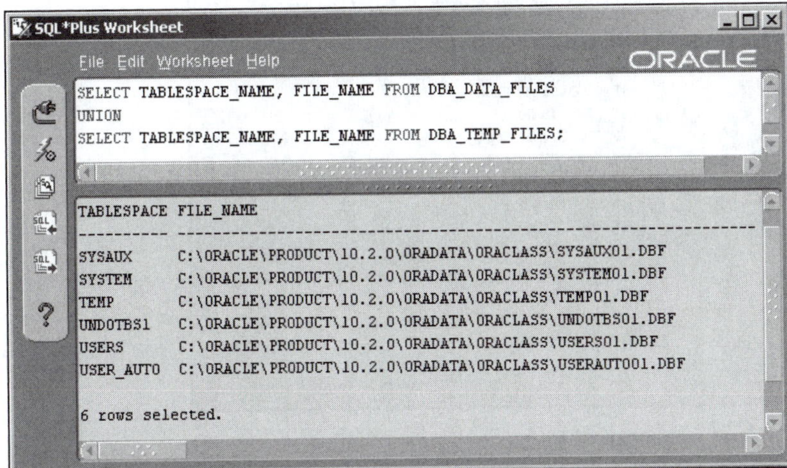

FIGURE 6-25 UNION merges rows and columns from two queries

Other variations of set operators are as follows:

- **UNION ALL**: Similar to UNION but retrieves all rows from both queries including duplicates. Duplicate rows are rows returned by both queries.
- **INTERSECT**: Returns distinct rows from both queries. An intersection is similar to an inner join.
- **MINUS**: Returns one query less the other, similar to a left outer join where only distinct rows in the first query are returned.

Hierarchical Queries

A hierarchical query allows display of hierarchical data in a single table, using a specialized query type. A hierarchical query is effectively a more formalized and much more comprehensive form of a self-join, as shown in Figure 6-26, and the following query:

```
SELECT SUPERTYPE_NAME "Parent", TYPE_NAME "Child", LEVEL
FROM DBA_TYPES
START WITH SUPERTYPE_NAME = 'ST_GEOMETRY'
CONNECT BY PRIOR TYPE_NAME = SUPERTYPE_NAME
ORDER BY 3, 1;
```

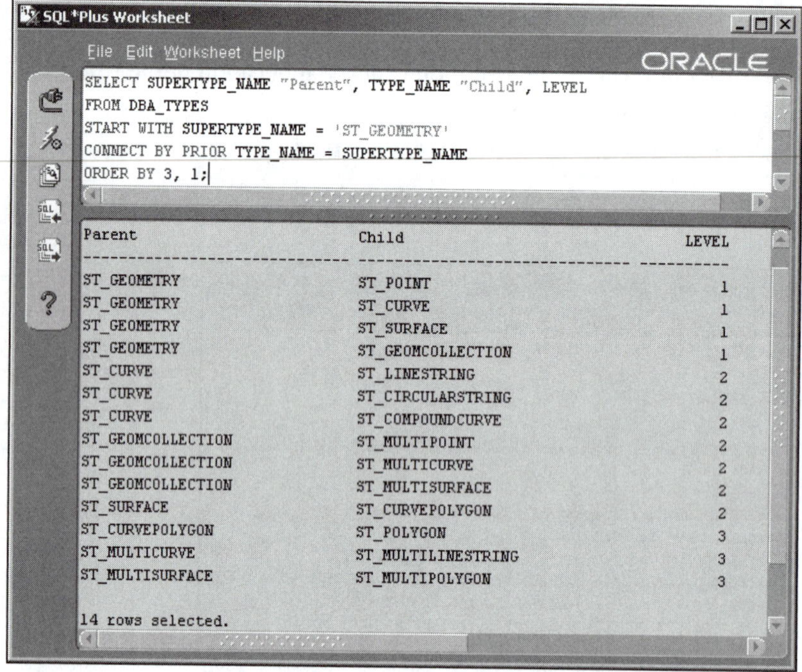

FIGURE 6-26 Hierarchical queries allow display of hierarchical data

The hierarchical query above introduces various new operators, clauses and pseudocolumns:

- **PRIOR operator**: Used with the CONNECT BY operator to match the specified column (PRIOR TYPE_NAME) in the parent row, with the parent column (SUPERTYPE_NAME) in the current row. In other words, in the fourth line of the example above PRIOR TYPE_NAME refers to the parent row, and SUPERTYPE_NAME refers to the current row (the child row).
- **CONNECT BY operator**: Without the PRIOR operator, CONNECT BY will perform a similar function to that of CONNECT BY PRIOR, except using the root row of the hierarchy as opposed to specific PRIOR the parent row. There are other more obscure variations of the CONNECT BY operator.
- **LEVEL pseudocolumn**: Used only in hierarchical queries, in tandem with the CONNECT BY operator, returning the level of a row in a hierarchy of rows, in a hierarchical query.
- **START WITH clause**: Allows a hierarchical query to begin displaying within a hierarchical structure, with a specified column and value.

Flashback Queries

A flashback query allows a flashback to, or snapshot of, data as it was at a previous point in time. AS OF flashback queries go back to a point in time using a timestamp or SCN. You can even execute flashback versions queries. A flashback versions query can be used to return more than one version of a single row both before and after a change.

NOTE

Manual rollback will not support flashback queries. Flashback queries must utilize automated undo.

General flashback query syntax is as follows:

```
SELECT ... FROM ...
[
   AS OF { SCN | TIMESTAMP } expression
  | VERSIONS BETWEEN { SCN | TIMESTAMP} { expression | MINVALUE }
    AND { expression | MAXVALUE }
]
[ WHERE ... ] [ ORDER BY ... ] [ GROUP BY ... ];
```

NOTE

The DBMS_FLASHBACK package can be used to create flashback queries at the session level.

There are also a number of pseudocolumns available for executing flashback queries, specifically for version flashback queries:

- **ORA_ROWSCN**. Returns a row SCN.
- **VERSIONS_{START|END}TIME**. First and last version timestamp.
- **VERSIONS_{START|END}SCN**. First and last version SCN.
- **VERSIONS_XID**. Transaction identifier.
- **VERSIONS_OPERATION**. Returns (I)nsert, (U)pdate or (D)elete.

An AS OF flashback query allows you to pull data from a database at a specific point in time, shown in the following query and in Figure 6-27:

```
SELECT * FROM CLASSMATE.CLIENT
AS OF TIMESTAMP(SYSTIMESTAMP - INTERVAL '1' DAY);
```

A flashback versions query can be used to shows multiple versions of rows, from the same table. You can change a row in the CLASSMATE user CLIENT table:

```
INSERT INTO CLASSMATE.CLIENT
VALUES(2000,'Bloggs','Joe','800-123-4567','joe@bloggs.net');
COMMIT;
```

Now delete the inserted row:

```
DELETE FROM CLASSMATE.CLIENT WHERE CLIENT_ID=2000;
COMMIT;
```

Execute a flashback versions query to find multiple versions of the same data, as shown in the following query and in Figure 6-28:

```
SELECT * FROM CLASSMATE.CLIENT
VERSIONS BETWEEN TIMESTAMP MINVALUE AND MAXVALUE;
```

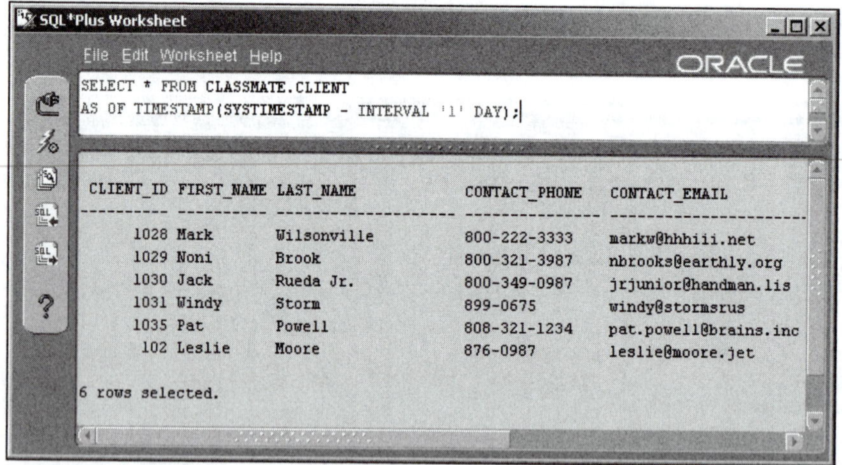

FIGURE 6-27 A flashback query can find data as of a point in time, or SCN, in the past

```
SELECT * FROM CLASSMATE.CLIENT
VERSIONS BETWEEN TIMESTAMP MINVALUE AND MAXVALUE;

CLIENT_ID FIRST_NAME LAST_NAME            CONTACT_PHONE   CONTACT_EMAIL
--------- ---------- -------------------- --------------- --------------------
     2000 Bloggs     Joe                  800-123-4567    joe@bloggs.net
     2000 Bloggs     Joe                  800-123-4567    joe@bloggs.net
     1028 Mark       Wilsonville          800-222-3333    markw@hhhiii.net
     1029 Noni       Brook                800-321-3987    nbrooks@earthly.org
     1030 Jack       Rueda Jr.            800-349-0987    jrjunior@handman.lis
     1031 Windy      Storm                899-0675        windy@stormsrus
     1035 Pat        Powell               808-321-1234    pat.powell@brains.inc
      102 Leslie     Moore                876-0987        leslie@moore.jet

8 rows selected.
```

FIGURE 6-28 A flashback versions query can return multiple versions of the same data, at different points in time

A single INSERT and a single DELETE were executed for the duplicated row shown in Figure 6-28. Thus one wonders why the row is duplicated; perhaps this is merely a "feature."

Flashback Database

Another feature of flashback technology is the ability to flashback an entire table or even a complete database using the following syntax:

```
FLASHBACK [ STANDBY ] DATABASE [ database ]
TO { SCN | TIMESTAMP } expression;

FLASHBACK TABLE { [ schema.]table , ... }
```

```
TO { SCN | TIMESTAMP } expression
[ { ENABLE | DISABLE } TRIGGERS ];
```

FLASHBACK DATABASE and FLASHBACK TABLE allow restore of either the entire database or a single table back to, and in the case of a table, even forward to a different SCN.

Parallel Queries

Parallel queries work best on multiple CPU platforms in tandem with partitioning, and particularly with multiple disks or RAID arrays. Additionally, parallel queries provide performance benefits for very large tables, or in very large databases, such as data warehouses. Using parallel queries on small highly active concurrent OLTP databases can cause more problems than they solve. The SQL queries and coding that can be executed in parallel are limited to the following types of functionality:

- Queries containing at least one table scan using SELECT, INSERT, UPDATE, and DELETE statements.
- CREATE INDEX and ALTER INDEX REBUILD statements.
- CREATE TABLE statements for generating a table from a SELECT command.
- Queries on partitions with local indexes. A local index is an index created on each separate partition.

You can cause parallel execution in two ways:

- The PARALLEL hint:

  ```
  SELECT /*+ PARALLEL(CLASSMATE.CLIENT, 2) */ *
  FROM CLASSMATE.CLIENT;
  ```

- CREATE TABLE and ALTER TABLE statements can include the PARALLEL clause using the following syntax:

  ```
  { CREATE | ALTER } TABLE ... [ NOPARALLEL | PARALLEL [n] ];
  ```

Chapter 7 will examine the basics of creating and managing tables in an Oracle database.

Chapter Summary

- The SELECT statement uses a list of columns and a FROM clause to retrieve rows from a table or view.
- The WHERE clause is a filter for removing rows from a query result, or to retain rows in a query result.
- The ORDER BY clause allows sorting of rows returned by a query, as one or more columns, expressions, or even by the position of columns in the SELECT statement column list.
- Aggregate (or summary) queries use the GROUP BY clause to group sets of rows into summarized rows, and the HAVING clause to filter in or out resulting grouped rows.
- Join queries allow merging of columns and rows from two or more tables, usually linking rows together, based on common column values in different tables.
- A subquery is a query embedded and executed within a calling query.
- A new table or view can be created from the result of a query.
- A composite query uses a set operator such as UNION to return rows from two separate queries, with rows in one table added to the rows of a second table.
- Oracle Database allows certain types of specialized queries for performing special case functionality, such as composite queries, hierarchical queries, flashback queries, and parallel execution queries.
- Oracle Database contains a large number of built-in functions for use in queries.
- Oracle built-in functions are generally classified as single row functions, data type conversions, or grouping functions for aggregate queries.
- Programmers can build custom functions for use in queries, using PL/SQL.
- NULL is a nothing value, meaning it is neither zero nor a space character.
- NULL values can be tested for using the IS [NOT] NULL conditional.
- The NVL function acts as a parse replace function for NULL values.
- The DUAL table is a dummy table, usually used to contain query results not in application tables or in metadata, such as the current system date and time.
- The DISTINCT clause is an SQL extension used to retrieve unique values from a set of values containing possible duplicates.
- The SELECT statement consists in its most basic form of a SELECT clause, a list of columns to be retrieved, followed by a FROM clause, typically containing a table or view, determining where columns are to be retrieved from.
- WHERE clauses, HAVING clauses, and any expression comparisons use comparison condition operators to compare two expressions.
- Multiple dual expression comparisons can be logically added together using AND, OR, and NOT logical operators.
- Comparison conditions include conditionals such as equality, inequality, LIKE pattern matches, and many others.
- A Top-N query allows the return of a small subset of rows from a very large query.

- Aliases can be used in queries to rename both tables and columns.
- Aliases must be used for tables and columns when there is some type of naming conflict, such as joining two tables on a column named the same in both tables.
- NULL values will always sort last by default, in an ascending sort, when an ascending sort, which is the default for an ORDER BY clause.
- A Cartesian product or cross join merges the results of two sources, without linking the two, resulting in every row in one table, joined to every row in the other table.
- A natural or inner join finds the intersecting rows between two tables, based on a common link column in both tables.
- An outer join finds the intersection of two tables, plus any rows in one table, and not in the other, or both.
- A left outer join returns intersecting rows, plus all those in the left table, and not in the right table.
- A right outer join returns intersecting rows, plus all those in the right table, and not in the left table.
- A full outer join returns intersecting rows, the left outer join, and the right outer join.
- A self-join is similar to a hierarchical query in that it joins rows in the same table, to each other, based on some kind of parent-child hierarchical relationship between different rows in that single table.
- A semi-join is similar to a join but often returns only rows from one of the tables in the join (using IN and EXISTS comparisons).
- Subqueries can find one or more rows, and one or more columns.
- Subqueries can allow a correlation between a calling query and a subquery, where values can be passed from calling query to subquery (NEVER from subquery to calling query).
- An inline view is a subquery, returning rows to the calling query, placed into the FROM clause of a SELECT statement.
- Subqueries can be used in INSERT, UPDATE and DELETE statements.
- A hierarchical query allows construction of hierarchical row-column maps from hierarchical data using specialized query clauses, operators, and pseudocolumns.
- A pseudocolumn is a dummy column, retrievable in a query, but not accessed from a table or view, or even metadata.
- Flashback queries snapshot data as it was at a previous point in time.
- Flashback versions queries can retrieve multiple versions of the same data over a period of time.
- Parallel operations can be performed on a limited set of query types, including queries containing full physical table scans, index creations and alterations, CREATE TABLE AS SELECT ... statements, and partitions using local partition indexing.

Review Questions

1. The WHERE clause filters rows in a query. True or False?
2. The ORDER BY clause is always required in a query. True or False?
3. Always place the HAVING clause before the WHERE clause in a query. True or False?
4. How many tables can be joined using a join query?
   a. 1 table and 3 views
   b. Only 2 tables
   c. 3 tables
   d. None of the above
5. NULL is equivalent to 0. True or False?
6. Is this a valid query? SELECT * FROM DUAL; True or False?
7. What type of query is this? SELECT * FROM CLIENT WHERE ROWNUM < 5;
   a. A small query
   b. An erroneous query
   c. A simple query
   d. A Top-X query
   e. None of the above
8. What type of query is the UNION keyword used in?
   a. Composite query
   b. Flashback query
   c. Hierarchical query
   d. Join query
   e. Parallel query
9. The HAVING clause filters aggregated rows and the WHERE clause selected rows. True or False?

Exam Review Questions—Oracle Database 10*g*: Administration (#1Z0-042)

There are no review questions required because this chapter is not included in the 1Z0-042 exam. This chapter is included to help you use your database a little more easily.

Hands-on Assignments

1. Create a SQL script that completes the following tasks successfully. Save the script in a file named **ho0601.sql** in the **Solutions\Chapter06** directory on your student disk.
   a. Find and write down two performance views used to read redo log files, in terms of their groups and members.
   b. Write a query, reading all columns, showing redo log groups excluding member details.

c. Write a query, reading all columns, showing redo log members as well as groups.

d. Write a query, reading useful columns only, linking all redo log members with their respective groups.

2. Find a view for displaying all tables and all views, within the CLASSMATE user, when logged in as SYS. Then return all tables and views as a single view, sorted by the name of the tables or view, regardless of the object being a table or view. Save the result in a file named **ho0602.txt** in the **Solutions\Chapter06** directory on your student disk.

3. Create a SQL script that completes the following tasks successfully, logged in as the SYS user. Save the script in a file named **ho0603.sql** in the **Solutions\Chapter06** directory on your student disk.

   a. Find all CLASSMATE user, CLIENT table, CLIENT_ID values between 1,000 and 2000.

   b. Find all CLASSMATE user CLIENT table rows, such that all phone numbers containing a 1-800 number are excluded.

   c. Find all CLASSMATE user, CLIENT table rows, such that the last names of Brook, Storm, and Moore are included.

   d. Find all datafiles that exist as tablespaces, using V$ performance views only. Use a single correlated subquery and a WHERE clause containing only an expression on the left of the WHERE clause comparison.

4. The ALL_OBJECTS metadata view contains far too many rows to return its result to a SQL*Plus tool window, without taking too much time. Find the first 10 rows in the ALL_OBJECTS view, regardless of content. Save the script in a file named **ho0604.sql** in the **Solutions\Chapter06** directory on your student disk.

5. Now do the same as in Assignment 4 except this time sort the result by OBJECT_NAME within OWNER column. Save the script in a file named **ho0605.sql** in the **Solutions\Chapter06** directory on your student disk.

Case Project

Your MIS manager wants to be able to view all tablespaces, datafiles, and redo logs periodically. Create a single view, using SELECT statements, in the SYS user allowing the MIS manager to view all required information, including current space occupied. Add your script to the file named **case0601.sql** in the **Solutions\Chapter06** directory on your student disk.

CHAPTER 7

BASIC TABLE MANAGEMENT

LEARNING OBJECTIVES

In this chapter, you:

- Describe the different types of tables and their storage methods
- Create relational and temporary tables
- Examine datatypes
- Create tables containing VARRAYs and nested tables
- Create object and partitioned tables
- View database object metadata in SQL*Plus, Enterprise Manager console, and the Database Control

INTRODUCTION

In this chapter, you dig deeper into the database structure by studying how tables are stored and managed. You will examine creating tables and, focusing on the storage options of the CREATE TABLE command. You find out how Oracle 10*g* keeps track of rows, blocks, and extents for a table and how to monitor and adjust the settings that affect table storage.

Before beginning this chapter, you should install some additional objects into your database that are needed for exercises in this chapter. Follow these steps to run a script that creates the objects you need.

1. Start up the Enterprise Manager console. In Windows, click **Start/All Programs/Oracle ... /Enterprise Manager Console**. In Unix or Linux, type **oemapp console** on the command line.
2. Click **Tools/Database Applications/SQL*Plus Worksheet** from the top menu. A background SQL*Plus process starts, and then the SQL*Plus Worksheet window appears.
3. Connect as the SYSTEM user. Click **File/Change Database Connection** in the menu bar. A logon window appears. Type **CLASSMATE** in the Username box, type **CLASSMATE** in the Password box, and **ORACLASS** in the Service box. Leave the connection type as "Normal." Click **OK** to continue.
4. Now that you are connected as the CLASSMATE user, you can run the script prepared for this chapter. Click **Worksheet/Run Local Script** from the top menu. This opens a window in which you select a file. Navigate to the **Data\Chapter07** directory on your student disk, and select the **setup.sql** file. Click **Open** to run the script in the worksheet.
5. The SQL*Plus Worksheet runs the script, and replies, "Type created" several times.
6. Exit the worksheet by clicking the **X** in the upper-right corner, or typing **EXIT**, clicking the **Execute** icon, and clicking **OK** in the Oracle Enterprise Manager dialog box that opens. This returns you to the console. Now, find the new object types and collection types created.
7. Exit the console by clicking the **X** in the upper-right corner.

The objects that were created are discussed later in this chapter.

INTRODUCTION TO TABLE STRUCTURES

As you know, a table is the basic storage unit for data in the Oracle 10g database. Tables are made up of rows and columns. When you create a table, you define the table's name as well as each column's name, datatype, and size. In addition, you can define the storage settings of the table. After the table is defined, you use SQL to add data to the table. One row in a table contains data for one record. One column in a row contains one field of data for one record.

There are several different kinds of tables you can create, depending on what you need to store:

- **Relational table:** This is the standard table traditionally used in a relational database. In Oracle 10g, it is referred to simply as a table. A **table** stores data of all types and is the most common form of storage in the database. Although it is not a strict requirement, most tables have a **primary key**, which is a column or set of columns that uniquely identify each row. A table resides in a single tablespace unless it is partitioned. Partitioned tables can store each partition in a separate tablespace. Tables can be partitioned to store large quantities of data and increase performance of access to that data.
- **Index-organized table:** This type of table is a table where the primary key contains both the index and the data (non-indexed columns in the table). In other words, the entire table is an index. There is no separate index for the primary key, as there is with a relational table. Use an index-organized table when the data is retrieved by primary key values or ranges of primary key values. Retrieval by primary key is faster because the database reads only the index, instead of reading the index and then reading the table. Index-organized tables can be partitioned just as relational tables can.
- **Object tables:** An **object table** holds objects and attributes of those objects. An object table is similar to a relational table, except that each row is a single unit of data defined by an object type. An **object type** is a set of column definitions that is defined ahead of time for use in objects. For example, before creating an object table that stores customer names and addresses, you create an object type that contains all the columns you plan to use. The object type's columns are called **attributes**. Some examples of attributes are customer name, street address, city, state, country, and zip code. After creating the object type, you use it to define an object table. The object table, like relational tables, holds the data in rows and columns. Object tables are used to store complex data that requires special handling whenever data is added or changed in the object table. That special handling can be programmed into **methods**, which make data maintenance simpler and more consistent. Methods are programmed processes stored with the object type's definition. You will not be working with methods in this book.

- **Temporary table:** These tables contain data like other tables; however, the data is privately retained for a connected user (not accessible by other users) and disappears when the user commits a transaction, or when a commit is forced, such as when a user session is disconnected from the database. The CREATE GLOBAL TEMPORARY TABLE command is used to create a temporary table. Temporary tables use temporary segments in either a temporary tablespace or a permanent tablespace.
- **External table:** These tables contain data in a file outside the database, such as in a text file. You define the table and its columns as usual, and then define where the external data resides and how each column maps to the external file's records. External tables are read-only, require read and write privileges in the directory created within, and can be created and loaded using an AS subquery clause.
- **Nested table:** Nested tables contain data that is stored within a single column of another table. The data in a nested table is actually stored in its own subsidiary table structure, whereas the main table contains an identifier that locates the associated nested table for that row. Define the subsidiary table structure as you would define a relational table, with a name and storage settings.
- **XML Table:** A table created with one column of the XMLtype datatype is an **XML table**. You can store XML interpretable and XML Document Object Model (DOM) accessible data in this new type of table. **XML** (eXtensible Markup Language) is a programming language for the Internet that provides both the data and the display details to a Web page or to an application. XML is also extremely versatile in that it can be used to contain both application and metadata, even to the point that an entire database can be built from XML alone. Additionally, XML is often used to transfer data between different computers because the format of XML documents is independent of the operating system, database vendor, platform, and essentially everything.
- **Cluster:** Technically, a **cluster** is not a table, but a group of tables stored together as if they were one table. Clusters store data of multiple tables in one segment. Clusters speed up the access to sets of tables that are frequently referenced together. For example, if your application always queries the CUSTOMER table joined with the CUSTOMER_ORDER table, creating a cluster of the two tables may speed up performance. A single I/O function can retrieve the combined data from both tables.
- **Partitions:** Oracle 10*g* allows the physical splitting of table data into separated sections called partitions. Partitions can be created, splitting rows in a table based on specific criteria. Those criteria can be ranges of values, lists of values, a hash value, or various combinations thereof. A range partition allows splitting of data based on ranges of values within a table's rows. A list allows splitting based on separate lists of values. Hash partitioning uses a hashing algorithm to spread table rows evenly across separate physical files in the database. The benefit of partitioning is one of potentially much improved performance, especially for large tables in large databases. A query can execute against a single partition. If a 1,000,000 row (1 million rows) table is split into 100 partitions of 10,000 rows each, then a query against a single partition can read 10,000 rows, instead of 1 million rows.

All these tables store data in tablespaces. As you saw in Chapter 5, tablespaces define how data is stored within them. In most cases, you can further define storage at the table level, effectively overriding storage structures defined at the tablespace level. Any storage settings not defined in a table simply default to those settings defined for the tablespace. To better understand storage, begin by looking into the lowest level of logical data storage: the data block.

Setting Block Space Usage

As illustrated in Chapter 5, a table's storage is contained in a single segment. The segment contains one or more extents. Each extent contains a contiguous set of data blocks. What is inside a data block?

Each block represents a group of bytes in the physical file. The default block size is 8192 bytes, determined on database creation. Oracle 10*g* manages space down to the last byte in each block.

TIP

In locally managed tablespaces, you don't need to worry as much about the way the blocks are managed. Even with locally managed tablespaces some block parameters are still adaptable. However, in dictionary-managed tablespaces, you must understand how the STORAGE parameters work. This way, you can match the settings with the type of activity in the table.

NOTE

The same STORAGE parameters discussed here for tables can be used in CREATE TABLE, CREATE INDEX, and other CREATE commands.

Recall that the DEFAULT STORAGE clause for a dictionary-managed tablespace has this syntax:

```
DEFAULT STORAGE (INITIAL <nn> NEXT <nn> PCTINCREASE <nn>
MINEXTENTS <nn> MAXEXTENTS <nn>)
```

The STORAGE clause syntax for a table or index or other object looks like this:

```
TABLESPACE <tablespace name>
STORAGE (INITIAL <nn> NEXT <nn> PCTINCREASE <nn>
MINEXTENTS <nn> MAXEXTENTS <nn>
FREELISTS <nn> FREELIST GROUPS <nn>
BUFFER_POOL KEEP|RECYCLE|DEFAULT
PCTFREE <nn> PCTUSED <nn> MINTRANS <nn>)
```

NOTE

The MAXTRANS option of previous versions of Oracle is deprecated in Oracle 10*g*.

The INITIAL, NEXT, PCTINCREASE, MINEXTENTS, and MAXEXTENTS parameters all pertain to the size, number, and management of extents and were discussed in Chapter 5. Tablespace parameters can be overridden by setting them when you create a table, index, and so on. This is not always strictly true. In particular, for a locally managed tablespace with uniform extent size, storage settings such as NEXT will be ignored when overridden for a table in that tablespace.

The PCTFREE, PCTUSED, and MINTRANS parameters help describe the way rows are stored and accessed inside each block in the extent. These three parameters are discussed later in this section.

The FREELIST and FREELIST GROUPS clauses tell Oracle 10*g* if you want more than one freelist for each datafile. Usually, you do not need to change these parameters. By default, there is one freelist and one freelist group per extent. The freelist keeps track of all the data blocks that allow rows to be inserted into the block.

The BUFFER POOL clause identifies for Oracle 10*g* the buffer for storing data blocks from the table when they are read from the disk and stored in the SGA. The default setting is BUFFER POOL DEFAULT and is appropriate for most tables. Specify BUFFER POOL KEEP for a table that is frequently used, so that data blocks stay resident inside the buffer longer than normal. The data blocks stay in the KEEP buffer until it is full and more data blocks need to be added. Then, the data blocks are flushed from the buffer with the least recently used blocks (**LRU blocks**) being replaced first. You can adjust the size of the KEEP buffer by modifying the DB_CACHE_KEEP_SIZE initialization parameter. Specify BUFFER POOL RECYCLE for tables that should be removed from the buffer as quickly as possible. This setting is best for infrequently used tables. Using this setting keeps the infrequently used table's data blocks from overwriting data blocks in the DEFAULT buffer. You can adjust the size of the RECYCLE buffer by modifying the DB_CACHE_RECYCLE_SIZE initialization parameter.

NOTE

The BUFFER_POOL_KEEP and BUFFER_POOL_RECYCLE parameters are still available (no longer listed in the manuals), but have been replaced with the DB_KEEP_CACHE_SIZE and DB_RECYCLE_CACHE_SIZE parameters respectively.

NOTE

Storage applies to both tables and indexes.

Tables store data in data blocks. Figure 7-1 shows the components that make up a single data block:

- **Common and variable header:** Identifying information, such as the type of block and block location.
- **Table directory:** Information about the table that has data in the block.
- **Row directory:** A list of row identifiers for rows stored in the block. This grows as more rows are inserted into the block. Even when a row is dropped, the row directory does not shrink.

- **Free space:** Bytes of storage space left unallocated. This shrinks as rows are inserted or updated with more data, causing both the row directory and the row data to consume more space. When a row is deleted or updated with less data, the free space grows, although the row directory never releases any of its space.
- **Row data:** Bytes of storage used for rows inserted or updated in the data block. Updates that make a row grow in size cause the row data to be shuffled so that the data for each row stays together. If a row is updated and requires more free space than the block contains, the entire row is **migrated**, that is, moved to another block. The original block has a pointer to the second block. If a row is too large to store in one block when it is inserted (for example, a row containing a 50Mb recorded song), the row is started in one block and the overflow is stored in another block. The original block has a pointer to the second block. If it needs a third block, the second block has a pointer to the third block, and so on. A row that spans multiple blocks is called a **chained row**.

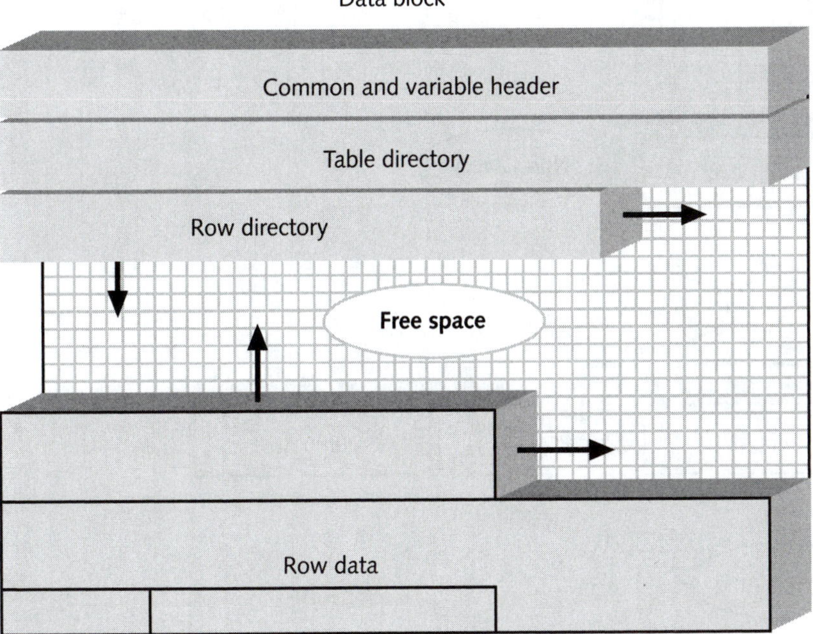

FIGURE 7-1 A data block's free space shrinks as overhead and row data grow

The first three data block components (common and variable header, table directory, and row directory) are called overhead because they store information about the maintenance and use of the block rather than actual data. The size is affected by changes in the data (more rows mean more bytes in the row directory) and by the setting of INITRANS in the STORAGE clause of a table.

INITRANS tells Oracle 10g how many concurrent transactions can access this block. The block header establishes a storage space inside the header for each transaction. The default value for INITRANS is 1 for tables and 2 for indexes. It is best not to modify INITRANS unless performing detailed physical tuning, especially in highly concurrent OLTP databases.

One of the key issues in good performance is to avoid the chaining and migrating of row data. A chained or migrated row requires additional work to retrieve.

NOTE

A chained row is a row split between multiple blocks. A migrated row is transferred in its entirety to another block, retaining only a pointer to the new block in the original block.

Oracle 10g has no direct way to reach the data stored in subsequent blocks. It must read the row header in the original block, locate the pointer to the second block, and read the second block. Figure 7-2 shows a normal row, a chained row, and a migrated row.

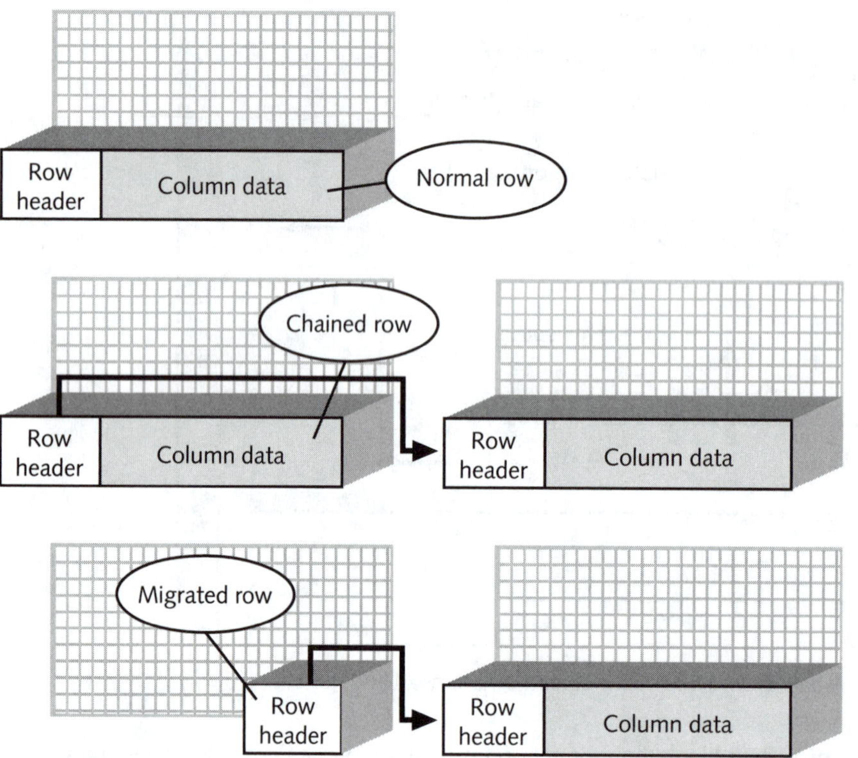

FIGURE 7-2 In some cases, chaining and migration allow one row to span multiple blocks

In the case of chained rows, the only solution is to increase the block size, and even that may not help in some cases. With large column data, such as those used by multimedia applications, some row chaining is inevitable.

To prevent row migration, you can adjust two storage parameters to suit the characteristics of your table's data. The two parameters are PCTFREE and PCTUSED. The two together must always add up to 100 or less.

PCTFREE tells Oracle 10g how much free space to reserve for updates to existing rows. For example, if PCTFREE is 20, rows can be inserted until the free space shrinks to 20 percent of the block. At this point, Oracle 10g allows no more inserting of rows. The remaining 20 percent of free space is reserved for the growth of the rows already in the block. For example, if you change the value in a column from "Hank" to "Henry-James," the size of that row grows, using some of the remaining free space. When a block stores rows that increase in size due to updates, a higher PCTFREE setting helps keep the rows from migrating.

PCTUSED tells Oracle 10g when to allow more rows to be inserted into the block after the deletion of rows from a block. If a row is deleted and the free space grows to more than 20 percent, the block still does not accept new rows (unless PCTUSED is set at 80 percent). The free space has to increase to at least the number in PCTUSED before new rows can be inserted. For example, if PCTUSED is 50 percent, the block initially is filled until only 20 percent is free (because PCTFREE is 20 percent). After this, if rows are deleted and the free space grows to 50 percent, rows can be inserted again. Rows can be inserted until the free space shrinks to 20 percent. This continues cycling as much as needed.

If your table data is not updated often, a lower PCTFREE allows you to use more space in the block. If your table data grows a great deal, a higher PCTFREE prevents row migration. A PCTFREE of zero packs rows in all the free space and works well for inserting masses of static data, such as a table of telephone area codes and their city, county, and state locations.

A higher PCTUSED allows Oracle 10g to reuse the space faster. The block is put back on the freelist that tracks block usage for the extent. However, because the row header keeps growing with every insert and does not shrink with deletes, it is conceivable that you could reach a point at which the row header consumes more room than the data.

The next section will cover details on how to set INITRANS, PCTFREE, and PCTUSED for a table.

Storage Methods

Storage methods for tables depend on the type of table you are creating and the characteristics of the tablespace in which you create the table. You can define a table with no storage settings, and Oracle 10g uses the default settings found in the tablespace, if they exist. If no storage defaults were set up for the tablespace, Oracle 10g uses its own standard default settings, which are as follows:

- PCTFREE=10
- PCTUSED=40
- INITRANS=1 (1 for tables and 2 for indexes)

The most important difference in storage methods is between locally managed and dictionary-managed tablespaces.

How to Set Storage for Locally Managed Tables

As you learned in Chapter 5, locally managed tablespaces handle free space using a bitmap within the tablespace. The DEFAULT STORAGE clause is not used with locally managed tablespaces. Similarly, only a few of the parameters of the STORAGE clause should be used in the tables created in a locally managed tablespace.

What storage information do you specify for a table in a locally managed tablespace?

- **TABLESPACE:** Name the tablespace in which the table is created. If you omit this parameter, the table is created in your default tablespace. The SYSTEM tablespace is used if the creator of the table does not have a default tablespace assigned. The SYSTEM tablespace contains metadata. Placing application tables into the SYSTEM tablespace creates conflict between Oracle Database system requirements and application requirements. This should be avoided.
- **STORAGE (INITIAL <nn>):** Oracle 10g looks at the tablespace attributes to determine extent size. If the tablespace specifies AUTOALLOCATE, Oracle can make the extent or extents any size that it determines is best. If the tablespace specifies UNIFORM, Oracle 10g allocates enough uniform extents to reach the INITIAL size. For example, if UNIFORM 1Mb set for the tablespace and INITIAL 4Mb is set for the table, Oracle 10g allocates 4 1Mb extents to the table initially. All the other parameters of the STORAGE clause are ignored, even if you include them.

Complete the following steps to create a table in a locally managed tablespace.

1. Start up SQL*Plus Worksheet. In Windows, select **Start/All Programs/Oracle ... /Application Development/SQLPlus Worksheet**. In Unix or Linux, type **oemapp worksheet** on a command line. The standard logon window for all Enterprise Manager tools appears.
2. Type **CLASSMATE** in the Username box, **CLASSPASS** in the Password box, and **ORACLASS** in the Service box. Leave the Connect set to "Normal" and click **OK**. The SQL*Plus Worksheet displays.
3. Type the following CREATE TABLE command in the top pane of the SQL*Plus Worksheet window. You learn about the other components of the command later in this chapter:

```
CREATE TABLE CH07BICYCLE
(BIKE_ID NUMBER(10) PRIMARY KEY,
BIKE_MAKER VARCHAR2(50) NOT NULL,
STYLE VARCHAR2(15))
TABLESPACE USERS
STORAGE (INITIAL 25M);
```

4. Execute the command by clicking the **Execute** icon on the left side of the worksheet. Oracle 10g creates the table and displays "Table created" in the SQL*Plus Worksheet window.
5. Exit SQL*Plus Worksheet by typing **EXIT**, clicking the **Execute** button, and clicking **OK** in the Oracle Enterprise Manager dialog box that opens.

TIP

You do not need to issue a COMMIT command after executing DDL (Data Definition Language) commands, because they are automatically committed.

If you omit the STORAGE clause altogether, Oracle 10*g* uses the tablespace settings, applying them to the table creation as a default for the table. If the tablespace has no MINEXTENTS setting, Oracle 10*g* creates the table with an initial extent of five data blocks (the default size) for AUTOALLOCATE, or one extent of whatever extent size is specified in UNIFORM.

Locally managed tables eliminate much of the work in managing storage space. However, it is still important to estimate the initial size of a table to keep its data in contiguous data blocks for faster data retrieval.

NOTE

Contiguous blocks are blocks that are located physically adjacent to each other on disk. This makes for more rapid accessing of similarly related data because the data is on disk in the same place.

How to Set Storage for Dictionary-Managed Tables

This section explores the details of specifying every possible storage parameter. INITIAL is the most important parameter because it defines the table's initial size. The next most important parameters in dictionary-managed tables are the PCTFREE and PCTUSED parameters.

CAUTION

Do not execute the example code in this section. The code will cause errors because your Oracle 10*g* database has no dictionary-managed tablespaces.

Consider the tablespace's DEFAULT storage settings before writing the STORAGE settings for a table. It is generally better to keep extent sizes the same or multiples of the same size within the tablespace so that any deallocated free space can be more readily used later on. If extent sizes are continually increased in size as they are added by the setting of PCTINCREASE, then any deleted extents will never be used by new allocations for a new extent because any new extent will never fit into the smaller empty extents. The result will be tremendous amounts of wasted space. All full table scans by queries read the entire table, including any deleted extents, ultimately leading to poor performance.

Following is an example of a table created in a dictionary-managed tablespace, in which all the default storage parameters are used, including the tablespace name. In this case, the tablespace name is derived from the user's default tablespace setting. That tablespace is dictionary-managed and has default storage settings that are used to create the table:

```
CREATE TABLE BIKE_MAINTENANCE
(BIKE_ID NUMBER(10),
 REPAIR_DATE DATE,
 DESCRIPTION VARCHAR2(30));
```

In this example, imagine that the tablespace's default storage settings are like this:

```
DEFAULT STORAGE (INITIAL 10M NEXT 2M
PCTINCREASE 0 PCTFREE 10 PCTUSED 80)
```

The BIKE_MAINTENANCE table would inherit all these settings. By using the same settings as other tables in the tablespace, extents tend to be uniform in size and, therefore, more easily reused when a table is dropped or storage is released by a TRUNCATE command.

Here is another table in which the initial size, extent size, and other table parameters, will override tablespace settings:


```
CREATE TABLE TRUCK_MAINTENANCE
(TRUCK_ID NUMBER(10),
 REPAIR_DATE DATE,
 PROBLEM_DESCRIPTION VARCHAR2(2000),
 DIAGNOSIS VARCHAR2(2000),
 BILLING_DATE DATE,
 BILLING_AMT NUMBER (10,2))
TABLESPACE USER_DTAB
STORAGE (INITIAL 80M NEXT 40M PCT INCREASE 0
MINEXTENTS 2 MAXEXTENTS 25
PCTFREE 25 PCTUSED 50 MINTRANS 1 MAXTRANS 2)
```

The TRUCK_MAINTENANCE table sets its own storage settings because it anticipates a higher volume required for storing its data. It still uses a multiple of the tablespace's default setting for INITIAL and NEXT so that any deallocated extents are more easily used by other tables.

TIP

The user can create a table in his default tablespace. In addition, if the user has the UNLIMITED TABLESPACE system privilege, he can create a table in any tablespace. Without the UNLIMITED TABLESPACE system privilege, a user must be given a quota of space on any tablespace he uses to store the tables that he creates. See Chapter 11 for more information.

The next section contains one final aspect of data storage: the makeup of row data. After working with row structure, you begin to dig into the details of what designers see as the building blocks of tables: columns.

Row Structure and the ROWID

Rows are stored in a data block in a compact structure, so a data block can accommodate the maximum amount of data. Figure 7-3 shows the components of a row.

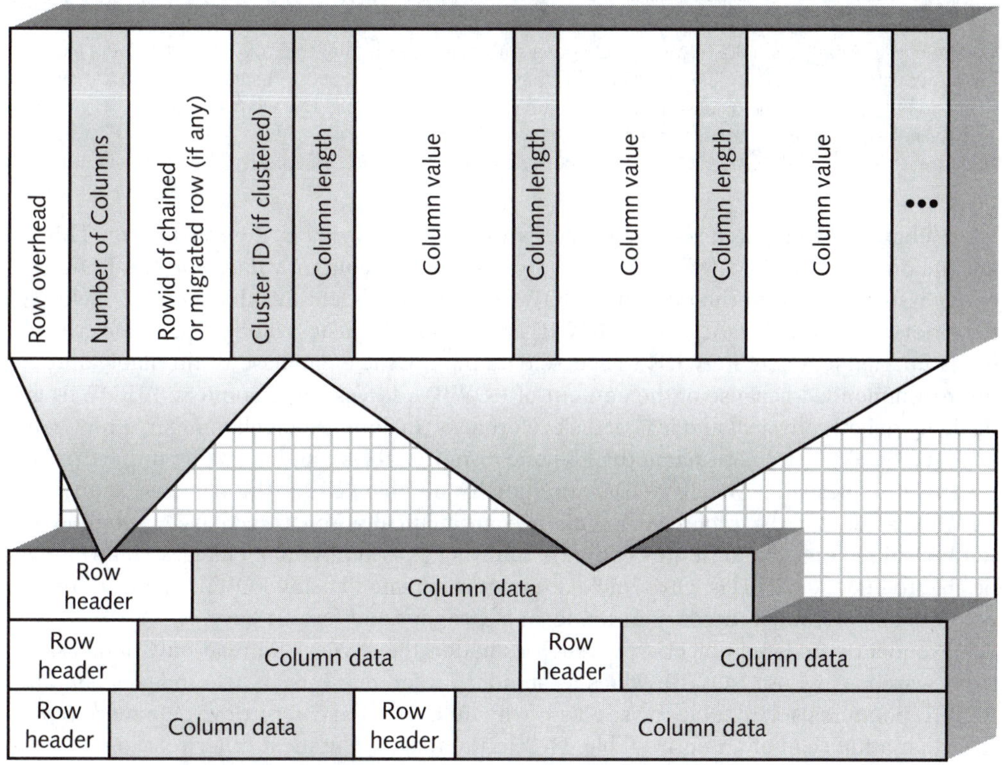

FIGURE 7-3 The row header points to a chained row if needed

As shown in Figure 7-3, a row is made up of two sections:

- **Row header:** This stores the number of columns contained in the column data area, some overhead, and the ROWID pointing to a chained or migrated row (if any). A **ROWID** contains the physical or logical address of the row. The row header is a minimum of three bytes long. If this is a row in a cluster, the cluster key ID is also included. A **cluster key ID** is a special type of ROWID for clustered tables. Oracle 10*g* stores only one **clustered row** (that is, the rows from each table in a cluster that correspond to one unique value in the cluster key) per block. Clustered tables are not covered in this book.
- **Column data:** The column data consists of a one- to three-byte section containing the length of the column's data followed by the actual data for the column. Notice that there is no column name or identifier in either the row header or the column data. Oracle 10*g* stores column data in the exact order as defined when the table was created. If a row's column has null values *and is at the end of the columns*, nothing is stored. A null column *followed by one or more columns with data* must have a placeholder, because Oracle 10*g* determines a column's name by its order in the list of column data. Therefore, Oracle 10*g* stores a column length of zero for the null column, followed immediately by the column length of the next column.

TIP

When creating the table, take advantage of space savings by placing the columns defined with the NOT NULL constraint first in the table, followed by columns that allow null values. Placing columns that never contain null values first helps save space by improving the odds that the row won't have extra placeholders for null columns that appear between columns with data.

A chained or migrated row is located by the ROWID stored in the row header. This is not the only use for the ROWID. Indexes store the index column values and the ROWID of the associated row in the table. The ROWID is a special identifier that allows Oracle 10*g* to retrieve a row quickly. Access by ROWID is probably the fastest method of data retrieval but is unreliable. Using ROWIDs to access data in an Oracle database is unreliable and not recommended because of the content of ROWID address value pointers. A ROWID is made up of both physical and logical disk references, including a block number, a row number (in a block), a segment name (can be one or more tables), and a datafile number relative to its tablespace. The ROWIDs can actually be changed by the database – any ROWID values stored in tables will not change automatically. You would have to rebuild your ROWIDs. In a large database this can be a daunting task. If anything changes you will not be able to find data. This is why Oracle does not recommend using ROWID value address pointers for referential integrity in the form of primary and foreign keys.

Frequently updated object types, such as materialized views, in read-only databases (data warehouses) use ROWID values as standard reference values. Also indexes contain ROWID pointers back to table rows. This is why indexes often can retrieve data more rapidly than a full scan of an entire table. Oracle Databases, like most other relational databases, have a special process called an optimizer. An optimizer process automatically calculates the best way to execute a query. It is this optimizer process that decides whether to use an index for a query, or if better efficiency is gained through reading the entire table and ignoring the index altogether.

Figure 7-4 shows a diagram of the two types of ROWIDs found in Oracle 10*g*.

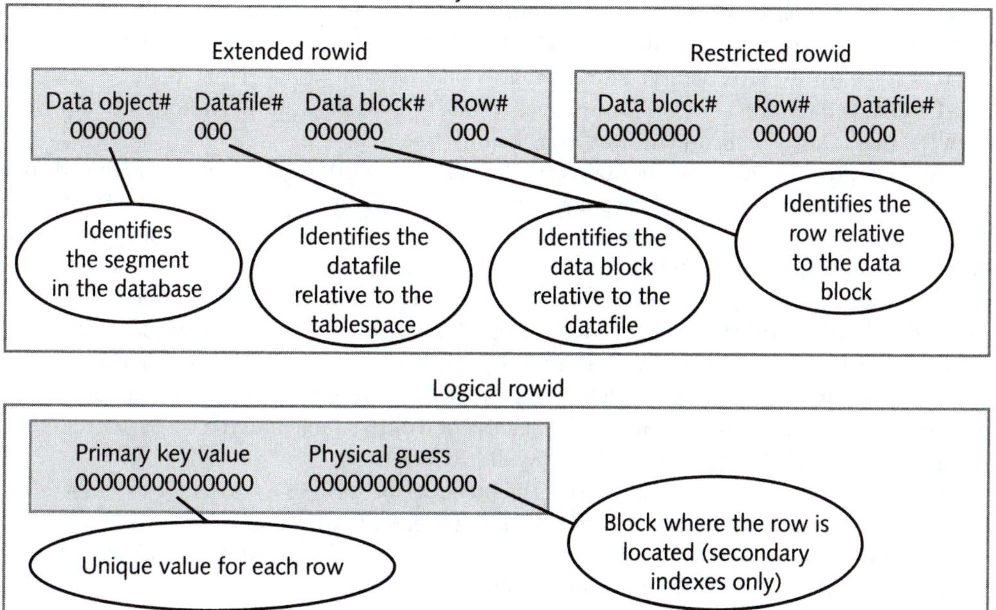

FIGURE 7-4 Physical and logical ROWIDs have different internal formats

As shown in Figure 7-4, the types of ROWIDs are:

- **Physical ROWID:** Identifies a row by its physical location in a datafile. The physical ROWID of a row never changes, unless the table is reorganized. There are two formats for the physical ROWID:
  - **Extended ROWID:** The default format for Oracle8*i* and above. This contains four values, as shown in Figure 7-4.
  - **Restricted ROWID:** This ROWID format was standard for Oracle7 and below and is preserved for backward compatibility. It is missing the object identifier found in the extended ROWID.
- **Logical ROWID:** Identifies a row by its primary key in an index-organized table. It may also contain the probable location of the block in which the row resides. A logical ROWID is used only for index-organized tables. The physical location of rows in an index-organized table changes as rows are added and deleted, so the physical ROWID is not accurate.

ROWIDs cannot be changed by any user. The ROWID is actually not stored with the row at all. A ROWID is only stored when an index on the table is created, or when you query the pseudocolumn ROWID, in which case the ROWID is constructed. A **pseudocolumn** acts like a column in a query, but is actually calculated by the database a query is executed. You can include a pseudocolumn when you query any table, and Oracle 10*g* delivers the appropriate data in the pseudocolumn. Another example of a pseudocolumn is SYSDATE, which returns the current date and time.

Indexes are automatically kept up-to-date as data in the table changes. An index is stored in the database as a table, and contains the values of the indexed columns and the ROWID of the corresponding table row. The ROWID is stored in a column with the UROWID datatype (or ROWID datatype for databases that are compatible with Oracle7 or lower). The **UROWID** datatype stands for universal ROWID datatype, and it can hold any type of ROWID, including a record identifier from a non-Oracle database.

You can create a table that contains the ROWIDs of another table's rows to take advantage of the high-speed retrieval rate for ROWIDs. For example, you have a CUSTOMER table with a primary key called CUST_ID. You build a CONTACT_LIST table that contains rows for each contact made to a customer. Ordinarily, you link the two tables together with a foreign key. However, you could create a column (with UROWID datatype) in the CONTACT_LIST table that stores the ROWID of the customer's row. Be cautious with this trick, because if a row in the CUSTOMER table is deleted, the corresponding ROWID found in any rows in the CONTACT_LIST for that customer are not automatically removed. Later, a query that joins the CONTACT_LIST table and the CUSTOMER table by the ROWID might fail because of the invalid ROWID.

You can use the ROWID to determine the physical location of a particular row in a table by looking at the portion of the ROWID storing the datafile information.

Now that you are familiar with the data block, how rows are stored, and what the ROWID contains, it is time to build some tables.

CREATING TABLES

You have already been exposed to the CREATE TABLE command earlier in this chapter. The command syntax varies according to what type of table you are creating. For relational tables (the most commonly used type of table), the syntax looks like this:

```
CREATE TABLE <schema>.<tablename>
(<column_name> <datatype> <size> NULL|NOT NULL
  DEFAULT <default_value> CHECK <check_constraint>,
... )
<constraints>
TABLESPACE <tablespace name>
STORAGE (INITIAL <nn> NEXT <nn> PCTINCREASE <nn>
MINEXTENTS <nn> MAXEXTENTS <nn>
FREELISTS <nn> FREELIST GROUPS <nn>
BUFFER_POOL KEEP|RECYCLE|DEFAULT
PCTFREE <nn> PCTUSED <nn> MINTRANS <nn> MAXTRANS <nn>)
```

NOTE

The STORAGE clause is highlighted because it is generally only applicable to dictionary-managed tablespaces when creating a tablespace. The STORAGE clause is retained in Oracle 10*g* for backward compatibility with older versions of Oracle. In other words, when creating locally managed tablespaces, do not specify a default storage clause. When creating tables in a locally managed tablespace, a storage clause can be defined, but any irrelevant parameters, such as PCTUSED, will simply be ignored by the CREATE TABLE statement. In truth, the STORAGE clause is still usable in locally managed tablespaces but it is discouraged. PCTFREE is significant for manual segment space management, but automated segment space management is encouraged, and thus free space is managed by Oracle. Additionally, automatic allocation of extents manages the size of all extents, and thus INITITAL, NEXT, PCTUSED, MINEXTENTS, MAXEXTENTS, even UNIFORM sized extents in locally managed tablespaces, are also all discouraged, and thus irrelevant to the future of Oracle databases. However, note that backward compatibility is an issue to existing installations, but not an issue to Oracle 10*g* certification exams. It can not be ignored.

The syntax for storage was discussed earlier in the chapter and in Chapter 6. The next section primarily discusses the column definition section of the command.

NOTE

The syntax for other types of tables and an alternative syntax for relational tables are discussed later in this chapter.

Columns and Datatypes

A table must contain at least one column to be defined in the database. Columns have names and datatypes and can optionally have a default value and a restriction on null values. A **datatype** is a predefined form of data. Oracle has numerous standard datatypes. A **default value** is a value that is filled into the column when an inserted row does not specify the value for that column. To prevent inserted or updated rows from leaving a particular column null, add the NOT NULL constraint to the column definition. **Constraints** are rules that define data integrity for a column or group of columns. See Chapter 10 for a complete description of constraints.

Table 7-1 shows a categorized list of Oracle's predefined (built-in) datatypes. More details for some of the datatypes are listed following the table.

TABLE 7-1 Oracle 10*g* built-in datatypes

Classification	Datatype	Parameters	Example
Simple Datatypes	VARCHAR2(n)	n=1 to 4,000	VARCHAR2(25)
	Text string with variable length up to 4000 bytes. If the column data's length is shorter than n, Oracle adjusts the length of the column to the size of the data. Trailing blanks are truncated. Use VARCHAR2 in favor of CHAR to avoid wasting space. VARCHAR is still a valid datatype but is replaced in favor of VARCHAR2.		
	CHAR(n)	n=1 to 2000	CHAR(14)
	Same as VARCHAR2 except it holds up to 2000 bytes, and is a static (fixed length) text string, regardless of the length of the data. Trailing blanks are preserved. Shorter data is padded to right with blanks. CHAR is the same as CHAR(1). Use CHAR rather than VARCHAR2 for short strings of a semi-fixed length or precisely known number of characters.		
	NVARCHAR2(n)	n=1 to 4,000	NVARCHAR2(65)
	Same as VARCHAR2, except that it stores characters for any language (national character set) supported by Oracle.		
	NCHAR(n)	n=1 to 2,000	NCHAR(30)
	Same as CHAR, except that the characters stored depend on a national character set (Chinese characters, for example).		
	NUMBER(p,s)	p=1 to 38, s= -84 to 127	NUMBER(10,2)
	Precision (p) is the total number of digits and the scale (s) is the number of digits to the right of the decimal. Oracle rounds data you insert if it has too many decimal places.		
	INTEGER		INTEGER
	Creates the same datatype as NUMBER(38).		
	SMALLINT		SMALLINT
	Creates the same datatype as NUMBER(38).		
	FLOAT(p)	p=1 to 126	FLOAT(20)
	A floating point or real number.		
	DATE		DATE
	Valid dates range from January 1, 4712 B.C. to December 31, 9999 A.D. Oracle stores DATE datatype values internally as 7-byte numbers including the time in hours, minutes, and seconds. If no time is specified when inserting a date, the time is set to midnight.		
	TIMESTAMP(p)	p=fractions of a second	TIMESTAMP(3)
	Same range as a DATE datatype, except this contains fractions of a second. For example, TIMESTAMP(4) has precision to 1/1000th of a second.		

TABLE 7-1 Oracle 10*g* built-in datatypes (continued)

Classification	Datatype	Parameters	Example
	TIMESTAMP(p) WITH TIME ZONE	p=fractions of a second	TIMESTAMP(4) WITH TIME ZONE
	Same as TIMESTAMP except the value includes the time zone of the user that inserts or updates the value.		
	TIMESTAMP(p) WITH LOCAL TIME ZONE	p=fractions of a second	TIMESTAMP(4) WITH LOCAL TIME ZONE
	Same as TIMESTAMP except the value converts the date to the time zone of the database and displays the time in the local time zone for the viewer.		
	INTERVAL YEAR [(p)] TO MONTH	p=year precision	INTERVAL is a specialized literal value
	Stores a time interval as years and months only		
	INTERVAL DAY [(p)] TO SECOND	p=day precision	INTERVAL is a specialized literal value
	Stores a time interval as days, hours, minutes and seconds		
	ROWID		ROWID
	Internal Oracle datatype that stores the physical locator string or logical pointer for a row of data.		
	UROWID		UROWID
	Universal ROWID. Hexadecimal string containing ROWID values for an index-organized table, object table, or non-Oracle entity. Can be up to 4000 bytes.		
	BINARY_FLOAT		BINARY_FLOAT
	32-bit binary precision floating point number including values for infinity and NaN. NaN means "not a number".		
	BINARY_DOUBLE		BINARY_DOUBLE
	64-bit binary precision floating point number including values for infinity and NaN. NaN means "not a number".		
Binary Objects	BLOB		BLOB
	Stores unstructured data in binary format, up to 4GB.		
	CLOB		CLOB
	Character data up to 4GB. Used for high volume text data.		
	NCLOB		NCLOB
	Stores large (up to 4 GB) data in unicode or a National character set.		

TABLE 7-1 Oracle 10*g* built-in datatypes (continued)

Classification	Datatype	Parameters	Example
	LONG		LONG (Desupported)
	Maximum size is 2GB. Used for text data. You should use BLOB instead of LONG when creating new tables.		
	RAW(n)	n=1 to 2000	RAW(500) (Desupported)
	Raw binary data of variable length, up to 2000 characters. Use CLOB instead.		
	LONG RAW		LONG RAW (Desupported)
	Raw binary data of variable length. The maximum length is 2GB. Use BLOB instead.		
Reference Pointers	BFILE		BFILE
	Stores pointers to an external file, such as an audio track. Oracle provides pre-defined functions for reading, storing, and writing a BFILE column. Requires a directory object in order to function.		
	REF	REF schema.objname	REF MUSIC.INSTRUMENT_OBJ
	Reference object identifier. Used for object tables to define a referential or object-parent to another object table, similar to a foreign key.		
Object Collections	VARRAY	(subscript)	Object(subscript)
	Fixed length array or reserved chunk of memory for a fixed number of array elements. VARRAY collections can be resized and used in temporary tables.		
	Nested Table	TABLE(...)	TABLE (SELECT ...)
	Dynamic array or pointer to a variable number of array elements. Nested table columns can be divided into separate tablespaces.		
	Associative Array		
	Indexed dynamic array. Faster access than a nested table using an index.		

TABLE 7-1 Oracle 10*g* built-in datatypes (continued)

Classification	Datatype	Parameters	Example
Specialized Types	XML		
	XML documents can be stored, retrieved and manipulated as XML documents. See Chapter 17.		
	Spatial		
	Special spatially oriented datatypes allowing for multiple dimensions such as for graphical (geographical) information systems (maps) and architectural and construction design, amongst various other types of geometric modeling data.		
	Media		
	Special multimedia datatypes of which there are numerous different datatypes.		
	Any		
	Unknown or generic datatypes.		

NOTE

Both DATE and TIMESTAMP contain hours, minutes, and seconds, as well as the year, month, and day. Only TIMESTAMP can contain fractions of a second (up to .000000001 second precision).

The following list contains a few more notes on some of the datatypes found in Table 7-1.

- **NUMBER [(p, s)]:** Precision (p) is the total number of digits. Scale (s) is the number of digits to the right of the decimal. Oracle 10*g* rounds an inserted value if it has too many decimal places, including no decimal places at all. If you use NUMBER without any scale or precision, the number can be any supported size. If you omit the precision, such as NUMBER(10), the number has 10 digits and no decimal places. Oracle 10*g* rounds off numbers that are a different precision. For example, this column has a precision of two:

  ```
  PAYRATE NUMBER(5, 2)
  ```

  In this example, inserting the value 100.256 causes Oracle 10*g* to round up and store 100.26.

- **DATE:** The default display format for DATE is DD-MMM-YYYY (15-JAN-2003, for example). Use the TO_CHAR function to display times or other date formats. Use the TO_DATE function to insert any format of date and time you want. No matter what format you use to insert the date, the date displays in the default format unless you use the TO_CHAR function to specify how you want it displayed.
- **TIMESTAMP WITH LOCAL TIME ZONE:** This datatype is the same as TIMESTAMP, except that the database converts the time to the local time of the database. Oracle 10*g* automatically converts the time to the user's local time when it is displayed. In addition, the time is adjusted to the local time zone of the database when inserted into the database.

- **INTERVAL YEAR TO MONTH:** This datatype stores intervals of time containing only a year and a month. A specialized interval string is used to contain and structure year and month values. The following query will return the string +01-08 representing 1 year and 8 months, or 20 months as shown in the query:

  ```
  SELECT INTERVAL '20' MONTH FROM DUAL;
  ```

  The following query will return a year and month datatype string representation of +20-02:

  ```
  SELECT INTERVAL '20-2' YEAR TO MONTH FROM DUAL;
  ```

  The following query will return the string +00010-00, as in years only, and up to 5 digits of precision:

  ```
  SELECT INTERVAL '10' YEAR(5) FROM DUAL;
  ```

  Consequently the following query does not have enough precision for the year value to be returned, and will return a precision error:

  ```
  SELECT INTERVAL '100' YEAR(2) FROM DUAL;
  ```

- **INTERVAL DAY TO SECOND:** This datatype stores intervals of time containing only days, hours, minutes, and seconds (not years or months). A specialized interval string is used to contain and structure day, hour, minute, and second values. These datatypes are similar to INTERVAL YEAR TO_MONTH datatypes, except using a different format.
- **UROWID:** This stores ROWIDs, including physical ROWID, logical ROWID, and external (for locating data in a non-Oracle database) ROWIDs. It can store up to 4000 bytes.
- **Object collections datatype**: Collections are repeating data contained within one column. They can be defined as an array (called **VARRAY** in Oracle 10g) or as a nested table. See the section titled "Creating VARRAYs and Nested Tables" later in this chapter.
- **Reference pointers:** Object tables can use any of the datatypes listed previously. In addition, there are two more datatypes just for object tables: REF and user-defined. The REF datatype is similar to a foreign key and links related object tables together. The **user-defined datatype** can be any collection of the other datatypes you want assembled into its own datatype. See the section titled "Creating Object Tables" later in this chapter.

The next sections guide you through creating various types of tables beginning with a relational table.

Creating Relational Tables

To create a relational table, you must design the table, deciding these factors:

- Name of the table
- Name and datatype of all columns
- Estimated initial size and growth pattern
- Location of the table
- Constraints, relationships to other tables, default data

Names of tables and columns follow naming rules of Oracle 10g: They should be 1–30 characters long. If you enclose the name in double quotes, it is case sensitive and can begin with any letter, number, or symbol character, including spaces. If you do not enclose the name in double quotes, it is interpreted as uppercase. An unquoted column name must begin with a letter character, although it can contain any letter, number, and the symbols # (number sign), $ (dollar sign), and _ (underscore).

Estimating the initial size and growth pattern helps you specify the storage settings for the table. In essence, you determine the average length of each column's data, add some overhead space for each column, and add them to determine the average row length. Then, you estimate the number of rows the table will have initially and multiply that by the average row length. This gives you a rough estimate of the initial size of the table (INITIAL). Growth patterns, such as how often rows are inserted, updated, and deleted, help you determine the size of the extents (NEXT) and how many extents to use (MAXEXTENTS).

Complete the following steps to create a relational table. During this exercise, you use the SQL*Plus Worksheet, which has a Windows-like format that makes editing long commands easier.

The tables you create in this chapter are the first tables you create in the Global Globe database project that you are developing throughout the book and, in particular, in the end of chapter Case Projects.

First, create a table to keep information about the Global Globe employees. You need to identify editors so that you can assign them to specific sections of the classified ads. In addition, you need their names and phone numbers. The phone number can have an extension, so you want to allow both numbers and characters in the field.

1. Start up the Enterprise Manager console. In Windows, click **Start/All Programs/Oracle ... /Enterprise Manager Console**. In UNIX or Linux, type **oemapp console** on the command line. The Enterprise Manager Console login screen appears.
2. Click **Tools/Database Applications/SQL*Plus Worksheet** from the top menu, as shown in Figure 7-5.

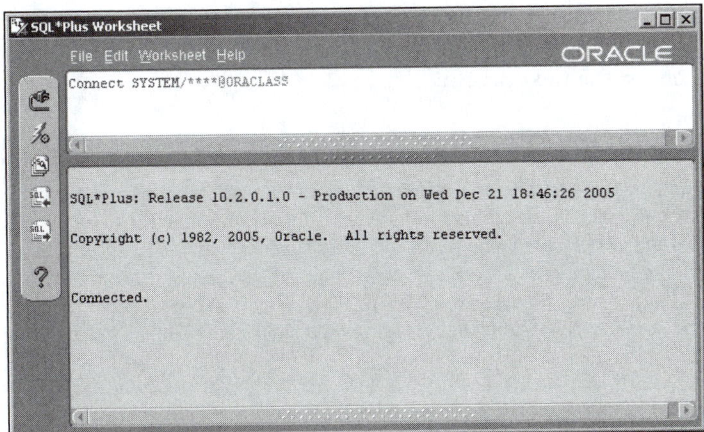

FIGURE 7-5 SQL*Plus Worksheet can be started from the Enterprise Manager console

3. Connect as the CLASSMATE user. Click **File/Change Database Connection** on the menu bar. A logon window appears. Type **CLASSMATE** in the Username box, **CLASSPASS** in the Password box, and **ORACLASS** in the Service box. Leave the connection type as "Normal." Click **OK** to continue.
4. Now that you are connected as the CLASSMATE user, you can create tables owned by CLASSMATE. (SYSTEM can also create objects owned by CLASSMATE; however, logging on as CLASSMATE uses this particular user's default tablespace and quotas.)
5. To begin creating a relational table, type this line in the top window of the worksheet:


```
CREATE TABLE EMPLOYEE
```

This line defines the table name as EMPLOYEE.
6. Now create the columns by typing these lines below the first line:

```
(EMPLOYEE_ID NUMBER (10),
JOB_TITLE VARCHAR2(45),
first_name varchar2(40),
last_name varchar2(40),
phone_number Varchar2(20)
```

Our example shows a mixture of uppercase and lowercase letters simply to illustrate that you can use either uppercase or lowercase letters in the column names as well as in the datatypes. As long as you don't use double quotation marks around the column name, Oracle 10g translates the name into all uppercase letters.
7. Complete the column list with a **closing parenthesis**. At this point, you can start defining storage settings. For this example, allow the tablespace defaults to be used by specifying no storage clause. Type a **semicolon** (;) to mark the end of the command. Figure 7-6 shows the entire command.
8. Click the **Execute** icon on the left side of the window. The icon looks like a lightning bolt. SQL*Plus Worksheet creates the table and displays "Table created" in the lower portion of the Worksheet.
9. Save your work in the **Solutions\Chapter07** directory of your student disk by clicking the **File/Save Input As** in the top menu and by selecting the directory. Name the file **employee.sql**, and click **Save** to save the file.
10. Remain logged on for the next practice.

NOTE

You can use the Enterprise Manager's Schema Manager to create tables. A Create Table Wizard guides you through all the steps, including calculating the storage requirements for your table. Alternatively, you can use the Create Table window and fill in a spreadsheet-like format for columns. Figure 7-7 shows the Create Table window of the Schema Manager. Use the Schema Manager when you want help calculating storage for a new table, or when you need help with the CREATE TABLE syntax. The Schema Manager has a feature with which you can view the DDL command that creates the table. This helps remind you of how to write a CREATE TABLE command.

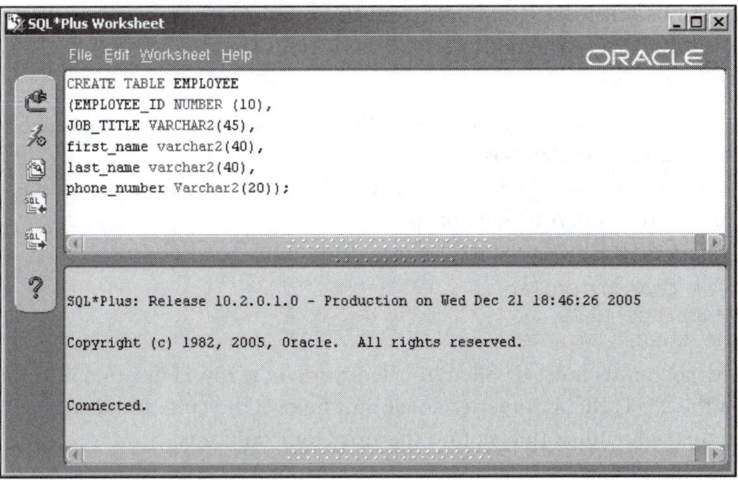

FIGURE 7-6 The completed command is ready to execute

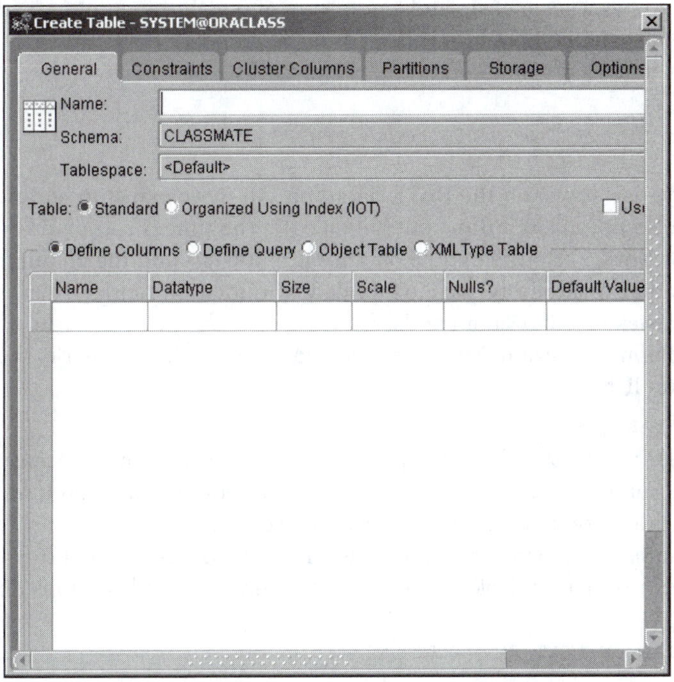

FIGURE 7-7 Creating a table in the Schema Manager is fast and easy

For the next table, you add a table that stores all the classified ads running in the paper. Ads are not removed after they run, so customers can renew ads if they would like to. You want to keep track of the editor that worked with the customer and entered the ad, plus the final price of the ad and its start date, end date, and the number of days it runs. You also add storage settings for this table. The table resides in a locally managed tablespace. Follow along with the steps.

1. If you are not already there, log on to SQL*Plus Worksheet as the user CLASSMATE in the ORACLASS database.
2. If a previous command is displayed, clear it by highlighting the text and clicking the **Delete** button on your keyboard.
3. Type in the first three lines to start defining the table. The lines are:

   ```
   CREATE TABLE CLASSIFIED_AD
   (AD_NO NUMBER NOT NULL,
    SECTION_NO NUMBER NOT NULL,
   ```

   These first two columns must be filled in whenever a new row is inserted into the database, so the NOT_NULL integrity constraint follows the column's datatype.
4. Type in the next four lines that define the next four columns:

   ```
   AD_TEXT VARCHAR2(1000),
   CUSTOMER_ID NUMBER,
   INTAKE_EDITOR_ID NUMBER,
   PRICE NUMBER (6,2),
   ```

   The price of the ad has a scale of six and a precision of two. If the number field is in units of dollars, the highest price allowed in the data is $9999.99.
5. Type in the next three lines that define three more columns:

   ```
   PLACED_DATE DATE,
   "Run Start Date" TIMESTAMP WITH LOCAL TIME ZONE,
   "Run End Date" TIMESTAMP WITH LOCAL TIME ZONE,
   ```

   All three lines are variations on the DATE datatype. All three can store dates and times. If you do not insert a time, but only a date, the time is set to midnight of that day. Notice also the double quotation marks around the column names. These names must always be surrounded by quotation marks when referenced in queries or other commands.
6. Type the final column definition. Notice the closing parenthesis that marks the end of the column list:

   ```
   RUN_DAYS INTERVAL DAY(3) TO SECOND(0))
   ```

   The INTERVAL DAY TO SECOND datatype is actually a DATE format. Because the ad runs for a number of days, you don't need any fractions of a second, so the precision of the second is zero, meaning whole seconds.
7. You add storage information after the column list. Type the tablespace and storage clauses and end with a semicolon to indicate that the command is complete:

   ```
   TABLESPACE USERS
   STORAGE (INITIAL 2M NEXT 2M MAXEXTENTS 10);
   ```

8. Click the **Execute** icon to create the table. SQL*Plus Worksheet replies "Table created" in the lower portion of the window.
9. Save the SQL command by selecting **File/Save Input As** in the menu bar and navigating to the **Solutions\Chapter07** directory. Type **classified_ad.sql** as the file name, and save by clicking the **Save** button.
10. Remain logged on for the next practice.

You have seen how to create a relational table with columns and storage settings. An alternative method of creating relational tables is illustrated in the next section.

Creating Temporary Tables

Temporary tables are tables created to store data for your session alone. The table remains after your session is finished; however, the data you added to the table is automatically deleted when you log off or execute a COMMIT command (directly or indirectly). The timing of data removal (on log off or commit) depends on the ON COMMIT clause that you used to create the temporary table. The syntax for creating a temporary table is:


```
CREATE GLOBAL TEMPORARY TABLE <tablename>
ON COMMIT DELETE ROWS|PRESERVE ROWS
<table specifications>
TABLESPACE <tablespace name>
STORAGE (<storage settings>);
```

The <table specifications> and <storage settings> can be the same as those for permanent tables. If you omit the TABLESPACE clause and you have a default temporary tablespace assigned to the user you are logged on as, all temporary tables are created in this tablespace. In previous versions of Oracle, the temporary table would have by default been created in the SYSTEM tablespace. This is unlikely to be the case in Oracle 10g as the CREATE DATABASE command includes the creation of a default temporary tablespace as a default for all users. If you name the tablespace, it must be the current temporary tablespace for the database or any other tablespace you have authority to use, preferably a locally managed tablespace.

Unlike working with permanent tables, adding or changing data in a temporary table does not generate redo log entries; however, it does generate undo log entries. Another difference between permanent tables and temporary tables is the allocation of segments. Temporary tables use temporary segments, and the segments are not allocated to the table until data is actually inserted into the table.

Follow along with this example to create a temporary table. In the example, you want to create a temporary table that contains data from both the CLASSIFIED_AD and the EMPLOYEE tables that you created earlier in the chapter. You use the subquery method to create the table that you used to create the other tables. A **subquery** is a query that is embedded in another SQL command, in this case, in the CREATE TABLE command:

1. If you are not already there, log on to the Enterprise Manager console, access the ORACLASS database, and start up the SQL*Plus Worksheet. Connect as the CLASSMATE user.
2. Clear any previous command by clicking **Edit/Clear All** in the menu bar.
3. Begin entering the temporary table command by typing this line:

   ```
   CREATE GLOBAL TEMPORARY TABLE EDITOR_REVENUE
   ```

4. Now specify how the rows are handled using the ON COMMIT clause. The default is ON COMMIT DELETE ROWS, which causes all the rows you inserted into the temporary table to be deleted when you issue a COMMIT command. In this case, you want to save the data for the duration of your session, even if you execute a COMMIT command, so you specify the ON COMMIT PRESERVE ROWS option:

```
ON COMMIT PRESERVE ROWS
```


5. The next line is an alternative method (the subquery method) of creating a table. You can use this method for any type of table, not just temporary tables. One advantage of this method is that it not only handles the DDL (by generating column details from the select statement), but also it inserts data into the table (with the results of the query) at the same time. Type **AS** followed by the first line of the subquery:

```
AS SELECT E.FIRST_NAME || ' ' || E.LAST_NAME EDITOR,
```

The double vertical bar (||) symbols are used to concatenate the left and right expressions together into a single string. In this case, the FIRST_NAME column is concatenated with a single blank space and then concatenated to the LAST_NAME column. The column that results is given the name EDITOR (which will be its column name in the temporary table). EDITOR is a column alias. A **column alias** is a short name for a results column and is used when generating tables or views to give the results column a valid column name. In addition, a column alias is used to provide a more readable column heading for query results.

6. Type the second line of the subquery:

```
SUM(CA.PRICE) ANNUAL_REVENUE
```

This is the final column in the SELECT statement, so there is no comma following the column alias. The SUM function is a group function. **Group functions** act on sets of column data. In this case, as you see when you complete the command, you are adding the price of every classified ad for the year 2003 for each employee. The employee's name and the sum of all that employee's ads are the two columns that are being created in the temporary table.

7. Type the FROM clause of the subquery:

```
FROM CLASSIFIED_AD CA JOIN EMPLOYEE E
```

As you can see, the subquery joins two tables. The tables have been assigned table aliases. A **table alias** is a shortcut name for a table used to prefix columns in an SQL command in place of using the entire table name. The table alias for CLASSIFIED_AD is CA and the table alias for EMPLOYEE is E. These aliases are used in the ON clause, which you type next.

8. Now, type in the criteria for joining the tables:

```
ON (CA.INTAKE_EDITOR_ID = E.EMPLOYEE_ID)
```

9. Besides the join criteria, you also include only the ads that were placed during the year 2003 by adding this WHERE clause:

```
WHERE TO_CHAR(CA.PLACED_DATE, 'YYYY') = '2003'
```

The previous line illustrates the TO_CHAR function, which you can use on DATE datatypes to extract any part of the date and time that is stored in the column.

10. The last line completes the subquery and the CREATE TABLE command:

    ```
    GROUP BY FIRST_NAME || ' ' || LAST_NAME;
    ```

    The GROUP BY clause is used only when you have a group function in the SELECT clause. It tells Oracle 10*g* how to collect the data into groups.
11. Click the **Execute** icon to run the command. Figure 7-8 shows the entire command as you see it in the SQL*Plus Worksheet and the "Table created" reply in the lower half of the screen that appears after you execute the command.

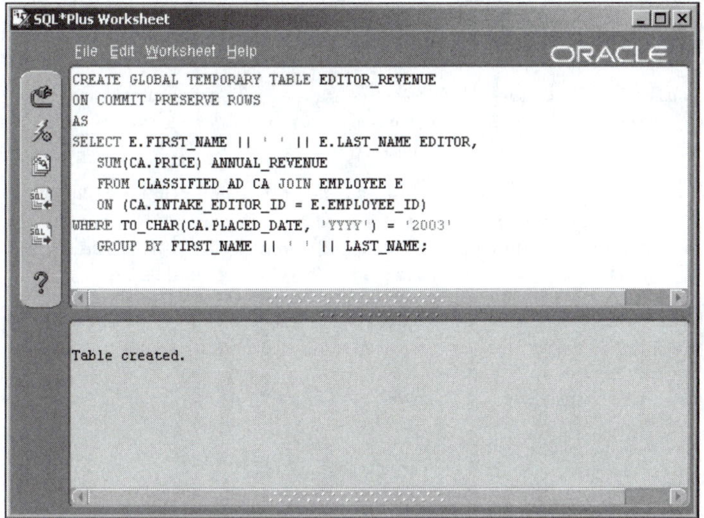

FIGURE 7-8 You must use column aliases when creating a table using a subquery if the columns include functions or calculations

12. Save the command by clicking **File/Save Input As** from the menu bar. Navigate to the **Solutions\Chapter07** directory. Type **editor_revenue.sql** as the file name, and click **Save** to save the file.
13. Remain logged on for the next practice.

TIP

The additional spaces in a SQL command are ignored by Oracle 10*g*, as are line breaks. The spaces and line breaks simply make the command more readable.

You can grant others access to a temporary table you create, but no two users ever share data. Every user session (or transaction, if you use the ON COMMIT DELETE ROWS option) has private rows. Two sessions can use the same temporary table at the same time; however, they do not see each other's data.

The following section describes tables within tables.

Creating VARRAYs and Nested Tables

When the capability to store a table within a column of another table was added to Oracle8, Oracle became an object-relational database. You can add a column that contains a table to a relational table. This column is most often used in object tables. The two types of embedded tables are called VARRAYs and nested tables. Table 7-2 shows the similarities and differences between the two structures.

TABLE 7-2 VARRAYs versus nested tables

Type	Nested table	VARRAY
Type of array (collection)	A dynamic length array or collection; meaning the number of elements is undetermined	A fixed length array, such that the number of elements is known, and relatively constant (VARRAYs can be increased in size)
Column restrictions	Must contain a single column of a table type data type	Must contain a single column of either a built-in data type, or a collection type data type
Storage	Stored in a table apart from the column (out of line) and can have its own storage settings	Stored inside the column (in line) along with the rest of the row; if large enough, Oracle 10g can move a VARRAY (internally) to a RAW or CLOB column
Order	Unordered: Access each row by data values in the row in the same manner as a relational table; the order of the rows is not preserved	Ordered: Access each row in order by its sequence number, not by values in the row; the order of the rows is not preserved
Indexing	Can be indexed	Cannot be indexed
Number of rows	Unlimited	Limited when the VARRAY is created

TIP

Use a VARRAY when you have a small number of items to store or you usually access the data by cycling through it (processing all elements in an array one by one and not referencing specific elements individually). Use a nested table when you have a large, variable number of items or unknown number of items, and you usually access the data by specific values. In programming terms VARRAYs are fixed-length arrays and nested tables are dynamic arrays.

Follow along with the next example to create one relational table with a nested table column and one relational table with a VARRAY column.

The first table, CLASSIFIED_SECTION, is used at the Global Globe to store information about each category of classified ads that the newspaper runs. In addition to the number and the name of the section, you also want to store a list of editors who specialize in helping customers place this type of ad. You build a nested table with a list of editors and mark one of those editors as the primary contact. This way, the telephone operator can direct calls to the primary editor, and if that editor is unavailable, the operator can go down the list of additional editors until a free editor is contacted for the customer.

The second table, CUSTOMER, stores details about a customer. Because many customers have several telephone numbers, you store the set of phone numbers in a VARRAY with a maximum of ten phone numbers per customer:

1. If you are not already there, log on to the Enterprise Manager console, access the ORACLASS database, and start up the SQL*Plus Worksheet. Connect as the SYSTEM user.
2. Go to the console window, double-click the **Databases** folder, and then double-click the **ORACLASS** database icon.
3. Double-click the **Schema** icon, and then double-click the **CLASSMATE** icon. The console displays objects owned by CLASSMATE in the right pane and categories of objects under the CLASSMATE icon in the left pane.
4. Double-click the **User Types** icon in the left pane, and then double-click the **Table Types** folder. The CONTACT_TABLE table type is displayed below the folder.
5. Double-click CONTACT_TABLE in the left pane. The property sheet for this table type is displayed as shown in Figure 7-9. As you can see, this table type is made up of rows of an object type named EDITOR_INFO.

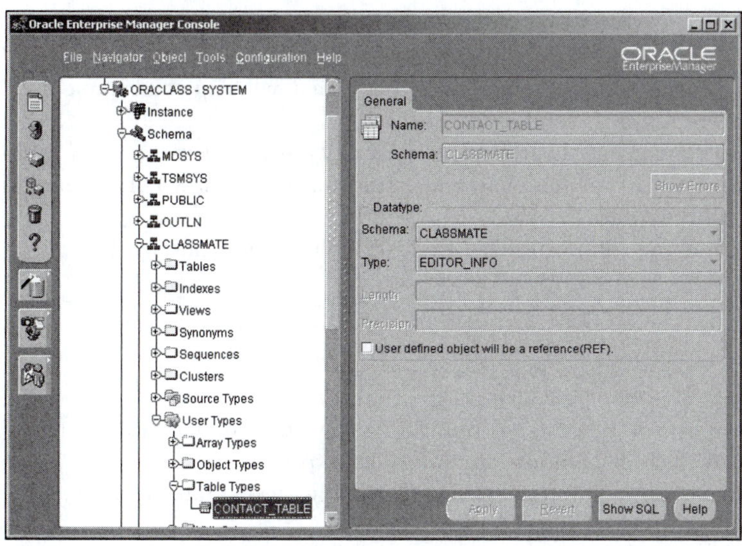

FIGURE 7-9 Object types, arrays, and table types used in this chapter appear under User Types

6. To view the attributes that will be stored in the CONTACT_TABLE table type, you must now look at the EDITOR_INFO object type. Double-click the **Object Types** folder to display a list of object types owned by CLASSMATE.
7. Click **EDITOR_INFO** to display the property sheet for this object type. As shown in Figure 7-10, the object type is made up of two attributes: PRIMARY_EDITOR and EDITOR_ID.

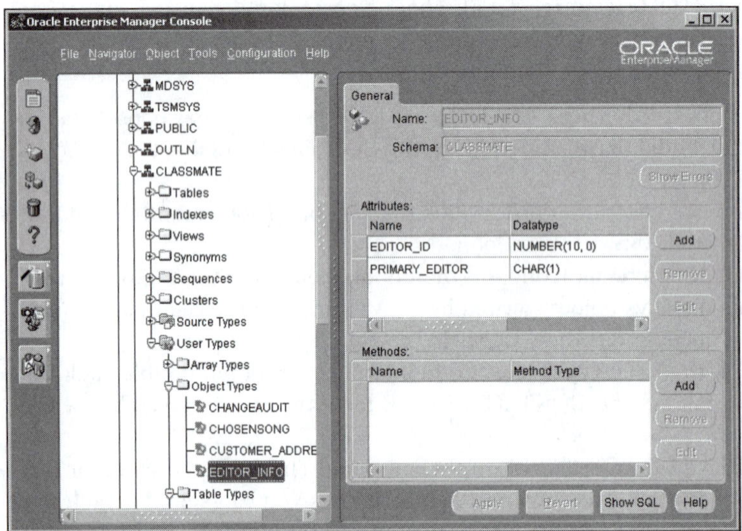

FIGURE 7-10 The EDITOR_INFO object has two attributes

Now you know how to look up object types in the console. You may want to practice this during the chapter exercises because it will help you complete the hands-on exercises at the end of the chapter.

8. Go to the SQL*Plus Worksheet window, clear any previous commands by clicking **Edit/Clear All** in the menu bar, and connect as the CLASSMATE user with password CLASSPASS.
9. Begin entering the CLASSIFIED_SECTION table command by typing these lines:

```
CREATE TABLE CLASSIFIED_SECTION
(SECTION_NO NUMBER NOT NULL,
 SECTION_TITLE VARCHAR2(50),
 BASE_RATE_PER_WORD NUMBER(4,3),
```

   These first lines look like the beginnings of a typical relational table you have seen before with three columns of built-in datatypes.

10. Now type in the last column definition. This is the nested table column:

```
CONTACT_EDITOR CLASSMATE.CONTACT_TABLE)
```

   The datatype of the CONTACT_EDITOR column is a user-defined table type. It was created as part of the initial setup for the chapter. The CONTACT_TABLE type defines a table with two columns: EDITOR_ID and PRIMARY_

CONTACT. The PRIMARY_CONTACT is a single character field and contains "Y" for the primary editor for that section and "N" for all other editors assigned to the section.

11. As described earlier, a nested table is stored in its own table. Although optional, it is advisable to define the name of the storage table so you can identify it more easily. Type the following line (the last line for the CREATE TABLE command) to define NESTED_EDITORS as the storage table for the nested table column data:

```
NESTED TABLE CONTACT_EDITOR
STORE AS NESTED_EDITORS;
```


12. Click the **Execute** icon to run the command. Figure 7-11 shows the entire command as you see it in the SQL*Plus Worksheet and the "Table created" reply in the lower half that appears after you execute the command.

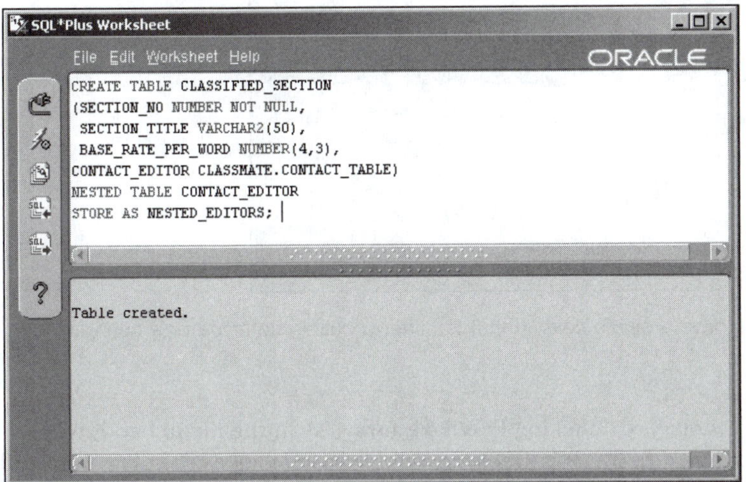

FIGURE 7-11 A nested table has its own name

13. Save the command by clicking **File/Save Input As** in the menu bar. Navigate to the **Solutions\Chapter07** directory. Type **classified_section.sql** as the file name and click **Save** to save the file.
14. Clear the command by clicking **Edit/Clear All** in the menu.
15. Next, begin entering the CUSTOMER table command by typing these lines:

```
CREATE TABLE CUSTOMER
(CUSTOMER_ID NUMBER NOT NULL,
 FULLNAME VARCHAR2(60),
 DISCOUNT_PERCENT NUMBER(3,1),
 EMAIL VARCHAR2(45),
```

Again, these are typical columns. The FULLNAME column has no underscore; simply to illustrate that, although it is often used to make column names readable, it is not required.

16. Now add the VARRAY column:

```
PHONE_LIST CLASSMATE.PHONE_ARRAY);
```

The collection type PHONE_ARRAY is a VARRAY that is predefined. It holds up to ten phone numbers in a repeating VARCHAR2(20) column.

17. Click the **Execute** icon to run the command. Figure 7-12 shows the entire command as you see it in the SQL*Plus Worksheet, and the "Table created" reply in the lower half that appears after you execute the command.

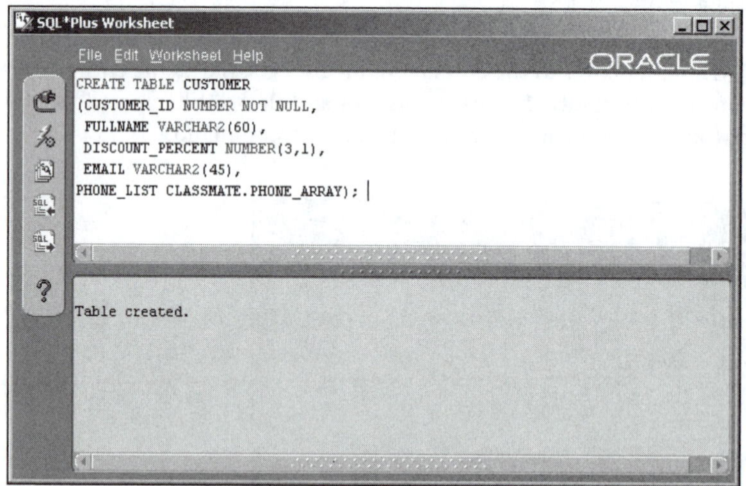

FIGURE 7-12 VARRAYs have a single column that is either a built-in datatype or a user-defined datatype

18. Save the command by clicking **File/Save Input As** in the menu bar. Navigate to the **Solutions\Chapter07** directory. Type **customer.sql** as the file name, and click **Save** to save the file.
19. Remain logged on for the final practice.

You now have two relational tables with collection datatypes in two of the columns. Figure 7-13 illustrates the layout of the CUSTOMER table that contains an array. Looking at the figure, the first row contains data for Mary Holman. She has three phone numbers, all recorded in the PHONE_LIST array. The second row has data for John Smith, who has five phone numbers in the PHONE_LIST array. A customer can have up to ten phone numbers.

Nested table rows are associated with a row in the parent table by the nested table identifier. This identifier is an added column in the nested table. In the case of a relational table with a nested table column, all the nested table rows are stored in a single nested table. If the table is an object table containing a nested table, the nested table is still stored out of line with the parent table; however, there might be separate table instances created for each row of the parent table instead of storing all the nested table data in a single table.

CUSTOMER

CUSTOMER _ID	FULLNAME	DISCOUNT _PERCENT	EMAIL	PHONE_LIST
1	Mary Holman	5	mholman@ebb.com	544-322-1234 544-322-5490 544-322-4499
2	John Smith	10	jsmith@hhy.com	334-998-3837 ext. 12 334-487-3895 334-449-3344 800-339-0987 334-444-5533 ext. 4327
3	Edward Jones		jonesey@hiya.com	872-334-1234 872-230-8017 ext. 100 800-555-3049 888-203-1234 504-308-2343

Arrays stored inline with rows

FIGURE 7-13 A VARRAY expands as new data is added to the array, but has a maximum number of entries

Figure 7-14 shows the CLASSIFIED_SECTION table and its associated nested table named NESTED_EDITORS. The CONTACT_EDITOR column contains a pointer, called a **Nested table ID**, that directs you to the corresponding rows in the NESTED_EDITOR table. For example, the second row is for the Personal section. The CONTACT_EDITOR column contains a pointer with the value "9092af." Using this value, you can see that the Personal section has four contact editors and the editor with EDITOR_ID of 101 is the primary contact. The value shown is simply for illustration; the actual nested table ID value would not be displayed in the CONTACT_EDITOR column. The nested table ID value is stored internally and is not seen when you work with the table. When you use a nested table, an unlimited number of rows can be stored in the nested table that are associated with one row in the main table. In the example, one classified section could have an unlimited number of contact editors.

The following section discusses how to create object tables.

Creating Object Tables

Object tables are always created with one column that always has a datatype that is a user-defined object type. An object type contains one or more attributes, which can be similar to relational table columns or can be made up of another object type. Object tables, therefore, can be built with layers of object types in their definitions.

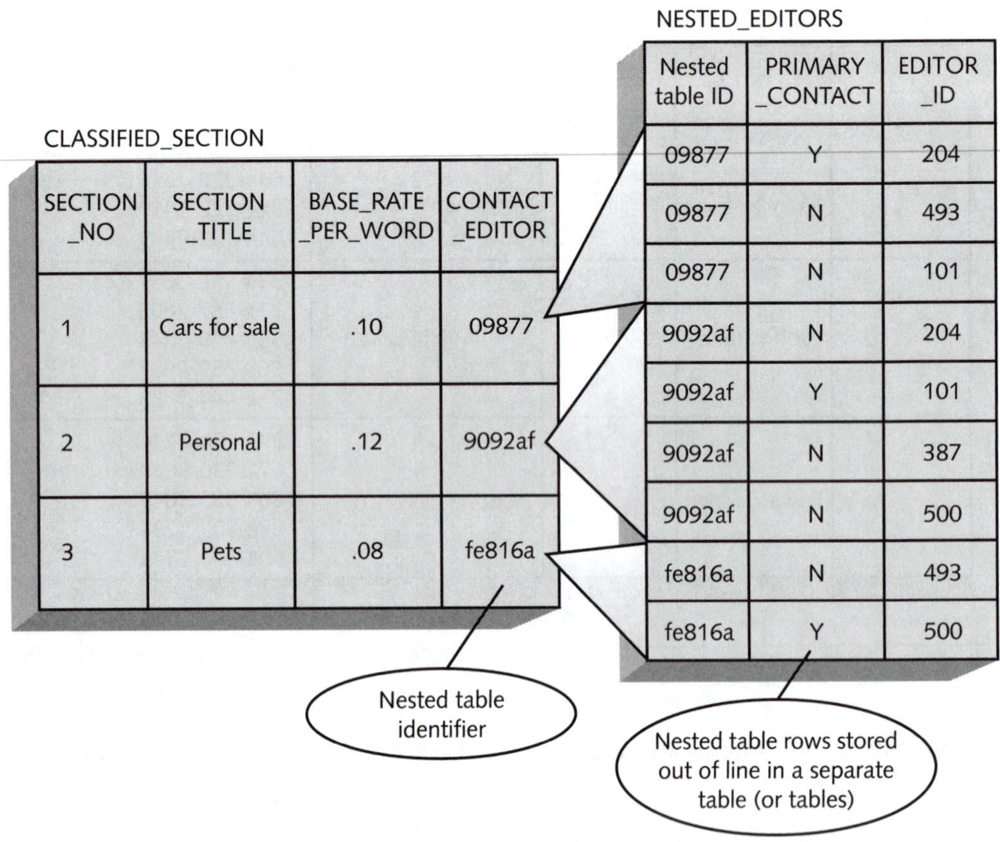

NESTED_EDITORS
Nested table ID
PRIMARY _CONTACT
EDITOR _ID
09877
Y
204
09877
N
493
09877
N
101
9092af
N
204
9092af
Y
101
9092af
N
387
9092af
N
500
fe816a
N
493
fe816a
Y
500
CLASSIFIED_SECTION
SECTION _NO
SECTION _TITLE
BASE_RATE _PER_WORD
CONTACT _EDITOR
1
Cars for sale
.10
09877
2
Personal
.12
9092af
3
Pets
.08
fe816a
Nested table identifier
Nested table rows stored out of line in a separate table (or tables)

FIGURE 7-14 The nested table column actually holds a pointer to the rows in the separate nested table

You must create an object type that has the design you want to use for your object table before actually creating the object table. Complete the following steps to view the predefined object type, and then create an object table.

In this example, you create an object table for storing customer addresses. A customer can have more than one address (such as a home and business address). The object type CUSTOMER_ADDRESS_TYPE was created for you to use when creating the object table.

1. If you are not already there, log on to the Enterprise Manager console, access the ORACLASS database, and start up the SQL*Plus Worksheet. Connect as the CLASSMATE user, with the password CLASSPASS.
2. Clear any previous command by clicking **Edit/Clear All** in the menu bar.
3. Type and execute this command to view the attributes defined in the CUSTOMER_ADDRESS_TYPE object type:

   ```
   DESC CUSTOMER_ADDRESS_TYPE
   ```

   This object type contains attributes for the CUSTOMER_ID, POBOX_SUITE, STREET_ADDRESS, CITY, STATE, and COUNTRY.

4. Type and execute the following command to create an object table that contains these attributes:

```
CREATE TABLE CUSTOMER_ADDRESS
OF CUSTOMER_ADDRESS_TYPE;
```

   You can add storage settings to an object table in the same way as you would a relational table.

5. Save the command by clicking **File/Save Input As** in the menu bar. Navigate to the **Solutions\Chapter07** directory. Type **customer_address.sql** as the file name, and click **Save** to save the file.
6. Exit SQL*Plus Worksheet by clicking the **X** in the top-right corner or type **EXIT** at the prompt, press the **Execute** icon, and click **OK** in the Oracle Enterprise Manager dialog box that opens.
7. Exit the console by clicking the **X** in the top right corner.

An object table is **instantiated** when you insert records into the table. Think of the object table you created as a template. Until the template is actually used (by inserting data), the object table is only a template and has no "real" existence. When you insert data (even if it is all null values) into the first row of an object table, the object table becomes instantiated. An object table row is also called an **object table instance**.

The next section describes how and why to create a partitioned table.

Creating Partitioned Tables

When a table grows very large, even an indexed search can become slow. To improve performance you can break a large table into separate sections, called **partitions**. You can partition any table except one that is part of a cluster. After a table is partitioned, queries that use the table access only the partition or partitions required to retrieve the data. Each partition is stored in a separate segment and can have its own storage settings. Partitions of the same table can even be stored in separate tablespaces, allowing you to divide the table across multiple disk drives if you want.

TIP

An application or SQL command does not have to be changed to access a partition. Oracle 10*g* automatically handles the translation between the table and its partitions.

It is important to understand how the table is queried and updated to best take advantage of partitioning. Some tables have no distinctive pattern to the queries, and so partitioning may not give you any performance improvements. An example is a table used by many diverse users to search for a variety of subjects or text strings in text columns that contain articles or descriptions. If, on the other hand, the table is consistently queried by its primary key, then partitioning the table according to that primary key improves performance. The idea is to determine the best partitioning method that enables queries (as well as updates, insertions, and deletions) to work within one partition rather than across multiple partitions.

Partitioning can be done in five ways:

- **Range:** To use range partitioning, define a set of dividing values so the table is stored in partitions according to a range of values in a set of columns (the **partitioning key**). You specify the upper boundary of each partition, aiming for an even distribution of the data volume in each partition. The last partition can be open-ended to allow for values higher than the current values in the table.
- **Hash:** With **hash partitioning**, you specify how many partitions to create and let Oracle 10g use a hash value (calculated on the partitioning key) to divide the data evenly among the set of partitions. This can be useful when you are not sure of the best range boundaries, but you want to spread the table across multiple disk drives.

NOTE

A hash value is a calculation producing an integer for a particular value of a string.

- **List:** To use **list partitioning**, set up a distinct list of partitioning key values and define which values go into each partition. This is useful for dividing by values, such as state or district office name, when the alphabetical order is an appropriate method of dividing the data. For example, you might divide a customer table into separate sales territories (locations, such as states), in which customers live.
- **Composite range-hash:** With this type of partitioning, you are actually partitioning the data and then partitioning the data within each partition (**subpartitioning**). Each subpartition is in a separate segment, so the subpartitions can reside in separate tablespaces. You specify range values (as with range partitioning), and then you specify the total number of subpartitions (as with hash partitioning) and the locations in which the subpartitions are to be stored.
- **Composite range-list:** This is another method for partitioning and subpartitioning, in which you again define a range of values for each partition. Then, you define subpartitions by lists of values on another set of columns in the table. For example, you may want to divide your historical sales data by ranges of sales dates, and then by sales territories within those ranges.

To create a partitioned table, you use the standard CREATE TABLE command as if you were creating a relational table, and add the PARTITION clause. Each type of partitioning has its own set of parameters. The following sections include examples for each type of partitioning.

Range Partitioning

The bank you work for has a massive transaction table. To manage it more effectively, you decide to create a table partitioned by ranges of account numbers. The tablespaces used do not exist in your database, so this command will not execute in SQL*Plus Worksheet:

```
CREATE TABLE TRANSACTION_RECORD
     ( ACCT_NO NUMBER NOT NULL,
       ACTION  VARCHAR2(5) NOT NULL,
       AMOUNT NUMBER(8,2) NOT NULL,
       SENDING_ACCT_NO NUMBER NOT NULL,
       ACTION_DATE DATE NOT NULL )
  PARTITION BY RANGE (ACCT_NO)
     ( PARTITION TRANS_P1 VALUES LESS THAN (2900999)
         TABLESPACE TBS1,
       PARTITION TRANS_P2 VALUES LESS THAN (5900999)
         TABLESPACE TBS2,
       PARTITION TRANS_P3 VALUES LESS THAN (9900999)
         TABLESPACE TBS3);
```


TIP

You can use multiple columns in the range by listing the columns and listing their range values within the parentheses, separated by commas. List ranges from lowest to highest.

Hash Partitioning

The bank has another large table that keeps track of mortgage loan history. You know that this historical table is nearly always accessed by date, and you want to allow the database to evenly divide the rows by hash value rather than determining a date range. Here is the CREATE TABLE command for a new, hash-partitioned history table. Once again, the tablespace used does not exist in your database, so this command will not execute in SQL*Plus Worksheet:

```
CREATE TABLE MORTGAGE_HISTORY
      (LOAN_NO NUMBER,
       ACCT_NO NUMBER,
       DATE_CREATED DATE,
       MORTGAGE_AMOUNT NUMBER)
   PARTITION BY HASH (DATE_CREATED)
   PARTITIONS 3
   STORE IN (HISTORY_TBS1, HISTORY2, HISTORYEXTENDED);
```

The tablespace to use for each partition is named in the STORE IN clause.

List Partitioning

The bank has a transaction history table that carries summarized transactions by branch. The table has grown so large that you need to divide it, and the BRANCH_REGION column provides a good basis for partitioning because most reporting that is done with this table is by regional managers looking for their own historical data. Here is the CREATE TABLE command. Again, tablespaces are not in your database:

```
CREATE TABLE TRANS_BY_BRANCH_HISTORY
      (BRANCH_ID NUMBER(9,0),
       BRANCH_REGION VARCHAR2(10),
       TRANS_YEAR NUMBER(4,0),
       TRANS_MONTH NUMBER(2,0),
       TRANS_AMOUNT NUMBER(10,2))
   PARTITION BY LIST (BRANCH_REGION)
  (PARTITION WESTERN VALUES ('WESTCOAST', 'NORTHWEST')
        TABLESPACE TBS1,
   PARTITION MOUNTAIN VALUES ('ROCKIES', 'SOUTHWEST')
        TABLESPACE TBS2,
   PARTITION MIDWEST
       VALUES  ('MIDWEST', 'IL-METRO')
       STORAGE (INITIAL 20M NEXT 40M)
        TABLESPACE TBS2,
   PARTITION NORTHEAST VALUES ('NY-METRO', 'NE-STATES')
        TABLESPACE TBS3,
   PARTITION SOUTHEAST VALUES ('SOUTHEAST'))
         TABLESPACE TBS4;
```

The STORAGE clause was added for the MIDWEST partition as an example of how you add individual STORAGE clauses to each partition if needed. Add a STORAGE clause to any or all partitions, the way you would to a table.

Composite Range-Hash Partitioning

The bank's range partitioned table has grown to such huge proportions that you decide to create a new table for the data that is partitioned by range as before, and in addition, subpartitioned by a hash value, based on the transaction date. Here is the CREATE TABLE command. Once again, tablespaces are not in your database:

```
CREATE TABLE TRANSACTION_RECORD
    ( ACCT_NO NUMBER NOT NULL,
      ACTION  VARCHAR2(5) NOT NULL,
      AMOUNT NUMBER(8,2) NOT NULL,
      SENDING_ACCT_NO NUMBER NOT NULL,
      ACTION_DATE DATE NOT NULL )
PARTITION BY RANGE (ACCT_NO)
SUBPARTITION BY HASH (ACTION_DATE)
    SUBPARTITIONS 4 STORE IN
    (TBS1, TBS2, TBS3, TBS4)
PARTITION TRANS_P1 VALUES LESS THAN (2900999),
PARTITION TRANS_P2 VALUES LESS THAN (5900999),
PARTITION TRANS_P3 VALUES LESS THAN (9900999));
```

TIP

Because the subpartitions are divided with hash values, you do not specify any storage sizes.

Composite Range-List Partitioning

Another method of composite partitioning is the range-list partition. Here, the range partition is divided into subpartitions based on a list of column values. For example, instead of using hash values as in the previous example, let's say you decide that the TRANSACTION_RECORD table should be subpartitioned by values in the ACTION column. Here is the CREATE TABLE command that makes this type of table. Again, tablespaces are not in your database:

```
CREATE TABLE TRANSACTION_RECORD
     ( ACCT_NO NUMBER NOT NULL,
       ACTION  VARCHAR2(5) NOT NULL,
       AMOUNT NUMBER(8,2) NOT NULL,
       SENDING_ACCT_NO NUMBER NOT NULL,
       ACTION_DATE DATE NOT NULL )
PARTITION BY RANGE (ACCT_NO)
SUBPARTITION BY LIST (ACTION)
(PARTITION TRANS_P1 VALUES LESS THAN (9900999)
   TABLESPACE TBS1
  (SUBPARTITION WESTERN VALUES ('WESTCOAST', 'NORTHWEST'),
   SUBPARTITION MOUNTAIN VALUES ('ROCKIES', 'SOUTHWEST'),
   SUBPARTITION MIDWEST VALUES  ('MIDWEST', 'IL-METRO'),
   SUBPARTITION NORTHEAST VALUES ('NY-METRO', 'NE-STATES'),
   SUBPARTITION SOUTHEAST VALUES ('SOUTHEAST')),
PARTITION TRANS_P2 VALUES LESS THAN (5900999)
   TABLESPACE TBS2
  (SUBPARTITION WESTERN VALUES ('WESTCOAST', 'NORTHWEST'),
   SUBPARTITION MOUNTAIN VALUES ('ROCKIES', 'SOUTHWEST'),
   SUBPARTITION MIDWEST VALUES  ('MIDWEST', 'IL-METRO'),
   SUBPARTITION NORTHEAST VALUES ('NY-METRO', 'NE-STATES'),
   SUBPARTITION SOUTHEAST VALUES ('SOUTHEAST')),
PARTITION TRANS_P3 VALUES LESS THAN (2900999)
   TABLESPACE TBS3
  (SUBPARTITION WESTERN VALUES ('WESTCOAST', 'NORTHWEST'),
   SUBPARTITION MOUNTAIN VALUES ('ROCKIES', 'SOUTHWEST'),
   SUBPARTITION MIDWEST VALUES  ('MIDWEST', 'IL-METRO'),
   SUBPARTITION NORTHEAST VALUES ('NY-METRO', 'NE-STATES'),
   SUBPARTITION SOUTHEAST VALUES ('SOUTHEAST'));
```

As you can see, subpartitioning by list requires you to repeat the list criteria as many times as there are partitions. It is possible to create a **subpartition template** instead of repeating the list criteria (or the hash criteria, for range-hash partitioned tables). The subpartition template describes all the subpartitions once and then all the partitions that use that template. The template is actually part of the CREATE TABLE command, not a specific object that is created. The template simply provides an alternative way to write the command for creating composite partitions. Here is the range-list partitioned table shown previously, using a template instead of repeating the subpartition criteria for each partition:

```
CREATE TABLE TRANSACTION_RECORD
     ( ACCT_NO NUMBER NOT NULL,
       ACTION  VARCHAR2(5) NOT NULL,
       AMOUNT NUMBER(8,2) NOT NULL,
       SENDING_ACCT_NO NUMBER NOT NULL,
       ACTION_DATE DATE NOT NULL )
```

```
PARTITION BY RANGE (ACCT_NO)
SUBPARTITION BY LIST (ACTION)
SUBPARTITION TEMPLATE
  (SUBPARTITION WESTERN VALUES ('WESTCOAST', 'NORTHWEST'),
   SUBPARTITION MOUNTAIN VALUES ('ROCKIES', 'SOUTHWEST'),
   SUBPARTITION MIDWEST VALUES  ('MIDWEST', 'IL-METRO'),
   SUBPARTITION NORTHEAST VALUES ('NY-METRO', 'NE-STATES'),
   SUBPARTITION SOUTHEAST VALUES ('SOUTHEAST'))
(PARTITION TRANS_P1 VALUES LESS THAN (9900999)
    TABLESPACE TBS1,
PARTITION TRANS_P2 VALUES LESS THAN (5900999)
   TABLESPACE TBS2,
PARTITION TRANS_P3 VALUES LESS THAN (2900999)
   TABLESPACE TBS3);
```


Index-organized tables can also be partitioned with the following restrictions:

- The partition key must be all or a subset of the table's primary key.
- Only range and hash partitioning are allowed.
- Only range-partitioned, index-organized tables can contain large object (LOB) columns.

VIEWING DATABASE OBJECT ATTRIBUTES

In a previous chapter you examined the use of metadata and performance views. Like many types of objects in an Oracle database, tables, indexes, and their attributes can be examined easily using some specific metadata dictionary views. You can also look at metadata structures using tools, such as the Enterprise Manager console and the Database Control. You begin with metadata views, then the console, and finally the Database Control.

Viewing Object Metadata in SQL*Plus

The traditional way of looking at metadata is using metadata and data dictionary views. In some ways using a command line type interface through a tool like SQL*Plus is a little easier and faster than using the more GUI-oriented tools, such as the console and the Database Control web interface. Partially it depends on what you are used to, partially it is easier to utilize caution using SQL*Plus because scripts can be built and tested beforehand. Also, using a step-by-step command line process is important to making sure you do not make errors. And Oracle certification tests do not ask questions about using GUI interfaces. There isn't much to test you on apart from how to click buttons and where to find things. The GUI tools are largely intuitive and easy to use. Oracle certification tests do however ask a lot of conceptual questions, and questions about command line syntax.

CAUTION

Caution as a database administrator is essential to your success in a commercial environment. I have seen some interesting things happen in the world. For instance, a DBA dropped an entire database by destroying all database users with a generic form of the DROP USER command, destroying an entire 200Gb OLTP database. And they didn't have effective backups. Using GUI tools tends to make things a little easier but can in some respects make administration tasks a little too easy, thus discouraging planning and thought before pressing the Enter key or clicking a mouse button. Learning about the details using the metadata views first will give you a better understanding of what is in front of you on a screen in the console or the Database Control. Two other significant factors with using GUI tools are the security issue and frequent low-level command line requirements of operating systems, such as Unix and Linux. With respect to security, how do you prevent just anyone from running your Database Control web interface, from anywhere in the world, and "administering" your production database? Some operating systems, Unix for instance, can sometimes require access using GUI tools.

Most importantly, there are some things you can do using SQL*Plus that you cannot do with the GUI tools, which you will soon see. Begin with a quick look at some metadata views by running some simple commands and queries:

1. Start up the command line SQL*Plus tool in Windows by selecting **Start/All Programs/Oracle ... /Application Development/SQL*Plus** from an Oracle server or client installation. In Unix or Linux simply execute sqlplus. You could also use SQL*Plus Worksheet or iSQL*Plus, but in this case I am insisting on a basic approach using just SQL*Plus.
2. Connect to your ORACLASS database as the CLASSMATE user when the login prompt window appears by typing CLASSMATE into the User Name box, your password (CLASSPASS) into the Password box, and the database TNS name (ORACLASS) into the Host String box.
3. Now type the following command into the SQL*Plus window and hit the return key:

   ```
   SET LINES 132 WRAP OFF PAGES 40
   ```

   This just makes things a little more visible, and somewhat less confusing.
4. Type the following command in and hit the return key:

   ```
   DESC USER_OBJECTS
   ```

   The USER_OBJECTS view contains all objects in the CLASSMATE schema, such as tables, indexes, sequences, types, or whatever has been created.
5. The OBJECT_TYPE column in the USER_OBJECTS view contains all the object types in the database. Type this command to see all different object types created within the CLASSMATE schema and hit the return key:

   ```
   SELECT DISTINCT(OBJECT_TYPE) FROM USER_OBJECTS ORDER BY 1;
   ```

   Your screen should look something like that shown in Figure 7-15. Figure 7-15 shows that the CLASSMATE schema in my database contains four different types of objects: tables, indexes, types, and views.
6. Examine the tables view by executing the following command and hit the return key:

   ```
   DESC USER_TABLES
   ```

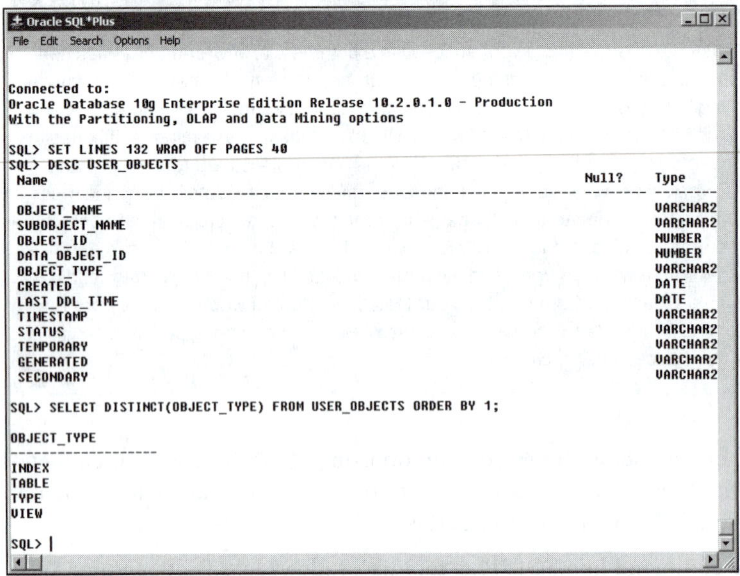

```
Connected to:
Oracle Database 10g Enterprise Edition Release 10.2.0.1.0 - Production
With the Partitioning, OLAP and Data Mining options

SQL> SET LINES 132 WRAP OFF PAGES 40
SQL> DESC USER_OBJECTS
 Name                                                                  Null?    Type
 --------------------------------------------------------------------- -------- --------
 OBJECT_NAME                                                                    VARCHAR2
 SUBOBJECT_NAME                                                                 VARCHAR2
 OBJECT_ID                                                                      NUMBER
 DATA_OBJECT_ID                                                                 NUMBER
 OBJECT_TYPE                                                                    VARCHAR2
 CREATED                                                                        DATE
 LAST_DDL_TIME                                                                  DATE
 TIMESTAMP                                                                      VARCHAR2
 STATUS                                                                         VARCHAR2
 TEMPORARY                                                                      VARCHAR2
 GENERATED                                                                      VARCHAR2
 SECONDARY                                                                      VARCHAR2

SQL> SELECT DISTINCT(OBJECT_TYPE) FROM USER_OBJECTS ORDER BY 1;

OBJECT_TYPE
-------------------
INDEX
TABLE
TYPE
VIEW

SQL> |
```

FIGURE 7-15 Discovering the different object types in a database schema

7. Now bring up some statistics for each table by executing the following query, whose result is shown in Figure 7-16:

```
SELECT TABLE_NAME, NUM_ROWS, CHAIN_CNT,
       SAMPLE_SIZE, LAST_ANALYZED
FROM USER_TABLES ORDER BY 1;
```

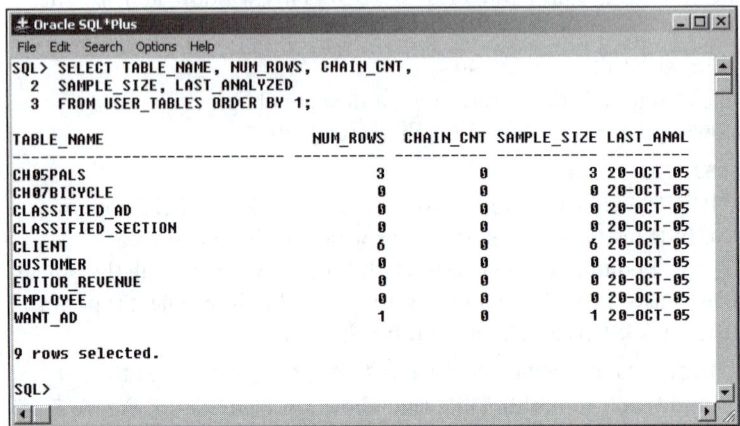

```
SQL> SELECT TABLE_NAME, NUM_ROWS, CHAIN_CNT,
  2  SAMPLE_SIZE, LAST_ANALYZED
  3  FROM USER_TABLES ORDER BY 1;

TABLE_NAME                       NUM_ROWS  CHAIN_CNT SAMPLE_SIZE LAST_ANAL
------------------------------ ---------- ---------- ----------- ---------
CH05PALS                                3          0           3 20-OCT-05
CH07BICYCLE                             0          0           0 20-OCT-05
CLASSIFIED_AD                           0          0           0 20-OCT-05
CLASSIFIED_SECTION                      0          0           0 20-OCT-05
CLIENT                                  6          0           6 20-OCT-05
CUSTOMER                                0          0           0 20-OCT-05
EDITOR_REVENUE                          0          0           0 20-OCT-05
EMPLOYEE                                0          0           0 20-OCT-05
WANT_AD                                 1          0           1 20-OCT-05

9 rows selected.

SQL>
```

FIGURE 7-16 The contents of the USER_TABLES metadata view

In Figure 7-16 you can see row counts for each table, chained rows, and statistical sample details. Statistics generation is somewhat automated in Oracle 10g so you should have values in the statistics column shown as in Figure 7-16. If not, then execute the following commands to generate statistics, and run the query in Figure 7-16 again (this script will only function on a Windows computer, assuming you have a C:\TEMP directory). For Unix or Linux change the paths:

```
SET HEAD OFF FEED OFF TERMOUT OFF
SPOOL C:\TEMP\stats.sql;
SELECT 'ANALYZE TABLE '||TABLE_NAME||' COMPUTE STATISTICS;' FROM USER_TABLES;
SPOOL OFF;
@C:\TEMP\stats.sql;
SET HEAD ON FEED ON TERMOUT ON;
```


8. Now run the following command to examine the columns in the USER_TAB_COLUMNS view:

```
DESC USER_TAB_COLUMNS
```

Where the USER_TABLES view contains tables in a schema, the USER_TAB_COLUMNS view contains all columns, in all tables, in a schema.

9. Here is something you cannot do in the console or Database Control GUI tools (only in a SQL*Plus tool). Execute the following query in your SQL*Plus window, joining all tables with all contained columns in the CLASSMATE schema. This script is contained in the **jointables.sql** script file in the **Data\Chapter07** directory on your student disk. Remain logged on as CLASSMATE to run the script:

```
SET WRAP OFF LINESIZE 132 PAGES 80
COLUMN TAB FORMAT A20
COLUMN COL FORMAT A15
COLUMN POS FORMAT 990
COLUMN TYP FORMAT A10
COLUMN TBS FORMAT A25
BREAK ON TAB NODUPLICATES SKIP 2 ON NAME NODUPLICATES
SELECT T.TABLE_NAME "TAB", C.COLUMN_NAME "COL"
,C.COLUMN_ID "POS", C.DATA_TYPE "TYP"
,DECODE(C.NULLABLE,'N','NOT NULL',NULL) "NULL"
,T.TABLESPACE_NAME "TBS"
FROM USER_TABLES T, USER_TAB_COLUMNS C
WHERE T.TABLE_NAME = C.TABLE_NAME
ORDER BY T.TABLE_NAME, C.COLUMN_ID;
```

The result of the above query for my database is shown in Figure 7-17.

There is much more you can do with SQL*Plus. In addition the above exercise can actually be performed in all of the SQL*Plus tools: SQL*Plus, iSQL*Plus, and SQL*Plus Worksheet. You can also execute SQL*Plus Worksheet from within the console, and even execute iSQL*Plus from with the Database Control.

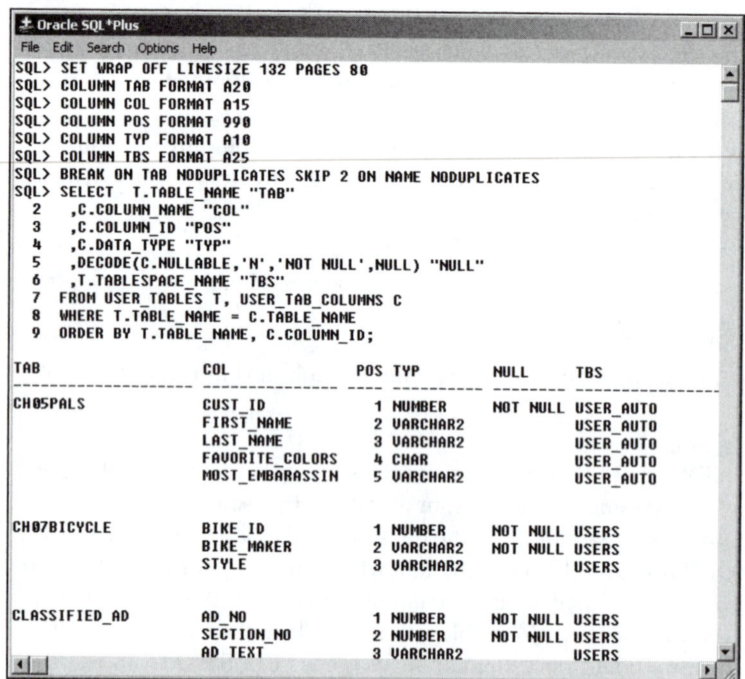

```
Oracle SQL*Plus
File Edit Search Options Help
SQL> SET WRAP OFF LINESIZE 132 PAGES 80
SQL> COLUMN TAB FORMAT A20
SQL> COLUMN COL FORMAT A15
SQL> COLUMN POS FORMAT 990
SQL> COLUMN TYP FORMAT A10
SQL> COLUMN TBS FORMAT A25
SQL> BREAK ON TAB NODUPLICATES SKIP 2 ON NAME NODUPLICATES
SQL> SELECT  T.TABLE_NAME "TAB"
  2    ,C.COLUMN_NAME "COL"
  3    ,C.COLUMN_ID "POS"
  4    ,C.DATA_TYPE "TYP"
  5    ,DECODE(C.NULLABLE,'N','NOT NULL',NULL) "NULL"
  6    ,T.TABLESPACE_NAME "TBS"
  7  FROM USER_TABLES T, USER_TAB_COLUMNS C
  8  WHERE T.TABLE_NAME = C.TABLE_NAME
  9  ORDER BY T.TABLE_NAME, C.COLUMN_ID;

TAB                  COL              POS TYP        NULL     TBS
-------------------- --------------- ---- ---------- -------- ----------------
CH05PALS             CUST_ID            1 NUMBER     NOT NULL USER_AUTO
                     FIRST_NAME         2 VARCHAR2            USER_AUTO
                     LAST_NAME          3 VARCHAR2            USER_AUTO
                     FAVORITE_COLORS    4 CHAR                USER_AUTO
                     MOST_EMBARASSIN    5 VARCHAR2            USER_AUTO


CH07BICYCLE          BIKE_ID            1 NUMBER     NOT NULL USERS
                     BIKE_MAKER         2 VARCHAR2   NOT NULL USERS
                     STYLE              3 VARCHAR2            USERS


CLASSIFIED_AD        AD_NO              1 NUMBER     NOT NULL USERS
                     SECTION_NO         2 NUMBER     NOT NULL USERS
                     AD_TEXT            3 VARCHAR2            USERS
```

FIGURE 7-17 Joining metadata views in SQL*Plus

Viewing Object Metadata in the Console

The Enterprise Manager console executes on a client machine; once loaded into memory it will execute efficiently. The console is limited in its ability to execute high-level database administration functionality, such as starting up and shutting down a database. This is actually a good thing. Even as a highly experienced database administrator, I prefer to exercise caution, particularly when maintaining a production database. I would prefer not to have access to functions such as database shut down unless I actually need to use it.

By now you should know how to open the Enterprise Manager console under an Oracle client installation.

NOTE

It is unlikely you will be able to log in to the console as the CLASSMATE user as that user should not have system level administration privileges. Remember that the console shows you all schemas, thus SYSTEM or SYS user database administrator access is required.

As an exercise, display the CLIENT table's fields on the screen, as shown in Figure 7-18. Note how you cannot display more than one table's fields at the same time, as you can in the previous section using SQL*Plus.

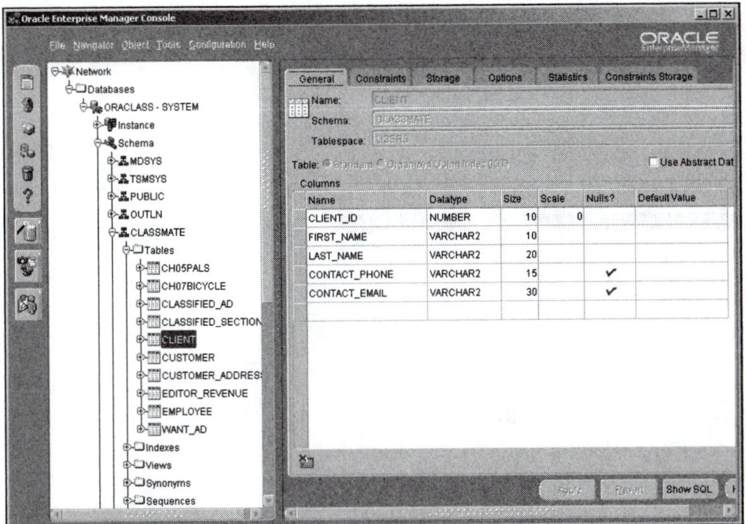

FIGURE 7-18 Displaying metadata in the Enterprise Manager console

Viewing Object Metadata in the Database Control

The Database Control is really *neat*! It gives you everything you need at your fingertips, and all in the same GUI tool. The Database Control runs on a database server through an HTTP protocol and could have security issues. In Oracle 10g release 2 (10.2) each database has it's own Database Control service running on a Windows computer (selected when creating the database using the Database Configuration Assistant or DBCA tool). The URL changes for each database using a separate port number (yours may be different), mine is as follows:

```
http://piii-450:1158/em
```

Go through the following steps:

1. On a client computer you can display the Database Control web interface by executing the URL http://<hostname>:<port number>/em in a browser, such as the Internet Explorer or Netscape. On a server computer (the computer on which your database is created), you can use the URL, or you can also use a menu option under **Start/All Programs/Oracle ... /Database Control – oraclass** (not in all versions of Oracle 10g).
2. You should get a login screen that looks something like that shown in Figure 7-19. Log in as SYSTEM.

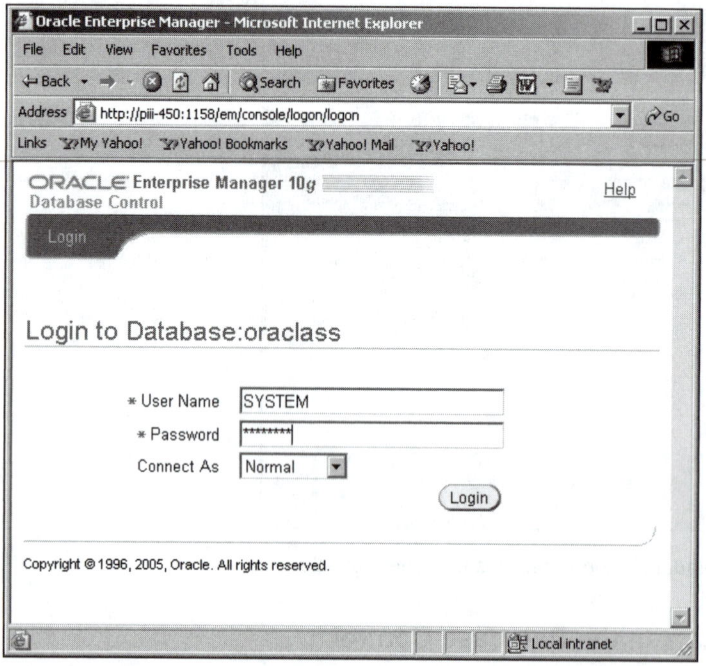

FIGURE 7-19 The Database Control interface Licensing screen.

3. You may get a licensing screen that looks something like that shown in Figure 7-20. This screen may not always appear. Scroll to the bottom of the screen and click the **I agree** button.

NOTE

Different releases of Oracle 10*g*, and even different patch sets of Oracle software, may display different screens.

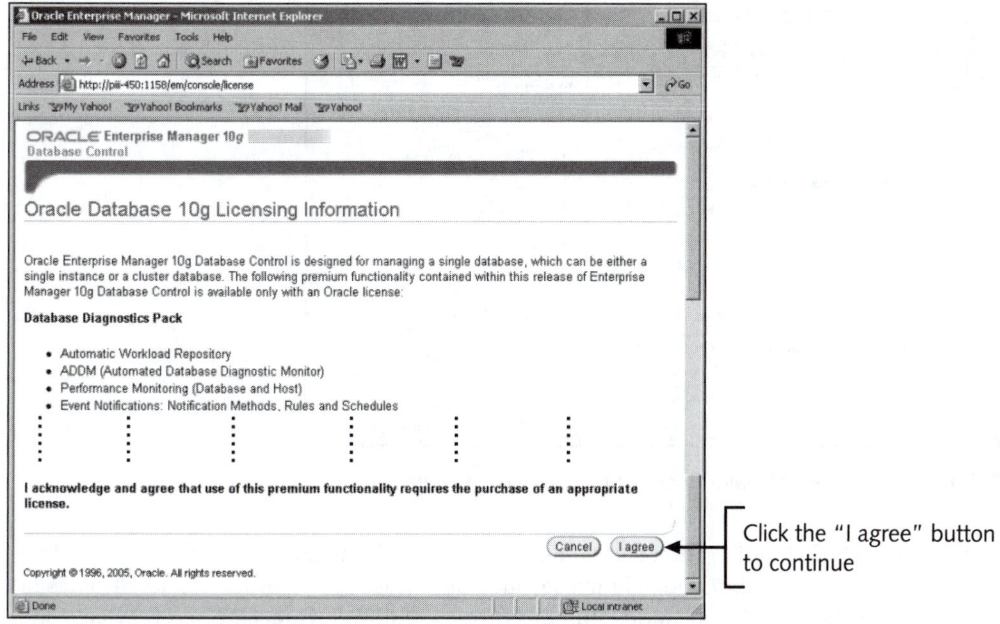

FIGURE 7-20 Connecting to the Database Control interface.

4. The next screen you get is the Home screen for the Database Control interface, as shown in Figure 7-21. There are four tabs across the top included with the Home tab, linking to other URLs: Home, Performance, Administration, and Maintenance. Click the Administration tab.

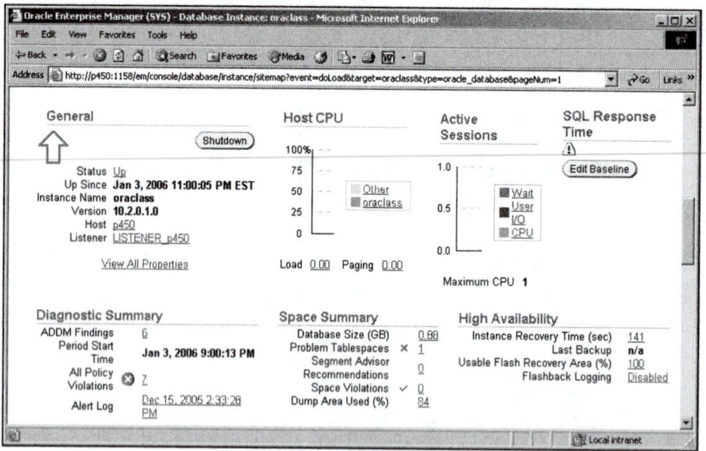

FIGURE 7-21 The Database Control interface Home screen

5. The Administration screen allows access to both examine and change various aspects of an Oracle database, such as storage, configuration, and all types of schema objects, among numerous other things. Figure 7-22 shows what you currently should be focused on under the header Database Objects under the Schema header. Also highlighted in Figure 7-22 is the Database Control interface access to physical database storage architectural facets, such as tablespaces and datafiles. Click on the Tables tab under the Database Objects section of the Schema section.

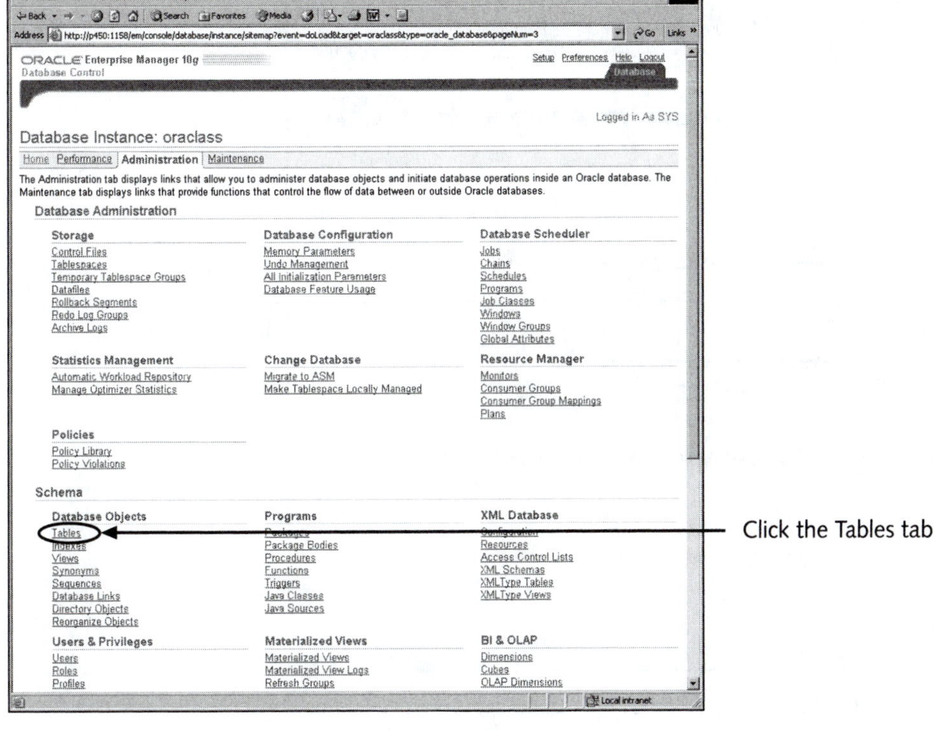

FIGURE 7-22 The Database Control interface Administration screen

6. Figure 7-23 shows a selection screen for table objects required for examination. Type the CLASSMATE schema name into the Schema box and leave the Object Name box empty. Click the Go button.

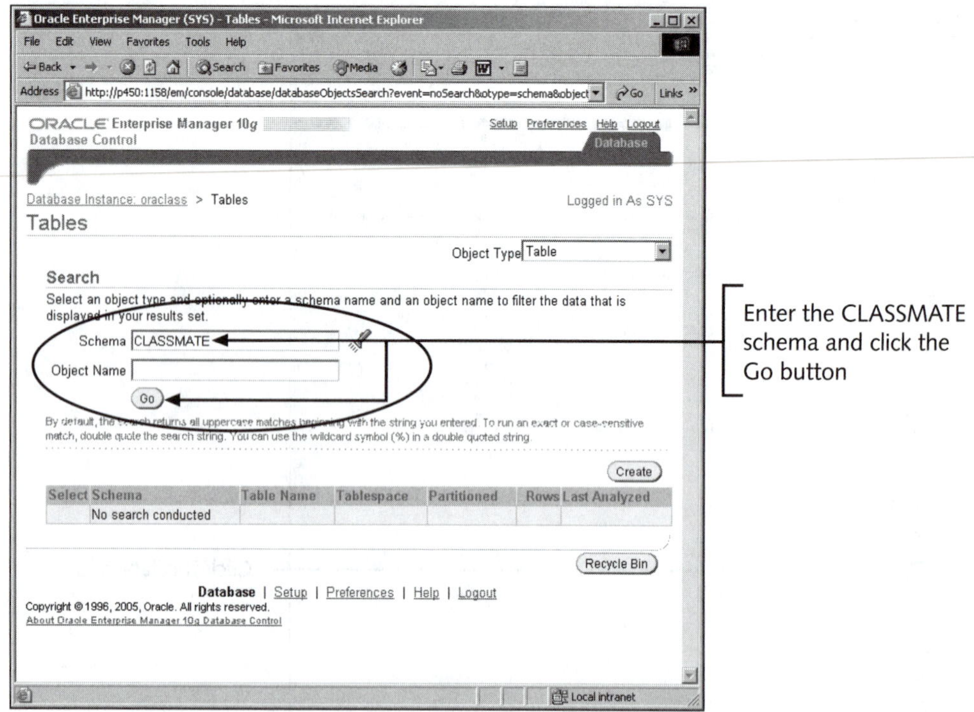

FIGURE 7-23 Selecting a schema in the Database Control interface

7. Figure 7-24 shows a list of all tables within the CLASSMATE schema. Select the radio button to the left of the CLIENT table. Click the View button. Be very careful not to click buttons such as Recycle Bin, which can potentially blow away your table.

NOTE

The recycle bin is a new Oracle 10*g* feature for deleting objects but allowing for later retrieval. The recycle bin can be purged to remove all deleted objects, and using command line syntax, objects can be destroyed without storing into the recycle bin for later possible undeletion.

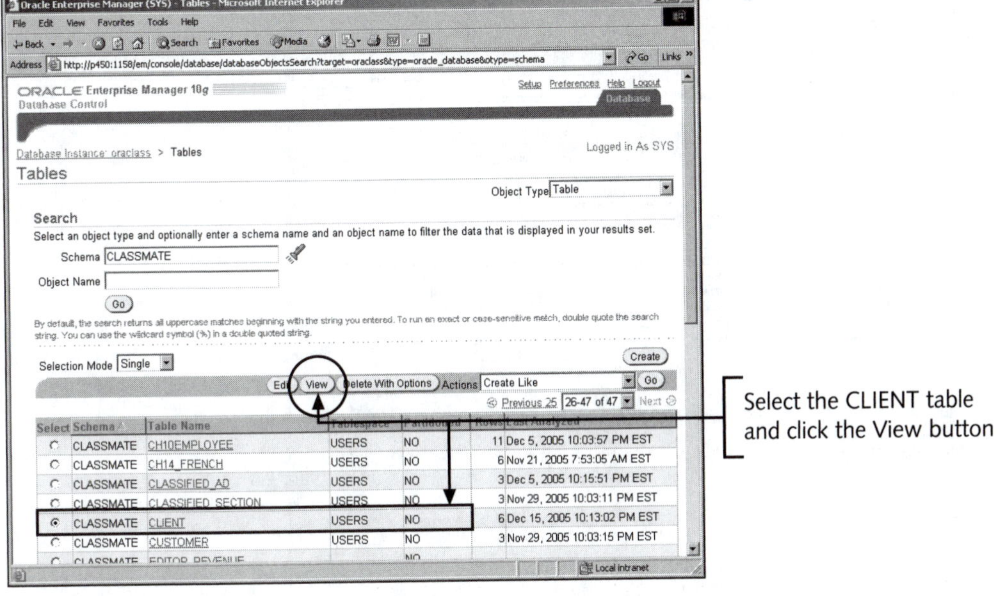
Select the CLIENT table
and click the View button

FIGURE 7-24 Selecting a table and its fields in the Database Control interface

8. Figure 7-25 shows all the metadata details of the CLASSMATE schema CLIENT table, including all sorts of other details as well.

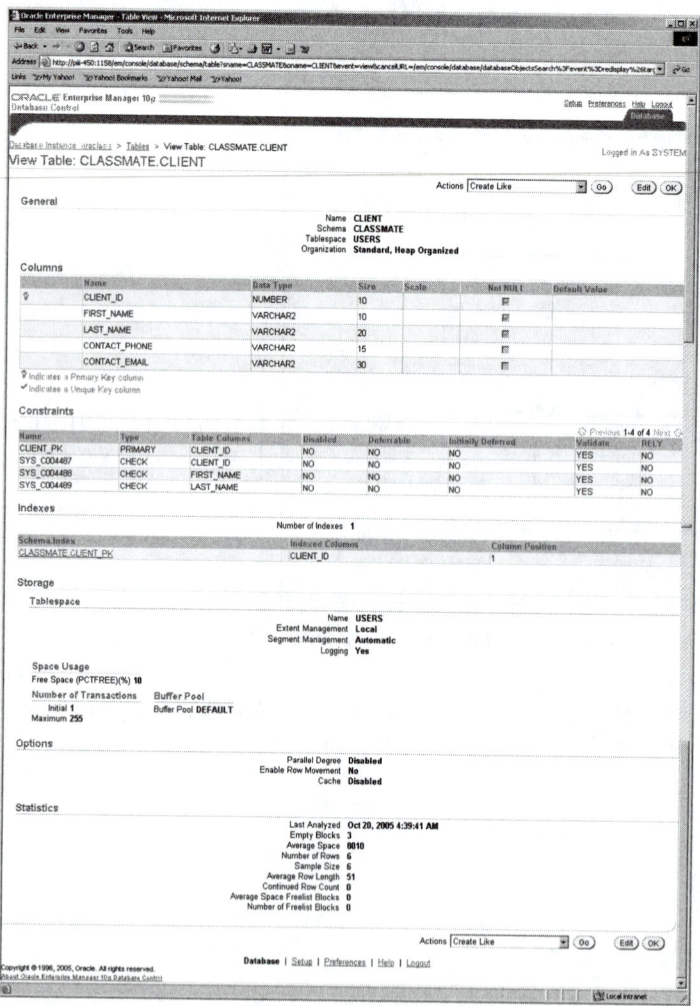

FIGURE 7-25 Viewing a table and its fields in the Database Control interface

9. Go back to the previous screen shown in Figure 7-24 using the browser back button, and click the Edit button. Figure 7-26 shows all the metadata details of the CLASSMATE schema CLIENT table. Many other details are also displayed.

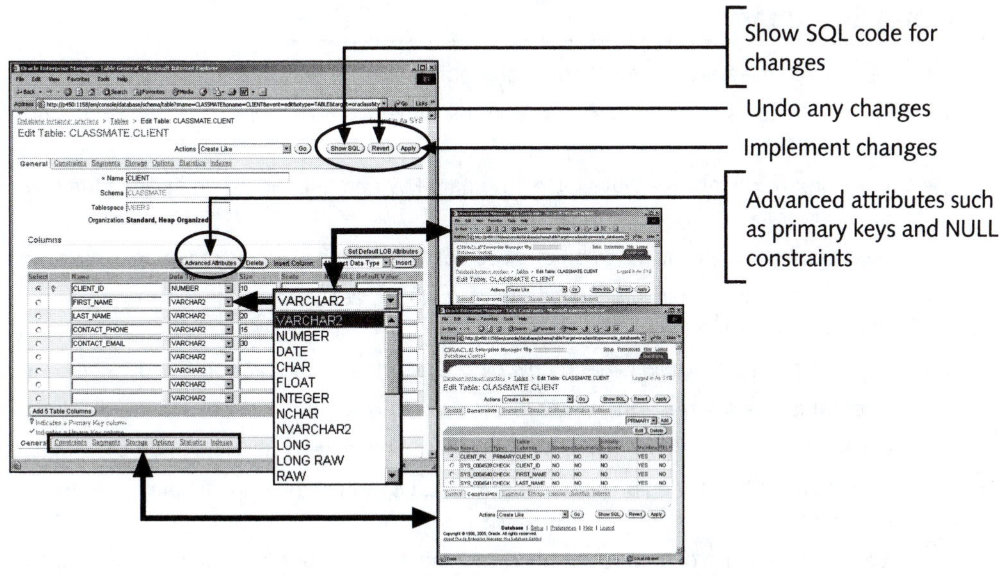

FIGURE 7-26 Editing a table and its fields in the Database Control interface

You have now examined and created a wide variety of tables. Chapter 8 looks at how to make structural changes to existing tables.

Chapter Summary

- Relational tables are the most common form of table in an Oracle database.
- Relational tables reside in a single segment within a single tablespace, unless the table is partitioned.
- Index-organized tables always have a primary key and use the key to arrange the physical order of the rows.
- Object tables have rows made up of an object type, which in turn is made up of attributes.
- In addition to data, object tables can have methods associated with them and can have other objects within them.
- A temporary table, which is seen only by one user, is created for private data.
- External tables are read-only and are used to retrieve data from outside the database.
- A nested table is a table within a single column or within an attribute of an object table.
- An XML table has a single column of the XML type datatype and contains XML formatted data.
- A table is made up of one segment, which contains one or more extents, each containing a group of data blocks.
- Locally managed tablespaces use the storage settings of a table but control the extent sizes and how rows are inserted and updated automatically.
- Database objects such as tables and indexes can use the default storage settings of tablespaces, augment them, or override tablespace settings entirely with storage settings specific to the object.
- A data block has these components: common and variable header, table directory, row directory, free space, and row data.
- A row that spans multiple blocks because it is too large to fit into one block is a chained row.
- A row that has been moved from one block to another (retaining only a ROWID pointer address in the original block) is known as a migrated row. Rows are migrated as a result of an update where the row's data no longer fits within the original block.
- INITRANS allocates space in a data block's header specifying how many slots are held for concurrent transactions accessing a data block.

Note: MAXTRANS is deprecated in Oracle 10*g*.

- Adjusting the PCTFREE and PCTUSED parameters can help avoid migrated rows.
- Chained rows and migrated rows can slow database performance.
- Tables in dictionary-managed tablespaces can use storage settings, such as PCTFREE and PCTUSED, plus settings that can also be set at the tablespace level. Locally managed tablespaces may ignore or prohibit storage setting parameters.
- A row is made up of a row header and column data.
- A row header has row overhead, the number of columns, a possible ROWID to a chained row, a possible ROWID to a migrated row (if any).

- Column data has the column data length followed by data, then the next column data length followed by data, and so on.
- A physical ROWID can be extended or restricted.
- A logical ROWID is used to locate rows in an index-organized table.
- You can query any row's ROWID value with the ROWID pseudocolumn.
- ROWIDs provide the fastest possible database access but they can change over time, resulting in potential data loss.
- Each column in a table has a datatype, both defining and restricting the format, structure, and content of data stored in a column.
- DATE datatypes contain a time and a date, unless restricted by an input mask.
- To adjust times for time zones, use the TIMESTAMP WITH LOCAL TIME ZONE datatype.
- Relational tables can be defined as a list of columns or as a subquery.
- Column names are case sensitive when enclosed within double quotes; otherwise, the names are interpreted as uppercase for all characters in the string.
- Temporary tables contain private data that can be viewed only by the user who inserted the data, and the data exists only for the duration of the user's session, or until the user commits the transaction or a commit occurs for any other reason.
- VARRAYs and nested tables are effectively tables built within a column of another table.
- A relational table or an object table can contain a VARRAY or nested table in its columns.
- When creating a VARRAY or nested table column, you must define a collection type (VARRAY type or table type) that is used as the column's datatype.
- Nested table data is stored out of line (in a separate location) from the primary table.
- A partition of a table resides in its own segment (can be physically separated, on disk, from other partitions).
- If a table has partitions and subpartitions, each subpartition resides in its own segment (can be physically separated, on disk, from other subpartitions).
- Partitioning can be done with range, list, hash, or composite methods.
- Composite partitioning can be range-hash or range-list.
- Range and hash partition keys can be made up of more than one column.
- List partition keys must contain only one column.
- To save typing, use composite range-list partitions with a subpartition template.
- Hash partitioning does not allow any storage settings other than what tablespace will contain the partition.

- Datatypes built-in to Oracle 10*g* can be divided into simple datatypes, binary objects, reference pointers, collections, and some specialized types such as XMLType for storing XML documents accessible as an XML DOM (Document Object Model).
- Use of the DOM for XMLType datatypes implies that an XMLType can be accessed and manipulated using an eXtensible Markup Language (XML) interpreter, and languages such as eXtensible Style Sheets (XSL).

Review Questions

1. Relational tables must have a primary key. True or False?
2. A non-partitioned table has one ______________ and one or more ______________ for storage.
3. You want to speed up queries on a table. Queries are always looking up rows based on the values of the row's primary key. What type of table would improve query speed?
4. Which of the following statements are true about both object tables and relational tables?
   a. Rows are stored in extents.
   b. Collection types can be used in a column.
   c. The table can be partitioned.
   d. Rows can be made up of one column.
   e. All of the above.
5. The PCTFREE storage setting can only be used at the table level. True or False?
6. Examine this SQL command, and determine which line has an error.

```
1 CREATE TABLE BEARTRAIL
2 (TRAIL_NO NUMBER,
3 "BEAR COUNT" NUMBER(10,0),
4 LASTCOUNT TIMESTAMP WITH LOCAL TIMEZONE);
```

   a. Line 1
   b. Line 2
   c. Line 3
   d. Line 4
   e. No errors exist in the SQL command.
7. Observe these two SQL commands.

```
CREATE TABLESPACE USER_SPACE
DATAFILE 'D:\oracle\product\10.2.0\oradata\mydb\user_space.dbf'
MINIMUM EXTENT 5M
DEFAULT STORAGE(INITIAL 25M);

CREATE TABLE ACCOUNT_LEDGER
(ACCOUNT_ID NUMBER, DEBIT NUMBER, CREDIT NUMBER)
TABLESPACE USER_SPACE
STORAGE (INITIAL 2M NEXT 10M);
```

What will the size of the table's first extent be?

a. 25 Mb
b. 2 Mb
c. 5 Mb
d. 12 Mb
e. 30 Mb

8. When you insert a row into a data block, and the free space drops below 20 percent, what happens?
   a. The data block is removed from the extent's freelist.
   b. The data block migrates the row to another block.
   c. The data block is not removed from the extent's freelist.
   d. There is not enough information to determine the action.

9. Look at the query and the first two rows of results below.

```
SELECT ROWID FROM STORE_INFO;
ROWID
----------------------
00000D45.0000.0021
00000D45.0001.0001
```

   What type of ROWID does the table contain?
   a. Universal ROWID
   b. Restricted ROWID
   c. Extended ROWID
   d. Logical ROWID
10. A temporary table is dropped when your session ends. True or False?
11. Which of the following is NOT a valid method of partitioning?
   a. List-hash
   b. Range
   c. Hash
   d. Range-hash
   e. None of the above
12. You want to create a database table that stores songs recorded for a new music CD. Which datatype is best suited for the songs?
   a. CLOB
   b. BFILE
   c. LONGRAW
   d. LOB
   e. Any of the above

Exam Review Questions—Oracle Database 10*g*: Administration (#1Z0-042)

1. Look at the following column definition.

```
COLUMN HATSIZE NUMBER(6,2)
```

Which of these statements are true?

a. The number 6.534 is rounded to 6.54 when inserted.

b. Nulls are not allowed.

c. The statement is invalid.

d. Numbers greater than 9999.99 return an error.

2. Look at this statement:

```
CREATE TABLE CAR_REPAIR (OWNER VARCHAR2(30),
REPAIR_DESC VARCHAR2(4000))
STORAGE (INITIAL 20M NEXT 5M MAXEXTENTS UNLIMITED
PCTFREE 20 PCTUSED 50 INITRANS 2)
TABLESPACE USERS;
```

How much of each data block must be filled with either overhead or row data before the block is taken off the freelist?

a. 20 percent

b. 50 percent

c. 80 percent

d. 21 percent

3. Look at this SQL command.

```
CREATE TABLESPACE USER_NEW
EXTENT MANAGEMENT LOCAL UNIFORM SIZE 5M;

CREATE TABLE INVENTORY_UPDATES
(INVENTORY_NO NUMBER, LAST_CHANGED DATE)
STORAGE (INITIAL 20M NEXT 5M
 MAXEXTENTS UNLIMITED PCTFREE 10)
TABLESPACE USERS_NEW;
```

Which of these statements is false with respect to the INVENTORY_UPDATES table?

a. The table's initial extent is 20Mb.

b. The INVENTORY_NO column can store numbers with decimals.

c. The table's next extent is 5Mb.

d. The PCTFREE setting is ignored.

e. None of the above.

4. Look at this table definition.

```
CREATE TABLE SALES_REPORT
(SALES_PERSON VARCHAR2(20), SALES_DATE DATE, SALES_AMOUNT
NUMBER(12,2), MANAGER_APPROVAL CHAR(3));
INSERT INTO SALES_REPORT VALUES ('Jerry Montell',
'15-JAN-2004');
```

Which statements are true about the inserted row?

a. The ROWID is stored in the row overhead.

b. The column length stored for the SALES_AMOUNT column is zero.

c. The column length for the MANAGER_APPROVAL column is three.

d. The SALES_AMOUNT is null.

e. The SALES_DATE has a time of 2 minutes past midnight.

f. None of the above.

5. Which of the following are valid datatypes? Choose three.

   a. YEAR TO MONTH

   b. TIMESTAMP

   c. NVARCHAR

   d. BFILE

   e. NCLOB

6. You want to create a temporary table that stores rows for your entire session. Arrange the command lines in the correct order.

```
1 (TARGET VARCHAR2(30), HITS NUMBER(5), MISSES NUMBER(4,3))
2 ON COMMIT PRESERVE ROWS
3 STORAGE (INITIAL 5M)
4 CREATE GLOBAL TEMPORARY TABLE MYROWS
```

   a. 2, 4, 1, 3

   b. 4, 1, 2, 3

   c. 4, 2, 1, 3

   d. 2, 4, 3, 1

7. Look at the following command that creates a table for storing flights from 1996 through 1999.

```
CREATE TABLE AIRPORT_HISTORY
     (CODE CHAR(4),
      FLIGHT NUMBER,
      FLIGHT_DATE DATE)
   PARTITION BY HASH (FLIGHT_DATE)
   PARTITIONS 4
   STORE IN (HISTORY1, HISTORY2, HISTORY3, HISTORY4);
```

Which partition stores flights from the year 1998?

   a. The third partition

   b. All the partitions

   c. Unknown

   d. None of the above

Hands-on Assignments

1. You have a small project to complete. You have almost finished an application that reads data from one table, performs a set of calculations, and inserts data into two output tables and commits its work. It reports errors on the screen, which you capture in a spooled file. You have to repeat your test of the application several times, checking the error report each time. You want to create a set of temporary tables for the output tables. Then, you can run the application repeatedly without clearing out the tables before repeating. Write the SQL command to create two temporary tables and explain how you can use them for testing your application. The first table should be named CH07TEMPCUSTORDER and should have columns for customer name and total order amount. The second table should be named CH07TEMPCUSTORDERTOTALS and should have columns for customer order number, total state tax, and total shipping charges. Save the commands and the explanation in a file named **ho0701.sql** in the **Solutions\Chapter07** directory on your student disk.

2. Recall that you created this table for Global Globe's employees:

```
CREATE TABLE EMPLOYEE
(EMPLOYEE_ID NUMBER(10),
JOB_TITLE VARCHAR2(45),
first_name varchar2(40),
Last_name varchar2(40),
phone_number Varchar2(20));
```

   Start with this CREATE TABLE command, and modify it to create a new range-partitioned table named CH07EMPLOYEE_RANGE that contains the same column as the employee table. The partition key is the EMPLOYEE_ID column, and there should be four partitions, all located in the USERS tablespace. Assume that the values in EMPLOYEE_ID start at one, reach 100,000, and are distributed evenly. Save the command in a file named **ho0702.sql** in the **Solutions\Chapter07** directory on your student disk.

3. Copy the range-partitioned table command from Assignment 2, and make these changes to it:
   - Name the table CH07EE_RANGEHASH
   - Keep the range partitions the same, and add hash subpartitioning on JOB_TITLE.
   - There are four subpartitions, all residing on the USERS tablespace.

   Run the command in your database so that the table is owned by CLASSMATE.

   Save your command in a file named **ho0703.sql** in the **Solutions/Chapter07** directory on your student disk.

4. You want to create a new table in the USER_DTAB tablespace. The table, named CH07GOLD, stores a history of the price of gold. A new row is added to the table approximately every two hours. Write the CREATE TABLE command for the table. Your command should address these points:
   - The columns are named PRICE, PRICE_DATETIME, and TIME_BETWEEN.
   - The PRICE can contain fractions of a penny as small as thousandths of a penny. The price is always under 1000 dollars.

- The TIME_BETWEEN column stores the number of days (up to 99 days), hours, minutes, and seconds (to the hundredth of a second) between this record and the previous record.
- Because the table is always receiving inserted data and is never updated, you want to minimize the storage space saved for updates.
- Assume that the average row length is 20 bytes, the table gets an average of 12 inserted records per day, and you want the table to be created with enough storage space for approximately 6 months.

Do not run the command in your database, because you do not have any dictionary-managed tablespaces, which this command requires. Save your command in a file named **ho0704.sql** in the **Solutions\Chapter07** directory on your student disk.

5. Your manager has just informed you that the GOLD table needs more columns. She asks you to create a new table called CH07GOLD_HISTORY that has the correct structure. Starting with the CREATE TABLE command you created in Assignment 4, modify the command to accommodate these changes:
   - Add a column called LOCATION that stores the city from which the price is quoted.
   - Add a column that contains multiple values for the price in five different currencies. The price is always added in this order: Francs, Yen, Pesos, Pounds, and Euros. (*Hint*: Use the Enterprise Manager console to find a collection type in the CLASSMATE schema that you can use.)
   - The added columns add an average of 30 bytes to each row, including overhead.
   - Adjust the storage settings so that the table resides in the locally managed tablespace, USERS.

   Run the command in your database as the CLASSMATE user. Save your command in a file named **ho0705.sql** in the **Solutions\Chapter07** directory on your student disk.
6. Once again, your manager says she wants a change to the GOLD table. It turns out that the table needs a few more columns and different storage settings. Instead of one price per record, each record holds ten prices. The date and time stamp columns are updated each time a new price is added to the table. The current record is updated until it is filled with ten prices. Then, a new record is inserted, and that becomes the current record to be updated.

   This time, she wants you to take the original table structure (from Assignment 4) and make these changes to the table:
   - Replace the price column with a column that holds up to 10 gold prices. (*Hint*: look for another collection datatype in the CLASSMATE schema.)
   - Recalculate the initial space requirements: The new average row length is 30 when it is inserted, growing to 50 bytes after the gold prices are filled in. The table gets an average of two records inserted per day, and you want the table to be created with enough storage space for approximately six months.
   - Adjust the storage settings to minimize row migration.

Name the table CH07GOLD_COLLECTION. Do not run the command, because it requires a dictionary-managed tablespace. Save your SQL in a file named **ho0706.sql** in the **Solutions\Chapter07** directory on your student disk.

7. Your new client has a collection of original music recordings in digital formats that he wants to store in his Oracle 10*g* database. He plans to give online customers the ability to select songs and assemble a customized CD that is then printed and mailed to them. Your job is to create the table to store the music. The client wants to use software other than Oracle 10*g* to update the music. Therefore, the music must reside in AUD format in a file that can be viewed by the database. (AUD format is a format used especially for music that can be played on the Internet or printed on a CD and played on a stereo like any music CD.) Create a table named CH07SONG that stores the song' s unique ID number, artist name, song title, song length (in minutes and seconds), and the song itself. Create another table named CH07SONGLIST that is used to assemble the customer's song list. This table should have the customer's name and address, CD name (the customer fills this in online, and it can be up to 40 characters long), and a list of song ID numbers (up to 15 songs). Once again, you find a collection appropriate for the list of songs in the Schema Manager.

   Create the tables in the CLASSMATE schema and save the SQL commands in a file named **ho0707.sql** in the **Solutions\Chapter07** directory on your student disk.

8. Look at the CUSTOMER table you created earlier in the chapter. Create a temporary table that has all the columns except the VARRAY and includes all the rows in the table. Name the table CH07TEMPCUSTOMER and rename all the columns with the prefix TEMP. The table should remove data rows after each commit.

   Execute the command as CLASSMATE in your database, and save your SQL command in a file named **ho0708.sql** in the **Solutions\Chapter07** directory on your student disk.

9. Create a table as described in Assignment 8, but make it a permanent table named ch07PERMCUSTOMER, and include the VARRAY. Execute the command as CLASSMATE in your database, and save your SQL command in a file named **ho0709.sql** in the **Solutions\Chapter07** directory of your student disk.

10. Create an object table named CH07NEW_ADDRESSES using the CUSTOMER_ADDRESS_TYPE as its object type and with storage in the USER_DTAB tablespace. The table needs 10Mb of storage and room for rows to double in size when they are updated.

    Do not create the table because it requires a dictionary-managed tablespace. Save the SQL command in a file named **ho0710.sql** in the **Solutions\Chapter07** directory on your student disk.

11. Find all the different objects in the entire ORACLASS database. Save the SQL command in a file named **ho0711.sql** in the **Solutions\Chapter07** directory on your student disk.

12. Use the sample query jointables.sql and do the same thing for indexes, returning table, name, index name, and all index columns (excluding table columns). Sort the result by index order, within index name, within table name. Save the SQL command in a file named **ho0712.sql** in the **Solutions\Chapter07** directory on your student disk.

Case Project

The Global Globe database now has tables for its classified ads (which were created in this chapter). Now, you turn your attention to the actual news articles. How will the articles be handled? You and your team of programmers and users decide that the articles should be moved from their current directory into the database, where they can be searched with Oracle's sophisticated text search tools. Here are the criteria for the table.

- The table is named NEWS_ARTICLE.
- One employee writes every article.
- Each article has a unique ID number, title, date it is run in the newspaper, and the article itself, which is a QuarkXPress document.
- The QuarkXPress document can be up to 500 Mb, because it sometimes contains digital images embedded within the document.
- The lead editor must approve an article before it is printed. Record the editor's employee ID and the date of approval.
- As an auditing and security feature, the table must store information about the last five times the article was modified. Store the employee's ID and the date and time of the modification. (Find an appropriate collection type owned by CLASSMATE.)
- Store the table in the USER_AUTO tablespace, and use the default storage settings.

After you have read the criteria, write and execute the command to create the table, and save your work in a file named **case0701.sql** in the **Solutions\Chapter07** directory on your student disk.

CHAPTER 8

ADVANCED TABLE MANAGEMENT

LEARNING OBJECTIVES

In this chapter, you:

- Create tables with large object (LOB) columns
- Create index-organized tables
- Analyze tables to collect statistics of data in a table
- Understand the tasks involved in table management including table storage structure
- Changing table columns
- Redefine a table when the table is online
- Specialized table changes including flashback and transparent encryption
- Understand the tasks involved in table management
- Use data dictionary views to find information about tables and underlying structures

INTRODUCTION

Table structures hold the vast majority of the data found in the database. You have practiced creating a variety of tables in the previous chapter. This chapter shows you how to create various other types of tables, including those containing large binary object (LOB) columns and index-organized tables. You then learn how to make many kinds of changes to the table structure, such as adjusting the storage settings, removing a column, and removing all the data. The final section of the chapter shows you how to query the data dictionary views related to tables, segments, and extents, and other table- or column-related elements.

ADVANCED TABLE STRUCTURES

Having seen so many ways to create tables, you are already familiar with the general syntax for creating a table. There still remain additional types of tables to learn to create. Two of these table types utilize unusual methods of storing data: tables with LOB columns and index-organized tables.

A table with LOB columns can store huge amounts of data in a single column. For example, the digital audio file for one song on a CD, or an entire CD of songs, can be stored in a single record containing one or more LOB columns. Other uses for LOB columns are large documents in PDF or other formats that contain special formatting symbols that must be preserved. These include images, such as satellite multi-spectrum photographs, high-resolution digital photographs, and scanned images of artwork. These large-sized files can be loaded into a LOB record in the database, where they are protected from unauthorized use and possible theft or damage.

Index-organized tables help you query table rows more quickly by reducing the number of times a process must read either memory or disk to retrieve the row data, when rows are read according to the primary key. A good use for index-organized tables is a table in which most of the columns are indexed within the primary key of the table and the data that is not part of the primary key is relatively small and static.

NOTE

It is not a prerequisite for efficiency that an index-organized table contain only static data. However, an index-organized table indexes all columns in a table. Subjected to constant DML change activity, an index-organized table will deteriorate more rapidly than a regular BTree index, simply because each index entry is larger.

An example of a semi-static table could be a table used to look up the population of a city by its state, county, and municipal code. The state, county, and municipal code are already indexed, so moving the population into the index only increases the size of the index slightly. By not having to read the ROWID from the index and then read the row from another block in the database, your queries perform better.

Tables with LOB Columns

As you know, a LOB column has one of four datatypes. These datatypes are BLOB (binary large object), CLOB (character large object – for storing text strings), NCLOB (as for CLOB but using Unicode), and BFILE (a pointer to an externally stored multimedia file). These four LOB datatypes are divided into two groups, according to where they are stored:

- **Internal LOB:** The BLOB, CLOB, and NCLOB datatypes all have their data stored inside the database, so they are called internal LOBs. Internal LOBs use copy semantics, which means that if you copy a LOB from one row or column to another, the entire LOB—including its data—is copied to the new location.

NOTE

LOB objects can also have a distinctive LOB storage clause. This means that a column containing a LOB object can be stored out of line from the table (not in the table). A LOB object can even be stored in a separate tablespace to that of the table, retaining only a pointer in the column.

- **External LOB:** The BFILE datatype is the only external LOB deliberately created as a column pointer, its data is stored outside the database in an operating system file. External LOBs use reference semantics, which means that when copying an external LOB, only the pointer to the location of the file is actually copied. The original file is not copied.

Here is an example of a table with two LOB columns, one for a 90-minute music track stored in a file outside the database and one for a high-resolution image of the band, for printing posters, stored inside the database.

```
CREATE TABLE SOUNDBYTES
(ALBUM_ID VARCHAR2(20),
 ARTIST VARCHAR2(40),
 CD_MUSIC BFILE,
 POSTER_SHOT BLOB);
```

There is also a special PL/SQL package called DBMS_LOB, used to simplify the manipulation of data in LOB columns. Procedures and functions in the DBMS_LOB package are all standard programming structures, used to open, close, read, and write files, along with various other activities. Some examples are as follows (all pseudo-coded, not necessarily functioning with this specific syntax, to provide a conceptual picture of functionality):

- **OPEN (<location>, <mode>).** Open a LOB object in a mode such as read, read-write, or read-only.
- **CLOSE (<location>).** Close an open LOB file.
- **ISOPEN (<location>).** Is a LOB file already open? If so, then you don't want to attempt to open it twice, returning an integer representing true or false:

  ```
  IF ISOPEN('file1') = TRUE THEN
        OPEN('file1', 'readwrite');
  END IF;
  ```

- **READ (<location>, <length>, <offset>, <output-buffer>).** Read data from a LOB, for a specified length, beginning from a specified offset, placing the result into an output buffer.
- **WRITE (<location>, <length>, <offset>, <output-buffer>).** Write data to a LOB, for a specified length, beginning at a specified offset, writing the result from an input buffer.
- **GETLENGTH (<location>).** Returns an integer value of the total length of a LOB. A function like this can be useful when reading an entire LOB element by element.
- **INSTR (<location>, <pattern>, <offset>, <n>).** Returns an integer value representing the nth position of a pattern in a LOB object.
- **SUBSTR (<location>, <length>, <offset>).** Returns a subset part of a LOB object for a specified length, from an offset position.

NOTE

Refer to the Oracle documentation for all the details of DBMS_LOB procedures and functions. The purpose of this book is not to simply duplicate Oracle documentation. As a database administrator you need to learn how to use Oracle documentation, rather than simply duplicate Oracle documentation. The Oracle documentation can be installed on your computer with your Oracle database server software, or your Oracle client software installation. If it is not installed your instructor may be able to provide it, and it is available on the Oracle website.

Oracle has consistently provided more tools for LOBs with each new release. Because LONG and LONGRAW are deprecated, all the capabilities and tools available for these old datatypes are being reintroduced and improved upon for the LOB datatypes.

LOB Storage

Because of their potentially huge size, Oracle 10*g* has special storage methods for LOBs.

Figure 8-1 shows the components of an internal and an external LOB. The data stored inside both inline and out of line LOB types is called its value. Sometimes, Oracle 10*g* stores an internal LOB's value within the row (inline). Unless you specify that all LOB values must be stored out of line (not within the row), Oracle 10*g* stores LOB values that are less than 4000-byte-long inline. The pointer that directs the database to the actual location of the value is called the LOB locator for internal LOBs, and a BFILE locator for external LOBs.

When an internal LOB is stored out of line, a separate LOB data segment is created. You can allow Oracle 10*g* to handle this itself, or you can specify a location and size for the LOB data segment when you create the table. You can also specify that all LOB data (even data that is less than 4000 bytes) is to be stored out of line. Here is brief syntax for creating a relational table with a BLOB column:

```
CREATE TABLE <tablename>
(<other_column_specs>,
 <LOBcolumnname> <LOBdatatype>)
LOB (<LOBcolumnname>) STORE AS <lobsegmentname>
 (TABLESPACE <tablespacename>
  ENABLE STORAGE IN ROW|DISABLE STORAGE IN ROW
  CHUNK <nn>
  STORAGE
     (INITIAL <nn> NEXT <nn> MAXEXTENTS UNLIMITED|<nn>)
  PCTVERSION <nn>|RETENTION
  LOGGING|NOLOGGING
   CACHE|NOCACHE);
```

Naming the LOB data segment (for example, STORE AS MOVIELOB) is optional; however, it makes queries on the data dictionary views more readable, because the name you choose appears rather than a system-generated name, such as SYS_103877.

The CHUNK parameter sets the number of bytes allocated for working with the LOB value. Oracle 10*g* writes and reads one chunk of the LOB value at a time. In addition, when storing LOB values out of line, Oracle 10*g* allocates space by chunks, rather than by data blocks. The chunk size must be a multiple of the database block size, and must be smaller than the INITIAL and the NEXT sizes (either in the tablespace or in the LOB storage clause). The maximum chunk size is 32 Kb. A larger chunk size is more efficient for the high

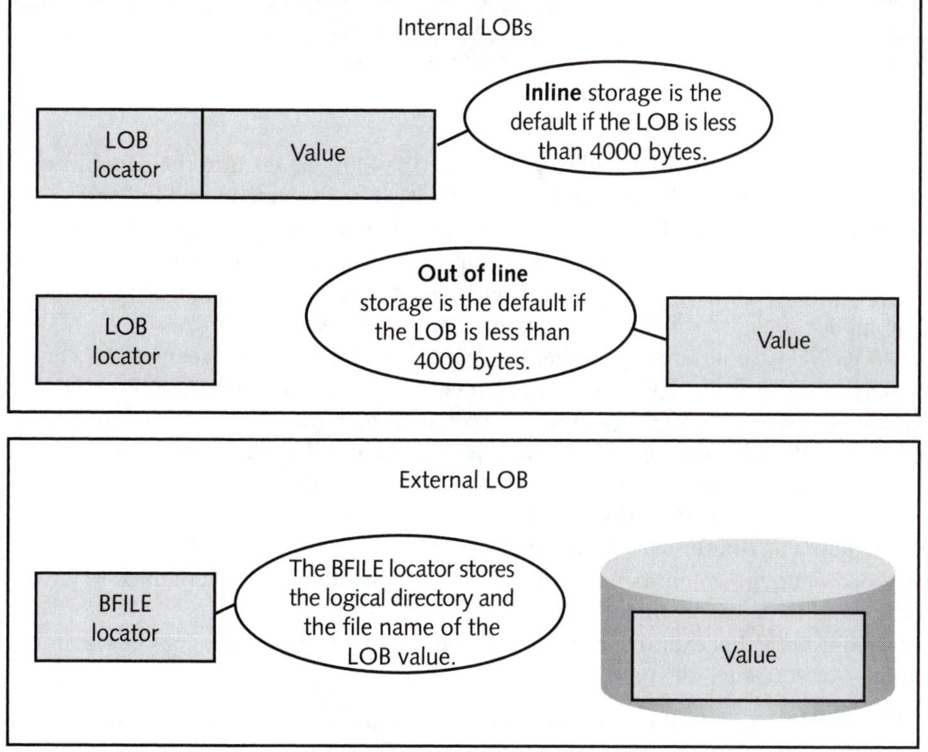

FIGURE 8-1 Even large objects stored inline with the row have a LOB locater (pointer)

volume I/O needed when manipulating very large LOB values. A smaller chunk size is better for LOBs, which have sizes that vary from row to row, because it reduces unused space allocated to the LOB storage. The default chunk size is one data block.

Specifying DISABLE STORAGE IN ROWS tells Oracle 10*g* to always use out of line storage for the LOB value. This might be valuable if you have many rows with LOB values that are close to the default threshold (4000 bytes). Storing these portions of the data out of line may prevent row chaining and allow faster access to the non-LOB columns. ENABLE STORAGE IN ROWS is the default, and it tells Oracle 10*g* to store small LOB values (less than 4000 bytes) inline and to store larger LOB values out of line.

The PCTVERSION <nn> parameter tells Oracle 10*g* to store old versions of the LOB value within the LOB itself, until the specified percent of storage is used up. After that, old versions are overwritten by newer versions. The RETENTION parameter is an alternative to the PCTVERSION and indicates that undo data should be generated as the method of retaining old versions of the LOB. Undo data is retained as long as it is specified in the UNDO_RETENTION initialization parameter. You can use one or the other, but not both of these parameters. The default depends on the database's undo mode. For automatic undo mode, RETENTION is the default. For manual undo mode, PCTVERSION is the default.

NOTE

Automated undo is now the default and acceptable mode of operation for an Oracle 10*g* database. Manual undo (otherwise known as rollback) is desupported, and should no longer be used.

The CACHE parameter tells Oracle 10g to place the LOB values into the data buffer for faster retrieval of frequently accessed data. NOCACHE is the default and indicates that the LOB values are placed at the end of the buffer, so they are first to be replaced when more space is needed in the buffer. From a performance tuning perspective, caching LOB objects is not sensible unless your application uses very small objects, such as iconic images, or applications are solely binary database storage in general.

The LOGGING parameter assures that the creation of the LOB data segment and mass inserts (such as those from SQL*Loader) are recorded in the redo logs. The NOLOGGING parameter suppresses logging of these two types of activities. The default depends on the CACHE or NOCACHE setting. If you use CACHE, LOGGING is the default. If you use NOCACHE, the default is whatever the tablespace default for logging specifies. You cannot specify CACHE and NOLOGGING together.

You can define attributes with LOB datatypes in object types for use in object tables or user-defined datatype columns in relational tables. You can use LOB columns in partitioned tables and in index-organized tables.

Follow along with this example to create a relational table with LOB columns that are stored in their own storage segment.

1. Start up the Enterprise Manager console. In Windows, click **Start/All Programs/Oracle ... /Enterprise Manager Console**. In Unix or Linux, type **oemapp console** on the command line. The Enterprise Manager Console login screen appears.
2. In the main window of the Enterprise Manager console, click **Tools/Database Applications/SQL*Plus Worksheet** from the top menu.
3. Connect as the CLASSMATE user. Click **File/Change Database Connection** on the menu. A logon window appears. Type **CLASSMATE** in the Username box, **CLASSPASS** in the Password box, and **ORACLASS** in the Service box. Leave the connection type as "Normal." Click **OK** to continue.
4. For this example, start with a prepared SQL command. Click **File/Open**. A window opens. Navigate to the **Data\Chapter08** directory, and select the **movie.sql** file. Click **Open**. The file appears in the SQL*Plus Worksheet window, as shown in Figure 8-2.
5. Replace the variables with actual names so that these characteristics are true:
   - The USERS tablespace stores the LOB data segment.
   - The LOB is stored in 32768 size chunks.
   - The initial extent is 64Kb, the next extent is 32Kb, and the maximum extent is unlimited.

   There is no LOB data segment name in this example because when you specify the LOB storage for two LOB columns, you cannot name the LOB data segment. The extent sizes used here are low for LOB data and are used only to conserve space for these examples. Normally, the LOB extent is 50Mb or more.

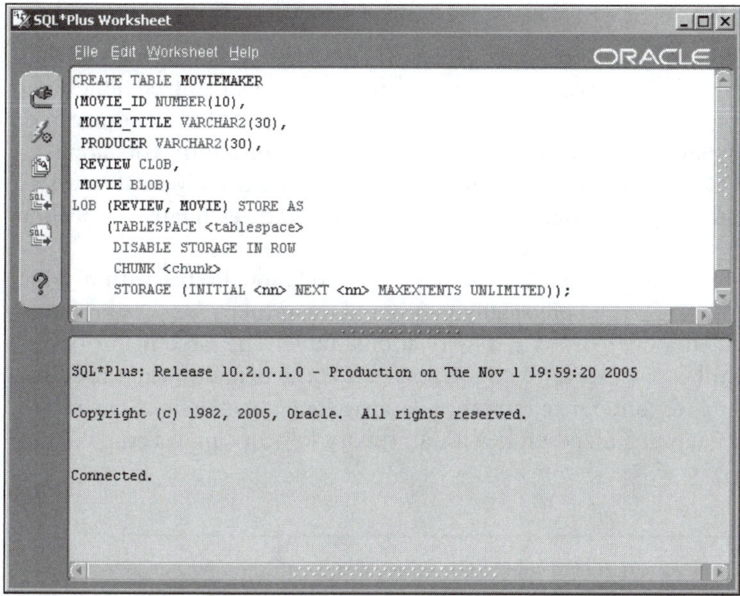
SQL*Plus Worksheet
File Edit Worksheet Help
ORACLE
CREATE TABLE MOVIEMAKER
(MOVIE_ID NUMBER(10),
MOVIE_TITLE VARCHAR2(30),
PRODUCER VARCHAR2(30),
REVIEW CLOB,
MOVIE BLOB)
LOB (REVIEW, MOVIE) STORE AS
(TABLESPACE <tablespace>
DISABLE STORAGE IN ROW
CHUNK <chunk>
STORAGE (INITIAL <nn> NEXT <nn> MAXEXTENTS UNLIMITED));
SQL*Plus: Release 10.2.0.1.0 - Production on Tue Nov 1 19:59:20 2005
Copyright (c) 1982, 2005, Oracle. All rights reserved.
Connected.

FIGURE 8-2 Load a file with File/Open and save a file with File/Save Input As

6. Run the command by clicking the **Execute** icon. The SQL*Plus Worksheet runs the command and responds "Table created."
7. Remain logged on for the next practice.

The file named **movie_final.sql** in the **Data\Chapter08** directory on your student disk contains the final version of the CREATE TABLE command.

LOB data segments can also be used to store a VARRAY. The settings for the LOB storage clause are the same as those you saw here for storing a LOB column, except the VARRAYs LOB cannot be in a different tablespace than the table. To specify that the VARRAY should use a LOB data segment, add the STORE AS LOB clause immediately after the VARRAY column name and datatype in the list of columns. For example, the following table has a VARRAY named RACE_LIST that is stored in a LOB data segment. (Create the VARRAY type first:

```
CREATE TYPE RACE_ARRAY AS VARRAY (1500) OF CHAR (25);
/
CREATE TABLE HORSERACE
(HORSE_NAME VARCHAR2(50),
 RACE_LIST RACE_ARRAY)
   VARRAY RACE_LIST STORE AS LOB RACEARRAYLOB
    (CHUNK 32768
     STORAGE (INITIAL 20M NEXT 40M MAXEXTENTS 100))
STORAGE (INITIAL 80M);
```

Use LOB storage for VARRAYs, when the VARRAY is intended to be very large.

Tables with LOBs are a type of table that requires special parameters in the CREATE TABLE command. The second type of table requiring special parameters is the index-organized table.

Index-Organized Tables

Tables with LOBs tend to have rows of extra-large size. They are so large that part of the row is often separated into its own storage segment. In contrast, index-organized tables tend to have smaller-sized rows that benefit from a consolidation of the table's index storage and the table's data storage. Whereas tables with LOBs are often used for multimedia applications, such as music, video, or images, index-organized tables are usually used in text-based or number-crunching applications, such as insurance rate estimates or airline reservations, in which speedy retrieval is key.

An index-organized table, as you saw in Chapter 7, is a relational table with a primary key, in which the rows are stored physically in order of the primary key. A non index-organized table is also called a heap-organized or relational table. The large majority of relational data is stored in heap-organized tables because it is the default format used by Oracle 10*g* for relational tables (and for many other database engines). Figure 8-3 shows the difference between a heap-organized table with a primary key and an index-organized table.

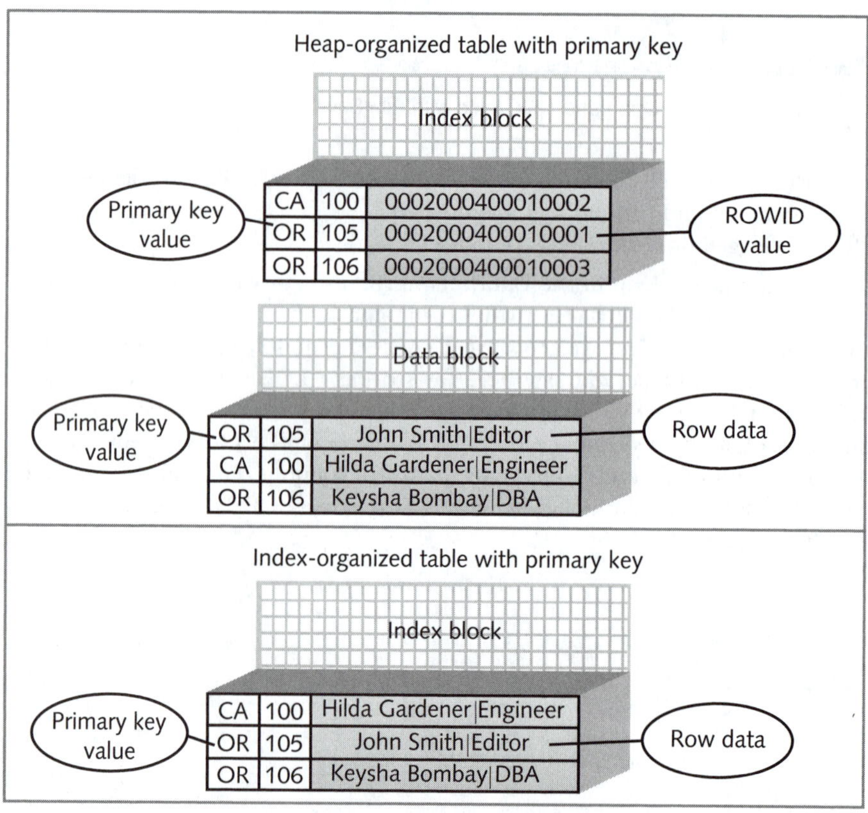

FIGURE 8-3 The index and the data are united in an index-organized table

A relational table with a primary key constraint automatically has a unique index associated with the primary key. That index is stored in primary key order in a BTree index structure, making queries by primary key very fast. A BTree index balances data in even branches that split into narrowing ranges of key values. The index stores the primary key columns and the ROWID of the associated row. The actual data is stored in the data block. When you execute a query looking for a particular primary key value, Oracle 10*g* searches the index, reads the ROWID, locates the row in the table, and reads the row. Using Figure 8-3 as an example, imagine that this query was issued and finds the row for John Smith:

```
SELECT NAME, JOB
FROM EMPLOYEE
WHERE STATE='CA' AND ID='105';
```


If the table is heap-organized (as in the top of Figure 8-3), Oracle 10*g* looks up the value of the ROWID for the matching primary key and then finds the row using the ROWID. It then retrieves the values of the NAME and JOB columns from the row.

An index-organized table does not have an index. Instead, the entire table becomes an index. Every row is stored in a BTree index structure. The primary key values are used to arrange the data into branches. When you execute a query looking for a particular primary key value, Oracle 10*g* searches the table, and retrieves the row. Continuing with the same example for Figure 8-3, Oracle 10*g* looks up the primary key in the index and finds the matching row. Then it uses this same row to retrieve the values of the NAME and JOB columns. By avoiding the second read of a data block, the index-organized table saves I/O time. The index-organized table also saves space because it does not store the primary key values in both the index and the data blocks and because it does not store the ROWID values at all.

The primary advantage of an index-organized table is that queries based on the primary key are faster than queries in heap-organized tables. The primary disadvantage is that inserts, updates, and deletes are slower, because they may cause an imbalance in the BTree structure, which requires Oracle 10*g* to shuffle rows into different index blocks to rebalance the structure. Additionally, continual change activity will eventually deteriorate the BTree structure to the point at which queries will actually be slower.

Follow along with this example to create an index-organized table:

1. If you are not already there, go to the SQL*Plus Worksheet, and log on as CLASSMATE/CLASSPASS in the ORACLASS database.
2. Start with a prepared SQL command found in the **Data\Chapter08** directory on your student disk. Click **File/Open**. A browse window opens. Navigate to the **Chapter08** directory, and select the **zip.sql** file. Click **Open**. The file appears in the SQL*Plus Worksheet window, as shown in Figure 8-4.
3. As you can see, the only new feature in this relational table so far is a primary key constraint. The first requirement of an index-organized table is that it must have a primary key on which the index is built. The second requirement is that you must include the ORGANIZATION INDEX clause. Type **ORGANIZATION INDEX** on the blank line below the command, and press **Enter**.

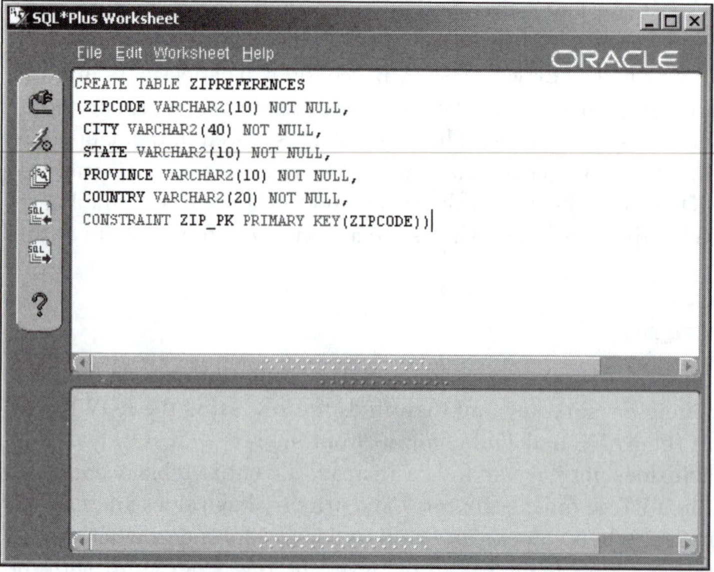

SQL*Plus Worksheet
File Edit Worksheet Help
ORACLE
CREATE TABLE ZIPREFERENCES
(ZIPCODE VARCHAR2(10) NOT NULL,
CITY VARCHAR2(40) NOT NULL,
STATE VARCHAR2(10) NOT NULL,
PROVINCE VARCHAR2(10) NOT NULL,
COUNTRY VARCHAR2(20) NOT NULL,
CONSTRAINT ZIP_PK PRIMARY KEY(ZIPCODE))

FIGURE 8-4 The relational table has a primary key constraint

4. "Optional additional phrases for index-organized tables are:
   - **OVERFLOW:** By restricting the length of each row within the table, adding this parameter helps keep the BTree index tightly packed into index blocks. For any row that exceeds the specified length, the balance of the row is stored in a separate segment. You can add storage specifications, such as the tablespace, extent size, and so on, along with the OVERFLOW parameter.
   - **PCTTHRESHOLD:** This tells Oracle 10g the maximum percentage of space in an index block that one row is allowed to use. If a row's total length exceeds this size, the row is divided, and the excess columns are stored in the overflow segment.
   - **INCLUDING <col>:** If you specify OVERFLOW, you specify the dividing line between which columns are stored in the main table and which are stored in the overflow segment. The primary key columns are always stored in the main table. When you specify INCLUDING, all columns after the named column are stored in the overflow segment. If you organize the list of columns in the table so that the most frequently accessed columns come first (after the primary key columns), this parameter can help reduce I/O by keeping the commonly accessed columns together.
   - **TABLESPACE** and **STORAGE:** Add storage and tablespace information for the overflow segment by specifying them after the OVERFLOW parameter.

   Type the following line to define the table's tablespace. Then press **Enter**:

   ```
   TABLESPACE USERS
   ```

5. Type the following two lines to indicate that a threshold of 20 percent per row will restrict the row size, and the remaining portion of the row (if any) will be stored in an overflow segment located in the USER_LOCAL tablespace. Press **Enter** after typing the lines:

```
PCTTHRESHOLD 20
OVERFLOW TABLESPACE USERS
```

6. Add storage parameters for the overflow segment by typing these two lines and pressing **Enter**:

```
STORAGE (INITIAL 64K NEXT 32K
MAXEXTENTS 50 PCTINCREASE 0);
```

7. Execute the command to create the table by clicking the **Execute** icon (the lightning bolt). The worksheet replies, "Table created" in the lower half of the screen.
8. Remain logged on for the next practice.

The complete command for the index-organized table named ZIPREFERENCES is in the **zip_final.sql** file in the **Data\Chapter08** directory on your student disk.

Here are some additional points about index-organized tables:

- You can include LOB columns.
- You can partition index-organized tables, but only using hash or range partitioning. In addition, the partition key must be all or a subset of the primary key.
- You can add secondary indexes; however, they should be rebuilt if the table is updated often, so that the physical guess stored in the index is accurate. Inaccurate physical guesses cause extra I/O.

This completes the details of procedures for creating complex table structures. You have practiced creating several tables in this chapter and in Chapter 7. What happens when you want to change something about a table you have created? For example, you may want to add a new column, change the datatype of a column, or add a default value to a column. You might want to give a table more storage space or reduce the storage it has been allocated but is not using. The next section shows you how to handle these kinds of tasks.

OVERVIEW OF TABLE MANAGEMENT

After you create a table, you can load data into it by using applications, or tools such as SQL*Loader or SQL*Plus. As time goes by, you may find your table structure needs adjustment. Here are the types of changes you can make to a table's structure:

- **Change the storage settings:** You can adjust the size of the next extent, maximum extents, percent free, and most of the other storage settings.
- **Reorganize the table online:** You can rearrange column order, add or remove columns, and change column names or datatypes while the table remains online.

NOTE

An online table is a table that is currently subject to concurrent access, involving both reading of data and changes to data in the table.

- **Drop columns:** You can mark a column so it is unavailable until it is dropped (saving valuable resources in actually destroying column data), or simply drop the column immediately.
- **Truncate or drop the table:** You can remove all the rows in a table without generating redo log entries by using the TRUNCATE command. Dropping the table using a DROP command removes the rows and the table structure as well.

Before making changes to a table's structure, it is always a good idea to analyze the data, so that you understand the table's makeup better. Perhaps even make a backup copy of the table first. In fact, you may discover that your table needs adjusting based on the outcome of the analysis. For example, you know a table needs better storage settings, because, after analyzing the table, you find it has 10 extents and 25 percent of the rows are migrated.

The next section describes how to analyze a table, or an entire schema. The following sections describe each of the table structure changes mentioned in the previous list.

Analyzing a Table

You should analyze the tables in a schema periodically to give the optimizer up-to-date information for optimizing queries and other SQL commands. If the table data changes (new rows, updated rows, or deleted rows) every day, you may want to analyze the table once a week, or even more frequently. If data in a table changes slowly, you could analyze the table on a monthly or quarterly basis. Analyzing a table also gives you up-to-date information so that you can make valid decisions regarding changes to the storage parameters of the table.

NOTE

Analyzing tables allows you to collect realistic and up-to-date statistics for tables. Even though statistics is, in effect, a performance tuning activity, it is essential to understand how storage structures can be changed. Therefore, it is necessary to present statistical analysis of tables. Oracle 10*g* can automate a certain amount of statistics gathering using the Automatic Workload Repository (AWR) and the Automatic Database Diagnostic Monitor (ADDM). Both of these tools are a part of the Database Control for a database. This will be covered in a later chapter dealing with performance monitoring issues. At this stage you will stick with simple ANALYZE commands and the DBMS_STATS built-in package.

To analyze a table, you issue a command that causes Oracle 10g to read the table's structure and update the table's metadata with current information about the size of the table, average row length, total number of rows, free space remaining in extents, and so on. There are two reasons for analyzing a table. First, analysis provides accurate statistics for the cost-based optimizer. Second, analysis gives you, the DBA, more in depth information, helping you decide which storage or column settings to change (if any).

The optimizer is a process used within Oracle 10*g* that decides the most efficient method of executing a query. The optimizer creates an execution plan for a query, commonly known as a **query plan**. A query plan is a list of steps taken to retrieve data for a query. For example, one query might require a full table scan followed by a sort operation. Another query might require an index lookup, followed by a merge with another table's data, followed by another index lookup on a third table. Queries are executed according to the plan selected by the optimizer, but not always. Why not always? Because sometimes the optimizer can get it all wrong, and during the execution process the database engine has the option of automatically changing a query plan. There are also various ways of changing query plans, such as using hints in SQL statements.

There is now only one type of optimizer available to Oracle 10*g*: cost-based optimization. Rule-based optimization is no longer available, even for backward compatibility. Rule-based optimization used static rules to rank possible access paths and select the best path. The cost-based optimizer is now the dominant and highly suggested method of optimization in Oracle 10*g*. Cost-based optimization uses statistics on the actual volume and distribution of table data to determine the best path to retrieve the data, taking into account the relative costs of I/O, CPU time, execution time, and other factors. The statistics used by the cost-based optimizer are stored in the table's metadata when you analyze the table.

NOTE

The only way to upgrade applications written as rule-based tuned SQL code (queries), in an older version of Oracle Database, is to use *stored outlines* (stored query plans). A stored outline allows you to freeze the method by which a query will be executed. *Plan stability* implies that a query is always executed in the way in which the outline (pre-tuned and stored query plan) has previously executed a query.

To analyze a table, you can use either the ANALYZE command or the DBMS_STATS predefined package. The ANALYZE command is easier to use but a little out of date. The DBMS_STATS package can be much more efficient, especially where large amounts are data are involved. Run the following examples in the ORACLASS database:

1. If you are not already there, go to the SQL*Plus Worksheet, and log on as CLASSMATE/CLASSPASS in the ORACLASS database.
2. Type the following line to enter the command to analyze the CUSTOMER table:

   ```
   ANALYZE TABLE CUSTOMER COMPUTE STATISTICS;
   ```

3. Click the **Execute** icon to run the command. The worksheet replies, "Table analyzed" when it completes its work. The command automatically gathers statistics on the table, all its columns, and all its indexes.
4. For very large tables, computing statistics takes more time. Therefore, Oracle 10*g* provides the ESTIMATE STATISTICS SAMPLE clause. You can specify a number of rows for the sample or a percentage of the table's rows. Two other clauses you can add to the ANALYZE TABLE command are the VALIDATE STRUCTURE (to validate the integrity of every data block and every row) and LIST CHAINED ROWS (to detect chained and migrated rows) options. Type the following command to experiment with the ANALYZE command:

```
ANALYZE TABLE CUSTOMER ESTIMATE STATISTICS
SAMPLE 1000 ROWS;
```

5. Click the **Execute** icon to run the command. The worksheet replies, "Table analyzed" when it completes its work.
6. Oracle recommends that you use the DBMS_STATS package instead of the ANALYZE command for gathering statistics. Oracle required in the past that you gathered statistics on at least one table in a query to enable cost-based optimization. If no tables have statistics, the rule-based optimizer is used. However, with the advent of substantial automation of statistics gathering using dynamic sampling (automated statistics gathering), and the Database Control features such as AWR and ADDM, this is now no longer necessary. Type in the following command to analyze the table with the DBMS_STATS package:

```
EXECUTE DBMS_STATS.GATHER_TABLE_STATS
('CLASSMATE','CUSTOMER');
```

7. Run the command by clicking the **Execute** icon (the lightning bolt). The worksheet replies, "PL/SQL procedure successfully completed" in the lower half of the screen.
8. An alternate method of running the procedure is to place it into a PL/SQL block and execute it. The advantage is that you don't have to type it all on one line, which makes the command more readable. Type and execute the following command, which has the same effect as the command in Step 6:

```
BEGIN
      DBMS_STATS.GATHER_TABLE_STATS
      ('CLASSMATE','CUSTOMER');
END;
```

9. You can also gather statistics on an entire schema at one time using the GATHER_SCHEMA_STATS procedure. To gather statistics for CLASSMATE, type and execute the following command:

```
EXECUTE DBMS_STATS.GATHER_SCHEMA_STATS('CLASSMATE');
```

10. This command takes longer to complete. After it is finished, the worksheet replies again with, "PL/SQL procedure successfully completed" in the lower half of the screen.
11. Remain logged on for the next practice.

Now that you have gathered statistics, you can query those statistics in the USER_, ALL_, and DBA_TABLES, in the USER_, ALL_, and DBA_INDEXES, data dictionary views, or you can see them in the Schema Manager within the Enterprise Manager console.

Adjusting Table Storage Structure

Many portions of the storage structure of a table can be modified after the table is created. For example, you can increase or decrease the size of the next extent.

All these changes are made by using various clauses within the ALTER TABLE command. The alter table syntax, with each clause listed, looks like this:

```
ALTER TABLE <schema>.<tablename>
  PCTFREE <nn> PCTUSED <nn>
```

```
  INITTRANS <nn> MAXTRANS <nn>
STORAGE (NEXT <nn> PCTINCREASE <nn>
        MAXEXTENTS <nn>|UNLIMITED)
  ALLOCATE EXTENT SIZE <nn> DATAFILE <filename>
  DEALLOCATE UNUSED KEEP <nn>
  COMPRESS|NOCOMPRESS
SHRINK SPACE [ COMPACT ] [ CASCADE ]MOVE TABLESPACE <tablespacename>
      STORAGE (INITIAL <nn> NEXT <nn> PCTINCREASE <nn>
               MAXEXTENTS <nn>|UNLIMITED)
      COMPRESS|NOCOMPRESS
      ONLINE
```

TIP

You do not have to specify the table schema when you are logged on as the owner of the table.

The following list describes each clause:

- **PCTFREE <nn> PCTUSED <nn>:** New settings affect all future rows inserted or updated (not existing blocks). You cannot specify these parameters when the table is in a locally managed tablespace.

NOTE

The STORAGE clause is only applicable to dictionary-managed tablespaces. Dictionary-managed tablespaces are not recommended for Oracle 10*g*. The STORAGE clause is retained in Oracle 10g for backward compatibility with older versions of Oracle. Do not specify a default storage clause for objects created in locally managed tablespaces. When creating tables in a locally managed tablespace, a storage clause can be defined but any irrelevant parameters, such as PCTUSED, will simply be ignored by statements such as CREATE TABLE or ALTER TABLE.

- **INITRANS <nn> MAXTRANS <nn>:** Changes the number of initial and maximum transactions allowed to concurrently access the data block. You cannot specify these for tables in locally managed tablespaces.

NOTE

MAXTRANS is obsolete

- **STORAGE (...):** Adjusts anything except the INITIAL extent of the table. Changes affect future extents, not current extents.
- **ALLOCATE EXTENT:** Explicitly adds an extent of a designated size and location to the table.

- **DEALLOCATE UNUSED:** Releases unused data blocks above the high watermark of the table. The high watermark is the boundary between used data blocks and unused data blocks in a table. There is often room in some of the blocks below the high watermark due to deleted rows or rows that were migrated. The boundary marks the last data block that is formatted for data. When deallocating space, the data blocks that are above the high watermark are released back to the database. Adding the KEEP <nn> clause tells Oracle 10*g* to keep that many bytes of storage above the high watermark within the table.
- **COMPRESS:** Changes a table from decompressed (the default) to compressed data storage. This saves space and memory; however, it is usually applied to tables that are primarily static. Updates and inserts take more time on compressed tables. You can reverse the process by specifying the NOCOMPRESS parameter for a table that is compressed.
- **SHRINK SPACE:** Manually shrink (decrease) the amount of space used by a table, and other database objects such as index-organized tables, partitions, LOB segments, and materialized views. You can use the shrink clause to compact space used by a table. You can adjust the high watermark and free space for use by other objects.

NOTE

The SHRINK SPACE clause applies only to tables in Automatic Segment Space Management (ASM), locally managed tablespaces.

Compaction of a table (segment) will likely require movement of rows and thus row movement must be enabled for the table. This will absolutely cause issues with ROWIDs. The COMPACT option of the SHRINK SPACE clause only defragments (places used blocks closer to each other). The COMPACT clause does not allow changing the high watermark. A later ALTER TABLE <table name> SHRINK SPACE will allow changing a table's high watermark. Using the CASCADE option cascades (passes on) the effects of the table SHRINK SPACE clause to dependent objects, such as indexes created for a particular table.

NOTE

Shrinking a table does not affect attached indexes directly. Shrink indexes using the ALTER INDEX command, which functions the same way as the SHRINK clause in the ALTER TABLE command.

- **MOVE:** Moves the table to another tablespace, although MOVE can also be used to change storage settings while keeping the table in the same tablespace. During a table move, you can adjust any or all storage settings, including INITIAL. You can also compress or decompress the data by adding the COMPRESS or NOCOMPRESS parameter.
- **ONLINE:** Allows users access to the table during the move process. This is only currently available for index-organized, non-partitioned tables.

NOTE

You cannot use all the clauses at once. Some clauses, such as DEALLOCATE and ALLOCATE, or COMPRESS and NOCOMPRESS are obviously quite contrary, and cannot be used together. Others clauses, such as MOVE, must be used alone.

Here is an example of modifying a table to release all unused space except 50Kb above the high watermark:

```
ALTER TABLE HORSERACE DEALLOCATE UNUSED KEEP 50K;
```

Modify the CUSTOMER table by moving it to a different tablespace and adjusting part of the storage settings.

1. If you are not already there, go to the SQL*Plus Worksheet, and log on as CLASSMATE/CLASSPASS in the ORACLASS database.
2. Let's assume that you have looked at the current storage values for the CLASSIFIED_AD table and have decided that the table will not need the 2Mb extents that you set when creating the table. The table appears to be using much less space. You cannot change the INITIAL storage allocation of 2Mb. However, you can change the size of subsequent extents by changing the NEXT parameter to a smaller size. Type and execute the following two lines to modify the CLASSIFIED_AD table:

   ```
   ALTER TABLE CLASSIFIED_AD MOVE TABLESPACE USERS
   STORAGE (NEXT 56K);
   ```

3. The worksheet replies "Table altered" when it completes its work.
4. Remain logged on for the next practice.

The next section describes how you can reorganize a table, while allowing users to continue to query and modify rows in the table.

Redefining Tables Online

Oracle 10*g* has a feature called online table redefinition, allowing you to make nearly any change to a table you need, while keeping the table available for inserts and updates most of the time. The phases in the online table redefinition process are creation of an interim table, redefinition of a table, and application of the redefinition back to the original table. The feature is implemented in a PL/SQL package called the DBMS_REDEFINITION package.

DBMS_REDEFINITION was created for high-availability applications, such as online airline reservations or online banking services. The package is intended for DBAs, and, therefore, you must have several DBA-level privileges to use the package. The package has been upgraded in Oracle 10*g* to allow online redefinition on many types of tables, including individual partitions, clusters, tables with object columns (VARRAYs and nested tables), table, check and NOT NULL constraints, plus a few other factors a little too specialized for this book. In addition, existing statistics are now preserved.

Here is an outline of the steps you follow to redefine a table online. See the Oracle 10g, Release 2 document named *Oracle Administrator's Guide* in the section titled *Managing Tables* for more details:

1. Let's say you want to use the package to modify the table structure of the CLASSIFIED_AD table. First, the CLASSMATE user does not have the privileges needed to run the package, so connect to the SYSTEM user by typing and executing the following command, replacing <*password*> with the actual password for SYSTEM:

   ```
   CONNECT SYSTEM/<password>@ORACLASS
   ```


2. Another minor detail to handle before you begin working with the package is that you must grant a privilege that is required for the package. Type and execute this command:

   ```
   GRANT EXECUTE_CATALOG_ROLE TO SYSTEM;
   ```

3. Verify that the table is a candidate for online redefinition by typing and executing these commands:

   ```
   BEGIN
   DBMS_REDEFINITION.CAN_REDEF_TABLE
   ('CLASSMATE','CLASSIFIED_AD');
   END;
   ```

   If an error message displays: ORA-12089: cannot online redefine table "CLASSMATE". "CLASSIFIED_AD" with no primary key. This will be because the statement failed due to the lack of a primary key. You will not be able to use the package to restructure the CLASSIFIED_AD table.
4. Try another table by typing and executing this command:

   ```
   BEGIN
   DBMS_REDEFINITION.CAN_REDEF_TABLE
   ('CLASSMATE','CLIENT');
   END;
   ```

   The statement should succeed. Thus the CLIENT table can be redefined using the package.
5. Create an interim table in the same schema to that of the original table, including all column changes. For example, let's add a new column FULLNAME just after the LAST_NAME column that contains the FIRST_NAME and LAST_NAME combined. Type and execute this command to create the new table:

   ```
   CREATE TABLE CLASSMATE.CLIENT1
   (CLIENT_ID NUMBER(10) NOT NULL,
    FIRST_NAME VARCHAR2(10) NOT NULL,
    LAST_NAME VARCHAR2(20) NOT NULL,
    FULLNAME VARCHAR2(32),
    CONTACT_PHONE VARCHAR2(15),
    CONTACT_EMAIL VARCHAR2(30),
    CONSTRAINT CLIENT_PK1 PRIMARY KEY(CLIENT_ID))
   TABLESPACE USERS PCTFREE 0
       STORAGE (INITIAL 64K NEXT 8K
       MINEXTENTS 1 MAXEXTENTS 10);
   ```

TIP

You can also include whatever storage parameters you want, regardless of what they are in the old table, assuming settings are appropriate for a locally managed tablespace. Notice that the table and primary key names are different from the original table and primary key names. This is needed because you cannot have two tables with the same name in the same schema. You also can't have two constraints with the same name in the same schema. You can however rename the table afterward. To rename the primary key constraint, you have to drop and re-create it, so you might want to leave it as it is.

6. Start the redefinition process by running this command:

```
BEGIN
DBMS_REDEFINITION.START_REDEF_TABLE ('CLASSMATE', 'CLIENT',
'CLIENT1',
 'CLIENT_ID CLIENT_ID,
  FIRST_NAME FIRST_NAME,
  LAST_NAME LAST_NAME,
  SUBSTR(CONCAT(RPAD(FIRST_NAME,LENGTH(FIRST_NAME)+1),
       LAST_NAME),1,32) FULLNAME,
  CONTACT_PHONE CONTACT_PHONE,
  CONTACT_EMAIL CONTACT_EMAIL');
END;
```

7. The above table redefinition command has parameter settings as the schema name, original table name, intermediate table name, and the *column mapping string*. The column mapping string contains a list of pairs for each column. The FULLNAME column is changed to contain a value consisting of the FIRST_NAME and LAST_NAME columns:

NOTE

If the START_REDEF_TABLE fails for any reason the ABORT_REDEF_TABLE procedure must be executed before executing the START_REDEF_TABLE procedure again:

```
BEGIN
      DBMS_REDEFINITION.ABORT_REDEF_TABLE
        ('CLASSMATE', 'CLIENT', 'CLIENT1');
END;
```

8. At this point, you can create any constraints, grants, triggers, or indexes you need on the interim table. These are transferred to the redefined table later. In this example, there is nothing that we need to create, so continue by running this command to complete the redefinition of the table:

```
BEGIN
      DBMS_REDEFINITION.FINISH_REDEF_TABLE
        ('CLASSMATE','CLIENT', 'CLIENT1');
END;
```

The procedure re-creates the table in the image of the interim table (including indexes, grants, and constraints), pours all the data in according to the column mapping, and then applies any data changes that took place between the beginning and end of the process to the redefined table. Constraints, indexes, and grants from the old table are transferred to the interim table and are dropped when you drop the interim table:

9. To view the results, query the restructured CLIENT table by typing and executing the following command. Figure 8-5 shows the results:

```
SELECT * FROM CLASSMATE.CLIENT;
```


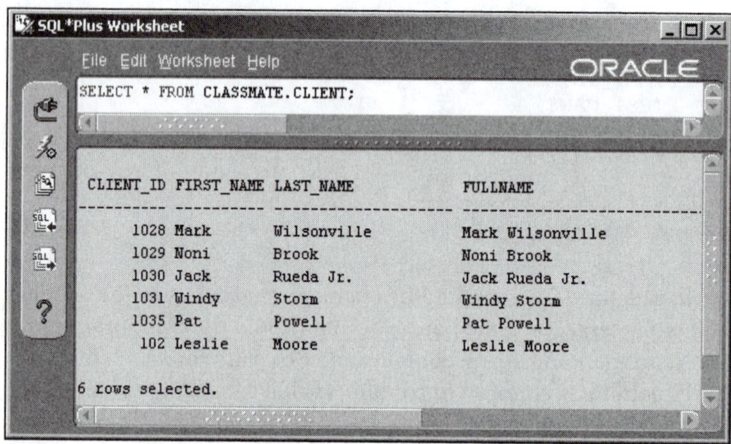

FIGURE 8-5 The redefinition procedure populated the new FULLNAME column

10. Remain logged on for the next practice.

TIP

If errors are encountered, there are a few other procedures included in the DBMS_REDEFINITION package for clean up and restoration. Refer to the Oracle 10*g*, Release 2, documentation in the book titled *PL/SQL Packages and Types Reference.*

When little or no down time is available for database maintenance, use the DBMS_REDEFINTION for tables in the high-demand system.

The next command is a form of the ALTER TABLE command that is strictly for object tables and relational tables, with object-type columns. Object tables and object-type columns are based on the definition of an object type. If the definition changes, the tables in which the object type is used do not change automatically. You must upgrade the table structure. Upgrading a table structure causes any object types used in the table to be updated with the most recent version of the object type definition. Here is an example of the command:

```
ALTER TABLE CUSTOMER UPGRADE INCLUDING DATA;
```

The INCLUDING DATA clause is the default, and data is modified to match the new type definition. You can use NOT INCLUDING DATA to leave data unchanged; however, this could cause problems in queries or other operations on the table's data.

You have seen how to make changes to a table's columns and to its storage settings. There are some other changes, such as changing the row movement mode that you can make on a table's structure. These will be described later in this chapter.

Dropping, Adding, or Modifying a Column in a Table

The previous section described a way to totally rework a table's structure using the DBMS_REDEFINITION package. Many times, you will not have to go through a total redefinition of a table. More often, you need a single column added, or the length of a column increased. These kinds of changes can be accomplished using the ALTER TABLE command and the clauses listed in the following syntax:


```
ALTER TABLE <schema>.<tablename>
  RENAME COLUMN <oldcolname> TO <newcolname>
  ADD (<colname> <datatype>, ... )
  MODIFY (<colname> <datatype>, ... )
  DROP (<colname>, <colname>,...)|COLUMN <colname>
        CASCADE CONSTRAINTS
  SET UNUSED (<colname>,<colname>,...)|COLUMN <colname>
      CASCADE CONSTRAINTS
  DROP UNUSED COLUMNS
```

TIP

Each of these clauses must be used alone in a single ALTER TABLE command. No other clauses can be added to the command. Notice, however, that the ADD, DROP, MODIFY, and SET UNUSED clauses allow you to list multiple columns.

The CASCADE CONSTRAINTS parameter can be added if there are any constraints in other tables that reference the column or columns you drop or set to unused status. Otherwise, Oracle 10*g* does not change the table and issues an error message. Be sure when using a CASCADE option that you actually intend to propagate changes throughout.

Follow along with these steps to use examples of each of these column clauses:

1. If you are not already there, go to the SQL*Plus Worksheet, and log on as CLASSMATE/CLASSPASS in the ORACLASS database.
2. Click **Worksheet/Run Local Script** in the Worksheet menu. Navigate to the file named **create_table.sql** in the **Data\Chapter08** directory on your student disk, and click **Open**. The commands are executed in the lower window as shown in Figure 8-6. This creates a table called CH08SURGERY and adds rows so you can experiment with various changes to the columns.

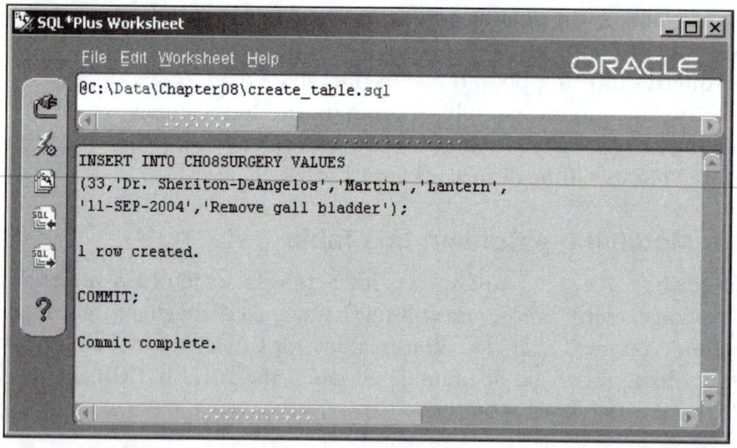

```
SQL*Plus Worksheet
File Edit Worksheet Help                                  ORACLE
@C:\Data\Chapter08\create_table.sql

INSERT INTO CH08SURGERY VALUES
(33,'Dr. Sheriton-DeAngelos','Martin','Lantern',
'11-SEP-2004','Remove gall bladder');

1 row created.

COMMIT;

Commit complete.
```

FIGURE 8-6 The commands run automatically when you select Run Local Script

3. Type **DESC CH08SURGERY** in the top window, and click the **Execute** icon. You will see a list of column names in the lower window. Notice that the third column name is misspelled. It is named PATIENT_FISRT_NAME.
4. Type and execute the following lines to enter the command to correct the column name:

   ```
   ALTER TABLE CH08SURGERY
   RENAME COLUMN PATIENT_FISRT_NAME TO PATIENT_FIRST_NAME;
   ```

   The worksheet replies, "Table altered" when it completes its work.
5. This table tracks surgeries in a hospital. The doctors need information about the outcome of the operation and the operating room used for the surgery. Type and execute the following lines to enter the command to add these two columns:

   ```
   ALTER TABLE CH08SURGERY
         ADD (OUTCOME VARCHAR2(40), OPERATING_ROOM_NO CHAR(4));
   ```

   The worksheet replies, "Table altered" when it completes its work.
6. You decide that the DOCTOR_NAME column is too long. It is currently defined as VARCHAR2(40). You cannot think of any doctor whose name is longer than 20 characters. Type and execute the following command to reduce the size of the column:

   ```
   ALTER TABLE CH08SURGERY
         MODIFY (DOCTOR_NAME VARCHAR2(20));
   ```

   The worksheet displays an error in the lower part of the window: "ORA-01441: cannot decrease column length because some value is too big." This is caused by the modification of the data in the column. At least one row has data that is longer than 20 characters, so the command failed.

7. Change the command so that the column is 25 rather than 20 characters long as shown below. Then click the **Execute** icon to run the command. The worksheet replies "Table altered" indicating that all the data fits within 25 characters or less:

```
ALTER TABLE CH08SURGERY
      MODIFY (DOCTOR_NAME VARCHAR2(25));
```

8. Imagine that time has gone by, and the doctors report to you that they are not using the PROCEDURES column or the OUTCOME column because the formats are incorrect. You decide the best course of action is to first render the two columns unusable and later, during an upcoming maintenance period, add replacement columns and drop the old columns. Type and execute the following command for making the two columns unusable:

```
ALTER TABLE CH08SURGERY
      SET UNUSED (PROCEDURES, OUTCOME);
```

   The worksheet replies, "Table altered" when it completes its work. The SET UNUSED clause does not fully remove the columns or the data in those columns. However, it does remove the columns from the Data Dictionary's information about the table and makes it impossible to update the data. You could execute the **DESC CH08SURGERY** command again to verify that the columns are no longer accessible, although the data still exists. The SET UNUSED clause is faster than the DROP COLUMN clause, because the data is not actually removed from each row in the table. You cannot undo an unused column.

9. Now it is time for maintenance. You have conferred with the doctors and determined the proper data types for the PROCEDURES and OUTCOME columns. Type and execute the following command to add the new definitions of the two columns:

```
ALTER TABLE CH08SURGERY
ADD (PROCEDURES PROCEDURES_ARRAY,
     OUTCOME NUMBER(2,0));
```

   The worksheet replies, "Table altered" when it completes its work. The added columns have the same names as the two columns you set to unused status. This is valid in Oracle 10g.

10. Now it is time to drop the two unused columns. Removing the two unused data columns is also known as restructuring the data to remove the unwanted data from each row in the table. Type and execute the following command to drop any unused columns:

```
ALTER TABLE CH08SURGERY
DROP UNUSED COLUMNS;
```

   The worksheet replies, "Table altered." Any column that was marked unused is dropped at this point.

11. Remain logged on for the next practice.

TIP

When you execute an ALTER TABLE DROP COLUMN command, all columns that are set to unused status are dropped, along with the column or columns you name in your DROP command.

There are a few restrictions when changing the data type or length of columns. If all the existing data cannot be translated into the intended datatype or length, then Oracle 10*g* cannot make the change. If you have special conversions that must be done to translate the data, you may be better off creating a new column and writing an UPDATE command to translate the column data from the old column into the new column. Then, you can drop the old column, and rename the new column, if you want.

Picture this scenario: You are running a program that inserts 10,000 rows into a table, committing its work each time it inserts 100 rows. You change the program, and then you want to run the program again. You issue a DELETE command and wait for a long time while Oracle 10*g* deletes 10,000 rows. Then, you issue a COMMIT command and wait again. How can you speed up this process? The next section shows how to drop the entire table, or remove all its contents quickly.

NOTE

The commands you ran in this section are all listed in the **alter_table.sql** file in the **Data\Chapter08** directory on your student disk.

Truncating and Dropping a Table

It is possible to remove all the rows from a table by typing:

```
DELETE FROM <tablename>;
```

This method of removing all the rows is slow, but can be reversed. Executing a DELETE command that removes all the table's rows causes Oracle 10*g* background processes to write undo records and redo log entries for each row deleted. This takes time, which might be a disadvantage. On the other hand, it also allows you to reverse the command (before committing) by executing the ROLLBACK command, which might be an advantage, allowing complete retrieval of all rows previously deleted.

If you are sure you do not need the data and want to remove thousands or millions of rows quickly, you are better off using the TRUNCATE command. The TRUNCATE command does not generate any undo records or redo log entries. This can improve the response time dramatically for removing all the rows in a table. The more rows you have to delete, the bigger the performance gain by using the TRUNCATE command. The syntax is:

```
TRUNCATE TABLE  <schema>.<tablename>
       DROP STORAGE|REUSE STORAGE
```

DROP STORAGE is the default setting and frees up all but the space allocated to the table except space required by the MINEXTENTS setting of the table or tablespace. In addition to the table's storage space, all indexes on the table are dropped, and all their space is released as well. This space returns to the tablespace and is available for creating new segments in the future.

Specify REUSE STORAGE to keep the space allocated to the table for future inserts and updates. This is appropriate if you are purging all the existing rows in preparation for inserting an equal number of new rows.

You can also remove all the table's rows along with the table structure by using the DROP TABLE command.

The primary difference between dropping a table and truncating a table is that truncating preserves a table's structure, whereas dropping a table does not. In addition, you cannot truncate a table that has constraints from other tables dependent on any of its columns, however you can drop a table that has constraints from other tables by adding the CASCADE CONSTRAINTS clause to the DROP TABLE command.

For example, if a table called COPCAR is related to a table called COP, given that there are one or more police officers in each cruiser, then the table with a foreign key (the COP table), should be dropped first, or dropped via the CASCADE CONSTRAINTS option in a drop command on the COPCAR table. To drop the COPCAR table, you must run this command:

```
DROP TABLE COPCAR CASCADE CONSTRAINTS;
```

The Recycle Bin

Oracle 10g has introduced a new toy into the mix of dropping tables, in the form of the recycle bin.

NOTE

The Oracle 10g recycle bin is not the same as the Windows operating system recycle bin but behaves in the same manner. Consequently you cannot recover Oracle 10g tables from the Windows recycle bin, only the recycle bin configured within the Oracle database.

There are a few other syntactical additions as a result of the introduction of the recycle bin, which you will see in this section. Thus the syntax for the DROP TABLE command is now changed to include a PURGE option as shown here:

```
DROP TABLE  <schema>.<tablename> [ CASCADE CONSTRAINTS ] [ PURGE ]
```

By default the CASCADE CONSTRAINTS and PURGE option are not included in a DROP TABLE. In other words, when dropping a table it will automatically be saved in the Oracle 10g database recycle bin, rather than purged from the recycle bin as well. It follows that a dropped table, saved to the recycle bin, must somehow be recoverable to its state prior to being dropped using a FLASHBACK TABLE command as shown in the following syntax:

```
FLASHBACK TABLE  <schema>.<tablename> TO BEFORE DROP [ RENAME
TO ,tablename> ]
```

You can even rename the table when recovering using flashback. Follow along with these steps to use examples of dropping, recovering (flashback), and purging tables from the recycle bin:

1. Once again, if you are not already there, go to the SQL*Plus Worksheet, and log on as CLASSMATE/CLASSPASS in the ORACLASS database.
2. Create a table called COPCAR using the following command:

```
CREATE TABLE COPCAR (VEHICLE_ID NUMBER,
 VEHICLE_MAKE VARCHAR2(32),
 VEHICLE_MODEL VARCHAR2(32),
 LAST_SERVICE DATE,
 MILEAGE FLOAT);
```


3. Now drop the table using the following command:

```
DROP TABLE COPCAR;
```

The worksheet replies, “Table dropped” when it completes its work.
4. Attempt to display the structure of the COPCAR table using the DESC COPCAR command, and you will find that the table is destroyed, indicated by the error: ERROR: “ORA-04043: object copcar does not exist”.

```
DESC COPCAR;
```

5. Recover the COPCAR table from the recycle bin, into a new table, called COPCARS, by executing the FLASHBACK TABLE command as shown here:

```
FLASHBACK TABLE COPCAR TO BEFORE DROP RENAME TO COPCARS;
```

The worksheet replies, “Flashback complete” when it finishes its work.
6. Use the DESC COPCARS command and you will see that the table has been recovered. The table has however been renamed during recovery, so the COPCAR table still does not exist.
7. Next drop the COPCARS table without the possibility for recovery using the following command:

```
DROP TABLE COPCARS PURGE;
```

8. Execute the flashback command again, in an attempt to recover the purged COPCARS table, and the error “ORA-38305: object not in RECYCLE BIN” will be displayed.
9. Remain logged on for the next practice.

NOTE

The DBA_RECYCLEBIN and USER_RECYCLEBIN metadata views can be used to display the contents of all recycle bins, for all users, or a specifically connected user recycle bin. The RECYCLEBIN view is often referred to and is simply a synonym for the USER_RECYCLEBIN view.

Next you will examine some specialized types of changes you can make to tables.

MAKING OTHER MORE SPECIALIZED TABLE CHANGES

As with most table changes, all the changes described in this section are made using the ALTER TABLE command. Here is the syntax examined in this section:

```
ALTER TABLE <schema>.<tablename>
  RENAME TO <newname>
  LOGGING|NOLOGGING
  ENABLE|DISABLE ROW MOVEMENT
  CACHE|NOCACHE
```

These parameters have the following effects:

- **RENAME TO:** This changes the name of the table. RENAME must be used alone as the only clause in the ALTER TABLE command.
- **LOGGING|NOLOGGING:** These change the logging mode of the table. The NOLOGGING clause suppresses the creation of redo log entries for some DDL commands, and some DML commands against the table, and for mass insert commands, such as those run by SQL*Loader.
- **ENABLE|DISABLE ROW MOVEMENT:** These allow or prevent Oracle 10g from moving rows during operations, such as data compression. Moving a row changes its ROWID, which could cause a problem, if you stored ROWIDs in a reference table for looking up data. It is usually not a problem, and the default setting is ENABLE ROW MOVEMENT.
- **CACHE|NOCACHE:** When you set CACHE for a table, data blocks read are placed in the "most recently used" section of the buffer. This keeps them stored in the buffer longer than the default NOCACHE setting. Use CACHE for frequently accessed tables.

Follow along with an example of how to modify the cache and monitoring of a table.

1. If you are not already there, go to the SQL*Plus Worksheet, and log on as CLASSMATE/CLASSPASS in the ORACLASS database.
2. Type and execute the following line to enter the command to modify the EMPLOYEE table:

   ```
   ALTER TABLE EMPLOYEE CACHE;
   ```

3. Note that the worksheet replies, "Table altered" when it completes its work.
4. Remain logged on for the next practice.

Table Flashback Recovery

You have already seen the FLASHBACK TABLE command used to retrieve tables dropped in to the recycle bin, as a result of using the DROP TABLE command, without using the PURGE optional clause. There is more to the FLASHBACK TABLE command than simply recovering tables from the recycle bin. This is the syntax for the FLASHBACK TABLE command:

```
FLASHBACK TABLE <schema>.<tablename>
  [ { TO SCN | TIMESTAMP } <expression> ] [ { ENABLE | DISABLE } TRIGGERS ]
  [ TO RESTORE POINT <expression> ] [ { ENABLE | DISABLE } TRIGGERS ]
  [ BEFORE DROP [ RENAME TO <tablename> ] ]
```

- **TO SCN | TIMESTAMP:** Restore a table to a previous time, as determined by a SCN (System Change Number) or a timestamp (specific point in time, in the past of course). The ENABLE TRIGGERS clause allows triggers to remain enabled during the flashback process, obviously causing possible problems as a result of referential integrity relationships.
- **TO RESTORE POINT:** Restore a table to a previously set restore point. A restore point must have been created with the CREATE RESTORE POINT command.
- **BEFORE DROP:** Restore a table into active use by recovering it from the recycle bin. Tables in the recycle bin have been previously dropped using the DROP TABLE command. The RENAME TO clause allows restoring of the previously dropped table to a table with another name. This can allow the recovery of a previously dropped table from the recycle bin, where a new copy of the table already exists.

NOTE

Table Flashback recovery is advanced and not included in the Oracle certification examination. However, it is interesting and thus included here for information purposes only.

Transparent Table Encryption

Oracle 10g has introduced encryption of data within the confines of an Oracle database, preventing potential snooping into datafiles from outside the confines of Oracle database software.

NOTE

Encryption requires the Oracle Advanced Security option and the setting of an encryption key.

The CREATE TABLE and ALTER TABLE commands can both be used to encrypt table data at the column level. The syntax is as follows:

```
{ CREATE | ALTER } TABLE <schema>.<tablename> ENCRYPT
  [ USING 'encryption algorithm' ]
  [ IDENTIFIED BY <password> ]
  [ [NO] SALT ]
```

All subsequent use of encryption is subsequently automated, or otherwise known as transparent, because there is no further intervention or action required. When authorized users store changes in a table their data is automatically encrypted. Similarly, whenever data is retrieved from a table by an authorized user, data selected is automatically decrypted. The various options shown in the syntax above can be described as follows:

- **USING ...:** A predetermined encryption algorithm is used as determined by various options available. The only prerequisite is that all columns across a single table must use the same algorithm.
- **IDENTIFIED BY ...:** The database will formulate a specific column key based on the password provided.
- **SALT ...:** A random string called a SALT string is added to the column in clear text prior to encryption.

NOTE

Transparent table encryption is advanced and not included in the Oracle certification examination. However, it is interesting and thus included here for information purposes only.

QUERYING DATA DICTIONARY VIEWS FOR TABLES AND OTHER OBJECTS

Because tables are so important in a database, there are many data dictionary views that contain information about them. The obvious views have either TAB or TABLE in their names, such as DBA_TABLES and USER_TAB_COLUMNS. The lesser known views report on segments, extents, and columns. Table 8-1 lists data dictionary views and dynamic performance views that contain information about tables. When the view begins with DBA_, remember there is almost always another view prefixed with USER_ and ALL_, with the same characteristics.

TABLE 8-1 Table-related data dictionary views

Name	Description
DBA_EXTERNAL_TABLES	List names of external tables and access methods used for retrieving data
DBA_LOBS	Lists LOB statistics, LOB storage settings, and so on for all but BFILE LOBS
DBA_NESTED_TABLES	List names and storage settings of nested tables
DBA_OBJECT_TABLES	List names and storage settings of object tables
DBA_SEGMENTS	List information about segments, such as the extents, blocks, freelists, and buffers used
DBA_TAB_COLUMNS	Lists all columns in the table, their datatypes, default values, and so on; also contains statistics gathered by ANALYZE and DBMS_STATS
DBA_TAB_PARTITIONS	List information about partition key, storage settings, and statistics on partitions
DBA_TAB_PRIVS	List table privileges granted on any table to any user

TABLE 8-1 Table-related data dictionary views (continued)

Name	Description
DBA_TABLES	Lists the name, storage, and other settings assigned when the table was created; also contains statistics gathered by ANALYZE and DBMS_STATS
DBA_UNUSED_COL_TABS	Lists tables that contain unused columns
USER_TAB_PRIVS	List table privileges granted to the current user on any table
USER_TAB_PRIVS_MADE	List table privileges granted by the current user on tables owned by the current user
V$OBJECT_USAGE	Lists results of monitoring tables

To sample one of these data dictionary views, follow these steps:

1. If you are not already there, go to the SQL*Plus Worksheet, and log on as CLASSMATE/CLASSPASS in the ORACLASS database.
2. Earlier in the chapter, you ran the DBMS_STATS package for the entire CLASSMATE schema. Look at the statistics now with this query. Type the query into the SQL*Plus Worksheet:

```
SELECT TABLE_NAME, TABLESPACE_NAME, MONITORING,
BLOCKS, EMPTY_BLOCKS
FROM USER_TABLES;
```

3. Click the **Execute** icon to run the command. The worksheet displays the list of tables and their information in the lower half of the window.
4. Log off of SQL*Plus Worksheet by clicking the **X** in the top-right corner.

You should run ANALYZE or DBMS_STATS on a regular basis. The time between runs depends on the volume of activity on your database. Keeping statistics current enables the optimizer to function more efficiently and predict more accurately which plan offers the fastest performance. In addition, the statistics are available through the data dictionary views to help determine whether you need to add storage space, move tables to different disk drives, and so on.

NOTE

Statistics are better automated in Oracle 10*g*.

Chapter Summary

- LOB datatypes are either internal LOBS, stored within the database, or external LOBS, stored outside the database.
- Oracle 10*g* contains a package called DBMS_LOB, used to manage and access LOB objects.
- By default, internal LOB values are stored inline, if they are less than 4000 bytes long.
- Internal LOB values larger than 4000 bytes are stored out of line in a LOB data segment.
- Oracle 10*g* works with LOBs by reading and writing one chunk at a time (a chunk can be from one data block that is up to 32 Kb in size).
- LOB segment storage settings are similar to settings for table storage.
- LOB datatypes can be part of partitioned tables, index-organized tables, and object tables.
- LOB data segments can also be used to store VARRAY data.
- Index-organized tables require a primary key and store data in order, as if the whole table were an index.
- You can split the column data in an index-organized table into the main segment and an overflow segment.
- The ANALYZE command gathers statistics on a table's size, number of rows, column distribution, and free space.
- You can use DBMS_STATS.GATHER_TABLE_STATS instead of ANALYZE.
- The DBMS_STATS package is recommended as a better option to that of the ANALYZE command.
- Changing a table's storage parameters involves using the ALTER TABLE command.
- DBMS_REDEFINITION is a package that can help you restructure a table while keeping the table online most of the time.
- Upgrading an object table or a relational table with object columns redefines their object types with the latest definition.
- Rename a table with the ALTER TABLE ... RENAME command.
- Modifying columns within a table also uses the ALTER TABLE command.
- To drop a column that is involved in another table's foreign key constraint, you must specify the CASCADE CONSTRAINTS clause.
- The SET UNUSED clause marks a column for dropping later.
- If you truncate a table, you cannot undo the transaction with a ROLLBACK command.
- Use the DROP TABLE ... CASCADE CONSTRAINTS command, if the table is named in foreign keys in any other related tables.
- Data dictionary views store table, segment, extent, column, and LOB segment information.

Review Questions

1. List all the internal LOB datatypes.
2. When a BFILE datatype column's value is copied into another row, the external file is copied into a new location. True or False?
3. LOBs can be stored on ________________ read-only devices such as a DVD.
4. A row of data has a BLOB column. The column's value is 4200 bytes long. Where is the value stored?
   a. Inline with the row data
   b. Out of line in the LOB data segment
   c. Depends on the LOB storage setting
   d. In an external file
5. Examine this LOB storage clause.

```
LOB (COMMERCIAL) STORE AS COMM_LOB
    (DISABLE STORAGE IN ROW
     CHUNK 32768
     STORAGE (INITIAL 5K NEXT 32K MAXEXTENTS UNLIMITED)
```

   Which statement is true about this clause?
   a. The clause is invalid, because you cannot name the LOB data segment.
   b. The clause is valid and will create a 5 Kb LOB data segment.
   c. The clause is invalid, because the initial extent must be a multiple of the chunk size.
   d. The clause is invalid, because the chunk size is not a multiple of the database block size.
6. Why is inserting a row in an index-organized table usually slower than inserting a row in a heap-organized table?
7. Index-organized tables can be partitioned using ________________ or ________________ partitioning.
8. Which of the following reasons are a valid cause for analyzing a table? Choose all that apply.
   a. To improve the accuracy of optimization
   b. To help you decide if storage settings fit the actual data
   c. To speed up inserts and updates
   d. To improve query performance
9. To release unused data blocks in a table, which of these commands can you use? Choose all that apply.
   a. TRUNCATE TABLE
   b. ALTER TABLE
   c. DROP TABLE
   d. MOVE TABLE
10. An object type has changed its definition. What should you do to a table containing a column with the changed object type?

11. Is the following command valid or invalid? Why?

```
ALTER TABLE CAR
DROP COLUMN CAR_OWNER
RENAME COLUMN CAR_MANUFACTURER TO CAR_MANUF;
```

Exam Review Questions—Oracle Database 10*g*: Administration (#1Z0-042)

1. The PARTS table has a foreign key referencing the CAR table. Which statement will remove all rows in the CAR table? Choose two.
   a. DROP TABLE CAR CASCADE CONSTRAINTS;
   b. TRUNCATE TABLE CAR;
   c. TRUNCATE TABLE CAR CASCADE CONSTRAINTS;
   d. DELETE FROM CAR;
2. You have a column called SNOW_DEPTH that you want to change from CHAR(4) to NUMBER(10, 2). The column contains numerical characters. What should you do? (Choose the best response.)
   a. Take the table offline, alter the column, update the data, and put the table back online again.
   b. Use the DBMS_REDEFINITION package to restructure the table with minimal down time.
   c. Execute an ALTER TABLE MODIFY command and allow data to be converted automatically.
   d. Create a new column, use an UPDATE command to load the data, and then drop the old column.
3. You are creating a table to keep digital copies of old movie clips. The table must meet the following requirements:
   - The table must use an internal LOB for the movie clips.
   - All LOB values must be stored out of line.
   - The LOB should store old versions of the value until 25 percent of the space is full.

   Which of the following CREATE TABLE commands meets all these criteria?

   a.

```
CREATE TABLE OLDMOVIE
(MOVIE_ID NUMBER(10),
 MOVIE_TITLE VARCHAR2(30),
 MOVIECLIP BLOB)
LOB (MOVIECLIP)  STORE  AS  MOVIELOB
    (TABLESPACE USER_LOCAL
     ENABLE STORAGE IN ROW
     PCTVERSION 25
     STORAGE (INITIAL 64M NEXT 32M));
```

   b.

```
CREATE TABLE OLDMOVIE
(MOVIE_ID NUMBER(10),
```

```
 MOVIE_TITLE VARCHAR2(30),
 MOVIECLIP BLOB)
LOB (MOVIECLIP)  STORE  AS  MOVIELOB
    (TABLESPACE USER_LOCAL
     DISABLE STORAGE IN ROW
     PCTVERSION 25
     STORAGE (INITIAL 64M NEXT 32M));
```

c.

```
CREATE TABLE OLDMOVIE
(MOVIE_ID NUMBER(10),
 MOVIE_TITLE VARCHAR2(30),
 MOVIECLIP BFILE)
LOB (MOVIECLIP)  STORE  AS  MOVIELOB
    (TABLESPACE USER_LOCAL
     DISABLE STORAGE IN ROW
     PCTVERSION 25
     STORAGE (INITIAL 64M NEXT 32M));
```


d.

```
CREATE TABLE OLDMOVIE
(MOVIE_ID NUMBER(10),
 MOVIE_TITLE VARCHAR2(30),
 MOVIECLIP BLOB)
LOB (MOVIECLIP)  STORE  AS  MOVIELOB
    (TABLESPACE USER_LOCAL
     DISABLE STORAGE IN ROW
     RETENTION 25
     STORAGE (INITIAL 64M NEXT 32M));
```

4. Which of the following statements are true about an index-organized table? Choose two.
   a. An index-organized table can contain LOB columns.
   b. Add the index-organized parameter when creating an index-organized table.
   c. Overflow segments always contain the same columns.
   d. Index-organized tables can be partitioned using hash or range partitioning.
5. Which of these dictionary views should you query to list all the columns in a table?
   a. USER_TABLES only
   b. USER_TABLES and USER_TAB_COLUMNS
   c. USER_TAB_COLUMNS and USER_UNUSED_COL_TABS
   d. USER_TAB_COLUMNS only
6. You want to move a table to a new tablespace and resize the initial extent, next extent, and turn on cache. Choose the best sequence of steps from this list.
   1. Create a new table with new extent sizes in the new tablespace.
   2. Drop the old table.
   3. Alter the table for new extent sizes.
   4. Alter the new table to turn on monitoring.
   5. Copy all data to the new table.
   6. Move the old table to the new tablespace.
   7. Move the old table to the new tablespace, and change extent sizes at the same time.

a. 1, 3, 4, 5, 2 (6 and 7 not needed)

b. 6, 3, 4 (1, 2, 5, 7 not needed)

c. 7, 4 (1, 2, 3, 5, 6 not needed)

d. 7, 4, 5, 2 (1, 3, 6 not needed)

7. Which of the following statements change(s) the size of the table's segment? Choose all that apply.

   a. ALTER TABLE CUSTOMER ALLOCATE EXTENT SIZE 55M

   b. ALTER TABLE CUSTOMER STORAGE (NEXT 55M)

   c. ALTER TABLE CUSTOMER DEALLOCATE UNUSED

   d. ALTER TABLE CUSTOMER COMPRESS

8. Examine the following SQL commands:

```
ALTER TABLE NEWSPAPER SET UNUSED COLUMN VOLUME_SOLD;
ALTER TABLE NEWSPAPER ADD (VOLUME_SOLD NUMBER(10,0));
ALTER TABLE NEWSPAPER DROP UNUSED COLUMNS;
```

   Which of the following statements is true?

   a. The second command will fail, and one column will be dropped.

   b. One column will be dropped, and one column will be added.

   c. The second command will fail, and no columns will be dropped.

   d. None of the above.

Hands-on Assignments

1. Write the SQL command to create a table with the following traits:

   - The table is named BEARS.
   - The table has nine columns: BEAR_TAG, BEAR_NAME, TAGGED_DATE, WEIGHT, HEIGHT, BIRTH_DATE, LAST_KNOWN_LOCATION (allow 2000 bytes), PHOTO (a BLOB), and MAP (a BFILE).
   - The table resides in the USERS tablespace and uses the default storage settings for the tablespace.
   - The table is index-organized on the BEAR_TAG column.
   - Overflow should begin after the BIRTH_DATE column.

   Create the table in the CLASSMATE schema in the ORACLASS database. Save the SQL command in the **Solutions\Chapter08** directory in a file named **ho0801.sql** on your student disk.

2. Copy the CREATE BEARS command you created in Hands-on Assignment 1. Change the name of the table to BEARS2 and define a LOB storage clause that places the BLOB column in a separate segment and requires all BLOB column values be stored in the LOB data segment. The chunk size is 8 Kb, and there should be an initial and next extent size of 16 Kb

(small because you are working on a test database), with unlimited extents. Allow 10 percent of the space to be used for old versions. Write the SQL command, execute it to create the table, and save the SQL command in the **Solutions\Chapter08** directory in a file named **ho0802.sql** on your student disk.

3. This assignment uses the CH08SURGERY table. If you did not create the table while working in this chapter, create it now by using SQL*Plus Worksheet to run the **create_table.sql** file in the **Data\Chapter08** directory on your student disk.

   Analyze the CH08SURGERY table.

   Write one or more queries on data dictionary view(s) to answer these questions:

   - How many rows are in the table?
   - What is the average length of a row in the table?
   - How many columns have no values in any rows?
   - What is the average length of the DOCTOR_NAME column?
   - How many distinct values are there in the PATIENT_FIRST_NAME and PATIENT_LAST_NAME columns?
   - What is the segment name and extent ID of the table's segment and extent?

   Save your SQL commands in a file named **ho0803.sql** in the **Solutions\Chapter08** directory on your student disk.

4. Copy the CREATE TABLE statement from the **create_table.sql** file in the **Data\Chapter08** directory on your student disk. Modify the statement to create a new table named CH08IOTSURGERY. Make it an index-organized table in the USERS tablespace. Insert the data from the CH08SURGERY table into the CH08IOTSURGERY table, except put a zero in every PROCEDURES column. Write a query to display the extent name and number of bytes for all segments associated with the CH08SURGERY table and the CH08IOTSURGERY table. (*Hint*: The primary key indexes have segments.) Which table uses more space? Save your SQL script and answer to the question in a file named **ho0804.sql** in the **Solutions\Chapter08** directory on your student disk.

5. Revise the BEARS table created in Hands-on Assignment 1. Make the following changes:

   - You have found that the bears' names are too long to fit into the column. The names sometimes are as much as 10 characters longer than the current maximum size.
   - In addition, the hand-held computer that loads data into the mainframe's database stores the TAGGED_DATE with a precision of hours, minutes, seconds, and hundredths of a second. You want your data to keep this level of accuracy.
   - The bear pictures are sometimes photos, but other times they are videos, so you want a different column name.
   - Save your script in a file named **ho0805.sql** in the **Solutions\Chapter08** directory on your student disk.

6. For this assignment, you should log on to SQL*Plus Worksheet as CLASSMATE and run the script called **ho0806setup.sql** in the **Data\Chapter08** directory. This script creates two related tables: CH08REPAIR_TYPE and CH08HOUSE_REPAIR. The CH08HOUSE_REPAIR table has a foreign key referencing the CH08REPAIR_TYPE table. You have discovered that the table CH08REPAIR_TYPE is loaded with invalid data, and that data has

been carried over into the CH08HOUSE_REPAIR table. You want to remove all the data from CH08REPAIR_TYPE while leaving the structure intact. You also want to remove the foreign key data in the CH08HOUSE_REPAIR table, so it can be reloaded later. Write two different SQL scripts to accomplish these tasks. One script should use TRUNCATE, and one should use DELETE. Save the script that uses TRUNCATE in a file named **ho0806a.sql** in the **Solutions\Chapter08** directory on your student disk. Save the script that uses DELETE in a file named **ho0806b.sql** in the **Solutions\Chapter08** directory on your student disk.

7. There are several tables owned by the CLASSMATE schema. Copy the file named **ho0807setup.sql** in the **Data\Chapter08** directory on your student disk. Use the file to build a file that tests all the tables owned by CLASSMATE. Test each one to see if it is eligible for the DBMS_REDEFINITION package. Which tables can use the package? Which ones cannot? Save the SQL commands in a file named **ho0807a.sql** in the **Solutions\Chapter08** directory on your student disk. Save the outcome of the SQL commands (showing the success or failure of the package) in a file named **ho0807b.txt** in the **Solutions\Chapter08** directory on your student disk. (*Hint*: connect as the SYS user for this assignment).
8. For this assignment, you should log on to SQL*Plus Worksheet as CLASSMATE and run the script called **ho0808setup.sql** in the **Data\Chapter08** directory on your student disk. This creates a table called CH08FAMILYTIES and an object type called CH08CHILDINFO that is used in the table in a column called CHILD_DATA. After creating the table, the script modifies the CH08CHILDINFO object type so that it contains an additional attribute. Write the SQL command needed to modify the attributes in the CHILD_DATA column in the CH08FAMILYTIES table. Save your command in a file named **ho0808.sql** in the **Solutions\Chapter08** directory on your student disk.
9. Using the table that you created in Hands-on Assignment 8, you now find that more changes are needed to the table's columns. Write the SQL commands to make these changes, and run them using SQL*Plus Worksheet.
    - The PARENT_ONE column needs to be split into two columns: PARENT1_FIRST_NAME and PARENT1_LAST_NAME. (*Hint*: use an UPDATE command with a SUBSTR function to load the data from the old column into the new columns.)

    The PARENT_TWO column should be split the same way that PARENT_ONE was split.

    Remove the old columns after the data has been transferred to the new columns. Save your script in a file named **ho0809.sql** in the **Solutions\Chapter08** directory on your student disk.

Case Project

The Global Globe's database system is developing well. You have a few changes to make as you learn more about the way the office runs. Here are the revisions you want to incorporate to improve the database. Use these specifications to create SQL scripts, then run the scripts, and save your work in a file named **case0801.sql** in the **Solutions\Chapter08** directory.

- There are many interactive applications that need the employee's ID and other data from the EMPLOYEE table. You have determined that the majority of queries read the table by primary key, so you decide this table should be index organized. Make the appropriate changes to the EMPLOYEE table. (*Hint*: You will have to re-create the table.)

- You discovered that many of the customers that place ads in the newspaper are businesses. You want the CUSTOMER table to have additional fields to track the business name, business Web site address, and business type. In addition, you want to rename the FULLNAME column to CONTACT_NAME. Make these changes to the CUSTOMER table.
- The newspaper has decided to add a new feature to its classified ads. For an additional fee, a customer's ad can be placed at the front of the section with a red star in front of it. You must add a new column to the CLASSIFIED_AD table to mark the ads that have priority placement. Also add a column to the CLASSIFIED_SECTION table, so the editors can set the priority placement pricing for each section. In addition, because it is so frequently used, you want the CLASSIFIED_AD table to stay cached in memory.

CHAPTER 9

CONSTRAINTS, INDEXES, AND OTHER SPECIALIZED OBJECTS

LEARNING OBJECTIVES

In this chapter, you will:

- Learn the types and the uses of constraints
- Examine the syntax and options for creating constraints
- Work with practical examples of creating, modifying, and dropping constraints
- Query database dictionary views to monitor constraints
- Learn the types of indexes Oracle offers and when to use each type
- Understand how to create each type of index
- Determine which data dictionary views contain information on indexes
- Find out how to monitor index usage and when to drop an index
- Learn how to modify, rebuild, and coalesce an index
- Learn the basics about views, sequences, and synonyms

INTRODUCTION

Defining the relationship of tables on paper or in a design tool is a good start, but the database system must incorporate the design and prevent invalid data from corrupting the design. Business rules are the statements defined during the design of a database system that inform both the database designer and the application programmer how data is used to support the business. For example, a business rule might state that every customer is assigned a unique customer identification number when he makes his first purchase. Applications were designed to handle all the business rules, including those at the database level (such as a primary key in a table) early in the development of relational database systems. Today, many of the business rules that apply to the database can be defined and enforced within the database itself, potentially removing the burden from applications. One of the easiest and most direct methods of defining business rules is the use of integrity constraints.

Constraints restrict values in a database. A constraint constrains data. Indexes on the other hand provide fast access to information in tables. An index on a table is like an index in the back of a book, allowing a rapid access method to items of information in large tables (rather than searching through the whole book, or searching an entire table in the case of a database). There are various types of indexes helping to improve performance. In this chapter you will create constraints and indexes. You learn to determine when each type of index is warranted, in what situations, and how to monitor the use of indexes. You will also discover how to modify and drop existing indexes.

The end of this chapter provides a brief look at some specialized database objects available in Oracle 10*g*. Specialized objects include views and materialized views, clusters, sequences, and synonyms.

WHAT ARE CONSTRAINTS?

Imagine that you are in charge of reservations for a theater. You and two assistants answer the telephone, making reservations as quickly as possible. You sell seats until the theater is filled. What would happen if two of you sold the same seat? An argument, a fight, perhaps even a refund. You could avoid this situation by using a method in which each of you knows when the other sells a seat. When ticket sellers share the same office, selling duplicates is less of a problem. However, if the reservations can be made online across the country, you need a database. The database can handle concurrent updates to theater seat reservations. You must tell the database not to sell a seat more than once. The rule, "Only one customer per seat" is called a constraint. Constraints are rules or restrictions that guide database inserts, updates, and deletions. Constraints keep invalid or erroneous data out of your database.

You can enforce integrity constraints in several ways. This chapter deals with the first method, declaring integrity constraints.

- **Declare an integrity constraint:** Integrity constraints are constraints defined at the database level on either a column or a table.
- **Write a database trigger:** A trigger is a program that runs when a certain event, such as inserting a row in the THEATER table, occurs. Like constraints, triggers are at the database level and provide more flexibility in designing complex constraints than integrity constraints, while keeping the constraint within the database.
- **Include constraints in an application:** This type of constraint is outside the database. Although application constraints are flexible and can be fine-tuned for each application, they enforce only those changes that are made within the application. Modifications to the database, which are made outside the application, are not affected by the constraint.

The integrity constraint method of enforcing constraints has several advantages over the other methods. First, integrity constraints are simple to create and maintain because they are centralized. Any constraint modifications occur only once. Second, integrity constraints are always enforced, no matter what application or tool is used to modify a table's data. Finally, integrity constraint checking performs faster than other methods of constraint enforcement, because the optimizer has been programmed to handle them efficiently.

Types of Integrity Constraints

Oracle 10g supports five types of integrity constraints. The constraint types handle all the basic requirements for relational database design and include the following:

- **PRIMARY KEY:** A primary key is the column or set of columns that define a unique identifying value for every row in the table. For example, the CLIENT table has a unique identifying column called CLIENT_ID. A primary key can contain a single column or multiple columns. When multiple columns make up the key, it is called a compound key or composite key. A primary key is often a system-generated number, guaranteeing uniqueness and stability. A PRIMARY KEY constraint has two rules underlying its enforcement: The column or columns must contain unique, non-null values. The PRIMARY KEY constraint is important in relational databases. Most tables have a primary key to comply with the standards of normalization of tables. A table can have no more than one PRIMARY KEY constraint. Normalizing tables is part of the design process, and one of the rules of normalization is that every table should have a key that contains a unique value for every row. In other words, every row in a table should be uniquely identifiable. In reality, this is not always strictly true of all databases. Primary keys maintain referential integrity between parent tables (containing a primary key), and child tables (containing foreign key copies of primary key values from the parent table). Referential integrity is quite literally the integrity, or soundness of data, maintained between two values, which reference each other (a primary key in one table and foreign key in another table).

- **UNIQUE:** This constraint is similar to the PRIMARY KEY constraint because both enforce unique values. The difference is that the UNIQUE key constraint allows null values in the column or columns that are named in the constraint. Columns with null values are ignored by the constraint. This constraint is useful for defining an alternate unique key for a table. A table can have more than one UNIQUE constraint.
- **FOREIGN KEY:** A foreign key establishes a relationship between two tables, in which one is the parent and the other is the child. The FOREIGN KEY constraint is placed in the child table, referencing the primary key of the parent table. Figure 9-1 shows two tables with a FOREIGN KEY constraint and two PRIMARY KEY constraints. The FOREIGN KEY constraint requires that every row in the child table (the ORDER table in Figure 9-1) contain either a value that matches a primary key value in the parent table or a null value in the CUST_ID column. The PRIMARY KEY constraint in each table requires that every row in the table should contain a unique value in the primary key column.

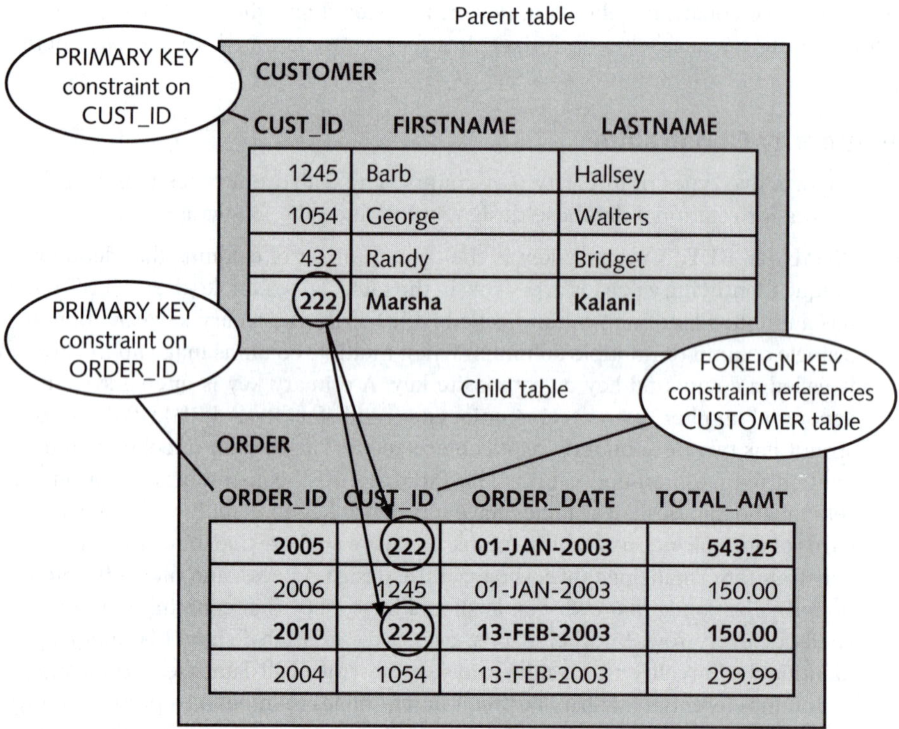

FIGURE 9-1 Primary and foreign keys relate tables to each other

- **NOT NULL:** Placing the NOT NULL constraint on a column indicates that all rows in the table must contain a value in the column. A null value is an empty column within a table row and usually means that the column value for that row is unknown or not applicable. For example, the column OCCUPATION can contain a null value for a row containing information about an unemployed person or a child. Use the NOT NULL constraint for columns holding essential information, such as the LAST_NAME column in a CUSTOMER table.
- **CHECK:** The CHECK constraint enforces a predefined list of values for a column. For example, the APPROVED column can have a CHECK constraint that requires the column to contain <YES>, <NO>, or a null value. Unless you also specify the NOT NULL constraint, a column with a CHECK constraint may contain a null value.

TIP

Creating a primary key in an Oracle 10*g* database automatically creates an internal unique index.

The next section describes how to create integrity constraints.

How To Create and Maintain Integrity Constraints

There are two ways to create integrity constraints on columns and tables:

1. Use the CREATE TABLE command to create the constraint while creating the table. This is the most common method for PRIMARY KEY and NOT NULL constraints, because these are essential parts of the table design and are usually known when the table is created.
2. Use the ALTER TABLE command later to create the constraint. Use this to add constraints that were either missed or added to the table design after table creation. Typically, all rows in the table must conform to the new constraint, although you can modify this default action if needed.

The next section describes how to add constraints to the CREATE TABLE command.

Creating Constraints Using the CREATE TABLE Command

Figure 9-2 shows the syntax of the CREATE TABLE command with notation pointing out the placement of constraints. Constraints that apply to a single column, such as the NOT NULL constraint, are called column constraints and appear inline with the column. An inline constraint appears immediately next to the column to which it applies. Constraints that apply to multiple columns, such as a constraint for a compound foreign key, are called table constraints and are placed immediately after the list of columns in the CREATE TABLE command. An out of line constraint appears after the full list of columns in the CREATE TABLE command and is usually used to define constraints to multiple columns. You can place single column constraints out of line, if you choose. The NOT NULL constraint must always be defined inline, as it always applies to individual columns.

Some constraints can be defined either inline or out of line. Table 9-1 shows the ways each constraint can be defined.

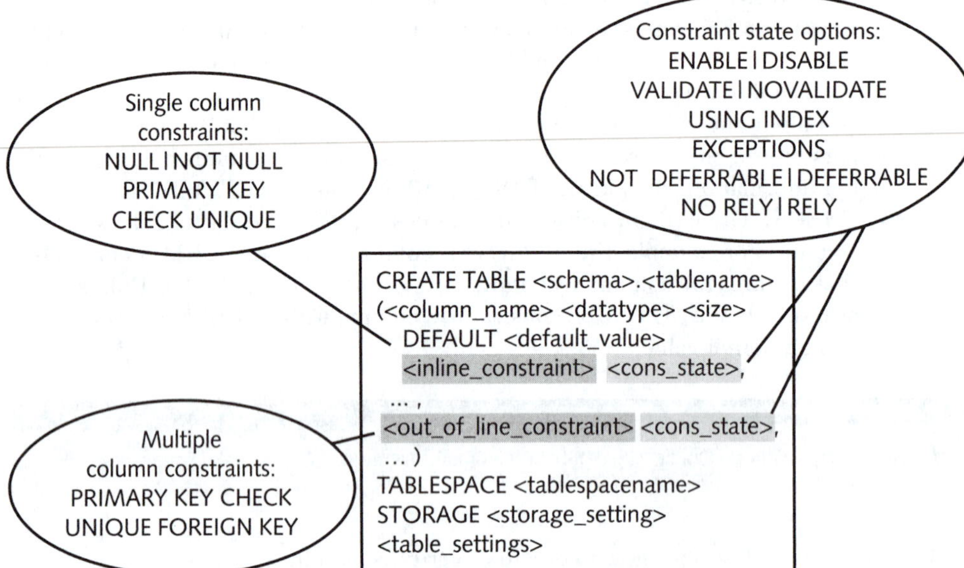

FIGURE 9-2 There are two valid locations for defining constraints, using the CREATE TABLE command

TABLE 9-1 Constraints - inline, out of line, or both?

Constraint	Inline in CREATE TABLE	Out of line in CREATE TABLE	Inline in ALTER TABLE	Out of line in ALTER TABLE
PRIMARY KEY	Yes, if single column	Yes	No	Yes
FOREIGN KEY	Yes, if single column and not named	Yes	No	Yes
NOT NULL	Yes	No	Yes	No
CHECK	Yes, if single column	Yes	No	Yes
UNIQUE KEY	Yes, if single column	Yes	No	Yes

Figure 9-3 shows an example of a table with several constraints in the CREATE TABLE command.

In the code example shown in Figure 9-3, you can see these constraints:

- **PRIMARY KEY:** The constraint is named PATIENT_PK; the PATIENT_ID column is the primary key.
- **NOT NULL:** The constraint is inline on the FULL_NAME column and has no name.

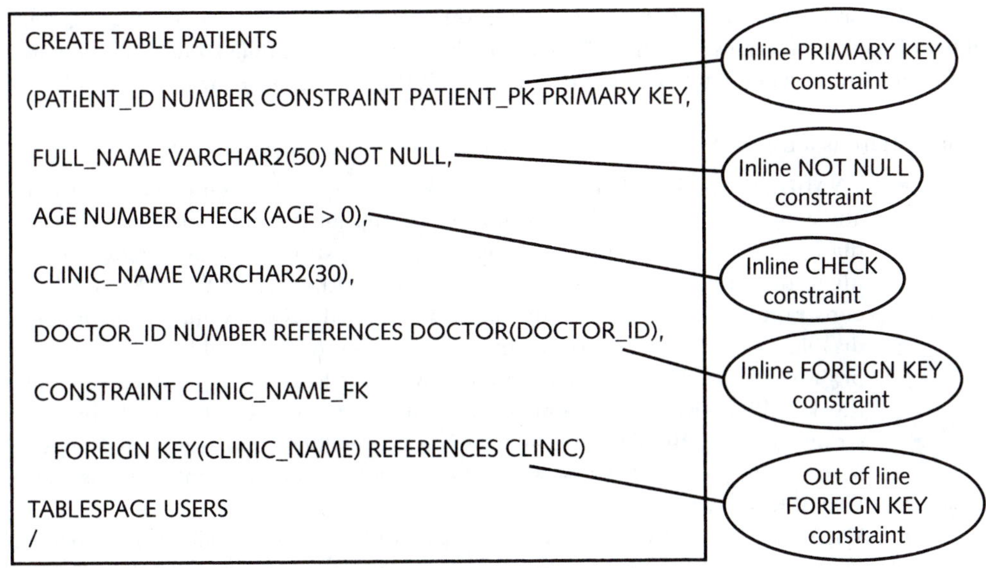

FIGURE 9-3 Constraints created both inline and out of line

- **CHECK:** The constraint is inline on the AGE column and has no name.
- **FOREIGN KEY:** The constraint is inline on the DOCTOR_ID column and has no name.
- **FOREIGN KEY:** The constraint is named CLINIC_NAME_FK and is out of line.

Frequently, NOT NULL and CHECK constraints are not named. Naming constraints is optional. When no name is specified, omit the CONSTRAINT keyword and the constraint name. When a constraint is not specifically named, Oracle 10*g* will automatically name a constraint with an internal name. A name can be useful, however, when you are rapidly reviewing constraints later to confirm that you have not missed anything in your schema. In fact, naming constraints can simplify later administration activities in general. In general, it is advisable to specifically name at least all PRIMARY KEY and FOREIGN KEY constraints, but it is not essential.

Although CHECK constraints can check multiple conditions, it is usually better to use multiple CHECK constraints instead of one complicated constraint. Name the CHECK constraints with names that describe their function, such as CHK_GRTR_THN_ZERO. The error message includes the check constraint name when the constraint is violated, and the user may have a better idea of how to correct the invalid data when she sees a meaningful constraint name.

TIP

A FOREIGN KEY constraint that is defined *inline* is not allowed to have a name. You must omit the CONSTRAINT keyword and the FOREIGN KEY keywords as well as the constraint name. Figure 9-3 shows both an inline FOREIGN KEY constraint and an out of line FOREIGN KEY constraint. A FOREIGN KEY constraint that is defined *out of line* can have a name, but is not required to have a name.

You can create or modify constraints in several different states. Figure 9-2 shows a partial list of constraint states. A constraint state is the attribute that tells Oracle 10g how to use the constraint when data is added to the table. If the state changes from DISABLE to ENABLE, the constraint state also tells Oracle 10g how to use the constraint on existing data. Here is a list of the constraint states and how they are used:

- **ENABLE|DISABLE:** Constraints are enabled by default when you create them. You can create a disabled constraint (or change an existing constraint) by adding DISABLE to the constraint. Disabling an existing constraint does nothing to the data, although if the constraint is a PRIMARY KEY or UNIQUE key constraint, the associated unique index is dropped. As an example, you might disable a FOREIGN KEY constraint to allow data to be loaded into a child table prior to loading data into a parent table. After the load is complete, you enable the FOREIGN KEY constraint again. You can re-enable a disabled constraint by using the ENABLE keyword. If you enable a constraint on a table with existing rows, all the rows are checked for compliance, unless you specify ENABLE NOVALIDATE.
- **VALIDATE|NOVALIDATE:** Constraints are validated (that is, all rows in the table must comply with the constraint) by default when they are created. If you are creating a constraint on a table that already has rows and use the NOVALIDATE keyword, the existing rows are not checked against the constraint, but new rows must comply (so long as the constraint is enabled). A typical use for the NOVALIDATE keyword is when you enable a constraint that you disabled temporarily to add several thousand rows. You know the rows comply with the disabled constraint, so you use ENABLE NOVALIDATE to switch the constraint to enabled because this saves time: Oracle 10g does not run through all the rows looking for rows that fail to comply with the constraint.
- **INITIALLY IMMEDIATE|INITIALLY DEFERRED:** In past versions of Oracle Database, all constraints were in the immediate state. A constraint in immediate state is validated as soon as a statement is executed. For example, you have a DOG_OWNER table and a DOG table. The DOG table has a FOREIGN KEY constraint referencing the DOG_OWNER table. You run an INSERT statement that has an invalid foreign key value. The INSERT statement fails immediately with an error caused by the failed constraint validation. A constraint in deferred state is validated only when you execute a COMMIT command. Continuing with the example, imagine that the FOREIGN KEY constraint on the DOG table is deferred. Now, you can insert a row in the DOG table first (without an error) and insert the dog's owner row into the DOG_OWNER table second. As long as you commit the rows after both are inserted, the COMMIT command succeeds. The INITIALLY IMMEDIATE state is the default state for constraints. Use INITIALLY DEFERRED to change the current state of the constraint to deferred state.
- **DEFERRABLE|NOT DEFERRABLE:** The default for all constraints is NOT DEFERRABLE. You must specify DEFERRABLE when you create the constraint to allow it to be deferred. *Furthermore, Oracle 10g does not allow you to use ALTER TABLE to change this constraint state.* Only constraints that are created as deferrable can be later modified to deferred state. You must drop and recreate the constraint to change this setting.

- **RELY|NORELY:** This state tells the optimizer to either use (RELY) or ignore (NORELY) disabled constraints when rewriting queries to better optimize the performance of complex queries. The RELY state is useful for data warehouses using materialized views and query rewrites. Otherwise, it is rarely used. A materialized view is a view that has been parsed and executed and its results stored in a table so that subsequent queries that use the materialized view do not have to repeat the parsing and executing of the view's SQL. Materialized views are efficient when the data in the underlying tables seldom changes. A query rewrite is the ability of Oracle 10*g* to modify the query you execute, changing it into an equivalent but more efficient statement before actually running the query. The query rewrite is especially useful when implementing materialized views, because the optimizer can rewrite your query so that a materialized view can be used to retrieve the data, making the query more efficient. Deferred constraints are normally ignored during the optimizer's rewrite activity. Therefore, the default state is NORELY, which tells the optimizer to ignore constraints that are disabled when rewriting queries or evaluating materialized views. RELY tells the optimizer to use disabled constraints.
- **USING INDEX <index>|<storage>:** You can specify this state only for the PRIMARY KEY and UNIQUE constraints. This tells Oracle 10*g* that you have made one of two choices: (1) You have an existing index and, therefore, no index needs to be created when creating or enabling the constraint, or (2) you are specifying the storage setting for the index that Oracle 10*g* creates for the constraint. For example, you can specify the tablespace in which to create the associated index when you add this state to a CREATE TABLE command. As another example, when you have a disabled PRIMARY KEY or UNIQUE constraint, you can create a unique index yourself. Then, when you enable the constraint, you specify that Oracle 10*g* use the existing index to enforce the constraint. You sample the USING INDEX state when you create a UNIQUE key constraint later in the chapter.
- **EXCEPTIONS|EXCEPTIONS INTO <tablename>:** This state is used only when enabling an existing constraint or adding a constraint to an existing table. The EXCEPTIONS keyword instructs Oracle 10*g* to list the ROWIDs of rows that fail validation in a table. If you don't specify a table, Oracle 10*g* inserts the exception information into a table, named EXCEPTIONS, that you own or to which you have access. Create this table by running the script: **utlexcpt.sql** (for supporting physical ROWIDs) or **utlexpt1.sql** (for supporting universal ROWIDs, such as those in index-organized tables). The script is found in **ORACLE_HOME\rdbms\admin**. If you want to use a table with a different name, copy the script and use it to create your own table. Then use the EXCEPTIONS INTO <tablename> format, specifying the table you created.

Creating or Changing Constraints Using the ALTER TABLE Command

The syntax for constraints within the ALTER TABLE command varies according to the type of constraint. The NOT NULL constraint has one variation, adding constraints to existing columns has a second variation, and changing existing constraints has a third variation.

Let's look at adding or removing NOT NULL constraints on an existing column. The NOT NULL constraint is always associated with a single column. Use the following syntax to either add or remove the NOT NULL constraint on a column.

```
ALTER TABLE <tablename>
MODIFY(<columnname> NULL|NOT NULL);
```

Use NULL to remove the NOT NULL constraint. Use NOT NULL to add the constraint.

TIP

To add a NOT NULL constraint successfully, all rows in the table must contain values for the column.

Next, add a new constraint to an existing table. Regardless of whether the constraint is for a single column or multiple columns, use the out of line format you saw in the CREATE TABLE command, to add a new constraint to an existing table. Here is the syntax:

```
ALTER TABLE <tablename>
ADD CONSTRAINT <constraintname>
  PRIMARY KEY (<colname>, ...) |
  FOREIGN KEY (<colname>, ...)
      REFERENCES <schema>.<tablename> (<colname>, ...) |
  UNIQUE (<colname>, ...) |
  CHECK (<colname>, ...) (<check_list>);
```

Specific examples of this format appear in many sections in this chapter to show you how to create and modify each type of constraint. If you omit the CONSTRAINT keyword, you must omit the constraint name as well.

Finally, this is how you change or remove a constraint. There are only two things you can change on an existing constraint: the *name* and the *state* of the constraint. Here is an example in which you change the name of the FOREIGN KEY constraint:

```
ALTER TABLE CUSTOMER
RENAME CONSTRAINT CUST_FK TO CUST_ORDER_FK;
```

The general syntax for changing the constraint state is:

```
ALTER TABLE <tablename>
MODIFY CONSTRAINT <constraintname>
  <constraint_state> <constraint_state> ...;
```

Here is an example in which you change the state of a UNIQUE constraint. You enable the constraint, report invalid rows to the BADCUSTOMERS table, and use the existing index, CUST_UNQ_INDEX, rather than creating a new unique index.

```
ALTER TABLE CUSTOMER
ENABLE CONSTRAINT CUST_UNQ
EXCEPTIONS TO BADCUSTOMERS
USING CUST_UNQ_INDEX;
```

TIP

You cannot change the state of a NOT NULL constraint; a NOT NULL constraint is always enabled. Removing a NOT NULL constraint is the only way to "disable" it.

Working with Constraints

The next five sections provide step-by-step examples in which you create, modify, and drop constraints of all types. New details about constraints are explained as you work through the examples.

The first constraint you examine is the easiest one: the NOT NULL constraint.

Adding or Removing a NOT NULL Constraint

The NOT NULL constraint is often used when creating tables. Use the NOT NULL constraint to require a value in a column when a row is inserted or updated. Follow along with the example to see how the NOT NULL constraint is handled:

1. Start up the Enterprise Manager console. In Windows, on the Taskbar, click **Start/All Programs/Oracle ... /Enterprise Manager Console**. In Unix or Linux, type **oemapp console** on the command line. The Enterprise Manager console login screen appears.
2. Start up the SQL*Plus Worksheet by clicking **Tools/Database Applications/SQL*Plus Worksheet** from the top menu in the console. A background SQL*Plus process starts, and then the SQL*Plus Worksheet window appears. If the background process appears in front, simply minimize it (click the **minus sign** in the top-right corner), so you can see the worksheet.
3. Connect as the CLASSMATE user. Click **File/Change Database Connection** on the menu. A logon window appears. Type **CLASSMATE** in the Username box, **CLASSPASS** in the Password box, and **ORACLASS** in the Service box. Leave the connection type as "Normal." Click **OK** to continue.
4. You are creating a sample table to practice adding and modifying the NOT NULL constraint. This table is used later for other constraint examples. Type this command to create a small table in which the three columns have NOT NULL constraints and one has a DEFAULT value. Execute the command by clicking the **Execute** icon. The SQL*Plus Worksheet replies, "Table created.":

```
CREATE TABLE CH09DOGSHOW
(DOGSHOWID NUMBER NOT NULL,
 SHOW_NAME VARCHAR2(40) NOT NULL,
 DATE_ADDED DATE DEFAULT SYSDATE NOT NULL);
```

   SYSDATE is a pseudocolumn that represents the current date and time. A pseudocolumn is available for use within SQL*Plus anywhere you can use a column value or (as in this case) a static value. When a new row is inserted with null values in the DATE_ADDED column, Oracle 10g places the current date and time in the column as the default value.

5. Try inserting a row with null in the SHOW_NAME column by typing this command and executing it. When you leave out a column, it defaults to a null value unless the column was defined with a DEFAULT value. In that case, the column omitted is assigned the DEFAULT value. In this example, there is no DEFAULT value, so the SHOW_NAME column value is null:

```
INSERT INTO CH09DOGSHOW
(DOGSHOWID, DATE_ADDED) VALUES
(1, '11-MAY-03');
```

   Figure 9-4 shows the error message that is returned stating that the SHOW_NAME column cannot be null.

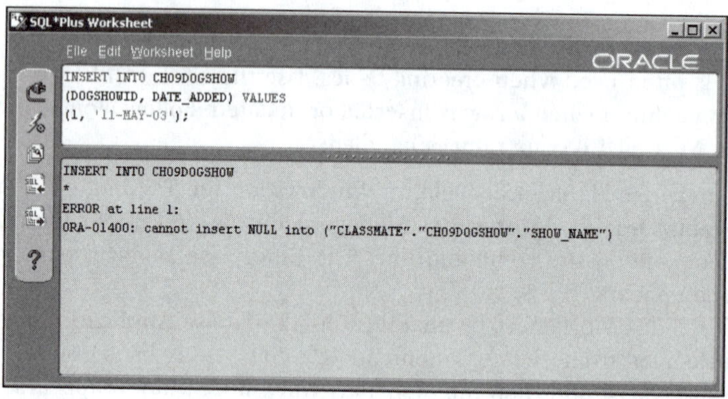

FIGURE 9-4 Omit the column to allow it to default to null (or its default value)

6. Now insert a row with null in the DATE_ADDED column. This is accepted, because the default value of SYSDATE is used. Type the following command and execute it:

```
INSERT INTO CH09DOGSHOW
(DOGSHOWID, SHOW_NAME) VALUES
(1, 'AKC Portland');
```

   The SQL*Plus Worksheet replies, "1 row created."

7. Remove the constraint from SHOW_NAME using the ALTER TABLE command. Type and execute this command:

```
ALTER TABLE CH09DOGSHOW
MODIFY (SHOW_NAME NULL);
```

   The SQL*Plus Worksheet replies, "Table altered." Note that the row inserted in Step 6 was automatically committed when you ran the ALTER TABLE command. Any DDL command causes Oracle 10g to commit previous changes that have not yet been committed.

8. To place the NOT NULL constraint back on the SHOW_NAME column, use the same format you used to remove it. Type the following command and execute it:

```
ALTER TABLE CH09DOGSHOW
MODIFY (SHOW_NAME NOT NULL);
```

   The SQL*Plus Worksheet replies, "Table altered."

9. Remain logged on for the next practice.

TIP

Including a DEFAULT setting on the column provides a fail-safe method of populating a NOT NULL column. Any row inserted with a null value in the column is automatically modified so that the column contains the default value instead.

Adding and Modifying a PRIMARY KEY Constraint

The PRIMARY KEY constraint is important for the integrity of the table. The primary key can be one column, such as the PATIENT_ID in the previous example, or it can be more than one column. The columns are usually placed at the beginning of the table, although this is not a requirement. Follow along with this example to create a new table with a compound primary key:

1. If you are not already logged on, start up the Enterprise Manager console and log on to the SQL*Plus Worksheet as CLASSMATE.
2. You are creating a sample table to practice adding and modifying the primary key. In this example, the primary key is the OWNER_ID column. Type the following two lines into SQL*Plus Worksheet to begin defining the table and to define the PRIMARY KEY constraint:

   ```
   CREATE TABLE CH09DOGOWNER
   (OWNER_ID NUMBER CONSTRAINT CH09_PK PRIMARY KEY,
   ```

   As you can see, the constraint is defined inline with the OWNER_ID column and is named CH09_PK.
3. Type the remaining lines and press the **Execute** icon to create the table:

   ```
   OWNER_NAME VARCHAR2(50),
   MEMBER_OF_AKC CHAR(3) DEFAULT 'NO',
   YEARS_EXPERIENCE NUMBER(2,0));
   ```

   The SQL*Plus Worksheet replies, "Table created."
4. The primary key has been created along with the table. In addition, an index named CH09_PK has been created in CLASSMATE's default tablespace. You decide the constraint should be named CH09_DOG_OWNER_PK, so you must modify the constraint. To modify a constraint, you must use the ALTER TABLE command. The NOT NULL constraint is the only constraint that can be modified using an inline constraint. All other constraints must be modified using an out of line constraint format. Type the following command to rename the constraint:

   ```
   ALTER TABLE CH09DOGOWNER
   RENAME CONSTRAINT CH09_PK TO CH09_DOG_OWNER_PK;
   ```

5. Run the command by clicking the **Execute** icon. SQL*Plus Worksheet replies, "Table altered."
6. Imagine that you decide you don't want the key after all. To remove the primary key, you must drop the constraint. Type the following command, and execute it by clicking the **Execute** icon:

   ```
   ALTER TABLE CH09DOGOWNER
   DROP CONSTRAINT CH09_DOG_OWNER_PK;
   ```

When you drop a PRIMARY KEY constraint, the associated index is also dropped.

NOTE

Any constraint other than NOT NULL can be dropped using the ALTER TABLE DROP CONSTRAINT command.

The SQL*Plus Worksheet replies, "Table altered."

7. As with all Oracle DDL commands, you cannot roll back (undo) the command after it has been executed. In this example, you discover you really need the PRIMARY KEY constraint that you just dropped. Place it back onto the table, this time using an out of line constraint and the ALTER TABLE command. Type the following command and execute it:

```
ALTER TABLE CH09DOGOWNER
ADD CONSTRAINT CH09_DOG_OWNER_PK PRIMARY KEY (OWNER_ID)
DISABLE;
```

   The SQL*PLUS Worksheet replies, "Table altered." The final line contains the DISABLE keyword. This means the constraint was created but is not currently enforced. A complete discussion of constraint states is presented earlier in this chapter.

8. Enforce the constraint by typing and executing this command:

```
ALTER TABLE CH09DOGOWNER
MODIFY CONSTRAINT CH09_DOG_OWNER_PK ENABLE;
```

   The SQL*Plus Worksheet replies, "Table altered."

9. Remain logged on for the next practice.

As you go through more examples, you practice using constraint states as well as other types of constraints.

Adding and Modifying a UNIQUE Key Constraint

As you read earlier, the UNIQUE key constraint and the PRIMARY KEY constraint are very similar. Both enforce a rule stating that every row in the table must contain a unique value within the column or columns in the constraint. The difference is that the PRIMARY KEY also enforces a NOT NULL constraint on the column or columns.

A UNIQUE constraint is often used in addition to a PRIMARY KEY rather than in place of it. For example, assume you are converting an old database system in which the primary key on the old table must be carried into the new system for reference. The old primary key does not conform to modern normalization rules because it is actually three distinct values combined into a single column. You want to break these distinct values into three columns and use the three new columns as the primary key. New rows inserted into the table do not carry the old key, so it must allow null values. You create a PRIMARY KEY constraint on the new key and a UNIQUE key constraint on the old key.

Follow these steps to create and modify a UNIQUE key constraint.

1. If you are not already logged on, start up the Enterprise Manager console, and log on to the SQL*Plus Worksheet as CLASSMATE.
2. You are creating a sample table to practice adding and modifying a UNIQUE key constraint. In this example, the table stores names of people from around the world. The primary key is made up of the COUNTRY and the PERSON_ID columns. The unique key is the US_TAX_ID column. Not every person has a U.S. tax number; however, every U.S. tax number is different. Type the following lines into SQL*Plus Worksheet to begin defining the table and to define the UNIQUE key constraint:

```
CREATE TABLE CH09WORLD
(COUNTRY VARCHAR2(10),
 PERSON_ID NUMBER,
 US_TAX_ID NUMBER(10) CONSTRAINT US_TAX_UNIQUE UNIQUE,
```


3. Finish the command by adding two more columns and the compound PRIMARY KEY constraint. Type these lines and execute the command:

```
FIRST_NAME VARCHAR2(10),
LAST_NAME VARCHAR2(20),
CONSTRAINT CH09WORLD_PK PRIMARY KEY (COUNTRY, PERSON_ID));
```

   The SQL*Plus Worksheet replies, "Table created."
4. A common event in the life of a database system is the loading of data from one source, such as an old file system, to a new table. Constraints are often disabled to prevent the data load from failing, because some rows do not comply with the new constraints. Therefore, to simulate a data load event, you first disable the UNIQUE key constraint. Type the following command and execute it:

```
ALTER TABLE CH09WORLD
MODIFY CONSTRAINT US_TAX_UNIQUE DISABLE;
```

   The SQL*Plus Worksheet replies, "Table altered."
5. Next, run a script that inserts rows into the table. Select **Worksheet/Run Local Script** and locate the file named **ch09data.sql** in the **Data\Chapter09** directory. Click **Open** to start the script. SQL*Plus Worksheet runs the script. You see that the phrase "1 row created" appears in the lower section of the screen several times, as shown in Figure 9-5.
6. Before enabling the constraint, you decide that if the constraint fails, you want to get a list of invalid rows. Using the EXCEPTIONS constraint state provides a list of invalid rows. However, you must create the EXCEPTIONS table before enabling the constraint. To do this, run the script that Oracle provides to create the table. Select **Worksheet/Run Local Script** and locate the file named **utlexcpt.sql** in the **ORACLE_HOME/rdbms/admin** directory. (Remember that ORACLE_HOME represents the full path of the directory that holds the Oracle 10g software. For example, on a Windows machine, ORACLE_HOME might be **C:\oracle\product\10.2.0\db_1**). Click **Open** to start the script. SQL*Plus Worksheet runs the script. You see the response, "Table created" in the lower part of the window. The table named EXCEPTIONS is created and owned by CLASSMATE (because you are logged on as CLASSMATE). The EXCEPTIONS table has these four columns:
   - **ROW_ID:** The ROWID of the row that failed the constraint rule
   - **OWNER:** The owner of the table where the row resides

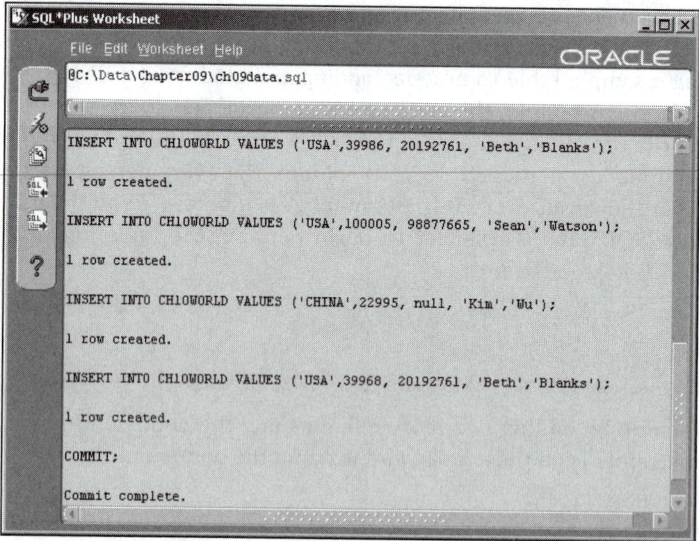

FIGURE 9-5 Running a script displays feedback, as the commands run

- **TABLE_NAME:** The name of the table where the row resides
- **CONSTRAINT:** The name of the constraint violated by the row

7. Now, enable the constraint by typing this command and executing it:

```
ALTER TABLE CH09WORLD
MODIFY CONSTRAINT US_TAX_UNIQUE ENABLE VALIDATE
EXCEPTIONS INTO EXCEPTIONS;
```

When you run the command, you see this error message: "ORA-02299: cannot validate (CLASSMATE.US_TAX_UNIQUE) - duplicate keys found." The constraint is not enabled.

The ENABLE VALIDATE option tells Oracle 10g to enable the constraint and validate all existing rows. (VALIDATE is the default when you specify ENABLE, so you can leave it out of the command and get the same results.) The last line tells Oracle 10g to load the ROWID of any rows that failed the validation into the table named EXCEPTIONS.

8. View the invalid records logged in the EXCEPTION table by typing and executing this query. The query joins the EXCEPTIONS table (alias of "E") with the CH09WORLD table (alias of "W") by matching the ROWID found in the EXCEPTIONS table with the actual ROWID of the rows in the CH09WORLD table. Matching rows are the ones that violated the constraint.

```
SELECT E.CONSTRAINT, W.*
FROM CH09WORLD W JOIN EXCEPTIONS E
ON (W.ROWID = E.ROW_ID)
ORDER BY W.US_TAX_ID;
```

Figure 9-6 shows the results. As you can see, there are two rows for Beth Blanks and two rows for Sean Watson. To successfully enable the UNIQUE constraint, the duplicate rows must be removed from the CH09WORLD table. For this exercise, you leave the constraints disabled.

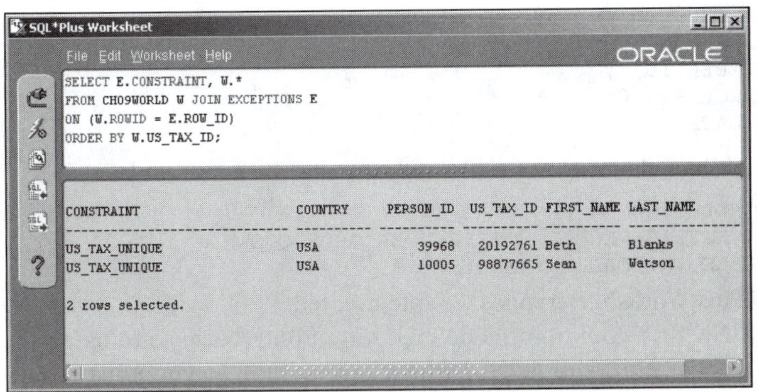

FIGURE 9-6 The constraint was violated 4 times

The query executed in this section is formatted in the new ANSI-standard JOIN syntax introduced in Oracle 10g.

TIP

For information on ANSI-standard JOIN syntax see the *Oracle 10g SQL Reference* document in the chapter titled *SQL Queries and Subqueries.*

The ANSI-standard JOIN equivalent query using the proprietary Oracle 10g format is:

```
SELECT E.CONSTRAINT, W.*
FROM CH09WORLD W, EXCEPTIONS E
WHERE W.ROWID = E.ROW_ID
ORDER BY W.US_TAX_ID;
```

9. Remain logged on for the next practice.

When you join tables together, you are nearly always using a column or set of columns in one table that corresponds to a column, or set of columns in another table. The corresponding columns are often connected with a constraint called the FOREIGN KEY constraint.

Working with a FOREIGN KEY Constraint

The FOREIGN KEY constraint requires a little more coding than the previous constraints. As you saw earlier, a foreign key references another table's primary key. Follow these examples to create and work with a FOREIGN KEY constraint:

1. If you are not already logged on, start up the Enterprise Manager console and log on to the SQL*Plus Worksheet as CLASSMATE.
2. You are creating a sample table to practice adding and modifying a FOREIGN KEY constraint. In this example, you are creating a table to register dogs for

a dog show. You already created a table for dog owners earlier in this chapter. The CH09DOG table uses a foreign key to connect the dogs to their owners. Type the following lines into SQL*Plus Worksheet to begin defining the table:

```
CREATE TABLE CH09DOG
(DOG_ID NUMBER,
 OWNER_ID NUMBER(10) ,
 DOG_NAME VARCHAR2(20),
 BIRTH_DATE DATE,
```

3. Type these final lines defining the table and execute the command:

```
CONSTRAINT CH09DOGOWNER_FK
     FOREIGN KEY (OWNER_ID) REFERENCES CH09DOGOWNER
     DEFERRABLE INITIALLY IMMEDIATE);
```

   The SQL*Plus Worksheet replies, "Table created."

   The FOREIGN KEY constraint is defined out of line (because to name the FOREIGN KEY constraint, you must define it out of line), so you must name the column in the table that contains the foreign key, in this case, the OWNER_ID column. The last line defines the constraint's state. DEFERRABLE means that at a later date, this constraint can be modified so that the constraint is validated only when inserts and updates are *committed* instead of immediately when the inserts or updates are *executed*. INITIALLY IMMEDIATE means that for now, the constraint is executed immediately when a row is inserted or updated.

4. Type and execute the following command to insert a row into the CH09DOG table. Because there are no rows in the CH09DOGOWNER table, the foreign key constraint will fail validation:

```
INSERT INTO CH09DOG VALUES
(1,2,'Chow Moon','15-JAN-02');
```

   Let's say you have a new owner to enter into the CH09DOGOWNER table along with this entry to the CH09DOG table. You have several choices to successfully insert both records:

   - Insert the owner, then the dog, and commit the transaction. The FOREIGN KEY constraint is not violated when you insert the owner row first.
   - Alter the CH09DOG table to disable the FOREIGN KEY constraint. Then, you can enter the rows in either order and commit the transaction. Then modify the table again to enable the constraint. This works only if you are the owner of the table, a DBA, or you have been granted the object privilege required to modify the table's constraints.
   - Use the SET CONSTRAINT command to adjust your *session* to defer all deferrable constraints. Now, you can enter the dog first, then the owner, then commit the transaction. This option provides a simplified method of deferring constraints without issuing any DDL commands. To defer a constraint using the SET CONSTRAINT command, you must have the SELECT privilege on the table, or be the owner. To reverse the setting, issue another SET CONSTRAINT command, or log off.

   The final selection (adjusting your session to defer constraints) is used for this example.

5. The SET CONSTRAINT command provides a way to change a named constraint or all constraints to a deferred state during your session. Of course, the only constraints that are affected were created with the DEFERRABLE setting. (Recall that by default, a constraint is not deferrable.) Type and execute the following command to defer all deferrable constraints. Remember, using this command does not change the constraint's state in the table; it only changes the constraint's state for your current session:

```
SET CONSTRAINTS ALL DEFERRED;
```

   The SQL*Plus Worksheet replies, "Constraint set."
6. Repeat the insert command from Step 4. SQL*Plus Worksheet replies, "1 row created."
7. Insert a row for the dog's owner by typing and executing the following command:

```
INSERT INTO CH09DOGOWNER VALUES
(2, 'Jack Maylew','YES', 3.5);
```

   SQL*Plus Worksheet replies, "1 row created."
8. Finally, commit both rows. When you execute the command, the constraint is validated. The FOREIGN KEY constraint is valid at this point. Type **COMMIT;** and click the **Execute** icon. SQL*Plus Worksheet replies, "Commit complete."
9. Reset your session by typing and executing this command:

```
SET CONSTRAINTS ALL IMMEDIATE;
```

   The SQL*Plus Worksheet replies, "Constraint set."
10. After a FOREIGN KEY constraint has been added, the parent table's PRIMARY KEY constraint cannot be removed without adding the CASCADE keyword that tells Oracle 10g to drop the PRIMARY KEY constraint and all FOREIGN KEY constraints that reference the table. Try this out with the CH09DOGOWNER table by typing and executing the following command:

```
ALTER TABLE CH09DOGOWNER
DROP CONSTRAINT CH09_DOG_OWNER_PK CASCADE;
```

   The SQL*Plus Worksheet replies, "Table altered."

   The CH09DOGOWNER no longer has a PRIMARY KEY constraint, and the CH09DOG table no longer has a FOREIGN KEY constraint.
11. Add the PRIMARY KEY constraint again by typing and executing this command:

```
ALTER TABLE CH09DOGOWNER
ADD CONSTRAINT CH09_DOG_OWNER_PK PRIMARY KEY(OWNER_ID);
```

   The SQL*Plus Worksheet replies, "Table altered."
12. Add the FOREIGN KEY constraint back to the CH09DOG table, but this time, use a new parameter. Type and execute the following command:

```
ALTER TABLE CH09DOG
ADD CONSTRAINT CH09DOGOWNER_FK FOREIGN KEY (OWNER_ID)
REFERENCES CH09DOGOWNER
ON DELETE CASCADE;
```

   The SQL*Plus Worksheet replies, "Table altered."

The ON DELETE CASCADE setting tells Oracle 10g that when a parent row is deleted, all the related child rows are also deleted. Another alternative to this is ON DELETE SET NULL, which instructs Oracle 10g to set the value of the foreign key column in the related child rows to null instead of deleting the rows. If you omit the ON DELETE parameter, you cannot delete a row from the parent table, if it has related children in other tables. Try this out by deleting the owner of the dog row now in the CH09DOG table.

13. Delete the row of the dog's owner by typing and executing this command:

```
DELETE FROM CH09DOGOWNER WHERE OWNER_ID = 2;
```

SQL*Plus Worksheet replies, "1 row deleted." But in fact, two rows were deleted. One row was deleted from the CH09DOGOWNER table because you executed the DELETE statement. Due to the cascading effect of deleting a parent row, one child row was also deleted from the CH09DOG table.

14. Execute this query to prove that the row was deleted from the CH09DOG table:

```
SELECT * FROM CH09DOG;
```

The SQL*Plus Worksheet replies, "no rows selected."

As you can see, no rows were retrieved. The one row that was inserted earlier has been deleted. This is useful when you know that the child rows are not needed after the parent row is deleted. For example, a CALENDAR_DAY table has a child table called DAY_APPOINTMENTS. When a day is deleted from the CALENDAR_DAY table, all the appointments for that day can be deleted automatically without issuing another DELETE command.

15. Roll back the deleted rows by typing **ROLLBACK;** and clicking the **Execute** icon.
16. The SQL*Plus Worksheet replies, "Rollback complete." Remain logged on for the next practice.

TIP

When using SET CONSTRAINTS, you can list one or more constraints by name, separated with commas, to defer only certain constraints. For example, to defer two constraints named DOG_FK and SHOW_NAME_FK, type and execute this command.

```
SET CONSTRAINTS DOG_FK, SHOW_NAME_FK DEFERRED;
```

The final constraint to examine helps validate data in one or more columns within the same row.

Creating and Changing a CHECK Constraint

The CHECK constraint helps validate the value within a column or a set of columns within one row. This is useful for columns with a fixed number of values, such as the days of the week or calendar months. You can use a few expressions to check against a range of values as well. For example, the number of wheels on a bus must be between 4 and 16 and divisible by 2.

You have already created several tables, so you are using the ALTER TABLE command to add CHECK constraints to columns that already exist. The syntax for adding a

CHECK constraint with CREATE TABLE and ALTER TABLE is identical. Like the other constraints, add the constraint inline with a new column if the constraint only refers to that column; add the constraint out of line when it refers to more than one column or when the column already exists.

Follow these steps to add CHECK constraints to some of the tables created in this chapter:

1. If you are not already logged on, start up the Enterprise Manager console, and log on to the SQL*Plus Worksheet as CLASSMATE.
2. Type and execute this command to create a CHECK constraint that requires the MEMBER_OF_AKC to have the values <YES>, <NO>, or null. (Null values automatically pass the constraint unless you specifically state the column cannot be null in the constraint):

```
ALTER TABLE CH09DOGOWNER ADD CONSTRAINT AKC_YN
CHECK (MEMBER_OF_AKC IN ('YES','NO'));
```

   The SQL*Plus Worksheet replies, "Table altered."
3. Add another constraint that requires that the names of all dog shows must be all uppercase letters. Because there is one row that violates the constraint, create the constraint in a disabled state. This gives you a chance to correct the data and enable the constraint. Type and execute this command:

```
ALTER TABLE CH09DOGSHOW ADD CONSTRAINT ALL_CAPS
CHECK (SHOW_NAME = UPPER(SHOW_NAME)) DISABLE;
```

   The SQL*Plus Worksheet replies, "Table altered."
4. Correct the data in the CH09DOGSHOW table to make all names uppercase by typing and executing this command:

```
UPDATE CH09DOGSHOW SET SHOW_NAME = UPPER(SHOW_NAME);
```

   The SQL*Plus Worksheet replies, "1 row updated."
5. Enable the constraint now by typing and executing the following command:

```
ALTER TABLE CH09DOGSHOW MODIFY CONSTRAINT ALL_CAPS ENABLE;
```

   The SQL*Plus Worksheet replies, "Table altered."
6. You can also use the CHECK constraint to compare one column against another column. Remember that the constraint can only use values within the current row. Add a constraint that rejects a row if both the first and last name of a person in the CH09WORLD table contains null values, or if the two names are identical. Type and execute the following command to accomplish this:

```
ALTER TABLE CH09WORLD ADD CONSTRAINT CHK_NAMES
CHECK ((FIRST_NAME IS NOT NULL OR LAST_NAME IS NOT NULL)
       AND(FIRST_NAME <> LAST_NAME));
```

   The SQL*Plus Worksheet replies, "Table altered."
7. Test the constraint by trying to insert a row where the first and last names match. Type and execute the following command:

```
INSERT INTO CH09WORLD VALUES
('USA', 1995, 99877689, 'Jeremy', 'Jeremy');
```

   SQL*Plus Worksheet replies, "ORA-02290: check constraint(CLASSMATE.CHK_NAMES) violated."

8. Test again by modifying the INSERT command so that the first and last names are both null values. The row is rejected with the same error message. Type and execute the following command:

```
INSERT INTO CH09WORLD VALUES
('USA', 1995, 99877689, NULL, NULL);
```

SQL*Plus Worksheet replies, "ORA-02290: check constraint (CLASSMATE.CHK_NAMES) violated."

9. For the last test, revise the INSERT command so that only the first name has a value. This row should pass the constraint. Type and execute the following command:

```
INSERT INTO CH09WORLD VALUES
('USA',1995, 99877689, 'Jeremy', NULL);
```

The SQL*Plus Worksheet replies, "1 row created."

10. Commit your work by typing **COMMIT;** and clicking the **Execute** icon. The SQL*Plus Worksheet replies, "Commit complete."
11. Remain logged on for the next practice.

TIP

CHECK constraints cannot contain queries or any references to other tables. They also cannot use pseudocolumns, such as SYSDATE and USER.

Now that you have created many constraints, the next section takes a look at the data dictionary views that display constraint information.

Data Dictionary Information on Constraints

There are only two constraint-related data dictionary views. They are:

- **ALL_CONSTRAINTS:** This view (and its corresponding views, DBA_CONSTRAINTS and USER_CONSTRAINTS) contains the definition of a constraint. The columns in this view include:
  - CONSTRAINT_NAME
  - CONSTRAINT_TYPE containing letters identifying each type, such as "P" for PRIMARY KEY, "C" for CHECK, and "R" for FOREIGN KEY (also called a referential constraint)
  - SEARCH_CONDITION (lists the condition for CHECK constraint)
  - STATUS (enabled or disabled)
- **ALL_CONS_COLUMNS:** This view lists the columns associated with each constraint. There is one row for each column in the constraint. For example, a primary key made up of two columns has two rows in this view.

Complete the following steps to examine the contents of the ALL_CONSTRAINTS view:

1. If you are not already logged on, start up the Enterprise Manager console, and log on to the SQL*Plus Worksheet as CLASSMATE.

2. Type and execute this command to see the constraints created by CLASSMATE:

```
SELECT CONSTRAINT_NAME, CONSTRAINT_TYPE, TABLE_NAME,
STATUS, SEARCH_CONDITION
FROM ALL_CONSTRAINTS
WHERE OWNER='CLASSMATE' ORDER BY TABLE_NAME;
```

Figure 9-7 shows the results of the query. Notice that the NOT NULL constraints are listed as CHECK constraints, and have system-assigned names. (You can name a NOT NULL constraint like any other constraint if you choose.)

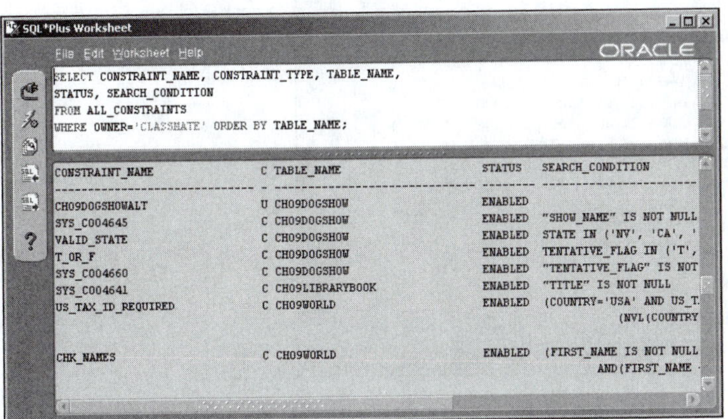

FIGURE 9-7 Foreign key constraints are indicated with the letter R (for "referential constraint")

3. This query goes one a step further, returning a formatted picture of all constraints, and all columns within all constraints:

```
SET WRAP OFF LINESIZE 132 PAGES 80;
COLUMN KEY FORMAT A10;
COLUMN POS FORMAT 990;
COLUMN COL FORMAT A10;
COLUMN CONS FORMAT A20;
COLUMN TAB FORMAT A20;
COLUMN OWN FORMAT A10;
SELECT DECODE(T.CONSTRAINT_TYPE, 'P','PRIMARY',
'R','FOREIGN', 'U','ALTERNATE', 'UNKNOWN') "KEY"
,T.TABLE_NAME "TAB"
,T.CONSTRAINT_NAME "CONS"
,C.COLUMN_NAME "COL"
,C.POSITION "POS"
FROM USER_CONSTRAINTS T, USER_CONS_COLUMNS C
WHERE T.CONSTRAINT_TYPE IN ('P','R','U')
AND T.TABLE_NAME = C.TABLE_NAME
AND T.CONSTRAINT_NAME = C.CONSTRAINT_NAME
ORDER BY T.TABLE_NAME, T.CONSTRAINT_TYPE, C.POSITION;
```

Figure 9-8 shows the results of the above query.

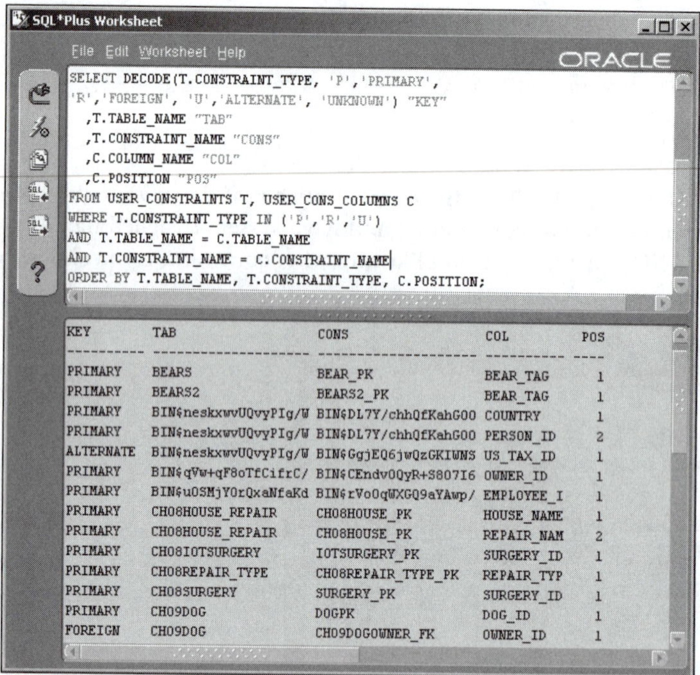

FIGURE 9-8 Reading both the constraints views

4. Log off SQL*Plus Worksheet by clicking the **X** in the top-right corner.

You have seen how to create your own tables and constraints. The next step is indexes.

WHAT ARE INDEXES?

An index is a database structure that is associated with a table or a cluster and speeds up data retrieval when the table or cluster is used in a query. An index has its own storage specification and does not have to be in the same tablespace as the associated table. You can create indexes on relational tables, object tables, nested tables, index-organized tables, and partitioned tables. When creating an index, you choose the type of index, the column or columns to be included in the index, and the location of the index. Oracle recommends that you separate the index and the table into separate tablespaces for better performance.

Oracle 10*g* automatically adds new rows to the index as new rows are added to the table. In addition, it updates or deletes index entries as the table rows are updated or deleted. Figure 9-9 shows a logical view diagram of a table and its index. In Figure 9-9, the rows in the CUSTOMER table are stored in two blocks within one table extent Block #000 and Block #001; rows are not ordered by the primary key, CUST_ID. The index is based on the primary key and is named CUST_PK. The index resides in its own segment and contains one block of index entries.

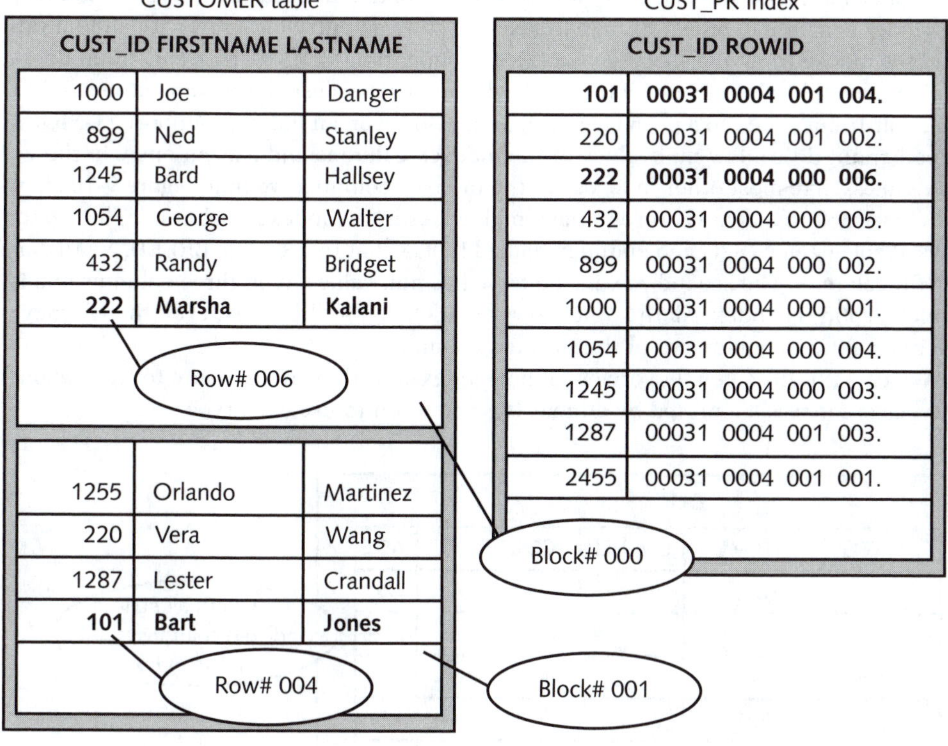

FIGURE 9-9 An example of a table and its index

Entries, on the other hand, are stored in order by primary key. Each entry stores the primary key value of one row and the ROWID for that row. For example, Marsha Kalani is in the sixth row in Block #000 and has a primary key of 222. In the index, the third row is hers and the last two numbers in the ROWID (representing the block and row numbers) are "000" and "006".

An index does not have to be based on a primary key. Any combination and order of columns can be used to create an index for a table. An index can have from one to 32 columns (only 30 columns if it is a bitmap index). An index containing multiple columns is called a composite index. Columns can be used in more than one index, although no two indexes on a table can contain the same combination and order of columns. For example, one index might use the LAST_NAME, FIRST_NAME, and ZIPCODE columns, whereas a second index might use the same three columns in a different order, and a third index might include only the ZIPCODE column.

In addition to providing faster access to data, certain indexes also have a second purpose: enforcing a PRIMARY KEY constraint or UNIQUE key constraint. The index for the PRIMARY KEY or UNIQUE key constraint is automatically created and has the same name as the constraint. Both constraints cause a unique index to be created. A unique index requires every row inserted into the table to have a unique value in the indexed column or columns.

An index does not store an entry for a row when all the indexed columns are null except if the index is a bitmap index. Bitmap indexes do store an entry for a row with all null values in the indexed columns. An index created to support a PRIMARY KEY constraint does not allow any null values in any of the indexed columns, because there is an automatic NOT NULL constraint added to a primary key column. However, an index to support a UNIQUE key constraint does allow null values in the indexed columns and lists an entry in the index, unless, like most other indexes, all the indexed columns are null. Figure 9-10 shows the behavior of an index when rows have null values in the indexed columns. The figure shows a table named DOG and an index named DOGX that indexes the BREED, NAME, and OWNER columns. The second row in the table has null values in all three columns and is not included in the index. All the other rows are included in the index, even though they have null values in one or two of the indexed columns.

Armed with this general overview of how indexes work, you are ready to learn about the types of indexes supported by Oracle 10*g* and when to use each type.

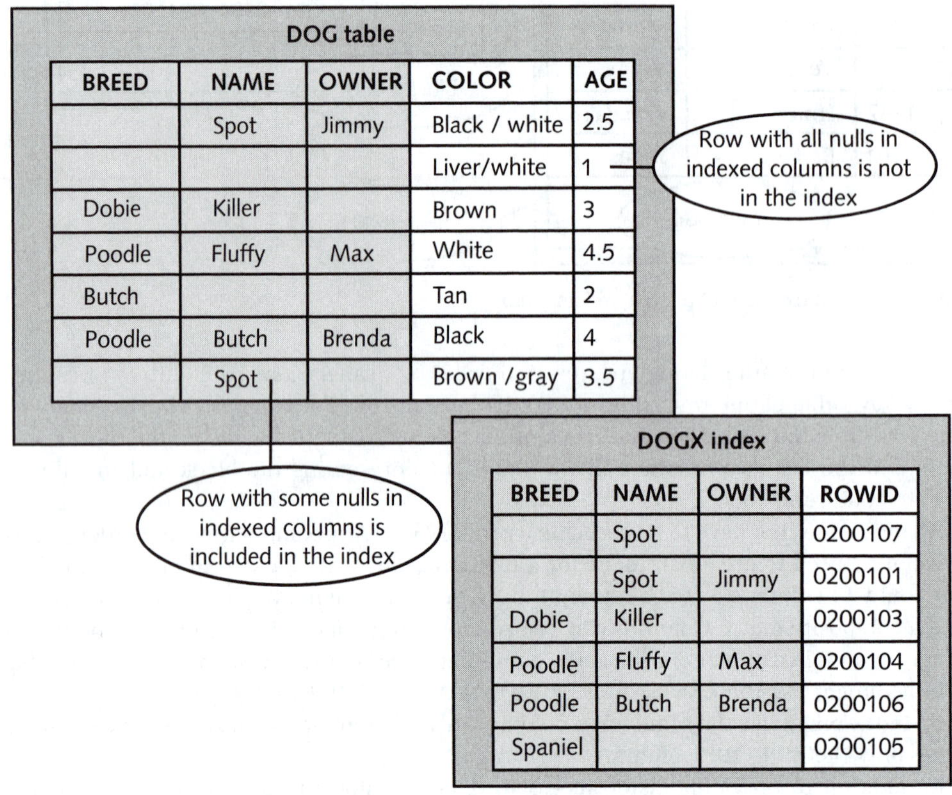

BREED	NAME	OWNER	COLOR	AGE
	Spot	Jimmy	Black / white	2.5
			Liver/white	1
Dobie	Killer		Brown	3
Poodle	Fluffy	Max	White	4.5
Butch			Tan	2
Poodle	Butch	Brenda	Black	4
	Spot		Brown /gray	3.5

BREED	NAME	OWNER	ROWID
	Spot		0200107
	Spot	Jimmy	0200101
Dobie	Killer		0200103
Poodle	Fluffy	Max	0200104
Poodle	Butch	Brenda	0200106
Spaniel			0200105

FIGURE 9-10 Indexes can include null values in some cases

Types and Uses of Indexes

There are numerous types of indexes you can create in Oracle 10*g*:

- **BTree index:** Used as the default type of index Oracle 10*g* builds for tables. There are also two specialized types of BTree indexes:
  - **Reverse key index:** Used to improve efficiency of Oracle Real Application Clusters. In general, a non-Oracle RAC database does not have a high enough level of concurrent insertion activity, to warrant use of reverse key indexes.
  - **Function-based index:** Used to store pre-computed expression values based on table columns.
- **Bitmap index:** Used to store index entries in a compact format and is used primarily for very large tables with a small number of distinct index values in data warehouses.
- **Local partitioned index:** Used on partitioned tables in which indexes are created for each individual partition. The index and the table are partitioned identically.
- **Global partitioned index:** Used on partitioned tables where the index is created on all partitions for the entire table, as a single index. The index is global to all partitions (encompasses all partitions). Oracle 10*g* now allows global hash-partitioned indexes, previous versions of Oracle allowed only local-partitioned hash indexes.
- **Cluster index:** Used specifically for creating indexes on clusters, by default a BTree but can also be a hash-clustered index.
- **Domain index:** Used for application- or cartridge-specific indexes.

All these index types except the last three are described in this chapter. The last two (cluster and domain indexes) are beyond the scope of this book.

Before moving into the BTree index section, take a look at the syntax for creating an index. This syntax is used for all types of indexes:

```
CREATE UNIQUE|BITMAP INDEX <schema>.<indexname>
ON <schema>.<tablename>
(<colname>|<expression> ASC|DESC,
 <colname>|<expression> ASC|DESC, ..)
TABLESPACE <tablespacename>
STORAGE (<storage_settings>)
LOGGING|NOLOGGING
ONLINE
COMPUTE STATISTICS
NOCOMPRESS|COMPRESS <nn>
NOSORT|REVERSE
NOPARALLEL|PARALLEL <nn>
PARTITION|GLOBAL PARTITION <partition_settings>
```

Many of these parameters have meanings that are identical to those described for the CREATE TABLE command. Refer back to Chapter 7 for a detailed description of the following parameters:

- TABLESPACE <tablespacename>
- STORAGE (<storage_settings>)
- LOGGING|NOLOGGING
- ONLINE

Other parameters are probably new to you because they are not used when creating tables. They are:

- **UNIQUE|BITMAP:** Specify UNIQUE to create an index that requires every row in a table to contain a unique value in the indexed column(s). Specify BITMAP to create a nonunique bitmap index (an index stored as a bitmap within the index segment). Omit them both to create a nonunique BTree index.
- **(<colname>|<expression> ASC|DESC, ...):** Use this phrase to list all the columns that are to be indexed. ASC is the default order, indicating that the column values should be arranged in ascending order. DESC signifies descending order. Separate columns by commas. An alternative to a column name is to list an expression. When an expression, such as UPPER(LAST_NAME), is listed within the list of columns included in the index, the index is called a function-based index.
- **COMPUTE STATISTICS:** Use COMPUTE STATISTICS to gather statistics while building the new index.
- **NOCOMPRESS|COMPRESS <nn>:** Use key compression to cause repeating values of a key column to be eliminated, saving space. Specify how many of the columns in the key you want compressed (1, 2, and so on). Key compression is used primarily for unique keys that have multiple columns. You cannot compress a bitmap index.
- **NOSORT|REVERSE:** Specify NOSORT to tell Oracle 10*g* to assume that the rows in the table are already in the same order as the index, and, therefore, no sorting needs to be done. The purpose of specifying NOSORT is to save time and temporary storage space when building the index. If the rows are not in order as expected, the CREATE INDEX command fails. The REVERSE parameter tells Oracle 10*g* to store the bytes of the index values in reverse order. This is primarily used with Real Application Clusters to improve performance on a distributed database. The NOSORT and REVERSE parameters are not allowed on bitmap indexes or on index-organized tables. In addition, REVERSE and NOSORT cannot be specified together.
- **NOPARALLEL|PARALLEL <nn>:** Use PARALLEL to indicate that <nn> number of threads is to be used in parallel to create the index. NOPARALLEL is the default.
- **PARTITION|NOPARTITION:** Partitioning an index can occur on partitioned tables as well as nonpartitioned tables and uses many of the same settings as those needed for partitioning a table.

This chapter contains examples of the parameters just listed. The next sections explain each type of index, beginning with the BTree index.

BTree Index

The BTree index is an index structure in which data is divided and subdivided based on the index key values to minimize the look-up time when searching for a key value. The BTree is a more complex variation on the binary tree. To understand the BTree, it is easiest to begin by learning about the binary tree.

A binary tree uses a branching formation that always divides the list of data into two halves. Each half is then divided in half again, and so on. Figure 9-11 shows two binary tree structures.

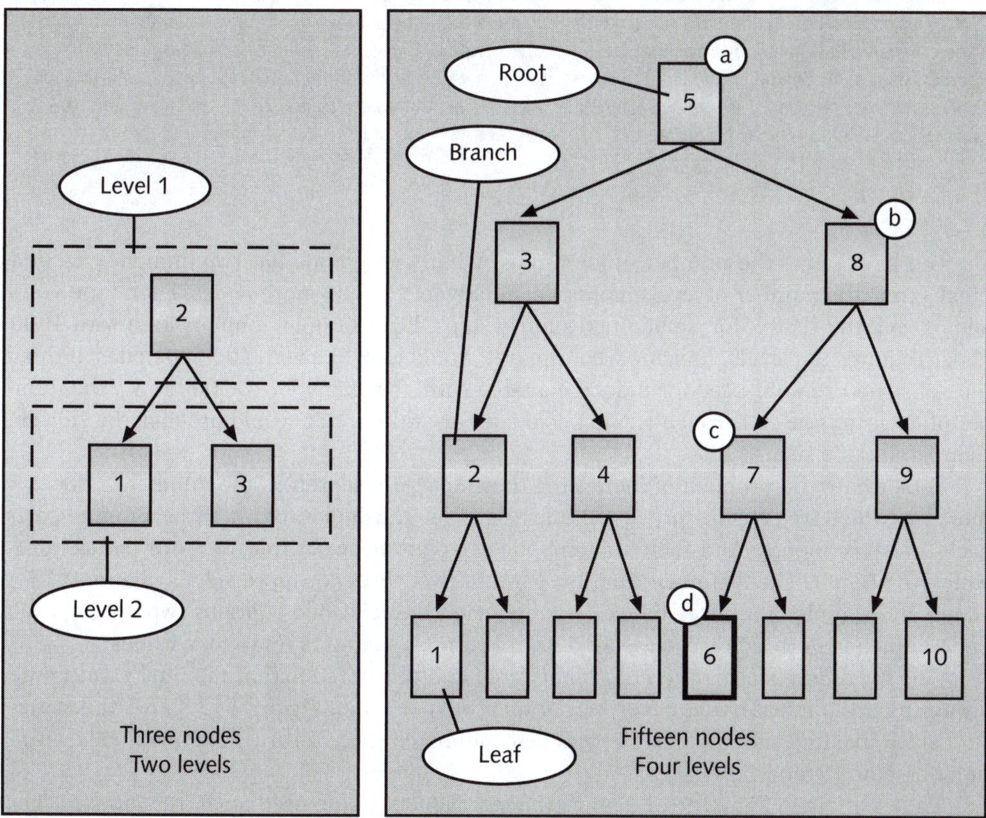

FIGURE 9-11 To find a value, follow the branches to the leaves

The tree on the left is the simplest example; one root has two leaves. The tree on the right has one root, six branches, and eight leaves. A root is the starting point for searching a binary tree or a BTree. A node is any point on the tree. A branch is a node that has more nodes below it. A leaf is the bottom level and is a node with no nodes below it. A leaf on the left has a value less than its parent branch, and a leaf on the right has a value greater than its parent branch. In the figure, five of the leaves are blank and available for additional data. Three of the leaves contain data. To find a value on the binary tree, you always start at the root. An example of a binary search is illustrated with the letters "a" through "d" in Figure 9-11, which shows how to search for the number six. Here are the steps:

a. Start with the root (5). Six is greater than the node (5), so go down the right branch to the next node (8).
b. Six is less than the node (8), so go down the left branch to the next node (7).

c. Six is less than the node (7), so go again down the left branch to the next node (6).
d. Six is equal to the node (6), so you have located the node.

NOTE

Oracle Database binary tree indexes have a maximum of three levels. They are not real binary trees by any mathematical definition, which would have an infinite number of levels. An Oracle Database binary tree consists of at most, one root node, a set of child branch nodes, and a set of child leaf nodes of those branch nodes. An Oracle Database binary tree allows a very fast depth-first traversal because the tree is only a minimal number of levels. The objective is to find a unique item by traversing as little of the tree as possible. Mathematically a binary tree expects a search from start to finish, throughout the entire tree, or until the required value is found.

You have found the number in four steps. When every node has two branches, as in a binary tree, the number of levels increases quickly as you add more nodes. Each time you add a level, the search for a node takes more steps. For example, a binary tree with 1000 nodes requires ten levels. Imagine what the tree would look like with 100,000 nodes. If this type of structure were used for indexing a table with, for example, 100,000 rows, the number of levels to search through would be so large that the index would probably be slower than a full table scan.

To alleviate the problem of increasing the steps for a search as the number of nodes increases, the BTree was invented. A BTree can divide the data into two, three, or more parts each time it branches. And each branch node can contain two, three, or more values from which to branch. The leaf node can also contain two, three, or more values. Figure 9-12 shows an example of a BTree index, in which each branch node contains two values, each node branches in three directions, and each leaf node contains up to four values.

One big difference between the binary tree and the BTree is that the final values are always found in a leaf node, never in a branch or root node. Figure 9-12 shows the structure of a BTree index for a table. The indexed column contains a three-character code with numbers and letters.

The example in Figure 9-12 depicts the storage technique used for BTree indexes. The approximate midpoint of the data is at the top of the tree. The key values or partial key values are listed in the branch nodes, and the complete key values with their associated ROWIDs are listed in the leaf nodes. With a BTree using three-branch nodes and four-value leaves, as in the example, your search algorithm is more complex than with a binary tree. An algorithm is a formula for solving a problem. For example, when guessing a number from 1 to 100, you might start at 50 and ask, "Is 50 higher or lower than the number"? If the answer is "higher," you divide 50 in half and ask, "Is 25 higher or lower than the number"? The pattern of dividing in half, asking a question, and repeating is an algorithm. In the case of an index, the algorithm tells you what to do at each branch node and at each leaf node. The goal of the BTree is to keep the leaf nodes from getting too far down the tree levels. The BTree accomplishes this by adding more values in each leaf node, adding more values in each branch node, and allowing each branch node to branch in more than two directions. All these changes help minimize the levels, which in turn shorten the

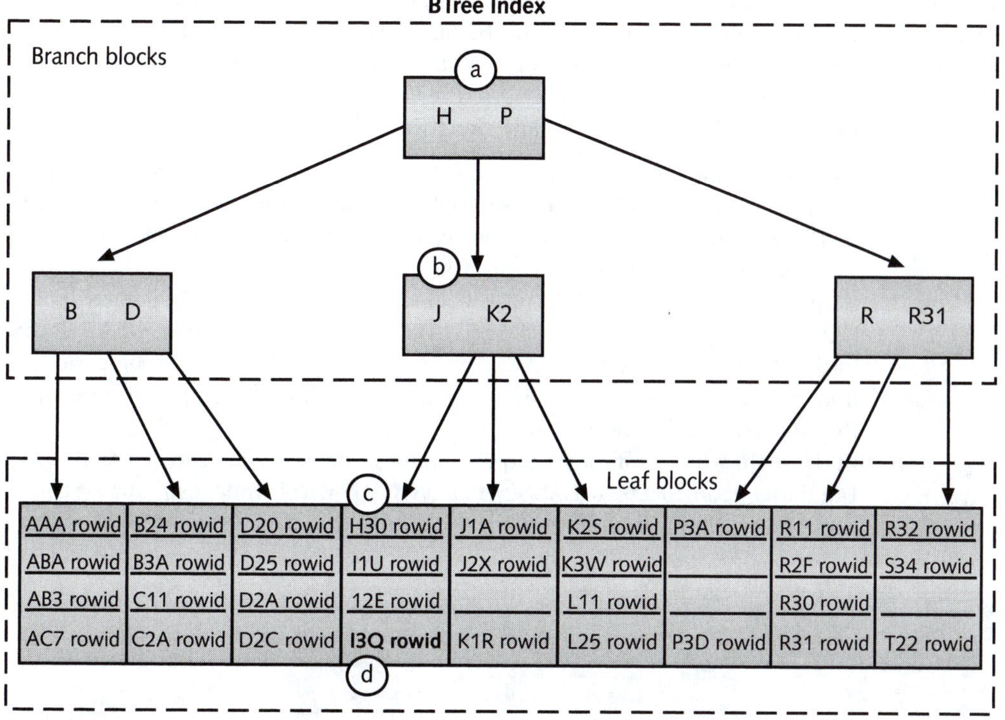

FIGURE 9-12 A BTree index has more nodes in fewer levels than a binary tree

number of steps needed to find any indexed value. For example, a four-level BTree with 50 values on each node can handle 10 million values.

To see how a BTree search works, follow along with these steps to find the ROWID of the row with an index key value of "I3Q." Each step letter is noted in Figure 9-12:

1. Start with the root (H P). "I3Q" is between H and P, so you take the center branch. (If the value were smaller than H, you would take the left branch, and if it were larger than P, you would take the right branch.)
2. The next branch node is "J K2." "I3Q" is less than "J," so you take the left branch.
3. You are now in the leaf node that contains "I3Q." Search through the values until you find "I3Q."
4. The fourth key is equal to "I3Q," so you read the ROWID and using this ROWID, you can quickly retrieve the actual data row.

Notice that the values in the branch nodes are sometimes a partial key value. This is a way for the BTree to save time. It stores only the minimum characters needed to make a decision on which branch to take. For example, look at the center branch node in Figure 9-12. The lower value is "J," whereas the upper value is "K2." There are leaf values that begin with "K" in both the center leaf and the right leaf below this branch node; therefore, a decision on which leaf to go to cannot be made using just the letter "K." By adding one more character, a valid decision point is made. "K2" is less than the first value in the right leaf ("K2S") and greater than or equal to the last value in the middle leaf ("K1R").

It is time to create a BTree index. Follow along with these steps:

1. Start up the Enterprise Manager console. In Windows, click **Start** on the Taskbar, and then click **All Programs/Oracle ... /Enterprise Manager Console**. In UNIX or Linux, type **oemapp console** on the command line. The Enterprise Manager Console login screen appears.
2. Connect as the CLASSMATE user. Connect to the ORACLASS database. A logon window appears. Type **CLASSMATE** in the Username box, **CLASSPASS** in the Password box, and **ORACLASS** in the Service box. Leave the connection type as "Normal." Click **OK** to continue. You should not need to connect as the SYS user in this case.
3. In the main window of the Enterprise Manager console, navigate starting at the **Databases** icon to **ORACLASS/Schema/CLASSMATE/Indexes**. Your screen should look similar to Figure 9-13.

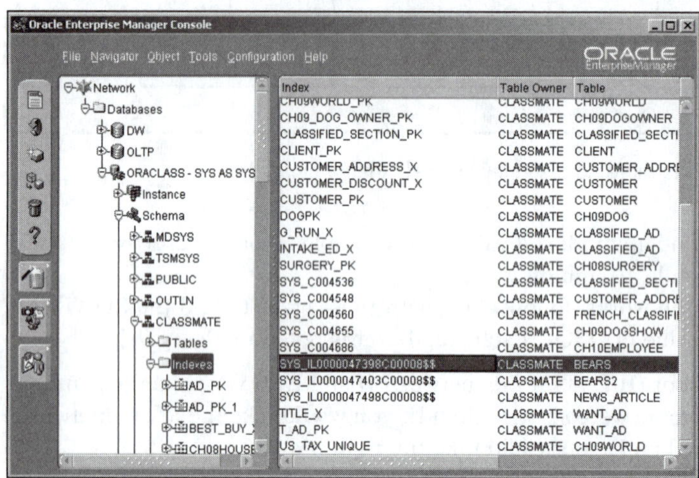

FIGURE 9-13 Use the console for a quick look at all the indexes owned by CLASSMATE

4. Notice that there are several indexes already in existence even though you have not specifically created any up to this point in the book. All these indexes were automatically created to support PRIMARY KEY constraints on tables created in earlier chapters. Some have readable names, such as SURGERY_PK, because the constraint was named when the table was created. Others have names that are system-generated such as "SYS_IL000047398C00008$$" for constraints that were not named when the table was created. All these indexes are unique BTree indexes.

NOTE

Index names may be different for your database.

5. To create a unique BTree index, start up the SQL*Plus Worksheet by clicking **Tools/Database Applications/SQL*Plus Worksheet** from the top menu in the console. A background SQL*Plus process starts, and then the SQL*Plus Worksheet window appears.
6. For this chapter, start with a prepared SQL command found in the directory named **Data\Chapter09** on your student disk. Click **Worksheet/Run Local Script**. A browse window opens. Navigate to the **Data\Chapter09** directory, and select the **ch09setup.sql** file. Click **Open**. The file runs in the SQL*Plus Worksheet window and creates tables needed for the examples in this chapter.
7. Clear the top part of the worksheet by selecting **Edit/Clear All** from the menu. Then type the first two lines of the command to create a unique index on the LIBRARYBOOK table:

   ```
   CREATE UNIQUE INDEX CLASSMATE.DEWEY_IX
       ON CLASSMATE.CH09LIBRARYBOOK
   ```

   The UNIQUE parameter is optional. Add it for unique indexes and omit it for nonunique indexes. If you are logged on as the owner of the index you plan to create, you can also omit the schema. It is possible to create an index owned by one user on a table owned by another user, provided that the creating user has either the CREATE ANY INDEX privilege or the INDEX privilege on the target table, or the user has been granted specific privileges for creating an index on that table.
8. Type the list of columns to be included in the index. For this example, one column, called DEWEY_DECIMAL, is included in the index:

   ```
   (DEWEY_DECIMAL)
   ```

9. Add other parameters, such as LOGGING, TABLESPACE, STORAGE, and so on. For this example, include the INITRANS, PCTFREE, and LOGGING parameters by typing these lines:

   ```
   INITRANS 2 PCTFREE 20 LOGGING
   ```

   These are all default settings. Although INITRANS is seldom changed for tables or indexes, Oracle recommends that you set the index's INITRANS to double the value of the table's INITRANS. You cannot set PCTUSED on an index, because when you update a column that is indexed, the index entry usually moves and is re-created anyway, rather than staying in the same location, as a table row does.
10. Add the parameter to have Oracle 10g gather statistics while building the index, and mark the end of the command by typing this line:

   ```
   COMPUTE STATISTICS;
   ```

11. Run the command by clicking the **Execute** icon. SQL*Plus Worksheet runs the command and responds, "Index created." Figure 9-14 shows the complete command in SQL*Plus Worksheet.

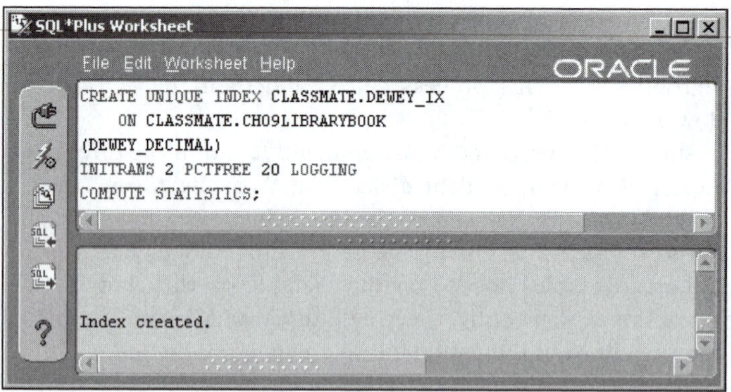

FIGURE 9-14 A unique index is created on the CH09LIBRARYBOOK table

12. Remain logged on for the next practice.

Unless you specifically instruct Oracle 10g to create a bitmap index, it always creates a BTree index. Bitmap indexes have a completely different search algorithm and storage method than the BTree index. You find out how they work in the next section.

Bitmap Index

A bitmap index is an index that does not use the BTree method to store index key values and ROWIDs. Instead, it uses a bitmap to store index key values and row locations. A bitmap implies a map of bits, a little like a 2-dimensional matrice, where each slot contains a 0 or a 1. If a slot is set to 1 then the value indexed exists, otherwise the value is something else or NULL. A bitmap index takes up a very small amount of storage space compared to a BTree index; however, the bitmap index is appropriate only for certain special circumstances:

- The indexed columns should have low cardinality. Low cardinality means that the number of distinct values in an indexed column should be low compared to the number of rows in the table. For example, the column MARITAL_STATUS might have three or four values over the entire table. Another example is a table with 100,000 rows and a column with 1000 distinct values. This column still qualifies as having low cardinality, because the ratio of distinct values to rows is 1 percent. Oracle advises that if the ratio of distinct values of a column to the total number of rows is less than 1 percent, or if the column has values that repeat more than 100 times, the column is a candidate for a bitmap index.

- The table should be used primarily for queries rather than updates, such as those in a data warehouse system. A data warehouse is a database used to collect and store information from many areas of a business. Usually, a data warehouse is used to perform queries for statistical applications, such as business trends and customer profiles.
- The majority of queries should use AND, OR, NOT, and "equal to" in the WHERE clause referencing the table's indexed columns. Queries that contain "greater than" or "less than" in the WHERE clause cannot use a bitmap index.
- The majority of queries include complex conditions in the WHERE clause, such as those used in statistical analysis of a data warehouse table.
- The table should not have concurrent updates, such as when a table is involved in an OLTP (online transaction processing) system.
- A bitmapped index cannot be a unique index.
- A bitmapped index cannot contain any DESC columns (columns in descending order in the index).

With all these qualifiers, you can see that the bitmap index has limited uses. However, for very large tables in a data warehouse system for which the types of queries described previously are used frequently, a bitmap index can save a significant amount of space and perform faster than an equivalent BTree index.

A bitmap index does not store ROWID values or key values. Figure 9-15 illustrates a conceptual view of a bitmap index and the corresponding table rows. In this example, the column BLOOD_TYPE and GENDER could be indexed using this command (*do not actually execute this* example as the PATIENT table does not exist in your database):

```
CREATE BITMAP INDEX PATIENT_BITMAP_X
ON PATIENT
(BLOOD_TYPE, GENDER);
```

The BITMAP parameter appears immediately after the word "CREATE" in the command. You cannot create a bitmap index that is also a UNIQUE index.

As a bitmap index is built, Oracle 10g builds a chart of each distinct value in each indexed column. There are four blood types and two genders in the PATIENT table in our example, so the bitmap has six values listed across the top. Oracle 10g walks through the table, from the first row to the last row building the bitmap index records. Each row is listed, in order of its appearance in the table, as a single row in the bitmap chart. Under each value column, a zero or a one is marked. A zero means, "No, this value is not in this row." A one means, "Yes, this value is in this row." You can see now why a small number of distinct values are important to a bitmap's efficiency, because every row gets a tick mark (0 or 1) for every value.

Look at Figure 9-15, and find the third row in the bitmap index. This row has a blood type of "B" and a gender of "F." Look at the third row in the table to see that it corresponds to these values. Similarly, the last row in the table has a null value in the BLOOD_TYPE column and "F" in the GENDER column. Looking at the last row in the bitmap index, you can see that all the blood type values are zero and the gender "F" has a 1 in the column.

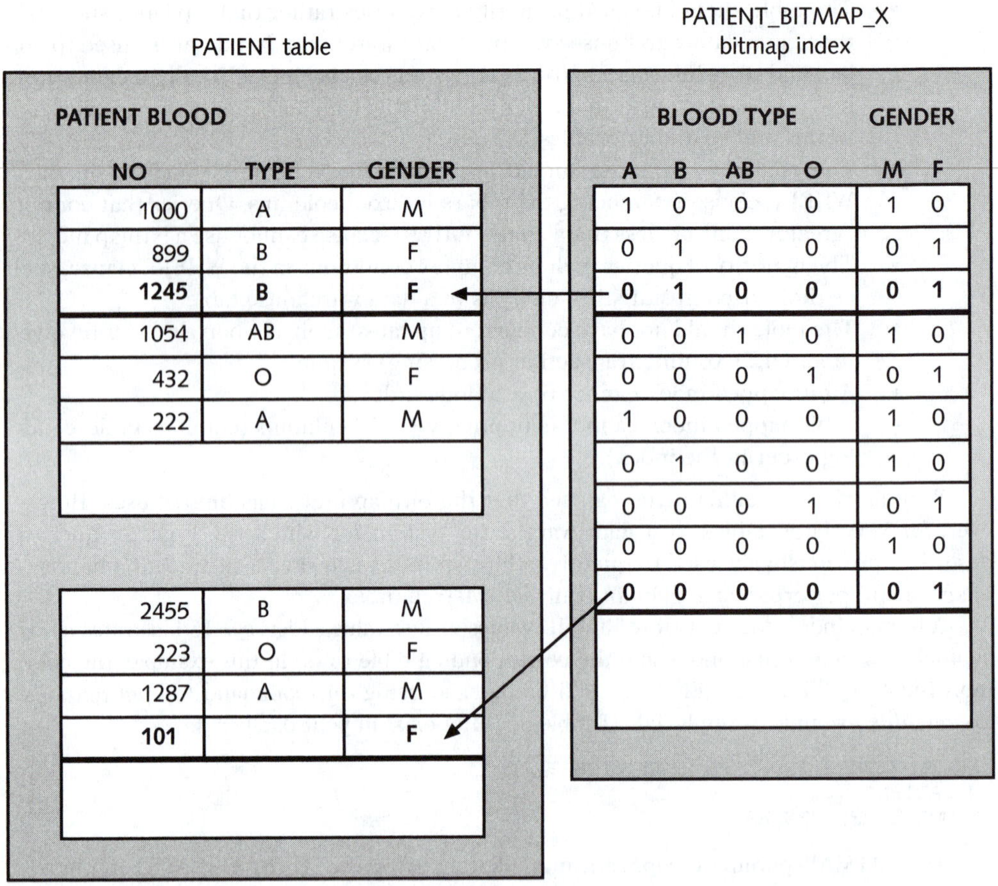

PATIENT table

PATIENT BLOOD

NO	TYPE	GENDER
1000	A	M
899	B	F
1245	**B**	**F**
1054	AB	M
432	O	F
222	A	M
2455	B	M
223	O	F
1287	A	M
101		**F**

PATIENT_BITMAP_X bitmap index

BLOOD TYPE A	B	AB	O	GENDER M	F
1	0	0	0	1	0
0	1	0	0	0	1
0	**1**	**0**	**0**	**0**	**1**
0	0	1	0	1	0
0	0	0	1	0	1
1	0	0	0	1	0
0	1	0	0	1	0
0	0	0	1	0	1
0	0	0	0	1	0
0	**0**	**0**	**0**	**0**	**1**

FIGURE 9-15 A bitmap index stores bits (1's and 0's), taking up very little physical space

NOTE

Unlike BTree indexes, bitmap indexes include table rows that have null values in all the indexed columns.

Imagine that four new rows are added to the PATIENT table in Figure 9-15. Recall from Chapter 7 that inserted rows are sometimes placed in data blocks that already have rows. If that happened here, then a new row might be added in the first data block. The bitmap index would require an inserted row in the same order as the table. Because each record in a bitmap index is so compact, inserting one record between the others causes much rewriting of data within a single index block. You can see how this could cause a problem and slow down performance. Slow performance on inserts (and updates) is why Oracle recommends using bitmaps for tables that are used almost entirely for queries.

To illustrate why bitmap indexes are faster than BTree indexes for certain types of queries and low-cardinality columns, imagine that you run the following query on the PATIENT table:

```
SELECT COUNT(*) FROM PATIENT
WHERE BLOOD_TYPE='A' and GENDER='M';
```

Looking again at Figure 9-15, you can quickly find the results by looking for the entries with "1" under the "A" column and the "M" column. If you were using a BTree index, you would be walking through several branches, finding the leaf nodes, and then counting the number of rows in each leaf node.

On the other hand, the following query simply cannot use a bitmap index, because there is not enough information stored in the bitmap index to determine an answer:

```
SELECT COUNT(*) FROM PATIENT
WHERE BLOOD_TYPE >'A' and GENDER='M';
```


This is why bitmap indexes are not appropriate for tables that are primarily queried with "greater than" or "less than" in the WHERE clause.

Local Partitioned Index

A local partitioned index is an index on a partitioned table, in which the index is partitioned in the same way and on the same columns as the table. Each partition of the local partitioned index has data from the associated table partition, which makes queries using the index faster than an index that contains values from the entire table. A local partitioned index is automatically updated if you update the partitioning of the table. For example, if you use the ALTER TABLE command to add a new partition, the local index automatically receives a new partition. Oracle 10g handles naming and locating the index partitions, unless you specifically name or locate the index partitions in the CREATE INDEX command.

Follow along with a few examples of partitioned indexes:

1. If you are not already there, start up the Enterprise Manager console and SQL*Plus Worksheet, logging on as CLASSMATE.
2. If you have already run the **ch09setup.sql** file earlier in the chapter, skip this step. Otherwise, click **Worksheet/Run Local Script**. A browse window opens. Navigate to the **Data\Chapter09** directory on your student disk, and select the **ch09setup.sql** file. Click **Open**. The file runs in the SQL*Plus Worksheet window and creates tables needed for the examples in this chapter.
3. Clear the top part of the worksheet by selecting **Edit/Clear All** from the menu. The CH09MORTGAGE_HISTORY table is a HASH-partitioned table with two partitions, one in the USERS tablespace and one in the USER_AUTO tablespace. You want to add a local partitioned index to speed up query-retrieval time. The table is partitioned on the DATE_CREATED column. To begin the command, type the following two lines:

   ```
   CREATE INDEX LOCAL_X ON CH09MORTGAGE_HISTORY(DATE_CREATED)
   LOCAL
   ```

   If you wanted Oracle 10g to do the rest (name the partitions and locate them in the corresponding tablespaces), you could run the command at this point. Instead, you will add details as to where each index partition is to be stored.

4. To specify the tablespaces for each index partition while allowing Oracle 10*g* to name the partitions, type the following line:

```
STORE IN (USERS, USER_AUTO);
```

5. Run the command by clicking the **Execute** icon. SQL*Plus Worksheet replies, "Index created." Other types of partitioned tables are handled the same way.
6. In this example, you are working with CH09MORTGAGE_CLIENT, which is RANGE partitioned by the LOAN_DATE. You are going to create a local partitioned index, in which you specify the partition names and the tablespace locations. In addition, this index is a bitmap index:

```
CREATE BITMAP INDEX MCLIENT_LOCAL_X
ON CH09MORTGAGE_CLIENT(LOAN_DATE)
LOCAL
```


7. Add details on each of the two partitions, including the index partition name, tablespace, and storage settings by typing these lines:

```
(PARTITION OLDER_X TABLESPACE USERS
           STORAGE (INITIAL 50K NEXT 10K),
 PARTITION NEWER_X TABLESPACE USER_AUTO
           STORAGE (INITIAL 40K NEXT 15K));
```

8. Run the command by clicking the **Execute icon**. SQL*Plus Worksheet replies, "Index created."
9. View the index partitions in the data dictionary view by querying the USER_IND_PARTITIONS. Type the following query to see the work you have done so far:

```
SELECT INDEX_NAME, PARTITION_POSITION POS,
PARTITION_NAME, TABLESPACE_NAME
FROM USER_IND_PARTITIONS
ORDER BY 1,2;
```

10. Run the query by clicking the **Execute** icon. SQL*Plus Worksheet displays the results, as shown in Figure 9-16. The system-generated index partition names may be different from the ones in the figure. The SYS_P25 and SYS_p26 partitions are internally generated by Oracle 10*g* and therefore may be different for your database.
11. Remain logged on for the next practice.

NOTE

If you accidentally specify the wrong number of partitions for a local partitioned index, Oracle 10*g* returns an error and does not create the index.

You cannot create a local partitioned index on a nonpartitioned table; however, you can create a different type of partitioned index—a global partitioned index—on a nonpartitioned table.

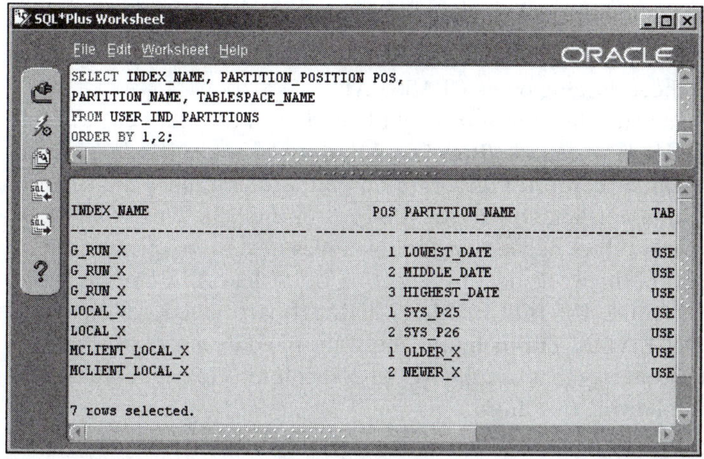

FIGURE 9-16 Oracle 10*g* named the SYS_P25 and SYS_P26 index partitions

Global Partitioned Index

A global partitioned index is an index that is partitioned when either the table is not partitioned or the table is partitioned in a different way than the index. Use a global partitioned index when you have queries on a table that are not using the partitions efficiently. For example, you have a list partitioned EMPLOYEE table that is partitioned by DEPARTMENT. You install a new online application that retrieves data by EMPLOYEE_NAME instead. So, you create a global partitioned index that is RANGE partitioned by EMPLOYEE_NAME to support efficient data retrieval for the new application.

NOTE

Global partitioned indexes can be RANGE or HASH partitioned.

An index can also be created, global to all partitions, but not using the partition key as the index. This type of index is a normal index (sometimes called a nonpartitioned global index), which is created on a partitioned table. The index contains data from all the partitions in the table. Use this type of global partitioned index for partitioned tables in which queries are imposed on columns that do not lend themselves to partitioning. For example, the list partitioned EMPLOYEE table (partitioned by DEPARTMENT) also has a column (RETIREMENT_ACCT_NO) that is unique. You want to create a unique global index on this column.

NOTE

Global indexes cannot be bitmap indexes; however, they can be unique indexes.

Follow along with these examples to add a global partitioned index to a table:

1. If you are not already there, start up the Enterprise Manager console and SQL*Plus Worksheet, logging on as CLASSMATE.
2. If you have already run the **ch09setup.sql** file earlier in the chapter, skip this step. Otherwise, click **Worksheet/Run Local Script**. A browse window opens. Navigate to the **Data\Chapter09** directory on your student disk, and select the **ch09setup.sql** file. Click **Open**. The file runs in the SQL*Plus Worksheet window and creates tables needed for the examples in this chapter.
3. Clear the top part of the worksheet by selecting **Edit/Clear All** from the menu. The CH09MORTGAGE_HISTORY table is a HASH-partitioned table partitioned on the DATE_CREATED column. You want to add a global partitioned index to speed up queries that use the ACCT_NO column. To begin the command, type the following two lines:

```
CREATE INDEX G_ACCT_X ON CH09MORTGAGE_HISTORY (ACCT_NO)
GLOBAL PARTITION BY RANGE (ACCT_NO)
```

4. The next part of the command looks just like the code to create a RANGE-partitioned table, in which you specify the partition details. In this example, you are dividing the table data into three partitions. Type the following lines to define the three partitions:

```
(PARTITION LOWEST_ACCT VALUES LESS THAN (5000),
 PARTITION MIDDLE_ACCT VALUES LESS THAN (10000),
 PARTITION HIGHEST_ACCT VALUES LESS THAN (MAXVALUE));
```

5. The final partition of a global partitioned index must contain MAXVALUE as the value. This ensures that all rows in the table fit into one of the partitions. (You can use MAXVALUE as the value in the highest range of a RANGE-partitioned table as well.)
6. Run the command by clicking the **Execute** icon. SQL*Plus Worksheet replies, "Index created."
7. Remain logged on for the next practice.

When you create a nonpartitioned global index on a partitioned table, the command looks like any other nonpartitioned index, except you add the GLOBAL parameter to the command. For example, a unique index on the LOAN_NO column of the CH09MORTGAGE_HISTORY partitioned table is created with this command:

```
CREATE UNIQUE INDEX G_LOAN_X
ON CH09MORTGAGE_HISTORY (LOAN_NO)
GLOBAL;
```

The global index can be useful in large tables. The next section describes an index type that can be used for large tables on a clustered database.

Reverse Key Index

A reverse key index creates an unusual form of BTree index. Every byte in the indexed column is reversed. The leaf blocks still include the ROWID (not reversed); however, the distribution of the data is changed.

A reverse key index is not the same as specifying DESC (for descending order) on an indexed column. The order is not reversed; the data is turned backwards and then put in order. For example, a nonreversed index on LAST_NAME would contain "Huxley," "Lopez," and "Madison" in alphabetical order. However, a reversed index on LAST_NAME would place them in this order: "Madison," "Huxley," and "Lopez," because the REVERSE function essentially spells each name backwards ("nosidaM," "yelxuH," and "zepoL"), and then alphabetizes the list.

What good is a list in reverse order? You cannot run a range search on the index, such as "LAST_NAME greater than 'Martin.'" In fact, the optimizer only uses a reverse key index when the query contains a single record search or a full index scan.

The real usefulness of the reverse key index is found when you have a clustered database using Real Application Clusters, in which many users are accessing the same section of data from a large table, causing I/O contention. I/O bottlenecks caused by too many users attempting to read or update the same group of data blocks can slow performance, because some users must wait for others to complete their work before accessing the data. For example, imagine that you have a large table called CURRENT_EVENTS that is indexed by CREATE_DATE. You have a Real Application Clusters database with an online application, from which all your clients want the last three days' events. This has caused a bottleneck in the database. Fix the problem by reversing the index. The reversed index distributes the three days across many have dropped the old index and are creating a new reverse key index with this command (*do not actually execute this example*):


```
CREATE INDEX EVENT_REVERSE ON CURRENT_EVENTS(CREATE_DATE)
         REVERSE;
```

NOTE

You can change an index from normal to reverse key and vice versa using the ALTER INDEX command.

Function-Based Index

Traditionally, both the cost-based and rule-based optimizer took full advantage of indexes when evaluating a query. However, only the cost-based optimizer is able to use the new function-based index type. A function-based index is an index with one or more columns indexed that are actually expressions or functions instead of columns.

Allowing functions and expressions within an index gives the optimizer the ability to search an index for values that otherwise would require a full table scan. For example, the CH09LIBRARYBOOK table has a column called PUB_DATE that contains the date a book was first published. Imagine that the online library search application has a box where the user types in a cutoff date for publication. For example, the user wants to search for books published after 1950, so he types *1950* in the box. The application generates a query that looks like this:

```
SELECT TITLE, AUTHOR, DEWEY_DECIMAL,
TO_CHAR(PUB_DATE,'YYYY') PUB_YEAR
FROM CH09LIBRARYBOOK
WHERE TO_CHAR(PUB_DATE,'YYYY') >= '1950';
```

Before Oracle9*i*, neither the cost-based nor the rule-based optimizer was able to use an index to resolve this query, *even when there was an index on the PUB_DATE column.* The index would be ignored because there is a function on the indexed column in the WHERE clause. Today, a function-based index solves this problem.

The user who creates a function-based index must have the usual privileges granted to a user (the RESOURCE role), plus the QUERY REWRITE system privilege. Steps 1 through 4 in the following list grant this privilege to the CLASSMATE user.

Follow along to create a function-based index for this table:

1. If you are not already there, start up the Enterprise Manager console and SQL*Plus Worksheet.
2. Connect as SYSTEM by clicking **File/Change Database Connection** on the menu and filling in the Database Connection Information window with **SYSTEM** in the Username box, the current password for SYSTEM in the Password box, and **ORACLASS** in the Service box. Leave the Connect as box set to "Normal." Click **OK**.
3. Clear the top part of the worksheet by selecting **Edit/Clear All** from the menu. To create a function-based index, the user must have the QUERY REWRITE system privilege, because in some cases the optimizer handles a function-based index by rewriting the query. Type the following command to grant the privilege to the CLASSMATE user:

```
GRANT QUERY REWRITE TO CLASSMATE;
```

4. Run the command by clicking the **Execute** icon. SQL*Plus Worksheet replies, "Grant succeeded."
5. Connect to CLASSMATE again by either clicking the **File/Change Database Connection** or by typing this line and pressing the **Execute** icon.

```
CONNECT CLASSMATE/CLASSPASS@ORACLASS
```

6. Now you are ready to create the function-based index. For this example, you will create an index on the PUB_DATE column so that the query you saw earlier can use the index. Type the command as follows, and click the **Execute** icon to run it:

```
CREATE INDEX PUB_YEAR_IX
ON CH09LIBRARYBOOK(TO_CHAR(PUB_DATE,'YYYY'));
```

   SQL*Plus Worksheet replies, "Index created."

7. Type the following query to look at the indexes owned by CLASSMATE. Click the **Execute** icon to run the query.

```
SELECT INDEX_NAME, INDEX_TYPE FROM USER_INDEXES;
```

8. Figure 9-17 shows the results in the lower half of SQL*Plus Worksheet. Your results should look similar. Notice that the index type for the function-based index you just created is "FUNCTION-BASED NORMAL."

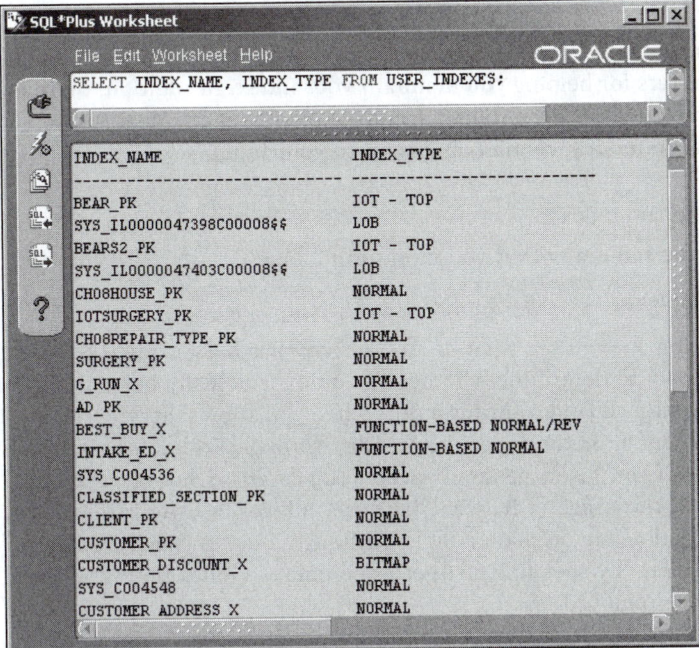

FIGURE 9-17 Function-based indexes are stored the same way as other indexes

9. You can use expressions in function-based indexes as well. For example, a complex mathematical calculation might work well for a function-based index, because the value of the calculation is stored, already calculated, in the index. This makes searches faster by shifting the computation time from the query to the building of the index (or to the insertion of new rows after the index is built). Type and execute this SQL command:

```
CREATE INDEX TOTAL_PAIDX
ON CH09MORTGAGE_CLIENT ((MORTGAGE_AMOUNT*NUMBER_OF_
YEARS*MORTGAGE_RATE)
      +MORTGAGE_AMOUNT)
TABLESPACE USERS
STORAGE (INITIAL 20K NEXT 10K) LOGGING COMPUTE STATISTICS;
```

10. Remain logged on for the next practice.

NOTE

As with other indexes, you can add storage settings, logging, and other parameters to the CREATE INDEX command of a function-based index.

Now let's examine managing indexes.

Managing Indexes

Like tables, your indexes may require some maintenance now and then. The ALTER INDEX command has many parameters for helping you maintain your index. In addition, there is the DROP INDEX command for removing an index from the database. The next few sections describe how to monitor, modify, rebuild, and remove your indexes.

Monitoring Indexes and Dropping Indexes

You can monitor an index the same way that you monitor a table:

```
ALTER INDEX <schema>.<indexname> MONITORING USAGE;
```


After you turn monitoring on, you see what, if any, activity has been using the index in question. This is a good way to determine whether the index is actually being used for queries. Be sure that you monitor the index during a time when you know that typical work is being done. For example, monitor the index on Monday through Friday for one week instead of over the weekend, if your users generally access the database during weekdays.

It is important to review the usage of indexes, because indexes create overhead that slows down insert, update, and delete tasks on a table. Although indexes usually make up for the slight lag in changing data by speeding up queries on data, an index that is not used has no benefits and should be removed.

To query the monitoring information, type the following query (*do not actually execute this code because we have not initiated any monitoring*):

```
SELECT * FROM V$OBJECT_USAGE;
```

If you wish, go into your SQL*Plus Worksheet and execute the query above, just for the sake of seeing what it retrieves.

If the index is being used for queries, the USED column contains "YES." If not, the USED column contains "NO." Before dropping the index that is not being used, you should review the queries and the index columns to see if you can modify one or the other, so the index can be used. For example, perhaps the queries add a function to the column in the WHERE clause. In this case, you might consider a function-based index, or a revision of the WHERE clause, so the function is not used on the indexed column.

If you determine that an index should be removed, because it is no longer being used, use the DROP INDEX command:

```
DROP INDEX <schema>.<indexname>;
```

Often, the change you want to make to cause a query to use an index involves reordering the columns used in the index. You must create a new index to accomplish this. You might also want to drop the old index, if it is not being used.

NOTE

You cannot change the column order of an existing index.

Reorganizing and Modifying Indexes

When reviewing an index's statistics, you may find that the index has much unused storage allocated or that the index has an incorrect setting for PCTINCREASE, NEXT, or other storage settings. In these cases, you can modify the index with the ALTER INDEX ... REBUILD command.

> **NOTE**
>
> Even indexes created by Oracle 10*g* to enforce a PRIMARY KEY constraint or UNIQUE key constraint can be modified to have different storage settings or to free up unused storage space.

Changing storage settings on an index is similar to changing the settings on a table. Like tables, you cannot modify the INITIAL and INITRANS settings of an existing index, but you can modify other storage settings, such as TABLESPACE, NEXT, MAXEXTENTS, and so on. The basic syntax for modifying storage and basic index settings is:

```
ALTER INDEX <schema>.<indexname>
REBUILD PARTITION|SUBPARTITION
REVERSE|NOREVERSE
TABLESPACE <tablespacename>
STORAGE (NEXT <nn> MAXEXTENTS <nn>)
PCTFREE <nn>
COMPUTE STATISTICS
COMPRESS|NOCOMPRESS
LOGGING|NOLOGGING
ONLINE;
```

All these settings have been discussed either earlier in this chapter (where the CREATE INDEX command is described) or in Chapter 7 (where the CREATE TABLE command is described). Refer to those areas for more information.

The ALTER INDEX ... REBUILD command also has these features:

- It automatically rebuilds the BTree structure of a normal index, which adjusts levels and leaves as needed.
- If successful, an index rebuild automatically corrects an index that has been marked "UNUSABLE" because a change was made to the structure of the underlying table or partition.
- An index rebuild can be performed on only one partition at a time for partitioned indexes.
- It can change a reverse key index to a normal index or vice versa.

Complete the following steps to experiment with an index rebuild:

1. If you are not already there, start up the Enterprise Manager console and SQL*Plus Worksheet, logging on as CLASSMATE.
2. Clear the top part of the worksheet by selecting **Edit/Clear All** from the menu. You want to move the PUB_YEAR_IX index to a different tablespace. This is the function-based index created earlier in the chapter. Type and execute the following command in SQL*Plus Worksheet.

```
ALTER INDEX PUB_YEAR_IX REBUILD
TABLESPACE USER_AUTO ONLINE;
```

SQL*Plus Worksheet replies, "Index altered." The ONLINE parameter tells Oracle 10g to keep the table and index available for queries during the operation.

3. Exit SQL*Plus Worksheet by clicking the **X** in the top-right corner.
4. Exit the console by clicking the **X** in the top-right corner.

You can change other index settings with the ALTER INDEX command as well. The following syntax shows additional settings that you can modify on indexes:


```
ALTER INDEX <schema>.<indexname>
COALESCE
UPDATE BLOCK REFERENCES
UNUSABLE
ONLINE
RENAME <oldindexname> TO <newindexname>
RENAME PARTITION <oldname> TO <newname>
DEALLOCATE UNUSED KEEP <nn>
LOGGING|NOLOGGING
NOPARALLEL|PARALLEL <nn>
MONITORING USAGE|NOMONITORING USAGE;
```

The first three parameters are unique to indexes, whereas the last four parameters should be familiar to you by now (refer to an earlier section of this chapter or to Chapter 7). Here is a description of the three new parameters:

- **COALESCE:** You coalesce an index to consolidate fragmented storage space in the leaf blocks. The excess space is kept unless you add DEALLOCATE UNUSED to the command. Coalescing an index is faster and takes less temporary storage space than rebuilding an index. Coalescing an index is useful to quickly compact the index without completely rebuilding it. You could coalesce or rebuild an index if performance begins to suffer. Figuring out a percentage of how many leaves (leaf index block – those with all the indexed values) contain unused space, as a result of deleted entries, is specific to an installation. Generally, this kind of thing is surmised when a database is performing poorly in general. You can use the INDEX_STATS view to check for details. This level of complexity is a little beyond the scope of this book. There is a chapter covering performance tuning later in this book. Query the INDEX_STATS view to check this condition. The COALESCE parameter must be used alone on either an entire index or on a partition.
- **UPDATE BLOCK REFERENCES:** The UPDATE BLOCK REFERENCES clause must be used alone. It is only for normal indexes on index-organized tables and is used to update the physical guesses of the indexed row's location stored in the index. Recall from Chapter 7 that the logical ROWID used for index-organized tables is a best physical guess of the location of a row. Refresh these guesses to aid performance of an index on the index-organized table after you have added or updated many rows of data on the table.
- **UNUSABLE:** The UNUSABLE clause tells Oracle 10g to mark the index unusable, which causes the optimizer to ignore the index when determining execution plans for queries. You might do this to experiment with performance time.

Here is an example using the COALESCE parameter:

```
ALTER INDEX DEWEY_IX COALESCE;
```

Partitioned indexes can be modified in most of the same ways that partitioned tables can be modified: You can add new partitions, split partitions, remove partitions, and rename partitions. There are a few restrictions on certain types of partitioned indexes. For example, you cannot add a new partition to a global partitioned index, because the highest partition must be set to MAXVALUE, so no additional partitions can fit above it. See the Administrator's Guide for details on maintenance of partitioned indexes and tables.

You have queried some of the data dictionary views that contain index information as you went through the practices in this chapter. The next section describes more data dictionary views that you can review when looking for information about indexes.

Data Dictionary Information on Indexes

Before looking at any specific data dictionary views for indexes, you should be sure to periodically run the DBMS_STATS command to analyze all your indexes. This supplies you with up-to-date statistics about the indexes and provides the cost-based optimizer with the information it needs to best utilize the indexes. Follow these steps to analyze all indexes in the CLASSMATE schema:

1. If you are not already there, start up the Enterprise Manager console and SQL*Plus Worksheet, logging on as CLASSMATE.
2. Clear the top part of the worksheet by selecting **Edit/Clear All** from the menu. Type and execute the following command to analyze all the tables and indexes in the CLASSMATE schema. Oracle recommends using the DBMS_STATS package to analyze database objects in general (rather than the ANALYZE command), because it gathers some unique statistics for the cost-based optimizer, it executes in parallel, and is faster (its syntax is however more complex):

   ```
   BEGIN
    DBMS_STATS.GATHER_SCHEMA_STATS
    (ownname=>'CLASSMATE',cascade=>TRUE);
   END;
   ```

   The process takes several seconds, and for schemas with more data or more tables, the process could take several minutes. When it is done, SQL*Plus Worksheet replies, "PL/SQL procedure successfully completed."

NOTE

When executing a package or a procedure, you must either wrap the command in a PL/SQL block by using the BEGIN and END; clauses on either side of the command, or type EXECUTE and the command all on a single line. Using the BEGIN and END; clauses allows you to type the command on more than one line.

3. One data dictionary view, INDEX_STATS, is updated with statistics that currently are only gathered by the ANALYZE command. Type and execute the following command to gather these statistics for the function-based index you

just created. The SQL*Plus Worksheet replies "Index analyzed." You can use the ANALYZE command. You don't have to use the DBMS_STATS package:

```
ANALYZE INDEX TOTAL_PAIDX VALIDATE STRUCTURE;
```

4. Now type and execute this query to view INDEX_STATS data:

```
SELECT NAME, BR_ROWS, BR_BLKS, LF_ROWS, DEL_LF_ROWS
FROM INDEX_STATS;
```

Figure 9-18 shows the results. This view shows the number of branches (BR_ROWS), leaves (LF_ROWS), and deleted leaves that still remain in the structure (DEL_LF_ROWS). If 30 percent or more of the leaves are deleted leaves, the index should be rebuilt. The next section describes how to rebuild an index. As Figure 9-18 shows, we don't need to rebuild the index at this time.

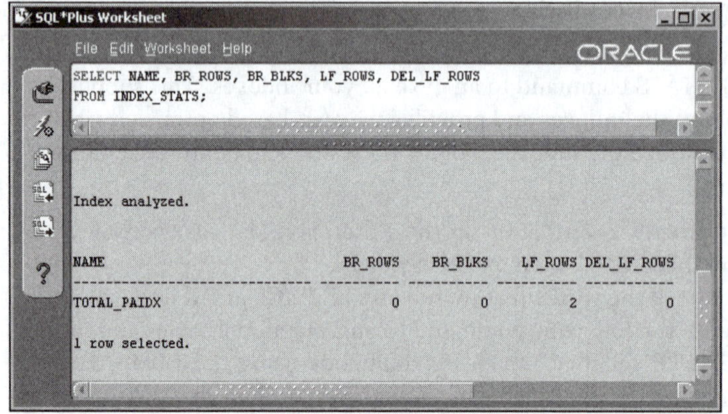

FIGURE 9-18 The leaves and branches of an index may need pruning

5. Remain logged on for the next practice.

Table 9-2 has a list of data dictionary views that display information about indexes. The views that begin with "DBA_" have similar views that begin with "USER_" and "ALL_". For example, the first view listed is the DBA_IND_COLUMNS that lists indexed column names and statistics for all indexes in the database. A second view named USER_IND_COLUMNS lists indexed column names and statistics owned by the user. A third view named ALL_IND_COLUMNS lists indexed column names and statistics either owned by the user or accessible to the user. Thus only ALL_ prefixed views are listed.

TABLE 9-2 Data dictionary views on indexes

View name	Description
ALL_IND_COLUMNS	Indexed column names and statistics
ALL_IND_EXPRESSIONS	Expressions found in function-based indexes
ALL_IND_PARTITIONS	Index partition names and statistics
ALL_IND_SUBPARTITIONS	Index subpartition names and statistics

TABLE 9-2 Data dictionary views on indexes (continued)

View name	Description
ALL_INDEXES	Index name, owner, type, and storage statistics
ALL_PART_INDEXES	High-level information on partitioned indexes, such as name, type of partitioning, storage settings, and statistics
INDEX_STATS	Statistics on leaf, branch, and deleted leaves gathered with ANALYZE ... VALIDATE STRUCTURE
V$OBJECT_USAGE	Details on index usage when you set MONITORING on for an index

The views that begin with "DBA_" have similar views that begin with "USER_" and "ALL_"

Now let's do the same thing with indexes and data dictionary views, as you saw when reading previously in this chapter, with respect to constraints. Consistency is always beneficial to understanding. The two index-related data dictionary views to examine are as follows:

- **ALL_INDEXES:** This view (and its corresponding views, DBA_INDEXES and USER_INDEXES) contains the definition of an index. The columns in this view include:
  - INDEX_NAME
  - INDEX_TYPE containing values such as IOT – TOP, LOB, FUNCTION-BASED NORMAL, NORMAL, CLUSTER (NORMAL represent a BTree index).
  - TABLE NAME is the name of the table the index is created on
  - UNIQUENESS containing UNIQUE or NONUNIQUE
- **ALL_IND_COLUMNS:** This view lists the columns associated with each index. There is one row for each column in an index. For example, a primary key made up of two columns has two rows in this view.

Complete the following steps to examine the contents of the ALL_INDEXES view:

1. If you are not already logged on, start up the Enterprise Manager console, and log on to the SQL*Plus Worksheet as CLASSMATE.
2. Type and execute this command to see the constraints created by CLASSMATE.

   ```
   SELECT INDEX_NAME, INDEX_TYPE, UNIQUENESS, TABLE_NAME
   FROM ALL_INDEXES
   WHERE OWNER='CLASSMATE' ORDER BY TABLE_NAME, INDEX_NAME;
   ```

   Figure 9-19 shows the results of the query.
3. Once again, as with constraints, the following query goes a step further than the last, returning a formatted picture of all indexes, and all columns within all indexes:

   ```
   SET WRAP OFF LINESIZE 132 PAGES 80;
   COLUMN POS FORMAT 990;
   COLUMN COMP FORMAT 90;
   COLUMN COL FORMAT A20;
   COLUMN IND FORMAT A25;
   COLUMN TAB FORMAT A25;
   ```

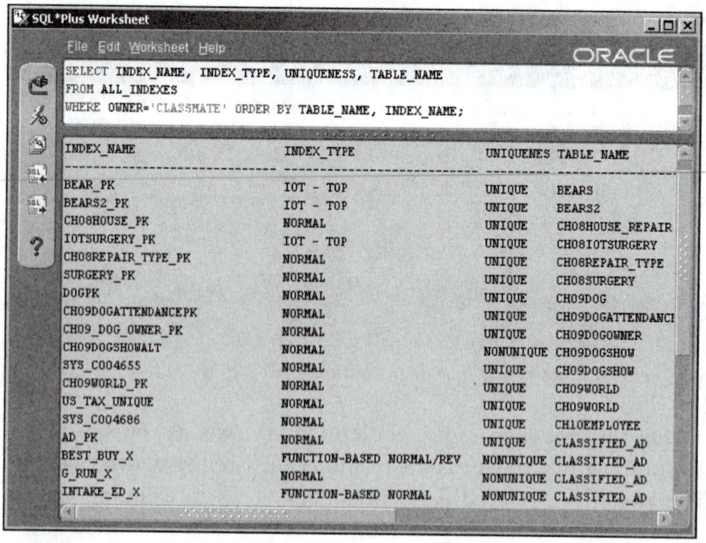

FIGURE 9-19 Indexes can be examined using the ALL_INDEXES data dictionary view

```
COLUMN TYP FORMAT A5;
COLUMN TBS FORMAT A25;
SELECT  T.TABLE_NAME "TAB",
DECODE(T.INDEX_TYPE,'NORMAL','BTREE',
'BITMAP','BITMAP',
'FUNCTION-BASED NORMAL',
'FUNCTION-BASED BTREE',
T.INDEX_TYPE) "TYP"
,T.INDEX_NAME "IND", C.COLUMN_NAME "COL"
,C.COLUMN_POSITION "POS", T.COMPRESSION "COMP"
,T.TABLESPACE_NAME "TBS"
FROM USER_INDEXES T, USER_IND_COLUMNS C
WHERE T.INDEX_NAME = C.INDEX_NAME
AND T.INDEX_TYPE NOT IN ('IOT - TOP','LOB')
ORDER BY T.TABLE_NAME, T.INDEX_NAME, C.COLUMN_POSITION;
```

Figure 9-20 shows the results of the above query.

4. Log off SQL*Plus Worksheet by clicking the **X** in the top-right corner.

You have seen how to create your own tables and constraints. The next step is indexes.

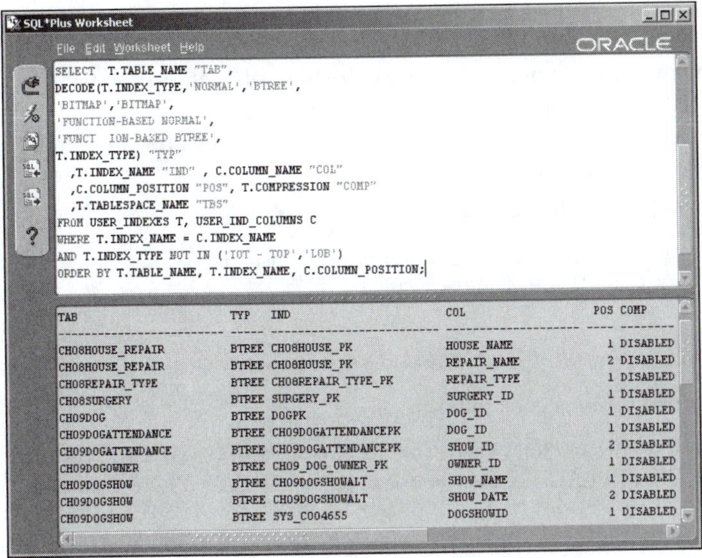

FIGURE 9-20 Reading both index views

OTHER SPECIALIZED DATABASE OBJECTS

There are numerous other specialized database objects available for creation in Oracle 10g. This section examines some of them briefly.

Views

A view is an overlay onto one or more tables. This overlay is built by creating a database object called a view, which contains no data, but only a SQL query SELECT statement, to retrieve data from underlying tables, whenever the view is queried.

Views can be categorized into two distinct forms:

- **Simple view:** Contains a query on one or more tables. Simple views can be used to narrow the focus or visible data window of a specific user from the entire table to a subset of the rows, or a subset of the columns.
- **Constraint view:** A view that can be used to insert a new row into the underlying table, as long as the row would be returned by a query, or the row exists for the view.

Common reasons for using views are as follows:

- **Security**: Create a view with a limited subset of the rows and/or columns in a table or tables and give the user permission to use the view, but not the base tables.
- **Simplicity:** Create a view that combines tables that have complex relationships so users writing queries do not need to understand the relationships.
- **Complex joins:** A small number of queries cannot be done without great difficulty unless you create a view similar to a temporary table first. For example,

you can create a view with a GROUP BY clause that summarizes data. You can join that summary data with other tables only by using a view.

- **Materialized views:** This is not a view as such since the data in the view is physically stored in the materialized view, thus the term materialized. Materialized views are a little too specialized for this book and are most commonly used in data warehouses.

Basic view creation syntax is as follows:

```
CREATE [ OR REPLACE ] [ [ NO ] FORCE ] VIEW <schema>.<viewname>
[ inline constraint ] [ out of line constraint ]
<columnname> [ , ...  ]
AS <subquery>
[ WITH { READ ONLY | CHECK OPTION [ CONSTRAINT <constraintname> ] } ];
```


The meaning of the various options is as follows:

- **OR REPLACE:** If a view exists then the existing view is overwritten. If the view exists, and the OR REPLACE option is not used, a CREATE VIEW command will fail.
- **FORCE:** Allows a view to be created, even if the subquery fails to execute properly. The view can be recompiled at a later point in time, when changes are made to allow the subquery to execute error free. For example, a view can be created to execute a table, which does not yet exist. This implies that an existing view can be changed without using an ALTER VIEW command, as opposed to a CREATE OR REPLACE VIEW command, as show in the following syntax:

  ```
  ALTER VIEW <schema>.<viewname> COMPILE;
  ```

NOTE

You can also modify and drop view constraints using the ALTER VIEW command as show in the following syntax:

- **<columnname> [, ...]:** Column names are only required if they are different from those in the subquery.
- **Constraints:** As for tables, constraints can be inline (defined for specific columns), or out of line (defined for more than one column, after all columns are listed).
- **AS <subquery>:** Can be any valid query with a few restrictions. This subquery is beyond the scope of this book.
- **READ ONLY:** Allows queries against the view only. No inserts, updates, or deletes are allowed directly through the view. The underlying tables must be accessed directly to change data.
- **WITH CHECK OPTION:** Used for constraint views to validate constraints. Constraints on underlying tables are complied within that views can actually allow changes to be made to underlying tables, by changing rows in the view. This option ensures that any changes comply with the restrictions on the view, such as if the view contains a WHERE clause filter, then any changes not matching that WHERE clause filter, will not be permitted.

Drop a view by using the DROP VIEW command as in the following syntax:

```
DROP VIEW <viewname>;
```

Follow these steps to practice with views:

1. If you are not already there, start up the Enterprise Manager console and SQL*Plus Worksheet, logging on as CLASSMATE.
2. For marketing and security purposes, you wish that all your clients, should be allowed to view all client names on your web site, but not contact details. As you can see from the DESC CLIENT command shown in Figure 9-21, there are a number of columns that need to be hidden from public view.

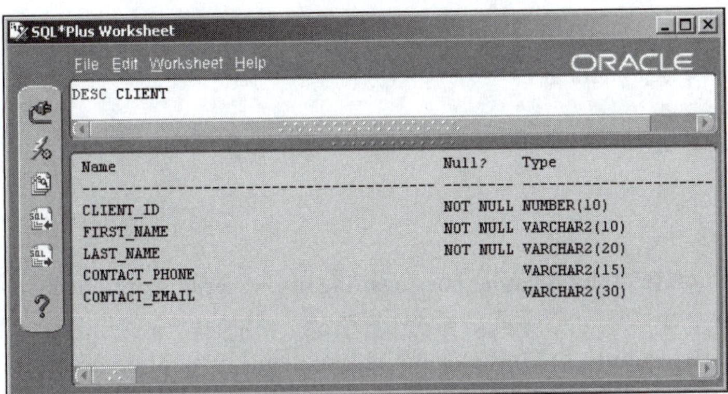

FIGURE 9-21 The CLIENT table has both names and contact information

3. Before you can create a view you may need a special privilege. Connect in SQL*Plus Worksheet as SYSTEM and execute the following command:

```
GRANT CREATE VIEW TO CLASSMATE;
```

4. Now connect back to the CLASSMATE user in SQL*Plus Worksheet. You can use a view like this one, to restrict column access. Type in this command and click the execute icon to view the results:

```
CREATE VIEW MARKETING
AS SELECT FIRST_NAME, LAST_NAME
FROM CLIENT;
```

5. If you query the contents of the view you have just created you will find that only 2 columns (not all the CLIENT table columns) are retrieved from the view. Query the view using a query like this:

```
SELECT * FROM MARKETING;
```

6. You are finished with the SQL*Plus Worksheet tool for this chapter so you can exit the tool.

Sequences

Sequences are covered briefly in Chapter 6. Here you will examine sequences in a little more depth. Sequences are used to contain automated counter values, most commonly used to retain counter values for integer-valued primary keys. The syntax for the CREATE SEQUENCE and ALTER SEQUENCE commands are shown below:

```
CREATE SEQUENCE <schema>.<sequencename>
[ START WITH <n> ]
[ INCREMENT BY <n> ]
[ [ NO ] MINVALUE <n> ]
[ [ NO ] MAXVALUE <n> ]
[ [ NO ] CYCLE ]
[ [ NO ] CACHE <n> ]
[ [ NO ] ORDER ];

ALTER SEQUENCE <schema>.<sequencename>
 [ INCREMENT BY <n> ]
[ [ NO ] MINVALUE <n> ]
[ [ NO ] MAXVALUE <n> ]
[ [ NO ] CYCLE ]
[ [ NO ] CACHE <n> ]
[ [ NO ] ORDER ];
```


All options are switched off (NO) by default. The meaning of the various options is as follows:

- **START WITH:** Determines the initial value of a sequence when it is first created, typically 0 or 1, and defaulted to 0.

NOTE

The START WITH option does not apply to the ALTER SEQUENCE command because the sequence is already created.

- **INCREMENT BY:** A sequence is by default increment by 1 but can be incremented by any value other than 0. A sequence set to increment by 0 would not be a sequence, as it would not count anything.
- **MINVALUE:** Sets a minimum value for a sequence. The default is NOMINVALUE. This is used for sequences that decrease rather than increase.
- **MAXVALUE:** Sets a maximum value for a sequence. The default is NOMAXVALUE. Be aware that a column datatype may cause an error if the number grows too large. For example, if the sequence is used to populate a column of NUMBER(5) datatype, once the sequence reaches 99999 then the next increment will cause an error.
- **CYCLE:** Causes a sequence to cycle around to its minimum when reaching its maximum for an ascending sequence, and to cycle around to its maximum when reaching its minimum for a descending sequence. The default is NOCYCLE. If you reach the maximum value on a sequence having NOCYCLE, you will get an error on the next query that tries to increment the sequence.

- **CACHE:** This option caches pre-calculated sequences into a buffer. If the database crashes then those sequence values will be lost. Unless it is absolutely imperative to maintain exact sequence counters then the default of CACHE 20 is best left as it is.
- **ORDER:** Ordering simply guarantees sequence numbers are created in precise sequential order. In other words with the NOORDER option sequence numbers can possibly be generated out of sequence sometimes, when there is excessive concurrent activity on the sequence.

The DROP SEQUENCE command is very simple:

```
DROP SEQUENCE <schema>.<sequencename>
```

As already stated, sequences are most often used to count out unique values for integer primary keys. Integer primary keys are called surrogate keys because they replace more meaningful, but less efficient columns in tables. A good example is CLIENT table, CLIENT_ID column, replacing the concatenation of the FIRST_NAME and LAST_NAME columns, as shown in Figure 9-21. Therefore the where and how of using sequences follows logically from what they are most often used for. A sequence allows for generation of unique, sequential values. Thus sequences are typically used in the types of SQL statements listed below:

- The VALUES clause of an INSERT statement.
- A subquery SELECT list contained within the VALUES clause of an INSERT statement.
- In the SET clause of an UPDATE statement.
- A query SELECT list.

Sequence values can be accessed using the CURRVAL and NEXTVAL pseudocolumns in the format as shown below:

- **<sequencename>.CURRVAL**. Returns the current value of the sequence. The sequence is not incremented by the CURRVAL pseudocolumn.
- **<sequencename>.NEXTVAL**. Returns the value of the sequence and increases the sequence one increment. Usually, sequences increase by increments of 1 each time, however, you can set a sequence to a different increment if needed.

Synonyms

A synonym can provide an alias to any object, in any schema in a database, assuming that the user has privileges to view the underlying objects. It makes an object appear as if you own it, because you do not have to use a schema prefix when querying or performing other tasks with the object. Synonyms can provide the following benefits:

- **Transparency:** A synonym masks the name of the schema owning the object. The object can even be in a remote database when you include a database link in the definition of the synonym. A database link is a direct gateway from one database to another database.
- **Simplified SQL code:** Code is simplified because schema names do not have to be included for table accesses where objects are in different schemas.
- **Easy Changes:** Moving objects to different schemas or databases in distributed environments do not require application changes since only synonyms need to be changed.

Syntax for synonyms is shown below:

```
CREATE [ OR REPLACE ] [ PUBLIC ] SYNONYM <schema>.<synonymname>
FOR [<schema>.]<objectname>[@<databaselinkname>];
DROP [ PUBLIC ] SYNONYM <schema>.<synonymname> [ FORCE ];
```

NOTE

There is no ALTER SYNONYM command. A change to a synonym requires dropping (DROP SYNONYM) and recreation (CREATE SYNONYM).

The various options can be described as follows:

- **OR REPLACE:** If a synonym exists then the existing synonym is overwritten. If the view exists, and the OR REPLACE option is not used, a CREATE SYNONYM command will fail.
- **PUBLIC:** A public synonym can make underlying objects (usually tables, but not always) available to all database users, regardless of whether the users are DBAs, power users, or even end users. Powerful privileges are required to create synonyms, particularly PUBLIC synonyms because they allow access to anything, for anyone. You would need to execute these two commands for the CLASSMATE user, to allow your user to create synonyms:

  ```
  GRANT CREATE SYNONYM TO CLASSMATE;
  GRANT CREATE PUBLIC SYNONYM TO CLASSMATE;
  ```
- **<objectname>:** The use of the term objectname implies that synonyms can be created for any directly accessible object in a schema, including tables, views, materialized views, and so on. An index would not be included in this list because indexes are not directly accessible by a query because a query always selects from tables. The optimizer can choose to read an index if appropriate but this is not under user control.
- **<@databaselinkname>:** A database link is a special type of link created between different databases, even databases residing on different computers located at far flung locations of the globe, and even databases from different manufacturers, such as Oracle, Sybase, DB2, and SQL-Server.

Data Dictionary Information on Views, Sequences, and Synonyms

Three views are of interest:

- **ALL_VIEWS:** This view (and its corresponding views, DBA_VIEWS and USER_VIEWS) contains the definition of a view. The columns in this view include:
  - VIEW_NAME
  - TEXT containing the text of the subquery
- **ALL_SEQUENCES:** This view (and its corresponding views, DBA_SEQUENCES and USER_SEQUENCES) contains the definition of a sequence.
- **ALL_SYNONYMS:** This view (and its corresponding views, DBA_ SYNONYMS and USER_ SYNONYMS) contains the definition of a synonym.

Chapter Summary

- Integrity constraints can be enforced using declared constraints, triggers, or application programming.
- A PRIMARY KEY constraint can cover one or more columns in a table and requires unique, non-null values in the primary key columns for every row.
- A UNIQUE constraint requires unique values or null values in the unique key columns.
- A FOREIGN KEY constraint enforces the integrity of a relationship between a parent and child table, and is defined on the child table.
- A NOT NULL constraint must always be on a single column and prevents null values in the column.
- A CHECK constraint compares the column to a list or range of values.
- Constraints can be created with the CREATE TABLE and the ALTER TABLE commands.
- Use the ALTER TABLE statement to rename or drop a constraint.
- Use the ALTER TABLE statement to change the constraint state.
- To remove the NOT NULL constraint, use ALTER TABLE MODIFY (column...) statement.
- The PRIMARY KEY constraint is crucial to a relational database design.
- When a PRIMARY KEY constraint is created (and not disabled), a unique index is created to help enforce the constraint.
- Dropping or disabling a PRIMARY KEY constraint causes the underlying index to be dropped.
- Use the NOVALIDATE constraint state when you do not want existing rows to be checked for compliance with a constraint.
- The default states of a constraint are ENABLE, VALIDATE, INITIALLY IMMEDIATE, NOT DEFERRABLE, and NORELY.
- USING INDEX applies only to PRIMARY KEY and UNIQUE key constraints in which you want to use a preexisting index or you want to specify storage settings for the index to be created when the constraint is created.
- EXCEPTIONS can be used after creating a table (usually called EXCEPTIONS) to hold the ROWID of rows that violate a constraint.
- Specifying DEFERRABLE INITIALLY IMMEDIATE specifies that the constraint is enforced after every statement (immediate), but can be deferred later if the user either alters the constraint or alters his session.
- SET CONSTRAINTS ALL DEFERRED switches all deferrable constraints to a deferred state during your current session only.
- A FOREIGN KEY constraint can define the behavior of the database when a parent row is deleted using the ON DELETE CASCADE or ON DELETE SET NULL parameter.
- The CHECK constraint can look for a specified list of values or other simple expressions.
- The ALL_CONSTRAINTS data dictionary view contains details on all five types of constraints.

- The ALL_CONS_COLUMNS data dictionary view lists all the columns involved in a constraint.
- An index is a structure that holds data and has its own storage in the database.
- An index contains columns and ROWIDs.
- A normal index stores rows in order by the index key values.
- An index can be based on a table's primary key or any combination and order of columns.
- An index can have up to 32 columns (30 for bitmap indexes).
- There are a number of types of indexes: BTree, bitmap, local partitioned, global partitioned, reverse key, function-based, BTree cluster, hash cluster, and domain indexes.
- Indexes are created using the CREATE INDEX command.
- BTree is Oracle's default type of index.
- BTree is a more complex variation of a binary tree structure.
- BTree has a root, branches, and leaves.
- Searches of BTrees take longer as more levels are added to the tree.
- BTree indexes are efficient even for very large amounts of data.
- Create a unique index with the CREATE UNIQUE INDEX command.
- Bitmap indexes consume far less storage space than BTree indexes.
- Bitmap indexes are best for columns with low cardinality and complex WHERE clauses involving AND, OR, or NOT, but not involving "greater than" or "less than."
- Bitmap indexes cannot be unique and are best suited for data warehouse systems.
- Local partitioned indexes mirror the partitioning of the partitioned table.
- The USER_IND_PARTITIONS data dictionary view lists partitioned indexes.
- Global partitioned indexes can be created on partitioned or nonpartitioned tables.
- Global partitioned indexes can only be RANGE partitioned.
- Global nonpartitioned indexes can be used on partitioned tables.
- Reverse key indexes contain columns that have been stored in reverse byte order.
- Reverse key indexes are usually only for Real Application Cluster systems.
- Function-based indexes substitute a function or expression for a column.
- You must run ANALYZE or DBMS_STATS to give the optimizer information about an index.
- The ALTER INDEX command allows you to make changes to the index.
- The ALTER INDEX REBUILD command lets you change storage parameters, change a reverse index to a normal index, and move the index to a different tablespace.
- The ALTER INDEX ... COALESCE command performs faster than REBUILD.
- REBUILD can consolidate free space, but not release it, whereas COALESCE consolidates and releases free space, if you specify the DEALLOCATE UNUSED parameter.
- You can only rebuild or coalesce one partition at a time of a partitioned index.

- Views can be used to restrict access to both columns and rows in one or more underlying tables.
- Sequences can be used to store automated sequential counters for surrogate primary key columns in tables.
- Synonyms allow easy access to database objects across schemas and databases.

Review Questions

1. In which of these situations would you enforce integrity constraints using database triggers?
   a. Your database contains many parent/child relationships.
   b. Your integrity constraints are complex and involve multiple table lookups.
   c. Your tables often use compound primary keys.
   d. All of the integrity constraints of your database system are enforced at the application level.
2. List the five types of integrity constraints.
3. Define an inline integrity constraint. When is it used?
4. Which constraints can be compound constraints?
5. You cannot specify storage information when creating a table that has integrity constraints. True or False?
6. The ________________ constraint allows nulls but not duplicate values, whereas the constraint does not allow nulls or duplicate values.
7. The following statement fails with a syntax error. True or False?

```
ALTER TABLE CLIENT
MODIFY(FIRST_NAME NOT NULLS);
```

8. A FOREIGN KEY constraint can reference columns in a UNIQUE key constraint. True or False?
9. The ________________ command changes a deferrable constraint to deferred state during your session.
10. Write the SQL statement to change the name of the KEY1 constraint (a PRIMARY KEY constraint) to PK1 on the MOUSETRAP table.
11. List the letters and what they stand for in the CONSTRAINT_TYPE column of the USER_CONSTRAINTS view.
12. The ________________ state tells Oracle 10*g* to check the constraint only when a transaction commits.
13. When you create an index, it is automatically stored in the same tablespace as the table. True or False?
14. What is the difference between a unique index that supports a PRIMARY KEY constraint and a unique index that supports a UNIQUE key constraint?
15. What is the difference between specifying REVERSE on an index and specifying DESC on an indexed column?

16. To create a BTree index named BOOK_X, the CREATE command should start like which of the following:
    a. CREATE INDEX BOOK_X
    b. CREATE BTREE INDEX BOOK_X
    c. CREATE UNIQUE BTREE INDEX BOOK_X
    d. CREATE BITMAP INDEX BOOK_X
17. Look at the following command:

```
1 CREATE INDEX CUST_IX ON
2 (CUST_ID ASC, BIRTH_DATE, AGE-18)
3 TABLESPACE USERS
4 LOGGING ONLINE;
```


    Which line has an error?
    a. Line 1
    b. Line 2
    c. Line 3
    d. Line 4
18. You have a table that has been loaded into the database in order by the CREATE_DATE column. You plan to create an index on the CREATE_DATE column. What parameter helps save time in creating the index?
    a. REVERSE
    b. NOSORT
    c. BITMAP
    d. NOCOMPRESS
19. You have a BTree index on a table with 1000 rows. How many levels will the index have?
    a. Four levels
    b. Ten levels
    c. Depends on the indexed columns
    d. A maximum of twenty levels
    e. None of the above
20. The CUSTOMER table has 100,000 rows and the FAVORITE_COLOR column has low cardinality. Queries usually look for an exact match on the color value. What kind of index would you use for the FAVORITE_COLOR column?
    a. Partitioned
    b. BTree
    c. Bitmap
    d. Function-based
21. A ________________ partitioned index is partitioned as a table is, but a ________________ partitioned index is partitioned differently to the table.

22. A global index must be partitioned. True or False?

23. The ________________ data dictionary view can tell you the name of a column in an index.

Exam Review Questions — Oracle Database 10*g*: Administration I (#1Z0-042)

1. The following constraints CANNOT be defined inline. Choose three.
   a. A compound primary key defined with ALTER TABLE
   b. A CHECK constraint on an existing column
   c. A NOT NULL constraint on an existing column
   d. A FOREIGN KEY constraint on an existing column
   e. A single column primary key defined with CREATE TABLE

2. The following statements were executed. What are the results?

```
CREATE TABLE MYCUSTOMER
  (ID NUMBER PRIMARY KEY, FIRSTNAME VARCHAR2(10),
   AGE NUMBER(4,0) CONSTRAINT CHK_AGE
   CHECK (AGE > 18) DEFERRABLE);
SET CONSTRAINT ALL DEFERRED;
INSERT INTO MYCUSTOMER VALUES (1, 'JANE',12);
INSERT INTO MYCUSTOMER VALUES (1, 'MARY',21);
UPDATE MYCUSTOMER SET AGE = 19 WHERE FIRSTNAME = 'JANE';
COMMIT;
```

   a. The first insert succeeds, the second insert succeeds, the update succeeds, and the commit fails.
   b. The first insert succeeds, the second insert fails, the update succeeds, and the commit fails.
   c. The first insert succeeds, the second insert succeeds, the update succeeds, and the commit succeeds.
   d. The first insert succeeds, the second insert fails, the update succeeds, and the commit succeeds.

3. The DEFERRABLE INITIALLY IMMEDIATE clause has what effect?
   a. Existing rows are validated, and then the constraint is deferred.
   b. Existing rows are not validated, and then the constraint is immediate.
   c. Existing rows are validated, and then the constraint is immediate unless later deferred.
   d. Existing rows are not validated, and the constraint is deferred unless later made immediate.
   e. None of the above.

4. Which command would you use to defer a constraint? Choose two.
   a. ALTER CONSTRAINT
   b. DEFER CONSTRAINT
   c. DROP CONSTRAINT
   d. ALTER TABLE
   e. SET CONSTRAINT
5. What does it mean when a constraint is created with ON DELETE SET NULL?
   a. All parent rows are deleted when a child row is deleted.
   b. All child rows have the column(s) set to null when a parent row is deleted.
   c. All child rows are deleted when a parent row is deleted.
   d. All parent rows have the column(s) set to null when a child is deleted.
   e. A single column primary key is defined with CREATE TABLE.
6. Which query displays whether a constraint is deferred?
   a. SELECT CONSTRAINT_NAME, STATUS FROM USER_CONS_COLUMNS;
   b. SELECT CONSTRAINT_NAME, DEFERRED FROM DBA_CONSTRAINTS;
   c. SELECT CONSTRAINT_NAME, DEFERRABLE FROM ALL_CONSTRAINTS;
   d. SELECT CONSTRAINT_NAME, DEFERRED FROM DBA_TAB_CONSTRAINTS;
7. You query USER_CONSTRAINTS and find three constraints on the CLIENT table with the constraint type 'R'. What can be said about these constraints? Choose two.
   a. Each constraint contains a different set or order of columns.
   b. This is invalid because there cannot be three constraints of this type on one table.
   c. All three constraints are FOREIGN KEY constraints.
   d. When enabled, the constraints created unique indexes.
8. What type of constraint can enforce this business rule: "An employee's salary may be no more than 10 percent higher than his boss's salary." Choose two.
   a. CHECK constraint
   b. NOT NULL constraint
   c. Application code
   d. Database trigger
9. Examine the following statement. Which statements are true regarding the constraint? Choose two.

```
ALTER TABLE FRIENDS
ADD CONSTRAINT FRIEND_ALT_KEY UNIQUE (PHONENO)
USING INDEX PHONEX;
```

   a. The statement fails if the PHONEX index does not exist.
   b. A new index named PHONEX is created.
   c. The PHONENO column cannot contain null values.
   d. Existing rows are validated immediately.

10. Examine the following statement. What might cause the statement to fail? Choose the best two.

```
ALTER TABLE CLIENT
ADD CONSTRAINT CLIENT_CO FOREIGN KEY (COMPANY_ID)
REFERENCES COMPANY (COMPANY_ID) ON DELETE CASCADE;
```

   a. The COMPANY table has a UNIQUE key constraint on COMPANY_ID.
   b. The COMPANY table has no rows.
   c. The CLIENT table has duplicate values in the COMPANY_ID column.
   d. The COMPANY table has no PRIMARY KEY constraint.
   e. The CLIENT table has all null values in the COMPANY_ID column.

11. Your index, called ACCT.PROCESS_X, has just been created. You run a query and look at the plan that the optimizer will use to execute the query. You expected to see the index used in the plan. Which command will help you get the results you want?

   a. BEGIN
      DBMS_STATS.GATHER_SCHEMA_STATS
      (ownname=>'ACCT',cascade=>TRUE);
      END;
   b. ANALYZE INDEX PROCESS_X VALIDATE STRUCTURE;
   c. GRANT REWRITE QUERY TO ACCT;
   d. ALTER SESSION SET OPTIMIZER_MODE='COST';

12. You can gather usage information by setting the index for MONITORING USAGE. Which view shows the details of the monitoring activity?
   a. DBA_INDEXES
   b. DBA_INDEX_STATS
   c. V$OBJECT_USAGE
   d. V$INDEX_USAGE

13. When would it be wise to partition an index?
   a. When the underlying table is partitioned on the same columns as the index
   b. When the table contains low cardinality of the column to be indexed
   c. When a new application in a remote system needs the index
   d. When the table has a primary key on the columns to be indexed

14. Which query shows you all your indexes in the USERS tablespace?
    a. SELECT INDEX_NAME FROM DBA_INDEXES
       WHERE TABLESPACE_NAME = 'USERS';
    b. SELECT INDEX_NAME FROM INDEX_STATS
       WHERE TABLESPACE_NAME = 'USERS'
    c. SELECT INDEX_NAME FROM USER_INDEXES
       WHERE TABLESPACE_NAME = 'USERS';
    d. SELECT INDEX_NAME FROM USER_IND_COLUMNS
       WHERE TABLESPACE_NAME = 'USERS';
15. Which of these tasks can be performed with the ALTER INDEX ... REBUILD command? Choose three.
    a. Changing a reverse index to a normal index
    b. Moving an index to another tablespace
    c. Adding a new column to the index
    d. Moving an index partition to another tablespace
    e. Changing a bitmap index to a normal index
16. Examine the command below.

    ```
    ALTER INDEX BOOK_IX COALESCE ONLINE DEALLOCATE UNUSED;
    ```

    What effect does this command have?
    a. Rebuilds the branches and leaves of the BTree
    b. Releases free space in leaves of the BTree
    c. Keeps users from using the index in the future
    d. Combines two partitions into one
    e. None of the above
17. The LIBRARY_FINES table needs an index. The majority of queries search for a range of the sum of FINE_TOTAL plus INTEREST_TOTAL. The table has 100,000 rows and is in the USERS tablespace. What type of index should you create?
    a. Reverse key index
    b. Bitmap index
    c. Function-based index
    d. BTree index
    e. None of the above

18. Examine the following command.

```
CREATE INDEX CLASSMATE.READER_IX
ON CLASSMATE.LIBRARYBOOK(READER_ID, BOOK_ID)
STORAGE (INITIAL 10M NEXT 5K)
TABLESPACE USERS NOLOGGING PARALLEL 3;
```

What type of index is created?

a. Partitioned index

b. Bitmap index

c. Function-based index

d. Composite index

e. None of the above

19. Which of these situations calls for a bitmap index?

a. Several columns with low cardinality need indexing; the table has 50 million rows and infrequent updates.

b. The columns have low cardinality; the table has 25 million rows and has frequent OLTP activity and few queries.

c. The low cardinality columns need a unique index; the table has 50 million rows and is rarely updated.

d. Several high cardinality columns need indexing; the table has 8 million rows and is updated frequently.

e. None of the above

20. You need to create an index on the CUST_ID and ORDER_NO columns in the BILLING table. The table is in CUST_ID order. The index must be unique and have 5 M extents. Which of these commands will create the index?

a. CREATE INDEX BILLX ON BILLING
(CUST_ID, ORDER_NO)
UNIQUE NOSORT STORAGE (INITIAL 5M NEXT 5M)
TABLESPACE INDEXES ONLINE;

b. CREATE INDEX BILLX ON BILLING
(CUST_ID, ORDER_NO)
UNIQUE STORAGE (INITIAL 5M NEXT 5M)
TABLESPACE INDEXES;

c. CREATE UNIQUE INDEX BILLX ON BILLING
(CUST_ID, ORDER_NO)
ONLINE STORAGE (INITIAL 5M NEXT 5M)
TABLESPACE INDEXES;

d. CREATE BITMAP INDEX BILLX ON BILLING
(CUST_ID, ORDER_NO)
UNIQUE STORAGE (INITIAL 5M NEXT 5M)

TABLESPACE INDEXES;

21. Examine the following command.

```
CREATE INDEX LOANX
ON ACTMGR.MORTGAGELOAN (LOANID, CUST_CODE)
GLOBAL PARTITION BY RANGE (LOANID, CUST_CODE)
(PARTITION L1 VALUES LESS THAN ('L005',001)
TABLESPACE USERS STORAGE (INITIAL 100M NEXT 10M),
PARTITION L2 VALUES LESS THAN ('L499',025),
PARTITION L3 VALUES LESS THAN (MAXVALUE, MAXVALUE));
```

Which of the following statements are true? Choose two.

a. All the partitions are located in the USERS tablespace.

b. A row with LOANID='M001' and CUST_CODE=004 will fall into the L3 partition.

c. The MORTGAGELOAN table is partitioned.

d. The L2 partition will be located in ACCTMGR's default tablespace.

22. Examine the following command.

```
ALTER INDEX LOANX
COALESCE PARTITION L1
RELEASE UNUSED KEEP 10K ONLINE;
```

Which of the following statements are true? Choose two.

a. The index's leaves and branches are rebalanced to improve performance.

b. Unused space is released except 10 K above the HWM.

c. Users can query the table while the command runs.

d. The index's leaves are adjusted to release unused space.

e. The partition's leaves are adjusted to release unused space.

Hands-on Assignments

The assignments in this section require that you have completed all the practices in the chapter and that you also run another script to add rows to some of the tables. Follow these steps to prepare for the hands-on assignments.

1. Start up the SQL*Plus Worksheet. In Windows, on the Taskbar, click **Start/All Programs/Oracle ... /Application Development/SQLPlus Worksheet**. In UNIX, type **oemapp worksheet** on a command line. A logon window displays.
2. Type **CLASSMATE** in the Username box, **CLASSPASS** in the Password box, and **ORACLASS** in the Service box. Leave the connection type as "Normal". Click **OK** to continue.
3. If you have not completed all the chapter exercises, on the SQL*Plus Worksheet menu, click **Worksheet/Run Local Script**. Then navigate to the **Data\Chapter9** directory and select the file named **ch09catchup.sql** on your student disk. Click **Open** to execute the script.
4. Run a script to add data to some of the tables. On the SQL*Plus Worksheet menu, click **Worksheet/Run Local Script**. Then navigate to the **Data\Chapter09** directory, and select the file named **ch09_handson_setup.sql** on your student disk. Click **Open** to execute the script.
5. Remain logged on for the next hands-on assignments.

6. Delete the two invalid rows in the CH09WORLD table and enable the US_TAX_UNIQUE constraint. Assume that the row with lower value in the PERSON_ID column is the correct one for each duplicate record. Save the SQL you use in your **Solutions\Chapter09** directory in a file named **h0901.sql**.
7. Add a PRIMARY KEY constraint to the CH09DOG table. The primary key column is named DOG_ID. The constraint should be nondeferrable, immediate, and enabled. Include the parameters for the constraint states even though they are the default settings and not technically required. Save the SQL you use in your **Solutions\Chapter09** directory in a file named **h0902.sql**.
8. Add a PRIMARY KEY constraint to the CH09DOGSHOW table. The primary key column is named DOGSHOWID. The constraint should be named by the system, and the index should reside in the USER_LOCAL tablespace. Save the SQL you use in your **Solutions\Chapter09** directory in a file named **h0903.sql**. (*Hint*: Omit the word CONSTRAINT and the constraint name to allow the system to name the constraint.)

9. Create a new table that keeps track of which dogs attend which dog shows. The table should contain the foreign key columns needed to reference the CH09DOG and CH09DOGSHOW tables, plus columns for the placement category (such as Best of Show or Best of Breed) and rank (first through fourth place). The table name should be CH09DOGATTENDANCE. Include a primary key that is made up of the columns from the two foreign keys. Include a CHECK constraint on the rank column so that only "First", "Second", "Third", "Fourth," and a null value are allowed. Save the SQL you use in your **Solutions\Chapter09** directory in a file named **h0904.sql**.
10. Add a new column to the CH09DOGSHOW table. The column tracks the show date. It does not allow null values; however, because there is one record in the dog show table, do not add a NOT NULL constraint when you create the column.

    Next, update the row with a show date. Finally, add the NOT NULL constraint to the new column. Save the SQL you use in your **Solutions\Chapter09** directory in a file named **h0905.sql**.
11. Create a unique key on the CH09DOGSHOW table. The key should contain the SHOW_NAME and the SHOW_DATE columns. The constraint is deferrable but currently set to immediate state. Save the SQL you use in your **Solutions\ Chapter09** directory in a file named **h0906.sql**.
12. Modify a constraint so that when a parent row in the CH09DOGSHOW table is deleted, all the related rows in the CH09DOGATTENDANCE table are deleted. Save the SQL you use in your **Solutions\Chapter09** directory in a file named **h0907.sql**.
13. Change the CH09DOGOWNER table so that these business rules are enforced:
    - A dog owner must have at least one year of experience.
    - If the actual number of years experience is not known, the dog owner is assigned one year of experience until the actual number is known.

    Save the SQL you use in your **Solutions\Chapter09** directory in a file named **h0908.sql**.

14. Add new columns to the CH09DOGSHOW table as follows:
    - The location of the dog show needs to be tracked, and you need one column for the city and one column for the state of the show. The state must be a two-character state code and the only values allowed are: NV, CA, OR, WA, and TX. These can be left out if the location has not been determined yet.
    - The dog show must have a sponsor, such as PETCO or Purina. Allow up to 50 characters in the sponsor name. Sometimes, a tentative sponsor is recorded, so add another column that is a "T" for "tentative" or an "F" for "finalized." This column must be filled in, and should default to "T" if no value is specified.

    Save the SQL you use in your **Solutions\Chapter09** directory in a file named **h0909.sql**.
15. Modify the CH09WORLD table so that if a person is in the United States, he or she must have a US tax ID and if the person is in any other country, he or she must NOT have a US tax ID. Preexisting data should not be checked. Save the SQL you use in your **Solutions\Chapter09** directory in a file named **h0910.sql**. (*Hint*: Use "IS NULL" and "IS NOT NULL" to check a column for null or non-null values in a CHECK constraint.)

16. Write and execute the command to create a BTree index on the INTAKE_EDITOR_ID of the CLASSIFIED_AD table. The index should contain a 5 K first extent and 2 K subsequent extents. Create the index with a compressed key and in descending order. Save your SQL in a file named **ho0911.sql** in the **Solutions\Chapter09** directory on your student disk.
17. Move the index you created in Hands-on Assignment 1 to the USERS tablespace. In the same command, change the PCTINCREASE to 0 and decompress the key. Save your SQL in a file named **ho0912.sql** in the **Solutions\Chapter09** directory on your student disk.
18. The newspaper has added a special deal for the classified ads online. For a small fee, the words "BEST BUY" are added to the text of the ad, and these ads are listed on a special "Best Buys" page online. The online classified section has a query to bring up all the ads that qualify, but it runs very slowly. Here is the query.

    ```
    SELECT AD_TEXT, AD_NO
    FROM CLASSMATE.CLASSIFIED_AD
    WHERE INSTR(AD_TEXT,'BEST BUY') > 0
    ```

    Create an index that will be used for this query and, therefore, speed up its response time. Save your SQL in a file named **ho0913.sql** in the **Solutions\Chapter09** directory on your student disk.
19. There is an object table named CUSTOMER_ADDRESS in the CLASSMATE database. Add a nonunique index that sorts on the CUSTOMER_ID attribute followed by the STATE and CITY attributes of the table. Save your SQL in a file named **ho0914.sql** in the **Solutions\Chapter09** directory on your student disk.
20. Modify the index created in Hands-on Assignment 4. Consolidate the index leaf blocks and release unused free space. Save your SQL in a file named **ho0915.sql** in the **Solutions\Chapter09** directory on your student disk.
21. The CUSTOMER table has a column called DISCOUNT_PERCENT that gives a number for the usual discount each customer receives (if any). You have found that the column has only ten distinct values. The table has 20,000 rows and grows slowly with few updates to the discount for each customer. Create a bitmap index on the column. Save your SQL in a file named **ho0916.sql** in the **Solutions\Chapter09** directory on your student disk.

22. Modify the bitmap index created in Hands-on Assignment 6 by changing the PCTFREE to 0 and moving the index to the USER_AUTO tablespace. The DDL command should run without logging. Save your SQL in a file named **ho0917.sql** in the **Solutions\Chapter09** directory on your student disk.
23. Modify the index you created in Hands-on Assignment 3, changing it to a reverse key index. Then coalesce the index. Save your SQL in a file named **ho0918.sql** in the **Solutions\Chapter09** directory on your student disk.
24. Write a query on the data dictionary views that shows these components of all the indexes owned by CLASSMATE:
    - Index name
    - Index type
    - Table name
    - Expression (if any)
    - Column name

    Save your SQL in a file named **ho0919.sql** in the **Solutions\Chapter09** directory on your student disk.
25. Write a query that generates a set of ANALYZE INDEX commands for every index owned by CLASSMATE. For a hint on getting started, look at this example of a query that generates a sentence made up of a combination of literals and column values.

```
SELECT 'I own the ' || UT.TABLE_NAME ||
' table in the '||UT.TABLESPACE_NAME ||
' tablespace.'
FROM USER_TABLES UT
ORDER BY TABLE_NAME;
```

    You will find the previous query in the **Data\Chapter09** data directory on your student disk in a file named **ho0920setup.sql**. Run the previous query to see how it works. Then create your own query. Save your SQL in a file named **ho0920.sql** in the **Solutions\Chapter09** directory on your student disk.

Case Projects

1. The Global Globe newspaper database now has several tables. Your mission for this project is to add integrity constraints to all the tables you have created to enforce the business rules. Documentation is important in any project, so in addition to the SQL commands, your boss wants a document describing the connection between the constraints you create and the business rules. The business rules are listed in a text file named **case0901.txt** in the **Data\Chapter09** directory. Open the file, complete the entries in the text file as you do your SQL work, and save the modified text file in a file named **case0901.txt** in the **Solutions\Chapter09** directory on your student disk. Save all your SQL work in a file named **case0901.sql** in the **Solutions\Chapter09** directory on your student disk. (*Hint*: to create a NOT NULL constraint with a name, use the following syntax:

```
ALTER TABLE <tablename>
MODIFY(<columnname> CONSTRAINT <constraintname> NOT NULL);
```

2. Global Globe has implemented an online classified ad web site. You have found that nearly every query on the CLASSIFIED_AD table for the online ad web site contains this clause in the WHERE clause:

```
AND PLACED_DATE > SYSDATE-7
```

You are sure an index would help, especially if the index could quickly eliminate older ads that have expired and are just stored for historical purposes. Create an index that is partitioned on the PLACED_DATE column. Decide what type of partitioning is best for the column. Although your test database has only a few sample rows, assume that the table contains 50,000 rows and the table grows by 2500 rows per month. The oldest want ad was run on March 15, 1998. The PLACED_DATE column is 7 bytes long. Estimate the INITIAL size to handle the current number of rows and NEXT extent size to handle one month of data. Make three partitions. Store two of the partitions in the USERS tablespace and one in the USER_AUTO tablespace. Use the same storage parameters for all three. Run the command to create the index. Save your SQL command and also include a few sentences on why you chose a particular type of partitioning for the index. (*Hint*: You must use the TO_DATE function to specify a literal date in a partition range.) Save your SQL and comments in a file named **case0902.sql** in the **Solutions\Chapter09** directory on your student disk.

CHAPTER **10**

BASIC DATA MANAGEMENT

LEARNING OBJECTIVES

In this chapter, you will:

- Discover changing data using Data Manipulation (DML) statements
- Describe syntax for, and use the INSERT statement
- Describe syntax for, and use the UPDATE statement
- Describe syntax for, and use the DELETE statement
- Describe and learn how to control transactions

INTRODUCTION

In this chapter, you discover how data is changed in an Oracle database. You start by studying the use and syntax of the INSERT, UPDATE, and DELETE Data Manipulation Language (DML) statements. Following that, you explore transaction control. Transactional control allows management of multiple sequential, sometimes dependent, DML statements.

INTRODUCTION TO BASIC DATA MANAGEMENT

As you already know from previous chapters, a table is the most basic storage unit for data in Oracle 10*g*. Tables are two-dimensional storage structures comprised of rows and columns. A column stores individual values such as a person's name. A row stores all the things about a person in separate columns such as their name, address, and phone number. A table stores multiple entries of different people, all those people having names, addresses, and phone numbers.

You have to have some way to add new people, change existing people, and remove people no longer required in your database. Oracle SQL statements to accomplish these tasks are the INSERT, UPDATE, and DELETE statements. Let's say you have a table called CH10EMPLOYEE as shown in Figure 10-1.

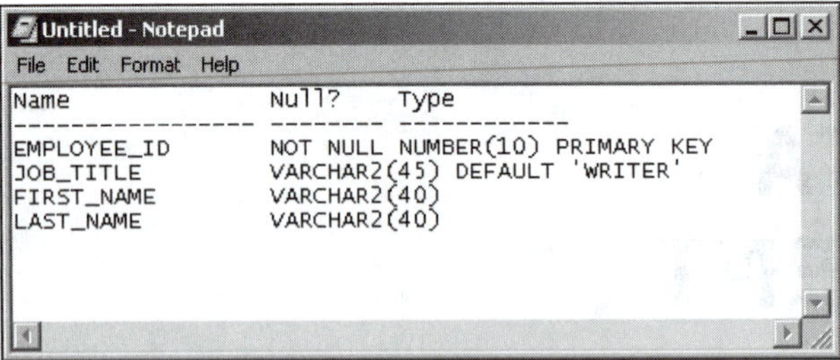

FIGURE 10-1 The CH10EMPLOYEE table

- The INSERT statement is used to add new rows to tables, as shown in Figure 10-2.

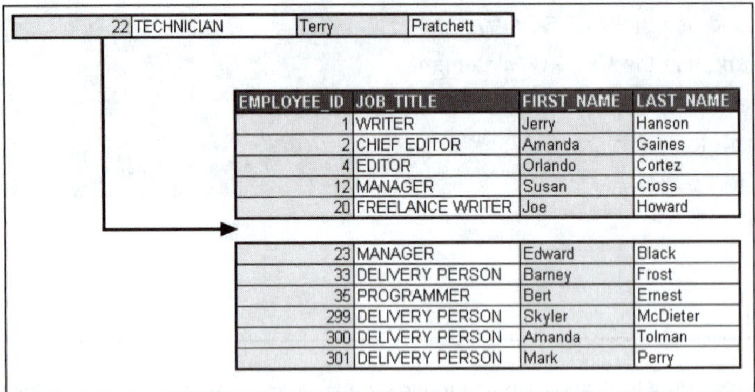

FIGURE 10-2 INSERT a new row into a table

- The UPDATE statement allows existing rows in a table to be changed, as shown in Figure 10-3.
- The DELETE statement lets you remove existing rows from tables as shown in Figure 10-4.

In a later chapter you will look into more advanced methods of data management including Programming Language for SQL (PL/SQL), plus various utilities. The utilities are Data Pump import and export utilities plus the SQL Loader utility. The export and import utilities can easily dump and import data from one or more tables, even entire schemas or a full database, including both data in tables and the table structures themselves. The SQL Loader utility, often called SQL*Loader or SQL*Ldr, can be used to rapidly load entire data sets into tables. SQL Loader can often be even as much as thousands of times faster in bulk loading data into a database, in comparison to executing a multitude of individually coded INSERT statements. This chapter will cover only DML, as DML is quite an extensive topic in itself.

WRITER

EMPLOYEE_ID	JOB_TITLE	FIRST_NAME	LAST_NAME
1	WRITER	Jerry	Hanson
2	CHIEF EDITOR	Amanda	Gaines
4	EDITOR	Orlando	Cortez
12	MANAGER	Susan	Cross
20	FREELANCE WRITER	Joe	Howard
22	TECHNICIAN	Terry	Pratchett
23	MANAGER	Edward	Black
33	DELIVERY PERSON	Barney	Frost
35	PROGRAMMER	Bert	Ernest
299	DELIVERY PERSON	Skyler	McDieter
300	DELIVERY PERSON	Amanda	Tolman
301	DELIVERY PERSON	Mark	Perry

FIGURE 10-3 Update an existing row in a table

EMPLOYEE_ID	JOB_TITLE	FIRST_NAME	LAST_NAME
1	WRITER	Jerry	Hanson
2	CHIEF EDITOR	Amanda	Gaines
4	EDITOR	Orlando	Cortez
12	MANAGER	Susan	Cross
20	FREELANCE WRITER	Joe	Howard
~~22~~	~~WRITER~~	~~Terry~~	~~Pratchett~~
23	MANAGER	Edward	Black
33	DELIVERY PERSON	Barney	Frost
35	PROGRAMMER	Bert	Ernest
299	DELIVERY PERSON	Skyler	McDieter
300	DELIVERY PERSON	Amanda	Tolman
301	DELIVERY PERSON	Mark	Perry

FIGURE 10-4 Delete an existing row from a table

So far in this book you have used the Data Definition Language (DDL) statement to create tables, create indexes, and apply constraints. DML statements are used to change the data content of tables (the row and column values). There are some essential differences between DDL and DML statements you need to understand before going any further.

What are DML and DDL?

You can change data in tables using Data Manipulation Language (DML) statements. As you already know the INSERT, UPDATE, and DELETE statements change data in a database and are thus all DML statements. DML statements change data in tables. On the other hand Data Definition Language (DDL) statements are used to change data structures such as tables—the objects containing data changed by DML statements. There is one significant difference between DML and DDL statements. DDL statements make permanent changes to the database and cannot be undone (rolled back). DML statements on the other hand make changes to a database, which can be undone, as long as they are not yet permanently stored in the database. Database changes are permanently stored to a database using the COMMIT statement and undone using a ROLLBACK statement. The COMMIT and ROLLBACK statements will be introduced shortly.

As stated in the previous paragraph DDL statements cannot be rolled back. This is because DDL statements inherently (automatically) execute a COMMIT statement, whether you choose to permanently commit data or not. For instance, if you were to execute three INSERT statements and no COMMIT statement those new rows can be removed from the database using a ROLLBACK statement. If a DDL statement such as a CREATE TABLE statement immediately follows the three INSERT statements, the CREATE TABLE statement will execute an automatic COMMIT statement, permanently storing changes to the database, made by the three INSERT statements. In other words, the new rows added cannot be removed from the database using a ROLLBACK statement because the CREATE TABLE DDL statement forced the execution of a COMMIT statement. This leads to the concept of a transaction.

What is a Transaction?

In commercial parlance, a transaction is a business agreement or exchange, usually involving money changing hands. In relational database terminology a transaction is a sequence of one or more DML statement-induced database changes, not yet permanently committed to that database. What is that in plain language? To use the three previous INSERT statements example, the three INSERT statements make up a single transaction, assuming no COMMIT or ROLLBACK statements are executed. So if the three INSERT statements are executed, then the first INSERT statement begins the transaction, and the transaction is not completed by either the second or third INSERT statements. The transaction is only complete when a COMMIT or ROLLBACK statement is executed or the CREATE TABLE DDL statement executes.

Let's take a brief look at the COMMIT and ROLLBACK statements.

COMMIT and ROLLBACK

It is important for database administrators to understand how COMMIT and ROLLBACK statements function. What happens to a database when DML changes are made? In other words what do COMMIT and ROLLBACK actually do when they are executed?

Changes to a database before COMMIT or ROLLBACK (shown in Figure 10-5):

1. Redo log entries representing changes to the database are written to redo logs.

NOTE

Redo log entries are written before anything else in order to maintain database recoverability.

2. Rollback entries are created for database changes to allow for possible future rollback.
3. Finally changes are made to database tables.

The COMMIT statement is shown in Figure 10-6. ROLLBACK entries made by database changes are simply removed. In reality rollback or undo segment entries are marked for reuse much like a file on disk is allocated as reusable when it is deleted.

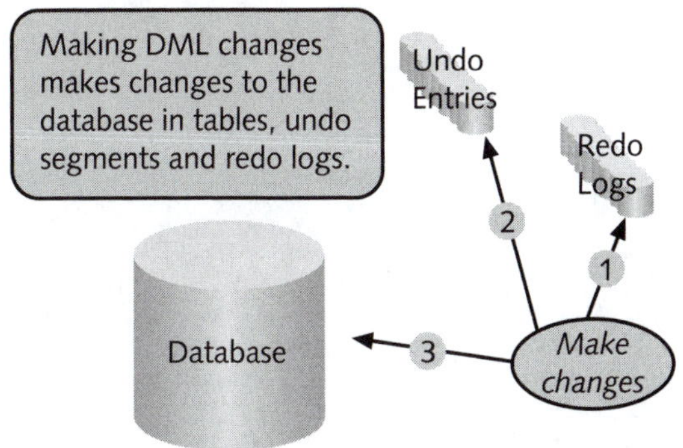

FIGURE 10-5 DML database changes prior to COMMIT or ROLLBACK

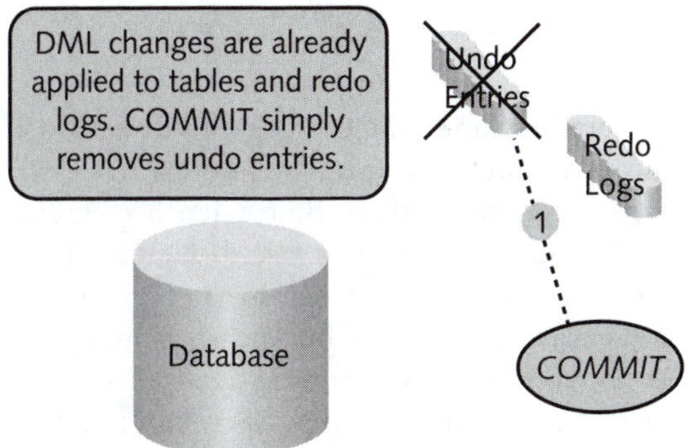

FIGURE 10-6 What does COMMIT do?

The ROLLBACK statement (shown in Figure 10-7):

1. Redo log entries representing undo changes to the database are written to redo logs. Remember that redo log entries are written before anything else in order to maintain database recoverability. In other words, if changes are made and subsequently rolled back, then there will be redo log entries for both the changes and the rollback of those changes.
2. Rollback entries are applied to tables to undo changes previously made to tables.
3. Rollback entries are removed because they are no longer required.

As we have already seen, there are three primary DML statements used to change data in an Oracle database. These statements are INSERT, UPDATE, and DELETE. There is an additional DML statement called MERGE, which is used to merge data from a source table into a target table. A MERGE statement is commonly known as an upsert in that rows already existing in the target table are updated by source table rows. Rows not in the target table are inserted from the source table.

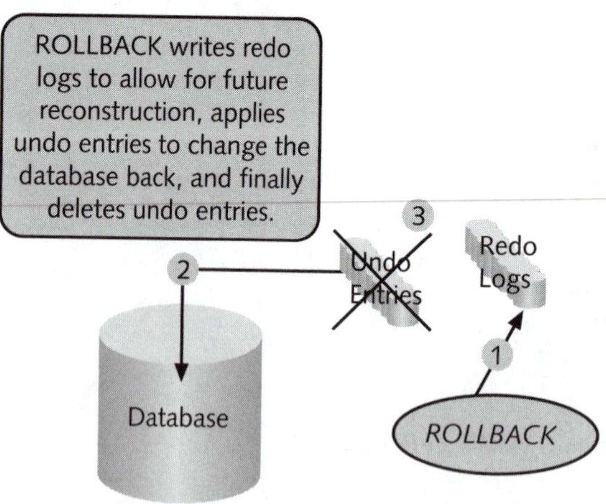

FIGURE 10-7 What does ROLLBACK do?

Why is it important to understand the internal workings of COMMIT and ROLLBACK statements? In general a properly coded application is likely to contain many more COMMIT than ROLLBACK statements. For example, it is more likely that when a new customer is added to a database, the customer will be saved to the database. It is much less likely that the clerk entering the new customer details will change his or her mind and abort the entry of the new customer. How often do you press the cancel button when using an ATM machine? Much less often than you press the enter key! Committing changes is a more frequent event than undoing changes because it is a more natural process. Therefore the execution of the COMMIT statement is much less complex than that of a ROLLBACK statement because it is naturally required to be so.

NOTE

The MERGE statement is not included in this exam (1Z0-042).

Additionally there are three types of INSERT statements:

- A single table INSERT statement adds to a single table.
- A non-conditional multiple-table INSERT statement adds to more than a single table, using a single INSERT statement.
- A conditional multiple table INSERT statement adds to more than a single table using the same INSERT statement, on a conditional basis. The term conditional implies that one, some, or all of the specified tables can be added to within a single INSERT statement.

Now you will dig deeply into the syntax and use of DML statements for changing data in a database.

USING DML STATEMENTS

This section will describe syntax for the INSERT, UPDATE, and DELETE DML statements. Syntax will be followed by examples of those DML statements.

The INSERT Statement (Adding New Data)

The INSERT statement allows you to add new rows to a table. You already know that the INSERT statement can have three forms: (1) a single table INSERT statement, (2) a non-conditional multiple table INSERT statement, (3) a conditional multiple table INSERT statement. Let's begin with a single table INSERT statement. This is the syntax for the single table INSERT statement:

```
INSERT INTO <table> [ (<column> [, ... ]) ]
{
      VALUES ( { <expression> | DEFAULT } [, ... ])
    | <subquery>
};
```


Parameters for the single table INSERT statement are as follows:

- **<table>**: The table can be any INSERT accessible table or updatable view.
- **(<column> [, ...])**: A single column or multiple columns (column list) contains column names for the table inserted into. A column list is entered as a comma-delimited list.
- **[(<column> [, ...])]**: If entries in the VALUES clause or a subquery exactly match both column order and/or datatypes for the table, then the column list is optional.
- **<expression>**: Any valid expression assuming the datatype is correct and no constraints are violated.
- **DEFAULT**: Sets a column in a table to a table column default value. The default value is created initially within the table such as with a CREATE TABLE statement.
- **| <subquery>**: The VALUES clause can be replaced with a subquery. The subquery must have exactly matching column order and/or datatypes for table columns.

Following are some example INSERT statements. This first example adds a row referencing all columns in both the column list and the VALUES clause:

```
INSERT INTO EMPLOYEE(EMPLOYEE_ID,JOB_TITLE,FIRST_NAME,LAST_NAME)
VALUES(100,'WRITER','Joe','Soap');
```

This example adds only the EMPLOYEE_ID column, the only nonnullable column. Because the JOB_TITLE column has a DEFAULT setting of *WRITER* the JOB_TITLE for this row will be *WRITER*:

```
INSERT INTO EMPLOYEE(EMPLOYEE_ID) VALUES(101);
```

This example adds all columns using the VALUES clause only:

```
INSERT INTO EMPLOYEE VALUES(102,'WRITER','Joe','Soap');
```

This example will produce an error as shown in Figure 10-8. Since the column list is not specified the SQL engine does not know which column to place the value *103* into:

```
INSERT INTO EMPLOYEE VALUES(103);
```

This example uses a subquery, selected from the DUAL table, to add a single row to the EMPLOYEE table:

```
INSERT INTO EMPLOYEE SELECT 104, 'WRITER', 'An', 'Author' FROM DUAL;
```

NOTE

The DUAL table is used in the above example because literal values are placed into the table. In other words, the values are not selected from a table.

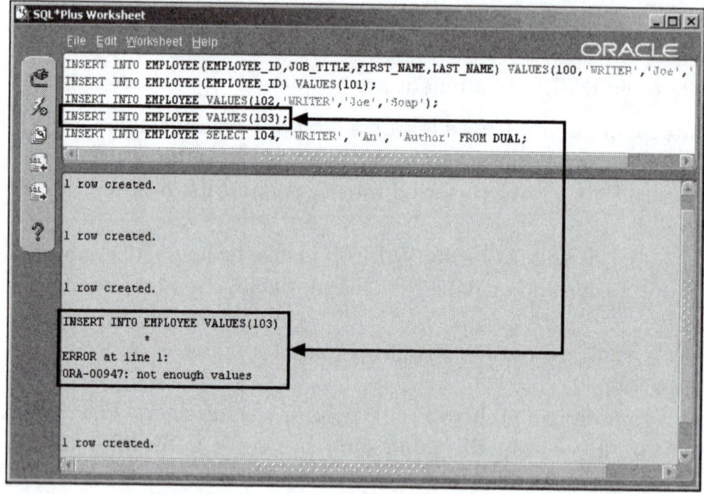

FIGURE 10-8 Some example INSERT statements

The non-conditional multiple-table INSERT statement is used to execute all INSERT statements for all rows selected by the subquery. This is the syntax for the non-conditional multiple table INSERT statement:

```
INSERT ALL <single-table-insert> [, ... ] <subquery>;
```

Parameters for the non-conditional multiple table INSERT statement are as follows:

- **ALL:** This option causes all contained INSERT statements to be executed for each row selected by the subquery.
- **<single-table-insert>:** Multiple single table INSERT statements are all executed.
- **subquery:** The subquery is required.

For the purposes of demonstration let's say you have two different versions of the EMPLOYEE table called EMPLOYEE1 and EMPLOYEE2:

```
CREATE TABLE EMPLOYEE1 AS SELECT * FROM EMPLOYEE WHERE ROWNUM < 1;
CREATE TABLE EMPLOYEE2 AS SELECT * FROM EMPLOYEE WHERE ROWNUM < 1;
```

Now you can use a non-conditional multiple table INSERT statement as shown below, adding rows from the original EMPLOYEE table in to both of the new tables EMPLOYEE1 and EMPLOYEE2 as shown below. The process is shown in Figure 10-9.

```
INSERT ALL
  INTO EMPLOYEE1(EMPLOYEE_ID,JOB_TITLE,FIRST_NAME,LAST_NAME)
  VALUES(EMPLOYEE_ID,JOB_TITLE,FIRST_NAME,LAST_NAME)
  INTO EMPLOYEE2(EMPLOYEE_ID,JOB_TITLE,FIRST_NAME,LAST_NAME)
  VALUES(EMPLOYEE_ID,JOB_TITLE,FIRST_NAME,LAST_NAME)
SELECT EMPLOYEE_ID,JOB_TITLE,FIRST_NAME,LAST_NAME FROM EMPLOYEE;
```


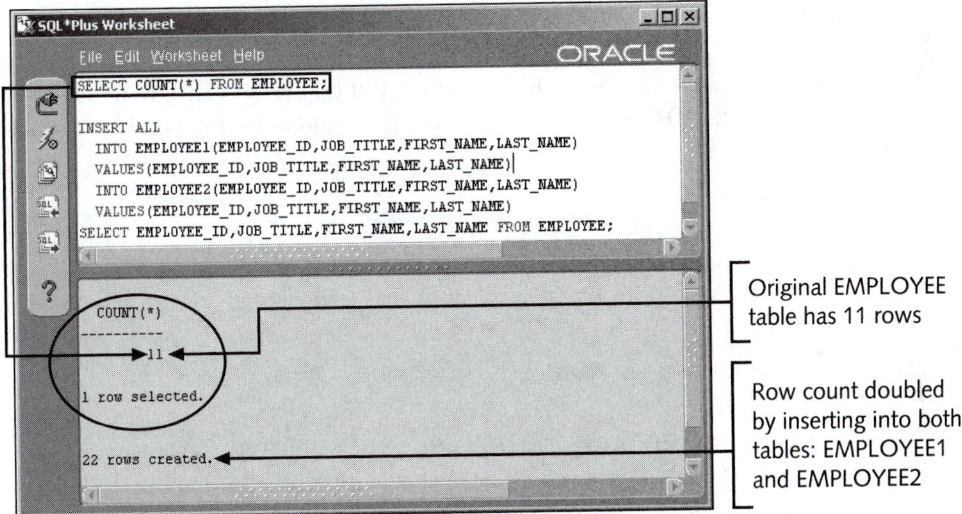

FIGURE 10-9 Executing a non-conditional INSERT statement

The conditional multiple table INSERT statement is used to insert rows into a single table, some tables, or all of a group of tables. This is the syntax for the conditional multiple table INSERT statement:

```
INSERT [ ALL | FIRST ]
{
    WHEN <condition> THEN { <single-table-insert > [, ... ] }
    WHEN <condition> THEN { <single-table-insert > [, ... ] }
    ...
    [ ELSE { <single-table-insert> [, ... ] } ]
}
<subquery>;
```

Parameters for the non-conditional multiple table INSERT statement are as follows:

- **[ALL | FIRST]:** FIRST executes only INSERT statements contained within the first successful WHEN clause condition for each row returned by the subquery. After one successful WHEN clause is encountered any subsequent WHEN clauses will not be executed even if the condition is met. ALL does completely the opposite of FIRST. Execute all successful WHEN condition INSERT statements regardless of previously successful WHEN clauses.
- **WHEN <condition>:** Causes one or more contained INSERT statements to be executed.
- ...: There can be any number of WHEN conditions.
- **THEN { <single-table-insert > }:** Causes contained single table INSERT statements to be executed.
- **<single-table-insert > [, ...]:** There can be one or many single table INSERT statements.
- **[ELSE] :** Optional and executed only if all WHEN conditions fail.
- **<subquery>:** The subquery is required.

To demonstrate let's say you have three tables, EMPLOYEE_WRITER, EMPLOYEE_EDITOR, and EMPLOYEE_OTHER. You create these three tables as empty tables with the following statements:

```
CREATE TABLE EMPLOYEE_WRITER AS
    SELECT EMPLOYEE_ID,FIRST_NAME,LAST_NAME FROM EMPLOYEE
    WHERE ROWNUM < 1;
CREATE TABLE EMPLOYEE_EDITOR AS
    SELECT EMPLOYEE_ID,FIRST_NAME,LAST_NAME FROM EMPLOYEE
    WHERE ROWNUM < 1;
CREATE TABLE EMPLOYEE_OTHER AS
    SELECT EMPLOYEE_ID,FIRST_NAME,LAST_NAME FROM EMPLOYEE
    WHERE ROWNUM < 1;
```

Now copy rows into the three separate tables from the EMPLOYEE table using the conditional multiple table INSERT statement shown below. The rows added to the three separate tables are shown in Figure 10-10:

```
INSERT FIRST
 WHEN JOB_TITLE LIKE '%WRITER%' THEN
  INTO EMPLOYEE_WRITER(EMPLOYEE_ID,FIRST_NAME,LAST_NAME)
  VALUES(EMPLOYEE_ID,FIRST_NAME,LAST_NAME)
 WHEN JOB_TITLE LIKE '%EDITOR%' THEN
  INTO EMPLOYEE_EDITOR(EMPLOYEE_ID,FIRST_NAME,LAST_NAME)
  VALUES(EMPLOYEE_ID,FIRST_NAME,LAST_NAME)
 ELSE
  INTO EMPLOYEE_OTHER(EMPLOYEE_ID,FIRST_NAME,LAST_NAME)
  VALUES(EMPLOYEE_ID,FIRST_NAME,LAST_NAME)
SELECT EMPLOYEE_ID,JOB_TITLE,FIRST_NAME,LAST_NAME FROM EMPLOYEE;
```

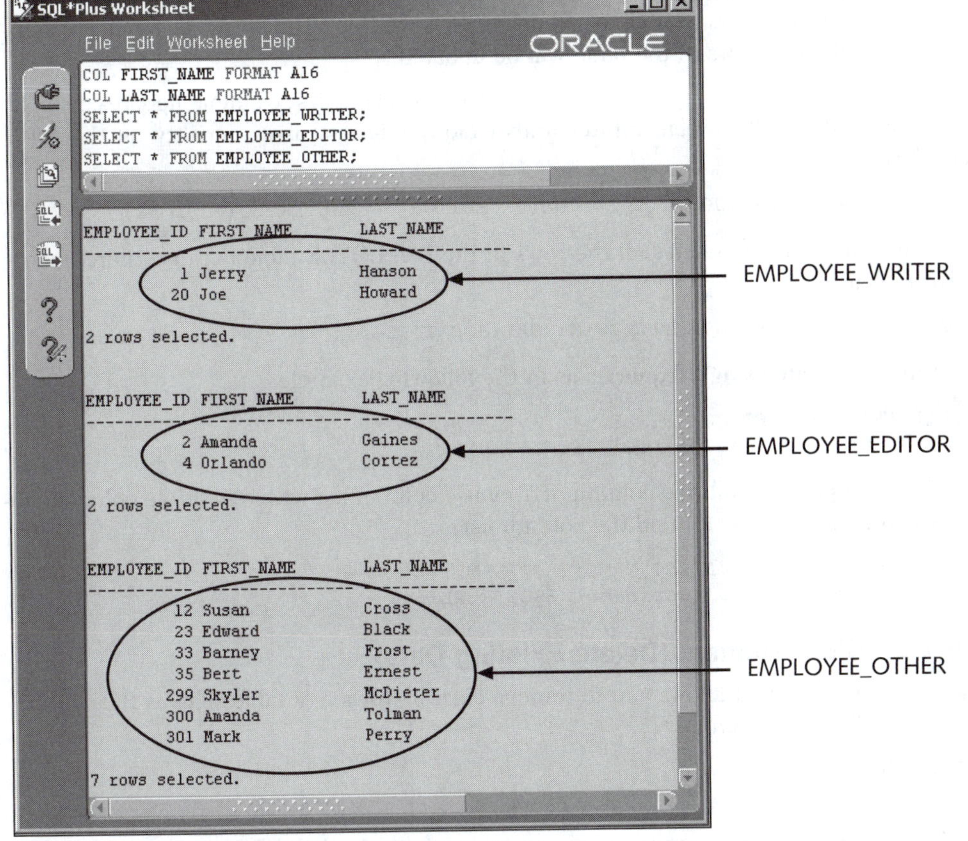

FIGURE 10-10 Conditional INSERT statement results

The UPDATE Statement (Changing Existing Data)

The UPDATE statement allows you to change column values in existing rows in a table. This is the syntax for the UPDATE statement:

```
UPDATE <table> SET
      { <column> = { <expression> | (<subquery>) | DEFAULT } [ , ... ] }
    | { { <column> [, ... ] } = (<subquery>) [, ... ] }
[ WHERE ... ];
```

- **SET:** This is the UPDATE SET clause, used to set columns to new values.
- **<column> = { <expression> | (<subquery>) | DEFAULT }:** A single column can be set to an expression, a scalar subquery (assuming the column is not a collection), and a table column default setting.
- **[, ...]:** Columns can be set in multiples as a comma-separated list.
- **<column> [, ...]:** Multiple columns can be set using multiple values such as multiple columns returned from a multiple column subquery.

- **[WHERE ...]:** An optional WHERE clause allows filtering to narrow down to specific rows to update, even a single row. If a WHERE clause is not present then all rows in the table will be updated.

Following are some example UPDATE statements. This first example updates a single row because the WHERE clause filter finds exactly one row using equality on the primary key:

```
UPDATE EMPLOYEE SET JOB_TITLE='JUNIOR EDITOR' WHERE EMPLOYEE_ID=4;
```

The next example updates all the rows in the EMPLOYEE table because there is no WHERE clause:

```
UPDATE EMPLOYEE SET JOB_TITLE=INITCAP(JOB_TITLE);
```

You can update using subqueries as in the following example:


```
UPDATE EMPLOYEE SET JOB_TITLE =
(SELECT 'This is a job title' FROM DUAL);
```

You can update a multiple column SET clause column list with a multiple column subquery (note the brackets around the column list):

```
UPDATE EMPLOYEE SET (FIRST_NAME,LAST_NAME) =
(SELECT 'First name','Last name' FROM DUAL);
```

The DELETE Statement (Delete Existing Data)

The DELETE statement allows you to remove existing rows in a table. This is the syntax for the DELETE statement:

```
DELETE FROM <table>
[ WHERE ... ];
```

The only explanation required here are the two following DELETE statements. The first deletes a single row and the second deletes many rows:

```
DELETE FROM EMPLOYEE WHERE EMPLOYEE_ID=1;
DELETE FROM EMPLOYEE WHERE EMPLOYEE_ID<5;
```

CONTROLLING TRANSACTIONS

We have already examined the COMMIT and ROLLBACK statements. To summarize and expand on transactional control, all relevant statements are described here as needed, both syntactically and in their use.

> **NOTE**
>
> A transaction starts whenever a DML statement is issued. That same transaction is terminated whenever a COMMIT or ROLLBACK statement is issued, or an implicit commit occurs. All DDL statements issue an implicit commit. An implicit commit can also occur under varying circumstances but generally when a session is cleanly terminated. For example, a database SHUTDOWN ABORT statement, or loss of power, do not perform a clean session termination.

The COMMIT Statement

The COMMIT statement makes pending changes permanent for an existing transaction in the current session. Why is the current session mentioned? If one user running a session changes a table at the same time as a different user in another session, the two sets of changes do not conflict with each other. COMMIT statement syntax is:

```
COMMIT;
```

The ROLLBACK Statement

The ROLLBACK statement removes pending changes (not yet committed) for an existing transaction in the current session. ROLLBACK statement syntax is:

```
ROLLBACK;
```

The SAVEPOINT Statement

The SAVEPOINT statement creates a label. That label can be used later in a transaction as a point to roll back to, reversing any pending changes up to and including changes made by the line of SQL code containing the SAVEPOINT statement. SAVEPOINT syntax is shown in the following pseudocode example:

```
... SQL code statements ...
SAVEPOINT <label>;
... SQL code statements ...
ROLLBACK TO <label>;
```

SET TRANSACTION Statement

The SET TRANSACTION statement permits control of a transaction as a whole from the first DML statement through to a transaction completion statement or event. The SET TRANSACTION statement is optional for any given transaction.

NOTE

A transaction is completed on execution of a COMMIT or ROLLBACK statement, or a disconnection from a session.

Syntax for the SET TRANSACTION statement is:

```
SET TRANSACTION
{
     [ READ WRITE | READ ONLY ]
   | [ ISOLATION LEVEL READ COMMITTED
       | ISOLATION LEVEL SERIALIZABLE ]
   | [ USE ROLLBACK SEGMENT <rollback segment> ]
}
[ NAME <transaction> ];
```

Default settings for the SET TRANSACTION statement are READ WRITE and ISOLATION LEVEL READ COMMITTED. Parameters for the SET TRANSACTION statement are as follows:

- **READ WRITE:** Statement-level read consistency. Forces all queries to see data as a snapshot at a single point in time regardless of any database changes after the start of execution of each query in a transaction.
- **READ ONLY:** Transaction-level read consistency. Forces all queries to see a snapshot of data prior to the start of a transaction, regardless of how many DML statements are within a transaction.
- **ISOLATION LEVEL READ COMMITTED:** DML statements within a transaction will wait for a row lock held by other transactions. This is the default.
- **ISOLATION LEVEL SERIALIZABLE:** DML statements within a transaction will fail immediately if a row lock held by another transaction is encountered. In other words, setting this option will basically remove any concurrency from the data involved, during the life of a transaction. Using this option is not recommended where any kind of concurrency is required for a database in general.
- **USE ROLLBACK SEGMENT <rollback segment>:** Manual rollback segment allocation. Manual rollback segments are deprecated in favor of automated undo as of Oracle 10*g*.
- **[NAME <transaction>]:** Optionally assigns a name to a transaction to monitor that transaction.

NOTE

Use of the SET TRANSACTION statement requires at least one of the parameters to be included.

The LOCK TABLE Statement

The LOCK TABLE statement places a lock on an entire table, prohibiting other transactions in other sessions from making changes to that table, until the lock is released by ending the transaction (COMMIT or ROLLBACK). Syntax for the LOCK TABLE statement is as follows:

```
LOCK TABLE { [<schema>.]<table> | [<schema>.]<view> } , ... IN
{
        [ ROW SHARE | SHARE UPDATE ]
      | [ ROW EXCLUSIVE ]
      | [ EXCLUSIVE ]
      | [ SHARE ]
      | [ SHARE ROW EXCLUSIVE ]
}
MODE [ NOWAIT ];
```

Default settings for the LOCK TABLE statement are ROW SHARE and MODE WAIT. Parameters for the LOCK TABLE statement are as follows:

- **[<schema>.]:** Optionally lock a table or view in any schema.
- **, ... IN:** Lock multiple tables and views with a single LOCK TABLE statement.

- Locking modes are listed as follows from the least prohibitive to the most prohibitive:
  - **ROW SHARE and SHARE UPDATE:** Prohibits locking of an entire table exclusively, allowing full concurrent multiple user access, with no restrictions on viewing and changing data in the table.
  - **ROW EXCLUSIVE:** As for ROW SHARE, additionally prohibiting a SHARE lock on the table.
  - **EXCLUSIVE:** Prohibits changes and allows queries only.
  - **SHARE:** Prohibits any changes to a locked table, permitting concurrent queries only.
  - **SHARE ROW EXCLUSIVE:** Lock a table (or an updatable view), disallowing locking of the table in SHARE mode, and prohibiting changes to the table.
- **MODE [NOWAIT]:** MODE NOWAIT will not wait for a table lock to be released if a lock is encountered. Without the NOWAIT option control is returned to the session when another lock is released, and the lock required by the executed LOCK TABLE statement is obtained.

DIRECTORY OBJECTS

A directory object is used to create a reference to a file, which is stored externally to an Oracle database. Typically, directory objects are used for external tables. An external table contains data in a file outside the database, such as in a text file. You can define the table and its columns as usual, and then define where the external data resides, and how each column maps to the external file's records. External tables are read-only, but requiring read and write privileges in the directory they are created in. They can be created and loaded using an AS subquery clause. They are writable during table creation but not after creation. In other words, external tables cannot have DML statements executed against them other than an initial subquery to insert the rows into the external table. Follow along with the steps to see how to create a directory object and an external table:

1. Start the Enterprise Manager console. In Windows, on the Taskbar, click **Start/All Programs/Oracle ... /Enterprise Manager Console**. In Unix or Linux, type **oemapp console** on the command line. The Enterprise Manager console login screen appears.
2. Start the SQL*Plus Worksheet by clicking **Tools/Database Applications/ SQL*Plus Worksheet** from the top menu in the console.
3. Connect as the CLASSMATE user. Click **File/Change Database Connection** on the menu. A logon window appears. Type **CLASSMATE** in the Username box, **CLASSPASS** in the Password box, and **ORACLASS** in the Service box. Leave the connection type as "Normal." Click **OK** to continue.
4. Create a directory called **tmp**. In Windows you may have to create the directory, create it on your C drive, in a shell, using a command like the following:

```
md C:\tmp
```

In Unix or Linux use a command like the following to create the directory in the root directory:

```
mkdir /tmp
```

5. Before creating a directory object you need the appropriate privileges. As an exercise, run the Oracle Enterprise Manager Console. Connect to the ORACLASS database as the SYS or SYSTEM user. Navigate the tree on the left of the Console, finding the **Security** option. Within the Security open the **Users** option. Then find the **CLASSMATE** schema and highlight the CLASSMATE schema. Click the System tab on the top right of the Console window. Scroll down the window below the tabs and find the **CREATE ANY DIRECTORY** privilege and highlight it with a click of the mouse. Then look for two little arrow icons between the two windows on the right of the Console window. Click the **down arrow** to add the CREATE ANY DIRECTORY privilege in the lower window, on the right side of the Console. At the bottom of the Console window on the right click the **Show SQL** button, showing you the script to grant the privilege. Then click the **Apply** button, and exit the Console.
6. Create a directory object in your tmp directory using the following command, executed from within SQL*Plus Worksheet:

```
CREATE OR REPLACE DIRECTORY DATA AS 'C:\tmp\oracleext';
```


NOTE

The REPLACE option can be used to update an already existing directory object. There is no ALTER DIRECTORY statement. You can drop directory objects using the DROP DIRECTORY statement.

7. Create an external table in SQL*Plus Worksheet with the following statement:

```
CREATE TABLE EMPLOYEES(
         EMPLOYEE_ID NUMBER(10)
       , JOB_TITLE VARCHAR2(45)
       , PHONE_NUMBER VARCHAR2(20)
)
ORGANIZATION EXTERNAL
(
       DEFAULT DIRECTORY data
       LOCATION ('empsext.ctl')
)
REJECT LIMIT UNLIMITED;
```

8. You cannot populate an external table using Oracle 10g DML commands, such as INSERT, UPDATE or DELETE. An external table is read only. You could create a file using a command line editor, as specified by the path and file name in the directory object. Then you could read data from the external table using a SELECT statement.

Possible benefits of using external tables are as read-only data, which is stored externally to the database structure. The metadata of an external table is stored in the database. The data is not stored in an Oracle database. Also data is read only. This makes external tables suitable for read-only data, which might benefit from being stored externally to the Oracle database SQL engine. Data warehouses might benefit from the use of external tables. Also perhaps some static data tables in OLTP databases – but I would not recommend it.

Chapter Summary

- DML or Data Manipulation Language statements are used to change data in tables in a database.
- DDL or Data Definition Language statements are used to change the structure of database objects such as tables, views, and indexes.
- DML statements can be rolled back using the ROLLBACK statement.
- DDL statements automatically execute a COMMIT statement, committing any pending DML statement changes.
- A transaction constitutes one or more database changes through DML statements.
- The COMMIT statement stores changes to the database.
- The ROLLBACK statement will undo or reverse any DML statement-induced changes not yet committed.
- The INSERT statement adds new rows to tables.
- The UPDATE statement changes rows in tables.
- The DELETE statement destroys rows in tables.
- The INSERT, UPDATE, and DELETE statements can be rolled back.
- INSERT statements can add rows to one or many tables, using a single INSERT statement or a multiple row table INSERT statement respectively.
- Multiple table INSERT statements can be non-conditional or conditional.
- A conditional multiple table INSERT statement allows addition of rows to each table in a set of many, dependant on specific conditions being met for each table to be added to.
- The SAVEPOINT statement creates a label allowing for partial transaction rollback.
- The SET TRANSACTION statement allows application of specific attributes for a transaction.
- The LOCK TABLE statement allows locking of tables during the processing of a transaction.
- Directory objects are used to create storage areas, outside of an Oracle database, for storing objects such as external table text files.

Review Questions

1. Which command is inherently faster? COMMIT or ROLLBACK?
2. How can the SET TRANSACTION command be used to force a transaction to fail when a lock is encountered?
   a. SET TRANSACTION NAME 'myTransaction';
   b. READ WRITE ISOLATION LEVEL SERIALIZABLE;
   c. READ WRITE ISOLATION READ COMMITTED;
   d. All of the above.
   e. None of the above.

3. Which of these commands are SQL transactional commands? Choose four.
   a. INSERT
   b. UPDATE
   c. ROLLBACK
   d. SET TRANSACTION
   e. COMMIT
   f. SAVEPOINT
4. What type of a DML statement is this?

```
INSERT ALL
WHEN id = (SELECT id FROM tableA WHERE name = 'X') THEN INTO table1
WHEN id = (SELECT id FROM tableA WHERE name = 'Y') THEN INTO table2
WHEN id = (SELECT id FROM tableA WHERE name = 'Z') THEN INTO table3
...
```

   a. A single table INSERT.
   b. A MERGE statement.
   c. A DELETE statement.
   d. A TRUNCATE statement.
   e. None of the above.
5. You can always undo changes made by both DML and DDL statements. True or False?

Exam Review Questions—Oracle Database 10*g*: Administration (#1Z0-042)

1. A table called RETAILERS containing three NUMBER datatype columns (col1, col2, col3), all containing integer values. All columns have a NOT NULL constraint. Which of these statements will fail?
   a. INSERT INTO VALUES(1,2,3);
   b. INSERT INTO RETAILERS VALUES(1,3);
   c. UPSERT INTO RETAILERS VALUES(1,2,3);
   d. INSERT INTO RETAILERS (col1, col2) VALUES(1,2,3);
   e. All will fail.
2. Again you have a RETAILERS table with three NOT NULL columns. Which of these statements will *not* produce an error?
   a. UPDATE RETAILERS SET col1=5;
   b. UPDATE RETAILERS SET (col1, col2) = (5, 2);
   c. UPDATE RETAILERS SET (col1, col2) = (SELECT 5, 2 FROM DUAL);
   d. UPDATE RETAILERS SET col1 = (col1 + col2) * col3;
   e. No errors will result.

3. When all these commands have executed, how many rows will be left in the table?

```
DELETE FROM RETAILERS;
ROLLBACK;
TRUNCATE TABLE RETAILERS;
ROLLBACK;
```

4. When all these commands have executed, how many rows will be left in the table, assuming the table is empty to begin with?

```
INSERT INTO RETAILERS VALUES(2,1,1);
INSERT INTO RETAILERS VALUES(3,1,1);
SAVEPOINT halthere;
INSERT INTO RETAILERS VALUES(3,1,1);
INSERT INTO RETAILERS VALUES(3,1,1);
ROLLBACK TO halthere;
```

Hands-on Assignments

1. Start up the SQL*Plus Worksheet. In Windows, on the Taskbar, click **Start/ All Programs/ Oracle ... /Application Development/SQLPlus Worksheet**. In UNIX, type **oemapp worksheet** on a command line. A logon window displays.
2. Type **CLASSMATE** in the Username box, **CLASSPASS** in the Password box, and **ORACLASS** in the Service box. Leave the connection type as "Normal". Click **OK** to continue.
3. Figure 10-2 shows a table called CH10EMPLOYEE. Write a script to create the table, and add all the rows shown in the Figure 10-2. Execute the script to make sure you have no errors. Save the script you use in your **Solutions\Chapter10** directory in a file named **h1001.sql**.
4. Figure 10-3 shows a change to a single row in the CH10EMPLOYEE table. Change the table to make the EMPLOYEE_ID column a primary key, and write a statement to change the value shown changed in Figure 10-3. Save the script you use in your **Solutions\Chapter10** directory in a file named **h1002.sql**.
5. Write a statement to delete the row changed in Figure 10-3, as shown in Figure 10-4. Save the script you use in your **Solutions\Chapter10** directory in a file named **h1003.sql.**

Case Project

1. The Global Globe newspaper database now has several tables. Your manager has noticed that end users have difficulty with the EMPLOYEE table FIRST_NAME and LAST_NAME columns. Change the table to concatenate all FIRST_NAME and LAST_NAME column values into a column called NAME on the EMPLOYEE table. As a last act drop the two FIRST_NAME and LAST_NAME columns from the EMPLOYEE table, such that only the NAME column remains. Save your SQL work in a file named **case1001.sql** in the **Solutions\Chapter10** directory on your student disk.

CHAPTER 11

ADVANCED DATA MANAGEMENT

LEARNING OBJECTIVES

In this chapter, you will:

- Examine the basics of PL/SQL
- Use the Data Pump export and import utilities
- Use the SQL Loader utility for rapid bulk data loads
- Briefly examine transportable tablespaces

INTRODUCTION

In this chapter, you discover how data is changed in an Oracle database by means other than using simple DML INSERT, UPDATE, and DELETE statements. PL/SQL can be used to manage data, and otherwise. Data Pump can be used to dump and reconstitute sets of metadata and data, such as entire schemas or tables. SQL Loader can be used to bulk upload very large amounts of data into a database.

NOTE

You are not required to know any SQL or PL/SQL for the first Oracle 10*g* certification exam. However, it will help you as a database administrator to know something about SQL and PL/SQL. Therefore, these topics are included in this book.

INTRODUCTION TO ADVANCED DATA MANAGEMENT

Chapter 10 covered basic data management including all DML commands, transaction control, and directory objects. This chapter builds on basic data management by exploring other methods used in Oracle databases to manage data. These advanced methods include Programming Language for SQL (PL/SQL), data pump technology, and SQL Loader.

PL/SQL allows construction of properly scripted intra-dependant SQL command structures. What does that mean? It means you can write proper programs with PL/SQL with things like control structures, strong datatype casting with variable scoping, and other tricks.

What are the components of PL/SQL?

- Modular block structures
- Anonymous blocks
- Exception trapping
- Stored (named) procedures:
    - Procedures
    - Functions
    - Procedure packaging using packages
    - Triggers
- Variables and datatypes
- Explicit and implicit cursors
- Control structures:
    - IF statement
    - CASE statement (both search condition and selector expression)
    - Looping constructs include the FOR loop, the WHILE loop, the LOOP... END (endless loop), and the FORALL statement.
    - Sequencing controls include the GOTO command and the NULL statement.

NOTE

There is a rich programming environment provided by packaged code and written in PL/SQL.

Data Pump (import and export) and SQL Loader are used to perform bulk loads and dumps into and out of an Oracle database. The original Import and Export utilities are called imp and exp. Data Pump import and export utilities are called expdp and impdp. Data Pump provides the following advantages over the original Import and Export utilities:

- Better handling of bulk loading and unloading, allowing complete restart or a temporary halt
- Parallel processing
- Better network support
- Metadata filters allowing inclusion and exclusion of specific metadata items (database objects such as tables)

- Interactive command mode allows monitoring of ongoing jobs
- Space estimations of a Data Pump export job
- Versions of objects can be specified
- Fine-grained object selection

The SQL Loader utility is generally the most efficient tool for the mass loading of data into an Oracle database. SQL Loader allows loading of operating system flat files into one or more tables at once. Additionally, SQL Loader can apply formatting to operating system flat files, SQL code parsing, and has numerous options applicable to various situations.

CODING SQL INTO PROGRAMS WITH PL/SQL

As you have already seen, a study of PL/SQL can be broken in specific parts. The topics required for the 1Z0-042 Oracle certification exam cover identification of PL/SQL objects, triggers, and the events that execute triggers. This text will cover a little more, even if briefly, because a little extra knowledge is always useful.

The order in which the facts of PL/SQL are presented is significant, because like any programming language, each part of a programming language builds on to the next. It makes sense to begin with the most fundamental—blocks of code.

Blocks, Exception Trapping, and Anonymous Blocks

What is a block? In programming terms a block is a block of code. A block of code is a self-contained sequence of one or more commands. In PL/SQL a block is a chunk of code containing both PL/SQL and SQL commands. PL/SQL is a programming form of SQL where SQL commands such as INSERT and SELECT can be contained with PL/SQL blocks. Let's begin with the first exercise for this chapter:

1. Start the Enterprise Manager console. In Windows, on the Taskbar, click **Start/All Programs/Oracle ... /Enterprise Manager Console**. In Unix or Linux, type **oemapp console** on the command line. The Enterprise Manager console login screen appears.
2. Start the SQL*Plus Worksheet by clicking **Tools/Database Applications/ SQL*Plus Worksheet** from the top menu in the console. A background SQL*Plus process starts, and then the SQL*Plus Worksheet window appears. If the background process appears in front, simply minimize it (click the **minus sign** in the top-right corner), so you can see the worksheet.
3. Connect as the CLASSMATE user. Click **File/Change Database Connection** on the menu. A logon window appears. Type **CLASSMATE** in the Username box, **CLASSPASS** in the Password box, and **ORACLASS** in the Service box. Leave the connection type as "Normal." Click **OK** to continue.

4. The simplest block of code you can create in PL/SQL is a block, enclosed between BEGIN and END commands, containing the command NULL. Type these commands on separate lines and click the **Execute** icon. SQL*Plus Worksheet replies, "PL/SQL procedure successfully completed.":

```
BEGIN
      NULL;
END;
```

NOTE

If using SQL*Plus and not SQL*Plus Worksheet, don't forget the / character on the last line as the code block will not execute without it. The script in step 6 shows the / character included because there are lines after the END command. The / character is required in step 6 for this reason.

5. Clear the top half of the screen by clicking on the top window, then select Edit, and Clear All from the menu.
6. Now let's build a simple block of code that does something. NULL does just as it states: nothing. Type these commands on separate lines and click the **Execute** icon. SQL*Plus Worksheet replies, "PL/SQL procedure successfully completed." Executing the ENABLE procedure, "This is a test" executing the PUT_LINE procedure, and "PL/SQL procedure successfully completed." Executing the block of code:

```
SET SERVEROUTPUT ON;
EXEC DBMS_OUTPUT.ENABLE(1000000);
BEGIN
      DBMS_OUTPUT.PUT_LINE('This is a test');
END;
/
EXEC DBMS_OUTPUT.DISABLE();
SET SERVEROUTPUT OFF;
```

NOTE

The first two lines and the last two lines in the script above allow for output in SQL*Plus Worksheet and within a procedure using the DBMS_OUTPUT package. They allow for output to the screen using the DBMS_OUTPUT procedure. Setting the value 1000000 sets an output buffer because the buffer is written out to the display (SQL*Plus) on completion of the procedure. If you want further explanation both can be found in Oracle documentation.

The result of the above script is shown in Figure 11-1.

7. Now let's add an exception trap. An exception trap is a block of code, within the main block, that will execute if a coding error occurs that is not trapped by the compiler. Type these commands on separate lines and click the **Execute** icon. In the following script the CLIENT_ID=500 filter returns no rows.

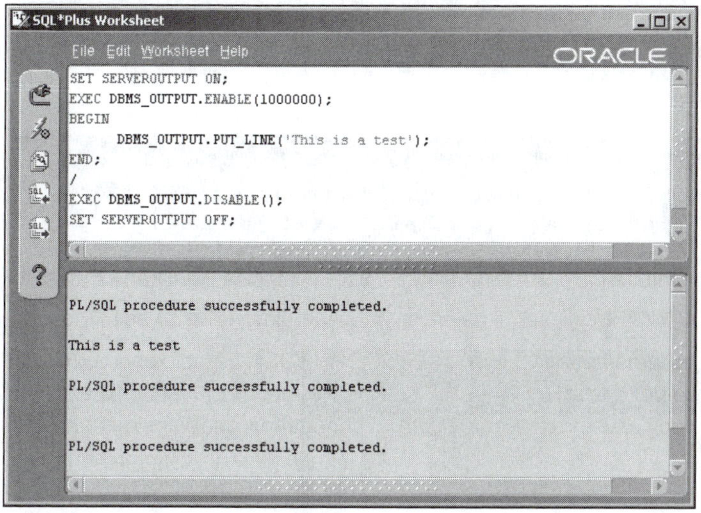

FIGURE 11-1 Executing an anonymous (unnamed) block in PL/SQL

Among other things, SQL*Plus Worksheet replies, "No data found". The result is shown in Figure 11-2. The exception trap section was executed, ignoring the PUT_LINE procedure after the query:

```
SET SERVEROUTPUT ON;
EXEC DBMS_OUTPUT.ENABLE(1000000);
DECLARE
     CLIENT_ID INTEGER;
BEGIN
     SELECT CLIENT_ID INTO CLIENT_ID
     FROM CLIENT WHERE CLIENT_ID=500;
     DBMS_OUTPUT.PUT_LINE('Client 500 found');
EXCEPTION WHEN OTHERS THEN
      RAISE;
END;
```

NOTE

The INTO keyword is required in PL/SQL in to copy unique values retrieved by a SELECT statement into program variables.

NOTE

The exception trap error in this case is WHEN OTHERS, implying any error at all. Other exceptions such as EXCEPTION WHEN NO_DATA_FOUND would cause a specific exception. Including the WHEN OTHERS exception in addition to any other exceptions, such as the NO_DATA_FOUND exception, will trap all runtime errors not catered for with specific exception traps.

NOTE

The RAISE keyword causes an error to be returned from the block of code.

NOTE

In an anonymous procedure or block any local variables must be declared using the DECLARE keyword, in the declaration section. Anonymous is a term given to a PL/SQL procedure or block of code that is not named, and is not stored in the database.

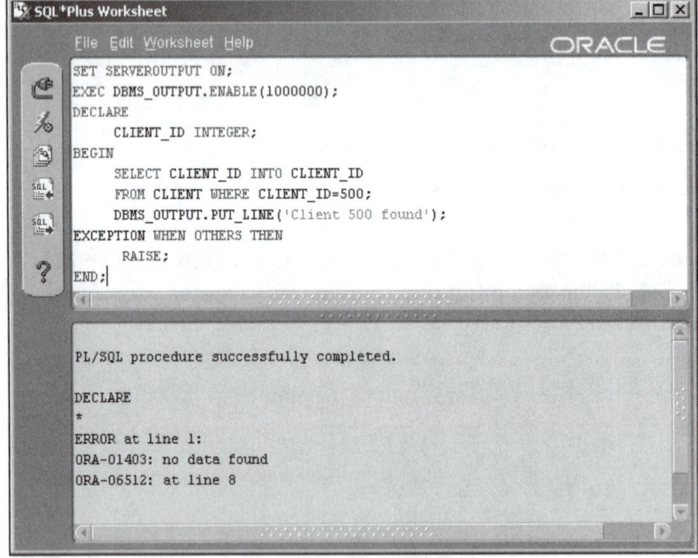

FIGURE 11-2 Executing an exception trap in PL/SQL

8. Remain logged on for the next practice.

So a block of PL/SQL code is enclosed with the BEGIN and END keywords. Also, exceptions or errors (not compilation errors) can be detected within the block using the EXCEPTION trapping command.

Named Blocks and Packages

We have seen an anonymous block in the previous section that is a block of code executed once, unless physically typed in again (copied and pasted in my case). A named block is a chunk of code given a label (a procedure name), so that block of code can be stored in the database. A named block of code is stored and can be recalled later and re-executed again, without having to retype the entire block of code.

Named blocks of code can be of various types with various restrictions, which are beyond the scope of this book:

- **Procedure**. Allows by value and by reference parameters with no return value.

NOTE

A by value parameter is a parameter passed into a procedure or function as a literal, meaning it cannot be changed within the procedure or function, and then passed back as a changed parameter. A by reference parameter passes a memory address, which is known as a pointer in programming parlance. The address points to an area of memory containing a stored value. The procedure or function can change the value of the pointed to area of memory, but not the pointer. So a by value parameter passes a value that cannot be changed. A by reference parameter passes a variable because the memory area pointed to can be altered.

- **Function**. Like a procedure except it allows a return value and can thus be embedded in an expression.

NOTE

The only difference between a function and a procedure is that a function returns a value, allowing a function to be included in an expression. For example, let's say the function ADDUP(<list of numbers>) adds numbers in a comma-delimited list. Thus (5 + (10 * ADDUP(1,2,3,4,5))) shows the ADDUP function embedded into a multiple hierarchical expression.

- **Trigger**. No transactional termination commands allowed and executed automatically by database event occurrences. Triggers are known as event-driven procedures.
- **Package**. A simple wrapper for grouping multiple procedures and functions together into a blocked (*packaged*) unit. Oracle Database comes with a multitude of pre-prepared packages. In the previous example you have seen use of the DBMS_OUTPUT package. The DBMS_OUTPUT package contains three procedures seen so far: ENABLE, DISABLE, and PUT_LINE.

NOTE

There are many built-in packages available with Oracle 10*g*. In fact, there are so many built-in packages, it is best to direct you to the PL/SQL Packages and Types Reference book in the Oracle documentation.

TIP

All PL/SQL named structures (procedures, functions, triggers, and packages) can be created in various interfaces including all SQL*Plus programs, as well as the Oracle Enterprise Manager console, and the Database Control. The latter two do not require demonstration because their interfaces are intuitive and easy to use. Generally, Oracle certification tests do not ask many questions about using their GUIs because they are so intuitive and easy to use. Few questions can be asked. On the other hand, Oracle certification tests consistently ask many questions about syntax. So you have to know things like the CREATE PROCEDURE command and the CREATE TABLE command, and their syntax, and how to execute them in programs like SQL*Plus and SQL*Plus Worksheet.

Let's do another exercise:

1. You should already be connected to SQL*Plus Worksheet. If not, go back to the first exercise in this chapter and follow the first three steps to reconnect.
2. We begin by creating a simple procedure. Type these commands on separate lines and click the **Execute** icon. Note how the procedure is no longer anonymous, but now a name. The OR REPLACE option allows the procedure to overwrite an existing occurrence of the named procedure. SQL*Plus Worksheet should respond, "Procedure created.":

```
CREATE OR REPLACE PROCEDURE GETPHONE
(pNAME IN VARCHAR2) AS
    vPHONE CLIENT.CONTACT_PHONE%TYPE;
BEGIN
    SELECT CONTACT_PHONE INTO vPHONE
    FROM CLIENT WHERE FIRST_NAME = pNAME;
    DBMS_OUTPUT.PUT_LINE(pNAME||': '||vPHONE);
EXCEPTION WHEN OTHERS THEN
    RAISE;
END;
/
```

NOTE

The %TYPE clause in the declaration section defines the local variable vPHONE as having the same datatype and size as that of the PHONE_NUMBER field in the CLIENT table.

NOTE

The || character sequence concatenates two strings together.

NOTE

If the response is not "Procedure created" then type in the command SHOW ERRORS, press the Return button, and fix the errors.

NOTE

A procedure without parameters is created without having to add the parentheses on the first line.

3. Type in the following command and then click the **Execute** icon:

```
EXEC GETPHONE('Mark');
```

You should get Mark's phone number, as shown in Figure 11-3.

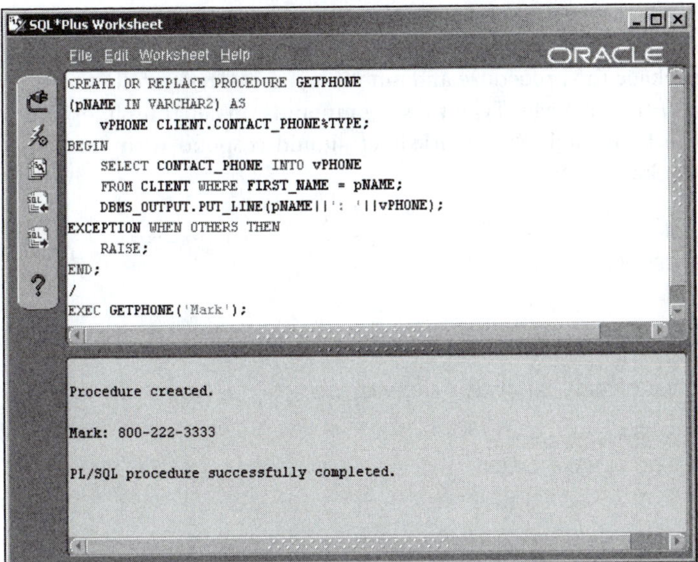

FIGURE 11-3 Executing a named procedure in PL/SQL

4. In step 2 you created a named procedure. Now create a named function to strip out the area code in the phone number. Type these commands on separate lines and click the **Execute** icon. SQL*Plus Worksheet should respond, "Function created.":

```
CREATE OR REPLACE FUNCTION STRIPCODE
(pPHONE IN VARCHAR2) RETURN VARCHAR2 IS
BEGIN
    RETURN SUBSTR(pPHONE,5,8);
EXCEPTION WHEN OTHERS THEN
    RAISE;
END;
/
```

5. Now embed the function into the procedure by re-creating the GETPHONE procedure. Type these commands on separate lines and click the **Execute** icon. SQL*Plus Worksheet should return a stripped phone number:

```
CREATE OR REPLACE PROCEDURE GETPHONE
(pNAME IN VARCHAR2) AS
```

```
    vPHONE CLIENT.CONTACT_PHONE%TYPE;
BEGIN
    SELECT CONTACT_PHONE INTO vPHONE
    FROM CLIENT WHERE FIRST_NAME = pNAME;
    DBMS_OUTPUT.PUT_LINE(STRIPCODE(vPHONE));
EXCEPTION WHEN OTHERS THEN
    RAISE;
END;
/
```

6. Type in the following command and then click the **Execute** icon:

```
EXEC GETPHONE('Mark');
```

You should get Mark's phone number but this time without the 800 prefix.

7. You could also package the procedure and function you created as follows—this is the definition of the package. Type these commands on separate lines and click the **Execute** icon. SQL*Plus Worksheet should respond with "Package created." and "Package body created.":


```
CREATE OR REPLACE PACKAGE PHONENUMBERS AS
    FUNCTION STRIPCODE(pPHONE IN VARCHAR2) RETURN VARCHAR2;
    PROCEDURE GETPHONE(pNAME IN VARCHAR2);
END;
/
CREATE OR REPLACE PACKAGE BODY PHONENUMBERS AS
   FUNCTION STRIPCODE
   (pPHONE IN VARCHAR2) RETURN VARCHAR2 IS
   BEGIN
       RETURN SUBSTR(pPHONE,5,8);
   EXCEPTION WHEN OTHERS THEN
       RAISE;
   END;
   PROCEDURE GETPHONE
   (pNAME IN VARCHAR2) AS
       vPHONE CLIENT.CONTACT_PHONE%TYPE;
   BEGIN
       SELECT CONTACT_PHONE INTO vPHONE
       FROM CLIENT WHERE FIRST_NAME = pNAME;
       DBMS_OUTPUT.PUT_LINE(STRIPCODE(vPHONE));
   EXCEPTION WHEN OTHERS THEN
       RAISE;
   END;
END;
/
```

NOTE

A package consists of a definitional section and a BODY section. The definitional section contains the name of the package and any variables used within the scope of the package. The package BODY section contains procedures and functions compiled within the package.

8. Now execute the procedure from within the package. Type this command and click the **Execute** icon. SQL*Plus Worksheet should respond again with Mark's phone number:

   ```
   EXEC PHONENUMBERS.GETPHONE('Mark');
   ```

9. Remain logged on for the next practice.

Triggers and Events Firing Triggers

A trigger is executed based on an event occurring in the database. The basic syntax for a trigger is as follows:

```
CREATE [ OR REPLACE ] TRIGGER [ <schemaname>.]<triggername>
   { BEFORE | AFTER | INSTEAD OF }
   {
         INSERT ON <tablename>
       | UPDATE ON <tablename> [ OF <columnname> ]
       | DELETE ON <tablename>
   }
BEGIN
   ... trigger code ...
END;
/
```


Examine the syntax above to observe various events that will cause a trigger to execute:

- BEFORE INSERT ON <tablename>
- BEFORE UPDATE ON <tablename> [OF <columnname>]
- BEFORE DELETE ON <tablename>
- AFTER INSERT ON <tablename>
- AFTER UPDATE ON <tablename> [OF <columnname>]
- AFTER DELETE ON <tablename>
- INSTEAD OF INSERT ON <tablename>
- INSTEAD OF UPDATE ON <tablename> [OF <columnname>]
- INSTEAD OF DELETE ON <tablename>

You can refer to column values in a row both prior to and after the DML command change; obviously new and old values exist depending on the DML command in question. In other words, INSERT has only new values, UPDATE has both old and new values, and DELETE has old values only. The following trigger executes when a row is updated in the CLIENT table (an UPDATE event):

```
CREATE OR REPLACE TRIGGER uCLIENT
    AFTER UPDATE ON CLIENT FOR EACH ROW
BEGIN
    DBMS_OUTPUT.PUT_LINE('Client''s name changed from '
    ||:old.FIRST_NAME||' to '||:new.FIRST_NAME);
EXCEPTION WHEN OTHERS THEN
    DBMS_OUTPUT.PUT_LINE(SQLERRM(SQLCODE));
    RAISE;
END;
/
```

NOTE

Every SQL statement issued stores its resulting state into a cursor—the most recent SQL statement executed. If that SQL command causes an error it will be accessible using the SQLERRM built-in procedure. SQLCODE is stored in memory automatically when an exception trap occurs. SQLERRM is a built-in procedure that returns an error message, determined by the value of SQLCODE.

The two single quotes are required in the string containing client names because Oracle Database requires a single quote to apply an escape sequence to the second, ignoring the quote as a string terminator. The following UPDATE statement will display the output within the trigger after the change has occurred:

```
UPDATE client SET first_name = 'Marcus' WHERE first_name = 'Mark';
```

The result is shown in Figure 11-4.

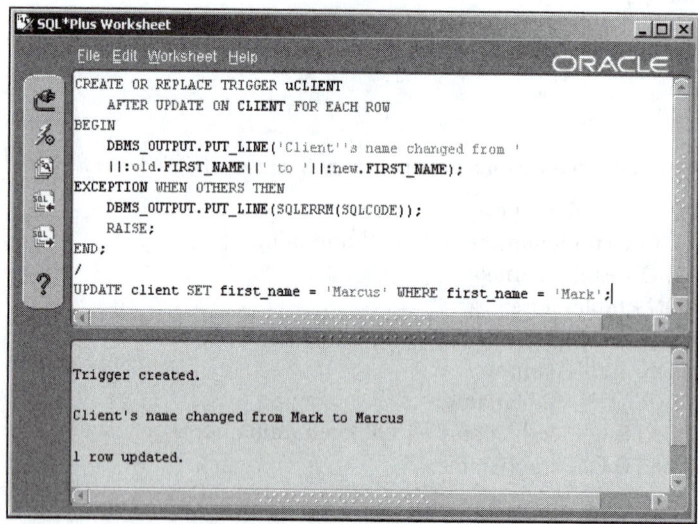

FIGURE 11-4 Executing a trigger in PL/SQL

. You can type all the trigger code and execute the UPDATE statement much as you did in the last two exercises, but it is not essential.

PL/SQL Variables and Datatypes

PL/SQL uses all the datatypes used in SQL as shown in Table 7-1. PL/SQL also uses some additional datatypes exclusive to PL/SQL, as listed in the following section:

- **NUMBER**: Different NUMBER datatypes such as FLOAT and SMALLINT.
- **BINARY_INTEGER**: Stores a signed integer value. There are various subtypes.
- **BOOLEAN**: Stores a TRUE, FALSE, or null value.

- **RECORD**: A composite structure similar to a VARRAY or TABLE datatype that allows the creation of a table row structure in memory. The following line uses ROWTYPE to duplicate the column structure of the CLIENT table row into a RECORD called RCLIENT:

```
RCLIENT CLIENT%ROWTYPE;
```

  In the next sample a new record structure is built using an added identifier field (ID), the name of the client and his phone number. A RECORD datatype called RCLIENT is declared as having the structure of the created type, allowing repetitions of the new structure to be stored in memory:

```
DECLARE
   TYPE TCLIENT IS RECORD
   (
        CLIENT_ID INTEGER
      , NAME CLIENT.FIRST_NAME%TYPE
      , PHONE CLIENT.CONTACT_PHONE%TYPE
   );
   RCLIENT TCLIENT;
BEGIN
   NULL;
END;
/
```

- **Reference Datatypes**: In addition to the REF object pointer type PL/SQL also includes a REF cursor. A REF cursor is a by reference cursor. By reference (BYREF) implies that a variable is a pointer and can be passed into, as well as out of, a procedure within procedure parameters, including returning any changes made to the REF cursor within the procedure.
- **Associative Arrays**: Associative arrays are currently only allowed in PL/SQL and not Oracle SQL. An associative array is a dynamic array much like a nested table object. The only difference is that an associative array is indexed and capable of much better performance than a nested table. The following script snippet shows how an associative array is declared in PL/SQL as opposed to VARRAYs and nested tables:

```
DECLARE
   TYPE TTABLE IS TABLE OF VARCHAR2(32);
   TYPE TVARRAY IS VARRAY(100) OF INTEGER;
   TYPE TITABLE IS TABLE OF VARCHAR2(32)
   INDEX BY BINARY_INTEGER;
   VPOINTER TTABLE;
   VARRAY TVARRAY;
   VINDEXEDPOINTER TITABLE;
BEGIN
   NULL;
END;
/
```

What Is a Cursor?

What is a cursor? A cursor is a temporary area in memory used to store the results of a query. Oracle Database calls the area of memory in which a cursor is temporarily placed a Work Area. In programming terms a cursor is a pointer to an address in memory, a chunk of memory. SQL statement results are processed in cursors during execution. In PL/SQL cursors can be created as programming structures for holding iterations of data (equivalent to rows in a table). Cursors can be used for queries returning one or many rows and can be of two types, implicit and explicit cursors. An implicit cursor is declared automatically by PL/SQL, an explicit cursor is declared by the programmer. An explicit cursor allows greater programming control.

Explicit cursors are operated on using the special cursor commands OPEN, FETCH, and CLOSE. The example below opens a cursor as containing all the rows in the CLIENT table:

```
DECLARE
   CURSOR CCLIENT IS SELECT * FROM CLIENT;
   RCLIENT CLIENT%ROWTYPE;
BEGIN
   OPEN CCLIENT;
   LOOP
      FETCH CCLIENT INTO RCLIENT;
      EXIT WHEN CCLIENT%NOTFOUND;
      DBMS_OUTPUT.PUT_LINE(RCLIENT.FIRST_NAME);
   END LOOP;
   CLOSE CCLIENT;
END;
/
```


An implicit cursor is automatically opened and closed by SQL or PL/SQL and is used to process INSERT, UPDATE, DELETE, and SELECT statements. A special type of implicit cursor exclusive to PL/SQL is called a cursor FOR loop. A cursor FOR loop is an implicit cursor on the basis that it does not require use of the OPEN, FETCH, and CLOSE statements:

```
DECLARE
    CURSOR CCLIENT IS SELECT * FROM CLIENT ORDER BY FIRST_NAME;
BEGIN
    FOR RCLIENT IN CCLIENT LOOP
        DBMS_OUTPUT.PUT_LINE(RCLIENT.FIRST_NAME);
    END LOOP;
END;
/
```

The result is shown in Figure 11-5.

There is no hard and fast rule about when to use an implicit cursor and when to use an explicit cursor. There are only two rules: (1) using an implicit cursor makes for easier programming, and (2) an explicit cursor allows for better programming control.

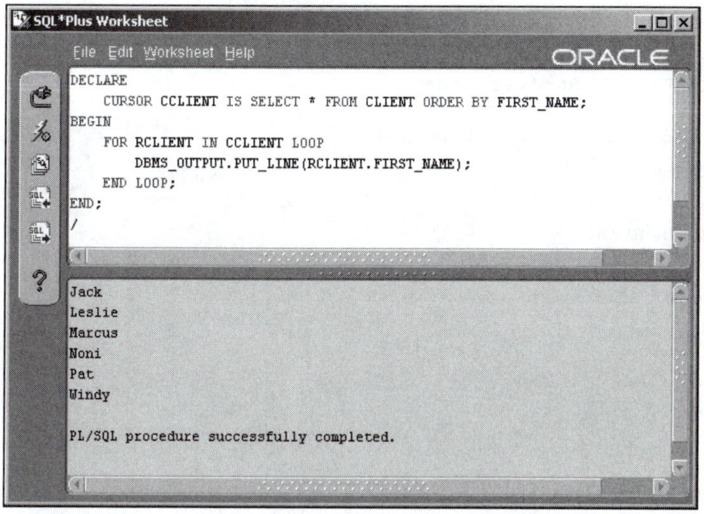
SQL*Plus Worksheet
File Edit Worksheet Help
ORACLE
DECLARE
CURSOR CCLIENT IS SELECT * FROM CLIENT ORDER BY FIRST_NAME;
BEGIN
FOR RCLIENT IN CCLIENT LOOP
DBMS_OUTPUT.PUT_LINE(RCLIENT.FIRST_NAME);
END LOOP;
END;
/
Jack
Leslie
Marcus
Noni
Pat
Windy
PL/SQL procedure successfully completed.

FIGURE 11-5 Executing an implicit cursor in PL/SQL

PL/SQL Programming Control Structures

A control structure in a programming language is a special type of keyword (or set of keywords) used to control the flow through a program. Most programming languages have largely the same control structures, with minor variations in syntax. PL/SQL control structures are:

The IF Statement

The IF statement allows for decisions:

```
IF condition THEN ...
ELSIF condition THEN ...
ELSE ...
END IF;
```

For example:

```
IF CLIENT_ID = 1 THEN RETURN 'one'
ELSIF CLIENT_ID = 2 THEN RETURN 'two'
ELSE RETURN 'neither'
END IF;
```

The CASE Statement

The CASE statement has both search condition:

```
CASE
CASE WHEN condition THEN ...
CASE WHEN condition THEN ...
ELSE ...
END CASE;
```

For example:

```
CASE
WHEN THEDATE < SYSDATE THEN RETURN 'yesterday';
WHEN THEDATE = SYSDATE THEN RETURN 'today';
WHEN THEDATE > SYSDATE THEN RETURN 'tomorrow';
ELSE ...
END CASE;
```

And selector-expression variations:

```
CASE selector
WHEN expression THEN ...
WHEN expression THEN ...
ELSE ...
END CASE;
```

For example:

```
CASE THEDATE
WHEN SYSDATE-1 THEN RETURN 'yesterday';
WHEN SYSDATE THEN RETURN 'today';
WHEN SYSDATE+1 THEN RETURN 'tomorrow';
ELSE ...
END CASE;
```


Loops in PL/SQL

Looping constructs include the FOR loop, the WHILE loop, and the LOOP...END (endless loop).

The FOR loop iterates through a known set of values:

```
FOR counter IN [REVERSE] lower-value .. higher-value LOOP
     EXIT WHEN expression
END LOOP;
```

For example:

```
DECLARE
     STEP INTEGER;
BEGIN
     FOR STEP IN 1..5 LOOP
          DBMS_OUTPUT.PUT_LINE(TO_CHAR(STEP));
     END LOOP;
END;
/
```

The WHILE loop continues as long as a condition holds true:

```
WHILE condition LOOP
     ...
     EXIT WHEN expression
END LOOP;
```

For example:

```
DECLARE
     STEP INTEGER DEFAULT 1;
BEGIN
     WHILE STEP < 10 LOOP
```

```
            DBMS_OUTPUT.PUT_LINE(TO_CHAR(STEP));
            STEP := STEP + 1;
      END LOOP;
END;
/
```

The LOOP...END LOOP loop allows for a potentially infinite loop, iterating without a condition on the loop statement. This loop can be used as a WHILE loop, exiting after the first iteration and a condition is met:

```
LOOP
      EXIT WHEN expression
      ...
END LOOP;
```

This LOOP...END construct can be used as an UNTIL loop (iterates until a condition is true) by placing an EXIT WHEN clause just before the END LOOP statement:

```
LOOP
      ...
      EXIT WHEN expression
END LOOP;
```

For example:

```
DECLARE
      STEP INTEGER DEFAULT 1;
BEGIN
      LOOP
            DBMS_OUTPUT.PUT_LINE(TO_CHAR(STEP));
            STEP := STEP + 1;
            EXIT WHEN STEP > 5;
      END LOOP;
END;
/
```

Sequencing Controls

Sequencing controls include the GOTO command and the NULL statement. A sequencing control is allowed to disrupt to logical flow of a program through its lines of code. The NULL statement has been used repeatedly in this book (it does nothing). The GOTO statement allows branching from one part of a program to another, either up or down the code:

```
 BEGIN
<<labelone>>
      ...
IF condition THEN GOTO labelone; END IF;
...
IF condition THEN GOTO labeltwo; END IF;
      ...
<<labeltwo>>
...
END;
```

That's PL/SQL. Next we will examine bulk importing and exporting using the export and import procedures, and Data Pump versions.

BULK IMPORTS AND EXPORTS USING DATA PUMP

In past versions of Oracle Database the export (exp) and import (imp) utilities were used to export data from a database as a dump, and import data back into a database from that dump. Data Pump export (expdp) and import (impdp) utilities are much the same as their older counterparts, except they are more versatile and much faster.

Exporting and importing is allowed at all logical layers such as export of individual tables, entire schemas, groups of objects, or even an entire database. Those same layers of object structure can again be imported. The idea of the export and import utility is to allow a copy from a database of a complete set of data and its metadata. In other words, when exporting a table you not only export the rows in the table but also the columns and datatypes of all columns in the table. This allows you to re-create the table in its entirety in a different schema or even different database. Database exports can even be used to migrate and upgrade between different versions of Oracle Database.

Some of the major differences between Data Pump exports/imports and the older versions are that Data Pump can be executed in parallel, failed or stopped jobs can be restarted, metadata can be filtered out (copying row and column values only—not their definitions), and version control, among other things.

Figure 11-6 highlights that all of Data Pump export, import, and SQL Loader utilities can be configured, managed, and executed from within the Database Control. This book will demonstrate command-line versions of these utilities for two reasons. First, the command-line versions are a better teaching tool. Second, with my version of Oracle 10g (10.2.0), running on a Win2K, I could not get any of these utilities to work properly running from within the Database Control.

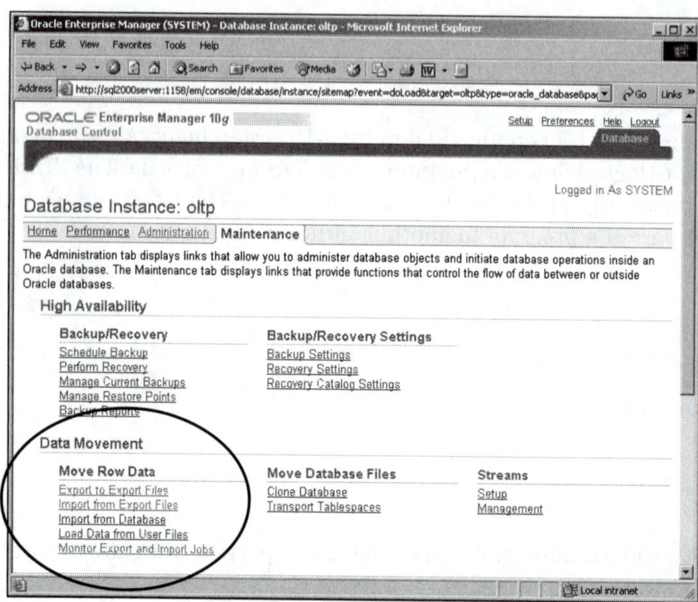

FIGURE 11-6 Data Pump exports and imports, plus SQL Loader in the Database Control

Exporting Using Data Pump

Whenever I use export or import utilities, the first thing I do is go into a shell and bring up the help screen for the utility by typing expdp help=y, and hitting the Enter key. Let's do that:

1. Start a shell. In Windows click **Start/Run**, type **cmd**, and click **OK**. For Unix or Linux you will have to ask your instructor how to get a shell because it depends on the interface you are using.
2. When a shell is on the screen, type **expdp help=y**. (Case is important in Unix or Linux.) Read and understand what all the options do. The options that describe various facets of the export utility should be obvious from the help screen, such as individual objects, starting and stopping jobs, direct path loads, parallelism, and so on.
3. Let's do a simple export. The minimum syntax to execute an entire schema (based on the connected user), would be a command like the following (do not execute this command yet):

```
expdp classmate/
classpass@oraclass DIRECTORY=dmpdir DUMPFILE=classmate.dmp
```

4. The DIRECTORY parameter requires a directory object to be created in the database. At the end of the previous chapter you learned how to create directory objects for external tables. Now you can create a directory object for Data Pump C/. Log in a sys:

```
CREATE OR REPLACE DIRECTORY DP AS 'C:\';
```


NOTE

The directory is created on the database server computer, not a client computer. The older EXP and IMP utilities allowed exports and imports across a network to and from whatever computer you were working on.

5. Now you can execute the expdp utility and export as shown below. The following command will actually export the SYSTEM user. You will likely have to log in to your computer as the Windows Administrator user (root on Unix or Linux). Your instructor may allow this, but if not don't be too concerned:

```
expdp system/password@oraclass DIRECTORY=dp DUMPFILE=system.
dmp NOLOGFILE=Y
```

NOTE

The NOLOGFILE option is specified as the utility requires a log to an administrative or superuser to write log files to the dump directory. In their current form Data Pump utilities are for administrative-level users only.

6. This command will export the CLASSMATE schema (overriding the database connection string). Substitute for the directory object you created previously. You will likely have to log in to your computer as the Windows Administrator user (root on Unix or Linux). Your instructor may allow this, but if not don't be too concerned:

```
expdp system/password@oraclass DIRECTORY=dp DUMPFILE=classmate2.
dmp SCHEMAS=('CLASSMATE') NOLOGFILE=Y
```

NOTE

The username is changed to the SYSTEM user (DBA user) because special privileges are involved for a non-DBA user to write to the disk. One factor is the setting of the UTL_FILE_DIR that allows any user to write to the disk structure of a database server computer. That is a serious security issue!

7. Developers should probably continue to use the older versions of the import (IMP) and export (EXP) utilities for development work as Data Pump technology appears to require high-level, superuser, even root (Unix and Linux) and Administrator (Windows) level access.
8. Stay logged on for the next exercise.

The usability of Data Pump technology is in question because it is only available to database- and operating-system level administrative personnel. The older import and export utilities serve to allow exports and imports across a network. Generally these utilities are secure because users have to have the schema password names to access schemas.

Importing Using Data Pump

Data Pump import is the opposite of Data Pump export, allowing imports into a database from Data Pump exported dump files. Try this brief exercise:

1. Start a shell. In Windows click **Start/Run**, type **cmd**, and click **OK**. For Unix or Linux you will have to ask your instructor how to get a shell because it depends on the interface you are using.
2. When a shell is on the screen type **impdp help=y** (case is important in Unix or Linux). Once again, read and understand what all the options do. The options that describe various facets of the import utility should be obvious from the help screen, such as individual objects, starting and stopping jobs, direct path loads, parallelism, and so on.

NOTE

Direct path loads are very important to performance as they allow much faster performance during imports, by appending directly to the ends of datafiles, without having to pass through the SQL engine.

The most likely use of these utilities are as fast backups and as an easy method of copying databases. That concludes our discussion of Data Pump export and import utilities. The older versions of Data Pump Import and Export utilities are the Import utility (IMP) and the Export utility (EXP). In general, IMP and EXP are easier to use than Data Pump, with the same options, just fewer options.

BULK DATA LOADS WITH SQL*LOADER

The most significant factor for SQL*Loader and performance is that it can perform magnificently in direct path, parallel mode, and using external tables. Direct path loads using SQL*Loader allow appending to tables. Appending implies that existing blocks are not considered when data is added and only new blocks are used. Some situations will cause single tables and even entire SQL*Loader executions to execute using a conventional path load. For example, not specifying DIRECT=Y PARALLEL=Y causes conventional path loading. Additionally, any kind of SQL transformation functionality placed into SQL*Loader control files, such as TO_CHAR conversion functions, will cause direct path loading to switch to conventional path loading.

Once again, Figure 11-6 highlights that all of Data Pump export, import, and SQL Loader utilities can be configured, managed, and executed from within the Database Control. This book demonstrates command-line versions of these utilities for the same reasons stated for the Data Pump export, import, and SQL Loader utilities.

Obviously, direct path loads appending new blocks is not necessarily the best option when rows are deleted from tables because existing blocks will never have rows slotted into them. This could cause uncontrollable growth.

SQL*Loader is not limited to individual table loads. It can load into more than one table at once, considering all constraints. SQL*Loader can also perform fast direct loads with external tables.

Figure 11-7 shows the basic architecture of SQL Loader. SQL Loader simply takes a data input file and applies rules stored in a control file to that data, which maps the data into tables in the database. Because of the loading of data into the database SQL Loader also generates a log file (a history of activity), a discard file (any rows discarded because they are not wanted in the database by the person loading the data), and finally a bad rows file (rows wanted but not allowed because they contain errors).

Before going any further, one very important aspect of using the SQL Loader utility is the difference between direct and conventional path loads.

Direct Versus Conventional Path Loads

A direct path load is potentially much faster than conventional path loads. A parallel direct load is even better. There is also an external table load. Let's look at these options and explain what each is, so that you can understand the difference:

- **Conventional path load**: Rows are parsed into an array structure according to field specification, as defined by SQL Loader and the table to be added to. Then all elements in the array are processed by a bulk SQL INSERT command, passing through the SQL engine, and being added to a table as any INSERT statement would do. In other words, rows are physically added to extents anywhere within the physical segment of the underlying datafiles.

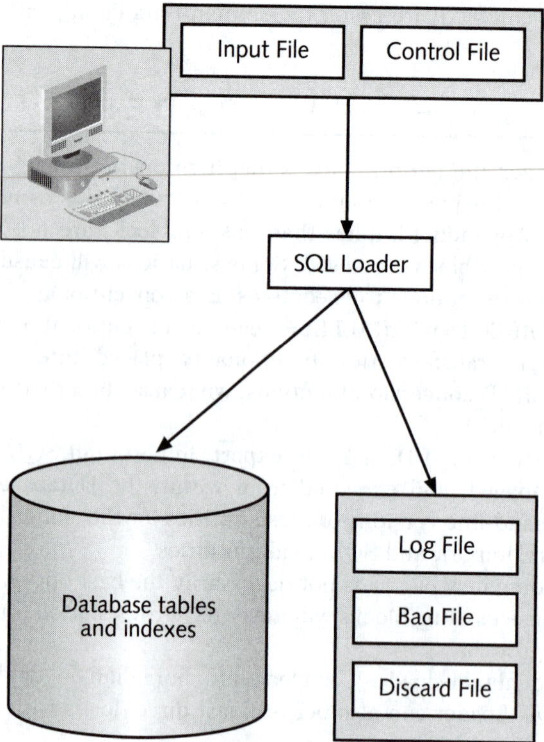

FIGURE 11-7 SQL Loader architecture

- **Direct path load**: A direct path is similar to a conventional path load. However, the load writes data to the end of a table into new block and extent structures, creating new blocks as it goes. Direct path loads bypass not only the regular SQL engine but also any existing physical blocks and extents that may have available space. In other words, there is little checking and validation with existing data, and thus little conflict with existing data. That's why a direct path load is so much faster than a conventional path load.
- **Parallel direct path load**: A parallel direct path load is a variation on a direct path load and is potentially faster than a direct path load. In this case multiple loading sessions can execute in parallel with each other, loading data into one or more physical areas of a datafile in parallel.
- **External table load**: This option creates and loads an external table.

Now let's examine each part of the SQL Loader architecture, beginning with the input data file.

SQL Loader Input Data Files

The input data files provide the data loaded in SQL*Loader. Input data file contents can be fixed-length rows, variable-length rows, or a single stream (string). Input data can consist of a single data file in the control file, or a separate data set in the input data file. Therefore, the input data file does not have to exist and that input data can all be included in the control file. To avoid confusion, it is best to divide things into the different appropriate files. In this section we gradually build up the various files for an example you can run at the end of this section:

1. Open the data file called **ch11.dat** in the **Data\Chapter11** directory on your student disk and look at it.
2. The data file you are looking at is a comma-delimited data input file you will ultimately load into the CLIENT table of the CLASSMATE schema, using SQL Loader.
3. The data file looks like the following list—comma delimited between columns and new line delimited between rows:

```
2001,Joe,Soap,123-4567,joesoap@yahoo.com
2002,Jim,Bloggs,324-3987,jimbloggs@aol.com
2003,Jimmy,Bean,459-0987,jimbean@aol.com
```

Next is the control file.

The SQL Loader Control File

One or more data files can be specified in the control file. In our case we will specify a single data file, the ch11.dat file C/:.

1. Open the control file called **ch11.ctl** in the **Data\Chapter11** directory on your student disk and examine it.
2. The control file you are looking at will load the data in the data file you looked at before called **ch11.dat**.
3. The control file looks like this:

```
LOAD DATA
   INFILE 'data\Chapter11\ch11.dat'
INTO TABLE client APPEND
FIELDS TERMINATED BY "," TRAILING NULLCOLS
(
    CLIENT_ID INTEGER,
    FIRST_NAME CHAR(10),
    LAST_NAME CHAR(20),
    CONTACT_PHONE CHAR(15),
    CONTACT_EMAIL CHAR(30)
)
```

4. SQL Loader datatypes are slightly different from Oracle Database tables in that they are somewhat simpler but also easier to define. Strings are fixed length and whole numbers are integers, as shown above.

5. The bad rows and discard rows can be defined after the line INFILE ... in the previous control file sample. This example will have no discards and no bad rows, but the control file could have lines like this added to it:

```
BADFILE 'solutions\Chapter11\bad.dat'
DISCARDFILE 'solutions\Chapter11\discard.dat'
```

6. The FIELDS TERMINATED BY option dictates that rows in the input file are comma delimited. In other words, each column's data is terminated by a comma, except for the last column, CONTACT_EMAIL, which is terminated by a new line character.
7. The TRAILING NULLCOLS option implies that any fields with whitespace will have trailing whitespace removed.
8. The APPEND option will cause rows to be added to the end of the CLIENT table.

The SQL*Loader control file contains a mapping between input data and table structures into which data is to be loaded. The control file can be simple or highly complex depending on requirements, and can contain details as described in the following sections.

Row Loading Options

Rows can be loaded as shown in the following syntax, INSERT being the default:

```
LOAD DATA INFILE ... BADFILE ... DISCARDFILE ...
INTO TABLE table1 [ INSERT | REPLACE | TRUNCATE | APPEND ]
FIELDS TERMINATED BY "," TRAILING NULLCOLS
(
    FIELD1 INTEGER EXTERNAL,
    FIELD2 INTEGER EXTERNAL,
     ...
    FIELDn ...
)
```

- **INSERT**: The default; requires an empty target table.
- **REPLACE**: Removes and replaces all rows.
- **TRUNCATE**: Truncates all rows first and can cause constraint issues. Truncation destroys all rows without the option to undo changes.
- **APPEND**: Adds to the end of files without removing existing data.

Column Delimiters

Field definitions in the control file can be defined using positioning or specific datatypes and sizes. The first example below splits data into separate columns using positions on each line, assuming fixed-length records:

```
LOAD DATA
INTO TABLE table1 TRUNCATE
(
   FIELD1POSITION(001:010) CHAR(10) TERMINATED BY WHITESPACE,
   FIELD2 POSITION(011:030) CHAR(20) TERMINATED BY WHITESPACE,
... ,
    FIELDn ...
)
```

As already seen in the control file example in the recent practice session, this example specifies values into specific places within each row, defined as having specific datatypes and lengths but separating fields in each row by using a comma as determined by the FIELDS TERMINATED BY clause:

```
LOAD DATA
INTO TABLE table1 TRUNCATE FIELDS TERMINATED BY "," TRAILING NULLCOLS
(
    FIELD1 INTEGER EXTERNAL,
    FIELD2 INTEGER EXTERNAL,
     ...
    FIELDn ...
)
```

Next let's look at load filters.

Load Filters

A load filter in SQL Loader uses a WHEN clause, which can be used to discard rows from the loading process, potentially placing discarded rows in the discard file and not loading them into target database tables. In the example below only rows with specified values are loaded, in this column 2 must be filled with the value "ABC". Others are ignored and placed into discards:

```
LOAD DATA INFILE ... INTO TABLE table1 TRUNCATE
FIELDS TERMINATED BY "," TRAILING NULLCOLS
WHEN (2) = 'ABC'
(
    FIELD1 INTEGER EXTERNAL,
    FIELD2 INTEGER EXTERNAL,
     ...
    FIELDn ...
)
```

Unwanted Columns

Unwanted columns can be removed from the loading process and replaced with NULL values by using the FILLER clause. Other column values in the same row are still loaded, as illustrated in the following example:

```
LOAD DATA INFILE ...
INTO TABLE table1(
col1 CHAR(10),
col2 FILLER CHAR(12),
col3 CHAR(1) TERMINATED BY WHITESPACE)
TRUNCATE FIELDS TERMINATED BY "" TRAILING NULLCOLS
...
```

Control File Datatypes

SQL*Loader has a limited datatype set with specific commands for handling different situations for each of the various datatypes, namely strings, dates, and numbers:

- **Strings**: CHAR[(n)].
    - DEFAULTIF col1=BLANKS sets to spaces or NULL.
    - NULLIF col1=BLANKS replace with NULL.

NOTE

The CHAR datatype in SQL*Loader is different from the CHAR datatype in an Oracle 10*g* table. SQL*Loader CHAR datatypes allow both fixed, and variable-length strings. The table column datatype controls how data is added to the table.

- **Dates**: DATE, TIME, TIMESTAMP and INTERVALs.
- **Numbers**: Numbers are externally or internally defined C/:.
  - EXTERNAL { INTEGER | FLOAT | DECIMAL } is defined by the number loaded. DEFAULTIF col1=BLANKS sets to 0. NULLIF col1=BLANKS replace with NULL.
  - Non EXTERNAL. INTEGER(n), SMALLINT, FLOAT, DOUBLE, BYTEINT, and DECIMAL(p,s).

NOTE

The clause BLANKS can be replaced with a literal string or number value.

Embedded SQL Statements

Including SQL statements in the control file, to be applied to every row of data input, disables APPEND loads. You will not get an error, only a message in SQL*Loader output indicating a switch from direct path to conventional path load. The following example contains a column reference for COL1, a literal SQL string enclosed in double quotes for COL1 and COL2, and finally embedded functionality and a custom written function for COL2 and the TEMPERATURE columns:

```
LOAD DATA INFILE 'input.txt' INTO TABLE table1 TRUNCATE
FIELDS TERMINATED BY "," TRAILING NULLCOLS
(
     col1 CHAR(10)"UPPER(:col1)"
    ,col2 CHAR(12)"UPPER(INITCAP(SUBSTR(:col2,4,20)))"
,temperature FLOAT"FToC(temperature)"
)
```

If you would like to know more about control file syntax, refer to the Oracle documentation in the Utilities manual.

The Parameter File

SQL*Loader can include a parameter file containing repeated settings, across multiple executions of SQL*Loader. Let's look at the parameter file you will use:

1. Open the data file called **ch11.par** in the **Data\Chapter11** directory on your student disk and look at it.
2. The file you are looking at is as shown here:

   ```
   USERID = CLASSMATE/CLASSPASS@ORACLASS
   DISCARDMAX = 2
   ERRORS = 1000000
   ```

NOTE

The PARALLEL = TRUE and DIRECT = TRUE options are omitted because constraints are enabled on the CLIENT table.

3. In the parameter file shown you connect SQL Loader to the database as the CLASSMATE user. The maximum number of rows that can be discarded in a discard file, using a WHEN clause filter, is two rows. The DISCARDFILE option is not used, but it can stay in the parameter file as it does no harm. Setting ERRORS to 1,000,000 is the maximum value for this parameter, allowing up to 1,000,000 row entries into the bad file (determined by the BADFILE option).

Load your new client rows. As with the Data Pump import and export. Now let's run SQL Loader:

1. Start a shell again (unless you already have the same one running). In Windows click **Start/Run**, type **cmd**, and click **OK**. For Unix or Linux you will have to ask your instructor how to get a shell because it depends on the interface you are using.
2. When a shell is on the screen type **sqlldr help=y** (case is important in Unix or Linux). Read and understand what all the options do. The options that describe various facets of the sqlldr utility should be obvious from the help screen.
3. You can execute your SQL Loader scripts using the command shown below. This command will only run on a database server. If you are on a client computer, the paths must all exist on the database server itself, not your client computer:

```
sqlldr control=<path>\data\Chapter11\ch11.
ctl log=<path>\data\Chapter11\ch11.
log parfile=<path>\data\Chapter11\ch11.par
```

4. Make sure the command above is typed into a single line, even if it wraps. Also replace the <path> variables with the appropriate path names on your database server computer.
5. Some of the most likely uses of SQL*Loader are to bulk load large amounts of data into a data warehouse, or when importing data from outside (another company, a legacy system), into an existing database.

Chapter Summary

- PL/SQL can be coded into named blocks or anonymous blocks that can be executed once.
- Named blocks are stored in the database and can be executed later.
- Named blocks can be procedures (accepting parameters), functions (accepting parameters and returning a single value), triggers (executed on a database event), or packages (grouping multiple named blocks into a package).
- A cursor is an area of memory reserved for a SQL statement to execute within or during its execution.
- PL/SQL allows standard programming control constructs such as IF statements, CASE statements, and loops.
- The Data Pump export utility is used to dump database objects such as tables and schemas to an output file.
- The Data Pump import utility is used to import files created by the Date Pump export utility into a database.
- SQL Loader uses a control file to map data input between the data input and the tables into which it is loaded.
- A client-side utility can be executed through an Oracle TNS network name, to a database. Both client-side and server-side (the server contains the database) computers can have a TNS configuration. Therefore, client-side utilities can be executed from both the server side and the client side.
- Most Oracle 10*g* Database utilities can execute as both client-side and server-side utilities, although some utilities will require that output files are written to the server-side computer.

Review Questions

1. Which PL/SQL command could you use to jump up a few lines to a label?
2. How can you change the following anonymous procedure to make it execute in SQL*Plus? It already runs in SQL*Plus Worksheet.

```
BEGIN
    NULL;
END;
```

3. Which of these are PL/SQL objects? (Choose four.)
   a. Procedure
   b. Package
   c. Function
   d. Event
   e. Trigger

4. Which of these commands belongs in the definition and management of an implicit cursor?
   a. OPEN
   b. CLOSE
   c. GOTO
   d. CURSORS
   e. None of the above
5. Data Pump technology is more efficient than older versions of import and export utilities. True or False?
6. Which is the fastest method of executing SQL Loader? Will it always be executed?
   a. Conventional path
   b. Direct path
   c. Parallel direct
   d. All the above perform the same

Exam Review Questions—Oracle Database 10*g*: Administration (#1Z0-042)

1. Which line in the trigger definition below has an error?

```
1. CREATE OR REPLACE TRIGGER uCLIENT
2.    AFTER UPDATE ON CLIENT
3. BEGIN
4.    DBMS_OUTPUT.PUT_LINE('Client''s name changed from '
    ||:old.FIRST_NAME||' to '||:new.FIRST_NAME);
5. EXCEPTION WHEN OTHERS THEN
6.    DBMS_OUTPUT.PUT_LINE(SQLERRM(SQLCODE));
7.    RAISE;
8. END;
9. /
```

2. Which of these are not valid syntax options for a trigger?
   a. BEFORE INSERT ON <tablename>
   b. AFTER INSERT ON <tablename>
   c. INSTEAD OF UPDATE ON <tablename> [OF <columnname>]
   d. AFTER SELECT ON <tablename>
   e. All are valid
3. What type of file is this?

```
LOAD DATA
   INFILE 'data\Chapter11\ch11.dat'
INTO TABLE client APPEND
FIELDS TERMINATED BY "," TRAILING NULLCOLS
(
    CLIENT_ID INTEGER,
    FIRST_NAME CHAR(10),
    LAST_NAME CHAR(20),
    CONTACT_PHONE CHAR(15),
    CONTACT_EMAIL CHAR(30)
)
```

4. What type of SQL Loader file would all these entries be stored in?

```
USERID = CLASSMATE/CLASSPASS@ORACLASS
DISCARDMAX = 2
ERRORS = 1000000
PARALLEL = TRUE
DIRECT = TRUE
```

Hands-on Assignments

1. Write and execute a stored procedure called GETEMPLOYEES that displays all employee names on the screen by concatenating all columns. Save your script in your **Solutions\Chapter11** directory in a file named **h1101.sql**. (*Hint*: Use an implicit cursor FOR loop.)
2. Write a stored function called GETINITIALS, which finds only the initials of each of the employees retrieved by the procedure called GETEMPLOYEES. Save the script you use in your **Solutions\Chapter11** directory in a file named **h1102.sql**.
3. Change the stored procedure created in step 1 to use the function created in step 2. Execute the changed stored procedure. Save the script you use in your **Solutions\Chapter11** directory in a file named **h1103.sql**.

Case Project

1. The Global Globe newspaper database now has several tables. Your manager has asked you to add 50 new employee names. Assume the input data file you receive is a comma-delimited file, containing all columns in the EMPLOYEE table, in the correct order, with all the correct datatypes. Save your work in a file called **case1101.sql** in the **Solutions\Chapter11** directory on your student disk, as a SQL Loader control file.

CHAPTER **12**

SECURITY MANAGEMENT

LEARNING OBJECTIVES

In this chapter, you will:

- Create, modify, and remove users
- Discover when and how to create, use, and drop profiles
- Manage passwords
- View information about users, profiles, passwords, and resources
- Identify and manage system and object privileges
- Grant and revoke privileges to users
- Understand auditing capabilities and practice using auditing commands
- Discover when and why to use roles
- Learn how to create, modify, and remove roles
- Learn how to assign roles
- Examine data dictionary views of roles
- Assign roles and privileges using the Enterprise Management console

INTRODUCTION

You have already been exposed to several users during previous exercises in this book. For example, the CLASSMATE user owns the sample data used in the book. You generally logged on as the CLASSMATE user to perform your exercises. Making changes to the database settings required you to log on as the SYSTEM or SYS user.

This chapter is divided into three distinct parts. The first part gives you in-depth coverage of how to create, modify, and remove database users. In addition, you learn how to set limits on the amount of space and other resources users may consume and query the data dictionary views for information about users.

In the second part you learn about how to allow users access to the database and its contents. You examine the types of privileges available and how to combine privileges into convenient groups. **Privileges** are tasks you are authorized to carry out in the database. Finally, in the third part of this chapter you learn about how to make groups of privileges, called roles. You'll find out how roles can simplify security and what predefined roles Oracle 10*g* provides.

USERS AND RESOURCE CONTROL

When a new database instance is created, at least two users are created. These two users are the SYS and the SYSTEM users. They are allowed to create the tables that form the backbone of the database. The SYS user owns most of the tables needed to run the database, as well as the data dictionary views. The SYS user also owns a host of packages and procedures built into the database both for your use, and internal use by the Oracle Database software. The SYS user can perform the highest-level tasks, such as starting up and shutting down a database instance, as well as performing backup and recovery tasks. You should log on as the SYS user to do these tasks. However, because the SYS user can drop the SYS tables (this could crash the entire system), you should log on as the SYS user for only the very highest level of tasks, and not for other, more routine tasks.

The SYSTEM user owns some tables, packages, and procedures. It is given the DBA role, which allows it to perform routine database administration tasks such as creating new users, monitoring database activities, and regulating database resources. You should log on as the SYSTEM user when you want to perform these routine tasks.

NOTE

A role is a specialized database object used to group privileges into a single convenient database object. A privilege is something that is granted to, or revoked from a user. Privileges allow a user to perform specific tasks, such reading rows from a table or creating a table.

During database creation, Oracle 10g creates other users to help it install some of the features of the database. For example, the MDSYS user is created to own functions, packages, views, and objects related to Oracle Spatial.

NOTE

Oracle Spatial is a feature for storing dimensional data in the database.

After database creation, these users are disabled to prevent anyone from logging on to the database with their accounts. You'll learn about disabling users later in the chapter.

After the database instance is up and running, you create users that own tables and other objects needed to implement whatever business system you plan to implement. Creating new users retains the system tables of the database, and your individual user tables in distinct logical groups. You can limit the ability of each user to create objects by setting storage limits on tablespaces. You can also create a special group of resource limits, called a profile, and assign the profile to any user. Profiles allow for sharing of resources between multiple tasks, allowing resource sharing, continually switching from one task to the next. In other words, a queue of tasks is continuously serviced by allocating specific amounts of processing time to each task, switching through the queue, returning repeatedly to service the same task.

After creating users to own the business tables, you need to create users who access these tables, but don't need to create any tables of their own. These are the users that the term end-user fits best: the people who use the tables created to support their business functions by adding, changing, removing, and viewing data. You can control how often the user's password expires and what resources are available to these kinds of users.

Begin exploring users by learning how to create new users in the next section.

Creating New Users

The two types of users (those who own tables and other objects and those who use tables and objects owned by other users) are created with the same command. Figure 12-1 shows the syntax of the CREATE USER command and outlines the commands that are explained in the first section of this chapter.

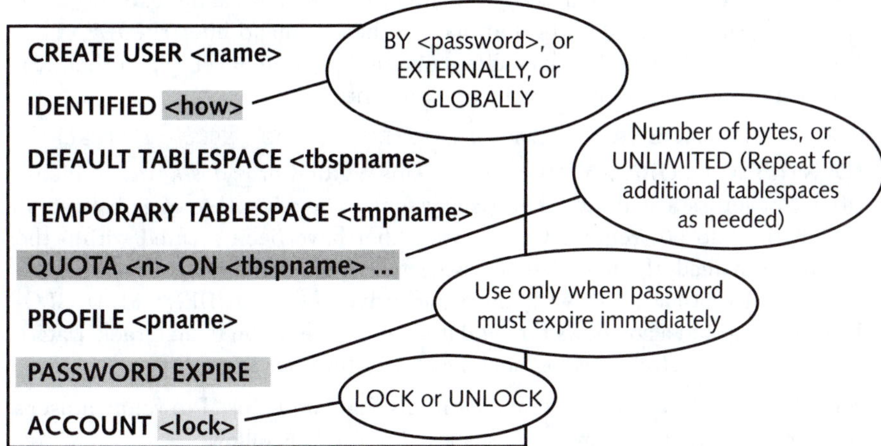

FIGURE 12-1 You must have the CREATE USER privilege to create a new user

The first two lines in the command syntax are actually the only required lines. The first line defines the new user's name. As you create users, keep in mind that user names are subject to the same naming conventions as tables and other database objects: a user name must be no more than 30 characters long, made up of alphanumeric characters (plus the

$, #, _ symbols), and cannot start with a number. In addition, to avoid confusion, it is recommended that a user name not be identical to a keyword, such as TABLE or INDEX.

The second line defines how the database confirms the user's identification. A user's identity can be validated in three ways:

- **IDENTIFIED BY <password>:** Oracle 10g is responsible for authenticating the user. The user must provide a password whenever logging on to the database. This type of user is called a local user. The password you provide must comply with the same naming conventions as other database objects. For example, you might use the password BLUESTARFISH. Many studies have shown, however, that passwords that contain only words are easier to break or guess than passwords that contain at least one number. If you require more complexity in passwords, Oracle 10g provides a utility for password management. See the section on managing passwords later in this chapter.
- **IDENTIFIED EXTERNALLY:** Oracle 10g allows a user whose user name matches his operating system logon name combined with a prefix such as OPS$ (the default). The prefix is determined by the OS_AUTHENT_PREFIX initialization parameter. The prefix can be set to null, so that the two user names match exactly. As an example, if the OS_AUTHENT_PREFIX is WIN, and the user's operating system logon name is SBAILEY, then the Oracle user name you create for SBAILEY should be named WIN_SBAILEY. Users identified externally do not have to supply their user names or passwords to log on; they simply type a slash in place of user name and password. Oracle Database retrieves the user's name from the operating system and assumes that the operating system has already validated the user. For example, to log on to SQL*Plus, the user could type **sqlplus /** on the command line. The IDENTIFIED EXTERNALLY option also allows operating system level SSL authentication, using a certificate handshake as shown here:

```
CREATE USER secureSSL IDENTIFIED EXTERNALLY AS 'SSLcertificate';
```

- **IDENTIFIED GLOBALLY AS <name>:** This method provides a way to identify users across distributed database systems. The advantage of global users is that they do not require Oracle users that have been created within the database. Instead, the user can be associated with a common user name that is shared by other users with the same privileges. The IDENTIFIED GLOBALLY AS option also allows for validation and connection to an Oracle Database using a directory service, such as Oracle Internet Directory.

The remaining syntax shows you optional additions that can be used to refine a user's capabilities in a database. These optional clauses are described as follows:

- **DEFAULT TABLESPACE:** This is only needed if the user is allowed to create database objects, such as tables or indexes. Otherwise, no default tablespace is needed. If you do not assign a default tablespace, the new user inherits the default tablespace from the default tablespace specified on database creation. Do not assign the SYSTEM or SYSAUX tablespaces as the default tablespace for any user. The SYSTEM and SYSAUX tablespaces contain database objects that are critical to the functioning of the database. You should especially avoid creating users with SYSTEM as their default tablespace.

- **TEMPORARY TABLESPACE:** Sometimes Oracle 10g needs temporary space on disk to sort query results, execute a join, or other tasks. These tasks are performed in the user's temporary tablespace. If you omit this clause, the user's temporary segments are stored in the default temporary tablespace named either in the CREATE DATABASE command, or in a subsequent ALTER DATABASE command. If no default temporary tablespace is designated, either in the CREATE USER or CREATE DATABASE commands, then the user's temporary segments are stored in the SYSTEM tablespace. Storing temporary segments in the SYSTEM tablespace is inadvisable because the SYSTEM tablespace houses so many critical tables for the operation of the database. When assigning a temporary tablespace to a user, the tablespace you specify must be a temporary tablespace.
- **QUOTA:** The default quota (the maximum amount of storage space a user can be allocated in the tablespace) is unlimited. Allowing unlimited storage space can sometimes lead to database errors because you might run out of disk space. For example, an inexperienced programmer may insert rows multiple times when testing an application and accidentally consume all the storage space available in the tablespace. In this case, other users who try to access the tablespace cannot create new tables or insert rows because they cause new extents to be allocated. Remember, even when a tablespace is set to automatically allocate new extents as needed (AUTOEXTEND ON), there is a physical limitation at the point when the tablespace cannot grow because its datafiles have reached the maximum storage capacity of the physical devices. Limiting each user's total storage space prevents uncontrolled growth of the database. Limiting the temporary storage space prevents one user from consuming so much of the temporary tablespace that other user operations are slowed or halted. However, limiting space may also cause a logical I/O error when Oracle 10g will not allow a datafile to be expanded anymore. This causes the same effective scenario as running out of disk space. Here are a few more points about quotas:
    - Define quotas in any tablespace except the undo tablespace, which is used only for undo records.
    - Define quotas to as many tablespaces as you want.
    - Set a tablespace quota to unlimited by using the UNLIMITED keyword.
    - Set a tablespace quota to zero by using the number zero as the quota.
- **PROFILE:** All users have a profile. If you do not name a profile when you create the user, Oracle 10g assigns DEFAULT to the user. The DEFAULT profile is a predefined profile created just after the database is created.
- **PASSWORD EXPIRE:** You can set the password to automatically expire by adding the PASSWORD EXPIRE clause to the statement. This optional clause tells Oracle 10g to create the user with the given password, then immediately cause the password to expire. An expired password makes the database prompt the user for a new password the next time he or she logs on to the database.

- **ACCOUNT:** This clause determines whether the user can log on to the database at all, even if he or she provides the correct password, or is validated externally by the operating system. Locking a user with the ACCOUNT LOCK clause completely locks this user out of the database. This is not usually done when a user is first created, although it is available. The default setting is ACCOUNT UNLOCK, which allows someone to log on with the new user name and password. The ACCOUNT LOCK option is most often used with the ALTER USER statement, preventing an existing user from accessing the database.

Now that you have examined the syntax, it is time to dive in and create some users. Complete the following steps to create new users and examine the components of the CREATE USER command:

1. To start the Enterprise Manager console in Windows, on the Taskbar, click **Start/All Programs/Oracle ... /Enterprise Manager Console**. In Unix or Linux, type **oemapp console** on the command line. The Enterprise Manager console login screen appears.
2. Start the SQL*Plus Worksheet by clicking **Tools/Database Applications/**SQL*Plus Worksheet from the top menu in the console.
3. Connect as the SYSTEM user. Click **File/Change Database Connection** on the menu. A logon window appears. Type **SYSTEM** in the Username box, the current password, *provided by your instructor*, in the Password box, and **ORACLASS** in the Service box. Leave the connection type as "Normal." Click **OK** to continue.
4. You are creating a new user named STUDENTA, who can create tables and other objects in the database. Begin by typing the first line of the command as shown here:

   ```
   CREATE USER STUDENTA
   ```

   Press the **Enter** key. Next, you assign a password, or define an alternative method of authentication. For this example, use the password TRUE#1 for the user by typing this line and pressing Enter, as you continue to code the CREATE USER statement:

   ```
   IDENTIFIED BY TRUE#1
   ```

5. To create a new user, you need to supply only the user name and password. However, in this example, you examine the remaining options of the CREATE USER statement. The first optional addition is assigning a default tablespace to the user. Assume that this user can create tables, and that you want the user to create tables in the USERS tablespace. Type this line and press **Enter** to continue:

   ```
   DEFAULT TABLESPACE USERS
   ```

6. Next, assign a default temporary tablespace. This is useful even when a user cannot create tables. Type this line, and press **Enter** to assign TEMP as the temporary tablespace for this user:

   ```
   TEMPORARY TABLESPACE TEMP
   ```

Now give the user 10 Mb in the USERS tablespace and 5 Mb in the USER_ AUTO tablespace. Type the following lines, and press **Enter** to continue to the next step:

```
QUOTA 10M ON USERS
QUOTA 5M ON USER_AUTO
```

NOTE

Oracle9*i* allowed a quota allocation on the TEMP tablespace. Oracle 10*g* does not.

7. The next line defines the user's profile. A later section in this chapter describes how to create profiles. For now, assign the user the default profile, which, conveniently, is named DEFAULT. Type the following line and press **Enter**:

   ```
   PROFILE DEFAULT
   ```

8. The final optional setting is the ACCOUNT clause, which can block a user from logging on to the database. For this user, type the **default setting**, the **semi-colon** (;), and then click the **Execute** icon to run the command.

   ```
   ACCOUNT UNLOCK;
   ```

   Figure 12-2 shows the complete command. SQL*Plus Worksheet replies, "User created."

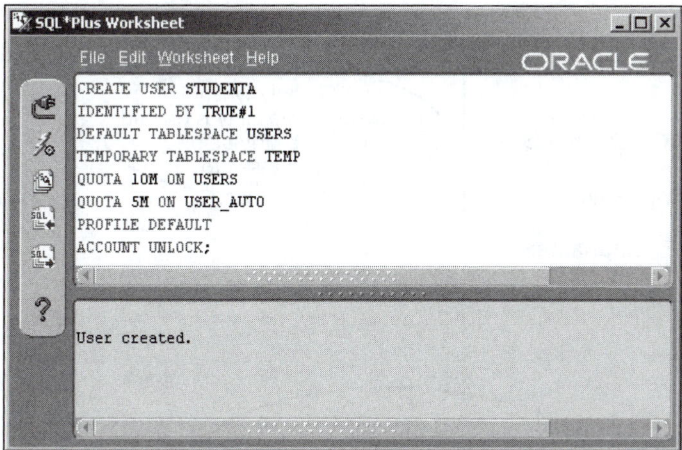

FIGURE 12-2 The new user STUDENTA has default tablespace and quotas allocated

9. Now make the following changes in your SQL*Plus Worksheet window: (1) change the user name to STUDENTB, and (2) change the password to TRUE#2. Execute the CREATE USER command again to create a second user named STUDENTB. The SQL*Plus Worksheet replies, "User created."

10. Type and execute one final command to give the users the system privilege needed to log on to the database:

```
GRANT CREATE SESSION TO STUDENTA, STUDENTB;
```

SQL*Plus Worksheet replies, "Grant succeeded."

11. Remain logged on for the next practice.

The new users you created have access to the database and quotas limiting their total storage usage. You might think that the users can begin creating a schema. However, these new users cannot create any tables until given more system privileges, which will be described in a subsequent section.

TIP

The default quota on all tablespaces for a user is zero unless you grant the RESOURCE role or the UNLIMITED TABLESPACE system privilege to the user. Use of the RESOURCE role is discouraged in Oracle 10*g*.

Modifying User Settings with the ALTER USER Statement

After a user is created, you can change the initial settings using the ALTER USER statement. Figure 12-3 shows the ALTER USER syntax. As you can see, all of the settings found in the CREATE USER statement are found here as well. In addition, a new clause, DEFAULT ROLE, is available. A later section in this chapter will describe roles, and when to modify a user's default roles.

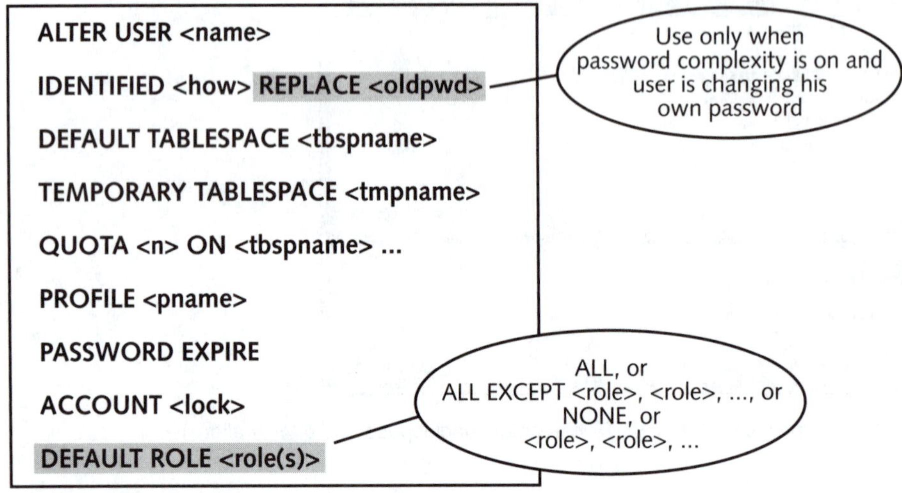

FIGURE 12-3 Modify most user settings with the ALTER USER statement

The EXTERNALLY AS 'SSLcertificate' and GLOBALLY AS 'DIRECTORYcertificate' options apply to the ALTER USER statement, as for the CREATE USER statement. You must have the ALTER USER system privilege to issue the ALTER USER statement. There is one exception: a user can run this command to change their own password. See the section on password management later in this chapter.

TIP

You cannot change a user's name; instead you must drop the user and create a new user with the new name.

Follow these steps to change some of the initial settings of the STUDENTA user, which you created in the previous section:

1. Currently, the STUDENTA user has quotas on three tablespaces. To view the settings, type the following query, and run it in SQL*Plus Worksheet. The results are shown in Figure 12-4:

```
SELECT USERNAME, TABLESPACE_NAME, MAX_BYTES, BYTES
FROM DBA_TS_QUOTAS
WHERE USERNAME = 'STUDENTA';
```


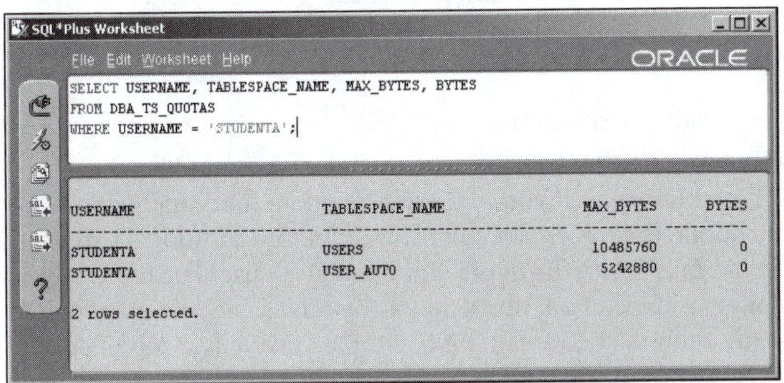

FIGURE 12-4 Quotas and the amount of space used is listed in the DBA_TS_QUOTAS metadata view

2. You want to allow STUDENTA to have unlimited storage on the USER_AUTO tablespace. Type and execute this command to modify the quota:

```
ALTER USER STUDENTA
QUOTA UNLIMITED ON USER_AUTO;
```

SQL*Plus Worksheet replies, "User altered."

3. In addition, you have changed plans, so you want to adjust the allocated 10 Mb for this user on the USERS tablespace to zero. This prevents STUDENTA from creating any new tables in the USERS tablespace. Any existing tables in the USERS tablespace would remain intact but are prevented from allocating additional extents. In other words, STUDENTA can only add rows until a new extent is required. Set the quota to zero by typing and running this command:

```
ALTER USER STUDENTA
QUOTA 0 ON USERS;
```

The SQL*Plus Worksheet replies, "User altered."

4. To confirm the settings, run the query you ran in Step 1 again. Figure 12-5 shows the new results. As you can see, the quota on USER_AUTO is now "-1." This is the number Oracle 10*g* uses to identify the quota as unlimited. The USERS tablespace no longer appears, which means the user has no quota in the USERS tablespace.

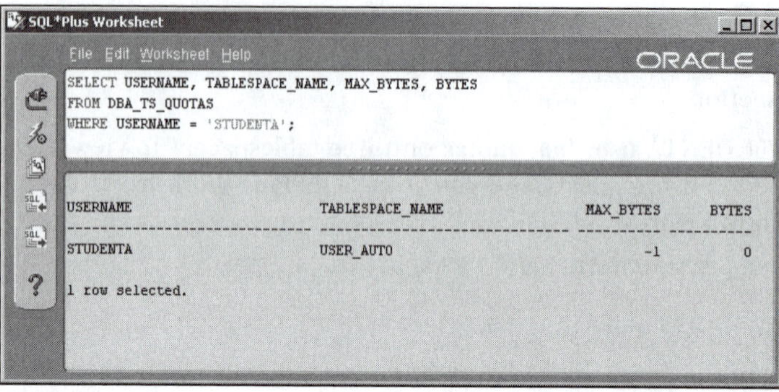

FIGURE 12-5 User STUDENTA has an unlimited quota setting on the USER_AUTO tablespace

5. Remain logged on for the next practice.

When you change a user's quota in a tablespace, what happens to the tables that the user has already created in the tablespace? If you increase the quota, nothing happens to the tables. If you decrease the quota, the tables can still be used, but no additional extents are allocated until the storage use falls below the quota. For example, a user has three tables using 5 Mb of space. The quota is reduced to 4 Mb. None of the tables can add an extent. Then, the user drops one of the tables. The remaining two tables use only 3 Mb of space. Now, either table can add an extent until the total storage reaches 4 Mb again.

If the quota on a tablespace with existing tables is reduced to zero, these tables remain but are never allowed to allocate more space. In other words, you can't add new rows, or change any rows that would require more physical space.

You have seen how to create and modify a user. Now, let's try removing a user.

Removing Users

Removing users requires the DROP USER system privilege, which the SYSTEM user has. The syntax for the command is simple:

```
DROP USER <user> CASCADE;
```

Use the CASCADE keyword if the user owns any tables or other database objects. For example, Figure 12-6 shows all the tables, indexes, and views owned by MELVIN.

NOTE

The user MELVIN does not exist in your database.

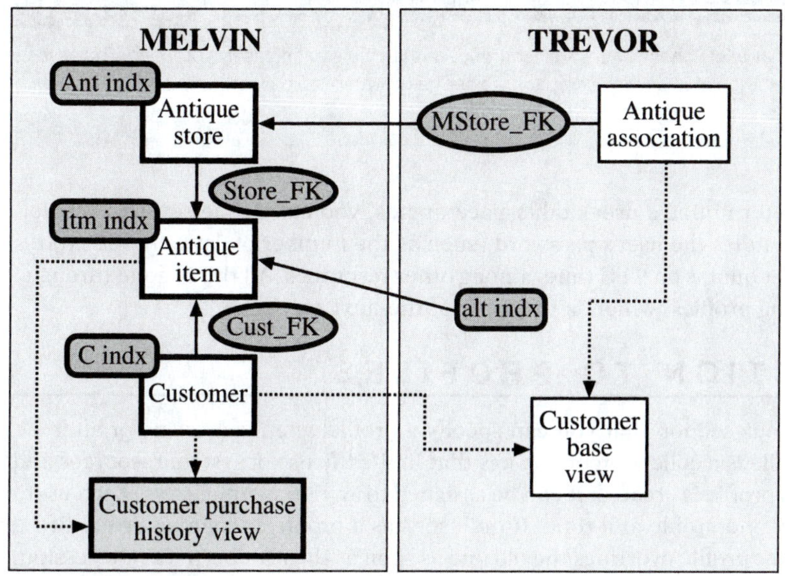

FIGURE 12-6 When the user MELVIN is dropped, some of the tables, indexes, and views owned by the user TREVOR are also affected

In addition, tables and views owned by TREVOR are shown. TREVOR has created a view that joins one of MELVIN's tables with his own. In addition, TREVOR has created a FOREIGN KEY constraint on one of his tables that references one of MELVIN's tables.

NOTE

The user TREVOR also does not exist in your database.

The items shown in Figure 12-6 with heavy outlines would be dropped when dropping the user MELVIN, with the CASCADE option. Views belonging to TREVOR would be marked invalid if they reference any of MELVIN's tables.

To drop a user named MELVIN, type the following command:

```
DROP USER MELVIN CASCADE;
```

To drop a user named TREVOR, who owns no tables or other database objects, type the following command:

```
DROP USER TREVOR;
```

Do not attempt to execute these above DROP TABLE commands because the users MELVIN and TREVOR are not in your database.

TIP

If a user has created other users, those users are not dropped when the creating user is dropped. The new users do not belong to the original user's schema. If a user has created tables you want to keep, do not drop the user. Instead, change the user account to LOCK status.

In addition to determining a user's tablespace quotas, you can define several aspects of how Oracle 10g handles the user's password (such as the number of days until it expires again) and you can set quotas on CPU time, among other resources. All this is done through creating and assigning profiles, which is the topic of the next section.

INTRODUCTION TO PROFILES

You saw in the previous section that you can specify a profile when you create or alter a database user. A profile is a collection of settings that limits the use of system resources and the database. After a profile is created, it can be assigned to any number of users. Each user can be assigned only one profile at a time. If the user has a profile and you assign a different profile, that new profile overrides the old one as soon as the user begins a new session. The user's current session is not affected by a profile change.

NOTE

Even after creating and assigning profiles to users, none of the limits are enforced until the database initialization parameter RESOURCE_LIMIT is set to TRUE. By default, the parameter is set to FALSE. You can change this parameter in the **init.ora** file, then restart the database instance. You can also modify this parameter dynamically using the ALTER SYSTEM command.

The default profile, named DEFAULT, has no limits on resources or database use. This works well for smaller systems in which resources are plentiful. However, as a system grows, resources may become stretched. This may cause some tasks to take longer to complete. For example, your system supports an online ordering system in which response time is critical. In addition, you have background jobs that process accounting and billing, which run at night. Now that the online ordering is international in scale, online users are complaining that response time during night hours is very slow. Your research shows that long-running queries from online users performing generalized searches on the database have caused other online users to wait for resources. To solve the problem, you can create a profile for online users to limit the amount of data blocks read and the amount of CPU time used in a single call. As a result queries that are too general and reading too much data at once, will be terminated.

A second use of profiles involves managing passwords. A profile does not contain a user's password. However, a profile determines how often the user must change his password, how many days must pass before an old password can be reused, and other important settings.

The next section shows how to create a profile.

Creating Profiles

The syntax for creating a profile looks like this:

```
CREATE PROFILE <profile> LIMIT
<password_setting> ...
<resource_setting> <limit> ...;
```

Replace <password_setting> with the setting you want to limit, replace <resource_setting> with the name of the resource, and replace <limit> with either UNLIMITED, DEFAULT, or a number.

There are seven password settings that you can set within a profile. If you do not name one of the password settings, Oracle 10*g* uses the value set for that password setting that it finds in the DEFAULT profile. If none is found then Oracle 10*g* assumes the password setting is unlimited. The password settings are:

- **FAILED_LOGIN_ATTEMPTS:** The maximum number of times the user can retry the password before the account is locked
- **PASSWORD_LIFE_TIME:** The maximum number of days the password can be used without changing
- **PASSWORD_REUSE_TIME:** The minimum number of days before the same password can be used again
- **PASSWORD_REUSE_MAX:** The minimum number of times the password must change before the same password can be used again
- **PASSWORD_LOCK_TIME:** The number of days (or a fraction of a day, such as 1/24 — one hour), that the account is locked after the maximum set by FAILED_LOGIN_ATTEMPTS has been reached
- **PASSWORD_GRACE_TIME:** The number of days after the password expires in which the user is given a warning that the password is expired, and is allowed to log on with the old password
- **PASSWORD_VERIFY_FUNCTION:** The database function called to verify the complexity of the password

TIP

PASSWORD_REUSE_TIME and PASSWORD_REUSE_MAX cannot be set at the same time. You must set one of them to UNLIMITED if you set the other to a number.

There are nine resources you can limit with a profile. If you do not name one of the resources, Oracle 10*g* uses the limit set for that resource that it finds in the DEFAULT profile. If none is found then Oracle 10*g* assumes this resource is unlimited. The resources on which you can set limits include:

- **SESSIONS_PER_USER:** Set the maximum number of concurrent sessions.
- **CPU_PER_SESSION:** Set the maximum CPU time (in hundredths of a second) for a user's entire session.
- **CPU_PER_CALL:** Set the maximum CPU time (in hundredths of a second) for any single call to the database. A call is a single task sent to the database, such as parsing a new SQL statement or executing a query.

- **CONNECT_TIME:** Set the maximum number of minutes a user's session can last.
- **IDLE_TIME:** Set the maximum number of minutes a user's session can remain inactive. A long-running query or other task in which the user is waiting for the database does not count toward idle time.
- **LOGICAL_READS_PER_SESSION:** Set the maximum number of data blocks read from either memory or disk during a user's session.
- **LOGICAL_READS_PER_CALL:** Set the maximum number of data blocks read during a single call.
- **PRIVATE_SGA:** Set the number of bytes that the user's session may allocate in the shared pool of the SGA. Set the value in bytes, kilobytes (Kb), or megabytes (Mb). This only applies to a shared server, in which part of the shared pool is allocated to user sessions. This setting is ignored on dedicated servers.
- **COMPOSITE_LIMIT:** Set the maximum resource cost (in service units) for a session. A service unit (defined in the "Controlling Resource Usage" section) calculates a resource cost based on a weighted sum of CPU_PER_SESSION, CONNECT_TIME, LOGICAL_READS_PER_SESSION, and PRIVATE_SGA. Oracle 10g monitors these values for each user session. To use the COMPOSITE_LIMIT setting, you must assign a weight to one or more of these four resources. By default, Oracle 10g weights them all at zero, meaning a user never reaches a composite limit, because all the values add up to zero. You work with an example of weighting resources later in the chapter.

TIP

You can modify the DEFAULT profile in the same way you modify profiles that you create, but you cannot drop the DEFAULT profile.

To help you experiment with profiles, begin by creating two profiles that you will use when setting both password and resource limits later in this chapter. You should still be logged on to the SQL*Plus Worksheet as the SYSTEM user. Follow these steps to create the two profiles:

1. You decide that you need two different profiles. The first profile is for programmers and the second for end users. Create the first profile, named PROGRAMMER, for the programmers by typing and executing this command:

   ```
   CREATE PROFILE PROGRAMMER LIMIT
   SESSIONS_PER_USER 2;
   ```

   The SQL*Plus Worksheet replies, "Profile created."

   The resource name, SESSIONS_PER_USER, limits the number of concurrent sessions a user is allowed to maintain. In this case you want programmers to avoid logging on with multiple sessions. You will add more limits to the profile later in this chapter.

2. Now, create the second profile and call it POWERUSER. You want the users to change passwords at least once every two months, so you create the profile with the password setting PASSWORD_LIFE_TIME of 60 days. Type and run the following command:

   ```
   CREATE PROFILE POWERUSER LIMIT
   PASSWORD_LIFE_TIME 60;
   ```

   The SQL*Plus Worksheet replies, "Profile created."
3. Remain logged on as you work with these profiles in the next sections.

The next section continues examining profiles, with emphasis on the password limits available.

Managing Passwords

You saw how to assign a password to a user when creating the user. Now, you look at methods for managing a user's password. There are three different areas to examine when working with passwords:

- **Changing a password and making it expire:** These are the only password properties that you can change specifically for a user by using the ALTER USER command.
- **Enforcing password time limits, history, and other settings:** All other password properties can be changed only through profiles. You create a profile, give the profile password properties, and then assign the profile to a user.
- **Enforcing password complexity:** This uses a combination of a function and a profile. You run a predefined SQL script (or one you create yourself), to verify the complexity of a password. Then adjust the PASSWORD_VERIFY_FUNCTION setting in a profile and assign that profile to a user.

Follow these steps to examine all three of these areas. Continue in SQL*Plus Worksheet, logged on as SYSTEM:

1. Imagine that the new user created earlier, named STUDENTA, has forgotten his password. You have forgotten it as well, so you attempt to find it in the database by typing and executing this query to look at the password in the DBA_USERS table:

   ```
   SELECT USERNAME, PASSWORD
   FROM DBA_USERS
   WHERE USERNAME = 'STUDENTA';
   ```

   Figure 12-7 shows the results of the query. As you can see, the password appears to be a string of letters and numbers, but this is not the password you assigned. The string is an encrypted form of the password that Oracle 10g uses to store all passwords. Unless you recorded the password, you cannot recover the actual value from the database, even if you are the DBA.

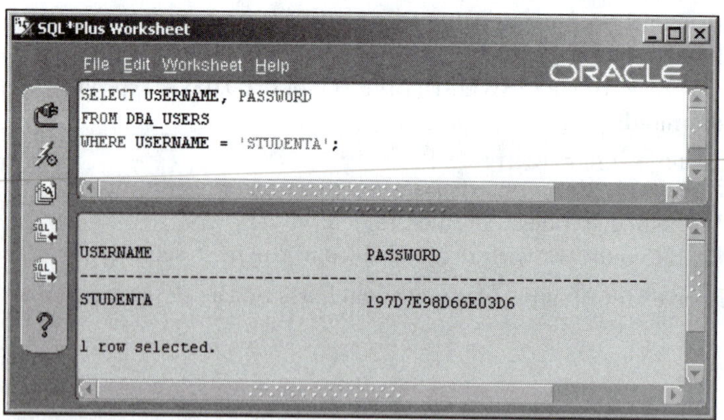

FIGURE 12-7 All passwords are encrypted in the database

2. Having determined that you cannot find the original password, you decide to change the password and inform the user of the new password. In addition, you make the password expire immediately so the user must choose a new password that he finds easier to remember. Type and execute this command:

```
ALTER USER STUDENTA
IDENTIFIED BY STUDENTA
PASSWORD EXPIRE;
```

NOTE

This is the first method of password management: changing the password and making it expire.

3. Remain logged on to the SQL*Plus Worksheet, and go to a command prompt to simulate a session for STUDENTA. In Windows, on the Taskbar, click **Start/All Programs/Accessories/Command Prompt**. In Unix or Linux go to the *$* prompt.
4. Type this command at the command prompt, and press **Enter** to start up SQL*Plus:

```
sqlplus studenta/studenta@oraclass
```

SQL*Plus starts and detects that the user's password has expired. SQL*Plus sends an error message and prompts for a new password.
5. Type **MYPASSWORD** as the new password, and press **Enter**. SQL*Plus prompts you to reenter the password.
6. Type **MYPASSWORD** again, and press **Enter**. SQL*Plus accepts the new password and starts a session for STUDENTA. Figure 12-8 shows the window at this point. This is how a user can change his own password using SQL*Plus.
7. Exit SQL*Plus by typing **EXIT** and pressing **Enter**.
8. If you are in Windows, close the command prompt window by clicking the **X** in the top-right corner. Return to the SQL*Plus Worksheet session you were using in Step 2.

```
Command Prompt (2) - sqlplus studenta/studenta@oraclass
Microsoft Windows 2000 [Version 5.00.2195]
(C) Copyright 1985-2000 Microsoft Corp.

C:\>sqlplus studenta/studenta@oraclass

SQL*Plus: Release 10.2.0.1.0 - Production on Fri Jan 6 15:41:16 2006

Copyright (c) 1982, 2005, Oracle.  All rights reserved.

ERROR:
ORA-28001: the password has expired


Changing password for studenta
New password:
Retype new password:
Password changed

Connected to:
Oracle Database 10g Enterprise Edition Release 10.2.0.1.0 - Production
With the Partitioning, OLAP and Data Mining options

SQL> _
```

FIGURE 12-8 SQL*Plus prompts for a new password when the old password expires

9. You can ensure that users don't reuse old passwords. This can only be done using profiles. Adjust the POWERUSER profile so that the user cannot reuse a password until he has used 10 other passwords, and require the account to be closed for one day if the user attempts and fails six times in a row to log on. Type and execute this command to modify the profile:

```
ALTER PROFILE POWERUSER LIMIT
PASSWORD_REUSE_MAX 10
FAILED_LOGIN_ATTEMPTS 6
PASSWORD_LOCK_TIME 1;
```

   The SQL*Worksheet replies, "Profile altered."

10. Assign the POWERUSER profile to the STUDENTA user by typing and running this command. The rules set for passwords in the POWERUSER profile will be imposed on STUDENTA immediately.

NOTE

This demonstrates the second method for password management: enforcing password time limits, history, and other settings.

```
ALTER USER STUDENTA
PROFILE POWERUSER;
```

   The SQL*Plus Worksheet replies, "User altered."

11. Later, you find out that STUDENTA has changed his password to "MYPASSWORD." This raises a flag in your mind, because a password like that is too common, making it very easy for an unauthorized user to guess. You ask a few other users what their passwords are, and discover that most of them use their user name or some simple word. You decide that this could be a

security issue and want to require users to choose more complex passwords, thus making it harder for unauthorized users to guess passwords. Oracle 10*g* provides a predefined function that you can create to specify these rules for all passwords:

- The password must contain at least four characters.
- The password cannot be identical to the user name (both user name and password are *not* case sensitive).
- The password must contain at least one alphabetic, one numeric, and one punctuation mark.
- The password must not be a simple word, such as user, database, or password.
- The password must differ from the previous password by at least three characters.

To use this password management function, log on as SYS and run the script that creates the password function and places it into the DEFAULT profile, as described in the next two steps.

12. Change your connection to SYS by selecting **File/Change Database Connection** in the SQL*Plus Worksheet. The Database Connect Information window appears. Type **SYS** in the Username box, the current password for SYS, *provided by your instructor*, in the Password box, and **ORACLASS** in the Service box. Select **SYSDBA** in the Connect As box. Click **OK** to log on.
13. SQL*Plus Workshop replies, "Connected" in the lower pane. Run the Oracle-provided, password management script by clicking **Worksheet/Run Local Script**. Navigate to the **ORACLE_HOME\rdbms\admin** directory and double-click on the **utlpwdmg.sql** file. Don't forget to replace ORACLE_HOME with the full path to the main directory containing the Oracle 10*g* software. The script runs and SQL*Plus Worksheet replies, "Profile altered." Figure 12-9 shows the worksheet at this point. The script created a function named VERIFY_FUNCTION and then modified the DEFAULT profile to adjust several password settings, including the use of the new function for verifying passwords.

    Even though the script changed the DEFAULT profile, and not the POWERUSER profile, it affects any user with the POWERUSER profile, because the setting associated with password complexity, PASSWORD_VERIFY_FUNCTION, has not been set in the POWERUSER profile. Only settings omitted from the POWERUSER profile fall back to the value in the DEFAULT profile. So, the STUDENTA user now has password complexity enforced the next time he changes his password.

NOTE

You have just used the third method of managing passwords: enforcing password complexity.

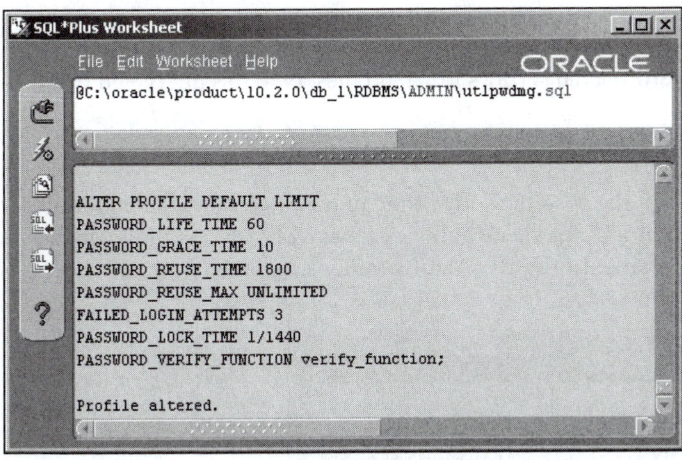

FIGURE 12-9 The script uses the ALTER PROFILE statement

14. Reconnect as the SYSTEM user by clicking **File/Change Database Connection** and logging on as SYSTEM.
15. Test the new password verification requirements by typing and executing this command, which should produce an error, because it changes STUDENTA's password to an invalid password based on the password complexity validation rules. Remember, either the DBA or the user is able to change the user's password:

```
ALTER USER STUDENTA IDENTIFIED BY A1PLUS;
```

Figure 12-10 shows the results: An error message was returned by the VERIFY_FUNCTION function.

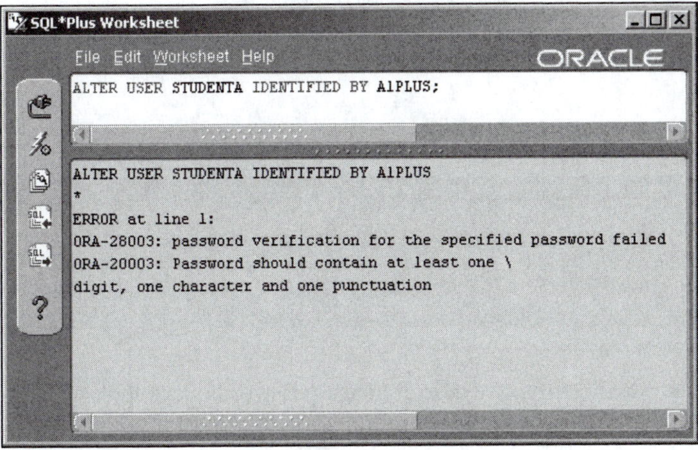

FIGURE 12-10 This password does not contain a symbol

16. Change the password to **A1PLU$** and try again. This time, the change succeeds. The SQL*Plus Worksheet replies, "User altered."
17. To see the password-related settings for all profiles, type this query and run it.

```
SELECT * FROM DBA_PROFILES
WHERE RESOURCE_TYPE='PASSWORD'
ORDER BY 1, 2;
```

Figure 12-11 shows the results. Notice that all parameters are listed for all the profiles. Those that are not specifically set have DEFAULT as their settings, meaning that the value in the DEFAULT profile is in force.

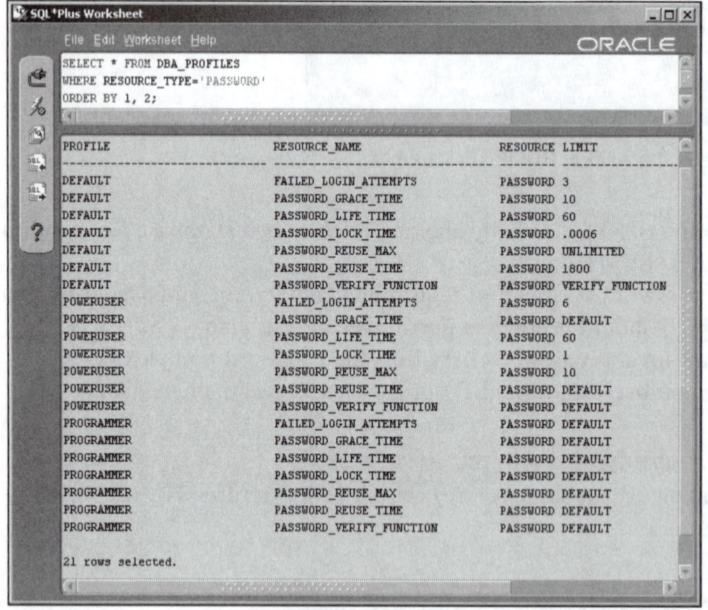

FIGURE 12-11 The POWERUSER profile has password-related settings

18. Remain logged on for the next practice.

TIP

Oracle documentation states that the **utlpwdmg.sql** file (used in Step 13) is provided more as an example than a real-world password validation tool. Use the source code as a model for writing your own password function.

Now that you have tried out the three ways to manage passwords, let's look at the resource limits that a profile can impose on a user session.

Controlling Resource Usage

You have already used the CREATE PROFILE and ALTER PROFILE commands. This section examines the profile settings that specifically control the use of computer resources. The syntax for ALTER PROFILE, with the resource clauses listed, looks like this:

```
ALTER PROFILE <profile> LIMIT
<password_setting> ...
SESSIONS_PER_USER <concurrent sessions>
CPU_PER_SESSION <hundredths of seconds>
CPU_PER_CALL <hundredths of seconds>
CONNECT_TIME <minutes>
IDLE_TIME <minutes>
LOGICAL_READS_PER_SESSION <data blocks>
LOGICAL_READS_PER_CALL <data blocks>
PRIVATE_SGA <bytes>
COMPOSITE_LIMIT <service units>
```

Follow these steps to try out resource settings for the PROGRAMMER profile that you created earlier in this chapter. You should still be logged on to SQL*Plus Worksheet as the SYSTEM user:

1. When you plan to adjust resource limits in any Oracle 10g database, you must first change the initialization parameter named RESOURCE_LIMITS. The default setting of RESOURCE_LIMITS is FALSE meaning that resource limits set in profiles are ignored. Changing the parameter to TRUE means that resource limits set in profiles are enforced. Type and execute this command:

   ```
   ALTER SYSTEM SET RESOURCE_LIMIT=TRUE;
   ```

   The SQL*Plus Worksheet replies, "System altered."
2. All the resource limits in the DEFAULT profile are set to UNLIMITED. Let's say you want to limit the users assigned to the PROGRAMMER profile to 15 minutes of idle time before the system ends their sessions. In addition, you want to limit the CPU_PER_CALL to 1 second, because your current system is running at 95 percent of its CPU usage. Type the command below and execute it to set these limits. Remember, the CPU_PER_CALL setting requires hundredths of a second, so 1 second is set by stating 100 hundredths of a second:

   ```
   ALTER PROFILE PROGRAMMER LIMIT
   IDLE_TIME 15
   CPU_PER_CALL 100;
   ```

   The SQL*Plus Worksheet replies, "Profile altered."
3. Next, you want to set a composite limit for the users. Composite limits require two steps. First, you set a relative weight for the resources you want to include when Oracle 10g calculates service units. Second, you set the maximum service units in the profile. Assume for now that you know that you want to weight the CPU_PER_SESSION at 1000 and the PRIVATE_SGA at 1. This means that 10 seconds of CPU (1000 hundredths of a second) have the same relative weight as one byte of storage in the private SGA memory. As you can tell, CPU

is not the most critical resource for this particular example. To set these resource weights, type and run the ALTER RESOURCE statement as follows:

```
ALTER RESOURCE COST
CPU_PER_SESSION 1000
PRIVATE_SGA 1;
```

The SQL*Plus Worksheet replies, “Resource cost altered.”

4. Add a composite limit to the PROGRAMMER profile by typing this command and executing it. You will find an explanation of how the composite limit is calculated for a user's session at the end of these steps:

```
ALTER PROFILE PROGRAMMER LIMIT
COMPOSITE_LIMIT 50000;
```

The SQL*Plus Worksheet replies, “Profile altered.”

5. Assign the PROGRAMMER profile to the STUDENTB user by typing and executing this command:

```
ALTER USER STUDENTB
PROFILE PROGRAMMER;
```

The SQL*Plus Worksheet replies “User altered.”

6. To view the resource settings for the system, type and run this query. Figure 12-12 shows the results.

```
SELECT * FROM RESOURCE_COST;
```

NOTE

Resources can be set to zero using the ALTER RESOURCE COST statement.

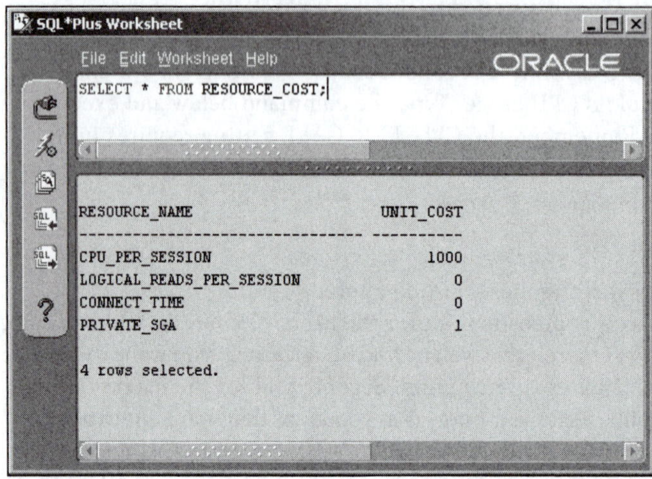

FIGURE 12-12 Resources can be set to zero using the ALTER RESOURCE COST statement

7. Remain logged on for the next practice.

TIP

To decide what type of weights to use, you may want to monitor your database's activity for a while.

In our example, the composite limit monitors two resources because only two of the four resources have been set to a nonzero cost. Let's imagine that the STUDENTB user is logged on to the database. STUDENTB has 3200 bytes of private SGA allocated. STUDENTB runs a complex query that requires 35/100 of a second to run. At this point, the composite cost of STUDENTB's session is calculated like this:

(3200 × 1) + (35 × 1000) = 37,000 service units

The user's session is below the maximum of 50,000 service units. A service unit is simply the name for the calculated value of the weighted sum of the resources consumed by the user session.

The STUDENTB user runs a series of updates, inserts, deletes, and queries that consume another 5,000 bytes of private SGA and increase the total CPU usage to one half second. Now the user session's total service units are:

(8200 × 1) + (50 × 1000) = 58200 service units

The session has exceeded the service unit cap of 50,000, so the session's transaction is rolled back and the user's session is ended with an error message.

Dropping a Profile

The syntax of DROP PROFILE is similar to the syntax for dropping a user in that it includes a CASCADE parameter:

```
DROP PROFILE <profile> CASCADE;
```

You must add CASCADE if any users have been assigned the profile being dropped. Oracle 10*g* automatically resets these users to the DEFAULT profile.

For example, if three users have been assigned to the ACCT_MGR profile, drop the profile like this:

```
DROP PROFILE ACCT_MGR CASCADE;
```

Next you need to know how users can view password and resource limits in their profiles.

OBTAINING USER, PROFILE, PASSWORD, AND RESOURCE DATA

You have already seen the following data dictionary views while going through the chapter:

- **DBA_USERS:** View user profile, expiration date of password, and account status.
- **DBA_TS_QUOTAS:** View the storage quotas of each user.
- **RESOURCE_COST:** View the weight setting for each resource used in calculating COMPOSITE_COST.
- **DBA_PROFILES:** View the settings for each profile.

As a user, your activity in the database is stopped if you reach a limit on a resource set in your profile. Follow these steps to run queries on the data dictionary views to learn about your profile settings:

1. Begin by changing your connection in SQL*Plus Worksheet, so that you are logged on as STUDENTA, one of the new users you created in this chapter. Click **File/Change Database Connection**. Type **STUDENTA** in the Username box, **A1PLU$** in the Password box, and **ORACLASS** in the Service box. Click **OK** to log on. The SQL*Plus Worksheet replies, "Connected."
2. Find out what your current password settings are by running this query:

```
SELECT * FROM USER_PASSWORD_LIMITS;
```

3. Figure 12-13 shows the results. The figure shows these characteristics of the password management for the STUDENTA user:
   - The password will expire in 60 days.
   - The user may continue to use the old password for 10 days after expiration.
   - The password cannot be reused until 10 other passwords have been used. (The PASSWORD_REUSE_MAX overrides the PASSWORD_REUSE_TIME, because the latter is defined in the DEFAULT profile, not the POWERUSER profile.)
   - The user is locked out for one day if he fails to log on six times in a row.
   - The password must be validated by the VERIFY_FUNCTION function.

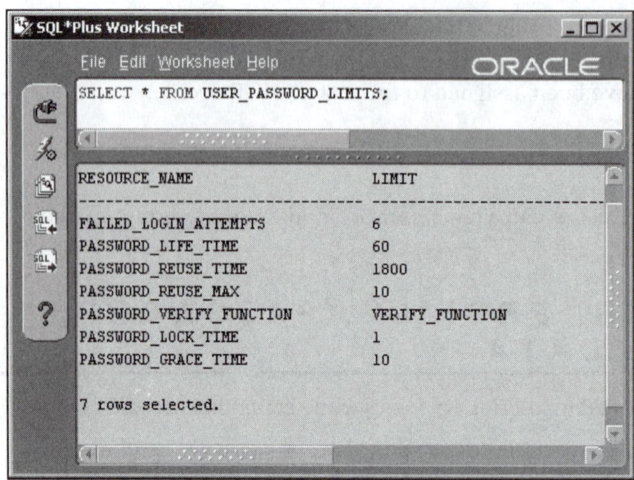

FIGURE 12-13 Users can view their own password limits with this view

4. Connect to STUDENTB now by clicking **File/Change Database Connection** in the menu. Type **STUDENTB** in the Username box, **TRUE#2** in the Password box, and **ORACLASS** in the Service box. Click **OK** to log on. The SQL*Plus Worksheet replies, "Connected."
5. Now query another view to see STUDENTB's resource limits:

   ```
   SELECT * FROM USER_RESOURCE_LIMITS;
   ```

   Figure 12-14 shows the results. From this, you know that STUDENTB has these resource limitations:

   - A composite limit of 50,000 service units
   - A maximum of 100 hundredths of a second (that is, one second) of CPU time per call
   - No more than 15 minutes of idle time
   - A limit of two sessions at a time
   - Unlimited resources otherwise

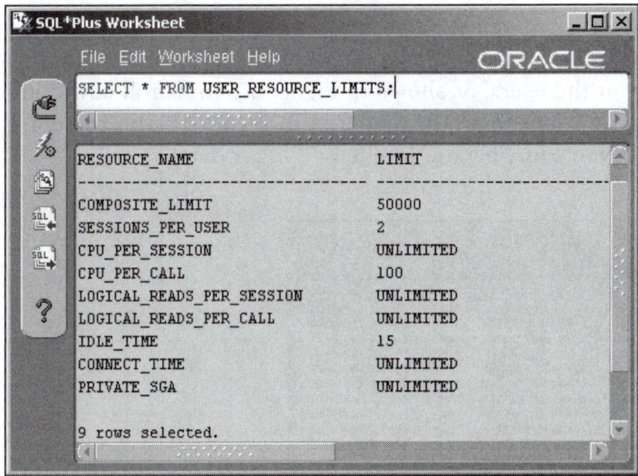

FIGURE 12-14 Resource limits exposed in the USER_RESOURCE_LIMITS view

6. Log off SQL*Plus Worksheet by clicking the **X** in the top-right corner.
7. You now see the Enterprise Manager console. Double-click the **ORACLASS** database, and then double-click the **Security** icon. This shows the Security Management main page, as shown in Figure 12-15.

NOTE

You may have to connect to the database at this stage. Connect as the **SYSTEM** user to the **ORACLASS** database.

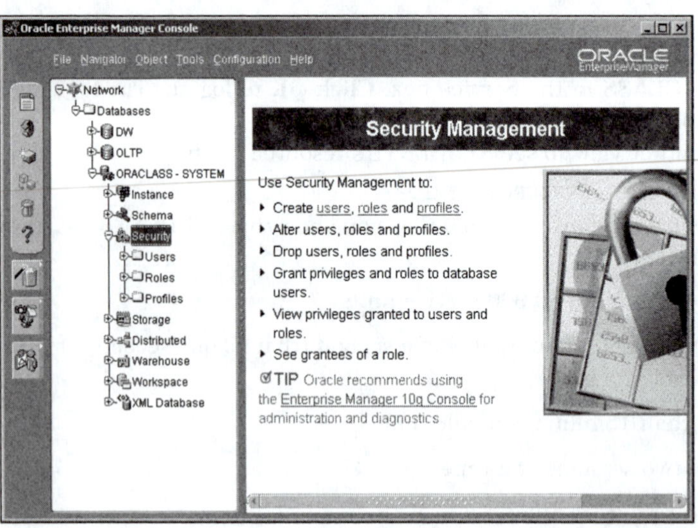

FIGURE 12-15 The security manager handles users, roles, and profiles

8. Click the **Users** folder. A list of users appears in the right pane along with high-level information about the users, as shown in Figure 12-16. Scroll down until you find STUDENTA, the line that is highlighted in Figure 12-16. The columns are adjusted so you can view the profile and other details easily.

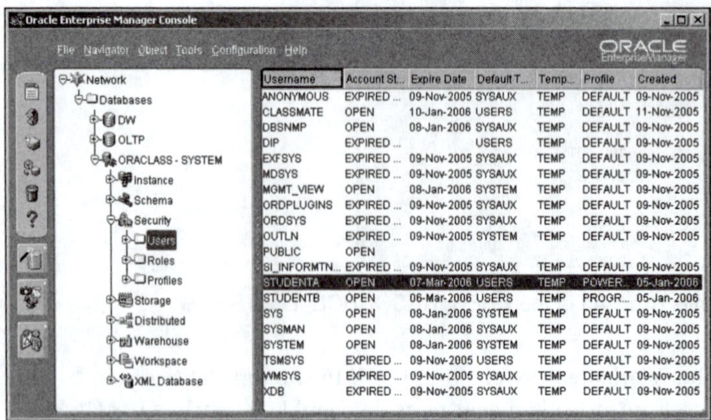

Username	Account St...	Expire Date	Default T...	Temp...	Profile	Created
ANONYMOUS	EXPIRED ...	09-Nov-2005	SYSAUX	TEMP	DEFAULT	09-Nov-2005
CLASSMATE	OPEN	10-Jan-2006	USERS	TEMP	DEFAULT	11-Nov-2005
DBSNMP	OPEN	08-Jan-2006	SYSAUX	TEMP	DEFAULT	09-Nov-2005
DIP	EXPIRED ...		USERS	TEMP	DEFAULT	09-Nov-2005
EXFSYS	EXPIRED ...	09-Nov-2005	SYSAUX	TEMP	DEFAULT	09-Nov-2005
MDSYS	EXPIRED ...	09-Nov-2005	SYSAUX	TEMP	DEFAULT	09-Nov-2005
MGMT_VIEW	OPEN	08-Jan-2006	SYSTEM	TEMP	DEFAULT	09-Nov-2005
ORDPLUGINS	EXPIRED ...	09-Nov-2005	SYSAUX	TEMP	DEFAULT	09-Nov-2005
ORDSYS	EXPIRED ...	09-Nov-2005	SYSAUX	TEMP	DEFAULT	09-Nov-2005
OUTLN	EXPIRED ...	09-Nov-2005	SYSTEM	TEMP	DEFAULT	09-Nov-2005
PUBLIC	OPEN					
SI_INFORMTN...	EXPIRED ...	09-Nov-2005	SYSAUX	TEMP	DEFAULT	09-Nov-2005
STUDENTA	OPEN	07-Mar-2006	USERS	TEMP	POWER...	05-Jan-2006
STUDENTB	OPEN	06-Mar-2006	USERS	TEMP	PROGR...	05-Jan-2006
SYS	OPEN	08-Jan-2006	SYSTEM	TEMP	DEFAULT	09-Nov-2005
SYSMAN	OPEN	08-Jan-2006	SYSAUX	TEMP	DEFAULT	09-Nov-2005
SYSTEM	OPEN	08-Jan-2006	SYSTEM	TEMP	DEFAULT	09-Nov-2005
TSMSYS	EXPIRED ...	09-Nov-2005	USERS	TEMP	DEFAULT	09-Nov-2005
WMSYS	EXPIRED ...	09-Nov-2005	SYSAUX	TEMP	DEFAULT	09-Nov-2005
XDB	EXPIRED ...	09-Nov-2005	SYSAUX	TEMP	DEFAULT	09-Nov-2005

FIGURE 12-16 STUDENTA has been assigned the POWERUSER profile

9. Double-click **STUDENTA** in the right pane. This brings up the property sheet for the user, as shown in Figure 12-17.

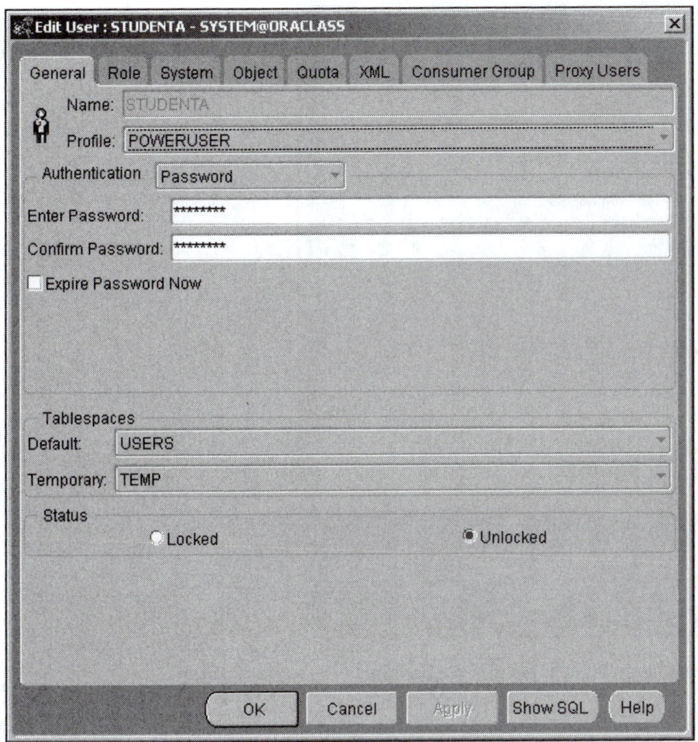

FIGURE 12-17 The user's property sheet displays many traits of the user

You can change the user's password, default tablespace, account status, and profile, in this window without writing any code.

10. Click the **Quota** tab. The quotas for each tablespace are listed here, as you see in Figure 12-18. Your list of tablespaces may be slightly different from those shown in the figure. You can even modify the quota by selecting a tablespace row and using the radio buttons and Value box at the bottom of the window.
11. Close the property sheet by clicking the **X** in the top-right corner. This returns you to the main console window.
12. Double-click the **Profiles** folder in the left pane. A list of profiles appears in the right pane, as you see in Figure 12-19.

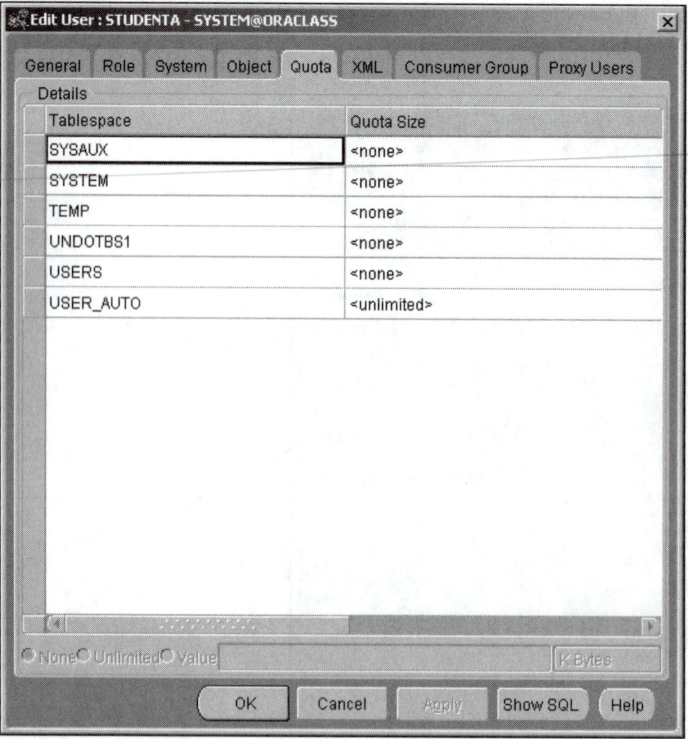

FIGURE 12-18 Quotas are set to unlimited, none, or a number

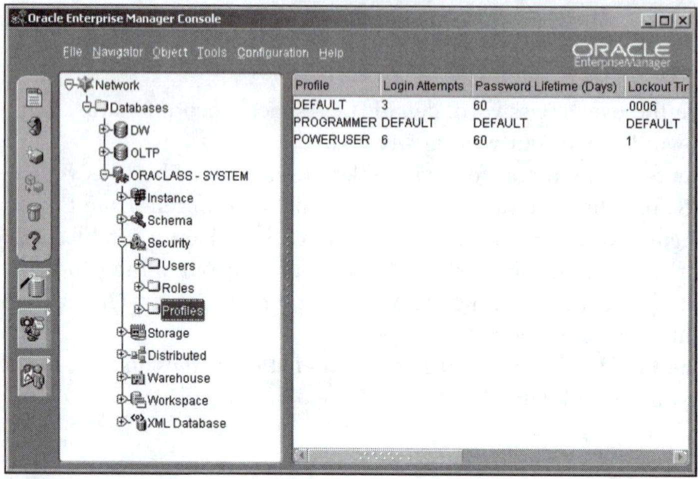

FIGURE 12-19 Password limits not set in a profile are marked DEFAULT

13. Double-click the profile named **POWERUSER** in the left pane. Figure 12-20 shows the property sheet that appears. You can click on any of these resource limits and choose from some suggested values in the drop-down lists, or type a value in the box.

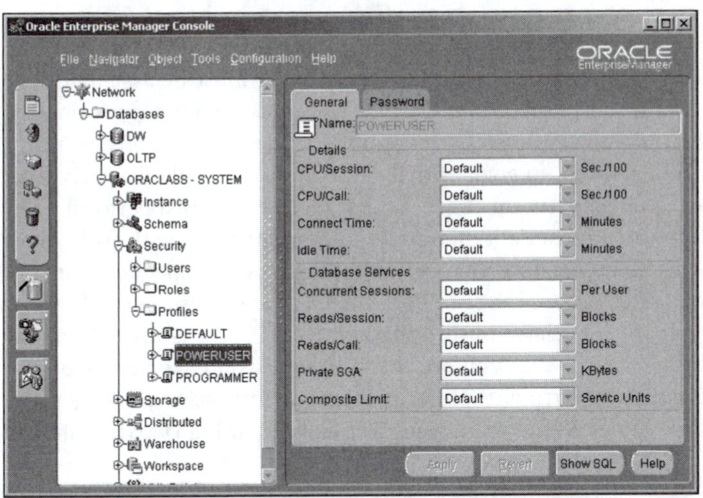

FIGURE 12-20 Any of the resource limits can be changed quickly

14. Click the **Password** tab. The password limits are now displayed, as shown in Figure 12-21. As you can see, you are provided with a simple click-and-type format for adjusting these settings. At times, you may find this an easier way to modify users and profiles.

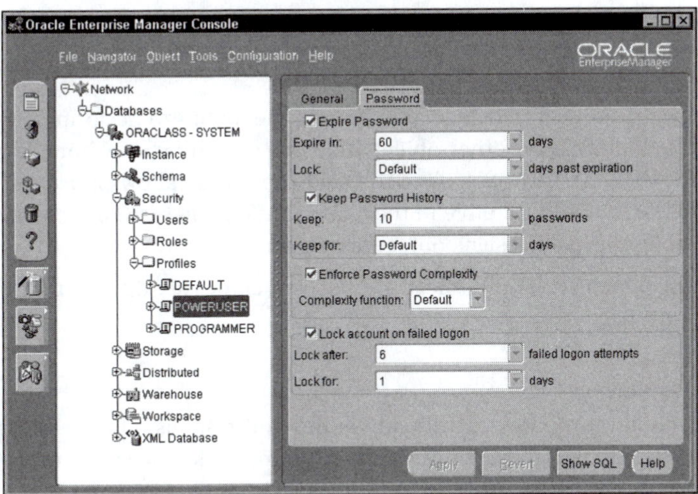

FIGURE 12-21 Anything marked Default reverts to the setting in the DEFAULT profile

15. Close the console by clicking the **X** in the top-right corner.

As you can see from performing these practices, the console gives you an excellent way to look at the settings for users and profiles, without having to write queries.

Next you will examine system and object privileges.

SYSTEM AND OBJECT PRIVILEGES

After a user has been created with a name and password, the user must be assigned the ability to log on to the database. Once logged on, however, the user cannot perform any other tasks unless given the privilege to do so. For example, a user must be given the privilege to create a table. It is possible to give a privilege to all users (a blanket privilege, you might say), and this can be convenient for the tasks that most users must perform. For example, the privileges of querying most data dictionary views that begin with ALL_ have been given to all users by Oracle by default. Most privileges are given to specific users or roles. A **role** is a named group of privileges that can be assigned to a user as a set rather than individually. Find out more about roles later in this chapter. Here is an example of how to use privileges in the database: Imagine that you are the DBA, and you have installed 10 tables to support the Accounting Department's new application. You assign privileges for querying and modifying these tables to the employees in the Accounting Department.

TIP

Assigning privileges only as they are needed is a form of security for the database. A database that allows all users to query or modify all database tables is the exception rather than the rule for Oracle databases.

There are two types of privileges that you can assign in the Oracle 10*g* database:

- **System privileges:** A **system privilege** gives a user the ability to manage some part of the database system. For example, the CREATE TABLE privilege allows a user to create a table, and the ALTER TABLESPACE privilege allows a user to modify an existing tablespace using the ALTER TABLESPACE command. The DBA usually assigns system privileges.
- **Object privileges:** An **object privilege** gives a user the ability to perform certain tasks on specific tables or other objects that are owned by a schema. For example, the INSERT ON CLASSMATE.EMPLOYEE privilege allows a user to insert rows into the EMPLOYEE table in the CLASSMATE schema. The schema that owns the object usually assigns object privileges.

The next sections describe each type of privilege so that you can easily differentiate between system and object privileges.

Identifying System Privileges

The position of database administrator comes with many duties and responsibilities. These include the tasks of creating the database, creating and maintaining tablespaces, monitoring resources, and backing up the database. Each of these tasks has its own system privilege, so that the DBA can perform the tasks or delegate them to an assistant. The predefined user, SYSTEM, has already been given the privileges needed for DBA activities. When you log on to the database as the SYSTEM user, you are able to perform DBA tasks. In addition, you are allowed to assign these privileges to other users.

There are over 100 system privileges in the Oracle 10g database. Fortunately, all of them are named very clearly, so it is usually easy to understand their use. Here are some of the most common and familiar system privileges:

- **SYSDBA:** You have already used this privilege when you logged on to the SQL*Plus Worksheet to create a database. This privilege allows the user to start up and shut down the database and to create an **spfile** (system initialization parameter file). The SYSTEM and SYS predefined users have this privilege.
- **SYSOPER:** This privilege is the same as SYSDBA, except it does not include the ability to create a database.
- **CREATE SESSION:** You assigned this privilege to a user in Chapter 11. This allows a user to log on to the database.
- **CREATE TABLE and CREATE VIEW:** These privileges allow users to create their own tables and views.
- **CREATE USER:** The DBA must have this privilege to create new users. You logged on as SYSTEM to create users in the previous chapter.
- **CREATE ANY TABLE:** This privilege is usually reserved for the DBA, allowing him or her to create a table in any schema. The CLASSMATE user has this privilege, which enabled you to create tables in previous chapters.
- **DROP ANY TABLE:** This privilege allows the user to drop any schema's tables, except those needed by the database system itself (the data dictionary tables).
- **SELECT ANY TABLE:** Again, this privilege is usually reserved for the DBA, because it allows the user to query any table on the database.
- **GRANT ANY PRIVILEGE, GRANT ANY OBJECT PRIVILEGE:** These allow the user (preferably the DBA) to assign any system privilege or any object privilege to other users. Later in the chapter, you see how to allow users to grant specific privileges.
- **BACKUP ANY TABLE:** This privilege allows the user to use the Export utility to export any table in the database. **Exporting** is a form of backup that can be used to back up specific tables or schemas, or to back up the entire database. An alternate way to back up the database is to copy the operating system files (the datafiles) to tape or CD. You do not need any Oracle-specific privileges to perform an operating system backup of Oracle files.

You can see a complete list of all the system privileges in the online documentation that Oracle provides in Oracle documentation (look up the GRANT statement in the SQL Reference manual).

You can also view and assign both system and object privileges by using the Security Manager in the Enterprise Management console. You will have some practice viewing system and object privileges in the console later in the chapter.

Using Object Privileges

Object privileges are more pinpointed than system privileges. That is, an object privilege has a much narrower focus. For example, the system privilege SELECT ANY TABLE gives the user the ability to query any table in any schema. On the other hand, the object privilege SELECT ON CUSTOMER gives the user the ability to query only the CUSTOMER table.

The user who owns a table or view is allowed by default to select, insert, update, and delete data in the table. Most of the time, users who actually log on and use the tables have their own user names and, therefore, must be assigned an object privilege for each of these tasks. The DBA user does not usually have to be assigned object privileges, because the DBA user (such as SYSTEM) already has system privileges that allow him to perform these tasks.

Object privileges always pertain to a table, function, procedure, or other object. There are several different object privileges, and some are available only for tables and views, whereas others are only available for functions, procedures, packages, or user-defined types. **Functions**, **procedures**, and **packages** are PL/SQL programs that reside in the database and can be called from SQL commands, such as SELECT and INSERT. **User-defined types** help define object columns or object tables. These types were discussed in a previous chapter, including object types, arrays, and nested tables. Table 12-1 shows the types of object privileges available for each type of object. This is not a complete list, but it covers both the objects that you study in this book and those usually covered by introductory Oracle 10g SQL classes or texts.

TABLE 12-1 Object privileges by object type

Object Privilege	Table	View	Sequence	Function, Procedure, Package	User-Defined Type
ALTER	Yes		Yes		
DELETE	Yes	Yes			
EXECUTE				Yes	Yes
DEBUG	Yes	Yes			
FLASHBACK	Yes	Yes			
INDEX	Yes				
INSERT	Yes	Yes			
REFERENCES	Yes	Yes			
SELECT	Yes	Yes	Yes		
UPDATE	Yes	Yes			

Some of these privileges have obvious meanings, such as SELECT, INSERT, and UPDATE. The following list defines privileges whose meanings are less obvious:

- **EXECUTE:** Call the function, procedure, or package while running an SQL query or other command.
- **DEBUG:** Run a debugging program that looks at triggers and SQL commands using a table or view.
- **FLASHBACK:** Run a flashback query on the table or view.
- **INDEX:** Create an index on the table.
- **REFERENCES:** Create FOREIGN KEY constraints that reference the table.

How do you assign a user an object privilege or a system privilege? The next section shows you what to do.

MANAGING SYSTEM AND OBJECT PRIVILEGES

When you **grant** a privilege, you assign a privilege to a user or a role, whether it is a system privilege or an object privilege. When you revoke a privilege, you take away the privilege. Granting privileges to roles is covered later in this chapter. For now, let's have you practice granting and revoking privileges to users.

Granting and Revoking System Privileges

The basic syntax of the GRANT command for system privileges is:

```
GRANT <systempriv>, <systempriv>,...|ALL PRIVILEGES
TO <user>,<user>...|PUBLIC
WITH ADMIN OPTION;
```

Here are some pointers about this command:

- List as many system privileges as you want, separating each with a comma. You can also substitute the phrase ALL PRIVILEGES for a list of privileges. Use ALL PRIVILEGES with caution, because it grants the user all available system privileges except the SELECT ANY DICTIONARY system privilege.
- Add the WITH ADMIN OPTION only when you want the user to be able to grant the same system privilege to other users. For example, this is appropriate when you hire a new assistant DBA and allow him to create new users.
- List all the users to whom you want to grant the same system privileges. Alternatively, use PUBLIC instead of a specific user name to grant the privilege to all users, including users created in the future.

Revoking a system privilege is simple:

```
REVOKE <systempriv>, <systempriv>,...|ALL PRIVILEGES
FROM <user>, <user>,...|PUBLIC;
```

Complete the following steps to practice granting and revoking privileges.

1. Start up the Enterprise Manager console. In Windows, click **Start** on the Taskbar, and then click **All Programs/Oracle ... /Enterprise Manager Console**. In Unix or Linux, type **oemapp console** on the command line. The Enterprise Manager console login screen appears.
2. The console appears. Before granting and revoking privileges, take a look at the privileges of the STUDENTA user, who was created in the previous chapter.
3. Double-click the **Databases** folder, and then double-click the **ORACLASS** database icon. The available tools are listed below the database.

NOTE

You may have to connect to the database at this stage. Connect as the **SYSTEM** user to the **ORACLASS** database.

4. Double-click the **Security** icon, and then double-click the **Users** folder. This displays the current users in the ORACLASS database, as shown in Figure 12-22.

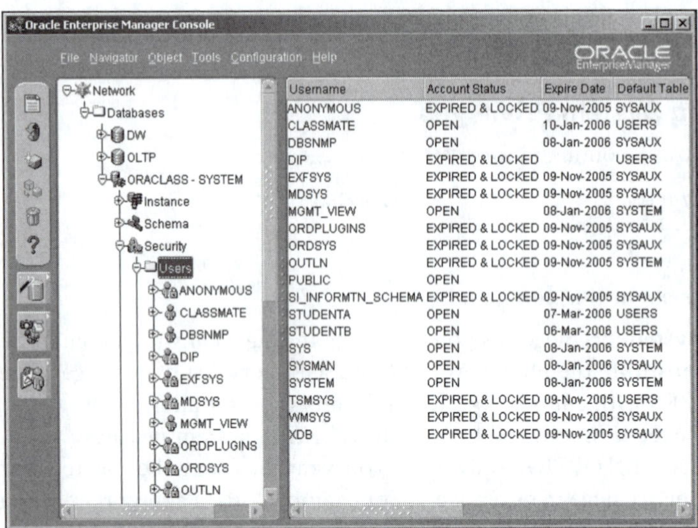

FIGURE 12-22 The Security Manager in the Enterprise Manager Console handles users

NOTE

The list of users on your screen may be slightly different from the list of users shown in the Figure 12-22.

5. Scroll down the left side to find and double-click on **STUDENTA**. The property sheet for STUDENTA appears on the right side of the console.

6. Click the **System** tab in the property sheet. This part of the property sheet displays the system privileges currently granted to the user. As you can see in Figure 12-23, CREATE SESSION is the only system privilege granted to STUDENTA. You could add more privileges right here; however, your goal is to become familiar with the GRANT command, so you will be adding privileges in SQL*Plus Worksheet instead.

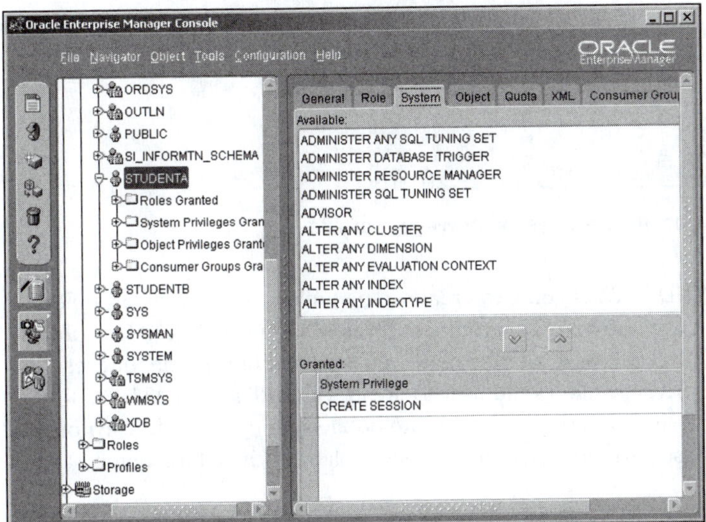

FIGURE 12-23 Available privileges are listed in the top panel, and granted privileges are listed in the lower panel

7. Start up the SQL*Plus Worksheet by clicking **Tools/Database Applications/**SQL*Plus Worksheet from the top menu in the console.

NOTE

You may have to connect to the database at this stage. Connect as the **SYSTEM** user to the **ORACLASS** database.

8. Notice that you are connected as SYSTEM. This is appropriate for granting system privileges, because the SYSTEM user has the privilege, GRANT ANY PRIVILEGE WITH ADMIN OPTION, meaning that SYSTEM can grant any system privilege to any other user. Display SYSTEM's privileges by executing this query:

```
SELECT * FROM SESSION_PRIVS
ORDER BY 1;
```

Figure 12-24 partially shows the results of the above query. Use this query for any user to display the currently enabled privileges.

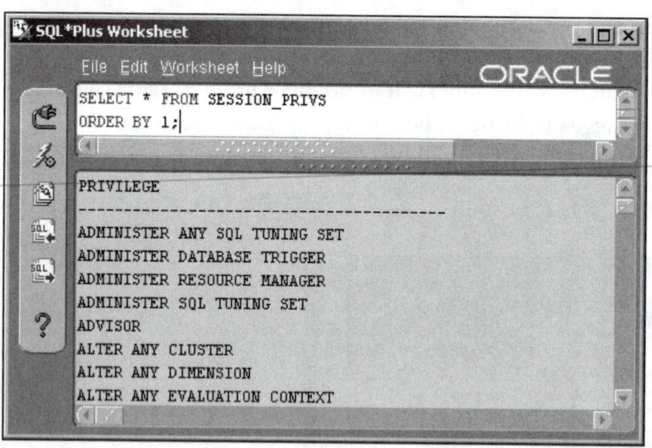

FIGURE 12-24 SYSTEM has nearly every system privilege available

9. Imagine that STUDENTA is going to create some tables, views, indexes, object tables, and public synonyms. This requires three system privileges. The CREATE TABLE privilege allows a user to create tables, object tables, and indexes. The CREATE VIEW privilege allows the user to create views. The CREATE PUBLIC SYNONYM privilege allows the user to create public synonyms. **Public synonyms** are like table aliases, which any user can use in place of the table's name.

NOTE

A public synonym does not give users the privilege to actually select from a table. That privilege must be explicitly granted.

Now, grant the system privileges to STUDENTA by executing these commands:

```
GRANT CREATE TABLE, CREATE VIEW, CREATE PUBLIC SYNONYM
TO STUDENTA;
```

10. The SQL*Plus Worksheet replies, "Grant succeeded." After granting this privilege, you find out that STUDENTA has an assistant (STUDENTB) who needs appropriate privileges.
Instead of granting the three system privileges, which you gave to STUDENTA, you decide to allow STUDENTA to grant these privileges (or only one or two of them) to STUDENTB. There is no command to alter a grant that has been given, so you must reissue the grant. You do not have to revoke the grant before reissuing it. Give STUDENTA the authority to grant the privileges by typing this command:

```
GRANT CREATE TABLE, CREATE VIEW, CREATE PUBLIC SYNONYM
TO STUDENTA WITH ADMIN OPTION;
```

The SQL*Plus Worksheet replies, "Grant succeeded."

11. STUDENTA is now allowed to grant these three system privileges to other users. Connect to STUDENTA to test this. The CONNECT command is an alternative to using the **File/Change Database Connection** menu selection. Type and execute this command:

```
CONNECT STUDENTA/A1PLU$@ORACLASS
```

12. The SQL*Plus Worksheet replies, "Connected." Now, as STUDENTA, you can look at your own privileges and find out which ones you have with the ADMIN option by typing and executing the following command. Figure 12-25 shows the results. As you can see, STUDENTA has the ADMIN option for all but the CREATE SESSION privilege:

```
SELECT * FROM USER_SYS_PRIVS;
```

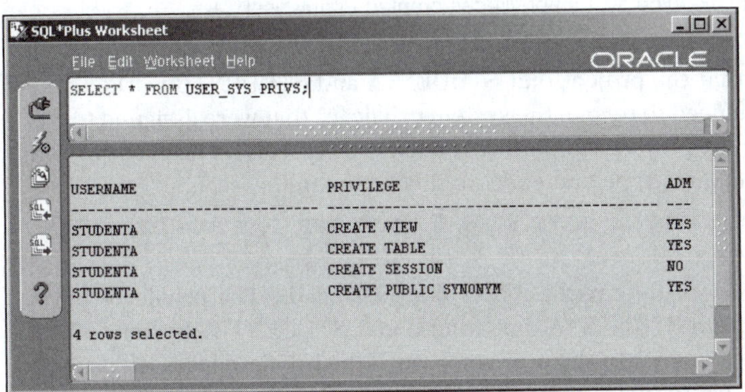

FIGURE 12-25 USER_SYS_PRIVS shows privileges belonging to the current user

13. The new user, STUDENTB, will be helping STUDENTA by creating tables and indexes. Grant STUDENTB the appropriate privilege by typing and executing this command:

```
GRANT CREATE TABLE TO STUDENTB;
```

14. The SQL*Plus Worksheet replies, "Grant succeeded." Now, connect to SYSTEM again by typing and executing this command, replacing <password> with the current password for the SYSTEM user:

```
CONNECT SYSTEM/<password>@ORACLASS;
```

The SQL*Plus Worksheet replies, "Connected."

15. Check the system privileges for the two users by typing and executing this query. Figure 12-26 shows the results:

```
SELECT * FROM DBA_SYS_PRIVS
WHERE GRANTEE LIKE 'STUDENT%';
```

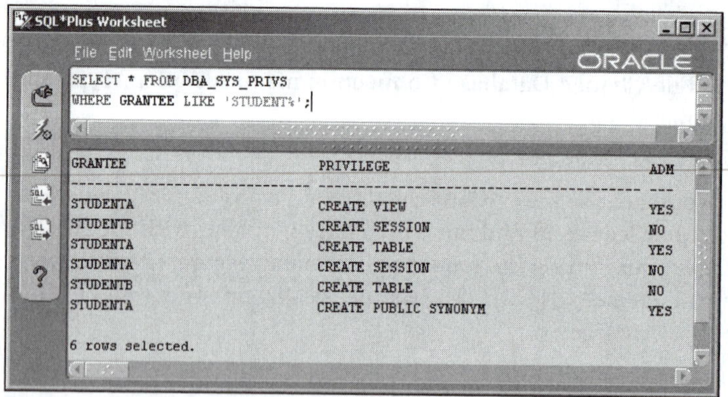
```
SELECT * FROM DBA_SYS_PRIVS
WHERE GRANTEE LIKE 'STUDENT%';

GRANTEE                        PRIVILEGE                                ADM
------------------------------ ---------------------------------------- ---
STUDENTA                       CREATE VIEW                              YES
STUDENTB                       CREATE SESSION                           NO
STUDENTA                       CREATE TABLE                             YES
STUDENTA                       CREATE SESSION                           NO
STUDENTB                       CREATE TABLE                             NO
STUDENTA                       CREATE PUBLIC SYNONYM                    YES

6 rows selected.
```

FIGURE 12-26 The DBA can examine system privileges granted to any user

16. Now, imagine that the project that STUDENTA and STUDENTB worked on is completed. You need to revoke the system privileges that were assigned to STUDENTA during the project. All that will be left is STUDENTA's privilege to connect to the database. Type and execute this command:

```
REVOKE CREATE TABLE, CREATE VIEW, CREATE PUBLIC SYNONYM
FROM STUDENTA;
```

17. The SQL*Plus Worksheet replies, "Revoke succeeded." The privileges have been revoked from STUDENTA, but what about STUDENTB? System privileges are not revoked when the user who issued the privilege loses his privilege. So, type and execute this command to revoke the privileges granted by STUDENTA to STUDENTB:

```
REVOKE CREATE TABLE FROM STUDENTB;
```

The SQL*Plus Worksheet replies, "Revoke succeeded."

18. Remain logged on for the next practice.

TIP

There is no distinct privilege for creating indexes on a table in your own schema. The CREATE ANY INDEX privilege allows you to create indexes on other schema's tables.

System privileges are needed to create tables in the database. The owner (schema) of a table can automatically select, insert, update, and delete data in his own tables. However, no other users (except the DBA) are allowed to even see the table's name without permission from the owner or the DBA. This is where object privileges come into play.

Granting and Revoking Object Privileges

The CLASSMATE schema has several tables that were created during previous chapters. So far, you have not worked directly with the tables, because this book focuses primarily on the DBA tasks involved in managing an Oracle 10g database. This section discusses tasks usually handled by the schema user who owns the tables. However, DBAs can also perform these tasks, as they commonly do in some companies.

The syntax for granting object privileges looks like this:

```
GRANT <objectpriv>, <objectpriv>,...|ALL
(<colname>,...) ON <schema>.<object>
TO <user>,...|PUBLIC
WITH GRANT OPTION
WITH HIERARCHY OPTION;
```

As you can see, the preceding syntax is similar to the syntax for granting system privileges. Here are some notes on the syntax:

- **Column list:** The column list (in the second line of the preceding syntax) is used only when you want to grant a privilege for specific columns in the table or view. Although this is not usually used, you could employ it as a security feature to restrict users from updating sensitive fields that they are allowed to query but not to update. The column list can only be used to grant UPDATE, REFERENCES, and DELETE privileges.
- **PUBLIC:** You can list object privileges for one object, and you can also list users who receive those privileges. PUBLIC is substituted for user names when you want to grant the privilege or privileges to all users.
- **WITH GRANT OPTION:** This is similar to the WITH ADMIN OPTION. Use this clause when you want the user to be able to issue grants to other users.
- **WITH HIERARCHY OPTION:** This clause is a special feature, which is used for objects that have contained objects (subobjects). A **subobject** is an object based on another object, for example an object type that is based on another object type. Although this book does not cover this option, the option is included for completeness of syntax. This clause instructs Oracle 10g to grant the object privilege to the user on the object and on all its contained objects.

Follow these steps to practice granting and revoking object privileges:

1. In your SQL*Plus Worksheet session, connect to the CLASSMATE user by typing and executing this command:

   ```
   CONNECT CLASSMATE/CLASSPASS@ORACLASS
   ```

   The SQL*Plus Worksheet replies, "Connected."
2. To see what kind of objects CLASSMATE owns, type and execute this query. Figure 12-27 partially shows the results:

   ```
   COLUMN OBJECT_NAME FORMAT A35
   SELECT OBJECT_TYPE, OBJECT_NAME
   FROM USER_OBJECTS
   ORDER BY 1;
   ```

   Refer to Table 12-1, which shows the types of object privileges available for different types of objects. Table 12-1 was presented previously in this chapter, in the section titled "Using Object Privileges." Some of the objects owned by CLASSMATE, such as individual table partitions and LOB segments, cannot be used with object privileges. The others, such as tables and views, can be used with object privileges.

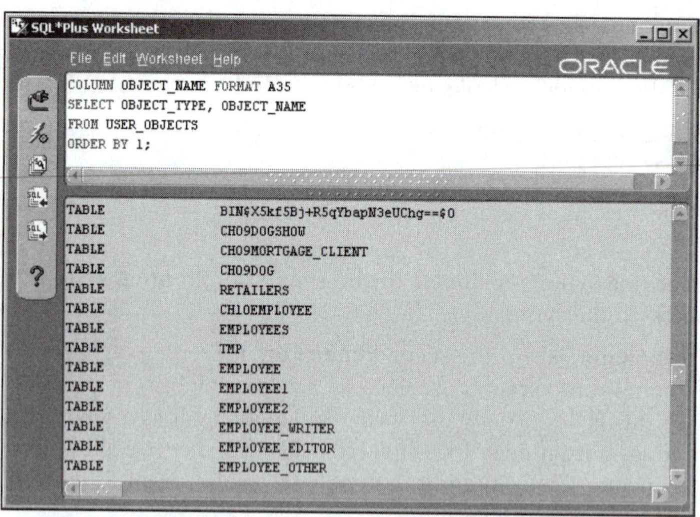
SQL*Plus Worksheet
File Edit Worksheet Help
ORACLE
COLUMN OBJECT_NAME FORMAT A35
SELECT OBJECT_TYPE, OBJECT_NAME
FROM USER_OBJECTS
ORDER BY 1;
TABLE BIN$X5kf5Bj+R5qYbapN3eUChg==$0
TABLE CH09DOGSHOW
TABLE CH09MORTGAGE_CLIENT
TABLE CH09DOG
TABLE RETAILERS
TABLE CH10EMPLOYEE
TABLE EMPLOYEES
TABLE TMP
TABLE EMPLOYEE
TABLE EMPLOYEE1
TABLE EMPLOYEE2
TABLE EMPLOYEE_WRITER
TABLE EMPLOYEE_EDITOR
TABLE EMPLOYEE_OTHER

FIGURE 12-27 CLASSMATE owns several different types of objects

3. Imagine that an application is about to be launched for the Sales Department of the Global Globe News Company. The application uses the WANT_AD_COMPLETE_VIEW view and the NEWS_ARTICLE table. For this imaginary example, there are only two users who need access to the tables when using the application: STUDENTA and STUDENTB. The application queries both the view and the table. Type and execute these commands to grant the appropriate object privileges:

```
GRANT SELECT ON WANT_AD_COMPLETE_VIEW
TO STUDENTA, STUDENTB;
GRANT SELECT ON NEWS_ARTICLE
TO STUDENTA, STUDENTB;
```

   The SQL*Plus Worksheet replies, "Grant succeeded."
4. The application also allows the users to create, modify, and remove records in the NEWS_ARTICLE table. Type and execute this command to grant the privileges needed:

```
GRANT INSERT, UPDATE, DELETE ON NEWS_ARTICLE
TO STUDENTA, STUDENTB;
```

   The SQL*Plus Worksheet replies, "Grant succeeded."
5. Type and execute this query to look over the grants you have created:

```
COLUMN PRIVILEGE FORMAT A10
COLUMN GRANTABLE FORMAT A10
SELECT GRANTEE, TABLE_NAME, PRIVILEGE, GRANTABLE
FROM USER_TAB_PRIVS_MADE
ORDER BY 1,2,3;
```

6. Figure 12-28 shows the results. A **grantee** is the user receiving a privilege. Each privilege for each user has a row in this table. Privileges for all types of objects are listed in this view, not just privileges for tables as its name implies. There

is a similar table, named USER_TAB_PRIVS_RECD that lists all the object privileges a user has received, rather than those the user has granted.

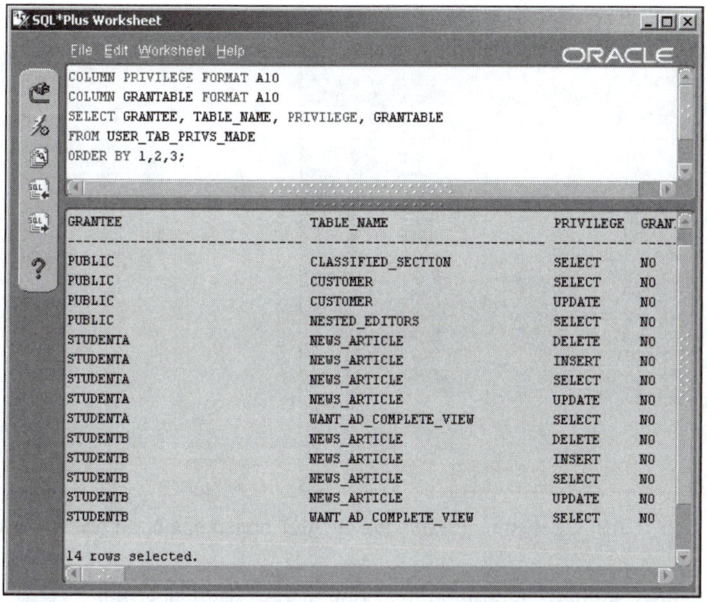

```
COLUMN PRIVILEGE FORMAT A10
COLUMN GRANTABLE FORMAT A10
SELECT GRANTEE, TABLE_NAME, PRIVILEGE, GRANTABLE
FROM USER_TAB_PRIVS_MADE
ORDER BY 1,2,3;
```

```
GRANTEE                        TABLE_NAME                     PRIVILEGE  GRAN
------------------------------ ------------------------------ ---------- ----
PUBLIC                         CLASSIFIED_SECTION             SELECT     NO
PUBLIC                         CUSTOMER                       SELECT     NO
PUBLIC                         CUSTOMER                       UPDATE     NO
PUBLIC                         NESTED_EDITORS                 SELECT     NO
STUDENTA                       NEWS_ARTICLE                   DELETE     NO
STUDENTA                       NEWS_ARTICLE                   INSERT     NO
STUDENTA                       NEWS_ARTICLE                   SELECT     NO
STUDENTA                       NEWS_ARTICLE                   UPDATE     NO
STUDENTA                       WANT_AD_COMPLETE_VIEW          SELECT     NO
STUDENTB                       NEWS_ARTICLE                   DELETE     NO
STUDENTB                       NEWS_ARTICLE                   INSERT     NO
STUDENTB                       NEWS_ARTICLE                   SELECT     NO
STUDENTB                       NEWS_ARTICLE                   UPDATE     NO
STUDENTB                       WANT_AD_COMPLETE_VIEW          SELECT     NO

14 rows selected.
```

FIGURE 12-28 This query shows object privileges granted to the current user

7. You decide that the STUDENTA user should be allowed to grant all his privileges on both tables to other users. Therefore, type and execute these commands to provide STUDENTA with the changed privileges:

   ```
   GRANT SELECT, INSERT, UPDATE, DELETE ON NEWS_ARTICLE
   TO STUDENTA WITH GRANT OPTION;
   GRANT SELECT ON WANT_AD_COMPLETE_VIEW TO STUDENTA
   WITH GRANT OPTION;
   ```

   The SQL*Plus Worksheet replies, "Grant succeeded."
8. Now you decide that STUDENTB should not have the ability to delete rows from the NEWS_ARTICLE table, so you modify his privileges by revoking that privilege. Type and execute this command:

   ```
   REVOKE DELETE ON NEWS_ARTICLE FROM STUDENTB;
   ```

   The SQL*Plus Worksheet replies, "Revoke succeeded."
9. Re-execute the query in Step 5 to confirm that STUDENTA has the GRANT option and STUDENTB has no DELETE privileges. To find the query, click the **Previous Command** icon on the left side of the SQL*Plus Worksheet. The previous command appears in the worksheet. After the correct command appears, click the **Execute** icon to execute it again. Figure 12-29 shows the results.
10. Switch now to the STUDENTA user by typing and executing this command.

    ```
    CONNECT STUDENTA/A1PLU$@ORACLASS
    ```

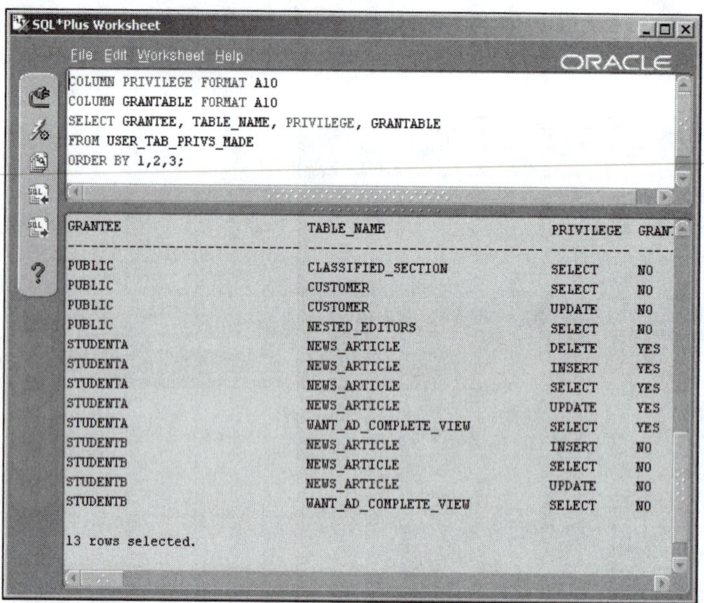

FIGURE 12-29 "YES" in the GRANTABLE column indicates the GRANT option has been given

11. SQL*Plus Worksheet replies, "Connected." Look at the privileges that STUDENTA has by typing and executing this query. The results show the privileges granted to STUDENTA:

```
SELECT GRANTOR, TABLE_NAME, PRIVILEGE, GRANTABLE
FROM USER_TAB_PRIVS_RECD
ORDER BY 1,2,3;
```

12. Try querying the NEWS_ARTICLE table by typing and executing this command. Two rows are returned. Remember that you must prefix the table with the schema when you don't own the table you are querying:

```
SELECT TITLE, RUN_DATE
FROM CLASSMATE.NEWS_ARTICLE;
```

13. Try querying the EMPLOYEE table, which you don't have authority to view. Type and execute this query. Figure 12-30 shows the resulting error message:

```
SELECT EMPLOYEE_ID, FIRST_NAME, LAST_NAME
FROM CLASSMATE.EMPLOYEE;
```

14. You decide that STUDENTB really does need the DELETE privilege on the NEWS_ARTICLE table. Give STUDENTB the privilege by typing and executing this command:

```
GRANT DELETE ON CLASSMATE.NEWS_ARTICLE
TO STUDENTB;
```

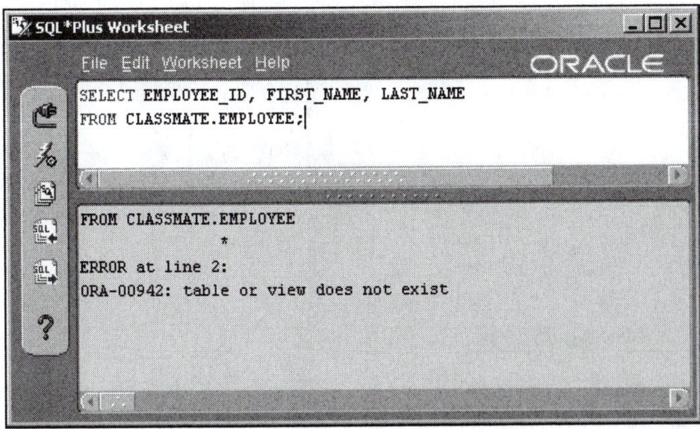

FIGURE 12-30 The error message is standard for object privilege violations

15. SQL*Plus Worksheet replies, "Grant succeeded." Confirm the existence of the privilege by typing and executing this command:

```
SELECT GRANTEE, TABLE_NAME, PRIVILEGE, GRANTABLE
FROM ALL_TAB_PRIVS_MADE;
```

16. The object privilege has been granted; however, the owner of the table (CLASSMATE) does not agree, and decides to remove STUDENTA's ability to grant privileges on the NEWS_ARTICLE table. Connect to CLASSMATE by typing and executing this command:

```
CONNECT CLASSMATE/CLASSPASS@ORACLASS
```

The SQL*Plus Worksheet replies, "Connected."

17. You cannot change an existing GRANT privilege; you must revoke the old one and grant the new one. Now, revoke the privileges from STUDENTA by typing and executing this command:

```
REVOKE SELECT, INSERT, UPDATE, DELETE ON NEWS_ARTICLE
FROM STUDENTA;
```

The SQL*Plus Worksheet replies, "Revoke succeeded."

18. Check the results by querying the USER_TAB_PRIVS_MADE view again using the following query:

```
COLUMN PRIVILEGE FORMAT A10
SELECT GRANTOR, TABLE_NAME, PRIVILEGE, GRANTABLE
FROM USER_TAB_PRIVS_RECD
ORDER BY 1,2,3;
```

Figure 12-31 shows the results.

Notice that STUDENTA no longer has any privileges on the NEWS_ARTICLE table, and STUDENTB no longer has the DELETE privilege that was granted to him by STUDENTA. When STUDENTA lost the privilege on the table, any object privileges created by STUDENTA were also lost. This is an important difference between system privileges and object privileges.

19. Remain logged on for the next practice.

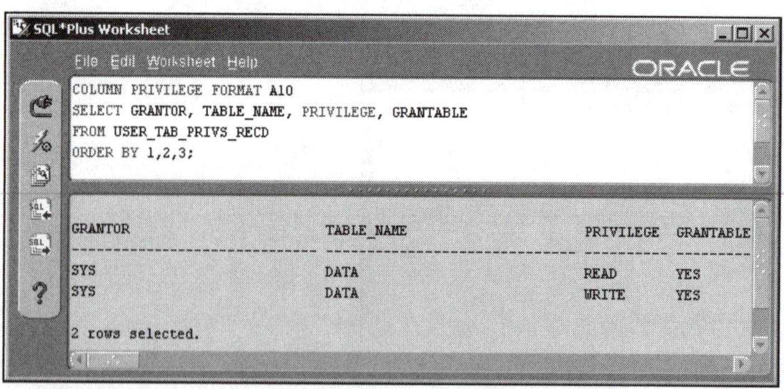

FIGURE 12-31 STUDENTB lost the DELETE privilege

TIP

The act of revoking an object privilege cascades downwards, whereas revoking a system privilege does not. The cascading of a revoked privilege occurs when you revoke a privilege from a user who has previously granted this privilege to other users. The privilege is revoked for the other users automatically.

You have examined how to grant and revoke system and object privileges. You have also queried several data dictionary views. Table 12-2 summarizes the data dictionary views and dynamic performance views related to system and object privileges. You have tried most of these views during the practice sessions in the chapter.

TABLE 12-2 Data dictionary views and dynamic performance views for privileges

Name	Description
DBA_SYS_PRIVS	All system privileges granted
DBA_TAB_PRIVS	All object privileges granted
USER_TAB_PRIVS_MADE	All object privileges granted by the user
USER_TAB_PRIVS_RECD	All object privileges when the user is the grantee
DBA_COL_PRIVS	All object privileges on column lists
SESSION_PRIVS	User's privileges currently enabled (see next chapter for more on enabling and disabling privileges through roles)

You have seen how Oracle 10*g* security works in giving out privileges only as they are needed. Very few privileges are given to PUBLIC (all users). You, as the DBA, control access to the database by restricting a user's authority to perform tasks, limiting users to only the tasks you assign. You also restrict a user's ability to query or modify data to only those objects required by the job. What if you have a large company in which many users have access to many tables? If a table's data appears to be corrupted by a malicious user, how do you identify the culprit? You can narrow it down by finding out which users are allowed to modify that table, but how can you pinpoint each user's activity? The next section shows how to use auditing to handle this situation.

DESCRIPTION OF AUDITING CAPABILITIES

Monitoring activity in a database is called **auditing**. Auditing is most frequently used to determine who is making unauthorized updates or deletions to sensitive data. There are three types of auditing that Oracle 10g can run automatically:

- **Statement auditing:** Audits types of SQL commands. You must have the AUDIT SYSTEM privilege to use this type of auditing. For example, an accountant with access to the CUSTOMER_ORDER table could use this access to adjust the balance due to zero for friends who have made orders. You can set up an audit that monitors all UPDATE commands performed by this accountant. The command looks like this (do not actually run this command):

```
AUDIT UPDATE TABLE BY JACK;
```


- **Privilege auditing:** Audits use of particular privileges. You must have the AUDIT SYSTEM privilege to use this type of auditing. As another example, imagine that your company works from 9:00 to 5:00, and you suspect that one of the programmers is moonlighting and using the database for personal projects. You cannot find any evidence left in the database (the programmer knows enough to drop any tables created for personal use). However, you can set up auditing to track anyone who creates tables and record what time this activity is happening. The command looks like this (do not actually run this command):

```
AUDIT CREATE TABLE;
```

- **Object auditing:** Audits activity on a certain object. You can use the AUDIT command to set up object auditing for any object you own. Otherwise, you must have the AUDIT ANY privilege to audit objects. For example, you have a table of employee pay rates and work history called EE_PRIVATE. Unauthorized access to the table would reveal confidential information. You set up auditing to monitor all queries of the table. The command looks like this (do not actually run this command):

```
AUDIT SELECT ON EE_PRIVATE;
```

TIP

Use auditing with caution, because it generates a high volume of data. It is best to start auditing when you have narrowed the area you want to monitor to a few tables, users, or privileges. This keeps the SYS.AUD$ table from growing too large too quickly.

Auditing commands have no effect until you set the AUDIT_TRAIL initialization parameter. You can set it to the values in the following list by modifying the **init.ora** file or the **spfile**. In either case, you must restart the database for the setting to take effect. Here are the valid settings for AUDIT_TRAIL:

- **TRUE** or **DB:** Starts auditing and places the audit trail records into the SYS.AUD$ table.
- **FALSE** or **NONE:** Turns off auditing. The default is NONE.
- **OS:** Starts auditing and places the audit trail records into an operating system file in the directory named in the AUDIT_FILE_DEST initialization parameter. The default setting for AUDIT_FILE_DEST is ORACLE_HOME\rdbms\audit. Recall that ORACLE_HOME stands for the full path on your computer where the Oracle Software is installed in your ORACLE_HOME directory on your database server computer.

Another initialization parameter, AUDIT_SYS_OPERATIONS, can be set to TRUE or FALSE to turn on or off monitoring of any activity of the SYS user and of other users logged on with SYSDBA privileges.

The syntax of the AUDIT command for object auditing is:

```
AUDIT <objpriv>,<objpriv>,...|ALL
ON <schema>.<object>|DEFAULT|NOT EXISTS
BY SESSION|BY ACCESS
WHENEVER SUCCESSFUL|WHENEVER NOT SUCCESSFUL;
```

To set this auditing for the automatic turn on of any new object that is created, substitute DEFAULT for an object name. This can save time later, if you decide to audit certain activities on every object. With DEFAULT, the auditing is automatically performed as new objects are created.

Substitute NOT EXISTS for an object name, and Oracle 10*g* creates an audit trail record for attempted actions that fail with the "object does not exist" error. This can be useful in determining who might be trying to access sensitive tables, because the "object does not exist" error is generated when you attempt to access an existing object on which you have no privileges.

The syntax for the AUDIT command for auditing SQL statements is:

```
AUDIT <priv>,<priv>,...|ALL PRIVILEGES|CONNECT|RESOURCE|DBA
BY <username>
BY SESSION|BY ACCESS
WHENEVER SUCCESSFUL|WHENEVER NOT SUCCESSFUL;
```

When you specify CONNECT, RESOURCE, or DBA instead of a list of privileges, Oracle 10*g* audits all privileges granted to that role.

The syntax for auditing SQL statements is:

```
AUDIT <sql>,<sql>...|ALL
BY <username>
BY SESSION|BY ACCESS
WHENEVER SUCCESSFUL|WHENEVER NOT SUCCESSFUL;
```

All three types of AUDIT commands are similar. You can narrow the focus of any AUDIT command by specifying any of these optional clauses:

- **BY SESSION:** This tells Oracle 10g to write one record to the audit trail for each session for the same SQL or privilege on the same object. This saves space in the audit trail.
- **BY ACCESS:** This tells Oracle 10g to write one record to the audit trail for every occurrence of the audited event. This is the default. You can specify either BY ACCESS or BY SESSION, but not both.
- **WHENEVER SUCCESSFUL:** This tells Oracle 10g to write a record to the audit trail only when the operation is successful.
- **WHENEVER NOT SUCCESSFUL:** This tells Oracle 10g to write a record to the audit trail only when the operation is not successful. If you don't specify this or the previous clause, Oracle 10g writes a record for the operation it is auditing regardless of whether it succeeds.

Follow these steps to practice creating some auditing commands:

1. Begin by connecting to the database as the SYSTEM user, because SYSTEM has the privileges needed for auditing. Type and execute this command, replacing <password> with the actual password:

```
CONNECT SYSTEM/<password>@ORACLASS
```

2. SQL*Plus Worksheet replies, "Connected." The first thing to check is the setting of the AUDIT_TRAIL initialization parameter. Type and execute this SQL*Plus command:

```
SHOW PARAMETERS AUDIT
```

NOTE

SQL*Plus that do not change metadata or data in the database are not required to be terminated with a colon (;).

Figure 12-32 shows the result of the above command.

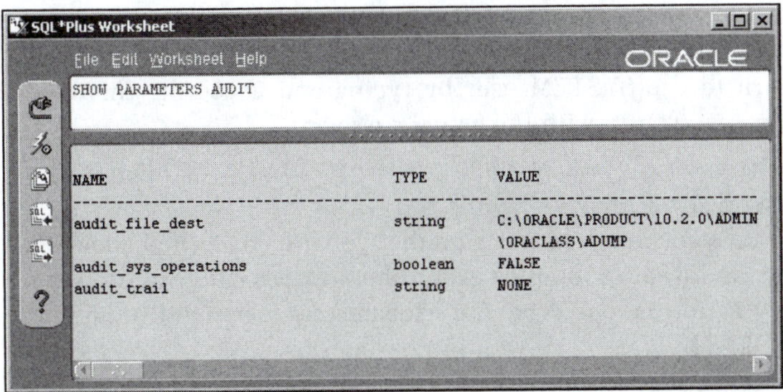

FIGURE 12-32 The TRANSACTION_AUDITING parameter is not involved in auditing

3. Auditing commands are accepted but have no effect unless the AUDIT_TRAIL parameter is set to a value other than "NONE." AUDIT_TRAIL is a static parameter, so you cannot change it in the current session. However, you can modify its value in the **spfile** and then restart the database to activate the parameter. Modify the **spfile** by typing and executing the following command. You are setting the parameter to "DB" to turn on auditing and to have Oracle 10g place audit trail records in the database:

```
ALTER SYSTEM SET AUDIT_TRAIL = 'DB' SCOPE=SPFILE;
```

The SQL*Plus Worksheet replies, "System altered."

4. Now, you must shut down and restart the database to put the parameter into effect. You must be logged on with SYSDBA privileges to shut down and start up the database, so connect to SYS by typing and executing the following command:

```
CONNECT SYS/<password>@ORACLASS AS SYSDBA
```

Replace <password> with the current password of SYS, provided by your instructor.

The SQL*Plus Worksheet replies, "Connected."

5. Shut down and start up the database by typing and executing the commands shown below. The **spfile** is automatically used for initialization parameters. This could take a minute or so. The last command displays the parameters again to confirm the change in the AUDIT_TRAIL parameter

```
SHUTDOWN IMMEDIATE
STARTUP
SHOW PARAMETERS AUDIT
```

NOTE

If your database does not restart it is very likely because of automated listener configuration. Ask your instructor for help. You could use the Net Manager to configure your listener properly. Make sure you restart your listener. You could also restart the database service (on Windows), or in desperation reboot your database server. Ask your instructor for help!

Figure 12-33 shows the results of the script above.

6. There is no need to be logged on as SYSDBA at this point, and Oracle recommends using SYS and SYSDBA privileges only when absolutely necessary. Therefore, return to the SYSTEM user by typing and executing this command, replacing <password> with the actual password:

```
CONNECT SYSTEM/<password>@ORACLASS
```

The SQL*Plus Worksheet replies, "Connected."

7. It is time to set up some auditing. Imagine that you are concerned about the STUDENTB user and want to monitor everything he does. You want to see a record for every action taken. Type and execute this command to audit all activity of STUDENTB:

```
AUDIT ALL BY STUDENTB;
```

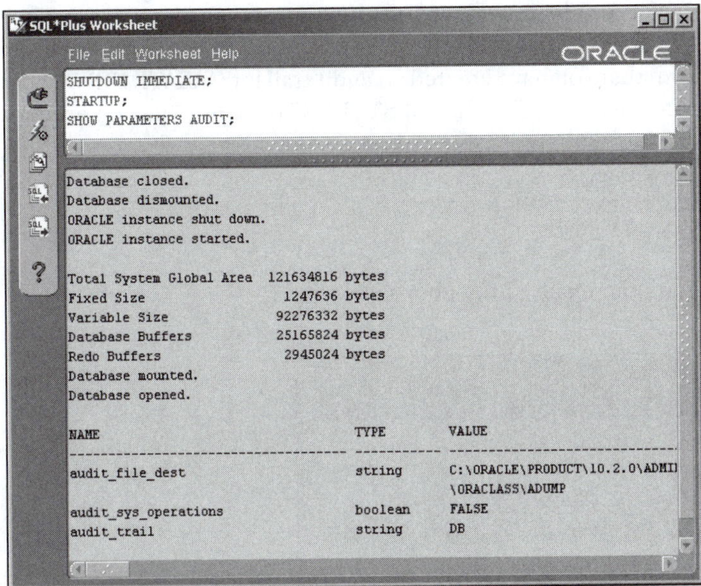

FIGURE 12-33 Restarting the database causes the new values in the spfile to take effect

The previous command is a statement auditing command. If it were a privilege auditing command, it would use "ALL PRIVILEGES" instead of "ALL." It does not name an object or display "ON DEFAULT," so it is not an object auditing command. Object auditing commands always contain the word "ON" followed by either "DEFAULT" or an object name.
The SQL*Plus Worksheet replies, "Audit succeeded."

8. In addition, you want to monitor all updates to the EMPLOYEE table that are made by anyone. You only need to see one record for a user's session, even if she updates the table 10 times in the session. This requires an object auditing command, so type and execute the following command:

```
AUDIT SELECT, UPDATE ON CLASSMATE.EMPLOYEE
BY SESSION;
```

The SQL*Plus Worksheet replies, "Audit succeeded."

9. And finally, you want to audit the CLASSMATE and STUDENTA users when they create a table or a view. You only want to see auditing records when the commands succeed. This requires a privilege auditing command. Type and execute this command to set up the audit:

```
AUDIT CREATE TABLE, CREATE VIEW
BY CLASSMATE, STUDENTA
WHENEVER SUCCESSFUL;
```

The SQL*Plus Worksheet replies, "Audit succeeded."

10. You can use data dictionary views to display the auditing settings you have created. There is a view for each type of audit statement: (1) DBA_OBJ_AUDIT_OPTS for object auditing, (2) DBA_PRIV_AUDIT_OPTS for privilege auditing, and (3) DBA_STMT_AUDIT_OPTS for statement auditing. There

is also a special view for object auditing in which you set the auditing as the default for all objects: ALL_DEF_AUDIT_OPTS. Query the view for privilege auditing to confirm that you have created an audit trail for CREATE VIEW and CREATE TABLE for the CLASSMATE and STUDENTB users. Type and execute this query:

```
COLUMN PRIVILEGE FORMAT A12
SELECT USER_NAME, PRIVILEGE, SUCCESS, FAILURE
FROM DBA_PRIV_AUDIT_OPTS
ORDER BY 1, 2;
```

Figure 12-34 shows the result of the above query.

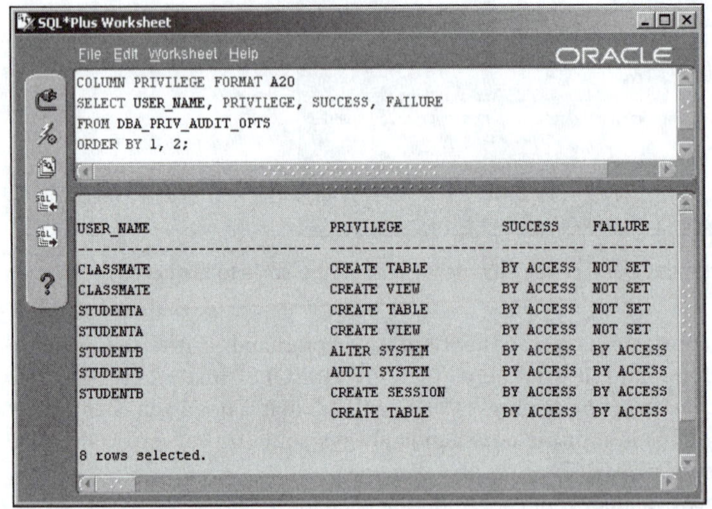

FIGURE 12-34 "NOT SET" in the FAILURE column indicates no auditing for failed commands

11. To simulate activity that would be audited, connect to the CLASSMATE user and run a script. Begin by typing and executing this command to switch to the CLASSMATE user:

    ```
    CONNECT CLASSMATE/CLASSPASS@ORACLASS
    ```

    The SQL*Plus Worksheet replies, "Connected."
12. Now, run the script that performs some updates on the EMPLOYEE table. Click **Worksheet/Run Local Script**. A window appears in which you can select a file from your computer.
13. Navigate to the **Data\Chapter12** directory, and select the **audit.sql** file. Click **Open** to retrieve the file, and run it in the worksheet.
14. Connect to SYSTEM again by typing and executing this command, replacing <password> with the actual password:

    ```
    CONNECT SYSTEM/<password>@ORACLASS
    ```

    The SQL*Plus Worksheet replies, "Connected."

15. The audit trail records are written to the SYS.AUD$ table. You can query this table directly, or you can use the data dictionary views to examine audit trail records. Query the object auditing records by typing and executing this query:

```
SELECT USERNAME, ACTION_NAME,
SES_ACTIONS FROM DBA_AUDIT_OBJECT
WHERE OBJ_NAME = 'EMPLOYEE';
```

Figure 12-35 shows the results.

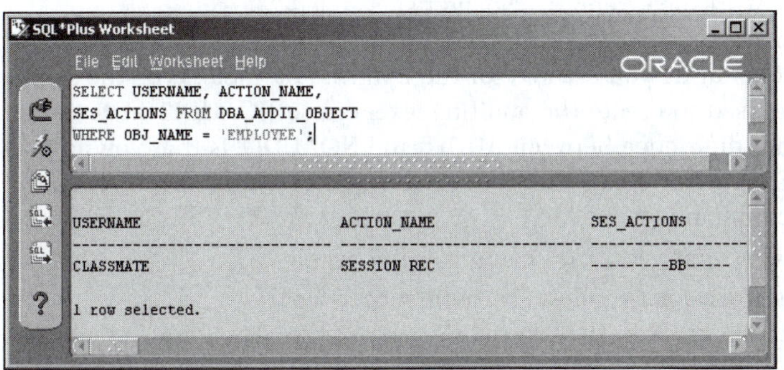

FIGURE 12-35 Activity on the EMPLOYEE table is consolidated into a single row

The SES_ACTIONS column has a set of codes. Each of the 16 positions represents one type of activity that can occur on the table. Your results may show a different combination of letters than the "B" and "S" shown in Figure 12-35, depending on other activities in the database. The tenth and eleventh positions are for SELECT and UPDATE. A dash (-) means no activity, a letter "S" means the action succeeded, a letter "F" means the action failed, and a letter "B" means that it failed and succeeded during this session. So, the CLASSMATE session included successful and failed SELECT actions and successful UPDATE actions. An update command has an implicit select that the audit trail records. One of the update commands failed because it attempted to update a nonexistent record. In this case, the implicit select failed.

16. Remain logged on for the next practice.

You used the DBA_AUDIT_OBJECT data dictionary view in the last practice. You can query several other data dictionary views for audit trail results. The complete list includes:

- **DBA_AUDIT_EXISTS:** Audit trail records generated by object auditing of nonexistent objects
- **DBA_AUDIT_OBJECT:** Audit trail records generated by object auditing
- **DBA_AUDIT_SESSION:** Audit trail records generated by session auditing
- **DBA_AUDIT_STATEMENT:** Audit trail records generated by statement auditing
- **DBA_AUDIT_TRAIL:** All audit trail records

All of the above metadata views have a corresponding USER_counterpart, except DBA_AUDIT_EXISTS.

You may want to turn off auditing or change what you are auditing. This is done with the NOAUDIT command. Its structure is exactly like the AUDIT command; it turns off the auditing it names. You can use it to turn off selective portions of the auditing you have set up. Follow these steps to explore the NOAUDIT command:

1. Recall that you have set up auditing of all statements made by STUDENTB. You can adjust this so that no audit records are generated by the SELECT command. Type and execute this command:

   ```
   NOAUDIT SELECT TABLE BY STUDENTB;
   ```

   The SQL*Plus Worksheet replies, "Noaudit succeeded."
   The other statements will still be audited.
2. To turn off the auditing you created for the EMPLOYEE table, type the same command you used to create the auditing, except use NOAUDIT instead of AUDIT. Another difference between AUDIT and NOAUDIT is that you do not specify BY SESSION or BY ACCESS in the NOAUDIT command. Type and execute this command:

   ```
   NOAUDIT SELECT, UPDATE ON CLASSMATE.EMPLOYEE;
   ```

   The SQL*Plus Worksheet replies, "Noaudit succeeded."
3. Log off by clicking the **X** in the top-right corner of the SQL*Plus Worksheet window.
4. Close the console by clicking the **X** in the top-right corner of the console window. You have seen how granting privileges and auditing database activity can be used to enhance security on the Oracle 10g database. Building on the base established in this chapter, the next chapter covers another security technique: using roles to manage privileges.

The next section covers database roles.

DATABASE ROLES

A role is a collection of privileges that is named and assigned to users or even to another role. A role can help you simplify database maintenance by giving you an easy way to assign a set of privileges to new users.

How to Use Roles

The primary use of roles is to help simplify security. For example, in a typical database system, you have a schema with 25 tables and a group of 50 users who need access to the schema. Figure 12-36 shows a miniature version of a database schema and its users, with the privileges they each require. As you can see, even scaled down to 5 tables and 6 users, you must grant 59 privileges. Even if you group them together in sets of users with the same sets of grants, you have at least 20 combinations to deal with.

Privileges	Tables				
	CUSTOMER	PAYMENT	ORDER	STORE	INVENTORY
SELECT	Joe Amy Henry Sarah John	Joe Amy Henry Sarah John	Joe Amy Henry Sarah John	Joe Amy Henry Sarah John Martin	Joe Amy Henry Sarah John
INSERT	Joe Amy	Joe Amy	Henry Sarah John	Martin	Henry Sarah John
UPDATE	Joe Amy	Joe Amy	Henry Sarah John	Martin	Henry Sarah John
DELETE	Joe Amy	Joe Amy	Henry Sarah John	Martin	Henry Sarah John

FIGURE 12-36 A matrix of privileges, tables, and users needing those privileges

If you look closely at the table in Figure 12-36, you can see a pattern of users who share the same set of grants. Figure 12-37 highlights the patterns, or groups of users who share the same set of privileges. This is where roles come into play. You can create a role for each group of users, grant the privileges to the role, and then grant the role to each user in each group.

Privileges	Tables				
	CUSTOMER	PAYMENT	ORDER	STORE	INVENTORY
SELECT	Joe Amy	Joe Amy	Joe Amy	Joe Amy	Joe Amy
	Henry Sarah John	Henry Sarah John	Henry Sarah John	Henry Sarah John Martin	Henry Sarah John
INSERT	Joe Amy	Joe Amy	Henry Sarah John	Martin	Henry Sarah John
UPDATE	Joe Amy	Joe Amy	Henry Sarah John	Martin	Henry Sarah John
DELETE	Joe Amy	Joe Amy	Henry Sarah John	Martin	Henry Sarah John

FIGURE 12-37 The 6 users fall into 3 groups

At first, creating roles may seem like an extra step. After you begin to add new users, however, you save time, because you must grant only roles to new users. In addition, you can revise the privileges in a role, and the change is automatically reflected for every user who has the role.

Using Predefined Roles

As stated previously in this chapter, there are over 100 system privileges, and as you create more tables, the number of possible object privileges grows as well. Even so, as you become familiar with these privileges, you quickly realize that many jobs require a typical set of privileges. The job of DBA, for example, requires nearly all of the system privileges. The job of applications developer requires the subset of the DBA privileges that enables the developer to create tables, procedures, and other database objects.

Oracle 10*g* provides predefined roles for common job titles to speed up your ability to get your users up and running with the privileges they require. Table 12-3 shows some of the commonly used predefined roles and what they are intended to handle.

TABLE 12-3 Predefined roles

Role Name	Description
CONNECT	Logs onto the database and performs limited activities within the user's own schema, such as creating tables, views, synonyms, and database links
DBA	Manages the database, including these tasks: creates users, profiles, and roles, and grants privileges; manages storage and security; starts up and shuts down the database
DELETE_CATALOG_ROLE	Gives the user the ability to delete from tables owned by SYS. This role was added because the system privilege DELETE ANY TABLE specifically excludes deleting from tables owned by SYS
EXECUTE_CATALOG_ROLE	Enables the user to execute any package supplied by Oracle that is owned by SYS. Most supplied packages are owned by SYS, and those most commonly used already allow users to execute them. If additional packages are needed, grant the user this role
EXP_FULL_DATABASE	Exports the database using the EXPORT utility
IMP_FULL_DATABASE	Imports the database using the IMPORT utility
RESOURCE	Provides more extensive abilities to create objects, such as procedures, triggers, and object types, for users who need to create their own objects. If you assign the CONNECT role to a user, you should grant a RESOURCE as well
SELECT_CATALOG_ROLE	Allows the user to query any data dictionary view or table owned by SYS. This can give a user more access to certain data dictionary views, although usually a user can already access those he needs, because the most common data dictionary views are viewable by all users

When you install and create the database, there are a dozen or so additional roles created that are for internal use by database tools and utilities. For example, there is a role called AQ_ADMINISTRATOR_ROLE, containing privileges specifically used for managing Oracle advanced queues, which is an Oracle feature for communicating between databases.

TIP

Oracle recommends that you create your own roles instead of using the CONNECT, RESOURCE, and DBA roles, because these three roles will be dropped in a future release. The intention is to minimize security risks.

Complete the following steps to query the data dictionary views and view the privileges that some of these roles include.

1. Start up the SQL*Plus Worksheet. In Windows, click **Start/All Programs/ Oracle ... /Application Development/SQLPlus Worksheet**. In Unix or Linux, type **oemapp worksheet** on the command line. The Oracle Enterprise Manager Login screen appears.
2. Login by typing **SYSTEM** in the Username box. In the Password box, type the current password for SYSTEM that is *provided by your instructor*. Type **ORACLASS** in the Service box, and leave the default selection of "Normal" in the Connect as box. Click **OK**. A background SQL*Plus process starts, and then the SQL*Plus Worksheet window appears.
3. Type and execute this query to list the system privileges granted to the IMP_FULL_DATABASE role. Figure 12-38 partially shows the results:

```
SET LINESIZE 100
SELECT * FROM ROLE_SYS_PRIVS
WHERE ROLE = 'IMP_FULL_DATABASE'
ORDER BY 1, 2;
```

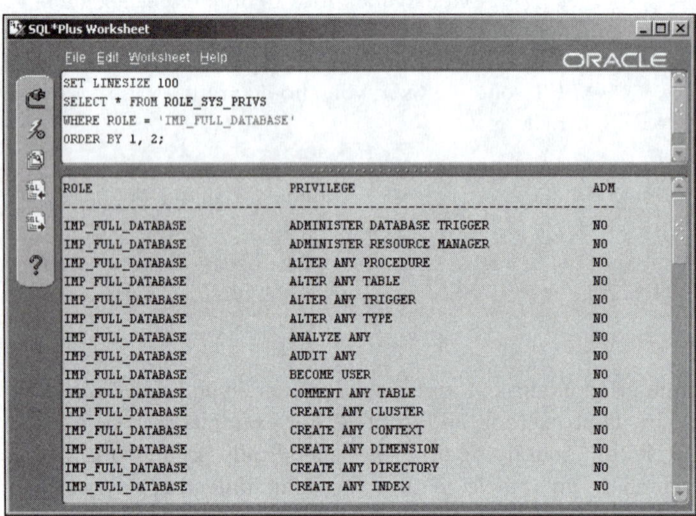

FIGURE 12-38 Roles are often granted system privileges

This role has over 60 privileges that are needed to perform a full database import. An import is a process that uses a previously created snapshot of the database and recreates all or part of it in the current database. An import can create users, tables, views, and triggers, add rows to existing tables, and much more.

4. Remain logged on for the next practice.

   You have seen Oracle's predefined roles, which give you an idea of how roles can be used. The next section shows you how to create your own roles.

CREATING AND MODIFYING ROLES

Roles are used to consolidate a group of system or object privileges so that you, the DBA, can assign the role to a user, rather than assigning all the underlying system and object privileges. You can also assign the role to another role.

The first step in this process is creating the role itself. The syntax is:

```
CREATE ROLE <name>
NOT IDENTIFIED|IDENTIFIED BY <password>
```

The NOT IDENTIFIED clause means that no additional authorization is required. This is the default, so omitting the clause is the same as including NOT IDENTIFIED. The alternative option is the IDENTIFIED BY <password> clause, which means that the user must provide the correct password to be able to use the privileges within that role.

The next step in working with roles is to assign privileges to the role. This is done with the GRANT command in exactly the same way as granting privileges to a user. Previously in this chapter you dealt with granting privileges to users. Now you'll see the same command, except that the user name is replaced by a role name. The syntax is:

```
GRANT <privilege> TO <role>;
```

TIP

You cannot grant a privilege and add WITH ADMIN OPTION or WITH GRANT OPTION when granting to a role.

The final step is assigning the role to a user. Again, this uses the GRANT command, except this time, you replace the privilege with the name of a role. You can also grant a role to another role with this command. The syntax is:

```
GRANT <role> TO <user>|<role>
WITH ADMIN OPTION;
```

Include the WITH ADMIN OPTION only when you want the user to be able to grant the role to other users. If you grant a role to a second role with the WITH ADMIN OPTION, any user who is granted the second role is allowed to grant the first role to others.

The only part of a role you can change is whether it uses a password. The syntax of the ALTER ROLE command is:

```
ALTER ROLE <name>
NOT IDENTIFIED|IDENTIFIED BY <password>
```

For example, to change the UPDATEALL role so that it requires a password, you use the following code:

```
ALTER ROLE UPDATEALL
IDENTIFIED BY U67DATR;
```

When a role switches to requiring a password, users currently logged on who are granted the role are unaffected until they log off and back on again.

Creating and Assigning Privileges to a Role

You have decided to create a role that allows a user to query the Global Globe tables that are needed to use an application that helps employees build their own queries. Follow along with these steps to create a new role and assign privileges to the role:

1. Create the role, named SELALL by typing and executing this command.

   ```
   CREATE ROLE SELALL;
   ```

2. SQL*Plus Worksheet replies, "Role created." Now, grant the SELECT object privilege to the role for the Global Globe tables by typing and executing these GRANT commands:

   ```
   GRANT SELECT ON CLASSMATE.CLASSIFIED_AD TO SELALL;
   GRANT SELECT ON CLASSMATE.CLASSIFIED_SECTION TO SELALL;
   GRANT SELECT ON CLASSMATE.CUSTOMER TO SELALL;
   GRANT SELECT ON CLASSMATE.CUSTOMER_ADDRESS TO SELALL;
   GRANT SELECT ON CLASSMATE.NEWS_ARTICLE TO SELALL;
   GRANT SELECT ON CLASSMATE.EMPLOYEE TO SELALL;
   ```

3. SQL*Plus Worksheet replies, "Grant succeeded" after it executes each command.
4. Remain logged on for the next practice.

TIP

Other users are allowed to grant privileges to a role created by the DBA. The user must, of course, have the authority to grant the privilege to others. For example, the CLASSMATE user could grant UPDATE ON EMPLOYEE to the SELALL role. If it makes sense to do so, users can grant privileges to the predefined roles as well.

Assigning Roles to Users and to Other Roles

Now that you have a role, you can assign (grant) the role to any user you want. In addition, a role can be granted to another role. The grantee role inherits all the privileges of the grantor role.

Continuing with the Global Globe example, follow these steps to grant SELALL to users and to another role:

1. For this exercise, you can use the two users created in the previous chapter, STUDENTA and STUDENTB. Type and execute this command to grant each of them the SELALL role:

   ```
   GRANT SELALL TO STUDENTA, STUDENTB;
   ```

2. SQL*Plus Worksheet replies, "Grant succeeded."
3. Connect to STUDENTA, and check to see what privileges he has. Type and execute these commands. Figure 12-39 shows the results:

```
CONNECT STUDENTA/A1PLU$@ORACLASS
SELECT TABLE_NAME, PRIVILEGE FROM USER_TAB_PRIVS;
```

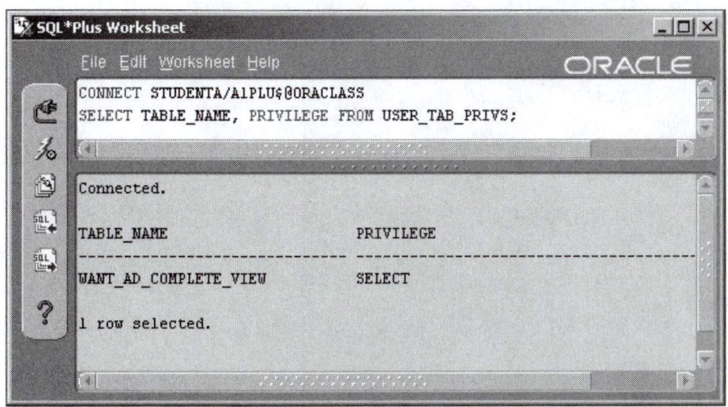

FIGURE 12-39 STUDENTA appears to have SELECT privileges on only one table

4. What went wrong? Nothing. However, when you grant a role, the user inherits the privileges granted to the role even though the privileges don't appear in the USER_TAB_PRIVS view. The following query shows that STUDENTA has been granted the SELALL role. Type and execute the query:

```
SELECT * FROM USER_ROLE_PRIVS;
```

5. Next is another query, in which the object privileges of the SELALL role and the object privileges of the STUDENTA user are combined to show a complete list of STUDENTA's privileges. Type and execute this query. Figure 12-40 shows the results:

```
COLUMN OWNER FORMAT A10
COLUMN PRIVILEGE FORMAT A10
SELECT OWNER, TABLE_NAME, PRIVILEGE FROM USER_TAB_PRIVS
UNION
SELECT OWNER, TABLE_NAME, PRIVILEGE FROM ROLE_TAB_PRIVS
WHERE ROLE IN (SELECT ROLE FROM USER_ROLE_PRIVS)
ORDER BY 1, 2;
```

6. The previous query works well, unless a role has been granted to another role. In that case you'd need a more complex query to derive the privileges of the underlying role. Reconnect to the SYSTEM user to create another role and grant one role to another. Type and execute this command, replacing <password> with the current password for SYSTEM.

```
CONNECT SYSTEM/<password>@ORACLASS
```

The SQL*Plus Worksheet replies, "Connected."

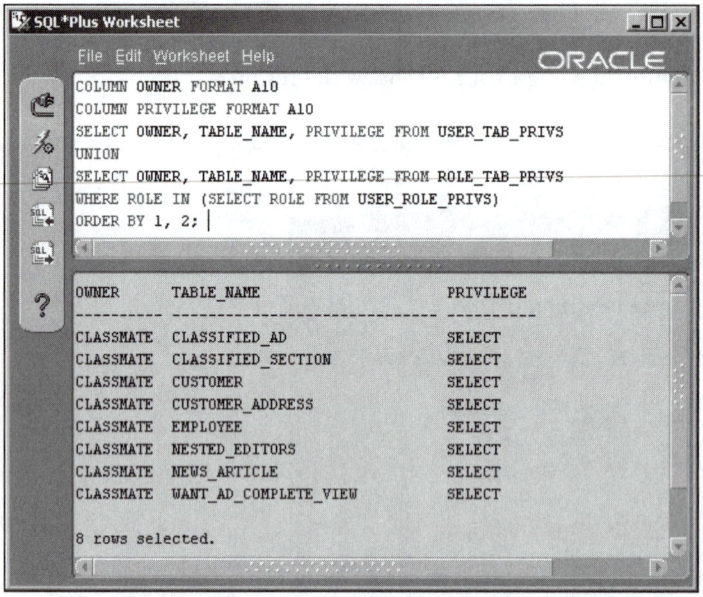

```
COLUMN OWNER FORMAT A10
COLUMN PRIVILEGE FORMAT A10
SELECT OWNER, TABLE_NAME, PRIVILEGE FROM USER_TAB_PRIVS
UNION
SELECT OWNER, TABLE_NAME, PRIVILEGE FROM ROLE_TAB_PRIVS
WHERE ROLE IN (SELECT ROLE FROM USER_ROLE_PRIVS)
ORDER BY 1, 2;

OWNER      TABLE_NAME                     PRIVILEGE
---------- ------------------------------ ----------
CLASSMATE  CLASSIFIED_AD                  SELECT
CLASSMATE  CLASSIFIED_SECTION             SELECT
CLASSMATE  CUSTOMER                       SELECT
CLASSMATE  CUSTOMER_ADDRESS               SELECT
CLASSMATE  EMPLOYEE                       SELECT
CLASSMATE  NESTED_EDITORS                 SELECT
CLASSMATE  NEWS_ARTICLE                   SELECT
CLASSMATE  WANT_AD_COMPLETE_VIEW          SELECT

8 rows selected.
```


FIGURE 12-40 Role-owned privileges combine with user-owned privileges in this query

7. You want another role for users who are allowed to update the EMPLOYEE and CUSTOMER tables. These users should also be able to query all the tables listed in the SELALL role. The new role, called UPDPEOPLE, is more sensitive, so you add a password requirement to the role. Type and execute this command:

   ```
   CREATE ROLE UPDPEOPLE IDENTIFIED BY M6XABC;
   ```

8. SQL*Plus replies, "Role created." Add the two privileges for the EMPLOYEE and CUSTOMER tables by typing and executing these commands:

   ```
   GRANT UPDATE ON CLASSMATE.CUSTOMER TO UPDPEOPLE;
   GRANT UPDATE ON CLASSMATE.EMPLOYEE TO UPDPEOPLE;
   ```

9. SQL*Plus replies, "Grant succeeded" after it executes each statement. To assign the privileges of the SELALL role to the UPDPEOPLE role, type and execute this command:

   ```
   GRANT SELALL TO UPDPEOPLE;
   ```

   The SQL*Plus Worksheet replies, "Grant Succeeded."

10. You can see which roles have been granted to another role by querying the ROLE_ROLE_PRIVS view. Type and execute this query. Figure 12-41 shows the results, in which you can see that the SELALL role was granted to the UPDPEOPLE role:

    ```
    SELECT ROLE, GRANTED_ROLE, ADMIN_OPTION
    FROM ROLE_ROLE_PRIVS ORDER BY 1,2;
    ```

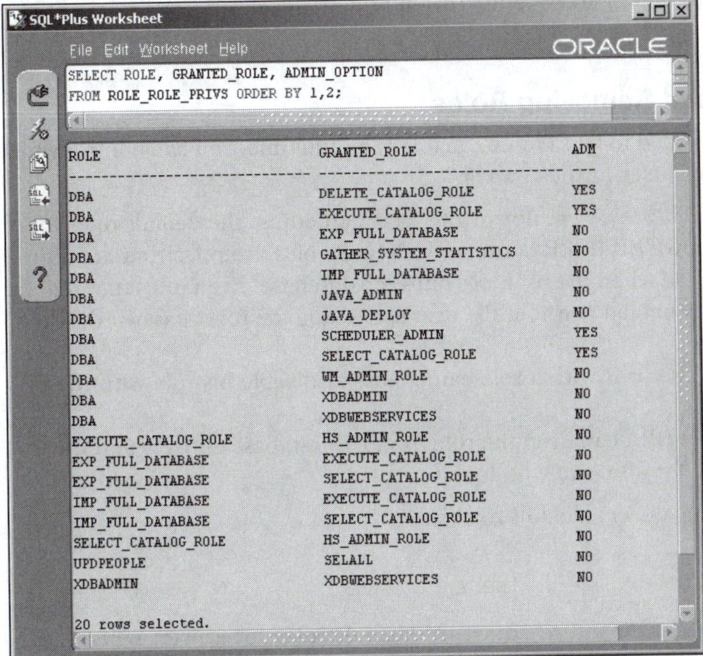

SQL*Plus Worksheet
File Edit Worksheet Help
ORACLE
SELECT ROLE, GRANTED_ROLE, ADMIN_OPTION
FROM ROLE_ROLE_PRIVS ORDER BY 1,2;
ROLE GRANTED_ROLE ADM
DBA DELETE_CATALOG_ROLE YES
DBA EXECUTE_CATALOG_ROLE YES
DBA EXP_FULL_DATABASE NO
DBA GATHER_SYSTEM_STATISTICS NO
DBA IMP_FULL_DATABASE NO
DBA JAVA_ADMIN NO
DBA JAVA_DEPLOY NO
DBA SCHEDULER_ADMIN YES
DBA SELECT_CATALOG_ROLE YES
DBA WM_ADMIN_ROLE NO
DBA XDBADMIN NO
DBA XDBWEBSERVICES NO
EXECUTE_CATALOG_ROLE HS_ADMIN_ROLE NO
EXP_FULL_DATABASE EXECUTE_CATALOG_ROLE NO
EXP_FULL_DATABASE SELECT_CATALOG_ROLE NO
IMP_FULL_DATABASE EXECUTE_CATALOG_ROLE NO
IMP_FULL_DATABASE SELECT_CATALOG_ROLE NO
SELECT_CATALOG_ROLE HS_ADMIN_ROLE NO
UPDPEOPLE SELALL NO
XDBADMIN XDBWEBSERVICES NO
20 rows selected.

FIGURE 12-41 Roles that contain granted roles are shown in the ROLE_ROLE_PRIVS view

11. You can grant the WITH ADMIN OPTION to either a role or a user. This allows the user (or a user with the role granted the WITH ADMIN OPTION) to grant that role to other users or roles. Change the UPDPEOPLE role so that it can administer the SELALL role by typing and executing this command:

```
GRANT SELALL TO UPDPEOPLE WITH ADMIN OPTION;
```

    The SQL*Plus Worksheet replies, "Grant succeeded."

12. Now, a user with the UPDPEOPLE role can assign the SELALL role to other users. Try this by granting the UPDPEOPLE role to Amanda Gaines, the chief editor whose user name is AGAINES. Type and execute these commands to create the user, and grant the role to her:

```
CREATE USER AGAINES
IDENTIFIED BY ABSOLUTE#1
DEFAULT TABLESPACE USERS
TEMPORARY TABLESPACE TEMP
QUOTA 10M ON USERS
QUOTA 5M ON USER_AUTO
PROFILE DEFAULT
ACCOUNT UNLOCK;
GRANT UPDPEOPLE TO AGAINES;
```

    The SQL*Plus Worksheet replies, "Grant succeeded."

13. Remain logged on for the next practice.

Roles are dynamic. After a user has a role, you can add more privileges to the role, and the user receives the privilege immediately.

The next session shows you how to enable (bring into effect) a role with a password, and how to control which roles are enabled.

Limiting Availability and Removing Roles

Sometimes, a role does not need to be in force (enabled) all the time. You can control when a role becomes enabled for a user in these ways:

- **Default roles:** The role's creator or the DBA can adjust the default roles for a user using the ALTER USER command. Default roles are roles that are automatically enabled when the user logs onto the database. Even roles that require a password are enabled (without the user specifying the role's password) when the user logs on.
- **Enable roles:** The user with a role can enable or disable his role with the SET ROLE command.
- **Drop roles:** The DBA can drop the role from the database entirely and thereby cancel the role for all users who had it.

The syntax for changing a user's default roles is:


```
ALTER USER <username> DEFAULT ROLE
<role>,...|ALL|ALL EXCEPT <role>,...|NONE
```

The DBA can issue this command to adjust the default roles for a user. When it is granted to a user, the role is automatically in the list of default roles. The only way to remove the role from the user's default roles is by issuing the ALTER USER command. To remove all the roles at once, use the NONE clause.

The syntax for the SET ROLE command is:

```
SET ROLE
<role> IDENTIFIED BY <password>,...|ALL|ALL EXCEPT|NONE|
```

The *user* can issue this command to adjust his enabled roles. To enable roles with passwords, include the IDENTIFIED BY <password> clause; otherwise, simply list the role. Any role not listed is disabled. Enable all roles by using ALL, and disable all roles by using NONE.

TIP

The roles remain enabled or disabled until the user issues another SET ROLE command, or until the user logs off. When the user logs on again, his roles are reset to the default roles dictated by the DBA.

Dropping a role revokes it from all users and roles, except those who are currently logged on and have the role enabled. The role is revoked for those users when their sessions end. The syntax of DROP ROLE is simply:

```
DROP ROLE <role>
```

Follow along to work with enabling, disabling, and dropping roles. Remember that in the previous practice, the UPDPEOPLE role was granted to AGAINES:

1. Add the CONNECT role and take the UPDPEOPLE role out of the default list for AGAINES by typing these commands:

```
GRANT CONNECT TO AGAINES;
ALTER USER AGAINES DEFAULT ROLE ALL EXCEPT UPDPEOPLE;
```

   The SQL*Plus Worksheet replies, "User altered."

2. Type and execute the following query to list the roles and to see whether the roles are default or not. Figure 12-42 shows the results. As you can see, the CONNECT role is a default role, and the UPDPEOPLE role is not a default role:

```
SELECT * FROM DBA_ROLE_PRIVS
WHERE GRANTEE = 'AGAINES';
```

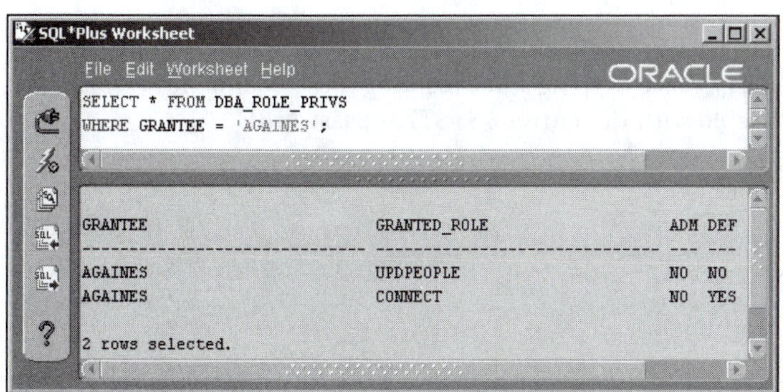

FIGURE 12-42 View roles assigned to users and the default setting of each role

3. Connect to AGAINES by typing and executing this command:

```
CONNECT AGAINES/ABSOLUTE#1@ORACLASS;
```

4. The only role currently enabled for AGAINES is the CONNECT role. To verify this, type and execute this query to see the currently enabled roles:

```
SELECT * FROM SESSION_ROLES;
```

5. When you enable it, the UPDPEOPLE role requires a password. Type and execute this command to enable the role by providing the correct password:

```
SET ROLE CONNECT, UPDPEOPLE IDENTIFIED BY M6XABC;
```

6. SQL*Plus Worksheet replies, "Role set." The CONNECT role was also listed to keep it enabled. Otherwise, it would have been disabled. Rerun this query to verify the enabled roles:

```
SELECT * FROM SESSION_ROLES;
```

   Figure 12-43 shows the results. The SELALL role appears, because it was granted to the UPDPEOPLE role.

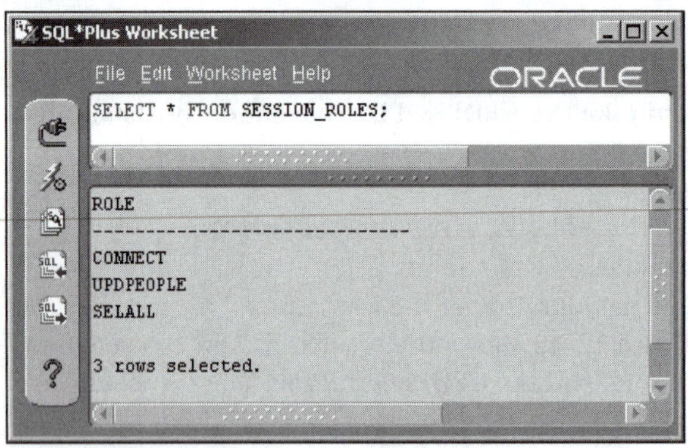
SQL*Plus Worksheet
File Edit Worksheet Help
ORACLE
SELECT * FROM SESSION_ROLES;
ROLE
CONNECT
UPDPEOPLE
SELALL
3 rows selected.

FIGURE 12-43 The current roles for the user include roles granted to other roles

7. While the UPDPEOPLE role is enabled, this user has authority to grant SELALL to another user. Try this by typing and executing these commands (replace the <password> variable with the current SYSTEM password):

```
CONNECT SYSTEM/<password>@ORACLASS;
CREATE USER JHANSON
IDENTIFIED BY TOTAL#1
DEFAULT TABLESPACE USERS
QUOTA 10M ON USERS
QUOTA 5M ON USER_AUTO
PROFILE DEFAULT
ACCOUNT UNLOCK;
CONNECT AGAINES/ABSOLUTE#1@ORACLASS;
GRANT SELALL TO JHANSON;
```

   The SQL*Plus Worksheet replies, "Grant succeeded."
8. Let's say that AGAINES is finished working on the EMPLOYEE table and wants to disable that role in her session. The SET ROLE command can be used to either enable or disable roles by including or excluding a role in the command. Type and execute this command to disable only the UPDPEOPLE role:

```
SET ROLE ALL EXCEPT UPDPEOPLE;
```

   The SQL*Plus Worksheet replies, "Role set."
9. Requery the SESSION_ROLE view, using the query in Step 6 above, to see which roles remain. Figure 12-44 displays the results, showing that only the CONNECT role is enabled now.
10. A role can be removed from the database by running the DROP ROLE command. This must be done by the DBA, so type and execute this command to connect as the SYSTEM user. Replace the <password> variable with the current SYSTEM password:

```
CONNECT SYSTEM/<password>@ORACLASS
```

   The SQL*Plus Worksheet replies "Connected."
11. Drop the UPDPEOPLE role by typing and executing this command:

```
DROP ROLE UPDPEOPLE;
```

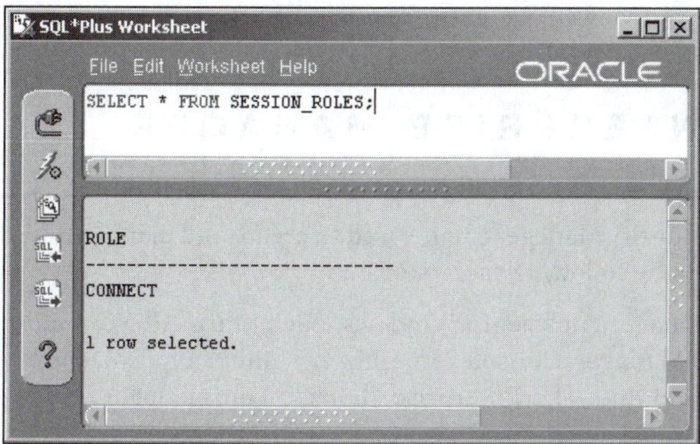
SQL*Plus Worksheet
File Edit Worksheet Help
ORACLE
SELECT * FROM SESSION_ROLES;
ROLE
CONNECT
1 row selected.

FIGURE 12-44 The SELALL role is disabled along with the UPDPEOPLE role

The SQL*Plus Worksheet replies, "Role dropped." The UPDPEOPLE role is thereby removed from the privileges of all users and roles that had been granted it. If a user has a session with the UPDPEOPLE role enabled when the DROP ROLE command is run, that role remains enabled for the user until her session ends.

12. Close your SQL*Plus Worksheet session by clicking the **X** in the top-right corner of the window.

TIP

Roles are similar to system privileges in that a role granted by a user is not revoked if the granting user loses the role.

DATA DICTIONARY INFORMATION ABOUT ROLES

You have used many of the data dictionary views while practicing the display of information about roles. Table 12-4 shows all the views, including those that you have already queried.

TABLE 12-4 Data dictionary views and dynamic performance views for privileges

Name	Description
DBA_SYS_PRIVS	All system privileges granted
DBA_TAB_PRIVS	All object privileges granted
USER_TAB_PRIVS_MADE	All object privileges granted by the user
USER_TAB_PRIVS_RECD	All object privileges when the user is the grantee
DBA_COL_PRIVS	All object privileges on column lists
SESSION_PRIVS	User's privileges currently enabled (see next chapter for more on enabling and disabling privileges through roles)

Remember that the DBA_ prefixed views have corresponding USER_ and ALL_ prefixed views.

ROLES IN THE ENTERPRISE MANAGER CONSOLE

The Enterprise Manager console's Security Manager can be used to create and manage roles very easily. Take a quick look at this by following these steps:

1. Start the Enterprise Manager console. In Windows, click **Start/ All Programs/ Oracle ... /Enterprise Manager Console**. In Unix or Linux, type **oemapp console** on the command line. The Enterprise Manager console login screen appears.
2. Connect as SYSTEM, with the appropriate password and database name.
3. In the main window of the Enterprise Manager console, navigate starting at the **Databases** icon to **ORACLASS/Security/Roles**. Your screen should look similar to Figure 12-45.

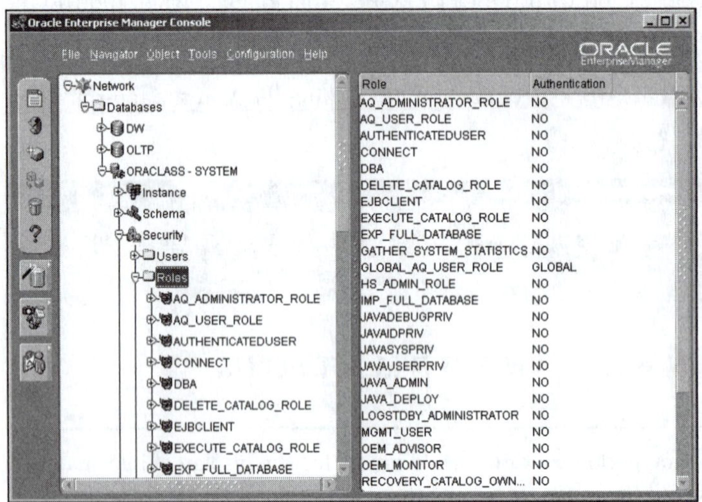

FIGURE 12-45 The Security Manager has an excellent tool for managing roles

4. Click the **CONNECT** role on the left side of the screen. This brings up the property sheet for the CONNECT role.
5. Click the **System** tab, which displays all the single system privileges assigned to the CONNECT role. Oracle 10g is updated to minimize on the initial power of privileges granted to user. The CONNECT role in past versions of Oracle had many powerful privileges granted to it. Now the CONNECT begins with the bare minimum, the CREATE SESSION privilege which allows a database connection. Adding privileges to users in Oracle 10g is now the purview of database administrators. This helps to maintain manageable security.

NOTE

The approach to security with roles, and particularly the CONNECT role, is known as a policy of least privilege.

Figure 12-46 shows the screen. You can add or remove privileges using this screen by highlighting a privilege. Then use the arrow buttons in the center to move the privilege up (revoking the privilege from the role) or down (granting the privilege to the role).

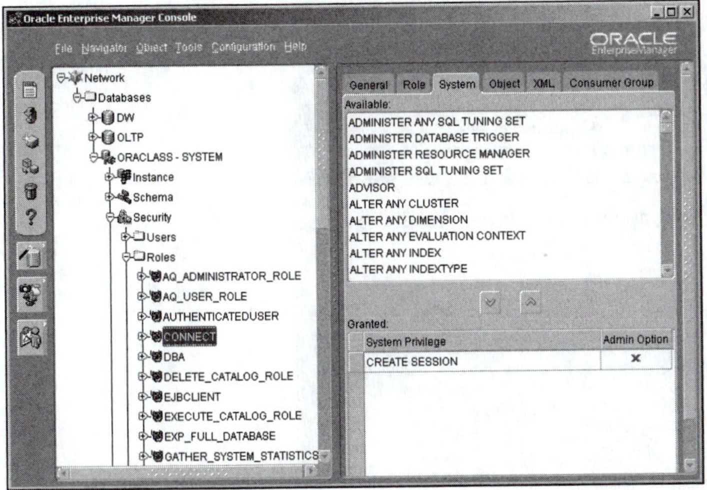

FIGURE 12-46 Available privileges are on the top, and granted privileges are on the bottom

6. Scroll down on the left and click the **SELALL** role. Then click the **Object** tab to display the object privileges assigned to this role, as shown in Figure 12-47. Notice the navigation tree in the top-right section of the screen. You can add new object privileges by navigating through this tree in the same way you use the Schema Manager.
7. Scroll up the navigation tree in the upper-right corner of the screen until you find the CLASSMATE schema. Double-click on **CLASSMATE**.
8. The list of available object types appears below the CLASSMATE schema. Double-click the **Tables** folder.
9. A list of tables owned by CLASSMATE is displayed below the Tables folder. Scroll down and select **WANT_AD**. A list of available privileges for the WANT_AD table appears in the box labeled "Available Privileges." This list changes according to the type of object you select. Figure 12-48 shows the screen at this point.

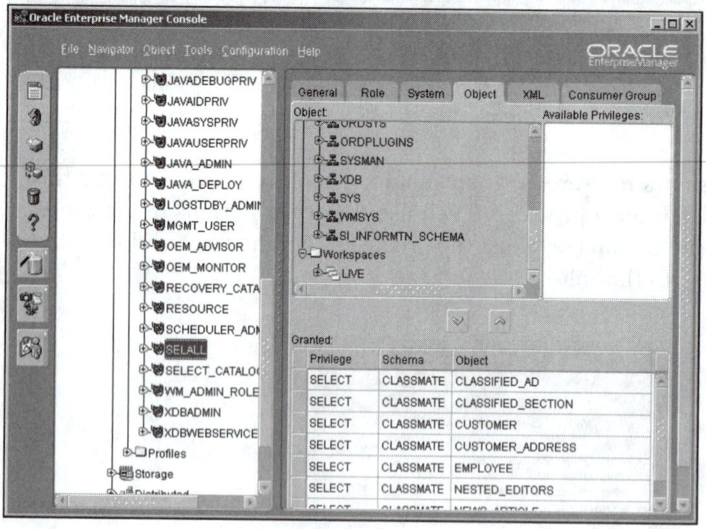

FIGURE 12-47 Object privileges appear in the lower half of the property sheet

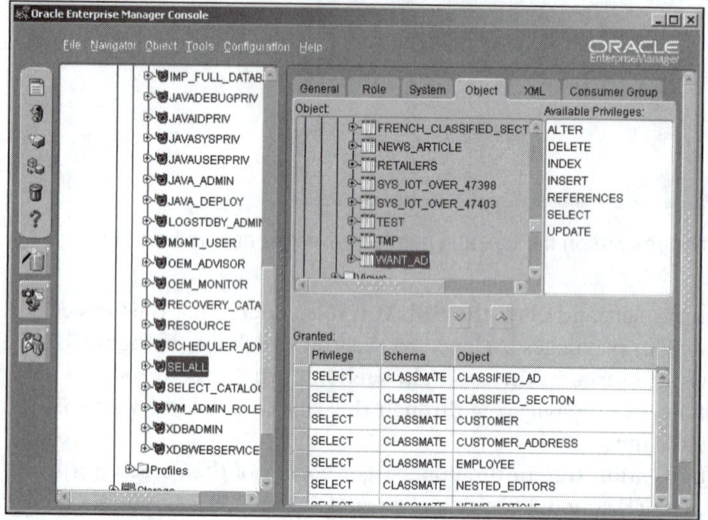

FIGURE 12-48 You have a choice of 7 object privileges for the WANT_AD table

10. Select the **SELECT** privilege, and click the **down arrow** button that is displayed in the middle of the property screen.
11. To see the change in the assigned privileges you have to scroll down to the bottom of the list, as shown in Figure 12-49. You see that the bottom line shows the new privilege for the WANT_AD table. This line has a plus sign in the far left column, meaning that this privilege has not yet been added.

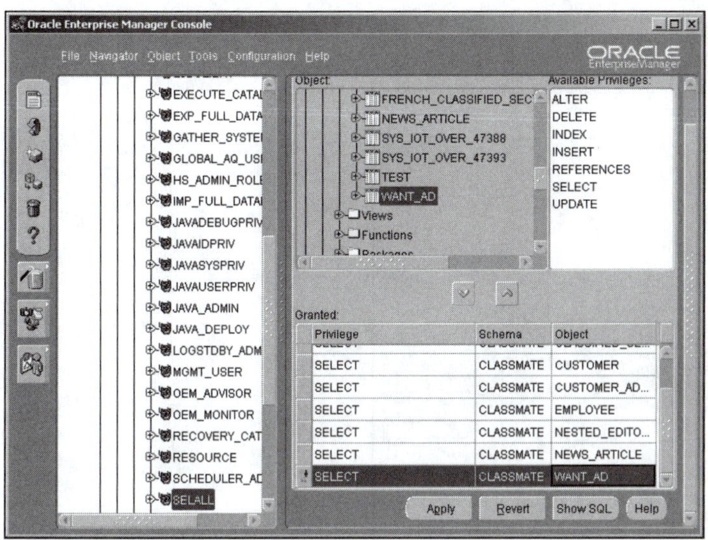

FIGURE 12-49 A privilege that has been selected but not yet been granted has a plus sign

12. Click the **Apply** button to grant the new privilege to the SELALL role. The plus sign disappears, and the screen reverts to its original display mode (similar to Figure 12-47). The SELECT on WANT_AD privilege appears in the list of granted object privileges.
13. You can also see which users have been assigned a role in the Security Manager. To find the list, right-click on the **SELALL** role on the left side of the console. A popup menu appears.
14. Select **Show Grantees** from the popup menu. A window displays the users who were granted this role, as shown in Figure 12-50. Notice that SYSTEM is shown with ADMIN OPTION for the role. This is because SYSTEM created the role. You could grant SELALL to more users (or roles) using this screen.
15. Click the **Cancel** button to close the additional screen.
16. Scroll up on the left side of the console, and double-click the **Users** folder. A list of all the users in the database appears below the folder.
17. Click the **AGAINES** user. A property window appears on the right showing the attributes of the user.
18. Click the **Role** tab. Figure 12-51 shows the screen. As you can see, the AGAINES user has the CONNECT role.
19. Scroll through the top right of the screen, until you find the SELALL role in the Available roles list. Then select the **SELALL** role, and click the **down arrow**, adding the SELALL role to the user AGAINES.

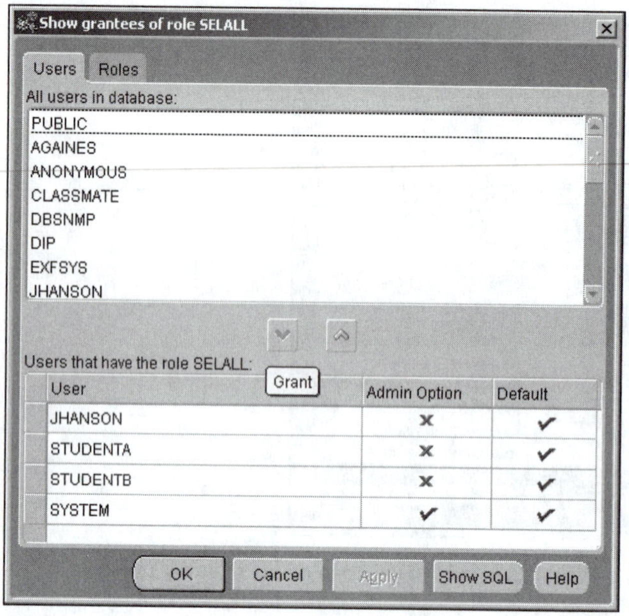

FIGURE 12-50 Users granted the SELALL role are listed in the lower half of the screen

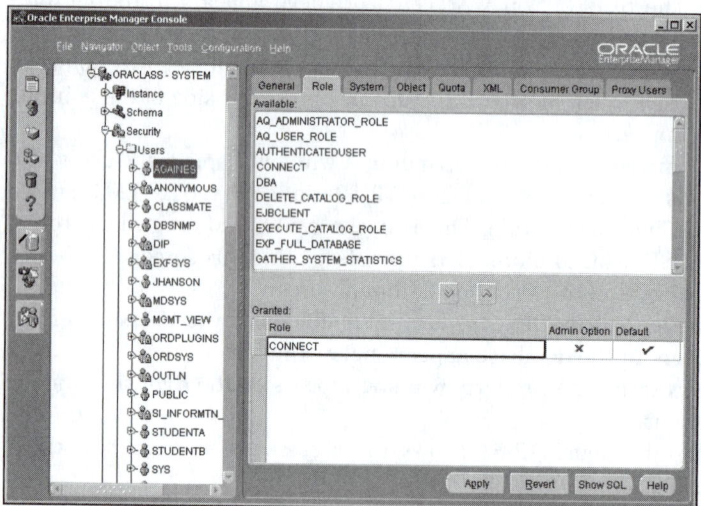

FIGURE 12-51 AGAINES has only one role assigned

20. The SELALL role is added to the list of granted roles. Click the **Apply** button to complete the process, and grant the role to AGAINES. Figure 12-52 shows the results.
21. Close the console by clicking the **X** in the top-right corner of the window.

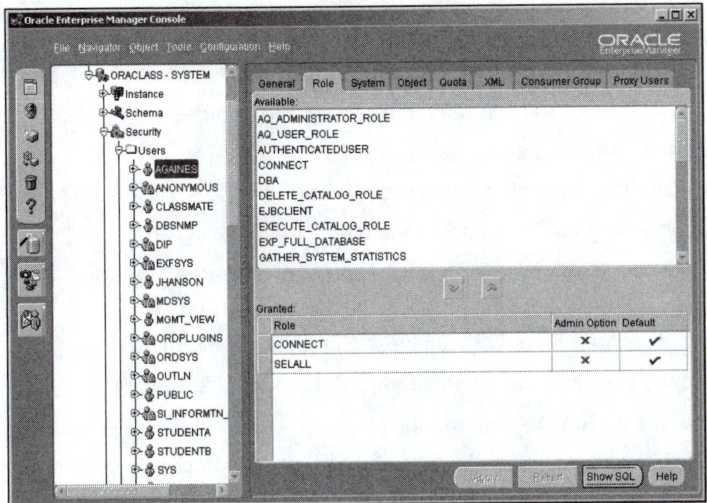

FIGURE 12-52 AGAINES now has two roles

You have seen how easy it is to grant privileges using the console. A real bonus of the console is the simple way it lists all the available system and object privileges for you. When they are so conveniently listed, you don't have to remember the exact names of the system or object privileges. The console can be a timesaving tool when you are working with roles.

Chapter Summary

- Users are created to either own a schema or access another user's schema.
- Users identified externally or globally are validated outside the database.
- Tablespace quotas limit a user's storage space.
- Profiles store a collection of password and resource limits.
- Passwords can be changed by the DBA and by the user.
- Profile password limits include how long a password can stay the same and when it can be reused.
- Use the profile as a method of limiting CPU usage, connect time, and more.
- System privileges allow a user to manage some part of the database system.
- Object privileges allow a user to work with an object.
- SYSDBA and SYSOPER are system privileges that allow a user to start up and shut down the database, as well as other high-level tasks.
- A grant made to PUBLIC gives all users the privilege.
- Revoked system privileges do not cascade to other users.
- Revoked object privileges cascade to other users.
- Object privileges can be granted on columns.
- The owner of a table can grant object privileges on that table.
- The grantor grants the privilege and the grantee receives the privilege.
- Querying an object without privileges to query causes an error stating that the object does not exist.
- Statement auditing is the monitoring of activity on a particular type of statement, such as SELECT.
- Privilege auditing audits any command that is authorized by the privilege, such as CREATE TABLE.
- Object auditing generates audit trail records as soon as the object is used, such as with SELECT or DELETE statements.
- A group of data dictionary views shows audit trail records for each type of auditing.
- Roles simplify security administration.
- Roles can be granted other roles, system privileges, and object privileges.
- Predefined roles help speed up administration by providing basic groupings of roles.
- Roles with passwords add security to the roles.
- After a user has been granted a role, subsequent grants to the role are effective immediately for the user.
- Default roles are roles enabled when you log on.
- Dropped roles are automatically revoked from users and other roles.
- The console displays roles and privileges within the Security Manager.

Review Questions

1. What two reasons justify creating new users?
2. The ________________ and ________________ methods of identifying a user do not require a password.
3. The OS_AUTHENT_PREFIX is null. You log on to your computer as JOEBANKS and then issue this command:

   ```
   sqlplus /
   ```

   What happens next?

   a. The database prompts you for a password.
   b. The database connects you as OPS$JOEBANKS.
   c. The database cannot connect you because no prefix is specified.
   d. The JOEBANKS user must be identified EXTERNALLY.
   e. None of the above.
4. The default tablespace for SARAH is USERS, where she has a quota of 10 Mb. She has no other quotas. SARAH creates a table in the USER_AUTO tablespace. What system privilege does SARAH have?
   a. CREATE TABLE
   b. CREATE USER
   c. UNLIMITED TABLESPACE
   d. UNLIMITED CREATE
5. The ________________ data dictionary view lists users.
6. What happens to existing tables when the table owner's quota is set to zero?
   a. The table remains but cannot be used.
   b. The table is dropped.
   c. The table remains but cannot increase in size.
   d. The table is marked invalid.
7. The ________________ data dictionary view lists users.
8. You are the DBA and you run the following SQL commands:

   ```
   CREATE USER MYEE IDENTIFIED BY MYEE
   DEFAULT TABLESPACE USERS
   PROFILE MANAGER_PROFILE
   PASSWORD EXPIRE;
   GRANT CREATE SESSION TO MYEE;
   CONNECT MYEE/MYEE;
   ```

   What happens next?

   a. You connect as the MYEE user.
   b. You are prompted to change the password for MYEE.
   c. You cannot log on because the password is expired.
   d. You cannot log on because MYEE does not have the RESOURCE role.

9. Examine the following SQL statement:

```
1 CREATE USER CARL
2 IDENTIFIED EXTERNALLY
3 DEFAULT TABLESPACE USERS
4 QUOTA 10M ON USERS
5 QUOTA 0 ON SYSTEM
6 ACCOUNT LOCK;
```

Which line could be removed without changing the outcome?

a. Line 6
b. Line 5
c. Line 3
d. Line 4
e. Line 2

10. You have a user named ALEX who owns several tables that are used by the Sales department. The user has left the company. You want to prevent the user from logging on to the database while keeping the tables intact. Which command(s) best handle this task?

a. ALTER USER ALEX ACCOUNT LOCK;
b. ALTER USER ALEX DROP ACCOUNT;
c. DROP USER ALEX CASCADE;
d. ALTER USER ALEX QUOTA 0 ON USERS; REVOKE CREATE SESSION FROM ALEX;

11. Which of these is not a system privilege?

a. SELECT TABLE
b. ALTER TABLESPACE
c. CREATE INDEX
d. SYSDBA

12. Object privileges must be granted by the DBA. True or False?

13. To create a table in another schema, a user must have the ________________ system privilege.

14. Which of the following object privileges *cannot* be granted on a view?

a. SELECT
b. UPDATE
c. FLASHBACK
d. EXECUTE

15. Explain why you would grant a system privilege to PUBLIC.

16. If user SMITH has the CREATE TABLE privilege and you want to revoke it, which statement would you use?

a. REVOKE CREATE TABLE FROM USER SMITH;
b. REVOKE CREATE TABLE FROM SMITH;
c. REVOKE ANY PRIVILEGE FROM SMITH;
d. REVOKE CREATE ANY TABLE FROM SMITH;

17. Which of the following is a valid statement to grant object privileges?
    a. GRANT SELECT ON CUSTOMER TO SMITH;
    b. GRANT SELECT ANY TABLE TO SMITH;
    c. GRANT ALL PRIVILEGES TO SMITH;
    d. GRANT ALL ON CUSTOMER;
18. You have a static table that is commonly queried by most users and applications. What is an appropriate action to handle privileges for the table?
    a. Grant all object privileges on the table to PUBLIC.
    b. Grant the SELECT object privilege to PUBLIC.
    c. Create a public synonym for the table.
    d. Grant the SELECT object to all users.
19. The ________________ clause of the GRANT command changes the audit trail so that it writes one record per session.
20. Which of the following data dictionary views should a user query to view any object privileges he has granted to other users?
    a. DBA_SYS_PRIVS
    b. USER_TAB_PRIVS_MADE
    c. USER_TAB_PRIVS_RECD
    d. USER_OBJ_PRIVS_MADE
21. List the three categories of privileges that a role may be granted.
22. Use the ________________ command to create a new role.
23. The ________________ clause of the CREATE ROLE command is optional, and is the default.
24. To give a user permission to enable a role, use the GRANT command. True or False?
25. Assuming that BOATROLE and SKIPPERROLE are both roles, which of the following commands is invalid?
    a. GRANT SELECT ON SHOWBOAT TO BOATROLE;
    b. GRANT CREATE TABLE TO BOATROLE;
    c. GRANT BOATROLE TO SKIPPERROLE WITH ADMIN OPTION;
    d. GRANT SKIPPERROLE TO BOATROLE IDENTIFIED BY CAPTAIN;
26. You want to rename a role. How do you do it?
    a. Use the ALTER ROLE command.
    b. Drop and recreate the role.
    c. Revoke and recreate the role.
    d. Use the CREATE ROLE command.

27. You want to enable all roles in your session. What command do you use?
    a. SET ROLE ALL;
    b. SET ROLE ENABLE ALL;
    c. ALTER SESSION ENABLE ALL ROLES;
    d. ALTER USER DEFAULT ROLE ALL;
28. What Data Dictionary role should you query to determine which roles are available to you, regardless of whether they are enabled?
29. You have granted the role ACCTMGR to many users. Which command revokes the role from all the users simultaneously and removes the role from the database?

Exam Review Questions—Oracle Database 10*g*: Administration (#1Z0-042)

1. You run this query with the results shown.

```
SELECT TABLESPACE_NAME, MAX_BYTES
FROM DBA_TS_QUOTAS
WHERE USERNAME = 'USER101';
TABLESPACE_NAME                         MAX_BYTES      BYTES
------------------------------ ---------- ----------
USER                                      5242880     395698
USER_AUTO                                      -1       5482
```

   Which statements are true? Choose two.
    a. USER101 had a quota on the USER_AUTO tablespace that was removed.
    b. USER101 has an unlimited quota on the USER_AUTO tablespace.
    c. USER101 has quotas on other tablespaces not shown.
    d. USER101 can use either USERS or USER_AUTO to create a new index.
    e. USER101 cannot use any more space in the USERS tablespace.
2. You issue the following SQL statements.

```
ALTER SYSTEM SET OS_AUTHENT_PREFIX = 'USA$';
CREATE USER STEVENS IDENTIFIED EXTERNALLY;
ALTER USER STEVENS QUOTA 10M ON USERS;
GRANT CONNECT TO STEVENS;
```

   Which statements below are true regarding the previous commands? Choose two.
    a. STEVENS is authenticated by Oracle Net.
    b. STEVENS cannot log on because he is missing a privilege.
    c. STEVENS is authenticated by the operating system.
    d. STEVENS logs on as USA$STEVENS.
    e. STEVENS can create tables in the USERS tablespace.

3. The table JOE.OLDCAR has 500 rows and uses 50 Mb of storage space. You issue the following SQL statements. Assume all work is done in the USERS tablespace.

```
ALTER USER FRANCIS QUOTA 100M ON USERS;
CREATE TABLE FRANCIS.NEWCAR AS SELECT * FROM JOE.OLDCAR;
ALTER USER FRANCIS QUOTA 20M ON USERS;
TRUNCATE FRANCIS.NEWCAR DROP STORAGE;
INSERT INTO FRANCIS.NEWCAR
SELECT * FROM JOE.OLDCAR WHERE ROWNUM < 10;
```

Which statements are true? (Choose two)

   a. All statements succeed.
   b. All but the last statement succeeds.
   c. FRANCIS.NEWCAR table has no rows.
   d. FRANCIS.NEWCAR table has 9 rows.
   e. The user FRANCIS has exceeded her quota.

4. The ALTER USER command **cannot** be used to accomplish which of these tasks?
   a. Change a user's password.
   b. Add quotas on two tablespaces to a user with one command.
   c. Revoke system privileges from a user.
   d. Change a user from a local user to a global user.

5. You can create a profile to limit which of these resources? Choose three.
   a. Complexity of passwords
   b. Days until password expires
   c. Hours after failed logon until a locked account unlocks
   d. Size of the SGA
   e. Number of users per session

6. You issued the following commands.

```
ALTER PROFILE DEFAULT LIMIT
PASSWORD_LIFE_TIME 90
CPU_PER_SESSION 100;
ALTER PROFILE ACCT_PR LIMIT
SESSIONS_PER_USER DEFAULT
PASSWORD_LIFE_TIME 10
CPU_PER_CALL 10;
ALTER USER KATE PROFILE ACCT_PR;
```

Which statements are true? Choose two.

   a. KATE's password expires in ten days.
   b. KATE can use up to 100 seconds of CPU time per session.
   c. KATE can run an unlimited number of concurrent sessions.
   d. KATE can use up to 10 seconds of CPU time per call.
   e. KATE cannot change her password for 10 days.

7. Observe this query and the results.

```
SELECT * FROM RESOURCE_COST;
RESOURCE_NAME                      UNIT_COST
-------------------------------- ----------
CPU_PER_SESSION                          100
LOGICAL_READS_PER_SESSION                 10
CONNECT_TIME                               0
PRIVATE_SGA                                0

4 rows selected.
ALTER PROFILE DEFAULT LIMIT
COMPOSITE_LIMIT 10000;
```

Here are the statistics for three user sessions:

JOE: CPU=10, Connect time=50, Logical reads=1000, Private SGA=2500

RICH: CPU=15, Connect time=100, Logical reads=100, Private SGA=5000

BETTY: CPU=100, Connect time=200, Logical reads=40, Private SGA=2500

Which statements are true? (Choose two)

a. The RESOURCE_LIMIT parameter must be set to TRUE to enforce composite limits.
b. The user sessions for JOE and RICH have exceeded the composite limit.
c. The user session for RICH has exceeded the composite limit.
d. The user sessions for JOE and BETTY have exceeded the composite limit.
e. The user sessions for all three users have exceeded the composite limit.

8. You have created a profile called ACCOUNTING and assigned the profile to all 10 accountants. Later, you create another profile called ACCT_MGR and assign this profile to five of the accountants. You then execute this command:

```
DROP PROFILE ACCOUNTING CASCADE;
```

What is the result of this command?

a. The five accountants with the ACCOUNTING profile switch to the DEFAULT profile.
b. The five accountants with the ACCOUNTING profile switch to the ACCT_MGR profile.
c. The statement fails.
d. The five accountants with the ACCOUNTING profile are dropped.

9. You want to enforce complexity rules on users' passwords. Which script do you run?

a. verify_function.sql
b. utlpwdmg.sql
c. utlpwmgt.sql
d. utlprofile.sql

10. You just fired a programmer who was using the database for his private consulting work. He has created numerous tables that you want to remove from the database. What is the best way to do this?

a. Lock the user account, write a query that generates a set of 'DROP TABLE' statements, and run the script.
b. Lock the user account, drop each table, and drop the user.
c. Drop the user with the CASCADE parameter.

d. Drop the tablespace containing all the user's tables.

11. Examine the following SQL statement.

```
1 GRANT CREATE USER, BECOME ANY USER,
2 CREATE TABLE, CREATE SESSION
3 TO NEWDBA, ADMINDBA
4 WITH GRANT OPTION;
```

Which line causes the statement to fail?

a. Line 1

b. Line 2

c. Line 3

d. Line 4

e. None of the lines cause an error

12. You are the DBA at a large company. Your assistant DBA is taking over for you while you are on vacation. You want your assistant to be able to create the schema needed for a new application system and grant the appropriate privileges to the end users. Which privileges will your assistant need? (Choose three)

a. CREATE SCHEMA

b. CREATE USER

c. CREATE ANY TABLE

d. GRANT ANY OBJECT PRIVILEGE

e. GRANT ANY PRIVILEGE

13. Examine the following SQL script.

```
SQL> GRANT SELECT, INSERT, UPDATE
2    ON  CLASSMATE.CUSTOMER
3    TO JIM
4    WITH GRANT OPTION;
Grant succeeded.
SQL> CONNECT JIM/jimmy@ORACLASS
Connected.
SQL> GRANT SELECT ON CLASSMATE.CUSTOMER
2  TO SMITH;
SQL> CONNECT SYSTEM/MANAGER@ORACLASS
Connected.
SQL> REVOKE SELECT ON CLASSMATE.CUSTOMER
2  FROM JIM;
Revoke succeeded.
```

What will happen if SMITH queries the CUSTOMER table?

a. The query fails with the error: "Insufficient privileges."

b. The query fails with the error: "Table does not exist."

c. The query succeeds.

d. SMITH is logged off immediately.

14. Which of the following queries displays the system privileges given out by the ORADBA user?
    a. SELECT * FROM DBA_SYS_PRIVS WHERE GRANTEE = 'ORADBA';
    b. SELECT * FROM DBA_TAB_PRIVS WHERE GRANTEE = 'ORADBA';
    c. SELECT * FROM DBA_SYS_PRIVS WHERE GRANTOR = 'ORADBA';
    d. SELECT * FROM SESSION_PRIVS WHERE GRANTOR = 'ORADBA';
15. You logged on as SYSTEM and granted CREATE TABLE WITH ADMIN OPTION to Susan. Susan then granted CREATE TABLE WITH ADMIN OPTION to Henry and Albert. You then revoked the CREATE TABLE privilege from Susan. Who has the CREATE TABLE privilege now?
    a. Henry and Albert
    b. SYSTEM, Henry, and Albert
    c. Only SYSTEM
    d. SYSTEM and Susan

16. You grant the SYSDBA privilege to your new DBA coworker. What tasks can the new DBA perform? (Choose two)
    a. Query any table
    b. Create new users
    c. Shut down the database
    d. Create an **spfile**
    e. Grant SYSDBA to another user
17. You plan to start auditing the database to investigate some unusual changes to the data. You change the AUDIT_TRAIL parameter with this command:
    ```
    ALTER SYSTEM SET AUDIT_TRAIL = 'DB' SCOPE=SPFILE;
    ```
    What is your next step to implement auditing?
    a. Execute the AUDIT command
    b. Shut down and restart the database
    c. Set the AUDIT_FILE_DEST parameter
    d. Back up the database
18. You are suspicious that the user named JOEY has been snooping and attempting to query sensitive tables that he is unauthorized to view. You want to monitor this type of activity. Which of these AUDIT commands is appropriate?
    a. AUDIT SELECT ON NOT EXISTS BY JOEY;
    b. AUDIT SELECT BY JOEY WHENEVER NOT SUCCESSFUL;
    c. AUDIT SELECT TABLE BY JOEY WHENEVER SUCCESSFUL;
    d. AUDIT SELECT ON JOEY WHENEVER NOT SUCCESSFUL;

19. Examine the following query.

```
SELECT USERNAME, ACTION_NAME,
SES_ACTIONS FROM DBA_AUDIT_STATEMENT
ORDER BY 1;
```

Which of the following AUDIT commands will generate records seen by the query?

a. AUDIT INSERT ON CUSTOMER WHENEVER SUCCESSFUL;

b. AUDIT SELECT TABLE BY STUDENTA WHENEVER NOT SUCCESSFUL;

c. AUDIT SELECT ANY TABLE BY ACCESS;

d. AUDIT UPDATE ON CUSTOMER BY SESSION;

20. Examine the following command.

```
GRANT SELECT, INSERT, UPDATE, DELETE
ON CLASSMATE.CUSTOMER TO JOEY, SAMUEL;
GRANT ALL ON CLASSMATE.CUSTOMER TO MARTIN;
REVOKE SELECT, DELETE
ON CLASSMATE.CUSTOMER FROM SAMUEL, MARTIN;
GRANT SELECT ON CLASSMATE.CUSTOMER TO PUBLIC;
```

Which of the users (JOEY, MARTIN, and SAMUEL) can successfully execute this query?

```
DELETE FROM CLASSMATE.CUSTOMER
WHERE FIRST_NAME = 'Mark';
```

a. JOEY and MARTIN

b. SAMUEL and JOEY

c. JOEY

d. JOEY, MARTIN, and SAMUEL

21. Which of the following roles are predefined? (Choose two)

a. IMP_FULL_DATABASE

b. POWER_USER

c. CONNECT

d. SYSOPER

e. CREATE_SESSION

22. You want to find out which object privileges have been granted to the ADMINACT role. Which query is the most useful?

a. SELECT * FROM ROLE_SYS_PRIVS
WHERE ROLE = 'ADMINACT';

b. SELECT * FROM SESSION_ROLES
WHERE ROLE = 'ADMINACT';

c. SELECT * FROM DBA_TAB_PRIVS
WHERE GRANTEE = 'ADMINACT';

d. SELECT * FROM ROLE_OBJ_PRIVS
WHERE ROLE = 'ADMINACT';

23. You plan to create a role named SALESREAD that does not require a password and gives read-only privileges to users. Which of the following commands creates the role?
    a. CREATE ROLE SALESREAD IDENTIFIED BY READ ONLY;
    b. CREATE ROLE SALESREAD NOT IDENTIFIED;
    c. CREATE READ ONLY ROLE SALESREAD;
    d. CREATE ROLE SALESREAD IDENTIFIED BY S123ALES;
24. Look at the following SQL script output.
```
GRANT SUPERUSER TO JONES WITH ADMIN OPTION;
Grant succeeded.
CONNECT JONES/XXX
Connected.
GRANT SUPERUSER TO HAROLD;
Grant succeeded.
CONNECT SYSTEM/XXX
Connected.
REVOKE SUPERUSER FROM JONES;
```
    Assuming that HAROLD is logged on when the last command is executed, what happens to HAROLD?
    a. HAROLD loses the SUPERUSER role immediately.
    b. HAROLD loses the SUPERUSER role when his session ends.
    c. HAROLD keeps the SUPERUSER role.
    d. HAROLD loses the SUPERUSER if it is enabled.
25. Which commands are valid, assuming that ROLEA and ROLEB are roles and USER1 is a user? (Choose two)
    a. GRANT SELECT ON CUSTOMERS TO ROLEA WITH GRANT OPTION;
    b. GRANT INSERT ON CUSTOMERS TO ROLEB, ROLEC, USER1;
    c. GRANT ROLEA TO ROLEB WITH ADMIN OPTION;
    d. GRANT CREATE VIEW TO ROLEA WITH ADMIN OPTION;
    e. GRANT CREATE VIEW TO USER1 WITH GRANT OPTION;
26. The TOPDOG role currently requires a password. The user JAMESK is logged on and has the TOPDOG role enabled. You execute the following command.
```
ALTER ROLE TOPDOG NOT IDENTIFIED;
```
    What happens in JAMESK's session?
    a. The TOPDOG role becomes disabled.
    b. The TOPDOG role becomes his default role.
    c. The TOPDOG role stays enabled.
    d. The TOPDOG role asks for a password.
    e. CREATE_SESSION

27. You execute the following commands.

```
GRANT DEALERROLE TO JOHNQ;
GRANT COLLECTORROLE TO JOHNQ;
ALTER USER JOHNQ DEFAULT ROLES NONE;
GRANT MANAGERROLE TO JOHNQ;
```

Which of the three roles are default roles for JOHNQ?

   a. DEALERROLE and COLLECTORROLE
   b. None of the roles
   c. MANAGERROLE only
   d. All three of the roles
   e. CREATE_SESSION

28. You are logged on as the user SMITH, and two roles you have are currently disabled. They were created with these commands:

```
CREATE ROLE MGR IDENTIFIED BY MGR123;
CREATE ROLE SALES IDENTIFIED BY SALES789;
```

Which of the following commands successfully enables both roles?

   a. SET ROLE MGR, SALES IDENTIFIED BY MGR123, SALES789;
   b. SET DEFAULT ROLE MGR, SALES;
   c. GRANT MGR IDENTIFIED BY MGR123; GRANT SALES IDENTIFIED BY SALES789 TO SMITH;
   d. SET ROLE MGR IDENTIFIED BY MGR123, SALES IDENTIFIED BY SALES789;

29. The user MARTHA logs on to the database. You, as the DBA, create a role called ADM and grant the CREATE TABLE privilege to the role. You grant the role to MARTHA. Then, you grant the CREATE VIEW privilege to the TOPDOG role. What happens to MARTHA?

   a. MARTHA can now create tables and views.
   b. MARTHA cannot create tables or views until she starts a new session.
   c. MARTHA can create tables but not views.
   d. MARTHA cannot create tables or views even when she starts a new session.

30. What are the advantages of granting privileges to roles rather than to users? Choose three.

   a. Better security on the database
   b. Easier to grant additional privileges to users
   c. Simplifies security administration
   d. Provides additional security through extra passwords
   e. Easier for users

Hands-on Assignments

Before working on the hands-on assignments, log on to SQL*Plus Worksheet as SYSTEM, and run the script named **setup.sql** found in the **Data\Chapter12** directory on your student disk:

1. You have three new consultants arriving to work for your company as database programmers. You want them to be able to create tables in the USER_AUTO tablespace that use up to a maximum of 10 Mb of space each. In addition, you want them to change their passwords every week, and you want the system to log them off the database if their sessions are idle for more than 10 minutes. The three user names to create are CONS01, CONS02, and CONS03. The profile to create for the consultants is named TEMPCONSULT. Write and execute the commands needed to accomplish these goals. Save the SQL script in a file named **ho1201.sql** in the **Solutions\Chapter12** directory on your student disk.
2. The three consultants (from Assignment 1) complain that they cannot work efficiently when the system keeps logging them off. They ask you to let them have 30 minutes before timeout. You agree to take care of it. You decide that you want the consultants to stop using the USER_AUTO tablespace and start using the USERS tablespace instead. You allow them a maximum of 10 Mb each on the USERS tablespace. However, the tables they have already created can remain on USER_AUTO. Write and execute the script to handle this, and save it in a file named **ho1202.sql** in the **Solutions\Chapter12** directory.
3. The three consultants who arrived in the office earlier (refer to Assignment 1) have left. You review the database and determine that CONS01 and CONS02 have created tables that can be safely removed. The third consultant, CONS03, however, has created some useful documentation tables that you are saving. Write and execute the script to remove all three users, so that the tables of CONS03 are preserved while the remaining tables are dropped. Save the script in a file named **ho1203.sql** in the **Solutions\Chapter12** directory on your student disk.
4. Write and execute a query that lists users whose passwords have expired or whose accounts are locked. Sort the users by create date (most recent date to oldest date). Save the script in a file named **ho1204.sql** in the **Solutions\Chapter12** directory on your student disk.
5. Your job as DBA in Local Locale Company requires you to help the staff of application programmers who work with the database. One common complaint from the programmers is: "I keep getting the error message, 'Table does not exist' when I know it is just a problem with permissions." The error does not appear for the programmer, because she works on development while logged onto the database as the table owner. The error shows up when the user logs on and runs the application. The application queries the CLASSMATE.CLASSIFIED_SECTION and CLASSMATE.CLASSIFIED_AD tables. Write an auditing statement that sets up auditing on these two tables, so that you can find out which users get the "Table does not exist" error message. Then write a query on the appropriate data dictionary view to list records in the audit trail.

   Save your work in a script named **ho1205.sql** in the **Solutions\Chapter12** directory on your student disk.
6. Look at Table 12-1. Create a list of these CLASSMATE schema objects and the privileges you can grant to them: CUSTOMER, EDITOR_INFO, CLIENT_VIEW, and CUSTOMER_ADDRESS.

Save your work in a file named **ho1206.txt** in the **Solutions\Chapter12** directory on your student disk.

7. There is an employee who has given notice and is going to another job in two weeks. You are worried that he might try to damage some of the database data. He has access to several different user names in the database, so you want to audit all those user names.

   Write an audit command to audit these three user names: STUDENTA, STUDENTB, and CLASSMATE. The audit should write an audit trail record every time one of these users attempts an update or delete on any table.

   Write a query on the audit trail records that will show not only the user name, but also the operating system name and the terminal from which the commands originated. (*Hint*: Find all the columns in a table or view by using the DESCRIBE command in SQL*Plus, which can be shortened to DESC. For example, DESC DBA_AUDIT_SESSION lists all the columns in the DBA_AUDIT_SESSION table.)

   Save your work in a file named **ho1207.sql** in the **Solutions\Chapter12** directory on your student disk.

8. Create a role called MAKEINDEX, and grant it the privileges needed to create indexes or views that query these CLASSMATE tables: CUSTOMER, CUSTOMER_ADDRESS, and WANT_AD. Grant the role to STUDENTB. Also grant the CONNECT role to STUDENTB, which enables the user to create various objects in the database. Connect to STUDENTB (password STUDENTB), and create a simple view on the CUSTOMER table. Save your work in the **Solutions\Chapter12** directory in a file named **ho1208.sql**.

Case Project

1. The Global Globe has a table of employees. Each employee needs a user name and password for the database. The employee's user name is the first name plus the first three letters of the last name. There are three categories of employees who need access to the database: managers, editors, and writers. Each of these needs resource limits, so you must set up three profiles.

   The MANAGERS profile has these traits: password expires after 30 days; maximum of 10 minutes idle time; reuse of password after 300 passwords; default resources otherwise.

   The WRITERS profile has these traits: password expires after 45 days; maximum of 15 minutes idle time; maximum session logical reads: 1000; maximum private SGA: 256 Kb; default resources otherwise.

   The EDITORS profile has these traits: password expires after 60 days; maximum of 45 minutes idle time; maximum time for one session: 8 hours; default resources otherwise.

   Create the three profiles. Save your work in a file named **case1201.sql** in the **Solutions\Chapter12** directory on your student disk.

   You can identify the category of each employee by his or her job title.

Write a query that generates the CREATE USER commands for all the managers, writers (including freelance writers), and editors. The CREATE USER command must include the appropriate profile. All the users have a password of TEMPPASS, and the password expires when the user is created.

Write another query that generates the GRANT CREATE SESSION command for all users. Run the queries, and then edit the script (if needed); run it to create all the users, and grant them the CREATE SESSION role. Save the queries in a file named **case1202.sql** in the **Solutions\Chapter12** directory on your student disk.

CHAPTER **13**

PERFORMANCE MONITORING

LEARNING OBJECTIVES

In this chapter, you will:

- Discover performance and performance monitoring
- Learn about different tools used for performance monitoring
- Learn about gathering statistics
- Discover invalidated database objects
- Find out about performance metrics
- Examine and change performance metrics
- React to performance issues
- Learn about locks

INTRODUCTION

Previous chapters in this book have focused on the architecture of a database, and how to examine and change that architecture. This chapter deals exclusively with trying to assess how well that architecture is functioning. How fast is the database performing? Is the database performing well? Is there anything that can be done about a poorly performing database?

Unlike previous chapters, this chapter does not focus on tables, tablespaces, security, and all the other physical and logical material in an Oracle database. This chapter begins the process of actually managing an Oracle database on a day-to-day basis. The performance of a database can be crucial to the success of an application, and sometimes even an entire organization. Performance is probably one of the most difficult jobs of a database administrator. Why is this so? In production environments, database administrators will usually be reacting to problems as they occur. However, there are ways that one can prepare for performance issues during the initial construction of the database architecture. Performance is about tuning a database to run faster. Preparing for fast performance is about planning, and is often a development (programming and design issue) rather than a database administration problem. However, performance often becomes a database administration problem.

In past versions of Oracle Database, performance tuning was a difficult, hands-on task. Oracle8*i* and Oracle9*i* Database introduced a tool called Oracle Enterprise Manager. In Oracle9*i* Database this tool allowed easy access to performance data. This easy access included many useful tools allowing drill-down through Oracle9*i* metadata. This permitted rapid and easy isolation of a problem as it was occurring. Oracle 10*g* Database has improved on the Console (Oracle9*i*) by providing the Database Control. The Database Control is a more *networkable* version of all the tuning tools found in the Oracle9*i* Database Health Check section of the Console.

This chapter will use the Database Control to explain how an Oracle 10*g* database can be monitored for performance problems. Also you will see how the tools can be utilized to help resolve performance problems as they arise. I have written two books on Oracle database performance tuning. Both are more than 700 pages in length. Performance tuning Oracle databases is an intense and broad topic. This chapter will cover only a small percentage of all there is to know about performance tuning, with a focus on detecting problems rather than solving problems. All you need to know at this stage is how to find the problems, and perhaps even go a small distance in isolating what might be causing a particular problem. At this early stage in your career as a database administrator, you do not have to know how to resolve all potential performance problems that can occur with as complex a beast as Oracle 10*g* Database.

INTRODUCTION TO PERFORMANCE MONITORING

What is performance? In terms of an Oracle 10g database, performance is a measure of how fast the database reacts to access data in the database. That access could be simple queries submitted by a custom built OLTP application or high-concurrency DML commands. Database access can even be long-running queries, such as in a data warehouse where queries can take hours to run. So, performance is really a measure of how quickly a database reacts to at a given request or activity. Ultimately the end users should be happy with the response of a web page or an application installed on their desktop or notebook computer. If end users are not happy, they might seek a similar service with a competing company, perhaps a company that offers a web interface that takes only a few seconds to update a web page, every time they select an item from a list.

Performance monitoring is the action of monitoring the speed of a database. This process of monitoring is most easily performed with software tools custom built for the job. There are numerous tools available with Oracle Database software, at no extra cost. These tools, along with more complex methods using command-line interfaces, allow access to underlying metadata and statistics within an Oracle database. That metadata and statistical record can be used to describe, and thus also monitor, how well an Oracle database is performing.

So let's discuss some of the tools available to monitor the performance of an Oracle database.

Different Tools for Performance Monitoring

There are many different tools that can be used to monitor the performance of an Oracle database, many of which are contained with Oracle Database itself, and there are numerous third-party tools offered by companies other than Oracle Corporation. This chapter focuses on using the Database Control to isolate performance problems. The Database Control provides an easy-to-use interface into performance data stored in an Oracle database.

It is useful to have some knowledge of all the tools that Oracle software can use for performance monitoring. This is a brief summary of some of those tools:

- **EXPLAIN PLAN**: The EXPLAIN PLAN command is used to examine potential query plans that the optimizer may utilize, when executing a particular query of DML statement.
- **Autotrace in SQL*Plus**: Autotrace displays a query plan in SQL*Plus every time a query is executed. Autotrace can be enabled using the SET AUTOTRACE ON EXPLAIN command, executed from within SQL*Plus.
- **SQL Trace and TKPROF**: SQL Trace allows for generation of trace files for Oracle Database. These trace files are like audit log trails of activities occurring in a database. Trace files are produced in 3 layers: (1) the alert log contains a record all of the highly significant activities including critical errors, (2) background process trace files contain details of the operation of each background process as it executes, and (3) each session has it's own trace file containing session activity. In the case of a shared server configuration, the

audit trail of activity for each session could be spread among multiple separate trace files. SQL Trace controls how much information is generated into trace files (BASIC, TYPICAL, or ALL). TKPROF is simply a formatting tool used to make the content of trace files readable to the human eye.

- **End-to-End tracing with TRCSESS**: The TRCSESS utility can be used to join the trace activity of a single session, from multiple trace files, when using a shared server configuration.
- **STATSPACK**: This is a set of scripts used to take a snapshot of a database. A snapshot is the state of all data, in an entire database, at a specific point in time. Each snapshot is then stored so that future snapshots can be compared. The objective is to compare acceptably performing snapshots with poorly performing snapshots to help to assess where a performance bottleneck may lie. This indicates what should be tuned.
- **V$ Views**: These views are also known as the Oracle database performance views. Database performance views are also known as dynamic performance views because they change constantly, depending on the state of a database.
- **The Wait Event Interface**: From a purely performance perspective V$ views provide a picture of statistics, and the ability to drill-down into those statistics. Drill-down effectively digs into statistical data allowing isolation of what specific thing is causing a performance problem or bottleneck.
- **Drill-Down in the Wait Event Interface**: The Wait Event Interface is a method of using the V$ performance views, to drill down into database statistics, available in the views.
- **Top*NNNN***: TopSQL, TopSessions, TopClient, TopConsumers, among others. These tools direct you to the things that are consuming the most resources. Let's say something is consuming your CPU time and RAM and performing a lot of I/O. It is not necessarily a problem unless the device or program is not supposed to be using these resources. Some things are supposed to be busy; some things are not.
- **Third-Party tools**: Spotlight is a particular tool that is very good for monitoring performance of busy production databases.
- **Operating system tools**: Generally these are tools provided by the different operating systems. These tools will provide information about the resources of an underlying Oracle database. Resources are CPU use, memory use, and I/O activity. These tools display things like processor use, I/O activity, memory usage, network activity, swapping, and so on.
- **Windows Performance Monitor**: This tool displays graphical representations of resource usage.
- **Unix utilities**: Solaris utilities such as sar, vmstat, mpstat, and iostat show CPU usage, I/O activity, and memory usage.

NOTE

For a database administrator who is more of a developer than a systems administrator, operating system level tools can be a little uninformative. Using these types of tools requires a good understanding of how an operating system works in general, in relation to resource usage.

The remaining sections of this chapter will mostly use the Database Control and its components. The Database Control is the primary Oracle 10*g* Database tool used for tuning. Also provided where appropriate to reinforce explanation of key concepts is information about commands, such as accessible V$ performance views.

STATISTICS

In this section you will learn about the importance of statistics, how to gather statistics manually, and how to gather statistics automatically using the Database Control.

What are statistics in an Oracle database? Statistics occur in various forms:

- **Object Statistics**: Database objects like tables and indexes can have statistics generated for them. Generally, object statistics record the number of rows in tables and the physical size of data. This allows the optimizer to have a more realistic picture of data. You can even create a histogram for a specific column in a table where the distribution of values in that column is skewed. A histogram allows the grouping of values into sets. The optimizer can utilize the set as a whole before searching for an individual element within that set.
- **System Statistics**: The database performs numerous activities. These activities are events occurring in a database. Whenever an event occurs it gets added to the sum for that event already recorded. The result is that if a particular event occurs too often, especially an event that can harm performance, you can track down the frequency of that event. Similarly, system statistics also store wait events, or waits for events. Obviously when one process is waiting for another process to complete, the result is a delay, wasted time, and wasted resources. A wait event that harms performance is effectively a bottleneck to performance because speed will be affected during the time in which the waiting event is waiting for the conflicting process to complete its task. Timed statistics parameters generate system statistics essential to tuning. System statistics gathering is controlled by the TIMED_STATISTICS and TIMED_OS_STATISTICS configuration parameters. Setting this parameter has negligible effect on performance and is set by default in Oracle 10*g*.

The Importance of Statistics

Why are statistics so important to the performance of an Oracle database? Statistics are very important to the way in which the optimizer assesses the best way to execute a query. The more up to date the statistics, the more accurately the optimizer can predict the best way to execute a SQL statement. Statistics are a computation or estimation of the exact size and placement of data in tables and indexes. Statistics are used by the optimizer to more effectively assess data in the database. This can produce a better query plan and potentially provide a more accurate match of the data in the database.

There are two general methods of gathering statistics in Oracle 10*g* Database: (1) a manual method, and (2) an automated process. Let's begin with the manual process.

Gathering Statistics Manually

Statistics can be gathered manually using a command-line tool such as SQL*Plus, with the ANALYZE command or the DBMS_STATS package. The DBMS_STATS package is the Oracle recommended method because it is parallel executable. The ANALYZE command is simple and easy to use. You can also manually gather statistics for snapshot comparison use, using both STATSPACK and the Database Control. From a manual perspective it is only necessary to examine the ANALYZE command and the DBMS_STATS package. Let's begin with the ANALYZE command.

NOTE

STATSPACK is beyond the scope of this book and also out of date for Oracle 10*g*. The Database Control is the recommended tool for statistics generation in Oracle 10*g*.

Using the ANALYZE Command

Let's look at syntax first. The following syntax diagram shows the basics of the ANALYZE command:


```
ANALYZE { TABLE | INDEX }
   COMPUTE [ SYSTEM ] STATISTICS [ FOR ... object specifics ... ]
      ESTIMATE [ SYSTEM ] STATISTICS [ FOR ... object specifics ... ]
      [ SAMPLE n { ROWS | PERCENT } ]
   | DELETE [ SYSTEM ] STATISTICS;
```

The meaning of the various options is as follows:

- **COMPUTE:** Calculates statistics for all rows and columns in an object, an object being a table or index. Precise results are produced.

NOTE

Whenever executing the ANALYZE command on a table, all indexes created on the table are also analyzed.

  COMPUTE can be time consuming because it reads an entire object.

- **ESTIMATE:** Calculates statistics on a sample number or percentage of rows, defaulted to 1,064 rows. This allows a scan of a small part of a very large table, helping to execute the ANALYZE command much faster.
- **DELETE:** Clears statistics.
- **SYSTEM:** Collects only system statistics, not user schema statistics.
- **FOR ...:** Allows the creation of a histogram for a specific column or columns. Histograms can help to logically normalize a skewed distribution.

Let's use the ANALYZE command:

1. To start the Enterprise Manager console in Windows, on the Taskbar, click **Start/All Programs/Oracle ... /Enterprise Manager console**. In Unix or Linux, type **oemapp console** on the command line. The Enterprise Manager console login screen appears.

2. Start the SQL*Plus Worksheet by clicking **Tools/Database Applications/**SQL*Plus Worksheet from the top menu in the console.
3. Connect as the CLASSMATE user. Click **File/Change Database Connection** on the menu. A login window appears. Type **CLASSMATE** in the Username box, type **CLASSPASS** (the current password), in the Password box, and **ORACLASS** in the Service box. Leave the connection type as "Normal." Click **OK** to continue.
4. Delete statistics from the EMPLOYEE table using the ANALYZE command with the DELETE option:

```
ANALYZE TABLE EMPLOYEE DELETE STATISTICS;
```

   SQL*Plus Worksheet should return "Table analyzed."
5. Go back to the Console. Connect as the SYSTEM user. In the tree structure on the left, double-click Schema, double-click the CLASSMATE user, double-click Tables, double click the EMPLOYEE table, and click the Statistics tab at the top right of the EMPLOYEE table details window. There should be no statistics for the EMPLOYEE table.
6. Go back to the SQL*Plus Worksheet and generate statistics for the EMPLOYEE table using the following command:

```
ANALYZE TABLE EMPLOYEE COMPUTE STATISTICS;
```

   SQL*Plus Worksheet should return "Table analyzed."
7. Now go back to the Console again, and with the EMPLOYEE table detail screen still up, and the Statistics tab highlighted, click the refresh button at the top-left corner of the tool bar. The refresh button is an arrow in a vertical circle. Refreshing should fill the EMPLOYEE table detail screen with statistical values, as shown in Figure 13-1.

NOTE

You might have to close the detail screen to get the refresh to work.

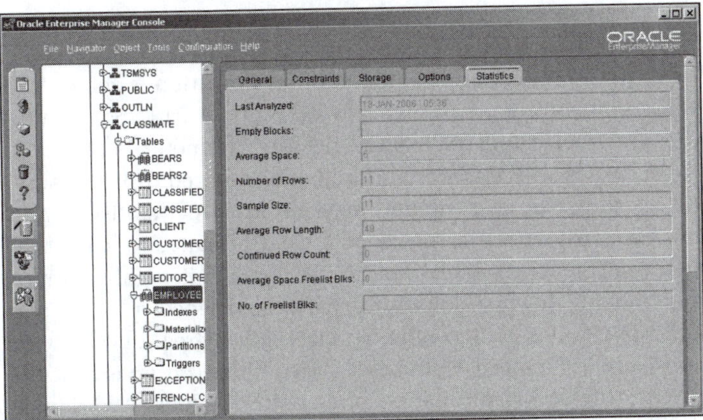

FIGURE 13-1 Viewing statistics in the client-side Oracle Enterprise Manager Console

8. Remain logged on for the next practice.

Using the DBMS_STATS Package

The DBMS_STATS package is more versatile and potentially faster than using the ANALYZE command to generate statistics. It is a little more complex to use. There are some non-optimizer statistics that can only be gathered with the ANALYZE command but the DBMS_STATS has parallel execution, Oracle Partitioning benefits, and performance tuning capabilities that the ANALYZE command does not provide.

The DBMS_STATS package can even be used to copy statistics between databases, allowing accurate simulated tuning between testing and production databases. The DBMS_STATS package can be used to gather statistics at all object layers of an Oracle database, such as for tables, indexes, even for a whole schema or an entire database. The following command can be used to gather statistics for a single table:

```
EXEC DBMS_STATS.GATHER_TABLE_STATS('<owner>', '<table>');
```

And the following command for a single index:

```
EXEC DBMS_STATS.GATHER_INDEX_STATS('<owner>', '<index>');
```


Let's use the DBMS_STATS package:

1. Go back to your SQL*Plus Worksheet and type in the following command to regenerate statistics for the EMPLOYEE table, except this time using the DBMS_STATS package:

   ```
   EXEC DBMS_STATS.GATHER_TABLE_STATS('CLASSMATE', 'EMPLOYEE');
   ```

2. SQL*Plus Worksheet replies "PL/SQL procedure successfully completed."
3. Go back to the Console again, and with the EMPLOYEE table detail screen still up and the Statistics tab highlighted, click the refresh button at the top-left corner of the tool bar. Refreshing should update the Last Analyzed date and time on the screen in front of you, and as shown in Figure 13-1.
4. Remain logged on for the next practice.

Automated Statistics Gathering

In Oracle 10g the default for statistics collection is automation. Automation is achieved by setting the STATISTICS_LEVEL configuration parameter to TYPICAL or ALL. Some automated statistics gathering occurs when the STATISTICS_LEVEL parameter is set to BASIC but stale (out of date) statistics will not be monitored and gathered. It is not sensible to ignore stale statistics as they can give the optimizer an incorrect statistical picture of data in your database. Similarly, it is also not good for performance to leave the STATISTICS_LEVEL parameter permanently set to ALL. However, ALL is highly comprehensive and can be useful for tracking down serious performance bottlenecks.

Unlike Oracle 9*i*, automated statistics gathering in Oracle 10g is a great development. Rule-based optimization is obsolete in Oracle 10g and cost-based optimization is now the only option. There must be some form of automation of statistical values. Additionally, dynamic sampling, the SAMPLE clause in queries, and CPU costing for the optimizer is now much more important. The primary objective of automation in statistics gathering for Oracle 10g, is to avoid situations of stale or non-generated statistics. Poor maintenance of statistics is one of the primary issues causing performance problems for Oracle database installations.

Automated statistics works (by default on database creation), using the GATHER_STATS_JOB job.

NOTE

Oracle Database has a job scheduling procedure allowing for jobs to be scheduled for execution at specified times, once, or repetitively.

The internal Oracle 10g Scheduler executes the GATHER_STATS_JOB job on a daily basis, by default, at default times which are generally overnight and on weekends. The job is executed for a set period of time and halted when required, regardless of completion or not. The default execution windows can be altered.

NOTE

Oracle 10*g* contains a scheduler that allows for better management of job scheduling. The Scheduler is part of the second Oracle 10*g* certification exam and is not covered in this book.

The GATHER_STATS_JOB job will find database objects with missing or stale statistics (more than 10% of rows have changed) and generate those statistics for you. The most needed statistics are generated first because this procedure has a time limit on its execution. The GATHER STATS_JOB job calls a DBMS_STATS procedure to generate statistics. You can verify the GATHER_STATS_JOB using the query as shown in Figure 13-2, using the following command:

```
COL OWNER FORMAT a16
COL JOB_NAME FORMAT a32
SELECT OWNER, JOB_NAME, JOB_TYPE FROM DBA_SCHEDULER_JOBS;
```

You must be connected as a DBA user to see the result of the above query, such as SYS or SYSTEM.

You can disable automated statistics gathering using the following command, logged in as SYS or SYSTEM of course:

```
EXEC DBMS_SCHEDULER.DISABLE('GATHER_STATS_JOB');
```

Some databases do not benefit from automation, such as those with a very large data warehouse, where either the available statistics-gathering window of time is unavailable, or the database is too large to fit processing into a reasonable amount of time. Constantly gathering statistics in a data warehouse that changes little is a waste of computer resources that might be better used on a more critical task.

It is also possible to force manual statistics generation on specific objects in a database—those you do not wish to have automated—using two procedures in the DBMS_STATS package. These procedures are called LOCK_TABLE_STATS, which can lock a specific table, and LOCK_SCHEMA_STATS, which exclude an entire schema. Some tables can

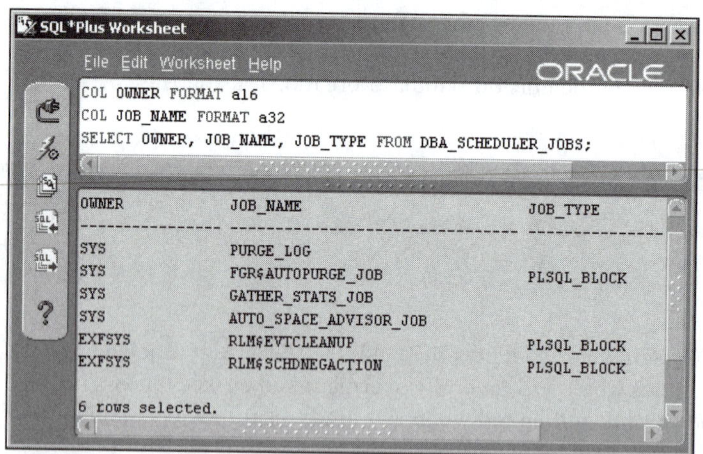

FIGURE 13-2 Verifying the automated statistics gathering procedure in Oracle 10*g*

be much larger than others and can inhibit performance by extending beyond the downtime window for automated statistics generation. Some tables may be read-only, or static, meaning the data is rarely altered. Dynamic sampling can be a good substitute for tables excluded from automated statistics gathering.

Dynamic Sampling

For statistics to be realistic and effective they must be frequently generated. In an OLTP database, data is changing all the time. Statistics can rapidly become stale. Oracle 10*g* query optimization (all cost based on statistics), against redundant statistics, can sometimes cause as big a problem as not having those statistics at all, depending on how out of date statistics are. If statistics are out of date or not present, then dynamic sampling may be used. Dynamic sampling reads a small number of blocks in a table to make a best guess at the statistical picture of data in a table.

Setting dynamic sampling to 0 switches it off, which is definitely not a recommended practice in Oracle 10*g*. The simplest difference between setting 1 and 2 is as follows:

- **OPTIMIZER_DYNAMIC_SAMPLING = 1**: Causes dynamic sampling of tables in queries, if certain criteria are met.
- **OPTIMIZER_DYNAMIC_SAMPLING = 2**: Causes dynamic sampling of all tables that have not been analyzed regardless of any rules.
- **OPTIMIZER_DYNAMIC_SAMPLING > 2**: The setting goes all the way up to 10. Setting to 10 uses everything and reads everything for an object.

NOTE

In Oracle 10*g* the OPTIMIZER_DYNAMIC_SAMPLING parameter is defaulted to 2, raised from 1 in Oracle 9*i*. This reinforces the heavier reliance and emphasis on automation of statistical gathering in Oracle 10*g*. Cost-based optimization requires this approach.

Leave the setting of the OPTIMIZER_DYNAMIC_SAMPLING parameter at its default level of 2, unless you have a very good reason.

The SAMPLE Clause

The SAMPLE clause (or SAMPLE BLOCK clause) is not really a part of automation. However, like dynamic sampling it can be used as a substitute semi-automated statistical generator. For example, when removing statistics from a specific object. The SAMPLE is added to the FROM clause of a query as a part of a SELECT statement. The SAMPLE clause causes the reading of a percentage of rows or blocks, in order to gather a statistical picture of data. That picture is passed to the optimizer allowing a better cost-based guess at how data should be accessed. The SAMPLE clause has the following syntax:

```
SELECT * FROM <table> SAMPLE(n) ...
```

The syntax will read a sample of n% of the rows in the table, when generating a query plan for reading the table in a query.

Automated Statistics and the Database Control

The heart of automated statistics gathering and management for Oracle 10*g* is in the Database Control browser interface. Much of that automation process is about automatic SQL tuning. Automatic SQL tuning is new to Oracle 10*g* Database where the optimizer can be switched from normal mode to a tuning mode. Tuning mode can consume large amounts of resources and is largely intended for use with complicated and long-running queries only. Automated SQL tuning can be performed manually using commands executed from within SQL*Plus but it is recommended to use only the Database Control. Using the Database Control is so much easier. Automated SQL tuning involves the following parts:

- **The AWR**: Automatic collection of statistics using the Automatic Workload Repository (AWR).
- **The ADDM**: Automatic performance diagnostics using the Automatic Database Diagnostic Monitor (ADDM).

NOTE

The ADDM has effectively replaced STATSPACK in Oracle 10*g* Database.

- **Automatic SQL Tuning**: Automatic SQL tuning using the SQL tuning advisor, SQL tuning sets and SQL profiles.
- **SQL Access Advisor**: Data warehouses and materialized view analysis using SQLAccess Advisor.

AWR executes statistical snapshots of the database. A snapshot takes a mathematical picture of the state of a database at a specific point in time.

NOTE

STATSPACK also executes snapshots, and compares one snapshot with another. STATSPACK is superseded by the AWR and the ADDM.

The AWR can be found in the Database Control, under the Administration tab, under Statistics Management. Let's find this page in the Database Control:

1. To start the Database Control, open an Internet browser.
2. Expand your browser up to full screen. The Database Control is BIG!!!
3. At the top of your browser window you should see an address line. Type the URL of your database server on this line, and execute the Database Control. It should look something like this:

   ```
   http://<hostname>:1158/em
   ```

   The <hostname> must be substituted with the hostname of your database server computer. Ask your instructor for the hostname. The port number may also be different depending on the release of Oracle 10g you are using, and if there is one or more databases created on the database server computer. Other ports used are 5500 and 5600. If in doubt ask your instructor.

TIP

The file called **portlist.ini** configuration file contains port number settings for various tools, including iSQL*Plus and the Database Control. Depending on the version of Oracle 10*g*, the file should be located in the following directory: $ORACLE_HOME/install. If the file cannot be found in that directory, then search your $ORACLE_HOME directory. The configuration file should look similar to this: iSQL*Plus HTTP port number =5560, Enterprise Manager Console HTTP Port (oraclass) = 1158, Enterprise Manager Agent Port (oraclass) = 3938

   Additionally, to execute the Database Control there needs to be a background process (a service on a Windows computer) running on the database server that executes the Database Control over the network. This service should be installed and started by the Oracle installation process on the database server. If it is not there, then ask your instructor.

4. Once again, a number of different screens may appear, depending on your release of Oracle 10g. The important screen is the login screen. Type **SYS** in the Username box, type the appropriate password (ask your instructor) in the Password box, and select **SYSDBA** from the Connect As box. Click **Login** to continue.
5. The screen that appears will look something like the one shown in Figure 13-3, where you will be at the Home portion for the Database Control of your database.

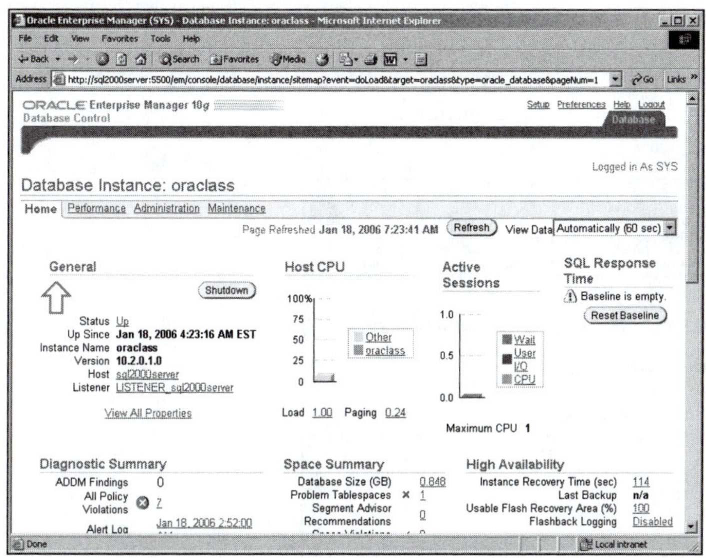

FIGURE 13-3 The Database Control home page

6. Next click the **Administration** tab and you will see the next screen as shown in Figure 13-4. What we are interested in at this stage is the **Statistics Management** section, as shown in Figure 13-4.

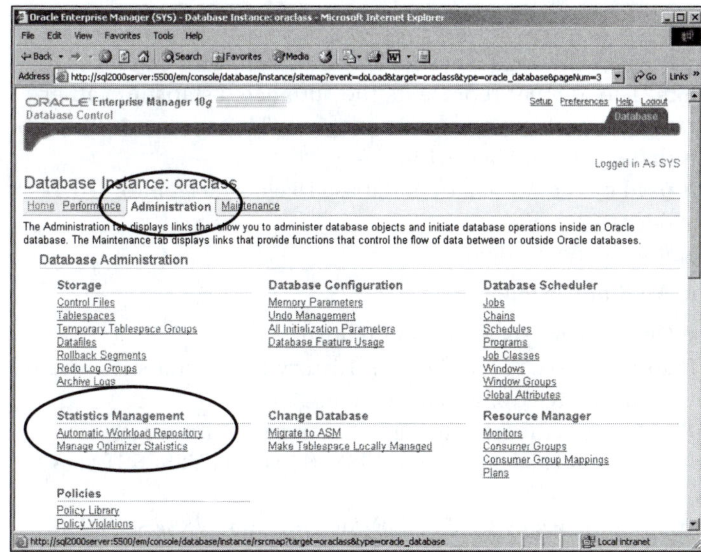

FIGURE 13-4 The Database Control administration page

7. Select the **Automatic Workload Repository** link under the **Statistics Management** section. This allows for configuration of the AWR, which as you can see in Figure 13-5, is fairly easy to do. Remember that the AWR controls how automated statistics are gathered.

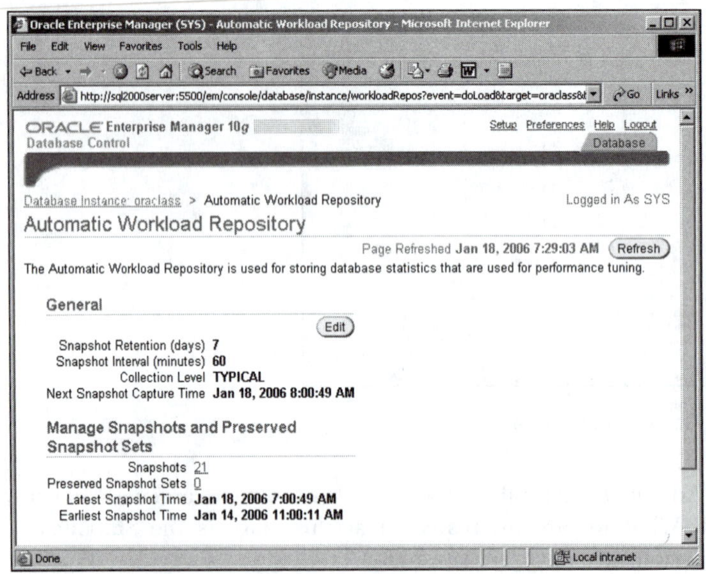

FIGURE 13-5 Configuring the Automatic Workload Repository (AWR)

Toward the bottom left of the page you should see an item with a number that is called **Snapshots**. The number represents the automated statistics gathering events that have occurred on your database server. This number is likely to be different from that shown in Figure 13-5.

8. Snapshots will be used by the Automatic Database Diagnostic Monitor (ADDM) to analyze and make assessments.
9. Remain logged on for the next practice.

Statistics gathered by the AWR are as follows:

- **Object statistics**: Objects such as tables and indexes
- **Active Session History (ASH)**: Active Session History (ASH) statistics where the word history means activity of recent sessions
- **High impact SQL**: SQL statements causing high impact on resources
- **System statistics**: System-level statistics in V$SYSSTAT and V$SESSTAT performance views
- **Time model system statistics**: V$SYS_TIME_MODEL and V$SESS_TIME_MODEL performance views

The next step is the Automatic Database Diagnostic Monitor (ADDM), which uses snapshots taken by the AWR, to make statistical comparisons and reports.

NOTE

STATSPACK does the same thing except using a command line interface. The Database Control is better.

Just like STATSPACK, the ADDM uses multiple snapshots taken by the AWR, and then performs an analysis between two snapshots. The following are common issues that the ADDM is searching for:

- **Over extended use of CPU time**: This can be for all sorts of reasons.
- **Sizing of memory structures**: It is rarely sensible to continue increasing buffers, such as the SGA and the database buffer cache. This tends to hide problems temporarily.
- **Heavy I/O usage**: For a data warehouse this is normal.
- **High consumption SQL statements**: These are either SQL statements causing problems or those doing a lot of work. Some SQL statements are supposed to do a lot of work. Also included here would be PL/SQL and Java code consuming excessive resources.
- **Configuration issues**: This involves file sizing, archives, inappropriate parameter settings, concurrency issues, hot blocking, and locking contention.
- **Anything busy**: Generally anything that is very busy.

Let's find the ADDM in the Database Control:

1. You should still be in the Automatic Workload Repository (AWR) page of the Database Control. Click the back button in your browser to return to the main screen containing the Home, Administration, Performance, and Maintenance tabs across the top-left corner of the page.
2. Click the **Performance** tab. The page should resemble Figure 13-6.

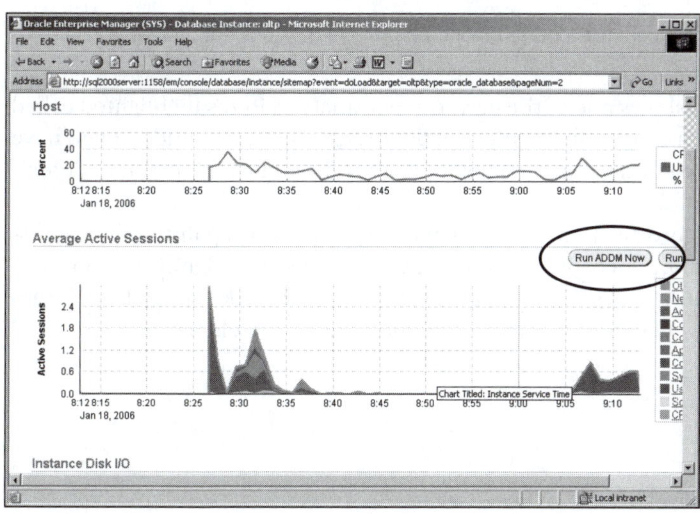

FIGURE 13-6 The Performance screen in the Database Control

3. To compare snapshots, you must execute the ADDM as highlighted in Figure 13-6. Click the **Run ADDM Now** button.
4. The next screen gives you a prompt to compare two snapshots. Follow the prompt by clicking the **Yes** button. Wait for processing to complete. This could take some time.
5. The screen that appears will look something like the one shown in Figure 13-7, where you will see a performance analysis and any significant errors. This database in particular is a busy database. Your screen is unlikely to look nearly as busy, and will probably contain no errors.

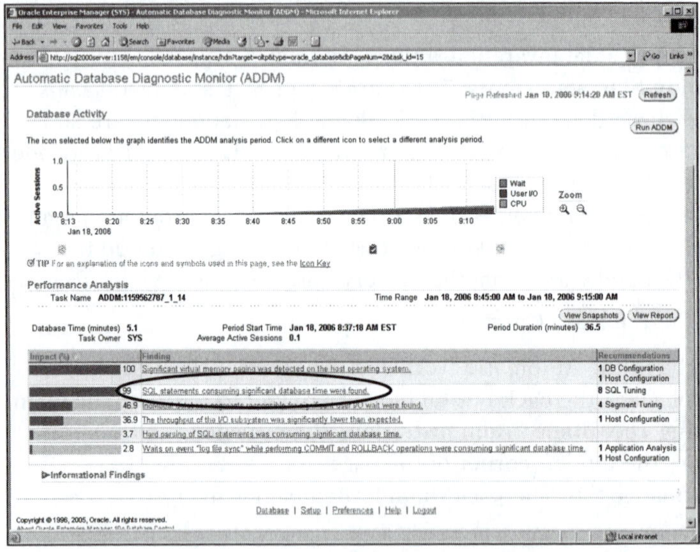

FIGURE 13-7 Running the ADDM shows a performance analysis and significant problems

6. Figure 13-7 also shows one of the performance analysis items highlighted called "SQL statements consuming significant database time were found." If you have a link in this area of the screen, click on it in the Database Control, to determine the problem.
7. As shown on Figure 13-8, the drill-down from the previous point is shown, displaying information for a SQL statement that is causing a significant performance hit. If there are no links in this section, then drill-down by clicking into whatever links you can and explore the functions of the elements.

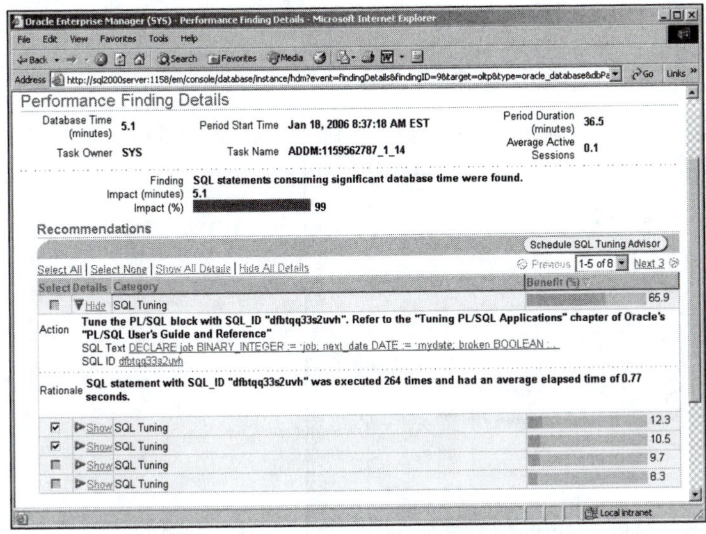

FIGURE 13-8 The ADDM allows drill-down into significant issues

8. Remain logged on for the next practice.

PERFORMANCE METRICS

What is a metric? A metric in Oracle Database is a special type of statistic, defined as a measure of a rate of change on a cumulative statistic. So a metric is a measure of a rate of change. Metrics can be found in many of the V$ performance views. The easiest way to access performance metrics is to administer and analyze them using the Database Control.

Let's look at all the performance metrics in the Database Control:

1. You should still be in the Database Control. If you are not at the Home page, click the back button on your browser until you return to the home page.
2. Click the **Performance** tab and scroll down to the bottom part of the performance page, using the scrollbar for your browser window. Figure 13-9 shows a picture of the lower half of the performance page of the Database Control.

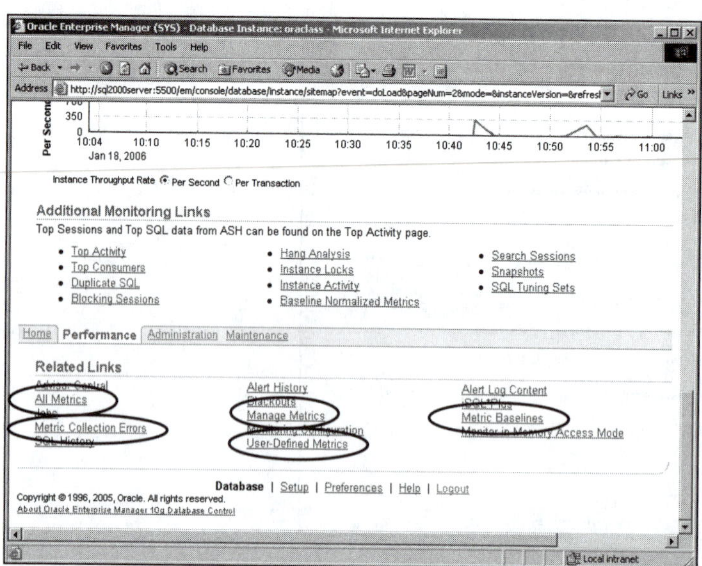

618 **FIGURE 13-9** Finding performance metrics in the Database Control

NOTE

If you find that the browser keeps refreshing and going back to the top of the page, perform the following steps. Scroll to the top-right corner of the browser page to the spin control called **View Data**. Click the default value in the spin control and change it to **Real Time: Manual Refresh**. That will stop the page from refreshing so frequently and give you a chance to examine the page.

The various metric links are highlighted in Figure 13-9.

3. Click the **All Metrics** link. You will see a list of many metrics. Click a few of the metric links and examine them. If you get a page with no information, then statistics are not available for that particular metric. All you need to do is look around and get a general idea as to how to navigate the interface. When a specific metric threshold is exceeded, the Database Control will return a warning or an error, similar to that shown in Figure 13-7.
4. Click the browser back button, and click the **Metric Collection Errors** link. If there is a record of any statistics collection (metrics) activity, and there were any problems on metric collection, then errors will be shown on this screen.
5. Click the browser back button again, and click the **Manage Metrics** link. This page allows you to change the metric structure in your database, such as changing thresholds. Change a threshold when the database returns a warning to the Database Control (shown in Figure 13-7). Do not change any metrics unless you have good reason to do so.
6. Click the browser back button again. There is also an option to create **User-Defined Metrics**. You can create your own metrics.

7. There is another link option allowing you to examine **Metric Baselines**. A baseline is a term applied to an established value, coagulated over a period of time, fixing an acceptable benchmark or value, for a specific metric. If a database suddenly begins to behave erratically then that unusual behavior can be detected by comparison against the baselines. In other words, a baseline value is an acceptable or expected mode of operation for a database. When things start going wrong, the Database Control will inform you, hopefully in time enough before things really become problematic. That's the idea of all this wonderful stuff.
8. The metric baselines page should show you a page including an **Enable Metric Baselines** button and a message stating this is not enabled. Click the **Enable Metric Baselines** button. Follow the prompts and click the **Yes** button to any questions. (Some of this detail will be covered in Chapter 14 but you might find it interesting to browse and read at this stage.) As with the Oracle Enterprise Manager Console in Oracle 9*i*, the Database Control in Oracle 10*g* is an excellent source for self-educating yourself about some of the more obscure details of Oracle Database software.
9. Remain logged on for the next practice.

Changing Performance Metric Parameters

To change performance metric parameters you must go back to the Performance page and click the Manage Metrics link. The resulting page is shown in Figure 13-10.

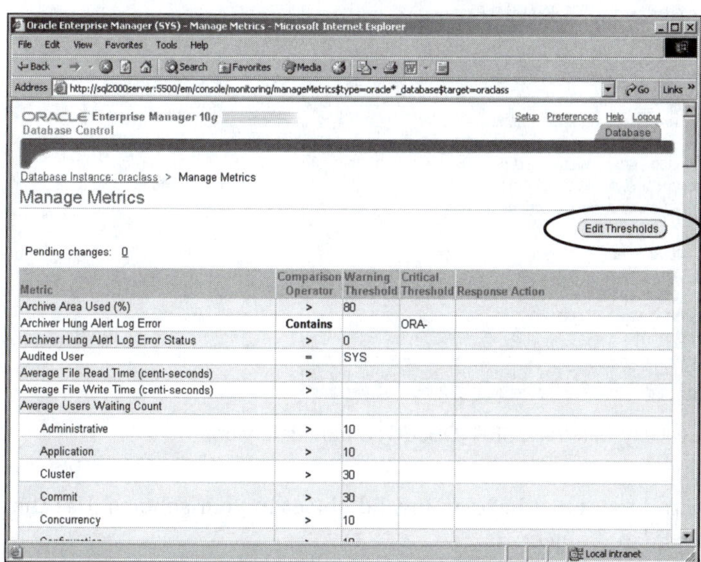

FIGURE 13-10 Changing performance metric parameters in the Database Control

If you click the **Edit Thresholds** button you will be taken to a page allowing changes to metrics. Just don't forget to click the OK button if you change anything. And make sure you understand what you are changing, and its effect.

SEARCHING FOR INVALIDATED OBJECTS

There are two easy ways to find invalid objects in Oracle 10*g*: (1) use the USER_OBJECTS metadata view and check the STATUS column (ALL_OBJECTS and DBA_OBJECTS can also be used), and (2) the metrics contained within the Database Control for finding invalid objects.

Let's begin with the USER_OBJECTS metadata view. You could use the following query to access the status of all objects inside the CLASSMATE schema, when connected as the CLASSMATE user:

```
COL OBJECT_TYPE FORMAT A16
SELECT OBJECT_TYPE, STATUS, OBJECT_NAME
FROM USER_OBJECTS ORDER BY 1,2;
```

A partial result for the above query is shown in Figure 13-11.

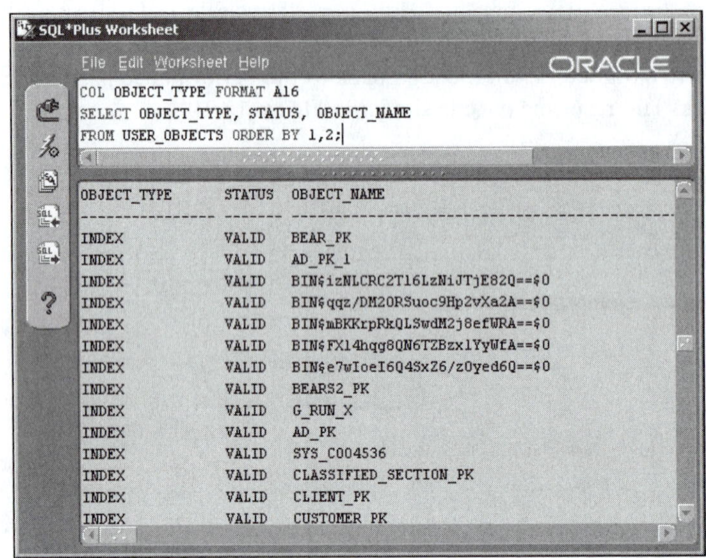

FIGURE 13-11 Finding object validity with the USER_OBJECTS metadata view

To use the Database Control you must have an invalid object in your schema. Use the following steps to find an invalidated object:

1. Click the **Manage Metrics** link from the home page. The page as shown in Figure 13-10, will display on the screen.
2. Click the **Edit Thresholds** button. This brings up the metric parameters editing screen.

3. Scroll down the browser page until you find the metric called **Owner's Invalid Object Count**. You can select this metric by selecting the radio button on the left.
4. Now change the default value to 0. This prohibits any invalid objects in a schema. The reason it is not set to zero is that any PL/SQL procedure compilication that fails because of something trivial like a typing error will cause a warning response in the Database Control.
5. Scroll down to the bottom of the browser page and click the **Cancel** button. You don't want to save that particular change.

LOCK CONFLICTS

NOTE

Chapter 10 introduced locking with respect to transactions and the LOCK TABLE statement. A transaction lock can be maintained for the duration of a transaction. This locks access (partially or completely) to data for the life of that transaction. A transaction is terminated on the execution of a COMMIT command (explicit or implicit), or when a ROLLBACK command is issued.

Additionally, in Chapter 10 you also learned about different types of locks. Basically, locks can be either shared or exclusive, in various forms. A share lock partially locks data where there is still partial access allowed to data by other sessions. An exclusive lock completely prohibits changes to data, but still allows read access.

You might be wondering why some locking information was introduced in Chapter 10 and other detail about locking is included here. The reason is as follows. Locking of transactions and tables using transactional control commands (COMMIT and ROLLBACK), as well as the LOCK TABLE command, are both manual intervention commands that create locking situations manually. In this chapter you examine how locks can occur as a result of Oracle 10g internally creating locks. Also certain things can happen when data is locked by Oracle 10g, such as deadlocks.

NOTE

A deadlock is where two interdependent rows are both locked and both rows require the release of the lock on the other row to continue processing. The result is a stalemate between the two rows—nothing can be done until the deadlock is released.

This section examines locking as it occurs automatically in Oracle 10g, within the database itself. In other words, if too many people access the same data at once, then some form of internal locking will result. Then again, locking is also a function of the type of access undertaken by a user causing that lock to occur. This chapter is about monitoring performance, so in this situation you need to understand how to detect and resolve locks.

Oracle 10g Database uses row locks instead of escalating locks. Row locks allow a certain amount of sharing at the block level. When a threshold is reached it does not escalate locks from a row up to an entire table, but rather it simply runs a little slower, being unable to service as many sessions at once. What you really need to know is what can you do when these situations occur.

NOTE

Oracle 10*g* has two distinctly different types of locks. The first type of lock functions for data in the database and is called a lock. The second type of lock is known as a latch. A latch manages sharing in memory buffers such as the database buffer cache and the shared pool. Latches are beyond the scope of this book.

How to Detect Lock Conflicts

There are two ways to detect locks. The first and more difficult method is to use the V$ performance views and the Oracle Wait Event Interface, using V$ performance views. The better way is to use the Database Control interface.

First you will examine lock conflict detection using V$ performance views, then you will do it the easy way using the Database Control. The most significant view is the V$LOCK view. Find sessions hold locks with the query:


```
SELECT * FROM V$LOCK;
```

NOTE

All examples in this section cannot be duplicated because I am using a busy database. So, you cannot duplicate these in your ORACLASS database.

The result of the above query is shown in Figure 13-12.

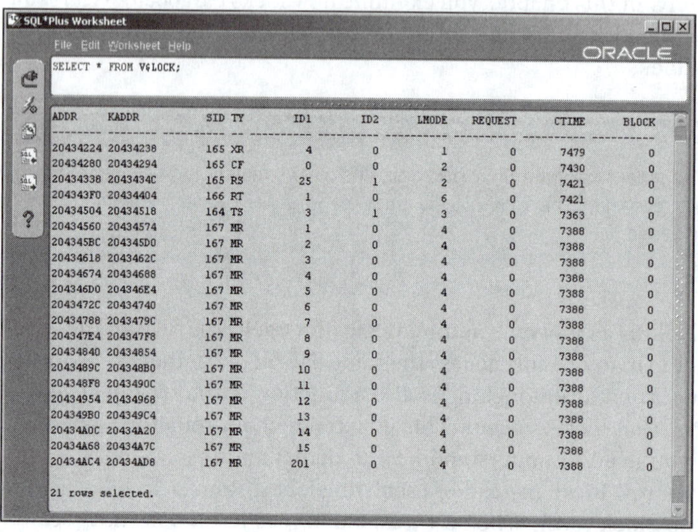

ADDR	KADDR	SID	TY	ID1	ID2	LMODE	REQUEST	CTIME	BLOCK
20434224	20434238	165	XR	4	0	1	0	7479	0
20434280	20434294	165	CF	0	0	2	0	7430	0
20434338	2043434C	165	RS	25	1	2	0	7421	0
204343F0	20434404	166	RT	1	0	6	0	7421	0
20434504	20434518	164	TS	3	1	3	0	7363	0
20434560	20434574	167	MR	1	0	4	0	7388	0
204345BC	204345D0	167	MR	2	0	4	0	7388	0
20434618	2043462C	167	MR	3	0	4	0	7388	0
20434674	20434688	167	MR	4	0	4	0	7388	0
204346D0	204346E4	167	MR	5	0	4	0	7388	0
2043472C	20434740	167	MR	6	0	4	0	7388	0
20434788	2043479C	167	MR	7	0	4	0	7388	0
204347E4	204347F8	167	MR	8	0	4	0	7388	0
20434840	20434854	167	MR	9	0	4	0	7388	0
2043489C	204348B0	167	MR	10	0	4	0	7388	0
204348F8	2043490C	167	MR	11	0	4	0	7388	0
20434954	20434968	167	MR	12	0	4	0	7388	0
204349B0	204349C4	167	MR	13	0	4	0	7388	0
20434A0C	20434A20	167	MR	14	0	4	0	7388	0
20434A68	20434A7C	167	MR	15	0	4	0	7388	0
20434AC4	20434AD8	167	MR	201	0	4	0	7388	0

21 rows selected.

FIGURE 13-12 Querying the V$LOCK performance view

Table 13-1 shows all the metadata and V$ performance views that can be used for tracking locks in an Oracle database.

TABLE 13-1 Oracle Database metadata and V$ performance views for tracking locks

View	Description
V$LOCK	Currently held locks, plus lock and latch requests
DBA_BLOCKERS	A blocking session holds a lock that another session is waiting for
DBA_WAITERS	A waiting session waits for the release of a lock held by another session
DBA_DDL_LOCKS	DDL commands can hold a lock on all the objects in a database. It contains current locks and lock requests. A lock request is a lock waiting to be placed, that cannot be placed because something has locked the required data
DBA_DML_LOCKS	Currently held DML locks, plus outstanding requests or locks
DBA_LOCK	All types of locks, plus outstanding requests for locks
DBA_LOCK_INTERNAL	All locks and latches, including currently held locks and latches, plus outstanding results for them

Figure 13-12 is meaningless because it is just numbers, codes, and addresses. The next query uses the DBA_LOCK metadata view to make the existing locks a more meaningful:

```
COL LOCK_TYPE FORMAT a20
COL MODE_HELD FORMAT a10
COL MODE_REQUESTED FORMAT a5
COL LOCK_ID1 FORMAT a10
COL LOCK_ID2 FORMAT a10
SELECT * FROM DBA_LOCK;
```

The query shown in Figure 13-12 is a little more informative but a lot of that information contained in the query result is beyond the scope of this book. There is an easier way.

Next, let's examine how to detect lock conflicts using the Database Control.

If you don't already have the Database Control on your screen execute it again:

1. The URL for the Database Control is as follows:

   http://<hostname>:1158/em

   Once again, the <hostname> must be substituted with the hostname of your database server computer. And again, the port number may be different depending on the release of Oracle 10g you are using, and if there are one or more databases created on your database server computer. Other ports used are 5500 and 5600.

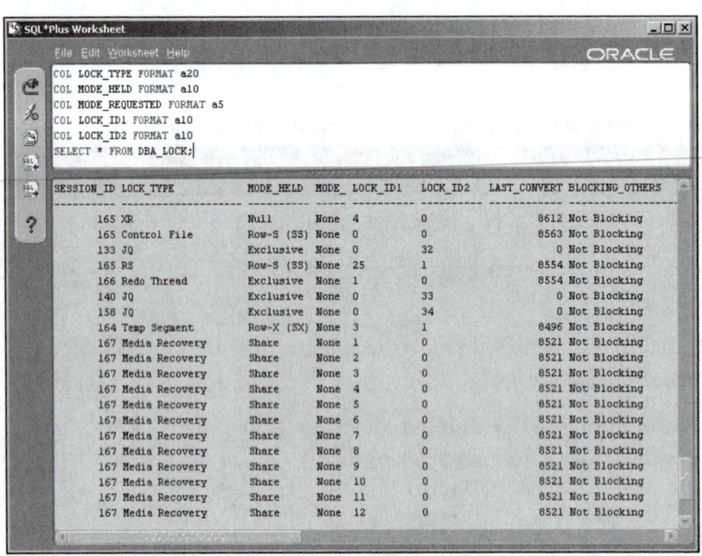

FIGURE 13-13 Querying the DBA_LOCK performance view

2. And as before, expand your browser up to full screen. The Database Control is BIG!!!
3. A number of different screens may appear, depending on your release of Oracle 10*g*. The important screen is the login screen. Type **SYS** in the Username box, type the appropriate password (ask your instructor), in the Password box, and select **SYSDBA** from the Connect As box. Click **OK** to continue.
4. The screen that appears will be the Home page. Click on the **Performance** tab.
5. The browser will probably keep refreshing and drive you completely nuts! Scroll the browser to the top of the page. At the top-right corner of the browser page is a spin control called **View Data**. Click the default value in the spin control and change it to **Real Time: Manual Refresh**. That will stop the page from refreshing so frequently to allow you to examine the content.
6. Now scroll the browser to the bottom of the page. You see a number of links at the bottom of the page. The links you are interested in for examining database locks are highlighted in Figure 13-14.

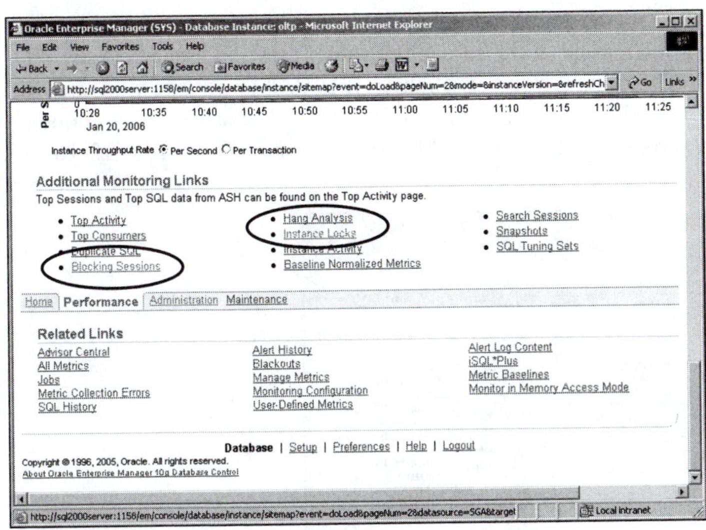

FIGURE 13-14 Finding lock conflicts in the Database Control

7. Click the link called **Instance Locks**. At the top-left corner of the page is a spin control called View. Click on the spin control and change its value to **All Instance Locks**. Then also select the check box called **Show MR (Media Recovery) Locks**. Your screen will now look something like that shown in Figure 13-15. You may also have some locks listed.

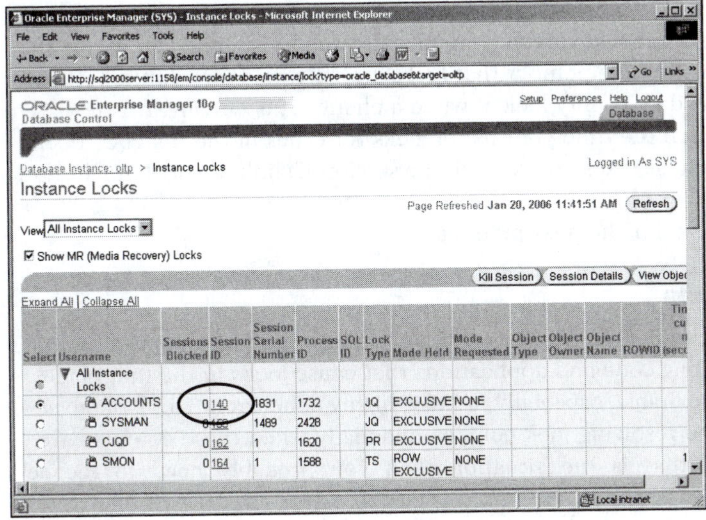

FIGURE 13-15 Viewing all instance locks using the Database Control

8. If the Database Control on your screen does show any locks you should see a column called Session ID. Each of the session ID values (session identifier) should appear as a link. This means you can click on it to drill-down into that specific lock. Click on one of the session ID links to drill-down. If you are quick enough you might see a screen that looks something like that shown in Figure 13-16.

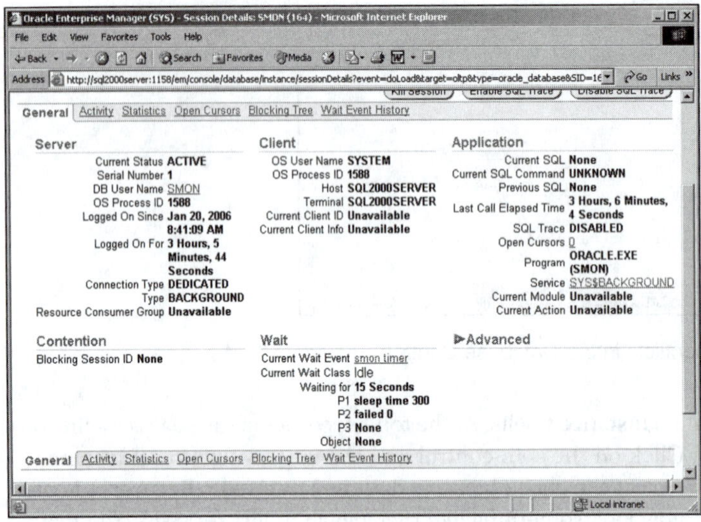

FIGURE 13-16 Drilling down into an instance lock using the Database Control

9. Bear in mind use of the word *might* in the previous point. What this means is that you may not get that screen. You may also get an error screen saying that the lock has expired. Remember that locks are only held for a limited period of time because data is only locked when a change is occurring. It is possible that between the time your list of locks refreshes in the Instance Locks page and the time it takes you to click the session ID link, the lock might no longer exist.
10. Remain logged on for the next practice.

Resolving Lock Conflicts

So how do we resolve lock conflicts? The best way from an architectural and programming perspective is to avoid building code and applications that cause locks in the first place. However, some locking is inevitable, especially in busy, highly concurrent OLTP databases. In general you should not be resolving lock conflicts manually when those conflicts are caused by applications. If that is the situation then you probably need to recode applications.

Lock conflicts that require manual resolution are usually caused in error, such as executing a DDL command on an active production database, and locking a busy table. Other potential problems are commonly caused by executing DML commands that are not committed or rolled back, or using the LOCK TABLE command, or changing all the rows in a busy table, and so on. Altering tables in a production database is always risky and should be avoided if possible.

The easiest way to demonstrate lock conflicts is to create a lock conflict and then try to resolve it as an exercise, such as the following:

1. Start up two sessions of SQL*Plus Worksheet. In Windows, click **Start/All Programs/Oracle ... /Application Development/SQL*Plus Worksheet**. That means start up SQL*Plus Worksheet twice from the menu.
2. In both instances of SQL*Plus Worksheet, connect as the CLASSMATE user when the logon window appears. Type **CLASSMATE** in the Username box, type **CLASSPASS** (the current password), in the Password box, and **ORACLASS** in the Service box. Leave the connection type as "Normal." Click **OK** to continue.
3. Now to create a lock, go to the first instance of SQL*Plus Worksheet and type the following UPDATE command, effectively locking all rows in the EMPLOYEE table (**DO NOT** include a COMMIT statement):

   ```
   UPDATE EMPLOYEE SET NAME = UPPER (FIRST_NAME);
   ```

   The response will be a number of rows updated.
4. Now go to the second instance of SQL*Plus Worksheet and type the following UPDATE command (again **DO NOT** include a COMMIT statement):

   ```
   UPDATE EMPLOYEE SET NAME = LOWER(LAST_NAME);
   ```

   There will be no response from your second instance of SQL*Plus Worksheet. You locked all the rows in the EMPLOYEE table with the first UPDATE statement in step 3 above. The second UPDATE statement in this point is waiting for the first UPDATE statement to be committed or rolled back.
5. If you don't have the Database Control still running, type the following into your browser:

   ```
   http://<hostname>:1158/em
   ```

   The <hostname> must be substituted for with the hostname of your database server computer. And the port number may be different.
6. Expand your browser up to full screen.
7. Connect as SYS. Click. Type **SYS** in the Username box, type the appropriate password (ask your instructor), in the Password box, and select **SYSDBA** from the Connect As box. Click **OK** to continue.
8. Fix the browser refresh problem. Scroll the browser to the top of the page. At the top-right corner of the browser page is a spin control called **View Data**. Click the default value in the spin control and change it to **Real Time: Manual Refresh**. That will stop the page from refreshing so frequently.
9. The screen that appears will be the Home page. Click on the **Performance** tab. Then click the **Instance Locks** link at the bottom of the performance page again.

10. You should get a screen similar to Figure 13-17. Figure 13-17 shows an exclusive blocking lock, held on the EMPLOYEE table in the CLASSMATE schema. Now you simply go to the first SQL*Plus Worksheet instance and execute a COMMIT or ROLLBACK command to clear the lock, allowing the UPDATE statement in the second SQL*Plus Worksheet to complete. But we won't do that.

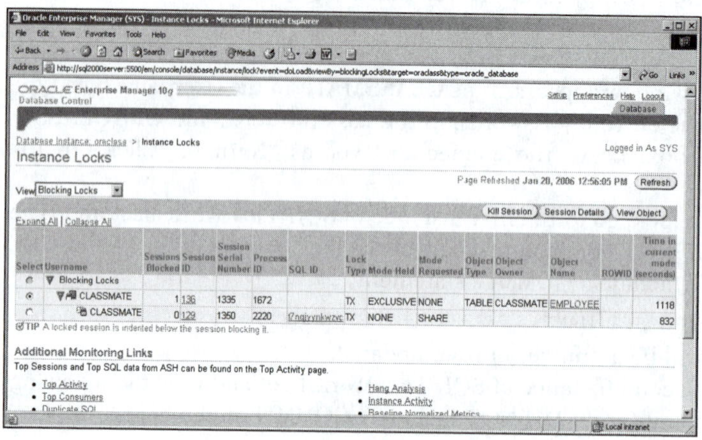

FIGURE 13-17 A blocking lock held on the CLASSMATE.EMPLOYEE table

In Figure 13-17 you are able to see that the first line in the list with the CLASSMATE username indicates an exclusive table lock. The second line with the CLASSMATE username indicates a second session requesting a lock on the same table, as a share lock. The second item is indicating that it is awaiting release of the first lock.

11. So how do you clear the lock using the Database Control? You can't commit or rollback SQL statements from within the Database Control but you can remove the blocking session. Make sure that the session holding the exclusive lock is selected. Click the **Kill Session** button at the top-right corner of the screen.
12. The next screen gives you a few options. Leave the **Kill Immediate** option selected and click the **Yes** button at the bottom right of the window.
13. You will be returned to the Instance Locks screen. Click the **Refresh** button at the top right of the window. The locks should disappear.
14. Now return to the second instance of SQL*Plus Worksheet. You will see a message indicating that the session that was previously blocked has performed its UPDATE statement successfully. Type the **COMMIT** command into the second instance of SQL*Plus Worksheet and execute it by clicking the little lighting bolt button in the top-left corner of the SQL*Plus Worksheet window.
15. Go to the first instance of SQL*Plus Worksheet and click the lighting bolt button to attempt to execute the UPDATE statement again. You will get an error indicating that this session was disconnected, as shown in Figure 13-18.

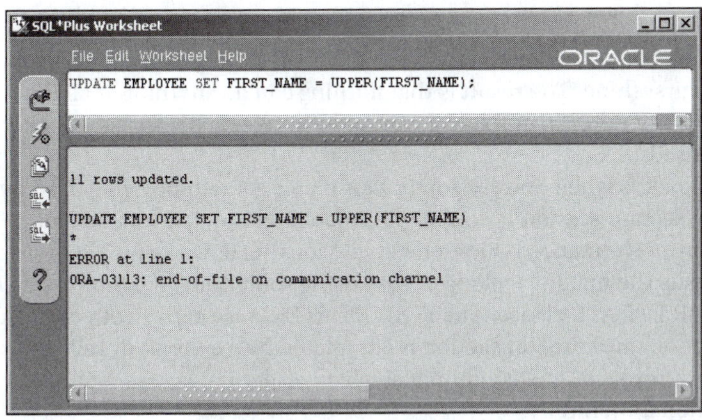

FIGURE 13-18 Clearing a lock in the Database Control kills the parent session

16. Return the second instance of SQL*Plus Worksheet again. Type the following query:

    ```
    SELECT FIRST_NAME, LAST_NAME FROM EMPLOYEE;
    ```

    The result of the above query is shown in Figure 13-19. As you can see the FIRST_NAME column is not upper case (was not changed), and the second column is changed to upper case. This indicates the first UPDATE statement failed and the second succeeded. This is because the parent session of the first UPDATE statement was killed in Database Control.

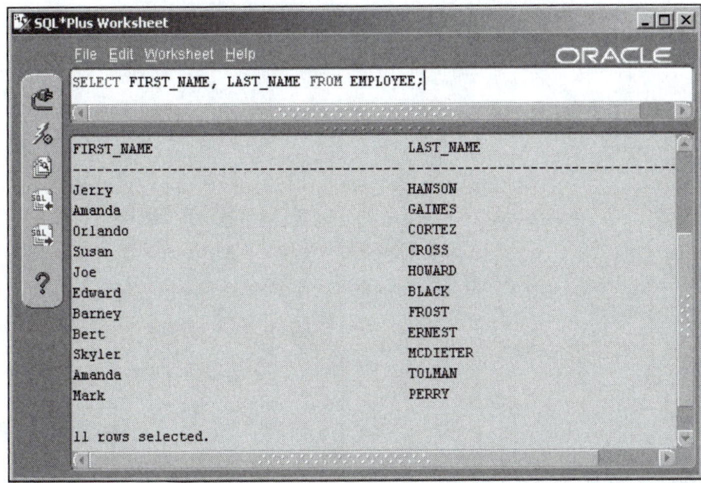

FIGURE 13-19 Clearing a lock in the Database Control aborts a pending transaction

17. This is the last exercise for this chapter. You can now shut down any instances of SQL*Plus Worksheet, the Oracle Enterprise Manager console, and the Database Control browser window.

What is a Deadlock?

A deadlock is when one thing is waiting for another, but that other thing is also waiting for something else, or even the first thing. The result is that nothing can do anything at all. The reason for a deadlock is that a pending lock request cannot be serviced, because the lock required will never be released.

The worst kind of deadlock is when one session is waiting for something being used by another session, where the second session is waiting for a lock held by the first session. Deadlocks can cause serious performance issues, effectively locking down data. The most common cause of deadlocks is the manual locking of data using something like the LOCK TABLE command or the FOR UPDATE clause. The FOR UPDATE clause can effectively lock everything within a transaction until the transaction is completed by a commit or rollback. This is especially true where rows are locked up in a single transaction, across multiple dependent tables in a relational hierarchy.

Usually deadlocked sessions must be rolled back manually. However, Oracle Database can sometimes automatically detect deadlocks, and even automatically select one of the offending transactions for automatic roll back, typically the transaction containing the command that detected the deadlock.

Two possible ways of resolving a deadlock manually are: (1) rollback one of the deadlocked transactions, or (2) kill one of the sessions executing one of the deadlocked transactions (perform an implicit roll back).

Chapter 14 will examine proactive maintenance and ways to counteract potential problems before they occur.

Chapter Summary

- Performance is the speed at which a database services its users and applications.
- Performance monitoring can be manually performed by database administrators or by software.
- Performance monitoring tools include EXPLAIN PLAN, SQL*Plus Autotrace, SQL Trace with TKPROF, STATSPACK, V$ performance views, the Wait Event Interface, and numerous tools contained within the Database Control.
- Statistics are important because they provide the optimizer a realistic and up to date mathematical picture of data.
- Statistics can be gathered manually using the ANALYZE command and the DBMS_STATS package.
- Statistics can be gathered in the Database Control using the AWR and ADDM.
- Automated statistics can be augmented using dynamic sampling and the SAMPLE BLOCK clause.
- The Database Control contains a plethora of performance metrics for all sorts of performance issues.
- Performance metrics operate against thresholds.
- When a performance metric threshold is exceeded, a warning is sent to the Database Control main screen.
- Performance metrics can be altered in the Database Control.
- Search for invalidated object using the USER_OBJECTS, DBA_OBJECTS, and ALL_OBJECTS.
- Invalidated objects can also be found using the Database Control.
- Locks can be both explicit and implicit.
- An explicit lock occurs when an administrator issues the LOCK TABLE command.
- Implicit locks can occur when there is too much competition for the same data.
- Implicit locks are either shared or exclusive.
- QA shared lock allows access to data.
- An exclusive lock allows only read access to data.

Review Questions

1. Which of these are Database Control tools?
   a. TopSQL
   b. TopSessions
   c. TopClient
   d. TopConsumers
   e. All of the above

2. Which command can be used to return a query plan in SQL*Plus?
   a. EXPLAIN PLANS
   b. EXPLAINED PLAN
   c. SET AUTOTRACE ON EXPLAIN
   d. EXPLAIN PLAN
   e. None of the above
3. Both system and object statistics are generated in Oracle 10*g* automatically, by default. True or False?
4. The STATISTICS_LEVEL parameter can be set to values, and still retain at least some statistics gathering automation in Oracle 10*g*?
   a. TYPICAL
   b. BASIC
   c. ALL
   d. All of the above
5. Which of these commands is recommended for Oracle 10*g*?
   a. ANALYZE TABLE EMPLOYEE COMPUTE STATISTICS;
   b. EXEC DBMS_STATS.GATHER_TABLE_STATS('CLASSMATE', 'EMPLOYEE');
   c. Neither
   d. Both
6. What does AWR stand for?
   a. `Automatic Workload Repository`
   b. `Automatic Wiring Repository`
   c. `Automatic Workload Recovery`
7. What is the ADDM searching for?
   a. Too much memory used
   b. Memory buffers too small
   c. Heavy CPU usage
   d. Too much I/O
   e. SQL statements using too many resources relative to other SQL statements
   f. Inappropriate configuration parameter settings
   g. Anything nasty causing performance problems
   h. All of the above
8. A lock can be cleared by killing the session containing a transaction that is not holding a lock. True or False?
9. Deadlocks are sometimes automatically detected and resolved by Oracle 10*g*. True or False?

Exam Review Questions—Oracle Database 10*g*: Administration (#1Z1-042)

1. What is the name of the configuration parameter explicitly causing Oracle 10*g* to generate system statistics?
2. What does the following command?

```
EXEC DBMS_SCHEDULER.DISABLE('GATHER_STATS_JOB');
```

3. The hostname of your computer is *mycomputer* and the port for the Database Control is 1158. Which of these URLs will execute the Database Control in your browser?
   a. http://<hostname>:1158/dbcontrol
   b. http://<hostname>:1158/databasecontrol
   c. http://<hostname>:1158/dbc
   d. http://<hostname>:1158/enterprisemanager
   e. None of the above
4. What is the result of this sequence of commands?

```
ANALYZE TABLE EMPLOYEE DELETE STATISTICS;
ANALYZE TABLE EMPLOYEE COMPUTE STATISTICS;
ANALYZE TABLE EMPLOYEE DELETE STATISTICS;
EXEC DBMS_SCHEDULER.DISABLE('GATHER_STATS_JOB');
```

5. Which of these options is true?
   a. The AWR manually collects statistics.
   b. The AWR automatically collects statistics.
   c. The ADDM compares three snapshots created by the AWR.
   d. The ADDM compares two snapshots created by the AWR.
   e. None of the above.
6. Which query will find invalid functions in Oracle 10*g*?
   a. SELECT * FROM DBA_OBJECTS WHERE STATUS = 'VALID';
   b. SELECT * FROM DBA_OBJECT WHERE STATUS = 'INVALID';
   c. SELECT * FROM USER_OBJECTS WHERE STATUS = 'VALID';
   d. SELECT * FROM USER_OBJECT WHERE STATUS = 'INVALID';
   e. SELECT * FROM ALL_OBJECTS WHERE STATUS = 'VALID';
   f. SELECT * FROM ALL_OBJECT WHERE STATUS = 'INVALID';
   g. None of the above
7. How is a metric in the Database Control exceeded?

Hands-on Assignments

1. Make sure statistics are automated for all tables by executing a single query against your database. Create a SQL script and save the script in a file named **ho1301.sql**, in the **Solutions\Chapter13** directory on your student disk.

2. Use a single command to switch off all automated statistics generation in your database. Save your commands in a file named **ho1302.sql** in the **Solutions\Chapter13** directory on your student disk.
3. You change your mind on a switch automated statistics generation for the entire database on again, but this time you switch off dynamic sampling. Do not shut down the database or recreate the configuration parameter file. Any changes must be made at the system level and not the session level. Save your commands in a file named **ho1303.sql** in the **Solutions\Chapter13** directory on your student disk.
4. Something is causing a lock on a table in your database. Use the Database Control to find the lock, and resolve the conflict. Describe how you find the lock and resolve the conflict. Save your answer in a file named **ho1304.txt** in the **Solutions\Chapter13** directory on your student disk.

Case Projects

1. Your MIS manager wants you to generate statistics manually for all tables in your CLASSMATE schema, using the ANALYZE command. Write a script to execute in the command-line version of SQL*Plus. Create a log file called **case1301.log**. Your manager wants to see something, specifically the commands executed against the database. The script should execute without any administrator intervention, and produce an audit of results. Store your script in a file named **case1301.sql** in the **Solutions\Chapter13** directory on your student disk. Generate the log file into the **Solutions\Chapter13** directory on your student disk as well.
2. You persuade your manager that the ANALYZE command is out of date. Use a single procedure execution command to do the same as above. Don't forget to create a log file because your manager wants to see results. Call the log file **case1302.log**. Store your script in a file named **case1302.sql** in the **Solutions\Chapter13** directory on your student disk. Make sure the log file is generated into the **Solutions\Chapter13** directory on your student disk as well.

CHAPTER **14**

PROACTIVE MAINTENANCE

LEARNING OBJECTIVES

In this chapter, you will:

- Learn more about the Database Control
- Discover how to use advice performance tools
- Learn about the segment advisor
- Learn about undo management and the undo management advisor
- Learn about the memory advisor
- Learn about the SQL access advisor
- Learn about the SQL tuning advisor
- Learn more about changing and using baseline metrics

INTRODUCTION

The previous chapter introduced various small parts of the Database Control. This included an introduction to performance metrics and baselines, along with some basics on use of the Automatic Workload Repository (AWR) and the Automatic Database Diagnostic Monitor (ADDM). Various parts of the Database Control have also been summarized throughout this book. This chapter concentrates on some of the more complex areas of Database Control functionality.

Topics to be covered in this chapter are thresholds and what to change, alerts and what do to about them, and the baseline metrics. Advisory tools present suggestions generally in the area of performance and are a large part of using the Database Control. In fact, this chapter, Proactive Maintenance deals mostly with performance. Proactive maintenance implies solving problems before they occur. Not as they occur or after they have occurred.

Chapter 13 showed how to find and set Database Control tools such as the advisors, metrics, alerts and warnings, and setting thresholds. This chapter describes some of those tools.

WHAT IS PROACTIVE MAINTENANCE?

Once again, proactive maintenance is really all about setting up the Database Control to predict and track performance problems. The objective is to prevent critical problems before they occur and damage a database. This is contrary to reactive maintenance, which is reacting to a problem after it has occurred. Much of proactive activity is related to performance tuning a database in one way or another. Proactive maintenance can be managed from within the Database Control including areas such as warnings and alerts, metrics for determining thresholds of when and how warnings and alerts occur, and a variety of advice from various Database Control tuning advisory tools. There is also a default configuration within the advisors and metrics provided when you create a database. For most databases it is unlikely that you will ever need to change any of the default setup in the Database Control. Let's begin by examining the various advice tools, under the Advisor Central section, within the Database Control.

ADVICE PERFORMANCE TOOLS

The advice performance tools were available in past versions of Oracle Database as specialized views. These views were accessible using one of the SQL*Plus tools. In Oracle10g you can access advisors and find plenty of good advice by using the Database Control.

Let's start up the Database Control:

1. To start the Database Control, open a browser, such as Internet Explorer in Windows. Ask your instructor if you don't know.
2. Expand your browser to full screen. The Database Control is BIG!!!
3. At the top of your browser window you should see an address line. Type the correct version of the URL shown below into the browser address entry. Replace the **<hostname>** value with the host name of your database server computer:

   `http://<hostname>:1158/em`

   Ask your instructor for the hostname if you don't have it. The port number may also be different depending on the release of Oracle 10g you are using, and if there is one or more databases created on the database server computer. Other ports that may be used are 5500 and 5600.

   Additionally, to execute the Database Control there needs to be a background process (a service on a Windows computer) running on the database server that executes the Database Control over the network. This service should be installed and started by the Oracle installation process on the database server. If it is not there, then ask your instructor.

4. Once again, a number of different screens may appear, depending on your release of Oracle10g. The important screen is the login screen. Connect as the SYS user. Type **SYS** in the Username box, type the appropriate password (*ask your instructor*), in the Password box, and select **SYSDBA** from the Connect As box. Click **Login** to continue.
5. The screen that appears will look something like the one shown in Figure 14-1, where you will be at the Home screen for the Database Control of your database.

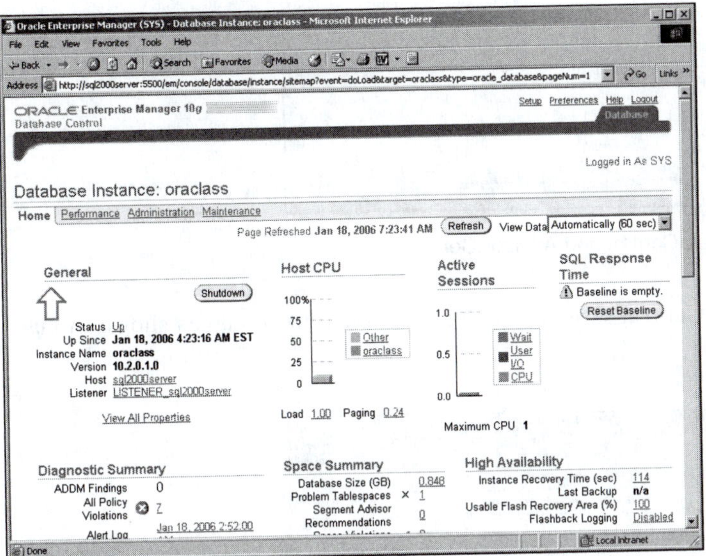

FIGURE 14-1 The Database Control home page

6. Before you do anything else, select the **View Data** spin control at the top-right corner of the screen, and select the **Manually** option.
7. Now scroll down to the bottom of the browser screen. You will see a link at the bottom-left corner of the Database Control window called **Advisor Central**, as shown in Figure 14-2. Click the **Advisor Central** link.

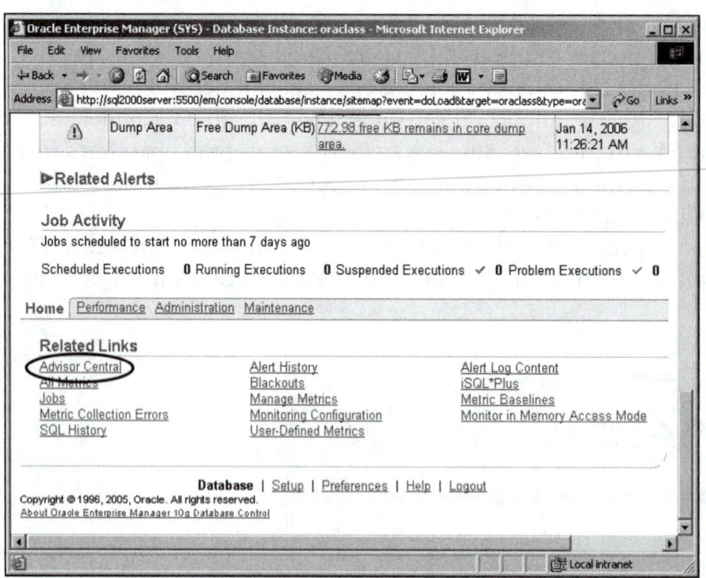

FIGURE 14-2 The Database Control and Advisor Central

8. The **Advisor Central** screen should appear on your screen as shown in Figure 14-3.

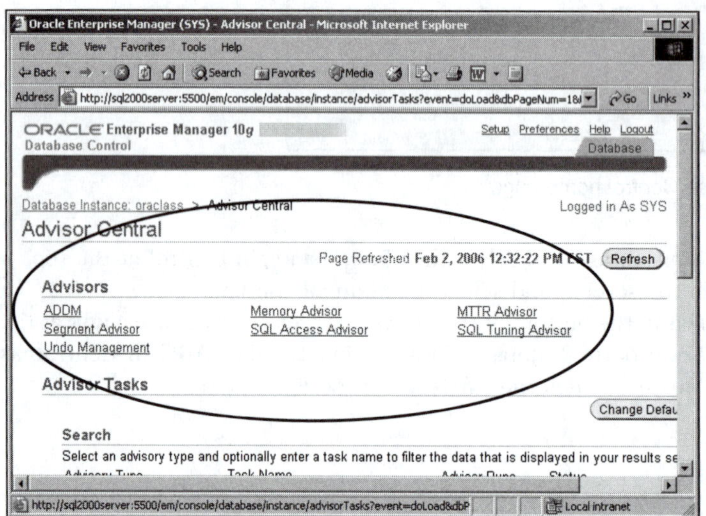

FIGURE 14-3 The Advisor Central screen in the Database Control

9. Remain logged on for the next practice.

The Different Tuning and Diagnostic Advisors

As you can see in Figure 14-3, Advisor Central is made up of a number of options. These options are as follows, some of which have been seen already in Chapter 13:

- **ADDM (Automatic Database Diagnostic Monitor)**: This advisor performs automated diagnostics, by comparing multiple statistical snapshots taken against the database. The AWR (Automatic Workload repository) collects those snapshots. Both the AWR and ADDM tools are covered in Chapter 13.
- **Segment Advisor**: This advisor searches for, automatically detects, and posts warnings and alerts for segment issues. A segment is essentially the purely physical I/O and disk aspects of a database object, such as a table.
- **Undo Management and Undo Advisor**: This tool advises as to the optimal configuration of automated undo.
- **Memory Advisor**: Warns and alerts with respect to potential instance recovery, potential media recovery, and potential flashback recovery.
- **MTTR (Mean Time To Recovery) Advisor**: Advises on the usage and sizing of the different memory buffers, including the entire SGA (System Global Area), and session connection memory—PGA (Program Global Area).
- **SQL Tuning Advisor**: Performs automated SQL tuning activities using the SQL Tuning Advisor, SQL tuning sets, and SQL profiles.
- **SQL Access Advisor**: This tool is generally used for materialized view analysis and usually in data warehouses.

Next you will get to examine and experiment with each of the advisor tools in turn. The ORACLASS database you have does not have any consistent activities. It is not constantly busy. I will be using another database, which I have used for writing various other books. This database has a large amount of activity because it executes some complex simulation coding. Therefore, everything that you will see in this section will be different from what you will see on your screens in terms of content. In other words, do not expect data on your screen to match the figures displayed in this chapter. The mere act of following through all these advisory tool exercises on your screen, and examining the figures in this book, will give you a comprehensive explanation of how the advisory tools function.

NOTE

The only tool to be omitted from the subsequent sections, covering all the Advisor Central tools, is the MTTR, which is covered in Chapter 15. Both the ADDM and AWR are covered in some detail in Chapter 13.

The Segment Advisor

The segment advisor tells you about disk space, how it is used, how it should not be used, and what you might want to alter to help everything run smoothly. Of particular interest are shrinkage, fragmentation, growth, capacity planning, row chaining, and row migration.

The primary focus is forecasting. Proactive maintenance is all about forecasting and perhaps preventing running out of disk space before it happens. Running out of disk space can crash a database because it has no space to write changes to disk.

The segment advisor consists of a number of separate sections:

- **Scope**: Scope allows definition of what you want advice on, such as tables or tablespaces.
- **Objects**: After selecting what kind of things to analyze (the scope of analysis), you can select exactly which tables to analyze. This of course assumes you selected tables as the scope of your selection.
- **Schedule**: You can schedule a segment advisory analysis to execute at a future point in time. You can also accept the default and execute the advisor immediately.
- **Review**: This step allows you to make sure that what you have selected in the previous three steps is what you actually want to do.

Let's practice using the segment advisor:

1. You should still be in the Database Control. If you are not, then go back to the first exercise in this chapter and navigate to the Advisor Central page of the Database Control.
2. Don't forget, when on the home page, to select the **View Data** spin control at the top-right corner of the screen, and select the **Manually** option.
3. Once in Advisor Central, click the **Segment Advisor** link, as shown in Figure 14-3.
4. As you can see in Figure 14-4, you can select the radio button options to analyze the tablespace of schema objects (tables, indexes, materialized views). Leave the option set to tablespaces and click the **Next** button to go on to the next step. Click the Add button on the following screen.

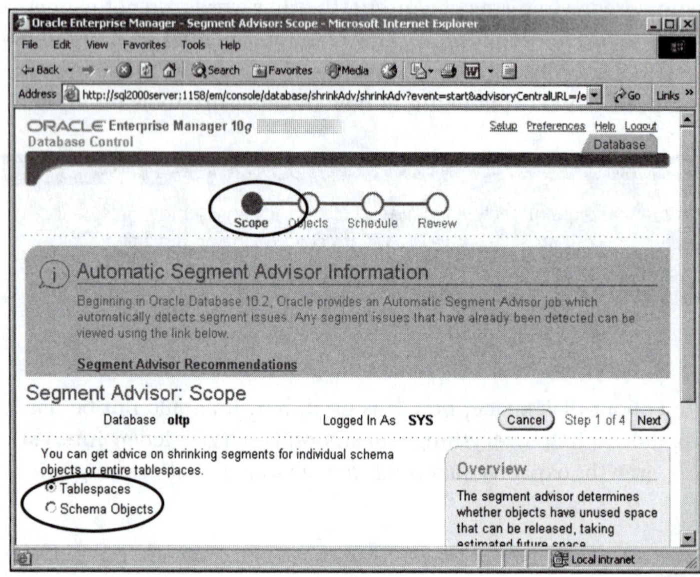

FIGURE 14-4 Executing the Segment Advisor

5. Now you have the object page in front of you, and a list of tablespaces you can select for analysis. As you can see in Figure 14-5, there are a number of tablespaces. Most of the application data in my database resides in my DATA and INDX tablespaces. So I select both DATA and INDX tablespaces (you can select the USERS tablespace for your ORACLASS database). Now click the **OK** button.

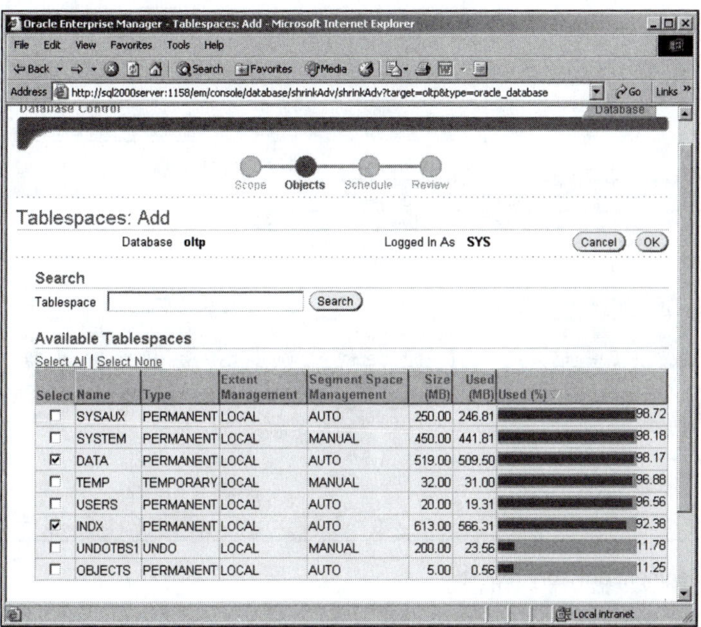

FIGURE 14-5 Selecting objects in the Segment Advisor

6. You will get another screen giving you the option to submit an analysis to the Segment Advisor, or to abort at this point. Click the **Submit** button to execute the analysis. If you click the **Next** button you will see various screens allowing scheduling of the segment analysis operation. You could also review what your analysis will do. Those screens are all straightforward and easy to use.
7. To examine your created segment analysis, click your browser back button until you get back to the Advisor Central page. If you miss the page, go back to the Database Control home page, scroll to the bottom, and click the **Advisor Central** link.
8. At the bottom of the Central Advisor page, you should see something like that shown in Figure 14-6. In Figure 14-6 there is a segment advisory job, shown as currently running in my database. Make sure that the **Segment Advisor** radio button is selected, and even though my job is still running, I can click the **View Result** button. Click the **View Result** button now.

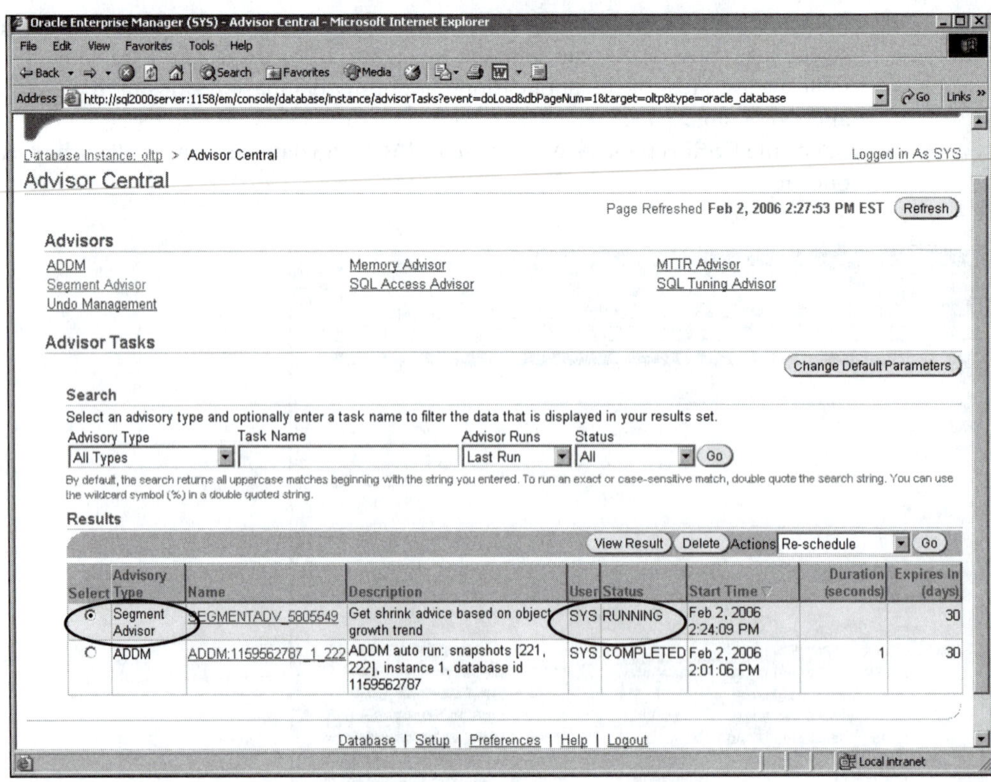

FIGURE 14-6 Executing a segment advisory report

9. The result is as shown in Figure 14-7. As you can see in Figure 14-7 no segments have problems. Can you guess why? The answer is because I am using locally managed tablespaces, with both automated extent and automated segment space management configuration, on all of those tablespaces.

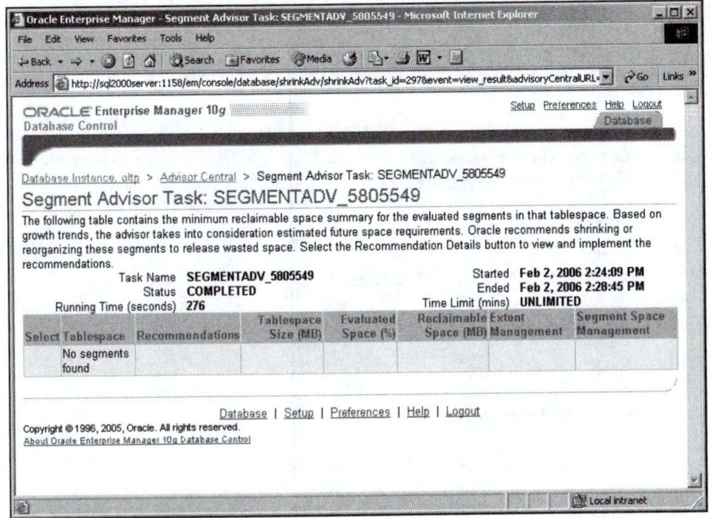

FIGURE 14-7 Segment Advisor problems

10. Remain logged on for the next practice.

Undo Management and the Undo Advisor

The undo advisor works best with automated undo. The undo advisor will help you with the settings for automated undo configuration parameters, including retention and tablespace sizing.

Automated undo has removed the need for complex maintenance of manual rollback segments. Automated undo allows rollback of transactions, recovery of aborted and partly completed transactions, and some recovery. Additionally, undo space allows for read consistency. Read consistency implies that if one user makes a change without committing or rolling back, then only the user who made the change can see that change. Other users will see data the way it was before the change was made, until the change is committed. Most database engines function by making changes physically in the database as soon as the change is made. The execution of a COMMIT or ROLLBACK command simply removes the potential for rollback. This is the most efficient method of physical database update because it is assumed that far more data is committed than is ever rolled back. This is a perfectly sensible assumption. Thus read consistency is very important to a relational database.

Let's practice with the undo advisor:

1. You should still be in the Database Control. If not then go back to the first exercise in this chapter, and navigate to the Advisor Central page of the Database Control.
2. Once again, don't forget that when on the home page, to select the **View Data** spin control at the top-right corner of the screen, and select the **Manually** option.
3. Once in Advisor Central, click the **Undo Management** link, as shown in Figure 14-3.

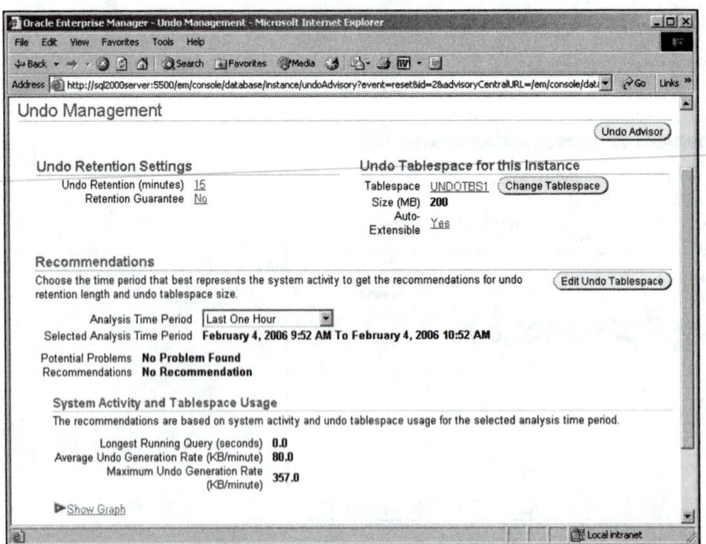

FIGURE 14-8 Executing Undo Management in the Database Control

4. To go into the undo advisor click the **Undo Advisor** button. Shown at the top-right of Figure 14-8.
5. A number of callouts appear on the graph shown on your screen, as shown in Figure 14-9. The entire graph is telling you is that there is a straight-line curve between automated (recommended minimum), and maximum possible retention (making undo space larger). If you want to make sure that you always retain changes for possible undo, then set undo retention to the maximum. Use the default in an OLTP database. The same concept of retention applies to sizing of the undo tablespace. The larger the undo tablespace, the larger the physical size of changes that can be undone.
6. The undo advisor screen does not reveal extensive information. Remain logged on for the next practice.

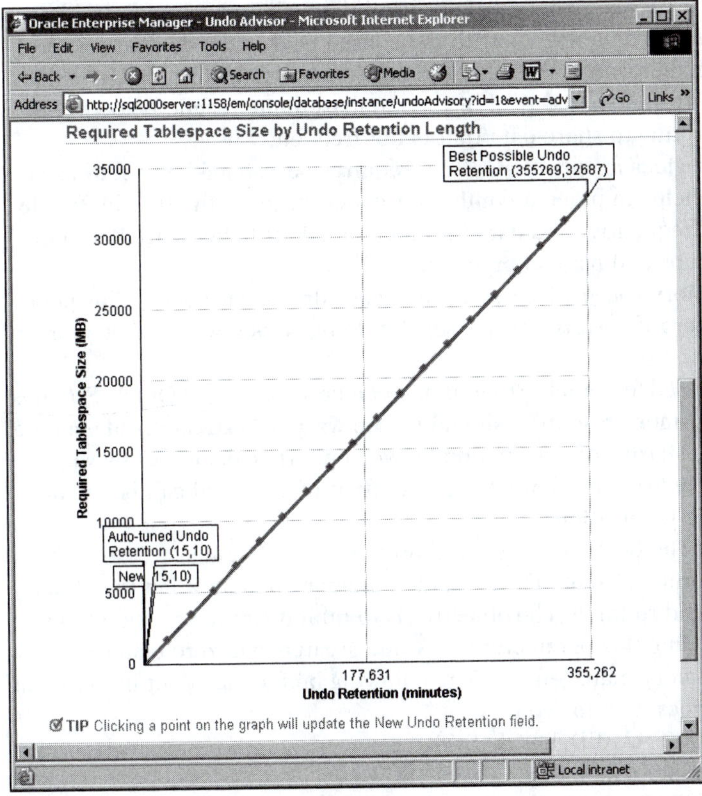

FIGURE 14-9 Examining undo advice

The Memory Advisor

The memory advisor page is split into advisors for both System Global Area (SGA) memory and Program Global Area (PGA) memory. The SGA is made up of various buffers including the shared pool and the database buffer cache. PGA is the session connection memory that is shared by connections to the database server. PGA manages memory operations on a database server, such as sorting, creating bitmaps, and creating hash tables.

The SGA consists of the following configuration parameters:

- **Shared pool**: Contains the dictionary cache and the library cache. The dictionary of a relational database contains all system metadata and application tables, indexes, materialized views, and so on. The library is all the preparsed SQL and PL/SQL code, which is reused when re-executed, within a reasonable period. The shared pool is essentially the most shareable memory buffer, providing concurrency for data and the database. The shared pool is usually the most critical buffer for any Oracle database. It cannot be too small because it simply won't share enough of the most frequently used data. Consequently, it also cannot be sized too large because it can become too complex for Oracle Database to manage.

- **Database buffer cache**: Contains application data (the rows in your tables, indexes, materialized views). As with the shared pool, if the database buffer cache is too small it won't provide fast enough access to highly shareable data. However, as with the shared pool, if the database buffer cache is too big it could be attempting to share data that is not frequently used.
- **Large pool**: Provides for processing that requires occasional large chunks of memory. This helps to prevent conflict for concurrency with other buffers by not pushing infrequently used data into memory. This buffer is used for odd tasks like backups and parallel processing.
- **Java pool**: Memory space allocated for database-driven Java code. This pool can be set to zero if Java is not in use. Not doing so is a waste of memory resources.
- **Streams pool**: Used for Oracle Streams and can be set to zero if Oracle Streams is not in use. Memory resources should not be wasted. A stream is like a pipe in that it allows sharing of information between multiple databases by way of messages. Oracle Streams allows for distribution of data, and can be an alternative to Oracle Replication.
- **SGA max size**: The SGA can be resized up to the maximum size setting. This limits the maximum size of SGA-contained memory buffers when the Oracle instance is up and running. The objective is to inhibit *run-away* operations.
- **SGA target**: Setting this parameter to a value greater than zero enables automated SGA memory management. The following buffers need not be set at all as Oracle 10g does it all for you:
  - Shared pool (SHARED_POOL_SIZE)
  - Database buffer cache (DB_CACHE_SIZE)
  - Large pool (LARGE_POOL_SIZE)
  - Java pool (JAVA_POOL_SIZE)
  - Streams pool (STREAMS_POOL_SIZE)
  - These parameters are excluded from automated SGA management and the setting of the SGA_TARGET parameter:
    - The log buffer (LOG_BUFFER_SIZE)
    - All the database buffer caches apart from the primary DB_CACHE_SIZE parameter. Excluded parameters are all the keep and recycle pools (DB_KEEP_CACHE_SIZE, DB_RECYCLE_CACHE_SIZE, BUFFER_POOL_KEEP, BUFFER_POOL_RECYCLE), plus all the non-default block size database buffer caches (DB_2K, 4K, 8K, 16K, 32K_CACHE_SIZE).

The PGA is controlled by the PGA_AGGREGATE_TARGET parameter. Setting this parameter to a value greater than zero enables automated PGA memory management. The PGA_AGGREGATE_TARGET parameter automates settings for all the *AREA_SIZE parameters, including the parameters BITMAP_MERGE_AREA_SIZE, CREATE_BITMAP_AREA_SIZE, HASH_AREA_SIZE, SORT_AREA_SIZE, and SORT_AREA_RETAINED_SIZE.

Now that you have read about memory architecture, let's practice with the memory advisor:

1. You should still be in the Database Control. If not then go back to the first exercise in this chapter, and navigate to the Advisor Central page of the Database Control.
2. Don't forget, when on the home page, to select the **View Data** spin control at the top-right corner of the screen, and select the **Manually** option.
3. Once in Advisor Central, click the **Memory Advisor** link, as shown in Figure 14-10.

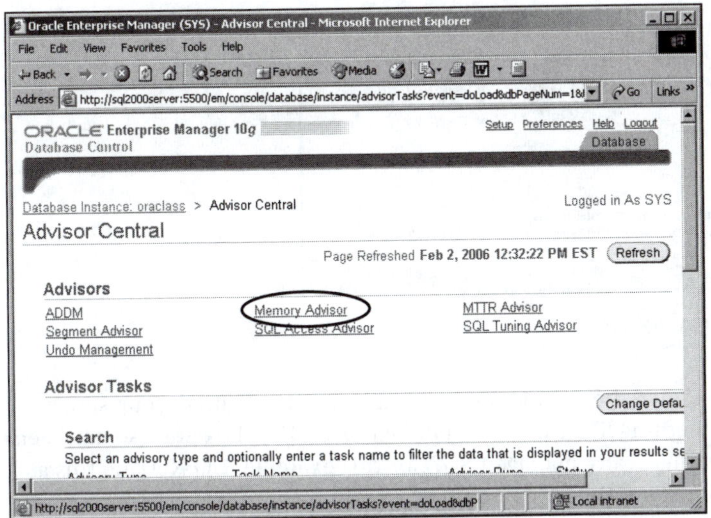

FIGURE 14-10 Executing the memory advisor in the Database Control

4. Figure 14-11 shows the SGA memory management screen, with manual SGA memory management configured. There is also a PGA memory management screen, which we will get to shortly. The SGA and PGA options are highlighted.

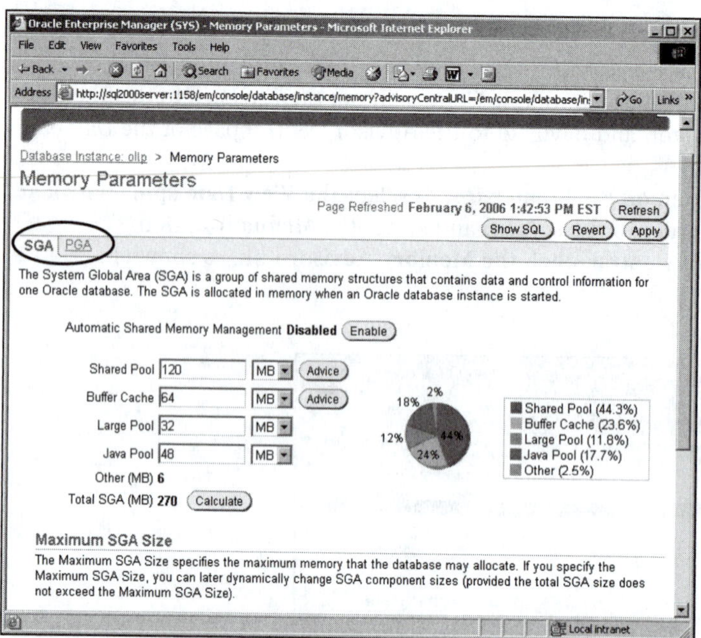

FIGURE 14-11 Managing SGA memory

If you have automated SGA memory management configured your screen should look something like that shown in Figure 14-12. The database has been switched to automatically managed SGA for this example, DO NOT do this as it can take quite some time.

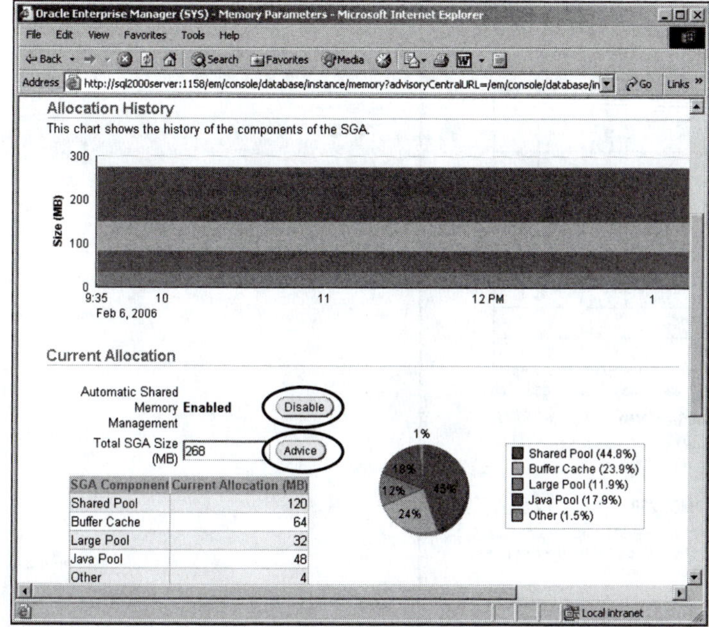

FIGURE 14-12 Managing automated SGA memory

5. Assuming you have automated SGA memory management installed, click the **Advice** button as highlighted in Figure 14-12. You should get a graph as shown in Figure 14-13, showing that any increase to the size of the total SGA is pointless. Therefore, simply click the Cancel button, and disable automated shared memory management by clicking the **Disable** button on the first screen and OK on the next screen, which is also highlighted in Figure 14-12.

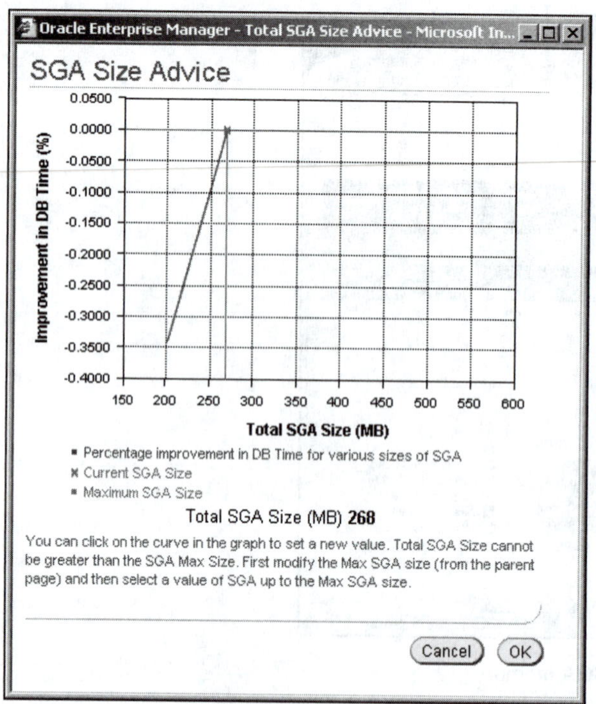

FIGURE 14-13 Getting advice on the configuration of automated SGA memory

6. This returns us to the screen as shown in Figure 14-11, with manual SGA memory management enabled. Click the **Advice** buttons for both the **Shared Pool** and **Buffer Cache** memory buffers in turn, and get graphs as shown in Figure 14-14.

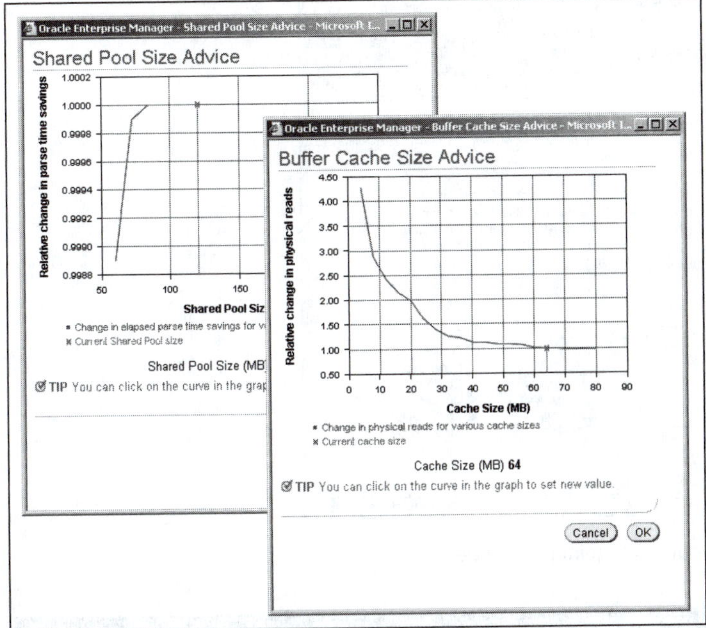

FIGURE 14-14 Advice for shared pool and buffer cache (database buffer cache) buffers

7. Press the **Cancel** button for each to remove the two graphs from the screen. Figure 14-14 illustrates that by increasing the shared pool there will be no difference in performance. However, a decrease in shared pool size may well harm performance. Additionally, decreasing the database buffer cache may harm performance, whereas an increase in database buffer cache size might help performance a little.
8. Change the manual SGA memory buffers as shown in Figure 14-15.
9. According to the graphs shown in Figure 14-14, it would make little sense to size the shared pool to anything other than a minimum value. The shared pool is the most highly concurrent section of memory for an Oracle database. The larger the shared pool, the more likely the database will be sharing data and structure that should not be shareable. The same general concept can be applied to the database buffer cache. Of course, these settings are both dependant on database type and applications. For example, a data warehouse does not need to utilize the database buffer cache nearly as much as an OLTP database so it can be set as small as software will allow. Why? Because a data warehouse spends 99% of its time reading and writing huge amounts of data to and from disk. No amount of RAM can handle data warehouse throughput requirements. Why waste time passing huge amounts of data in and out of the database buffer cache when that data is constantly found on disk anyway. The current requirements for the database are neither a large pool nor a Java pool, therefore, set these to zero. There is no point in wasting memory that can be used for other tasks.

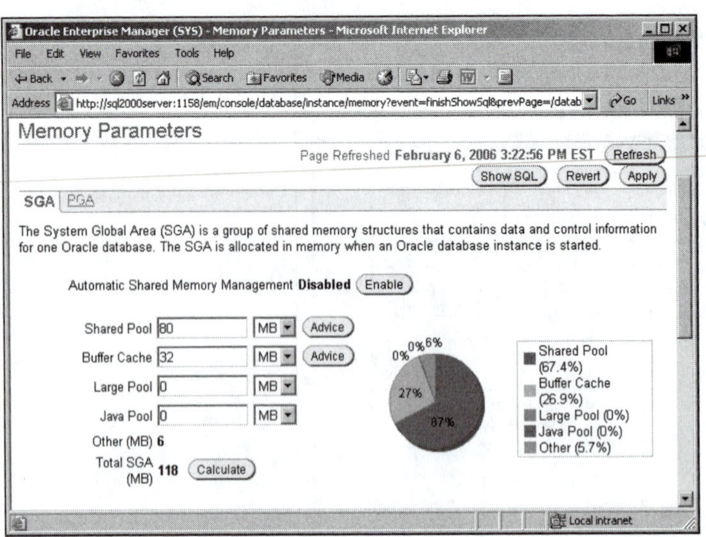

FIGURE 14-15 Setting SGA buffers to minimum values

NOTE

The Database Control will not allow the Java pool buffer size to be set to 0. Instead, make the change using an ALTER SYSTEM statement, or by changing the text parameter file, and then regenerating the binary parameter file.

10. Recall from Figure 14-11, there is a highlight at the top-left corner of the screen, for the **SGA** and the **PGA** links. You have seen how to manage SGA memory. Now click the **PGA** link. The screen should look as shown in Figure 14-16.

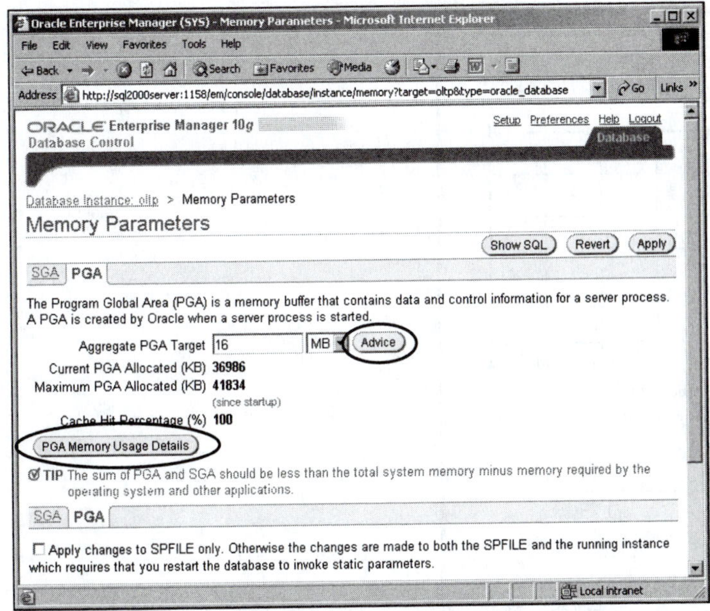

FIGURE 14-16 Setting the PGA buffer to a minimum value

11. PGA memory has been automatically managed since Oracle9*i*, and recommended as such. Automated SGA memory management has been introduced in Oracle10g. There is no option in the Database Control to alter PGA memory management into a manually managed mode. Click the **Advice** button as highlighted in Figure 14-16.
12. As shown in Figure 14-17, the PGA memory advisor does not indicate any recommended change to the PGA memory buffer configuration parameter (PGA_AGGREGATE_TARGET). Now click the **Cancel** button to remove the graph from your screen.

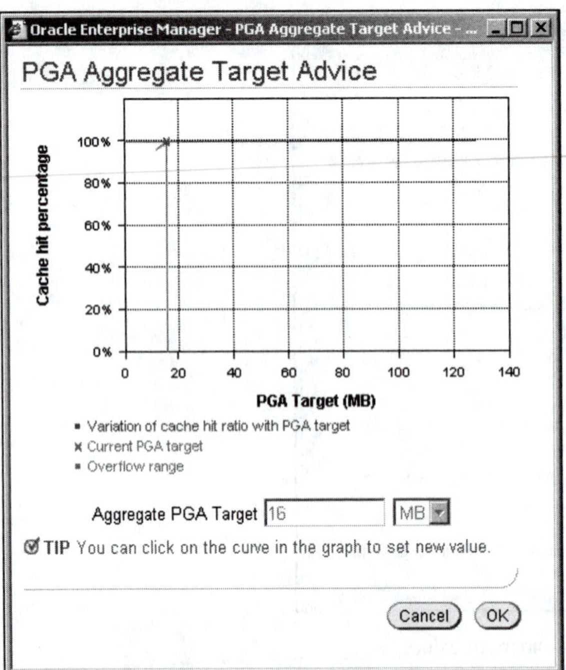

FIGURE 14-17 The PGA memory advisor

13. Also highlighted in Figure 14-16 is the **PGA Memory Usage Details** button. Click that button now. Figure 14-18 shows the result, which can display either **Execution Percentages** or **Number of Executions** (select using the radio button). Click the **OK** button to remove the graph from your screen.

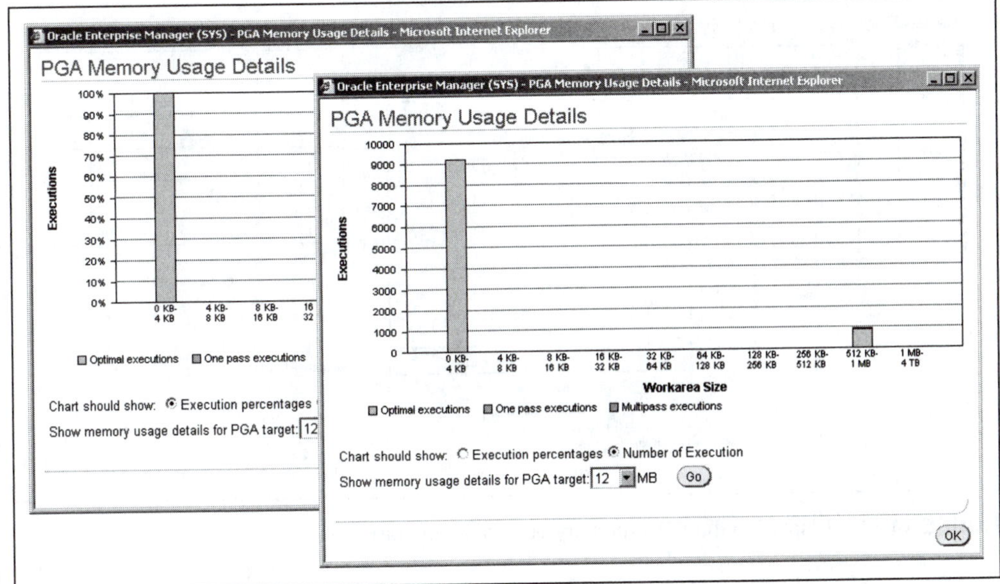

FIGURE 14-18 Displaying the PGA buffers as execution percentages and number of executions

14. Remain logged on for the next practice.

So far in this section covering the Memory Advisor in the Database Control, a static database has been used. Now, let's switch to a very busy simulated OLTP database, showing a few of the memory advice graphs to see if there is any difference. The result, as shown in Figure 14-19, is that changing the shared pool or the PGA buffer would make no difference, even when using a very busy database. However, Figure 14-19 does appear to indicate that the database buffer cache could be larger. This makes perfect sense because 32 Mb for a database buffer cache is a little too small. As a rule, for an OLTP database the database buffer cache should be similar in size to the shared pool. The PGA contains memory required for concurrent connections. If concurrency is very high (you have lots of users such as for an Internet site), the PGA can be set much higher than both the shared pool and database buffer cache combined. For a data warehouse the database buffer cache should be as small as possible, such as 16 Mb.

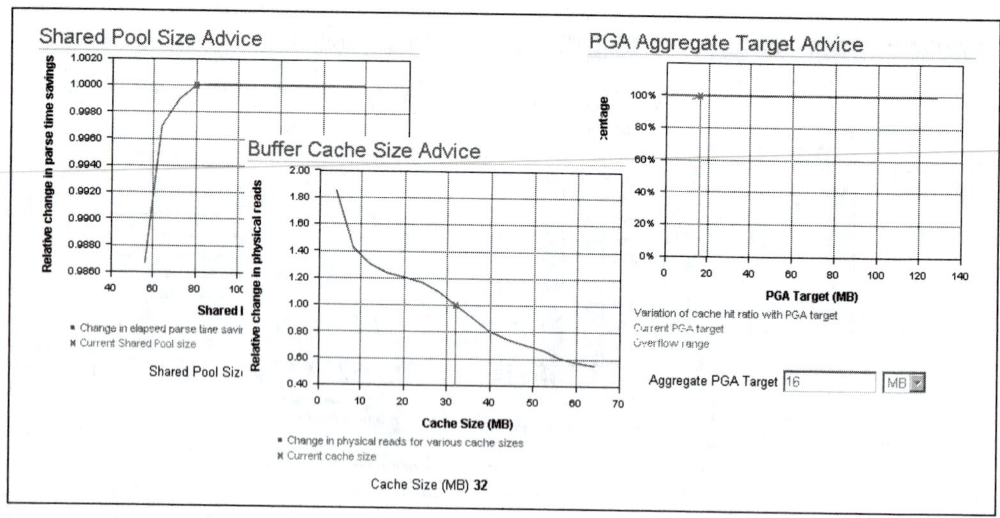

FIGURE 14-19 Displaying the SGA memory buffers under some stress

The SQL Access Advisor

The SQL access advisor allows you to analyze various database objects, such as indexes and materialized views, potentially making recommendations for improving performance. Let's do a brief overview of this advisor, as a full description is beyond the scope of this book:

1. You should still be in the Database Control. If not, then go back to the first exercise in this chapter, and navigate to the Advisor Central page of the Database Control.
2. Don't forget, when on the home page, to select the **View Data** spin control at the top-right corner of the screen, and select the **Manually** option.
3. Once in Advisor Central, click the **SQL Access Advisor** link.
4. Figure 14-20 shows the initial options for the SQL access advisor tool. Leave the **Use Default Options** set as it is, and click the **Continue** button.

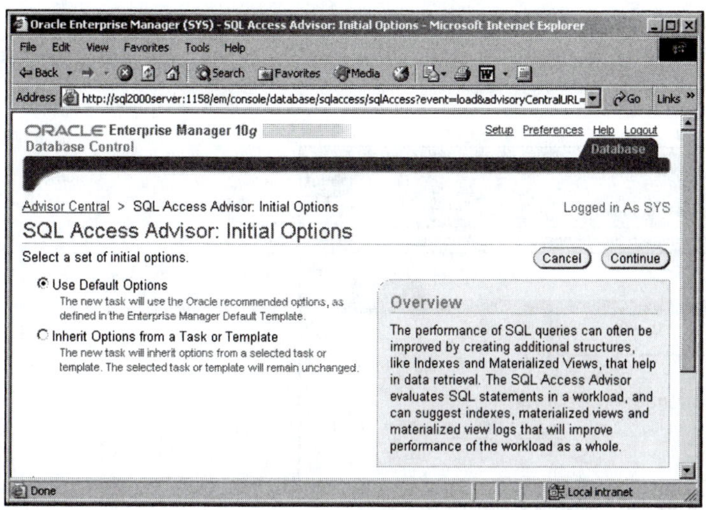

FIGURE 14-20 Initial options for the SQL access advisor

5. The next screen is shown in Figure 14-21, which is the workload screen. In other words, where is the SQL access advisor processing going to get information? Again, leave this option set as the default, which is **Current and Recent SQL Activity**. Click the **Next** button.

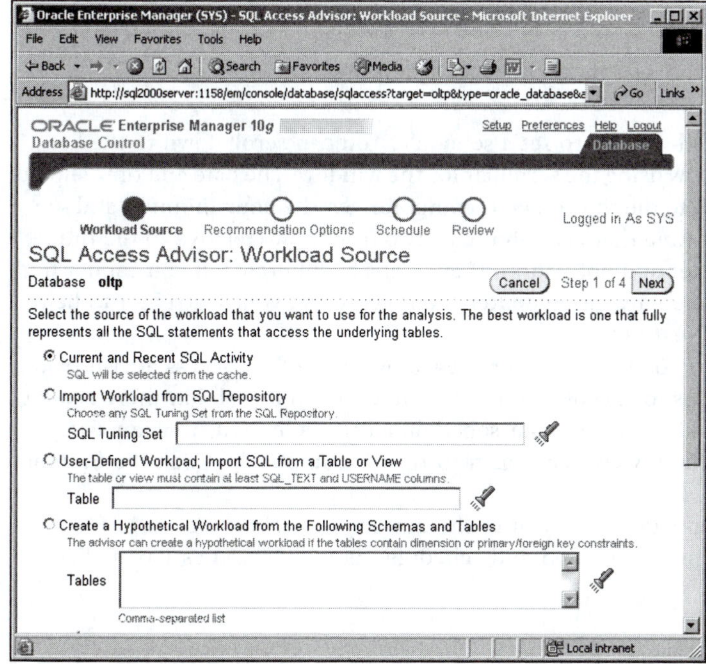

FIGURE 14-21 The SQL access advisor workload source screen

6. Figure 14-22 shows the screen containing recommendations. What exactly are you trying to analyze? Again, leave everything set to the default values and click the **Next** button.

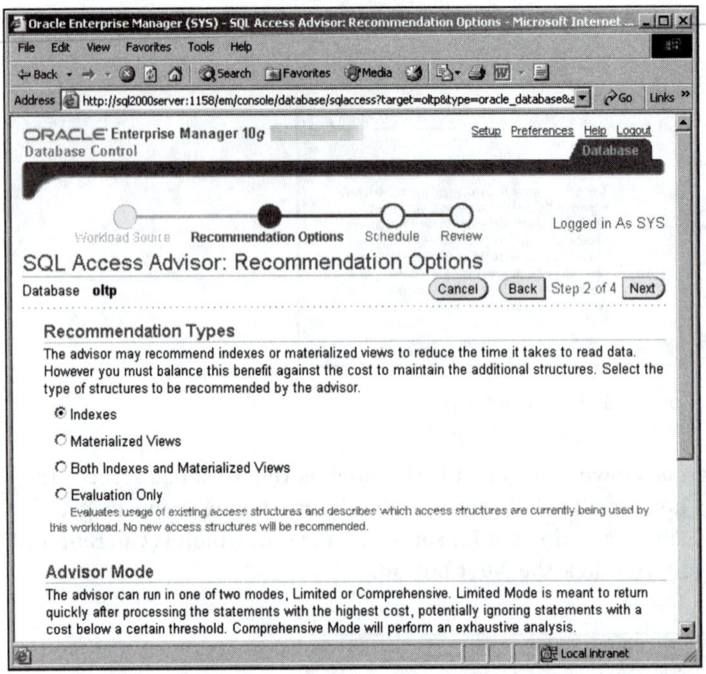

FIGURE 14-22 Recommendation options for SQL access advisor

7. The next page allows you to set a schedule. You can scroll down to the bottom of the window using the scrollbar for the window. The date and time should be set according to the time on your computer. Set the hour, minutes, and seconds fields to a time that is a minute or two from now (relative to the time on your computer and not yours—they might be different). If you cannot find the date and time on your computer then please ask your instructor to help you. Click the **Next** button.
8. The next screen you will see is a review of what the SQL access advisor will do when it begins to execute. You can read it all if you wish but there is a lot of complex detail there and your scheduling time is limited. You can always come back to the review page again to read it completely. Click the **Submit** button.
9. The next screen will return you to the advisor central screen, as in Figure 14-23, where the job created and scheduled is highlighted as having been **CREATED**.

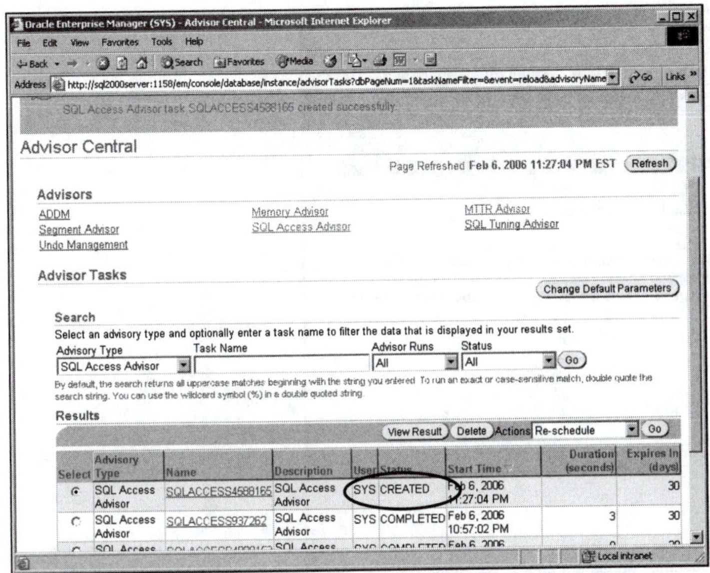

FIGURE 14-23 The SQL access advisor job has been created and scheduled

10. When the job has completed executing, as indicated by a status of **COMPLETED**, you can click on the link under the NAME column for useful recommendations. Again, this tool is beyond the scope of this book.
11. Remain logged on for the next practice.

The SQL Tuning Advisor

The SQL tuning advisor is similar to the SQL access advisor, but will analyze SQL code, perhaps making recommendations for improving performance. The following section provides a brief overview of the SQL access advisor:

1. You should still be in the Database Control. If not, then go back to the first exercise in this chapter, and navigate to the Advisor Central page of the Database Control.
2. Don't forget, when on the home page, to select the **View Data** spin control at the top-right corner of the screen, and select the **Manually** option.
3. Once in Advisor Central, click the **SQL Tuning Advisor** link.
4. Figure 14-24 shows the various options for the SQL tuning advisor tool. Click the **Top Activity** link.

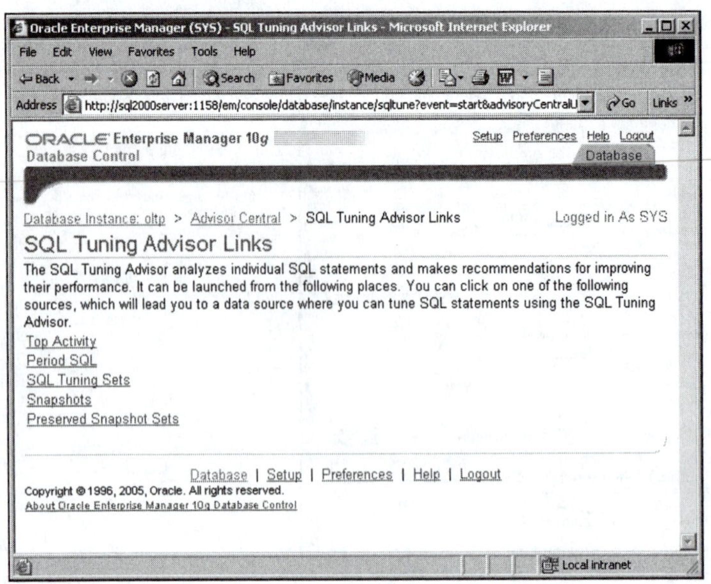

FIGURE 14-24 SQL tuning advisor links

5. The next screen is shown in Figure 14-25. This screen has a graph at the top, with various activities shown. The lower part of the screen shows different sessions that are potential problems. Your screen is unlikely to look anything like the picture shown in Figure 14-25.
6. The right side of the graph in Figure 14-25 shows a sharp increase in general activity because the simulation code was reinitiated to illustrate what a lot of activity looks like. Click the back button (the reverse or backwards pointing arrow) for your browser window, to return to the screen shown in Figure 14-24.

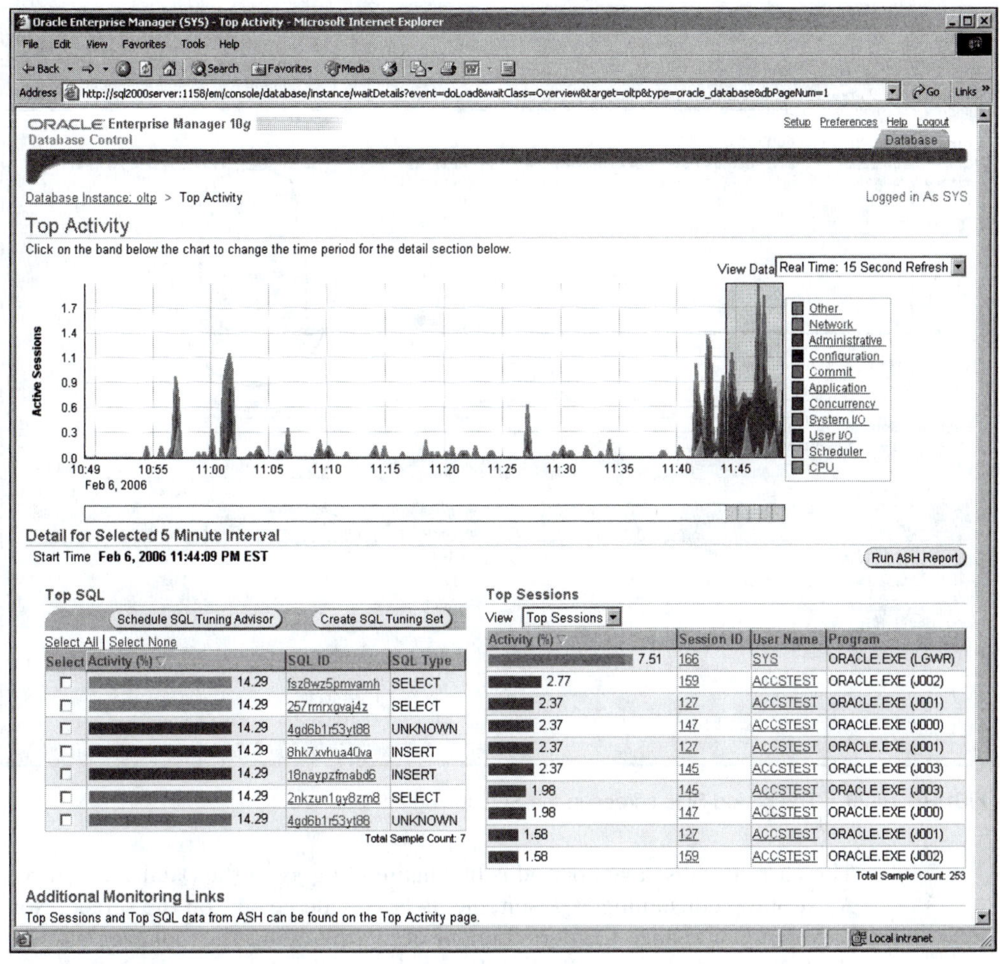

FIGURE 14-25 SQL tuning advisor top activities

7. Click the **Period SQL** link as shown in Figure 14-24. This tool allows you to look at potentially problematic SQL statements during a period. The Period SQL tool is shown in Figure 14-26.

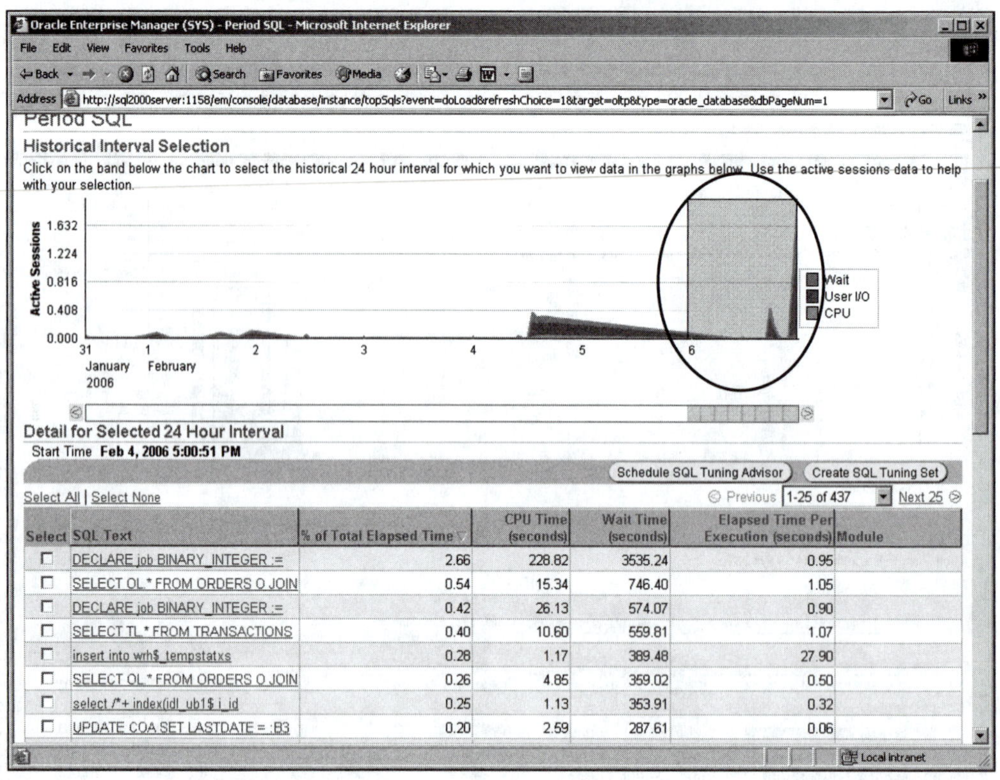

FIGURE 14-26 The period SQL screen in the SQL tuning advisor

The currently displayed period is highlighted in gray in the Database Control. A list of top consuming SQL statements is provided below on the screen. Not only can you change the period under observation, but the tool even alters the SQL code statement displayed according to the time period selected.

8. Click on the band (with the two little arrows at each end) underneath the graph. You will see the gray box shift (assuming you don't click under where the gray box is now). The SQL statements listed below on the screen are likely to change as well. Click the back button in your browser again.

9. Click on the **SQL Tuning Sets** link. A SQL tuning set is a set of SQL statements that are present in a database. Figure 14-27 shows the next page.

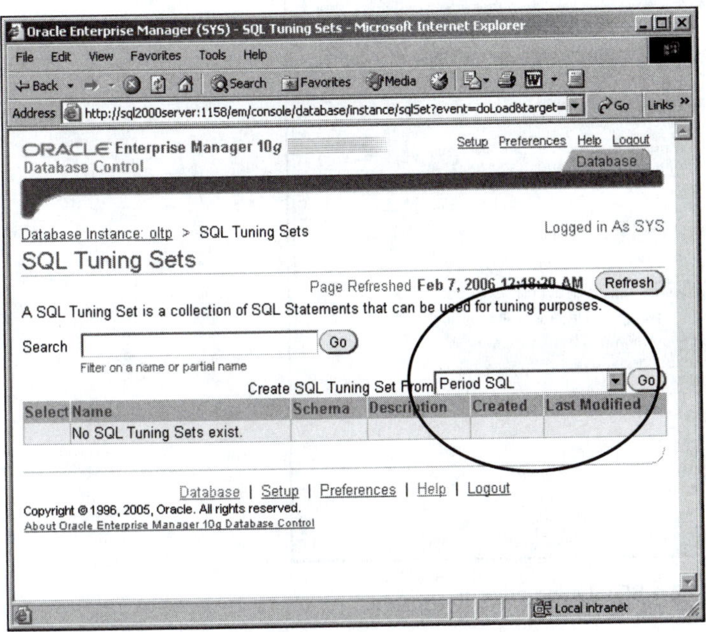

FIGURE 14-27 The SQL tuning sets screen in the SQL tuning advisor

10. Figure 14-27 highlights the spin control on the right. Click the spin control and you can see that you can create a SQL tuning set from existing data sets, either a period of time from a snapshot (which you saw in Figure 14-26), or a stored snapshot set. Set the spin control to **Period SQL** and click the **Go** button.
11. Scroll to the left of the screen and click the **Select All** link (this selects all SQL statements currently displayed on the screen). Go to the right of the screen, find the **Create SQL Tuning Set** button, and click it. You should get yet another screen. Click the **OK** button.
12. Now you should see the same screen as shown in Figure 14-27. However, this time you should see a tuning set listed, which should read **No SQL Tuning Sets exist**.
13. Now click the browser back button until you return to the SQL tuning advisor links screen, as shown in Figure 14-24.
14. Click the **Snapshots** link and you are taken to a screen, as shown in Figure 14-28, allowing you to create snapshots. A snapshot is a picture of a database at a specific point in time. In theory, multiple snapshots can be compared to allow for search and analysis of problems. The purpose is isolation of a root cause of a problem. Typically snapshots are created when the database is operating normally. When something goes wrong, a new snapshot can be taken. The two snapshots can then be compared to assist in finding out what is causing a database to perform poorly.

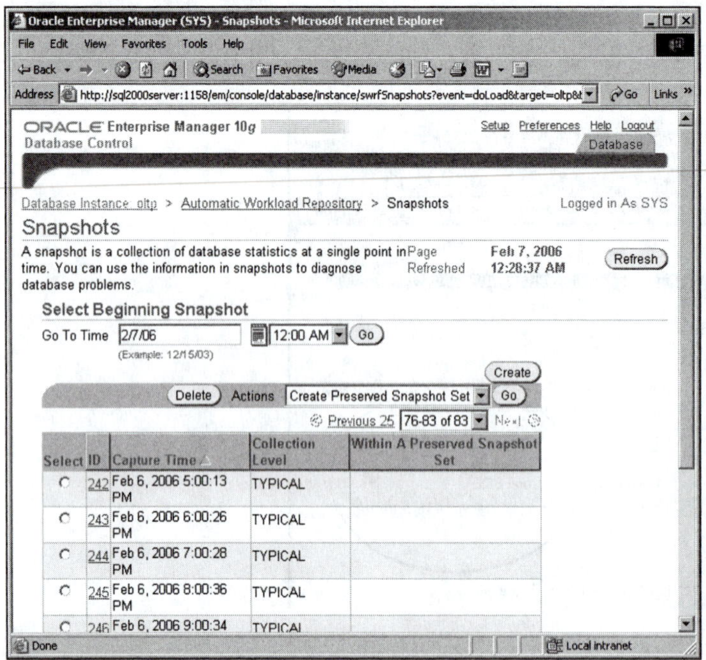

FIGURE 14-28 Snapshots in the SQL tuning advisor

15. The **Preserve Snapshots** link allows for creation and examination of preserved snapshots. In other words, you can permanently store older snapshots. Preserved snapshots are used in the same manner as snapshots, in that they can be used for comparison against other snapshots.
16. Remain logged on for the next practice.

That concludes the discussion of the Advisor Central screen. Next, we move on to baseline metrics.

BASELINE METRICS

In Chapter 13 you experimented with how to access and alter metrics. A metric is essentially a measure of a rate of change. A baseline metric is an established, expected, or even hoped for value. Anything not conforming to a baseline value indicates a problem or potential problem. Additionally, many of the Database Control metric links were examined during exercises in Chapter 13, as highlighted in Figure 14-29. Therefore, you already know how to find this information in the Database Control.

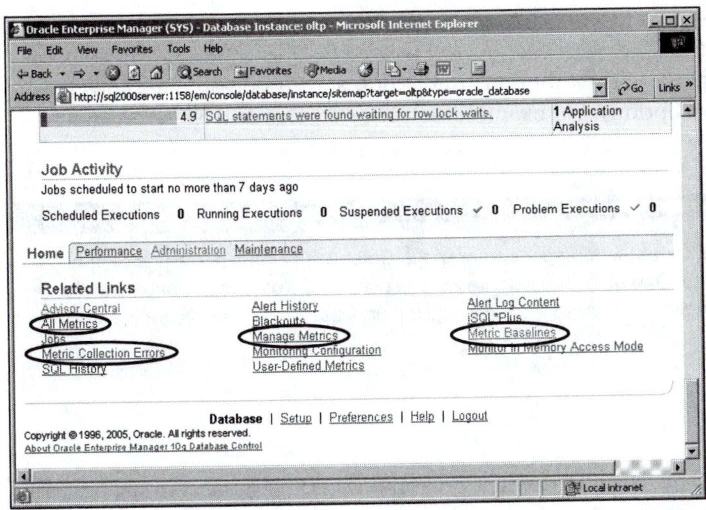

FIGURE 14-29 The different metric links on the Database Control home page

What are Baseline Metrics?

Metric baselines must first be enabled for a database if they are not already. To enable metric baselines, click the **Metric Baselines** link at the bottom of the Database Control home page or follow the prompts and selections through subsequent pages including clicking an **Enable Metric Baselines** button, and an **OK** button (this was covered in Chapter 13). Ultimately you get to the screen shown in Figure 14-30.

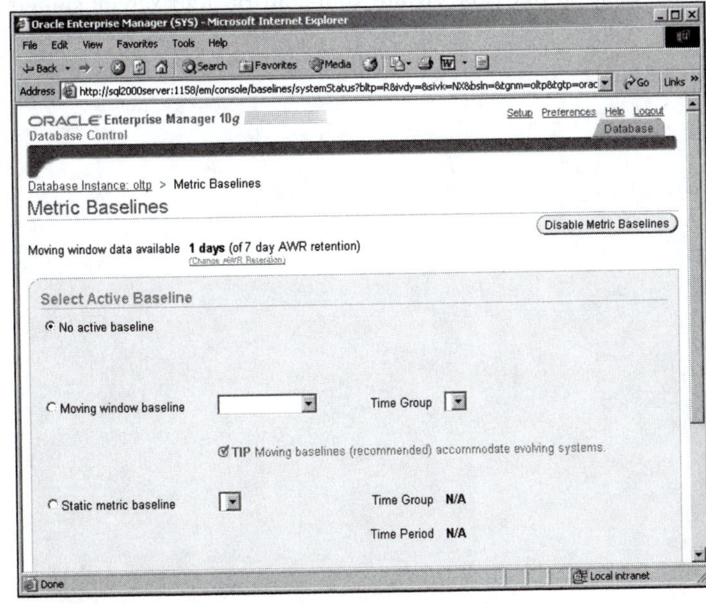

FIGURE 14-30 Managing metric baselines

A moving window baseline moves with the current date setting for your database server, from today backwards in time, for whatever period is specified. In a static window baseline you select a specific period. The different time group settings create baselines (regularly for the moving window option). For example, selecting **By Day and Night** creates two metric baseline groups: one during the day and one at night.

NOTE

You can always click the Help link at the top-right corner of the Database Control page, and read about the contents of the current page. One of the most informative uses for the Database Control, and Oracle Enterprise Manager in general, is the extensive content of its help files. Those help files are an excellent learning tool.

Let's set this up quickly:

1. You should still be in the Database Control. If not, then go back to the first exercise in this chapter, and navigate to the Advisor Central page of the Database Control.
2. Don't forget, when on the home page, to select the **View Data** spin control at the top-right corner of the screen, and select the **Manually** option.
3. Scroll the bottom of the home page in the Database Control. At the bottom-right corner of the page, click the **Metric Baselines** link. This should give you the window shown in Figure 14-30.
4. Now select the **Moving window baseline** option.
5. Click the spin control next to it and select **Trailing 7 Days**.
6. Give the browser some time to update. Select **By Day and Night, over Weekdays and Weekend** in the **Time Group** spin control. Your screen should now look similar to Figure 14-31.

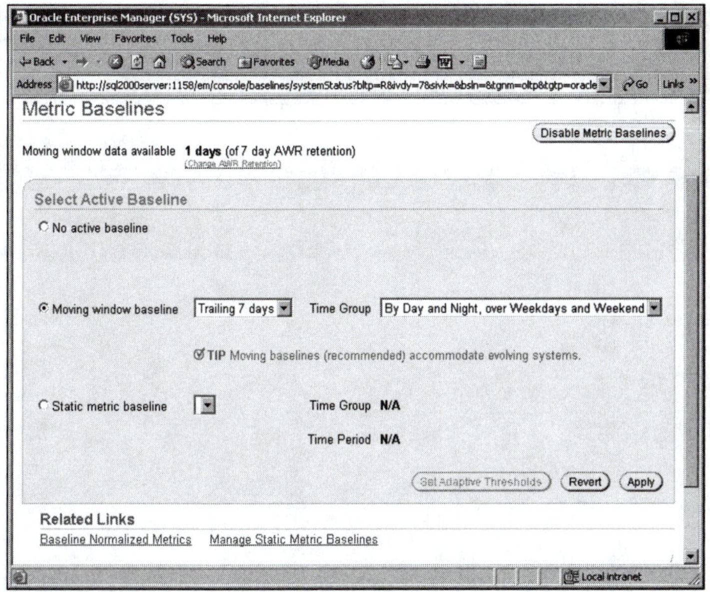

FIGURE 14-31 Setting up metric baselines

7. Click the **Apply** button, and select **Yes** to confirm on the subsequent screen. When you are returned to the screen shown in Figure 14-31, click the **Set Adaptive Thresholds** button at the bottom of the browser window.
8. This will display a list of baseline metrics. Click the plus signs next to both the **Workload Type Metrics** and **Workload Volume Metrics**, as shown in Figure 14-32. Then click the **Select All** link.
9. Click the **Edit** button at the top-right corner of the Database Control browser window.

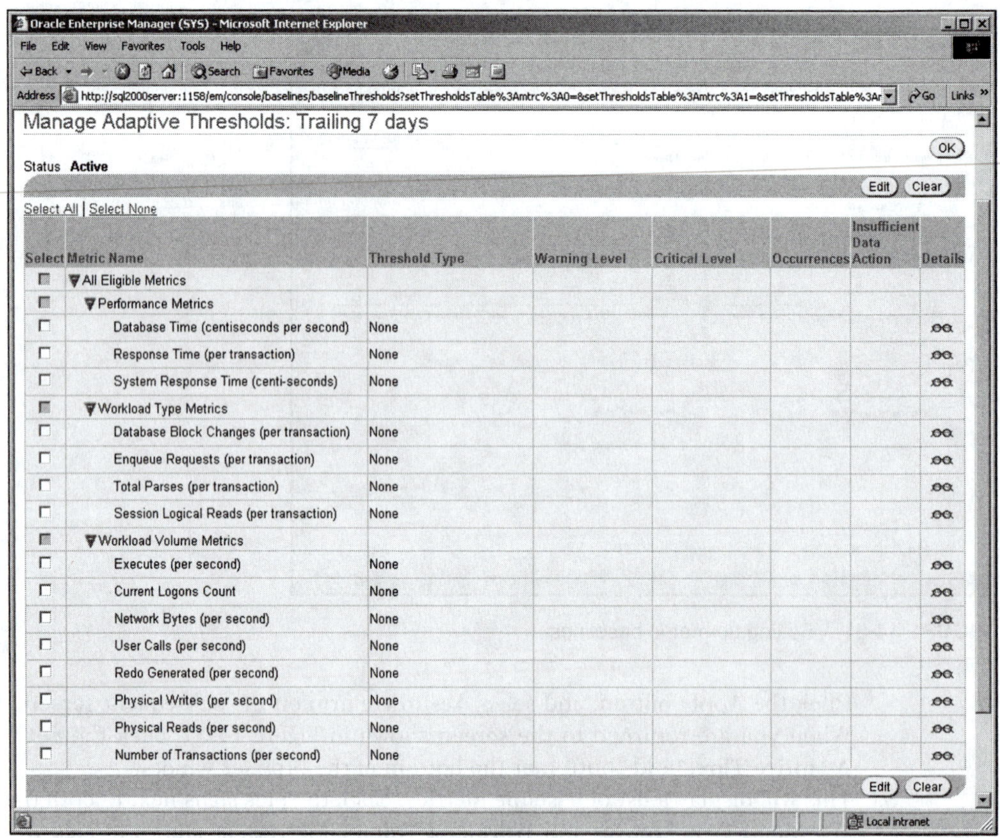

FIGURE 14-32 Editing metric baselines

10. Remain logged on, leaving the screen open for the next practice.

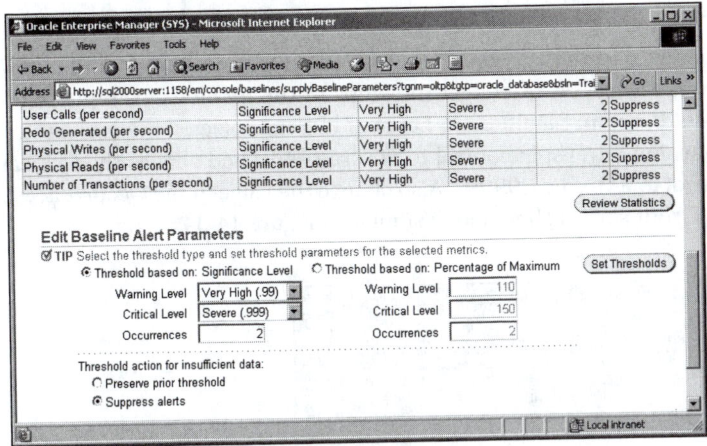

FIGURE 14-33 Editing baseline alert parameters

All the options shown in Figure 14-33 can be described as follows:

- **Threshold based on:**
  - **Significance level**. Alerts are created if anything is above the selected level:
    - **Warning level**: The alert is a warning
    - **Critical level**: The alert is a critical error
    - **Occurrences**: An alert can occur a specified number of times before it is created as a warning or error in the Database Control
  - **Percentage of maximum**: Alerts are created if anything is above the selected percentage:
    - **Warning level**: The alert is a warning
    - **Critical level**: The alert is a critical error
    - **Occurrences**: An alert can occur a specified number of times before it is created as a warning or error in the Database Control
- **Threshold action for insufficient data:**
  - **Preserve prior threshold**: Alerts can be created when there is insufficient data for an alert
  - **Suppress alerts**: Alerts are only created when there is an absolute reason for the alert

Now back to the exercise:

1. Resume the position in the Database Control. If you are no longer at the window as shown in Figure 14-33, repeat the steps in the previous example.
2. Leave **Threshold base on: Significance Level**. If it is not selected then select it. Change the **Warning Level** spin control to **Severe (.999)** and change the **Critical Level** spin control to **Extreme (.9999)**. Also change the value in the **Occurrences** entry field to 1. Your screen should look similar to Figure 14-34.

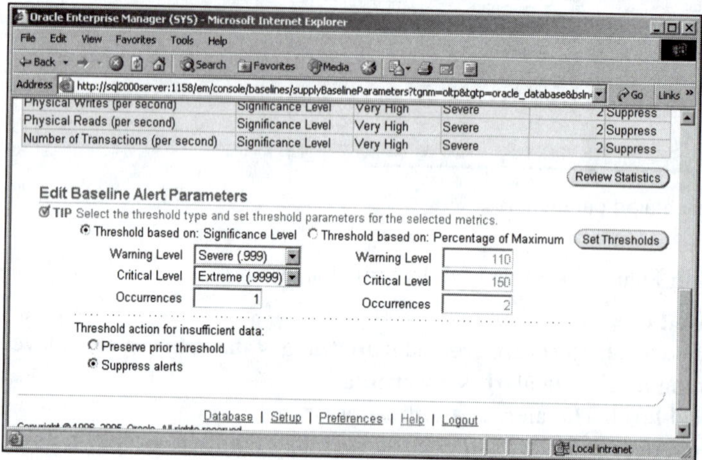

FIGURE 14-34 Editing baseline alert thresholds

3. Click the **Set Thresholds** button. You will return to a version of the screen similar to Figure 14-32. You can expand on the **Workload Type Metrics** and **Workload Volume Metrics** entries (the plus signs). As shown in Figure 14-35 you should see that all the **Warning Level**, **Critical Level**, and **Occurences** values have been changed, when compared with Figure 14-32.

4. Remain logged on for the next practice.

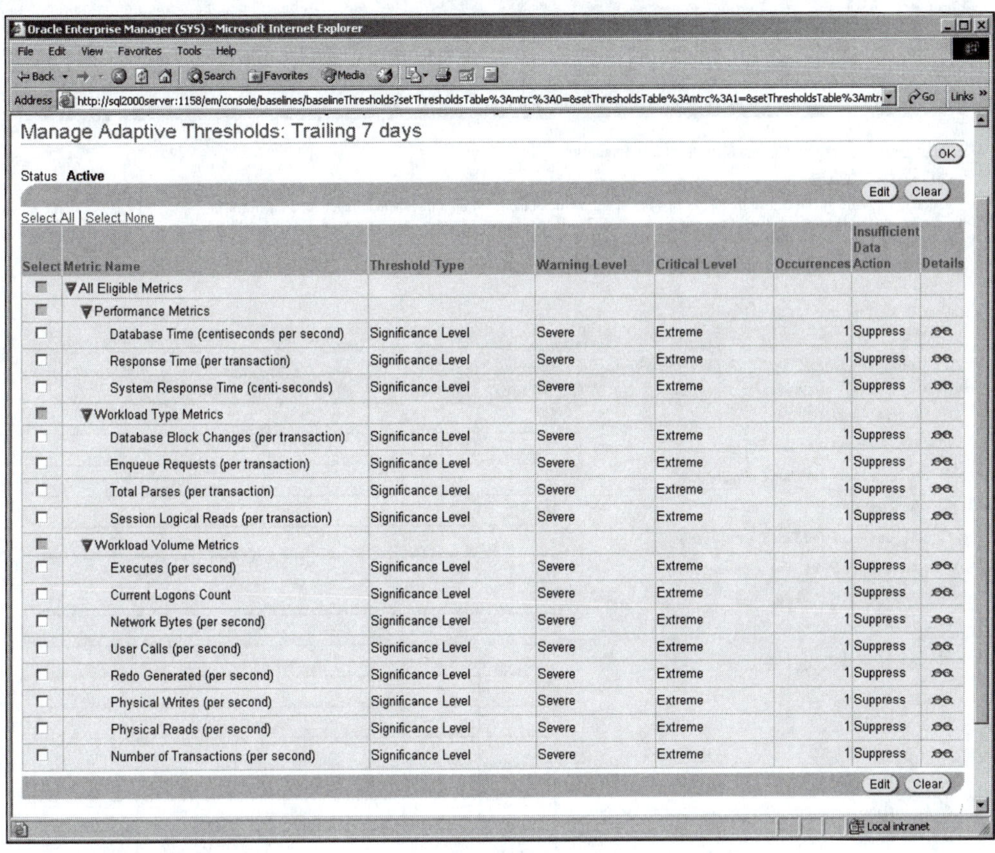

Manage Adaptive Thresholds: Trailing 7 days

Status **Active**

Select	Metric Name	Threshold Type	Warning Level	Critical Level	Occurrences	Insufficient Data Action	Details
	All Eligible Metrics						
	Performance Metrics						
	Database Time (centiseconds per second)	Significance Level	Severe	Extreme	1	Suppress	
	Response Time (per transaction)	Significance Level	Severe	Extreme	1	Suppress	
	System Response Time (centi-seconds)	Significance Level	Severe	Extreme	1	Suppress	
	Workload Type Metrics						
	Database Block Changes (per transaction)	Significance Level	Severe	Extreme	1	Suppress	
	Enqueue Requests (per transaction)	Significance Level	Severe	Extreme	1	Suppress	
	Total Parses (per transaction)	Significance Level	Severe	Extreme	1	Suppress	
	Session Logical Reads (per transaction)	Significance Level	Severe	Extreme	1	Suppress	
	Workload Volume Metrics						
	Executes (per second)	Significance Level	Severe	Extreme	1	Suppress	
	Current Logons Count	Significance Level	Severe	Extreme	1	Suppress	
	Network Bytes (per second)	Significance Level	Severe	Extreme	1	Suppress	
	User Calls (per second)	Significance Level	Severe	Extreme	1	Suppress	
	Redo Generated (per second)	Significance Level	Severe	Extreme	1	Suppress	
	Physical Writes (per second)	Significance Level	Severe	Extreme	1	Suppress	
	Physical Reads (per second)	Significance Level	Severe	Extreme	1	Suppress	
	Number of Transactions (per second)	Significance Level	Severe	Extreme	1	Suppress	

FIGURE 14-35 Viewing edits to baseline alert thresholds

Changing Baseline Metrics

Changing baseline metrics involves altering the threshold values. Changing those values is simple in the Database Control:

1. You should still be in the Database Control. If not, then go back to the first exercise in this chapter, and navigate to the home page of the Database Control.
2. Don't forget, when on the home page, to select the **View Data** spin control at the top-right corner of the screen, and select the **Manually** option.
3. Scroll the bottom of the home page in the Database Control and click the **Manage Metrics** link. The page you get was already shown in Figure 13-10. Click the **Edit Thresholds** button. The screen that appears is similar to Figure 14-36, which is the screen allowing an administrator to change the actual values that thresholds are compared against. Don't change anything!

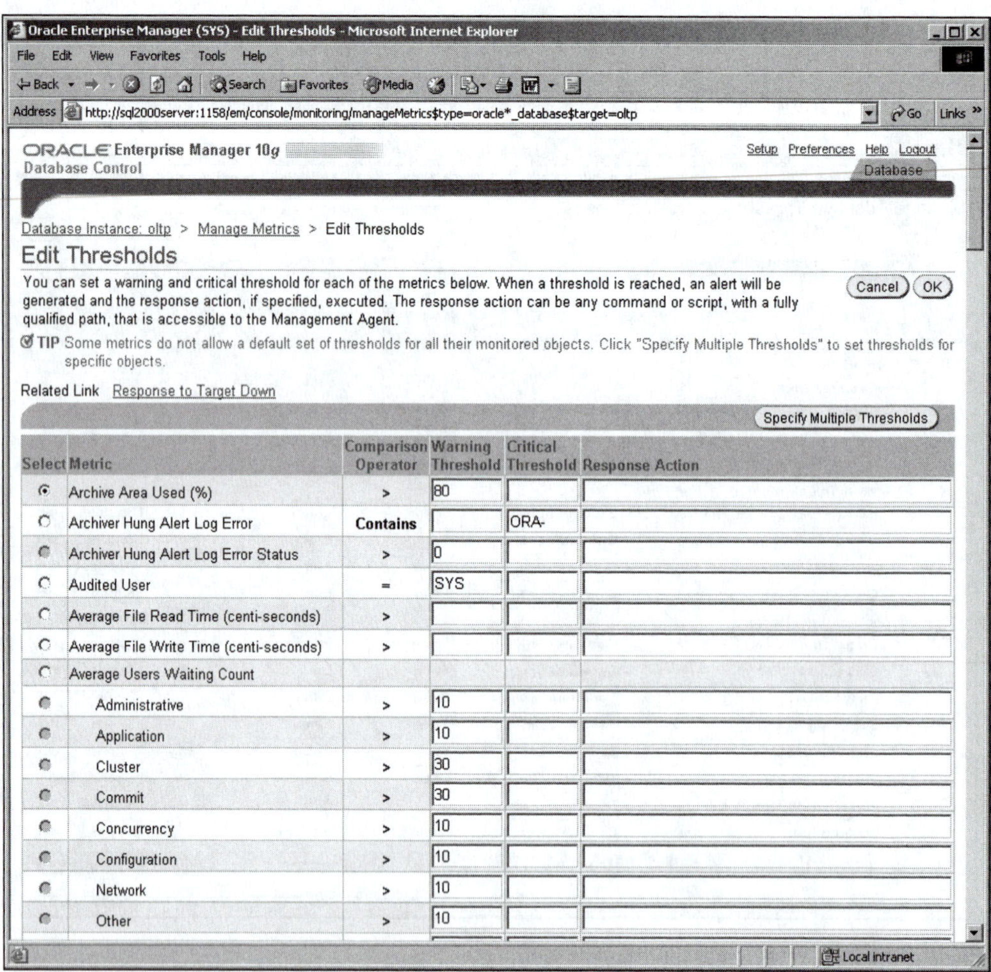

FIGURE 14-36 Changing threshold values

4. You can now close the Database Control.

Chapter 15 will examine the basics of backup and recovery.

Chapter Summary

- Proactive maintenance is all about trying to predict and track potential problems before they occur.
- The Database Control is an excellent graphical tool for managing and performing proactive maintenance.
- The segment advisor searches for, automatically detects, and posts warnings and alerts regarding segment issues. A segment is essentially the purely physical I/O and disk aspect of a database object, such as a table.
- The undo advisor advises as to optimal configuration of automated undo.
- The memory advisor warns and alerts with respect to potential instance recovery, potential media recovery, and potential flashback recovery.
- The SQL tuning advisor performs automated SQL tuning activities using the SQL Tuning Advisor, SQL tuning sets, and SQL profiles.
- The SQL access advisor allows for analysis of various database objects, such as indexes and materialized views, potentially making recommendations for improving performance.
- A baseline metric is an established, expected, or even hoped for value. Thus anything not conforming to a baseline value could possibly be an indicator of a potential problem.
- Baseline metrics establish an entire automated architecture, allowing for automated monitoring of a database in all facets and at all levels of detail.
- Baseline metrics can be managed using the Database Control.

Review Questions

1. Which of these are advisor tools?
   a. TopSQL
   b. TopSessions
   c. Memory Advisor
   d. TopConsumers
   e. All of the above
2. What is the name of the link to all the advisor tools?
   a. Advisor
   b. Advisor Central
   c. Advisor Control
   d. Advisor Control Panel
   e. None of the above
3. Which of these tools is used to examine the physical aspects of tables and indexes?
   a. SQL Access Advisor
   b. SQL Tuning Advisor
   c. MTTR Advisor

d. Memory Advisor

e. None of the above

4. Put these steps for the Segment Advisor into the correct order?

a. Objects

b. Scope

c. Review

d. Schedule

5. You can use the SQL Tuning Advisor to create SQL tuning sets. True or False?
6. The shared pool and database buffer cache are both part of the PGA. True or False?
7. All the *_AREA configuration parameters are catered to by the PGA_AGGREGATE_TARGET parameter. True of False?
8. Which of these buffers are catered to by the SGA_TARGET configuration parameter?

a. Shared pool

b. Large pool

c. Java pool

d. Database buffer cache

e. Streams pool

f. Log buffer

g. All of the above

Exam Review Questions—Oracle Database 10*g*: Administration (#1Z1-042)

1. What is the name of the configuration parameter used to automate SGA memory?
2. Which of these commands terminates a transaction?

a. INSERT

b. CREATE TABLE

c. COMMIT

d. ROLLBACK

e. Both a. and c. above

3. Which of these configuration parameters is included in the PGA?

a. SHARED_POOL_SIZE

b. DB_2K_CACHE_SIZE

c. DB_CACHE_SIZE

d. LOG_BUFFER

e. None of the above

4. What is the name of the configuration parameter used to automate PGA memory?

5. Which of these configuration parameters is included in SGA_TARGET?
   a. JAVAS_POOL_SIZE
   b. JAVA_POOL_SIZE
   c. SHARED_POOL_SIZE
   d. STREAMS_POOL_SIZE
   e. DB_KEEP_CACHE_SIZE
   f. None of the above

Hands-on Assignments

1. Start the Database Control. Find the warning threshold for the archive area used percentage and change it to 90%. Save your answer in a file named **ho1401.txt**, in the **Solutions\Chapter14** directory on your student disk.
2. Find the window in the Database Control listing metric collection errors? Save your answer in a file named **ho1402.txt**, in the **Solutions\Chapter14** directory on your student disk.

Case Project

1. Your MIS manager decides that the company's Oracle 10*g* database should be modernized. Make sure that the shared pool and PGA memory are both automatically managed by Oracle 10*g*. Also check to see that automated undo is configured, but don't change anything. Save your answer in a file named **case1401.txt**, in the **Solutions\Chapter14** directory on your student disk.

CHAPTER **15**

BACKUP AND RECOVERY

LEARNING OBJECTIVES

In this chapter, you will:

- Discover the difference between backup, restore, and recovery
- The difference between cold and hot backups
- Learn about different tools used for backup and recovery
- Learn about different types of failure that create a need to recover a database
- Learn about different backup strategies
- Learn about essential configuration as applied to making a database fully recoverable
- Learn about the specific backup and recovery scenarios

INTRODUCTION

The intention of this chapter is to describe the backup and recovery processes and tools involved in managing the recoverability for Oracle 10*g*. Most of the details of the Recovery Manager (RMAN) tool are not contained in the 1Z0-042 Oracle 10*g* certification exam, and thus details of RMAN are not covered in this book.

The primary focus of this chapter is to provide an overview of how backup and recovery function in an Oracle database, using datafiles, controlfiles, redo logs, and archive logs. Additionally, some other minor points are covered, including some use of the export utility and mention of other architectural database structures such as standby databases and replication.

Before beginning this chapter there are some points to note. First, there is not as much hands-on work in this chapter because that kind of activity can be brutal to a running database. You don't want to have to re-create your database as a result of a small mishap. Second, the first exam for Oracle 10*g* certification introduces backup and recovery from a conceptual point of view. Thus this chapter does not have to dig too deeply in the practical aspects of performing backup and recovery procedures.

INTRODUCTION TO BACKUP AND RECOVERY

What is backup and recovery? Backup is the process of making some kind of copies of parts of a database, or an entire database. Recovery is the process of rebuilding a database after some part of a database has been lost.

Another term commonly used by Oracle is restoration. Restoration is the process of copying files from a backup. Recovery is the process of executing procedures in Oracle Database to update the recovered backup files to an up to date state.

So Oracle backup and recovery is actually broken up into three separate actions—backup, restoration, and recovery.

What is Backup?

Backup copies files from an Oracle database and stores the resulting copies somewhere secure and inaccessible to the database. This process is shown in Figure 15-1 where datafiles, controlfiles, and archive logs can all be duplicated to a set of backup files.

The redo log files are typically covered for backups by simply duplexing redo log groups to have multiple members in each group. As shown in Figure 15-1, redo log files should be backed up for a cold backup, when the database is completely shut down to provide a snapshot of the entire database. Making a consistent copy of redo log files during a hot backup is pointless, as a consistent snapshot of inactive redo log files is not possible.

What is Restoration?

Restore copies files from a backup copy back into an Oracle database. Copying redo log files from a cold backup back into a database is only necessary when redo log files have been lost. In this situation, updating the datafiles past the time of the restored backup is not possible. As shown in Figure 15-2, all of datafiles, archive logs, and redo logs can be recovered from a backup, but preferably not all.

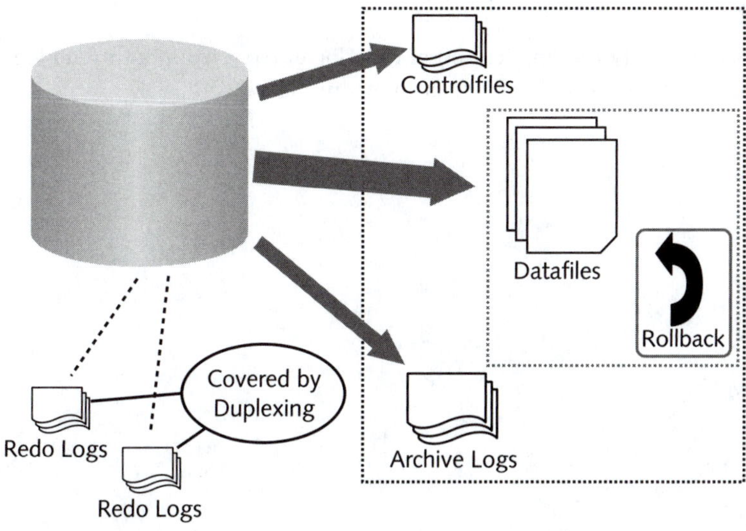

FIGURE 15-1 What is copied during an Oracle Database backup?

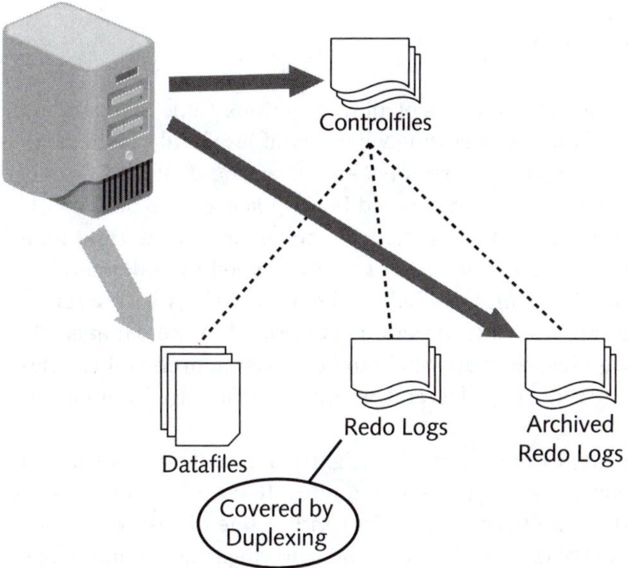

FIGURE 15-2 Which files can be recovered from a backup into an Oracle database?

What is Recovery?

The recovery process is essentially the application of redo log entries, from both redo log files and archive log files, back into the datafiles, as shown in Figure 15-3.

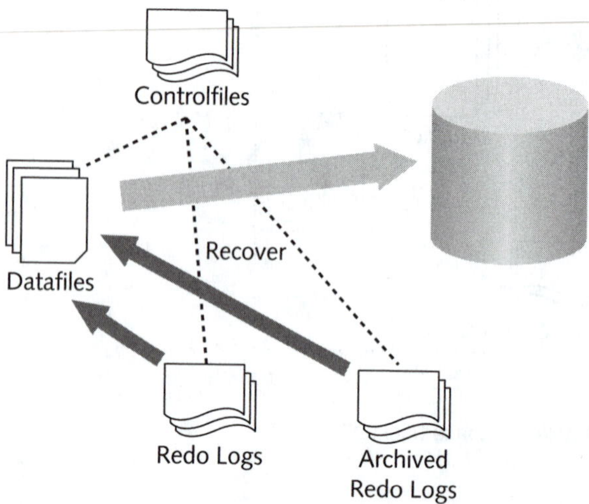

FIGURE 15-3 How the recovery process works

The redo logs and archive logs consist of records of all transactions made to a database. The controlfiles contain pointers to datafiles, dictating where datafiles should be in relation to redo log entries. If a datafile is restored from a backup, then the controlfile will be ahead of the datafile in time. Restoration of the recovered backup is a simple process of applying redo log entries to the datafile, until the datafile "catches up" to the time indicated by the controlfile. It's that simple! The only other thing you need to understand is that archive logs are copies of old redo log files, copied just before redo log files are reused. Thus archive log files contain older records of transactions performed in the database. If records are not available in redo log files, they are likely to be stored in archive logs. Restoration can utilize entries in both redo log files and archive log files to complete a recovery.

Obviously recovery of different types of files involve slightly different processes. The previous section described the recovery of one or more datafiles. Recovery of a controlfile does not actually involve any kind of recovery process, but rather one of restoration or rebuilding of the controlfile. Restoring of redo log files or archive log files means simply placing copies back into their respective subdirectories, within the database structure in the operating system.

NOTE

A point that a recovery is processed up to, is represented in the controlfile as a System Change Number (SCN). An SCN is a sequential number applied to every logged change to a database. Recovery can also be executed up to a point, but an automated recovery will use SCNs to determine what is updated, and up to when.

Methods of Backup and Recovery

The two basic methods of backup and recovery are cold backups and hot backups.

What is a Cold Backup?

A cold backup is the easiest method of backup and recovery. A cold backup or recovery requires a database to be completely shut down for the backup, and sometimes the recovery process as well.

NOTE

A cold backup takes a snapshot of the entire physical structure of a database (all the datafiles, controlfiles, redo logs, and archive logs). If the database is not shut down, and any changes are made during the copy process, an inconsistent backup will result. In other words, some files could have changes, and some could not because one file could have already been copied (before a change was made). Another file could be copied after a change. An inconsistent backup is not recoverable and is not a useful backup.

A cold backup is often made to make infrequent but extremely reliable copies of an Oracle database, most often when a database is periodically inactive and generally small in physical size. Parts of a cold backup can always be used to recover files into an active database, such as datafiles or archive log files, where the recovery process can update the datafiles with redo log entries (from both redo log files and archive log files), based on the latest SCN stored in the controlfile, for a restored datafile.

Figure 15-4 shows how a cold backup works, allowing all files to be copied, including datafiles, controlfiles, redo log files, and archive log files. A cold backup requires that a database be completely shut down. When a database is shut down, no activity can take place, and thus no changes can be made to the database. This allows for a consistent snapshot of the database to be copied for which all database files are copied at a state at a specific time.

NOTE

There are other methods of performing consistent backups, such as using export utilities, but by definition these are not cold backups because the database is not shut down during the backup process.

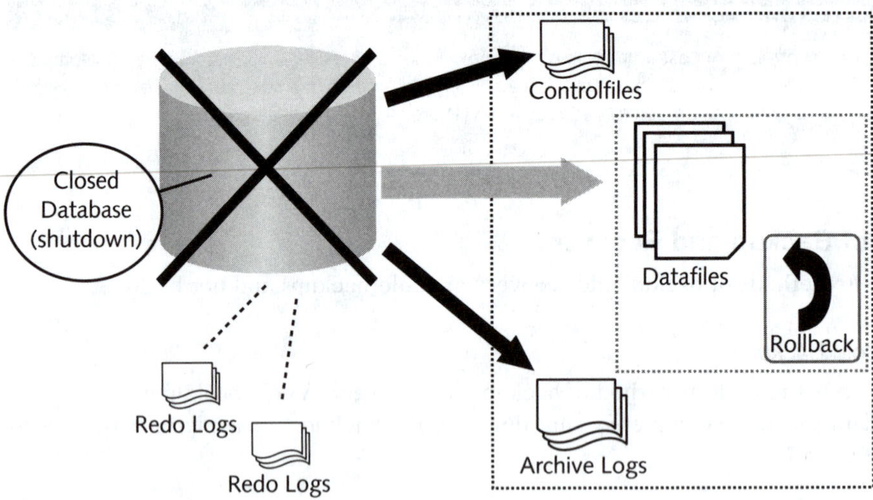

FIGURE 15-4 How does a cold backup function

What is a Hot Backup?

A hot backup is a backup performed when the database is online, active, and available for use. There are many different tools for performing hot backups, and many different methods of performing hot backups. Essentially, a hot backup takes a snapshot of a database, or preferably a part of a database, one file or type of file at a time. The result is that a hot backup is not necessarily a consistent backup across all the files in the backup (if the entire database is backed up in this manner). Essentially, a hot backup is made of pieces of a database (different files), where those files making up a complete database backup are not recoverable to a working database as a group. The individual files can be slotted into (restored into) a running database, and can be recovered individually, or as a group, depending on what types of files they are (datafiles, log files, controlfiles). You will learn about recovery of these different file types as you read through this chapter.

Tools for Backup and Recovery

Tools used for backup and recovery of an Oracle database are as follows:

- **Export and Import Utilities**: These utilities are superceded in Oracle 10g by the export and import data pump utilities.
- **Backup Mode Tablespace Copies**: Individual physical tablespace copies using tablespace backup mode.
- **RMAN (Recovery Manager)**: This tool allows both consistent and even incremental copies of parts of, or an entire database. RMAN is a complete backup and recovery management tool.
- **Oracle Enterprise Manager and the Database Control**: These tools include access to backup and recovery tools, generally executing one, some, or all of the three methods mentioned above.

These tools will all be discussed and demonstrated in this chapter.

Types of Failure

There are different types of failure that can occur in an Oracle database, ranging from the loss of a single file, through to a complete loss of an entire database server. It is important to understand what the various types of failure are so that you can be better prepared.

Media Failure

Media failure is storage device failure, such as when a disk fails, and denies your database access to one, some, or all of the files in your database. Media failure is fortunately rare because of many modern striping and mirroring utilities using specialized hardware such as RAID arrays. What can be done to ensure that media failure can be recovered from quickly? Always multiplex controlfiles and duplex redo logs. Use a RAID array for underlying disk storage, or some type of hardware and/or software architecture that allows for some type of mirroring of underlying file structures.

NOTE

A mirror is the automated maintenance of one or more copies of physical files.

User and Application Failure

User and application failure is much more likely than any other failure situation, such as dropping a table that should not be dropped. Also, applications can be improperly coded, causing errors to occur at the database level. What can be done to ensure that user and application failure will cause minimal disruption? User and application errors are usually object-centric, such as the dropping of a table. Sometimes those tables can be individually restored, particularly if a table contains static data. Export utility dump files can sometimes be used to recover from a situation like this. However, RMAN is capable of recovering individual objects using log entries, so the export utility is somewhat outdated.

Oracle Database-Induced Failure

Oracle Database-induced failure can sometimes be the result of a bug or an overload applied to the database itself. However, Oracle Database can also fail because of an administrator-induced problem, such as repetitive use of the SHUTDOWN ABORT command, or pulling the power plug out of the wall. An abort shutdown of a database, or a power failure, kills all processes and clears all memory buffers immediately, regardless of what is executing in the database. If a transaction is partially complete, that transaction may be aborted in its current state. Restarting the database can usually recover from a SHUTDOWN ABORT or a power failure, but not always.

What can be done to avoid failure at this level? (1) Use an uninterrupted power supply so that a clean shutdown can be performed in the event of a power failure; (2) don't pull the plug; and (3) never execute a SHUTDOWN ABORT, kill a process on a Unix or Linux box, stop and start the Windows Oracle Database service, or reboot your database server computer unless you have to. Always use SHUTDOWN IMMEDIATE rather than SHUTDOWN ABORT. The difference in speed between an aborted and a clean shutdown is minimal.

The processing that is performed by a clean shutdown to do things like terminate transactions and complete any writing of buffers to disk may be partially or even wholly performed automatically on database restart. Using SHUTDOWN ABORT rather than SHUTDOWN IMMEDIATE only gives an impression that a database restart is faster. Also, a SHUTDOWN ABORT can potentially lose information completely, causing a serious database failure, which could require recovery from a backup.

Backup Strategy

A backup strategy is required to plan what you might need. What types of backups should you use? Which tools should you use to backup, and what tools will you use in the event of failure? How often should you back up your database? Establish a plan for what you might need before implementing a backup plan. In other words, establish what strategy is needed to allow for a better selection of options when implementing backups. Selecting a backup strategy is dependent on factors such as the type of database, how much data can be lost, and available equipment.

Type of Database

An OLTP database can be large and active. Performing regular cold backups in this situation is generally unacceptable as it requires a complete database shutdown. OLTP databases are often required to be available on a 24-hour basis. Additionally, OLTP databases tend to change rapidly, in small chunks, in many different parts of the database, or all at once. Incremental backups using RMAN are often useful in this type of environment because they only copy what has changed since the previous backup. The same rule applies to data warehouses because the amount of regular updating is small in relation to the physical size of the entire data warehouse. Additionally, large parts of data warehouses are often static, and sometimes even read-only, and thus only need a single backup.

Database Availability (24x7x365)

As already mentioned, a database must be available globally without interruption. Hot backups are essential in this case. Additionally, hot backups, especially in the case of using RMAN, can allow for recovery of failure due to partial errors such as a single disk failure in a collection of disks, or the loss of a single table. There is no need to restore and recover an entire database of a few terabytes when only a single offending 1 Mb table is causing a problem. The essential requirement of this type of database is availability. Typically, this approach is more likely to apply to high-concurrency OLTP data, rather than to data warehouse data. Typically OLTP databases are customer facing, whereas a data warehouse contains historical information. Historical data is usually required internally within a company to track trends, such as what sold where, what sold the most often, and what made the most profit for the company.

Data Change Speed

How fast is data changing? Both an OLTP database and a data warehouse can change rapidly. However, where a data warehouse has new data added to the end of it (appended) at regular intervals, an OLTP database has small amounts of data changed in all parts of the database, around the clock. Thus, in terms of recoverability, a data warehouse could simply have batch processing re-executed. An OLTP database, on the other hand, has to recall all transactions from log files and essentially re-execute them on recovery. Additionally, during recovery some parts, or even all of a database, may be unavailable. Devise a backup strategy based on how long it will take to recover the database to an acceptable point, and perform the recovery as fast as possible. For example, if your database contains a very large table that is partitioned, back up each partition individually, and not the table as a single backup file. That way you could simply recover a single physical partition, which Oracle 10g recovery commands can recover, rather than set yourself up to recover an entire table, which could be hundreds of megabytes in size.

Acceptable Loss Upon Failure

Acceptable loss is how much data the company can afford to lose and still maintain usability and availability of data. Obviously, the less acceptable loss allowed, the more complex and longer backups will take to execute, and as a result the more time needed should restoration and recovery ever be required. In general, an OLTP database requires zero loss. A data warehouse can often be rebuilt by re-executing batch processing. Additionally, factors in the realm of acceptable loss include the amount of storage capacity available for backups, the type of media used for backups (disk is faster than tape storage), and even minor factors such as network bandwidth between backup media and active database storage.

Available Equipment

Are you allowed disk or tape backups? Can you use both? Backup to disk is much faster than backup to tape. However, if you need to retain backups for a number of years, using tape backups are much easier. Typically, many database installations will use a combination of both disk and tape backup storage. Recent backups will be stored on disk, allowing for rapid and specific recovery scenarios. After a period of time disk backups could be transferred to a sequential media such as tape, where recovery would be naturally slower and more cumbersome, but less expensive and easier to manage.

Planning for Potential Disaster and Recovery

Always plan for a potential disaster! It might happen to you, so be prepared. There is nothing worse than having a loss of a database, and then discovering that the people who administered the database before you did not back up the database properly, or with the correct methods and tools, or even at all. If you begin a new job, backup and recovery should be at the top of your list of priorities. Automate the backup process if possible.

Use scripting and scheduling to perform backups periodically and automatically. RMAN allows full automation of a backup strategy and even allows for embedded scripting, executed from within RMAN, stored in the RMAN catalog, and even executing backup processing in parallel or on a specific node in a clustered environment. Test the existing backup implementation if possible and always test anything you construct as new, preferably off the production server environment. Make sure it all works seamlessly. If you ever have to recover a production database, your customers, and of course your boss, will want the database server to be back online and usable as quickly as possible.

NOTE

All of the topic areas presented in this section are not included in the exam. However, there are many things that Oracle 10*g* Database can do that a database administrator should know.

Other Approaches to Backup and Recovery

Other approaches include **standby databases**, which can be simple or complex. A standby database is a backup database, which can be used to replace a primary customer facing database virtually in seconds, and even automatically, in the event of a primary database failure. Figure 15-5 shows a diagram of how a basic **physical standby database** functions. A physical standby database passes newly created archive log files to a standby database server computer. Those archive logs are then applied on the standby database computer to the standby database. The standby database is retained as an unchangeable but viewable read-only database that can be used for reporting and other read-only activities. However, its purpose is as a backup database that can be recovered within as little as a few seconds (even automatically), to the time of a primary database failure (by recovering any usable redo log files on the primary database).

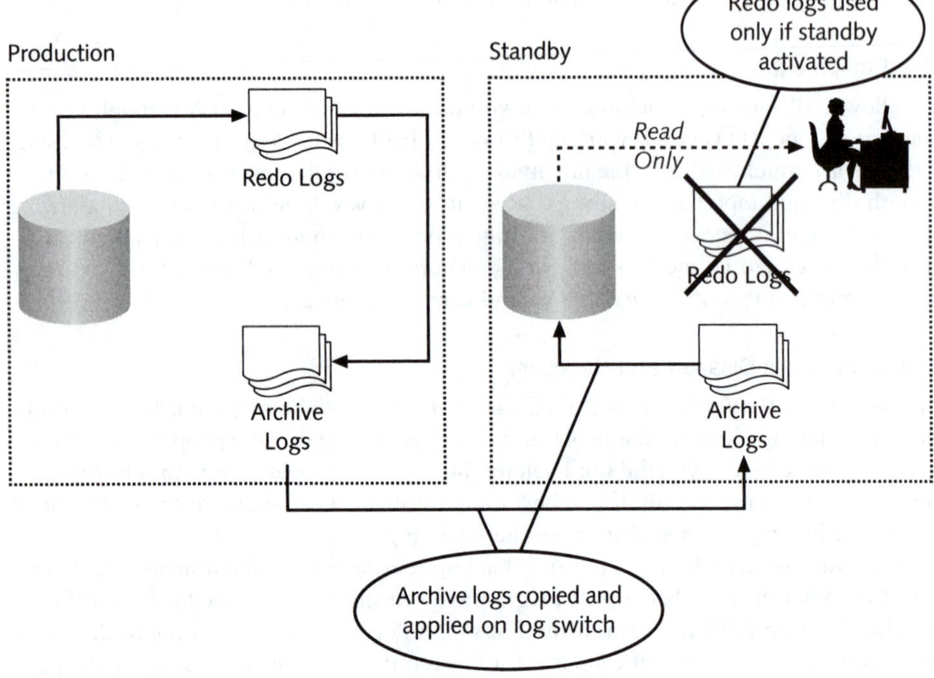

FIGURE 15-5 Physical standby architecture as a potential backup and recovery option

A **logical standby database**, as shown in Figure 15-6, can potentially pass log entries to a standby database server, log entry by log entry. In other words, each transaction is applied to the standby immediately after application to the primary database. Logical standby databases can be maintained as read-only databases or even read-write databases. In the case of a read-write logical standby database, existing structures from the primary database cannot be disturbed, but other additional structures, such as tables and materialized views, can be added and regularly updated.

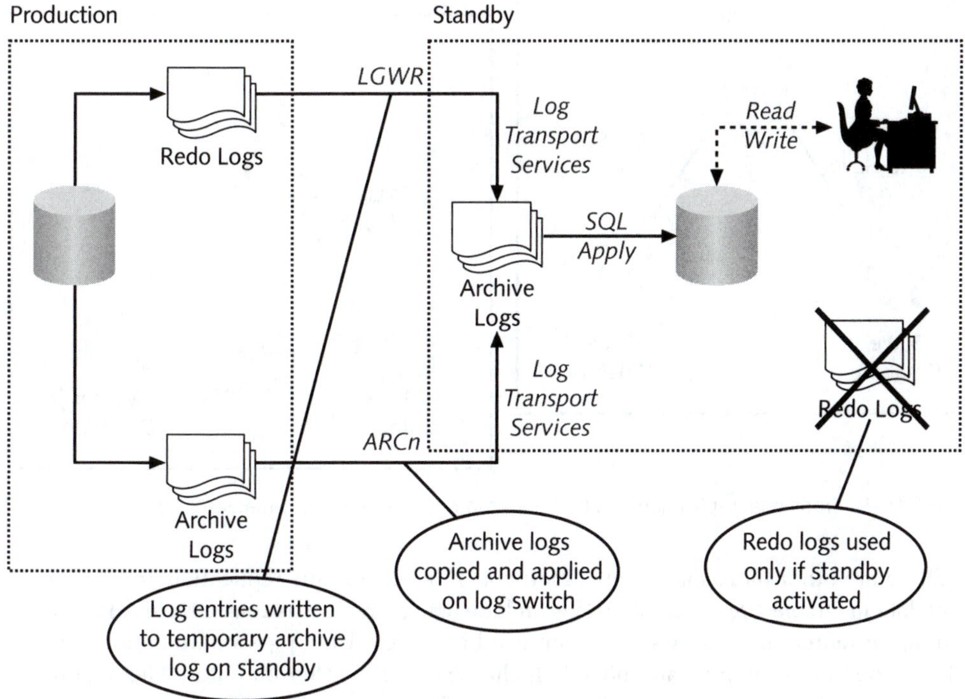

FIGURE 15-6 Logical standby architecture as a potential backup and recovery option

Replication is another option, but it is often misused where a standby database is sufficient or better than using replication. The primary function of replication is controlled distribution of data over a WAN.

NOTE

A WAN can be within a single building, across an entire city, entire country, or even global.

As shown at the top-left corner of Figure 15-7, a traditional replicated database accepts changes from a master database, effectively acting as a slave database to the master database. Changes are applied at the master database and then replicated out to one or more slave databases. The slave databases become mirrored copies of a single master database. On the right side of Figure 15-7 is a master-to-master database architecture. In a master-to-master replicated database architecture, any changes made to any one of the databases are replicated to all other databases in the replication architectural structure. It is also possible to mix replicated database architecture with more than one master database and more than one slave database.

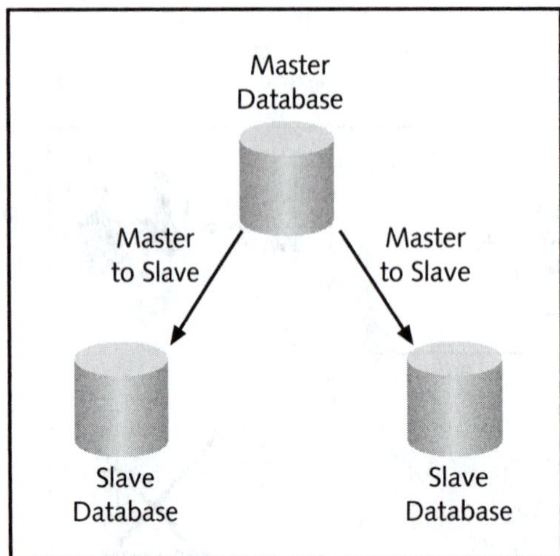

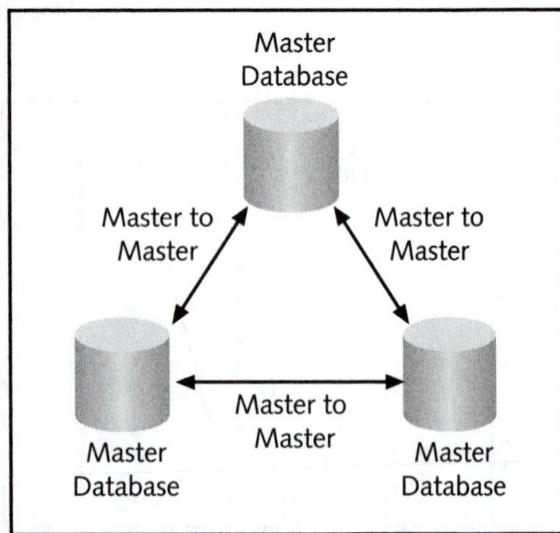

FIGURE 15-7 Traditional replication can be master-to-slave or master-to-master

There are other more manageable methods of implementing replication and distribution of data in Oracle 10g. These methods include using **Oracle Streams**, **Oracle Advanced Queuing**, or **materialized views**. A stream is a little like a Unix pipe where data can simply be pushed onto the pipe and pulled off the pipe at the other end of it. The queuing aspect implies that changes can be pushed and pulled at either end using a queue, which effectively applies a sequential order to the events onto the stream. Figure 15-8 shows a simplistic picture of using streams and queuing as a method of implementing replication.

As shown in Figure 15-8, either side of the stream (or pipe connection between two database servers) is a capture stream queue and an apply queue. The capture queue accepts changes from a master database. The apply queue applies changes to a slave database. The stream can also pass information in either direction, effectively allowing easy implementation of either master-to-slave replication or master-to-master replication.

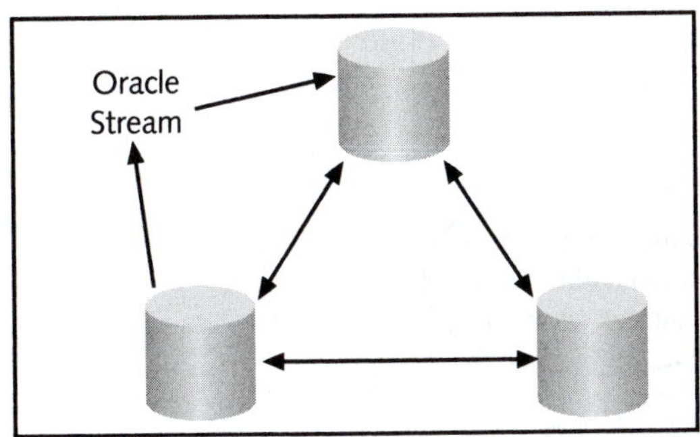

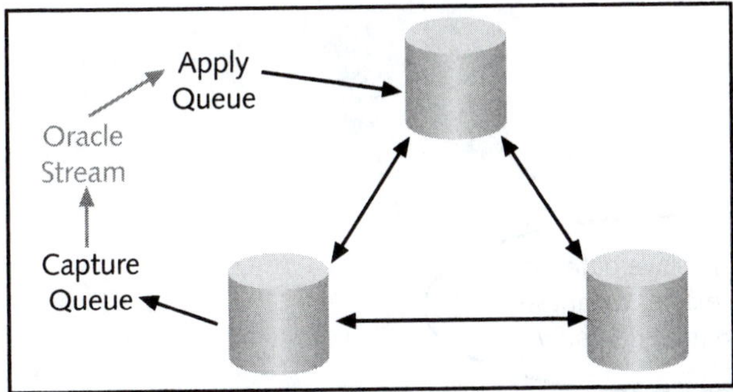

FIGURE 15-8 Implementing replication using streams and queuing

As shown in Figure 15-9, materialized views are another option for replication of data from database server to another, where the materialized views are placed on a slave database and updated regularly through database links.

NOTE

A database link is a direct, Oracle 10*g* managed network connection between two database servers.

Additional benefits of using a materialized views for replication include: (1) materialized views can be updated regularly and automatically, (2) the slave database can be used for reporting because the materialized views can be read on the slave database, and (3) other tables can be added to the slave database as long as the materialized views containing primary database information are not disturbed.

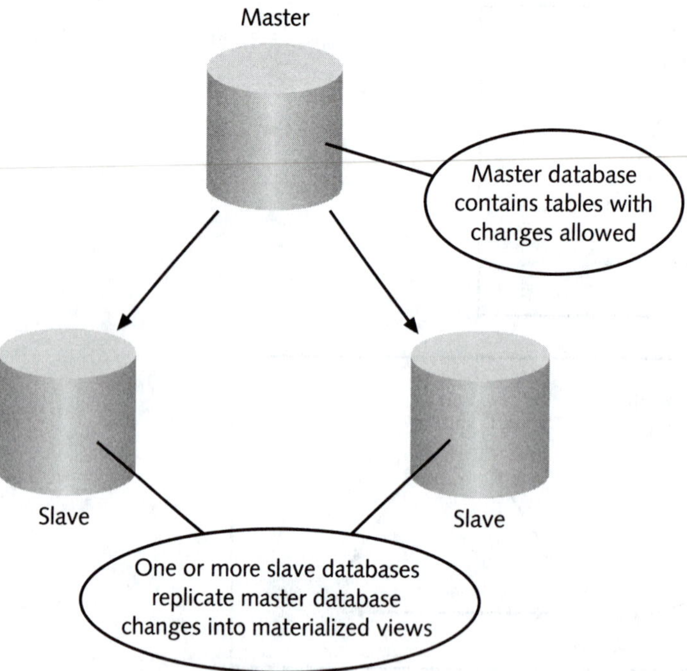
Master
Master database contains tables with changes allowed
Slave
Slave
One or more slave databases replicate master database changes into materialized views

FIGURE 15-9 Implementing master-to-slave replication using materialized views and a database link

Clustered environments consisting of multiple nodes can also be useful if one node fails, in which case another node assumes the node of both the failed node and of itself. Then the failed node in the cluster can be recovered easily and offline, with no disruption to service. In Oracle 10g a clustered environment consists of multiple instances (computers containing an Oracle installation) connected to a single, shared data repository (set of disks), as shown in Figure 15-10.

A **computer grid** differs from a cluster in that there is no common shared storage area. A grid simply links all the hardware resources of many computers together. The hardware resources of each computer include disk storage, on-board memory, and processors. The Oracle 10g database software is effectively operative on all of the computers in a grid such that the multitude of computers appears and behaves as if all computers were one computer. A computer grid is a little like the collective consciousness of the computing world.

Use of **transportable tablespaces** is another method of backup and recovery, as shown in Figure 15-11.

A transportable tablespace is a physical copy of a tablespace (containing data), plus an export utility dump of metadata (from the SYSTEM tablespace). Transportable tablespaces can be plugged in and out of any Oracle 10g database, regardless of the platform and operating system of the target database.

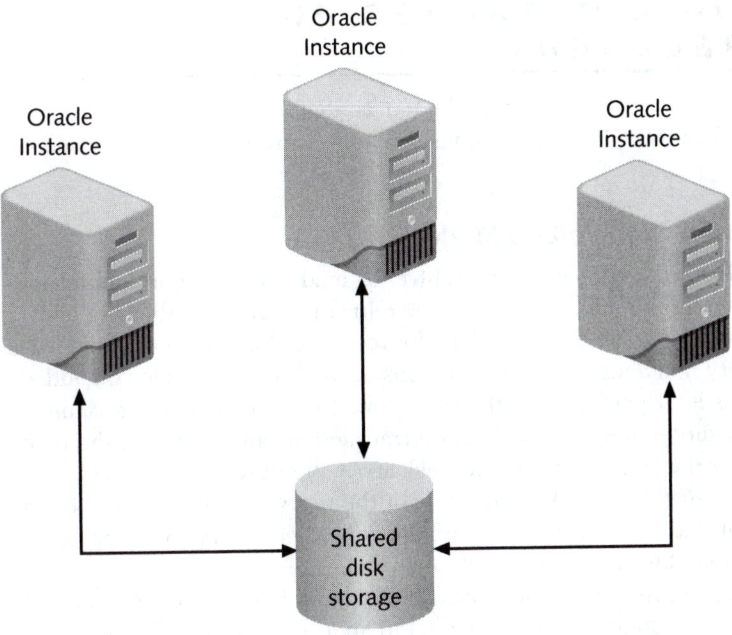

FIGURE 15-10 Implementing a cluster of database instances to a single shared storage area

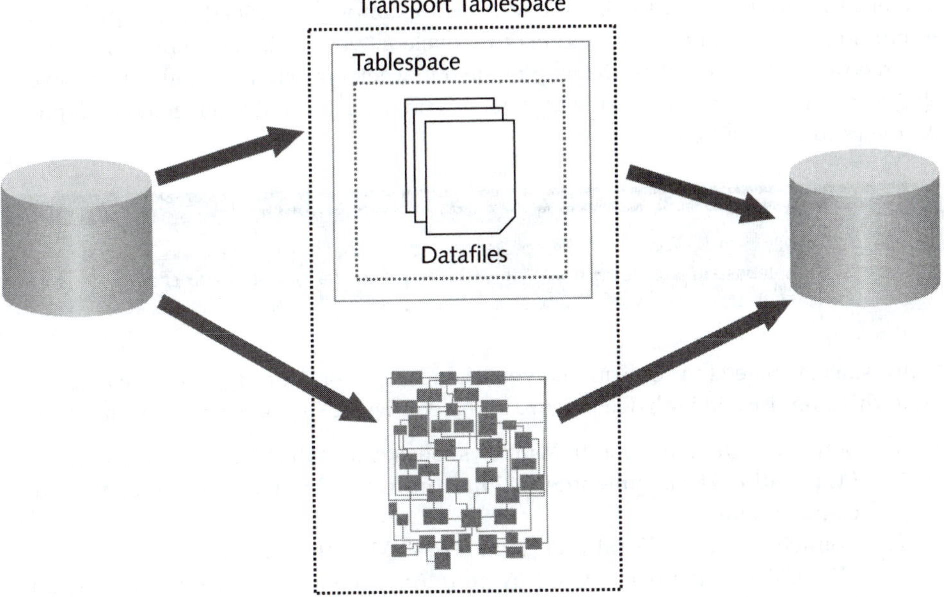

FIGURE 15-11 Transportable tablespaces are another possibility for backup and recovery

CONFIGURING A DATABASE FOR POSSIBLE RECOVERY

There are various things you can do with Oracle 10g configuration to ensure proper functioning of backups. The most important thing is making sure that your database is archived.

Setting the Database in Archive Log Mode

Why is it so important that your database is in archive log mode? Because, in archive log mode, the database will create archive logs for you. Archive logs are files that are copied from redo logs when a redo log file is switched out for recycling. Redo logs contain entries of all transactional activity in a database as transactions occur. Two weeks ago you added a row to a table. Two days later you updated the same row. Today you delete that same row. All of these three, addition, update and deletion transactions are written to the redo logs. If you have to recover the database from one week ago, without redo log entries your database will not have the row deleted. The objective of the recovery process, by applying part redo log entries, will find the row deletion and apply it when the recovery process reaches another week forward from the one-week-old backup.

Additionally, redo logs are recycled. There are a limited number of redo log groups. As those groups are filled, each redo log file is duplicated to an archive log file. The redo log group is then made available for new log entries. Thus the redo logs are recycled (overwritten). The result is that if a redo log is recycled, unless the redo log is copied to an archive log (history of log entries), all entries in that redo log group are effectively lost. Think about the two-day, one-week, and two-week examples described in the previous paragraph. If there are enough transactions to recycle all your redo log groups five times a day, any recovery from a database backup made one week ago will not be able to recover anything if archive logs are not created. A database must be in archive log mode to duplicate redo logs to archive logs.

NOTE

A database is generally regarded as being unrecoverable when it is not in archive log mode. Make sure you always create a database in archive log mode, unless perhaps you have serious disk space problems.

At this stage, you need to confirm that your database is configured for archive log mode. If it is not, this exercise will both deconfigure and reconfigure, so you can see how it's done:

1. Start the SQL*Plus tool. In Windows, click **Start/All Programs/Oracle ... /Application Development/SQL*Plus**. In Unix or Linux, type **sqlplusw** on the command line.
2. Connect as the SYSTEM user. Type **SYSTEM** in the Username box, *the SYSTEM password provided by your instructor* in the Password box, and **ORACLASS** in the Host String box. Click **OK** to continue.

3. Type the following line and hit the **Enter** key. The CONNECT command reconnects you as the SYS user (not possible from the initial logon screen). This command makes things easier to read and is covered in Chapter 14:

```
CONNECT SYS/<password>@ORACLASS AS SYSDBA
SET WRAP OFF LINES 132
```

   Don't forget to use the appropriate SYS user password in the CONNECT command above.

4. Is your database in archive log mode? Type the following command:

```
ARCHIVE LOG LIST;
```

   If your database is in archive log mode you will get a response like this:

```
SQL> ARCHIVE LOG LIST
Database log mode              Archive Mode
Automatic archival             Enabled
Archive destination            USE_DB_RECOVERY_FILE_DEST
Oldest online log sequence     24
Next log sequence to archive   26
Current log sequence           26
SQL>
```

   If your database is not in archive log mode you will get a response like this:

```
SQL> archive log list
Database log mode              No Archive Mode
Automatic archival             Disabled
Archive destination            USE_DB_RECOVERY_FILE_DEST
Oldest online log sequence     24
Current log sequence           26
SQL>
```

   If the response is that your database is not in archive log mode (the second result above), skip to step 7.

5. Assuming your database is in archive log mode, you will now switch your database out of archive log mode. Type in the following command and hit **Enter**:

```
ARCHIVE LOG STOP;
```

   The above command will prevent automated archiving. It will not take the database out of archive log mode, but will cease the production of any archive log copies.

6. To take your database out of archive log mode, execute the following commands:

```
SHUTDOWN IMMEDIATE;
STARTUP MOUNT;
ALTER DATABASE NOARCHIVELOG;
ALTER DATABASE OPEN;
```

   If you run the ARCHIVE LOG LIST command again you will see that your database is no longer in archive log mode.

NOTE

If at any stage during this process you do not know the open status of your database, you can execute the query SELECT OPEN_MODE, NAME FROM V$DATABASE; If the database is mounted or open (read write), you will get a response.

NOTE

If your database refuses to restart after shutting down it may be because of an issue with Oracle 10*g*. The automated listener configuration does not always function properly. The only solution is to manually create the listener.ora and tnsnames.ora configuration by using the Net Manager tool, or by manually editing the files. This is not required for this exam. If you have this problem, please consult your instructor. You will not be able to start the database in anything but open mode, if you restart your computer, or you restart the service on a Windows machine.

7. At this stage, your database is not in archive log mode. Let's switch your database to archive log mode. First, you need to restart your database in mounted mode by executing these commands in SQL*Plus:

```
SHUTDOWN IMMEDIATE;
STARTUP MOUNT;
```


NOTE

You may get a listener error. If your database does not restart it is because of automated listener startup in Oracle 10*g*. Ask your instructor for help.

8. Executing the ARCHIVE LOG LIST; command should list your database as not in archive log mode. To switch archive log mode on for your database, execute these two commands:

```
ALTER DATABASE ARCHIVELOG;
ALTER DATABASE OPEN;
```

9. Executing the ARCHIVE LOG LIST; command again should list your database as now in archive log mode. Automatic archiving should also now be enabled because the ALTER DATABASE ARCHIVELOG; command will inherently execute the ARCHIVE LOG START; command.
10. Leave the SQL*Plus window on the screen to use later in this chapter.

You should now understand the importance of maintaining a database in archive log mode, and how to switch a database in and out of archive log mode. You can also manage archive log mode by altering configuration parameter settings from within the Database Control.

Checkpoints, Redo Logs, Archive Logs, and Fast Starts

You already know what redo logs and archive logs are and how they are related to each other. Redo logs contain entries of transactions executed against a database. Archive logs contain copies of redo logs that have been recycled. The combination of archive logs and redo logs allows you to recover your database from a point in time, as long as you have retained the archive log files in your database structure.

What is a checkpoint? A checkpoint is a point in time where all buffers are flushed to disk. Those buffers include the redo log buffer and all database buffer caches, where all database changes are held before being written out to permanent storage on disk. If a database fails, any changes remaining in the database buffer cache that are not yet written to disk could be lost. The shorter the time between successive checkpoints, the fewer transactional changes to a database you could lose. This is why log entries are written out before database buffer cache entries: because redo log entries are used to recover in the event of failure, not database buffer cache entries. Similarly, the redo log buffer is always smaller than the database buffer cache because it should be written to disk more frequently. In a default database creation the log buffer is generally in kilobytes, and the database buffer cache is in megabytes. Sizing the redo log buffer to more than 1 Mb can be risky if recovery is required because the log buffer will not be written to permanent storage in redo log files as frequently as it should be. So this brings us back to the question of what is a checkpoint? Or moreover, what does a checkpoint do? A checkpoint writes all pending redo log buffer and database buffer cache changes to disk, writing the redo log buffer first, followed by the database buffer cache. By default a checkpoint is executed automatically when the log buffer is one-third full of pending changes, every three seconds, when a log switch occurs, or when COMMIT (explicit or implicit) or ROLLBACK commands are executed.

NOTE

A log switch is when one redo log file is filled. The current redo log group will be switched out as the active write log, the next redo log group in the redo logs becomes the active redo log group, and the switched out redo log file is copied to a new archive log file (if your database is in archive log mode).

You can alter checkpoint configuration for a database by altering the LOG_CHECKPOINT_INTERVAL and LOG_CHECKPOINT_TIMEOUT parameters. Let's examine management of checkpoint parameters in the Database Control interface.

Sacrificing Recoverability for Performance

Sacrificing recoverability for performance means that you can limit the number of checkpoints that occur, thereby potentially speeding up database performance. However, because checkpoints are not executed as frequently, your database becomes less recoverable because you could lose what has not been written to disk from buffers at the time of failure. This simply involves tweaking checkpoint parameter settings either directly in the parameter file or from within the Database Control.

FLASH RECOVERY AND BACKUPS

Oracle 10g includes a flash recovery capability, functioning in a similar fashion to flashback functionality, which has already been discussed in this book. Essentially, flashback recovery is an added option to that of standard log file recovery, allowing retention of potential flashback data for a specified time. Enabling flashback recovery does two things: (1) it simplifies backup and recovery management, and (2) it can potentially speed up performance in the event of a recovery, where the amount of information to be recovered falls within the flashback recovery retention period.

The difference between regular physical recovery and flashback recovery is a physical versus a logical one, respectively. Flashback recovery automatically passes through redo logs, archive logs, and even undo space, allowing recovery on specific parts of a database to be restored, or even locked during a period to logically restore a single table. Conversely, physical recovery requires restoration and recovery of physical objects such as entire datafiles. The difference between the methods is quite distinct.

Oracle flashback technology has been covered in various parts of this book already and includes capabilities such as the following:

- **Flashback Queries**: Execute a query on data at a specific time in the past, regardless of any changes to rows after the time specified.
- **Flashback Version Queries**: Multiple versions of the same rows at many times.
- **Flashback Transaction Queries**: Examine changes as an audit trail, made by one or more transactions in a record, or history of changes. For example, inserts, updates, and deletes of specific rows.
- **Flashback Database**: Even an entire database can be recovered using flashback capabilities.

Flashback technology relies generally on a combination of undo data (retained for a period of time depending on undo retention settings), and the recycle bin. When the retention period is exceeded, log files are used (both redo logs and archive logs). Additionally, flashback technology is potentially much faster in recovery times than traditional file restoration, combined with log entry records recovery. Flashback configuration can be maintained with various configuration parameters but it is more easily maintainable using the Database Control, highlighted in Figure 15-12, under the **Maintenance** screen.

The MTTR (Mean Time To Recovery) Advisor

The MTTR advisor in the Database Control allows you to monitor flash recovery. Let's begin with the Database Control:

1. To start the Database Control, open a browser such as Internet Explorer.
2. Expand your browser to full screen. The Database Control is BIG!
3. At the top of your browser window you should see an address line. Type a URL into that address line that goes to your database server and executes the Database Control. It should look something like this:

   ```
   http://<hostname>:1158/em
   ```

   The `<hostname>` must be substituted with the hostname of your database server computer. Ask your instructor for this. The port number may also be different depending on the release of Oracle 10g you are using, and if there is one or more databases created on the database server computer. Other ports used are 5500 and 5600. The port number depends on which release of

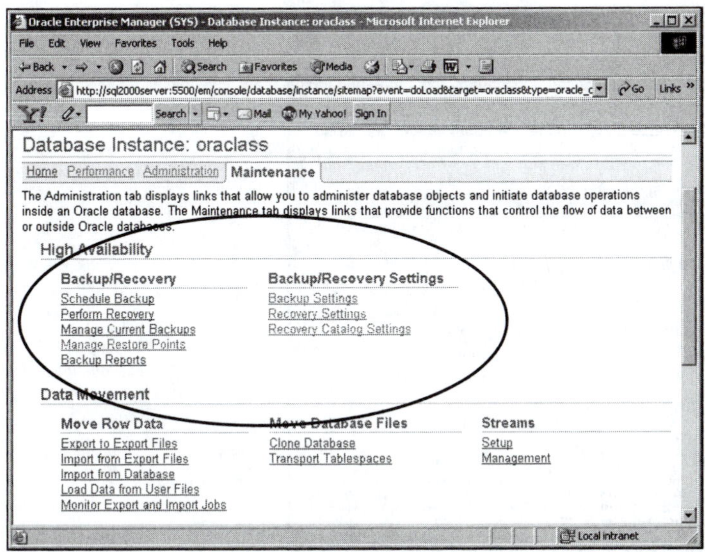

FIGURE 15-12 Managing flashback, backup, and recovery in the Database Control

Oracle 10*g* you have and how many Oracle databases exist on your database server computer. If in doubt ask your instructor.

Additionally, to execute the Database Control there needs to be a background process (a service on a Windows computer) running on the database server that executes the Database Control over the network. This service should be installed and started by the Oracle installation process on the database server. If it is not there, ask your instructor.

4. Once again, a number of different screens may appear, depending on your release of Oracle 10*g*. The important screen is the login screen. Connect as the SYS. Type **SYS** in the Username box, type the appropriate password (*ask your instructor*) into the Password box, and select **SYSDBA** from the **Connect As** box. Click **Login** to continue.
5. The screen that appears will look something like the one shown in Figure 15-13, where you will be at the Home portion for the Database Control of your database.
6. Before you do anything else, select the **View Data** spin control at the top-right corner of the screen, and select the **Manually** option.

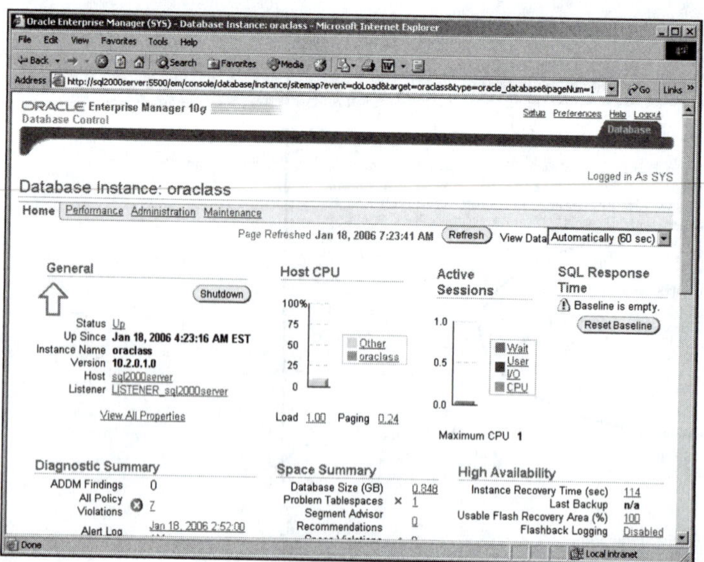

FIGURE 15-13 The Database Control home page

7. Now scroll to the bottom of the browser screen. You will see a link at the bottom-left corner of the Database Control window called **Advisor Central**, as shown in Figure 15-14. Click the **Advisor Central** link.

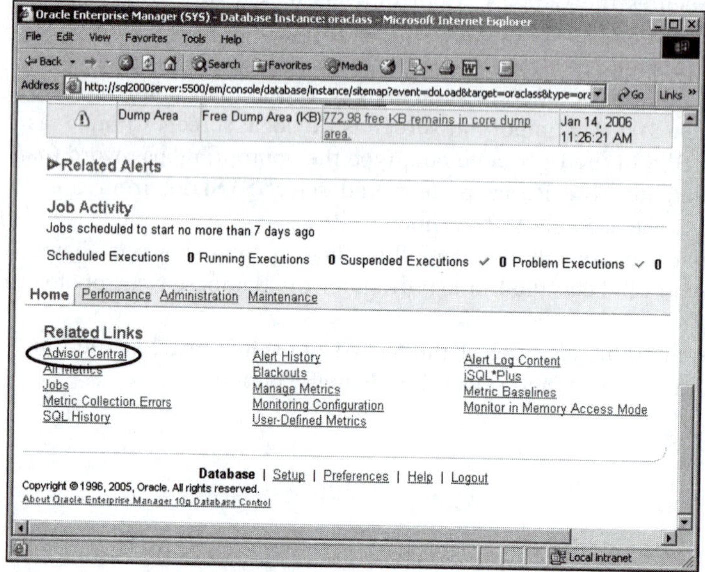

FIGURE 15-14 The Database Control and Advisor Central link

8. The **Advisor Central** screen should display, as shown in Figure 15-15.

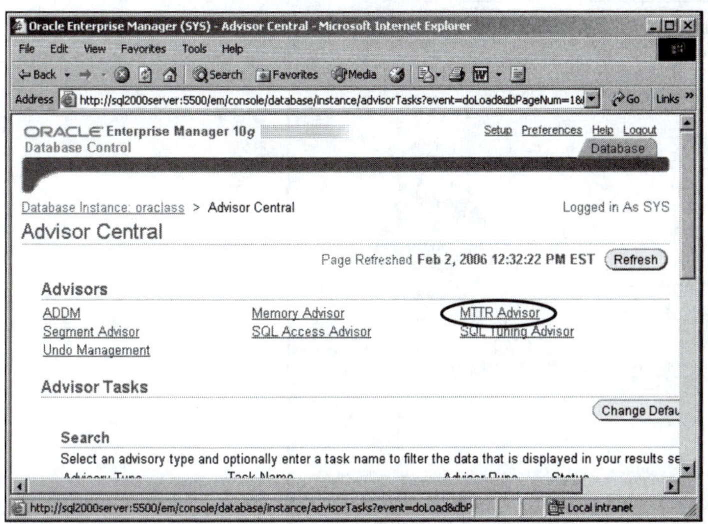

FIGURE 15-15 The Advisor Central screen in the Database Control

9. Once in the Advisor Central screen, click the **MTTR Advisor** link at the top-right corner of the page. The lower part of the window will look something like that shown in Figure 15-16, allowing monitoring of flash recovery, along with various configuration parameter settings.
10. You can shut down the Database Control.

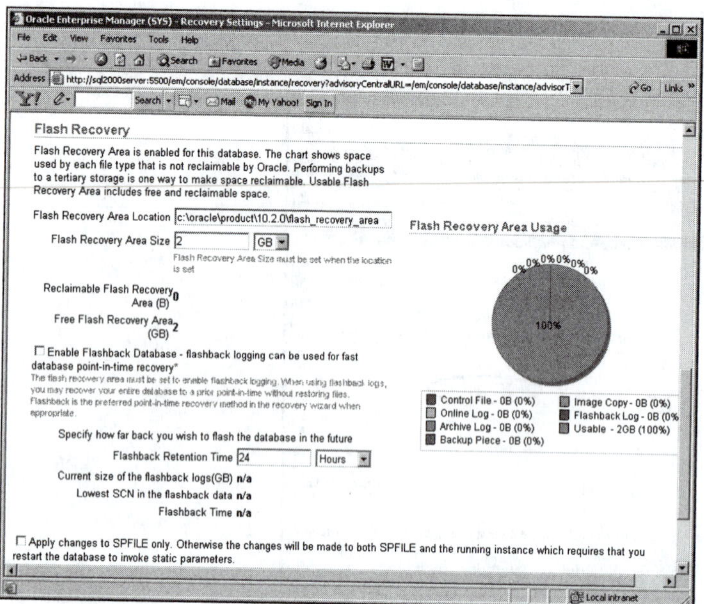

FIGURE 15-16 The MTTR Advisor in the Database Control

DATABASE BACKUP

Now let's experiment with executing some backups, using different methods.

NOTE

RMAN will not be covered as a backup tool in this book on an experimental basis because RMAN is in the second Oracle 10*g* database certification exam, and is thus not relevant to this book.

Cold Backups

The one thing you don't really need to have demonstrated is a cold backup. All you have to do for a cold backup is shut down the database completely and then copy all the files. The database is shut down and therefore all database files are in a static or frozen state, and will provide a consistent snapshot of a database. The files to back up are as follows:

- All datafiles
- All redo log files
- All archive log files
- All controlfiles
- Optionally you can also back up parameter files including the configuration parameter files (both text and binary versions), and any networking configuration files.

If you ever need to restore from a cold backup, you would need to restore at least all the datafiles and controlfiles. You could optionally use more current redo log files, archive log files, and controlfiles—allowing a recovery by applying redo log entries to datafiles.

Hot Backups

As you already know, a hot backup is a backup made with the database online, and generally available for use. The objective of a hot backup is to obtain a snapshot (consistent point in time) of all data in the database. Or you can create a backup that is reconstructable (recoverable) by applying log entries to datafiles, based on SCN matches between datafiles, controlfiles, and redo log entries. Different methods include export plus import utilities (or Data Pump technology), RMAN, or even traditional tablespace backups.

Consistent Backups Using Exports

The export utility can be used to create a consistent backup. Regardless of database activity, the export utility can scan through all types of data, including tables, redo and archive logs, and undo data. This allows for a consistent backup (export) based on a snapshot of data. In other words, regardless of any changes occurring to the database, during an export, the export utility will read datafiles and undo space to get a consistent backup. If a change is made to an object, then a previous incarnation of that object will be read from undo space.

NOTE

A consistent backup is the same as a snapshot. The only difference is that the export utility performs a logical backup, not a physical backup. A logical backup creates both DDL commands and DML commands, to re-create a database object (it uses the database SQL engine). A physical backup copies datafiles in the operating system. Therefore, exports cannot be used to recover any lost physical files, such as a controlfile or a datafile.

To run a simple export you need to open a shell:

1. In Windows, click **Start/Run**, type **CMD** into the **Open** entry, and click **OK**.
2. In the shell type **exp help=y**. Read all the information on the screen.
3. DO NOT type the following command. The command below could create a huge export file and you could run out of disk space. (I did!) The export command below will export the entire CLASSMATE schema. The export is very large because there are some large binary objects in the database:

```
exp system/<password>@oraclass file=classmate.
dat owner=classmate consistent=Y
```

NOTE

The consistent=Y setting ensures that a snapshot is obtained.

4. This command would export a copy of your CLASSMATE schema, as a file called classmate.dat, into whatever directory your shell is opened in.
5. You could also export individual tables, such as in the following export command, which exports the CLIENT and CUSTOMER tables to a single export dump file:

```
exp classmate/classpass@oraclass file=classmatetables.
dat tables=(client,customer)
```

NOTE

Notice that the individual tables export utility has been connected to the database as the CLASSMATE user, and not the SYSTEM user, as was previously done for the CLASSMATE a schema export.

6. Leave the shell open to use to experiment with recovering from the single table export you just created.

Tablespace Backups

Another type of hot backup is a tablespace backup, for which you switch a tablespace into a special mode called backup mode. This allows a physical operating system-level copy of that tablespace. In Oracle 10*g* this option is also improved by allowing a backup mode setting to be applied to all tablespaces at once.

The only problem with backup mode is a performance issue. Placing a tablespace into backup mode forces all change activity for that tablespace to be copied to the redo logs. If there are enough redo entries produced to recycle a redo log group, then the archive log files will be used as well. The longer a tablespace remains in backup mode, the more performance is affected, the longer any potential recovery is, and more redo log data is produced. Larger redo log entries cause redo logs to switch more often, producing more archive logs more often.

Let's make a backup of the USERS tablespace in the ORACLASS database, with the database remaining online:

1. Go back to your SQL*Plus window that you used earlier in the chapter.
2. Connect as the SYSTEM user. Type the following command:

```
CONNECT SYSTEM/<PASSWORD>@ORACLASS
```

3. Type the following command to place the USERS tablespace into backup mode:

```
ALTER TABLESPACE USERS BEGIN BACKUP;
```

Any changes made to any data (tables), contained within the USERS tablespace, will not update USERS tablespace datafiles.

4. Execute the following DML commands in the SQL*Plus to make a change to the USERS tablespace:

```
CREATE TABLE NOTBACKEDUP(ID INTEGER, STR VARCHAR2(16)) TABLESPACE USERS
INSERT INTO NOTBACKEDUP VALUES(1,'Row 1');
INSERT INTO NOTBACKEDUP VALUES(2,'Row 2');
COMMIT;
```

This creates a new table in the USERS tablespace. Add two rows to it and commit the changes. However, because the tablespace is currently in backup mode, the changes are stored as redo log entries, and not in the tablespace. The tablespace is effectively frozen. When the tablespace is switched out of backup mode the database will automatically recover all changes made (in the redo logs), and update the USERS tablespace to reflect the new table, with its new rows.

5. Before switching the tablespace out of backup mode again, you need to create a backup copy of all datafiles in the USERS tablespace. Execute the following query to display all datafiles contained within the USERS tablespace:

```
SELECT TABLESPACE_NAME,FILE_NAME
FROM DBA_DATA_FILES
WHERE TABLESPACE_NAME='USERS';
```

The result of the above query is shown in Figure 15-17.

```
Oracle SQL*Plus
File Edit Search Options Help
SQL> SELECT TABLESPACE_NAME,FILE_NAME FROM DBA_DATA_FILES WHERE TABLESPACE_NAME='USERS';

TABLESPACE_NAME
------------------------------
FILE_NAME
--------------------------------------------------------------------------------
USERS
C:\ORACLE\PRODUCT\10.2.0\ORADATA\ORACLASS\USERS01.DBF


SQL> |
```


FIGURE 15-17 Finding datafiles for a tablespace in backup mode

6. Go into the operating system on your database server by using a shell or a tool such as Windows Explorer and make a physical copy of the datafiles you found using the query shown in Figure 15-17. Make sure you copy all files in the USERS tablespace.
7. Now that you have a copy of the datafile, without any changes applied, you can switch the tablespace out of backup mode:

```
ALTER TABLESPACE USERS END BACKUP;
```

8. The END BACKUP command will recover any changes made during the BACKUP mode setting, for the tablespace, in the underlying datafiles. You don't see anything happen, but that is how it works.
9. Leave the SQL*Plus shell open for now; you will use it to experiment with recovering the tablespace.

Now let's move on to recovery.

DATABASE RECOVERY

There are various scenarios that can occur with different database failures because there are different types of failure, as discussed in this chapter. With an up-and-running Oracle 10g database, the following errors can occur:

- Losing a controlfile
- Losing a redo log file
- Losing the SYSTEM or SYSAUX tablespace datafile
- Losing an application tablespace datafile

The question is, how do you respond to these different failure scenarios?

Losing a Controlfile

The easiest way to resolve this problem is to shut down the database and copy back in a current controlfile copy. If this can't be done, one can rebuild the controlfile from a trace copy (see Chapter 4).

You should always multiplex controlfile so you have a copy that can simply be plugged into a shutdown database. There is no danger of the control file being mismatched in time with the other datafiles and redo logs in the database, because loss of the controlfile does not allow any change activity to continue, until that controlfile is repaired or replaced.

NOTE

Always multiplex the controlfile.

Losing a Redo Log File

Once again, redo logs should always be duplexed and replacing a redo log file involves shutting down the database and restoring an exact copy into the location of the erroneous redo log file, from a duplexed copy. Other methods of recovering redo log files, when duplexed copies do not exist, involve restarting the database but with cleared redo log files. This is a problem, and undesirable, because you will lose whatever is cleared from the redo log files.

NOTE

Always duplex your redo log files.

Losing an Individual Table

Loss of an individual table is simple to recover if you have an export. Let's say you lost the CUSTOMER or the CLIENT table from your CLASSMATE schema. You could easily replace them using the export dump file of those two tables, which you created earlier, using the import utility something like this:

```
imp classmate/classpass file=classmatetables.dat tables=customer
```

Losing an Application Datafile

Loss of an application datafile includes any non-system tablespace, excluding the SYSTEM and the SYSAUX tablespace.

NOTE

The UNDO and TEMP tablespaces are usually only backed up during a cold backup because of the flexible nature of their contents.

You created a tablespace backup mode copy of the USERS tablespace in a previous section. If you lost one or more datafiles in a tablespace such as the USERS tablespace, you would have to temporarily take the offending tablespace offline:

```
ALTER TABLESPACE USERS OFFLINE;
```

Then copy the erroneous datafiles from a backup in the operating system. Then bring the USERS tablespace back online:

```
ALTER TABLESPACE USERS ONLINE;
```

You should get a message citing that media recovery is required. Media recovery implies that a file (the datafile) is out of date for that tablespace. You will not be able to bring the table back online until you have recovered the tablespace. You can recover at multiple levels using the RECOVER command. In this case you would use this command:

```
RECOVER TABLESPACE USERS;
```

When media recovery is complete you can attempt to place the tablespace back online, error free, with this command:

```
ALTER TABLESPACE USERS ONLINE;
```

Losing SYSTEM or SYSAUX Tablespaces

If a SYSTEM of SYSAUX tablespace datafile is lost you are actually forced to shut down the database to copy the backup file back into the database architectural structure because these tablespaces contain the metadata for the database, including all the catalogs, built-in stored procedures, and many other critical features. These two tablespaces run the database. The system-level tablespaces contain everything that is the Oracle database. They have to be permanently online when the database is open for use. Therefore they can only be retrieved with the database shutdown, and recovered when the database is not open.

This completes this chapter and this book. I hope you have enjoyed reading it as much as I enjoyed writing it.

Chapter Summary

- Backup of a database is the process of taking copies of part or all of a database.
- The parts of a database that are backed up can help to restore the database to a specific time.
- Recovery of a database can be performed using parts of a previously created backup or an entire backup.
- A cold backup is a snapshot of a database when the database is completely shut down, inactive, and inaccessible.
- A hot backup is a backup that can be taken of a database when the database is running and fully available for use.
- A hot backup can be either a consistent snapshot of an entire database or inconsistent copies of individual physical and logical parts of a database.
- Tools used for backup and recovery include export and import utilities (plus Data Pump variations), RAM, and operating system-level copies of various forms.
- An Oracle database can fail in numerous ways, including media (a corrupt disk), user error, and even an Oracle software error.
- It is important to isolate the type of failure that has occurred to assist in deciding how a database should be recovered.
- A backup strategy is largely determined by how a database is used, what it is used for, how much downtime is acceptable, and if any loss of data is acceptable.
- A database is generally regarded as nearly 100 percent fully recoverable using an archive log database configuration.
- Other factors such as duplexed redo logs and multiplexed controlfiles can help achieve 100% recoverability.

Review Questions

1. Which of these can be used for recovery of part or all of a database?
   a. TopSQL
   b. TopSessions
   c. Memory Advisor
   d. TopConsumers
   e. None of the above
2. Which of these backups can be used for recovery using the RECOVER command?
   a. Export utility dump files
   b. Files containing lots of INSERT statements
   c. Tablespace copies from a running database with the tablespace completely available fort all activities
   d. Datafiles contained with a tablespace placed into backup mode
   e. None of the above

3. Where can SCNs be found?
   a. Redo log files
   b. Archive log files
   c. Controlfiles
   d. Datafiles
   e. All of the above
4. The ________________ Control interface can be used to manage, schedule, and perform backups.
5. The ________________ advisor in the Database Control is used to manage and tune backups for potential recovery.
6. What is the process of bringing a restored datafile back up to the latest point in time, contained within the controlfile?
   a. Backup
   b. Restoring
   c. Restoration
   d. Updating
   e. None of the above
7. A datafile that is copied without the parent tablespace in backup mode can be used for recovery? True or False.
8. Redo log files are copies of recycled archive log files. True or False?

Exam Review Questions—Oracle Database 10*g*: Administration (#1Z1-042)

1. What is the SQL*Plus command for displaying the current archive log mode configuration for a database?
2. Which of these command sequences will restart the database faster?
   a. SHUTDOWN ABORT;
   b. SHUTDOWN IMMEDIATE;
   c. Neither
3. Fill in the blanks in the following sequence of commands:

```
SHUTDOWN IMMEDIATE;
STARTUP ________________ ;
ALTER DATABASE ARCHIVELOG;
ALTER DATABASE ________________ ;
```

4. The LOG_CHECKPOINT ________________ and LOG_CHECKPOINT ________________ parameters can be used to control checkpointing in a database.

Hands-on Assignments

1. Write a sequence of steps to perform a cold backup of a database. Do not execute the script. Any non-SQL*Plus commands should be pseudocoded and commented out. Save your answer in a file named **ho1501.txt**, in the **Solutions\Chapter15** directory on your student disk.
2. Write a sequence of steps used to perform a hot backup of a database. Write the scripts containing all the tablespaces explicitly where each tablespace has a command of its own. Do not execute the script. Any non-SQL*Plus commands should be pseudocoded and commented out. Save your answer in a file named **ho1502.txt**, in the **Solutions\Chapter15** directory on your student disk.

Case Project

1. Your MIS manager decides that your Oracle 10*g* database should be backed up using an inconsistent backup and a consistent backup using two different backup methods or tools. Also assume there are tablespaces called SYSTEM, SYSAUX, USERS, TEMP, and UNDO. Do not execute anything. Write your answer as a sequence of steps. Save your answer in a file named **case1501.txt**, in the **Solutions\Chapter15.**

NEW FEATURES OF ORACLE 10g (for 1Z0-042)

The following is a list of the new features of Oracle 10*g*:

- Bigfile tablespaces that can hold as much as 4Gb blocks.
- Multiple default temporary tablespaces that are contained in a tablespace group. The temporary tablespace group can then be assigned to individual users for sorting purposes.
- A tablespace can be renamed.
- The SYSAUX tablespace is introduced, which is the auxiliary system tablespace. The SYSAUX tablespace contains application metadata, and the SYSTEM tablespace contains only database metadata.
- Automatic Storage Management (ASM) provides volume management of physical space contained within Oracle Database. ASM utilizes Oracle Managed Files (OMF), disk groups, mirroring, and striping.
- The DROP DATABASE statement can be used to drop an entire database.
- Flashback allows for features, such as rapid recovery on a small scale, plus flashback versions queries.

- The FLASHBACK TABLE statement allows for rapid recovery of a single table using undo space.
- The DROP option of the FLASHBACK TABLE statement allows for rapid recovery of a mistakenly dropped table.
- The flash recovery area contains files that can be used for potential recovery.
- Online table redefinition is now less restrictive allowing for online redefinition of more types of tables. For instance, individual partitions, clusters, object datatypes, some constraints, or statistics.
- Statistics are automatically gathered based on the setting value of the STATISTICS_LEVEL parameter.
- Shared server configuration is automated and only the SHARED_SERVERS parameter must be explicitly set.
- Automated memory management is improved by introducing automated management of the System Global Area (SGA). The SGA includes the main database buffer cache, the shared pool, the large pool, the java pool, and the streams pool.
- The COMPATIBILITY parameter cannot be retrogressed back to a previous version of Oracle Database, unless done by full database recovery back to a previous version database backup.
- A sorted hash cluster allows a hash cluster to be sorted. Thus faster retrieval can be achieved based on the sorting specification.
- The MAXTRANS parameter is removed. This value is now automated as 255 transactions per block.
- Manual rollback is replaced with the far more efficient method of automated undo. Additionally, any SQL coding containing manual rollback statements will not cause an error. Any manual rollback instructions will simply be ignored by Oracle Database.

- The UNDO_SUPPRESS_ERRORS parameter is no longer used.
- Automated tuning of undo retention makes snapshot too old errors less likely. A snapshot too old error occurs when a query is executed against undo data that has been cycled out of undo space.
- Transparent data encryption allows data in specific table columns to be encrypted. Encrypted data cannot be viewed without decrypting it first and is thus secure from snooping by outsiders.
- Partitioned tables and indexes can now be partitioned to a far greater degree.
- The Database Control Segment Advisor tool includes reporting on tables with excessive amounts of row chaining. Additionally, the Segment Advisory tool executes automatically as a scheduled job in the default maintenance window for a database.
- Online segment shrinking includes large binary objects (LOBs) and index-organized tables (IOTs).
- Some configuration parameters no longer require recreation of the control file for a database, such as MAXLOGFILES, MAXLOGMEMBERS, MAXLOGHISTORY, and MAXINSTANCES.

GLOSSARY

The terms in this glossary are defined in the context of the Oracle 10*g* database structures, rather than in the context of general computer technology. This approach should make the glossary more useful to you as a student of Oracle 10*g*.

add-on Components available in Oracle that can enhance the database system.

algorithm A set of steps, or mathematical calculation, used for solving a problem.

ANALYZE To issue a command that causes Oracle 10*g* to read the table's structure and update the table's metadata with current information about the size of the table, average row length, total number of rows, and free space remaining in extents. Much statistics generation in Oracle 10*g* is automated. *See* DBMS_STATS package.

ARCHIVELOG mode A database in ARCHIVELOG mode automatically archives (makes copies of) recycled redo log files. The opposite is NOARCHIVELOG mode.

ARCn Archiver background process. **n** indicates that multiple archiver processes can run concurrently. Used only when database is in ARCHIVELOG mode.

automated memory management Includes both automated manage of the System Global Area (SGA) and the Program Global Area (PGA). The SGA includes the main database buffer cache, the shared pool, the large pool, the java pool, and the streams pool. The PGA is essentially session connection memory.

automated statistics gathering Statistics are automatically gathered by default in Oracle 10*g*. Automation is based on the setting value of the STATISTICS_LEVEL parameter.

Automatic Storage Management (ASM) Provides volume management of physical space, contained within Oracle Database. ASM utilizes Oracle Managed Files (OMF), disk groups, mirroring and striping.

automatic undo management mode Automated method of managing rollback segments in which Oracle 10*g* automatically handles undo records, using an undo tablespace. Manual undo management of rollback space is obsolete in Oracle 10*g*.

BIGFILE — Tablespaces that can hold as much as 4G blocks. A SMALLFILE tablespace is the default.

BFILE locator A pointer directing the database to the physical disk location of an external LOB file.

binary tree A type of data structure that uses a branching formation, always dividing a list of data into two halves. Each half is then divided in half again, and so on. A binary tree contains a root node (at the top of the tree), branch nodes (between the root and the leaves), and leaf nodes (at the bottom of the tree).

bitmap index An index that stores index key values as a bitmap of 0's and 1's, pointing to row locations using ROWIDs.

branch node An element of a data structure. A branch is the node of a binary tree that has other nodes below it and above it. *See* binary tree.

BTree index An index structure in which data is divided and subdivided based on the index key values to minimize the look-up time required when searching for a key value. The BTree is so named, because it is a variation of a binary tree.

business rules Statements defined during the design of a database system that inform both the database designer, and the application programmer, of how data is used to support the business. Business rules are applied using database design and various other objects and substructures within a database architecture.

chained row A table row that spans multiple blocks. Access of chained rows introduces avoidable inefficiency.

change vector Component of a redo log entry. Describes a change made to a data block. Many change vectors can be contained within a single redo entry.

character set A standardized definition of the numeric bit values used to represent each character in a language. For example, in the American character set, the capital letter Q is stored as the bit value 81.

checkpoint A moment in time when the checkpoint (CKPT) background process signals that all used memory buffers are to be written to disk.

chunk Oracle 10g writes and reads one chunk from a large object (LOB) value at a time. A chunk can be from one data block that is up to 32K in size.

clearing (a group) Removes all entries from all online redo log groups. This can occur when a database is restored and recovered, to a point in time before any active redo logs are applied (all changes are applied from archive redo log files), or when a database is simply unrecoverable.

cluster A collection of information that is organized for efficient retrieval, for example name, address, and phone number. Technically, a cluster is not a table, but is a group of tables stored together as if they were one table. Clusters store data of multiple tables in one segment.

cluster key The set of columns from each table in a cluster that make up the index for the cluster.

clustered rows Rows from each table in a cluster that correspond to one unique value in the cluster key per block.

coalesce The combining of multiple adjacent free extents into a single contiguous (adjacent) free extent. For example, an index is coalesced to consolidate fragmented storage space in its leaf blocks.

collection data type A datatype that repeats, and is contained within, a single column. A collection can be defined as an array, called a VARRAY, as a nested table, or an associative array.

column constraint A restriction that applies to a single column and appears inline with the column.

column data Consists of a 1- to 3-byte section of a table containing the length of the column's data, followed by the actual data for the column.

common and variable header The section of a table that contains identifying information, such as the type of block and block location.

composite index An index that is made up of more than one column. No two indexes on a table can contain the same combination and ordered sequence of columns.

composite partition A method of table partitioning that is used when the data within a table is being partitioned, and each partition is partitioned in turn, as subpartitions.

composite range-hash partitioning A method of table partitioning where hash subpartitions are created within range partitions.

composite range-list partitioning A method used to partition where list subpartitions are contained within range partitions. *See* range partitioning and list partitioning.

compound key A field used to sort data that is composed of multiple columns.

connection The link from the user session, through the server session, to the database instance.

contiguous A term applied to blocks located adjacently on disk. Scanning contiguous blocks is better for performance than searching for blocks distributed randomly across disk storage space.

constraint A rule that defines data integrity for a column, or group of columns, in a table.

constraint state An attribute that tells s Oracle 10*g* how to use a constraint when data is added to a table. A constraint state can also tell Oracle 10*g* how to use a constraint on existing data, when a constraint is changed from an inactive state (DISABLE), to an active state (ENABLE).

control file A file containing the names, locations of, and most recent changes to all the datafiles and redo log files belonging to a database instance.

copy semantics Internal large objects (LOBs) use copy semantics, which means that the LOB value is copied like other data when you copy the LOB into a new location. *See* reference semantics.

optimizer Optimization uses statistics on the actual volume and distribution of table data, in order to help determine the best path when retrieving data. Optimization takes certain factors into account, such as the relative costs of I/O, CPU time, and execution time.

data block A data block is made up of a set number of contiguous physical bytes in a physical file. A data block is the smallest size of a logical unit in the database.

data dictionary view A window into the metadata and system data in the database. These views query the database's internal management tables, presenting data as a view.

data segment An area used to store data for objects such as tables, object tables, and materialized views.

data type A predefined form of data. Oracle has numerous standard data types, including numbers and dates.

database A group of files containing data, metadata, undo information and redo log entries. Accessed through a database instance. *See* database instance and database server.

database character set Each database has a base language, such as English, that is called its database character set. *See* character set.

database instance A running database consists of memory allocations and background processes. The memory buffers and processes make up the instance or instantiation of Oracle Database software running on a database server. *See* database and database server.

database object Any database item that can be individually set and manipulated. A database object is not only an object table, but also any other structure held in the database, such as a relational table, an index, a PL/SQL procedure, or even a customized attribute with methods and rules attached.

database — The combination of a database (the files) and a database instance (the SGA and the background processes).

Database Control The Oracle 10*g* incarnation of the Oracle Enterprise Manager software (OEM). OEM in previous versions of Oracle Database was written and executed using Java. The Database Control is written and executed in a browser. *See* Oracle Enterprise Manager (OEM).

datafile A physical file that is located on an operating system and contains data. It is the primary physical structure that contains all the data stored in the database.

DBA DBA stands for both database administration and database administrator.

DBA authentication method A method used to validate the login of users with SYSDBA (for SYSDBA privileges) or SYSOPER (for SYSOPER privileges).

DBMS_STATS — A built-in package allowing efficient database statistics generation. Much statistics generation in Oracle 10*g* is automated. *See* ANALYZE.

deadlock Occurs when two transactions cannot resolve their simultaneous requirements for the same data.

dedicated server A mode of a database connecting one user process with one server process. *See* shared server.

default role Any role assigned to a user that is automatically enabled when the user logs onto a database. *See* role.

default value A value that Oracle inserts into a column when an inserted row does not specify a value for that column.

deferred state A type of constraint state in which the constraint is validated only when a COMMIT command is executed. *See* immediate state.

denormalize Decomposition of normalized relations, reintroducing duplication to tables, usually employed in data warehouse star schema, but sometimes in Inter database for the purposes of improving performance, by making SQL coding less complex. *See* normalize.

dictionary-managed tablespace A tablespace whose free space and storage allocation information is stored inside the data dictionary tables in the SYSTEM tablespace. *See* locally managed tablespace.

dirty buffer Used memory blocks in cache or buffer blocks containing data changes, not yet written out to disk. These blocks are contained in the System Global Area (SGA).

disabled Disabled means switched off. For example, a disabled role allows a user no privileges. A disabled constraint does not enforce its data integrity rule. *See* enabled.

distributed database system A distributed database system has multiple instances that are used as if they were one instance. Data on any of the databases can be modified by any of the instances.

duplexing The process of making duplicated, parallel write copies of redo log files. This helps to guarantee that if any redo log files is damaged, another can take its place immediately. Redo log files are usually duplexed and control files are usually multiplexed. *See* multiplexing.

enabled Enabled means in force. For example, a database role does not have to be enabled all the time. A constraint can also be enabled or disabled. *See* disabled.

exporting A form of backup that can be used on specific tables or schemas, or to back up the entire database. Creates a file in Oracle-proprietary format that can only be used with Oracle's importing process. A database export contains DDL and DML commands to create both database objects and their contents, such as a table and its rows. *See* importing.

eXtensible markup language (XML) A programming language for the Internet that provides both the data and the display details to a web page or to an application. Oracle supports storing XML formatted data in tables.

extent A contiguous group of data blocks that are assigned to a segment. The number of data blocks in one extent is determined by the storage parameters of the object or the tablespace.

external LOB The data of an external large object (LOB) is stored outside the database in an operating system file. Commonly used for multimedia. The data type of an external LOB column is a BFILE datatype. *See* internal LOB.

external table A table that is defined in the database but whose data is stored in a file that is outside the database. Commonly used to copy external data into a table inside the database.

external user An Oracle user whose name and password are authenticated outside the database by the operating system or a third-party service. *See* local user and global user.

fast commit Oracle uses a fast commit mechanism to speed up performance. When a transaction is committed, the changed data blocks are kept in the buffer cache and are not written to the datafiles until later. Oracle 10*g* has introduced specialized optional clauses for the COMMIT statement, where a COMMIT is forced to wait for dirty log buffer entries to write to disk.

fixed-length A multibyte character set can be either fixed-length or variable-length. A fixed-length, multibyte character set uses a fixed number of bytes for each character in the character set. This simplifies retrieval but uses more space to store data. *See* variable-length and multibyte character set.

flashback — Allows for features, such as rapid recovery on a small scale, plus flashback versions queries.

flashback recovery area — Contains files that can be used for potential recovery.

flushing the buffer The process of copying the memory buffer contents to the appropriate datafile, or to the archive log file, and then deleting the memory buffer's contents.

foreign key The name of a relationship between two tables, in which one is the parent and the other is the child. Usually enforced with a foreign key constraint defined on the child table and a primary key constraint defined on the parent table. *See* primary key.

free space The contiguous blocks of storage space that are unallocated in a tablespace.

freelist A list of individual blocks within an extent that have room available for inserting new rows. To be included in the freelist, the percentage of available space in a block must be greater than the PCTFREE value for the table.

function-based index An index that uses the value of an expression or function instead of the value of a column as the indexed column. For example, the index uses TO_CHAR(SALES_DATE, 'MONTH') as one of its indexed column values, where the index contained the converted values containing months.

functions, procedures, and packages PL/SQL programs that reside in the database and can be called from SQL commands, such as SELECT and INSERT. Oracle 10*g* provides many pre-defined functions, procedures, and packages. You can also create your own. *See* PL/SQL.

global non-partitioned index A normal index that is created on a partitioned table and contains data from all the partitions in the table. *See* global partitioned index and local partitioned index.

global partitioned index An index that is partitioned. Can be created on a non-partitioned or partitioned table. When created on a partitioned table, the index is partitioned differently than the table, such that the index is created for the entire table, regardless of partitions. Global partitioned indexes can be created as hash key indexes in Oracle 10*g*. *See* local partitioned index.

global user An Oracle user whose name and password are authenticated outside the database by a directory service such as Light Weight Directory Access (LDAP) server. *See* local user and external user.

grant When a database administrator grants a privilege, he assigns the ability to access an object or use a system privilege to a user or to a role. *See* object privilege and system privilege.

grantee A user or role that receives (is granted) an object privilege, system privilege, or a role.

grantor A user who executes the GRANT command, granting a privilege or role to a user or role.

group function A function that acts on sets of column data in rows. For example, SUM(SALES_AMOUNT) adds the value of SALES_AMOUNT into a set, which aggregates into a group of rows.

hash partitioning A method of table partitioning, in which the database administrator specifies the partitioning key and the number of partitions. Oracle uses a hash value (calculated on the partitioning key), to divide the data evenly among the set of partitions. Hash algorithms are used to divide partitions into similarly sized groups.

heap-organized table A table that is not stored in index order and is literally stored in an unorganized pile or heap. This is the normal method of storing tables. *See* index-organized table.

high watermark (HWM) The boundary between used data blocks (blocks formatted for data), and unused data blocks (blocks allocated but not yet formatted for data) in a table. Used blocks can contain no data if the rows they previously stored were deleted. You cannot deallocate blocks below the HWM.

immediate state A type of constraint state in which the constraint is validated as soon as an insert, update, or delete statement is executed. *See* deferred state.

importing A process that uses a previously created snapshot of a database (or part of a database such as schema or a table), and recreates all or part of it in the current database. An import can perform many functions including adding rows to existing tables, as well as creating users, tables, views, and triggers. *See* exporting.

index A database structure that is associated with a table or a cluster and speeds up data retrieval when the table or cluster is used in a query. *See* bitmap index and BTree index.

index segment The logical structure that stores all the index data, except when the index is partitioned. For partitioned indexes, there is one index segment per partition.

index-organized table A relational table with rows stored in the physical order of the primary key values, in the same way an index is stored. Can improve performance for tables that are always queried by a primary key. *See* hash-organized table.

inheritance The act of one object type obtaining characteristics from another object type. For example, an animal object type has attributes inherited by a mammal object type, which is a specialized type of animal.

inline constraint An integrity rule (constraint) that, when defined, appears immediately next to the column to which it applies. *See* out of line constraint.

integrity constraint A rule that defines restrictions or relationship requirements on a column or a set of columns in a table. *See* constraint, foreign key, and primary key.

internal LOB A large object (LOB) column with data stored inside the database. The column's datatype is BLOB, CLOB, or NCLOB. *See* external LOB.

Java Virtual Machine (JVM) An internal component of the database that enables the storing, parsing, and executing of Java procedures within a database.

leaf node The lowest level of a binary tree, which has no nodes below it. The leaf blocks contain the actual index values and ROWIDs pointing to rows in tables. *See* binary tree, BTree index and branch.

list partitioning A method of table partitioning in which a distinct list of partitioning key values is set up, and values are defined to go into each partition. Useful when the partitioning key has a small number of distinct values, such as state abbreviations.

listener A database service that waits for incoming requests for the database server and responds to them.

LOB A value stored in a column that can be up to 4Gb in size. LOB stands for *large binary object*. LOBs usually contain multimedia information. *See* LOB value, internal LOB and external LOB.

LOB locator The pointer that directs the database to the actual location of the value is called the large object (LOB) locator or internal LOBs and the BFILE locator for external LOBs. The LOB locator resides in the table, while the LOB value resides in a separate location (either inside or outside the database.) Some LOB values are stored inside the table (no pointer required).

LOB value The data stored in a column with one of the LOB data types. LOB values can be stored with the rest of the table data (inline) or in a separate segment (out of line). *See* LOB pointer. LOB data types are BLOB, CLOB, NCLOB, and a BFILE pointer.

local partitioned index An index on a partitioned table in which the index is partitioned in the same way and on the same columns as the table.

local user A user who must enter a password whenever logging onto a database. User name and password are managed inside the database. *See* external user, and global user.

locally managed tablespace A tablespace that contains an internal bitmap that stores all the details about free space, used space, and the location of extents. Locally managed tablespace is the default setting for tablespaces. *See* dictionary-managed tablespace.

log switch An event in which the log writer (LGWR) process stops writing to one log group, and begins writing to another log group. *See* redo log group.

logging A parameter specified during the creation of a tablespace that sets the default action for objects in the tablespace so that all transactions that change the objects are recorded automatically. Logging is the opposite of nologging.

logical structure A structure that is composed of orderly groupings of information that allow for the manipulation and access of related data. A table is a logical structure. *See* physical structure.

low cardinality Low cardinality means that the number of distinct values in an indexed column is low compared to the number of rows in the table. For example, the CUSTOMER table has 50,000 rows and the STATE column has 50 distinct values. Remember that cardinality is relative. In other words, a column with values of M for male, and F for female, and only two rows in the table does not have low cardinality.

manual rollback management mode Older method of managing rollback segments. Typically used before Oracle9*i*, in which the DBA establishes files for storing undo records. Manual rollback has been replaced with automatic undo management, which is much more efficient.

materialized view — This type of view is termed materialized because it is materialized, or a copy of data from underlying tables. A view, on the other hand, contains a query that re-executes for every access to the view. Materialized views have traditionally been utilized in data warehouses. However, in Oracle 10*g* query rewrite is enabled by default. Thus materialized views are slowly coming into use, even in OLTP databases. Materialized views are not included in this exam.

media recovery Media recovery implies that some media (hard disk drive stored files such as a datafile), requires updating using redo log entries, and/or undo space entries. A recovery that requires restoring the database from a backup, rolling forward through archived redo logs, and finally rolling forward through online redo logs. It is called media recovery because the most common reason for this type of recovery is the failure of a disk drive or some other storage media.

member Each file in a redo log group is considered a member of that group.

metadata Data about a specific set of data in the database including how, when, and by whom it was collected, and how the data is formatted. For example, data dictionary views display metadata about tables.

methods Sets of predefined code segments that perform tasks on the data in an object. For example, a GIVE_RAISE method might update a record with a certain percentage increase in pay, perhaps based on length of service.

migrated If a row is updated and requires more free space than the block contains, then the entire row is migrated, that is, moved to another block. A pointer remains in the original row location to redirect a process to the correct location.

multibyte character set A character set that is complex and requires multiple bytes to represent a single character. Chinese is a multibyte character set. *See* character set.

multiple instance server With a multiple instance server installation, one computer has multiple instances, each instance with its own database files. In Oracle8*i* this is known as a parallel server environment. In Oracle9*i* this is a clustered server environment (Oracle Real Application Clusters – RAC). In Oracle 10*g* this is known as a clustered environment or even a computer grid.

multiplexing The process of making redundant copies of control files to guarantee that if the original control file is damaged, another can take its place immediately. Control files are usually multiplexed and redo log groups are usually duplexed. *See* duplexing.

nested table An object type that allows one column to store a table of data.

NOARCHIVELOG mode A database in NOARCHIVELOG mode does not archive redo log files. The opposite is ARCHIVELOG mode.

node Any point in a binary tree where two or more lines meet. *See* branch, leaf, binary tree, and BTree.

nologging A parameter specified during the creation of a tablespace, or overridden when creating a table, which sets the default action for objects in the tablespace. For example, transactions performing mass inserts do not have to be recorded. nologging is the opposite of logging.

normalize Normalizing tables is part of the design process in which this and other normalizing rules are followed: every table should have a key that contains a unique value for every row. *See* denormalize.

object privilege Gives a user or role the ability to perform certain tasks on specific tables or other objects that are owned by a schema. For example, the SELECT object privilege allows a user to query a specific table. *See* system privilege.

object table A table that holds objects and attributes. An object table is similar to a relational table, except that each row is a single unit of data defined by an object type.

object type An object type has the attributes (an attribute is the object equivalent term for a column of field), and aspects of an object-oriented databases, such as methods and inheritance.

offline tablespace A tablespace that is not available for use. Switching a tablespace offline switches all datafiles within that tablespace offline. A single datafile within a tablespace can also be switched offline.

online backup A backup of the tablespace that is made while the tablespace is available for use. Also known as an open database backup, or backup mode tablespace backup.

online tablespace A tablespace that is available for use.

online table redefinition Allows for online redefinition of tables.

operating system (OS) authentication When using operating system (OS) authentication, the user logs in without specifying an Oracle user name and password. Oracle derives the Oracle user name from the operating system user name.

optimizer A process used within Oracle 10g that selects the fastest access path for the execution of a SQL command. *See* rule-based optimizer and cost-based optimizer.

Oracle Advanced Security A feature that encrypts outgoing data before it goes onto the network or across the Internet. It also supports special methods of user authentication, such as the programs used with automated tellers that require a user possess a valid bank card, and know the personal identification number (PIN), in order to access account data.

Oracle Enterprise Manager (OEM) A Database Administrator tool with a Windows-like interface, which was first introduced with Oracle7. The tool integrates many utilities and monitoring tools into a single interface. This book makes extensive use of the OEM Console that can be accessed using the Oracle client software only. Typically, cautious DBAs will utilize client software as much as possible, not utilizing the database sever computer for administration whenever possible. *See* Database Control.

Oracle Financials A set of software functions designed for use in the fields of bookkeeping, accounting, inventory, and sales.

Oracle JDeveloper An application builder that writes Java code using a Windows-like interface.

Oracle Label Security A tool for restricting access to rows within a table or within a view. Useful when high security standards must be met. It labels individual rows with a security profile and matches that row with a user's security profile that is stored in the database.

Oracle On-line Analytical Processing (OLAP) Services Services that make it possible for standard file directory support to be combined with database delivery of tables or views.

Oracle Partitioning A feature that enables the process of dividing tables across multiple tablespaces and datafiles. High volume tables, such as historical records in data warehouses, benefit from partitioning because it speeds up data retrieval for queries. Partitioning is also highly applicable to very busy, high concurrency level OTLP databases in some circumstances.

Oracle Real Application Clusters (RAC) Provide management tools to support database clusters.

Oracle Spatial A feature that adds programmed packages to the database to handle spatial objects. Geographic mapping is a common use for Oracle Spatial.

ORACLE_BASE A root directory that stores all Oracle-related files, including database datafiles and configuration files. Database datafiles reside in subdirectories under ORACLE_BASE.

ORACLE_HOME The directory tree in which Oracle executable files are stored. This is by default a subdirectory within ORACLE_BASE.

out-of-line constraint A restriction (or rule) that appears after the full list of columns in the CREATE TABLE statement, usually applying to multiple columns. Constraints can also be applied in the ALTER TABLE statement. *See* inline constraint.

partitioning key A range of values in a set of columns, determining how a table or index is stored in partitions.

permanent tablespace Stores permanent objects such as tables and indexes. A permanent tablespace, as opposed to a temporary tablespace, is the default setting for tablespaces.

physical structure A structure that is composed of operating system components and has a physical name and location. Physical structures can be seen and manipulated in a computer operating system. In Oracle 10g, a datafile is a physical structure. *See* logical structure.

PL/SQL A procedural language that Oracle provides to write simple programs in the database. Uses SQL commands along with variable definitions, loops, if-then-else logic, and so on.

precompiler Precompilers (such as Pro*COBOL and Pro*C) support embedded SQL commands within programs in C, C++, or COBOL. The precompiler translates the SQL command into the appropriate set of commands for the program, which is then compiled and ready for executing.

primary key The column or set of columns defining a unique identifying value for every row in a table. For example, the CLIENT table has a unique identifying column called CLIENT_ID. A primary key is usually enforced by defining a primary key constraint on a table.

privilege A capability, such as the ability to create new users, or an authorization, such as the authority to SELECT on a table. A user or role can be given a privilege. *See* object privilege and system privilege.

profile A collection of settings that limit the use of system resources and the database. All users have one the DEFAULT profile allocated to them automatically on creation of the user.

pseudocolumn Acts like a column in a query, but actually is calculated by the database for the query. For example, ROWNUM is a pseudocolumn containing the sequence number of each row returned in a query.

public synonym A unique name for an object that allows any user to use the object without prefixing it with the owner name. A public synonym does not give users the privilege to actually select from the table. That privilege must be granted.

query plan A list of steps developed by the optimizer that will be taken to retrieve data for a query. Query plans can be accessed in various ways including the EXPLAIN PLAN statement and the Database Control interface.

query rewrite The ability of Oracle 10g to modify a query that is executed and change it into an equivalent but more efficient statement before actually running the query. This ability is enabled or disabled with the QUERY_REWRITE_ENABLED initialization parameter, which is enabled by default for Oracle 10g. Query rewrite allows automated switching of query plan executions to read data from materialized views, rather than underlying tables. *See* materialized view.

quota The maximum amount of storage space a user can be allocated in a tablespace. Set with the CREATE USER or ALTER USER statements.

range partitioning A method of table partitioning in which the table is stored in partitions according to a specified set of ranges of values defined by the partitioning key. For example, the first range is all rows with partitioning key values less than 100, the second range is all values less than 1000, and so on.

read consistency A type of status that allows only the user who issued an update command to see the changes to the data in queries, until such time as that user commits the change. Other users see the data as if it had not been changed until the commit occurs.

read-only tablespace A tablespace mode in which objects in the tablespace can be queried but not changed. No user can execute an INSERT, UPDATE, or DELETE statement on objects in a read-only tablespace.

recover This term applies to the process of bringing restored datafiles up to date with other datafiles, and/or the control, in a database by application of more recent redo log entries. Redo log entries can be found in both redo log files and archive log files. *See* restore.

record sections Lists of information by category within a control file. For example, one record section stores data file names and locations, whereas another record section stores recovery information, including the date of the most recent archive.

Recovery Manager (RMAN) A utility that helps to automate and improve database backup and recovery. RMAN is not covered in this exam in any particular detail, other than just conceptually.

redo entry Redo entries are essentially a record of all changes made to a database. For example, let's say a datafile is lost. That same datafile is subsequently restored from an older copy of the datafile. Recovery procedures will apply all new redo log entries to the datafile, bringing the datafile up to date with all other datafiles in the database. The redo log file contains redo entries that are made up of a related group of change vectors that record a description of the changes to a single block in the database. Redo log entries can be found in both redo log files and archive log files. *See* recovery and restore.

redo log file A file containing redo log entries. Redo log files store information critical to the recovery of changed data when the database has lost data as a result of a failure of some kind.

redo log group A file or set of files (members) that store redo entry data. A database must have at least two redo log groups containing at least one file each. Oracle 10*g* by default creates three redo log groups containing one redo log file each. *See* member, log switch, and duplexing.

reference semantics External large objects (LOBs) use reference semantics, which means that when copying an external LOB, only the pointer to the location of the external LOB is actually copied. *See* copy semantics.

relational table The type of table traditionally used in a relational database as well as in Oracle 10*g*. A relational table is usually referred to simply as a table. A table stores data of all types and is the most common form of storage in a database.

restore This term applies to the process of copying a backup file back into a database structure, on disk, when a file such as a datafile, redo log file, or control file has been lost of damaged. *See* recover.

role A collection of privileges that are given a name and can be assigned to users or even to another role.

rollback segment Created automatically by Oracle in an undo tablespace when using the Automatic Undo Management feature. Stores information that allows data changes to be undone (rolled back) if the change is not committed.

root node The starting point for searching a binary tree or BTree. The root has branch or leaf nodes below it, but no nodes above it.

row data A table component that consists of bytes of storage used for rows inserted or updated in the data block.

row directory A group of table components that consist of a list of row identifiers for rows stored in the block.

row header Stores the number of columns contained in the column data area, as well as some overhead, and the ROWID, pointing to a chained or migrated row (if any).

ROWID Contains the physical or logical address of a row.

rule-based optimizer A component of the database still available for backward compatibility that uses static rules to rank possible access paths and select the best path. Rule-based optimization is obsolete for Oracle 10g.

schema The collection of database objects created by one user, such as a table, index, user-defined attribute, an integrity constraint, or a procedure. A schema has the same name as the user who created the objects. The term schema applies more to the database objects and their inter-related structures. A user applies to any type of user. For example, a developer is allowed to create new tables and thus has a schema. On the other hand, an end-user who is not allowed to create tables does not have a schema of his own. However, an end-user can probably access schemas created by other users, either directly, or indirectly through the use of applications.

schema object An object created by a user (schema) in a database.

segment The set of extents that make up one schema object within a tablespace. For example, a table has one segment containing all its extents. A partitioned table has one segment per partition.

server process On the database side of a transaction, the process that interacts with a user process is called a server process.

service name A set of information that Oracle Net Services uses to locate and communicate between client computers and an Oracle database.

service unit The name for the calculated value of the weighted sum of the resources consumed by the user session. Used when setting the COMPOSITE_LIMIT parameter of a profile.

session A period of computer use that lasts from the time a user makes a connection to the database until the user ends the connection.

shared server A mode of database server that connects multiple user processes with one server process. *See* dedicated server.

single instance server The typical type of installation for Oracle. A single instance server installation involves one computer with one set of database files accessed by that one instance.

sorted hash cluster Allows a hash cluster to be sorted. Thus faster retrieval can be achieved based on the sorting specification.

standby database A clone of the current working database that is kept current with the existing database by applying changes stored in the archived redo logs. If the primary database fails, the standby database replaces it automatically.

status (of a tablespace) Defines the availability of a tablespace to end-users and also defines how the tablespace is handled during backup and recovery.

store table The data in a nested table is actually stored in its own table (called the store table), whereas the main table contains an identifier that locates the associated nested table for that row.

subobject An object that is based on another object. For example, an object type can be based on another object type. Also called a subtype.

subpartition template Describes all the subpartitions in a table once, and then all the partitions that use that template. Used as a shorter method of defining subpartitions in a CREATE TABLE statement.

subpartitioning The partitioning of data within each partition of a partitioned table.

subquery A query embedded in another SQL command.

SYSAUX The auxiliary SYSTEM tablespace. The SYSAUX tablespace contains application metadata, and the SYSTEM tablespace only database metadata.

system change number (SCN) A sequential number that is incremented for each change that modifies Oracle datafiles.

System Global Area (SGA) The portion of computer memory (RAM) allocated for database memory by a database instance.

system privilege Represents the ability to manage some part of the database system. Can be assigned to users or roles. For example, the ALTER TABLESPACE privilege allows a user to modify an existing tablespace using the ALTER TABLESPACE statement. *See* object privilege.

table alias A shortcut name for a table that is used to prefix a column name in an SQL command in place of using the entire table name.

table constraints Restrictions (rules) that apply to multiple columns, such as a constraint for a compound foreign key. Table constraints are placed immediately after the list of columns in the CREATE TABLE statement. Table constraints can also appear in the ALTER TABLE statement. *See* out-of-line constraints.

table directory A data block component that consists of information about which table has data in the block.

tablespace A logical data storage space that maps directly to one or more datafiles. The storage capacity of a tablespace is the sum of the size of all the datafiles assigned to that tablespace.

tablespace group Multiple default temporary tablespaces that are contained in a tablespace group. The temporary tablespace group can then be assigned to individual users for sorting purposes.

temporary segment Created during execution of a SQL statement that creates a need for space to perform sorting or other operations. Temporary segments must be stored in a temporary tablespace.

temporary table As with standard tables, a temporary table contains data. However, temporary tables are filled and cleared on a temporary basis, and data contained within them is private to the session that created the data (not seen by other users). Thus rows disappear at the end of a user's session, or when the user commits a transaction. Multiple users can store data in one temporary tablespace, but each user sees only his or her own data.

temporary tablespace Stores objects, such as temporary segments, for the duration of a session.

trigger A program that runs whenever a certain event occurs in a table on which the trigger is defined, such as the insertion of a row into a table.

undo data Made up of undo blocks. Each undo block contains the before image of the data in the block. Usually stored in an undo tablespace and used to restore changed data to its original state if the change is not committed.

undo data retention The retention of data for a short time after the data has actually been rolled back or committed. Used for read consistency and flashback queries.

undo extents The data in the undo tablespace that is added in the form of extents.

undo tablespace An entire tablespace that is reserved for undo data.

unique index A unique index requires that every row inserted into a table has a unique value in the indexed column or columns.

user process Whenever a user runs an application that uses the database, the application creates a user process that controls the connection to the database process.

user-defined data type A datatype defined by a user. Must be an object type, array type and table type. User-defined types help define object columns or object tables. For example, the user-defined data type ADDRESS_TYPE is made up of three attributes: STREET, CITY, and COUNTRY, which are all VARCHAR2(30) data types.

utilities Programs that handle backup, migration, recovery, and various other functions..

variable-length A multibyte character set can be either fixed-length or variable-length. A variable-length, multibyte character set uses a variable number of bytes for each character in the character set. This saves space but requires more complex retrieval routines that take more time. *See* character set.

XML table A table that is created with one column of the XML type data type is an XML table.

INDEX

Special Characters

A

B

C

D

E

F

G

H

I

J

L

M

N

O

P

Q

R

S

T

U

V

W

X